Lecture Notes in Computer Science 9719

Commenced Publication in 1973
Founding and Former Series Editors:
Gerhard Goos, Juris Hartmanis, and Jan van Leeuwen

More information about this series at http://www.springer.com/series/7407

Long Cheng · Qingshan Liu
Andrey Ronzhin (Eds.)

Advances in
Neural Networks – ISNN 2016

13th International Symposium
on Neural Networks, ISNN 2016
St. Petersburg, Russia, July 6–8, 2016
Proceedings

 Springer

Editors
Long Cheng
The Chinese Academy of Sciences
Beijing
China

Andrey Ronzhin
SPIIRAS
St. Petersburg
Russia

Qingshan Liu
Huazhong University of Science
  and Technology
Wuhan, Hubei
China

ISSN 0302-9743 ISSN 1611-3349 (electronic)
Lecture Notes in Computer Science
ISBN 978-3-319-40662-6 ISBN 978-3-319-40663-3 (eBook)
DOI 10.1007/978-3-319-40663-3

Library of Congress Control Number: 2016941302

LNCS Sublibrary: SL1 – Theoretical Computer Science and General Issues

Printed on acid-free paper

This Springer imprint is published by Springer Nature
The registered company is Springer International Publishing AG Switzerland

Preface

This volume of *Lecture Notes in Computer Science* constitutes the proceedings of the 13th International Symposium on Neural Networks (ISNN 2016) held during July 6–8, 2016, in Saint Petersburg, Russia. Building on the success of the previous events, ISNN has become a well-established series of popular and high-quality conferences on neural networks and their applications. This year the symposium was held for the second time outside China, in Saint Petersburg, a very beautiful city in Russia. As usual, it achieved great success.

ISNN aims at providing a high-level international forum for scientists, engineers, educators, as well as students to gather so as to present and discuss the latest progresses in neural network research and applications in diverse areas. It encouraged open discussion, disagreement, criticism, and debate, and we think this is the right way to push the field forward.

This year, we received 104 submissions from about 291 authors in 40 countries and regions. Based on the rigorous peer-reviews by the Program Committee members and reviewers, 84 high-quality papers were selected for publication in the LNCS proceedings. These papers cover many topics of neural network-related research including intelligent control, neurodynamic analysis, memristive neurodynamics, computer vision, signal processing, machine learning, optimization etc.

Many organizations and volunteers made great contributions toward the success of this symposium. We would like to express our sincere gratitude to the City University of Hong Kong, the St. Petersburg Institute for Informatics and Automation, Russian Academy of Sciences, the IEEE Hong Kong Section (CIS Chapter), the International Neural Network Society, the Asia Pacific Neural Network Society, and the Russian Neural Networks Society for their technical co-sponsorship. We would also like to sincerely thank all the committee members for all their great efforts and time in organizing the symposium. Special thanks go to the Program Committee members and reviewers whose insightful reviews and timely feedback ensured the high quality of the accepted papers and the smooth flow of the symposium. We would also like to thank Springer for their cooperation in publishing the proceedings in the prestigious *Lecture Notes in Computer Science* series. Finally, we would like to thank all the speakers, authors, and participants for their support.

April 2016

Long Cheng
Qingshan Liu
Andrey Ronzhin

Organization

General Chairs

Tatiana V. Chernigovskaya   St. Petersburg State University, Saint Petersburg, Russia

Jun Wang   City University of Hong Kong, Hong Kong, SAR China

Rafael Yusupov   St. Petersburg Institute for Informatics and Automation, Russian Academy of Sciences, Saint Petersburg, Russia

Advisory Chairs

Vladimir Cherkassky   University of Minnesota, Minneapolis, USA

Boris Kryzhanovsky   Russian Academy of Sciences, Moscow, Russia

Danil Prokhorov   Toyota Motor Corporation, Ann Arbor, USA

Steering Chairs

Haibo He   University of Rhode Island, Kingston, USA

Derong Liu   University of Illinois-Chicago, Chicago, USA

Organizing Committee Chair

Alexey Potapov   St. Petersburg State University, Saint Petersburg, Russia

Program Chairs

Long Cheng   Institute of Automation, Chinese Academy of Sciences, Beijing, China

Qingshan Liu   Huazhong University of Science and Technology, Wuhan, China

Andrey Ronzhin   St. Petersburg Institute for Informatics and Automation, Russian Academy of Sciences, Saint Petersburg, Russia

Special Sessions Chairs

Jinde Cao   Southeast University, Nanjing, China

Min Han   Dalian University of Technology, Dalian, China

Xiaolin Hu   Tsinghua University, Beijing, China

Publicity Chairs

Huaguang Zhang	Northeastern University, Shenyang, China
Jun Zhang	Sun Yat-sen University, Guangzhou, China
Li Zhang	Chinese University of Hong Kong, Hong Kong, SAR China

Publications Chairs

Zhenyuan Guo	Hunan University, Changsha, China
Biao Luo	Institute of Automation, Chinese Academy of Sciences, Beijing, China
Sitian Qin	Harbin Institute of Technology at Weihai, Weihai, China

Registration Chairs

Yiu-Ming Cheung	Hong Kong Baptist University, Hong Kong, SAR China
Shenshen Gu	Shanghai University, Shanghai, China
Daniel W.C. Ho	City University of Hong Kong, Hong Kong, SAR China

Local Arrangements Chairs

Alexey Karpov	St. Petersburg Institute for Informatics and Automation, Russian Academy of Sciences, Saint Petersburg, Russia
Natalia P. Nesmeyanova	St. Petersburg State University, Saint Petersburg, Russia

Program Committee

Jose Aguilar	Universidad de Los Andes, Venezuela
Plamen Angelov	Lancaster University, UK
Daniel Araújo	Universidade Federal do Rio Grande do Norte, Brazil
Laxmidhar Behera	Indian Institute of Technology Kanpur, India
Pascal Bouvry	University of Luxembourg, Luxembourg
Salim Bouzerdoum	University of Wollongong, Australia
Chien-Lung Chan	Yuan Ze University, Taiwan
Jonathan Chan	King Mongkut's University of Technology Thonburi, Thailand
Andrey Chechulin	St. Petersburg Institute for Informatics and Automation, Russia
Ke Chen	Tampere University of Technology, Finland
Yuehui Chen	University of Jinan, China
Zengqiang Chen	Nankai University, China
Long Cheng	Chinese Academy of Sciences, China
Peng Cheng	Zhejiang University, China
Zhenbo Cheng	ZheJiang University of Technology, China

Vladimir Cherkassky	University of Minnesota, USA
Zheru Chi	Hong Kong Polytechnic University, Hong Kong, SAR China
Fengyu Cong	Dalian University of Technology, China
José Alfredo Ferreira Costa	Universidade Federal do Rio Grande do Norte, Brazil
Ruxandra Liana Costea	Polytechnic University of Bucharest, Romania
Francisco De A.T. De Carvalho	Federal University of Pernambuco, Brazil
Mingcong Deng	Tokyo University of Agriculture and Technology, Japan
Habib Dhahri	University of Sfax, Tunisia
Qiulei Dong	Chinese Academy of Sciences, China
Andries Engelbrecht	University of Pretoria, South Africa
Jianchao Fan	National Marine Environmental Monitoring Center, China
Mauro Forti	Universita di Siena, Italy
Wai-Keung Fung	Robert Gordon University, UK
Wenyin Gong	China University of Geosciences, China
Shenshen Gu	Shanghai University, China
Chengan Guo	Dalian University of Technology, China
Ping Guo	Beijing Normal University, China
Zhishan Guo	University of North Carolina at Chapel Hill, USA
Honggui Han	Beijing University of Technology, China
Ran He	National Laboratory of Pattern Recognition, China
Xing He	Chongqing University, China
Iliya Hodashinsky	TUSUR University, Russia
Tzung-Pei Hong	National Univesity of Kaohsiung, Taiwan
Zhongsheng Hou	Beijing Jiaotong University, China
Bill Howell	Natural Resources Canada, Canada
Jinglu Hu	Waseda University, Japan
Danchi Jiang	University of Tasmania, Australia
Haijun Jiang	Xinjiang University, China
Min Jiang	Xiamen University, China
Yaochu Jin	University of Surrey, UK
Kostas Karatzas	Aristotle University, Greece
Alexey Karpov	St. Petersburg Institute for Informatics and Automation, Russia
Sungshin Kim	Pusan National University, Korea
Irina Kipyatkova	St. Petersburg Institute for Informatics and Automation, Russia
Evgeniy Kostyuchenko	Tomsk State University of Control Systems and Radioelectronics, Russia
Georgios Kouroupetroglou	National and Kapodistrian University of Athens, Greece
Chiman Kwan	Signal Processing, Inc., USA
Chengdong Li	Shandong Jianzhu University, China
Chuandong Li	Chongqing University, China

Gang Li	Deakin University, Australia
Hongyi Li	Bohai University, China
Jianmin Li	Tsinghua University, China
Miqing Li	Brunel University, UK
Shuai Li	Hong Kong Polytechnic University, Hong Kong, SAR China
Tieshan Li	Dalian Maritime University, China
Yangmin Li	University of Macau, Macau, SAR China
Yongjie Li	University of Electronic Science and Technology of China, China
Jie Lian	Dalian University of Technology, China
Hualou Liang	Drexel University, USA
Alan Wee-Chung Liew	Griffith University, Australia
Chih-Min Lin	Yuan Ze University, Taiwan
Huaping Liu	Tsinghua University, China
Ju Liu	Shandong University, China
Lianqing Liu	Chinese Academy of Sciences, China
Qingshan Liu	Huazhong University of Science and Technology, China
Xingwen Liu	Southwest University for Nationalities of China, China
Wenlian Lu	Fudan University, China
Jinwen Ma	Peking University, China
Deyuan Meng	Beihang University, China
Felix Pasila	Petra Christian University, Indonesia
Zhouhua Peng	Dalian Maritime University, China
Irina Perfilieva	University of Ostrava, Czech
Leonid Perlovsky	US Air Force Research Laboratory, USA
Vladimir Red'Ko	Russian Academy of Sciences, Russia
Andrey Ronzhin	St. Petersburg Institute for Informatics and Automation, Russia
Manuel Roveri	Politecnico di Milano, Italy
Juha Röning	University of Oulu, Finland
Alireza Sadeghian	Ryerson University, Canada
Igor Saenko	St. Petersburg Institute for Information and Automation, Russia
Md. Shahjahan	Khulna University of Engineering and Technology, Bangladesh
Bo Shen	Donghua University, China
Qiankun Song	Chongqing Jiaotong University, China
Stefano Squartini	Università Politecnica delle Marche, Italy
Lev Stankevich	St. Peterburg Polytechnic University, Russia
Jian Sun	Beijing Institute of Technology, China
Ning Sun	Nankai University, China
Zhanquan Sun	Shandong Computer Science Center, China
Manchun Tan	Jinan University, China
Qing Tao	Chinese Academy of Sciences, China
Christos Tjortjis	International Hellenic University, Greece

Contents

Signal and Image Processing

Large Scale Image Steganalysis Based on MapReduce.   3
    Zhanquan Sun, Huifen Huang, and Feng Li

Edge Detection Using Convolutional Neural Network   12
    Ruohui Wang

Spectral-spatial Classification of Hyperspectral Image Based on Locality
Preserving Discriminant Analysis   21
    Min Han, Chengkun Zhang, and Jun Wang

Individual Independent Component Analysis on EEG: Event-Related
Responses Vs. Difference Wave of Deviant and Standard Responses.   30
    Tiantian Yang, Fengyu Cong, Zheng Chang, Youyi Liu,
    Tapani Ristainiemi, and Hong Li

Parallel Classification of Large Aerospace Images by the Multi-alternative
Discrete Accumulation Method.   40
    Vladimir I. Vorobiev, Elena L. Evnevich, and Dmitriy K. Levonevskiy

Instantaneous Wavelet Correlation of Spectral Integrals Related to Records
from Different *EEG* Channels.   49
    Sergey V. Bozhokin and Irina B. Suslova

The Theory of Information Images: Modeling of Communicative
Interactions of Individuals   56
    Alexandr Y. Petukhov and Sofia A. Polevaya

A Two-Stage Channel Selection Model for Classifying EEG Activities
of Young Adults with Internet Addiction   66
    Wenjie Li, Ling Zou, Tiantong Zhou, Changming Wang,
    and Jiongru Zhou

Text-independent Speaker Recognition Using Radial Basis Function
Network.   74
    Anton Yakovenko and Galina Malychina

Usage of DNN in Speaker Recognition: Advantages and Problems   82
    Oleg Kudashev, Sergey Novoselov, Timur Pekhovsky,
    Konstantin Simonchik, and Galina Lavrentyeva

Boosted Inductive Matrix Completion for Image Tagging.   92
    Yuqing Hou

Neurological Classifier Committee Based on Artificial Neural Networks
and Support Vector Machine for Single-Trial EEG Signal Decoding   100
    Konstantin Sonkin, Lev Stankevich, Yulia Khomenko,
    Zhanna Nagornova, Natalia Shemyakina, Alexandra Koval,
    and Dmitry Perets

Calculation of Analogs for the Largest Lyapunov Exponents for Acoustic
Data by Means of Artificial Neural Networks.   108
    German A. Chernykh, Yuri A. Kuperin, Ludmila A. Dmitrieva,
    and Angelina A. Navleva

Robust Acoustic Emotion Recognition Based on Cascaded Normalization
and Extreme Learning Machines.   115
    Heysem Kaya, Alexey A. Karpov, and Albert Ali Salah

Dynamical Behaviors of Recurrent Neural Networks

Matrix-Valued Hopfield Neural Networks   127
    Călin-Adrian Popa

Synchronization of Coupled Neural Networks with Nodes of Different
Dimensions   135
    Manchun Tan and Desheng Xu

Asymptotic Behaviors for Non-autonomous Difference Neural Networks
with Impulses and Delays   143
    Shujun Long and Bing Li

Optimal Real-Time Price in Smart Grid via Recurrent Neural Network   152
    Haisha Niu, Zhanshan Wang, Zhenwei Liu, and Yingwei Zhang

Exponential Stability of Anti-periodic Solution of Cohen-Grossberg Neural
Networks with Mixed Delays.   160
    Sitian Qin, Yongyi Tan, and Fuqiang Wang

Stability of Complex-Valued Cohen-Grossberg Neural Networks
with Time-Varying Delays.   168
    Zhenjiang Zhao and Qiankun Song

Space-Time Structures of Recurrent Neural Networks with
Controlled Synapses   177
    Vasiliy Osipov

A Practical Simulator of Associative Intellectual Machine   185
    Sergey Baranov

Hopfield Network with Interneuronal Connections Based on
Memristor Bridges   196
   Mikhail S. Tarkov

Two-Dimensional Fast Orthogonal Neural Networks   204
   A. Yu. Dorogov

Existence of Periodic Solutions to Non-autonomous Delay
Cohen-Grossberg Neural Networks with Impulses on Time Scales   211
   Zhouhong Li

Intelligent Control

Improved Direct Finite-control-set Model Predictive Control Strategy with
Delay Compensation and Simplified Computational Approach for Active
Front-end Rectifiers.   223
   Xing Liu, Dan Wang, and Zhouhua Peng

Distributed Tracking Control of Uncertain Multiple Manipulators Under
Switching Topologies Using Neural Networks   233
   Long Cheng, Ming Cheng, Hongnian Yu, Lu Deng,
   and Zeng-Guang Hou

A Novel Emergency Braking Method with Payload Swing Suppression
for Overhead Crane Systems   242
   He Chen, Yongchun Fang, and Ning Sun

Neural Network Approximation Based Multi-dimensional Active Control
of Regenerative Chatter in Micro-milling   250
   Xiaoli Liu, Chun-Yi Su, and Zhijun Li

A Distributed Delay Consensus of Multi-Agent Systems with Nonlinear
Dynamics in Directed Networks   260
   Li Qiu, Liuxiao Guo, Jia Liu, and Yongqing Yang

Discrete-Time Two-Player Zero-Sum Games for Nonlinear Systems
Using Iterative Adaptive Dynamic Programming.   269
   Qinglai Wei and Derong Liu

Neural Network Technique in Boundary Value Problems for Ordinary
Differential Equations   277
   Elena M. Budkina, Evgenii B. Kuznetsov, Tatiana V. Lazovskaya,
   Sergey S. Leonov, Dmitriy A. Tarkhov, and Alexander N. Vasilyev

Transmission Synchronization Control of Multiple Non-identical Coupled
Chaotic Systems   284
   Xiangyong Chen, Jinde Cao, Jianlong Qiu, and Chengdong Yang

Pneumatic Manipulator with Neural Network Control   292
    Anton Aliseychik, Igor Orlov, Vladimir Pavlovsky,
    Alexey Podoprosvetov, Marina Shishova, and Vladimir Smolin

Dynamic Noise Reduction in the System Measuring Efficiency of Light
Emitting Diodes   302
    Galina Malykhina and Yuri Grodetskiy

Neural Network Technique in Some Inverse Problems of
Mathematical Physics   310
    Vladimir I. Gorbachenko, Tatiana V. Lazovskaya, Dmitriy A. Tarkhov,
    Alexander N. Vasilyev, and Maxim V. Zhukov

The Model of the Robot's Hierarchical Behavioral Control System   317
    A.V. Bakhshiev and F.V. Gundelakh

Object Trajectory Association Rules for Tracking Trailer Boat
in Low-frame-rate Videos.   328
    Jing Zhao, Shaoning Pang, Bruce Hartill,
    and Abdolhossein Sarrafzadeh

Learning Time-optimal Anti-swing Trajectories for Overhead
Crane Systems   338
    Xuebo Zhang, Ruijie Xue, Yimin Yang, Long Cheng, and Yongchun Fang

Attitude Estimation for UAV with Low-Cost IMU/ADS Based on
Adaptive-Gain Complementary Filter.   346
    Lingling Wang, Li Fu, Xiaoguang Hu, and Guofeng Zhang

Hot-Redundancy CPCI Measurement and Control System Based on
Probabilistic Neural Networks.   356
    Dan Li, Xiaoguang Hu, Guofeng Zhang, and Haibin Duan

Individually Adapted Neural Network for Pilot's Final Approach
Actions Modeling   365
    Veniamin Evdokimenkov, Roman Kim, Mikhail Krasilshchikov,
    and German Sebrjakov

Clustering, Classification, Modeling, and Forecasting

On Neurochemical Aspects of Agent-Based Memory Model.   375
    Alexandr A. Ezhov, Andrei G. Khromov, and Svetlana S. Terentyeva

Intelligent Route Choice Model for Passengers' Movement
in Subway Stations   385
    Eric Wai Ming Lee and Michelle Ching Wa Li

A Gaussian Kernel-based Clustering Algorithm with Automatic
Hyper-parameters Computation   393
   Francisco de A.T. de Carvalho, Marcelo R.P. Ferreira,
   and Eduardo C. Simões

Network Intrusion Detection with Bat Algorithm for Synchronization
of Feature Selection and Support Vector Machines   401
   Chunying Cheng, Lanying Bao, and Chunhua Bao

Motion Detection in Asymmetric Neural Networks   409
   Naohiro Ishii, Toshinori Deguchi, Masashi Kawaguchi,
   and Hiroshi Sasaki

Language Models with RNNs for Rescoring Hypotheses of Russian ASR . . .   418
   Irina Kipyatkova and Alexey Karpov

User-Level Twitter Sentiment Analysis with a Hybrid Approach   426
   Meng Joo Er, Fan Liu, Ning Wang, Yong Zhang,
   and Mahardhika Pratama

The Development of a Nonlinear Curve Fitter Using RBF Neural Networks
with Hybrid Neurons   434
   Michael M. Li

Networks of Coupled Oscillators for Cluster Analysis: Overview
and Application Prospects   444
   Andrei Novikov and Elena Benderskaya

Day-Ahead Electricity Price Forecasting Using WT, MI and LSSVM
Optimized by Modified ABC Algorithm   454
   H. Shayeghi, A. Ghasemi, and M. Moradzadeh

Categorization in Intentional Theory of Concepts   465
   Dmitry Zaitsev and Natalia Zaitseva

A Novel Incremental Class Learning Technique for Multi-class
Classification   474
   Meng Joo Er, Vijaya Krishna Yalavarthi, Ning Wang,
   and Rajasekar Venkatesan

Basis Functions Comparative Analysis in Consecutive Data Smoothing
Algorithms   482
   F.D. Tarasenko and D.A. Tarkhov

FIR as Classifier in the Presence of Imbalanced Data   490
   Solmaz Bagherpour, Àngela Nebot, and Francisco Mugica

Neural Network System for Monitoring State of a High-Speed
Fiber-Optical Linear Path   497
  I.A. Saitov, O.O. Basov, A.I. Motienko, S.I. Saitov, M.M. Bizin,
  and V. Yu. Budkov

Pattern Classification with the Probabilistic Neural Networks Based
on Orthogonal Series Kernel.   505
  Andrey V. Savchenko

Neural Network Methods for Construction of Sociodynamic
Models Hierarchy   513
  Ekaterina A. Blagoveshchenskaya, Aleksandra I. Dashkina,
  Tatiana V. Lazovskaya, Viktoria V. Ryabukhina, and Dmitriy A. Tarkhov

Application of Hybrid Neural Networks for Monitoring and Forecasting
Computer Networks States   521
  Igor Saenko, Fadey Skorik, and Igor Kotenko

Multiclass Ensemble of One-against-all SVM Classifiers   531
  Catarina Silva and Bernardete Ribeiro

Long Exposure Point Spread Function Modeling with Gaussian Processes . . .   540
  Ping Guo, Jian Yu, and Qian Yin

Neural Network Technique for Processes Modeling in Porous Catalyst
and Chemical Reactor   547
  Tatiana A. Shemyakina, Dmitriy A. Tarkhov, and Alexander N. Vasilyev

Fine-Grained Real Estate Estimation Based on Mixture Models   555
  Peng Ji, Xin Xin, and Ping Guo

Intellectual Analysis System of Big Industrial Data for Quality
Management of Polymer Films   565
  Tamara Chistyakova, Mikhail Teterin, Alexander Razygraev,
  and Christian Kohlert

Some Ideas of Informational Deep Neural Networks Structural
Organization.   573
  Vladimir Smolin

Meshfree Computational Algorithms Based on Normalized Radial
Basis Functions.   583
  Alexander N. Vasilyev, Ilya S. Kolbin, and Dmitry L. Reviznikov

Evolutionary Computation

An Experimental Assessment of Hybrid Genetic-Simulated Annealing
Algorithm   595
  Cong Jin and Jinan Liu

When Neural Network Computation Meets Evolutionary Computation:
A Survey   603
  Zonggan Chen, Zhihui Zhan, Wen Shi, Weineng Chen, and Jun Zhang

New Adaptive Feature Vector Construction Procedure for Speaker Emotion
Recognition Based on Wavelet Transform and Genetic Algorithm   613
  Alexander M. Soroka, Pavel E. Kovalets, and Igor E. Kheidorov

Integration of Bayesian Classifier and Perceptron for Problem Identification
on Dynamics Signature Using a Genetic Algorithm for the Identification
Threshold Selection   620
  Evgeny Kostyuchenko, Mihail Gurakov, Egor Krivonosov,
  Maxim Tomyshev, Roman Mescheryakov, and Ilya Hodashinskiy

Cognition Computation and Spiking Neural Networks

Tracking Based on Unit-Linking Pulse Coupled Neural Network Image
Icon and Particle Filter   631
  Hang Liu and Xiaodong Gu

Quaternion Spike Neural Networks   640
  Luis Lechuga-Gutiérrez and Eduardo Bayro-Corrochano

Vector-Matrix Models of Pulse Neuron for Digital Signal Processing   647
  Vladimir Bondarev

About $\Sigma\Pi$-neuron Models of Aggregating Type   657
  Zaur Shibzukhov and Denis Cherednikov

Conversion from Rate Code to Temporal Code – Crucial Role of Inhibition . . .   665
  Mikhail V. Kiselev

Analysis of Oscillations in the Brain During Sensory Stimulation:
Cross-Frequency Relations   673
  Elena Astasheva, Maksim Astashev, and Valentina Kitchigina

Memristor-Based Neuromorphic System with Content Addressable
Memory Structure   681
  Yidong Zhu, Xiao Wang, Tingwen Huang, and Zhigang Zeng

Detailed Structure of the Cortical Magnetic Response to Words   691
  V.L. Vvedensky and A. Yu. Nikolayeva

Pre-coding & Testing Technique for Interfacing Neural Networks
Associative Memory   698
  Fayçal Saffih, Wan Abdulllah, and Zainol Ibrahim

The Peculiarities of Perceptual Set in Sensorimotor Illusions   706
  *Valeria Karpinskaia, Vsevolod Lyakhovetskii, Viktor Allakhverdov,
  and Yuri Shilov*

Generalized Truth Values: From Logic to the Applications in Cognitive
Sciences.   712
  Oleg Grigoriev

Modeling of Cognitive Evolution: Agent-Based Investigations in
Cognitive Science   720
  Vladimir G. Red'ko

Usage of Language Particularities for Semantic Map Construction: Affixes
in Russian Language   731
  *Anita Balandina, Artyom Chernyshov, Valentin Klimov,
  and Anastasiya Kostkina*

Author Index   739

Signal and Image Processing

Large Scale Image Steganalysis
Based on MapReduce

Zhanquan Sun[1,2(✉)], Huifen Huang[2], and Feng Li[3]

[1] Shandong Provincial Key Laboratory of Computer Network,
Shandong Computer Science Center (National Supercomputing Center in Jinan),
Jinan, Shandong, China
sunzhq@sdas.org
[2] Shandong Yingcai University, Jinan, Shandong, China
[3] Department of History, College of Liberal Arts, Shanghai University,
Shanghai, China

Abstract. Steganalysis is the opposite art to steganography, whose goal is to detect whether or not the seemly innocent objects like image hiding message. Image steganalysis is important research issue of information security field. With the development of steganography technology, steganalysis becomes more and more difficult. Regarding the problem of improving the performance of image steganalysis, many research work have been done. Based on current research, large scale training set will be the feasible means to improve the steganalysis performance. Classic classifier is out of work to deal with large scale images steganalysis. In this paper, a parallel Support Vector Machines based on MapReduce is used to build the steganalysis classifier according to large scale training samples. The efficiency of the proposed method is illustrated with an experiment analysis.

Keywords: Image steganalysis · MapReduce · Support vector machines · Feature extraction

1 Introduction

Steganography and steganalysis can be considered as opposite sides of the game, they are progressing during the competition between each other. Steganalysis is the opposite art to steganography, whose goal is to detect whether or not the seemly innocent objects like image hiding message. Many steganography models have been proposed, such as LSB (least significant bit) replace and LSB matching, STC (Space-time-coding), HUGO (Highly Undetectable Steganography), JSteg, OutGuess, F5, nsF5, MBS, YASS, MOD and so on [1–3]. Steganalysis aiming at a special Steganography method has lots research work [4, 5]. Universal steganalysis is a difficult problems. Some research has been done on the topic [6–8]. But how to improve the steganalysis performance is still a difficult issue.

With the development of information technology, data scale becomes larger and larger. Big data technology develops quickly in recent years. Taking full use of large scale training data's information to improve steganalysis performance is the

© Springer International Publishing Switzerland 2016
L. Cheng et al. (Eds.): ISNN 2016, LNCS 9719, pp. 3–11, 2016.
DOI: 10.1007/978-3-319-40663-3_1

develop trend [9]. Most classic classification methods are out of reach in practice in face of big data. Efficient parallel algorithms and implementation techniques are the key to meeting the scalability and performance requirements entailed in such large scale data mining analyses. Many parallel algorithms are implemented using different parallelization techniques such as threads, MPI, MapReduce, and mash-up or workflow technologies yielding different performance and usability characteristics [9]. MapReduce is a cloud technology developed from the data analysis model of the information retrieval field. Several MapReduce architectures are developed now. Iterative MapReduce based on memory computing is the most efficient architecture to deal with large scale computing problems. Commonly used iterative MapReduce software are Spark and Twister [9, 10]. In this paper, parallel SVM model is proposed to deal with large scale image steganalysis problem. The efficiency is illustrated with a practical experience.

The rest of this paper is organized as follows. In Sect. 2, image feature extraction based on wavelet transform is introduced. Iterative MapReduce architecture is present in Sect. 3. Parallel SVM model based on iterative MapReduce is present in Sect. 4. Image steganalysis procedure based on MapReduce is described in detail in Sect. 5. A practical example is analyzed in Sect. 6 to illustrate the efficiency of the proposed method.

1.1 Discrete Wavelet Transform

A wavelet series is a representation of a square-integrable function by a certain orthonormal series generated by a wavelet. It is a multi-resolution analysis tool. Apply L levels wavelet decomposition on an image, it will produce multi sub-band images. Let the initial image be denoted C_0. The coefficients matrix of wavelet coefficients $\psi(x)$ is denoted by and the coefficients matrix of scaling coefficients $\Phi(X)$ is denoted by H. The wavelet decomposition algorithm is as follows.

$$\begin{cases} C_{j+1} = HC_jH_T \\ D_{j+1}^h = GC_jH_T \\ D_{j+1}^v = HC_jG_T \\ D_{j+1}^d = GC_jG_T \end{cases} \tag{1}$$

where j denote the level of decomposition, denote horizontal, vertical, and diagonal component respectively. H_T and G_T denote the conjugate transpose matrix of H and G.

1.2 Histogram Moment in Frequency Domain

The histogram moment is defined as the following equation.

$$M_n = \sum_{k=-N/2}^{N/2} |f_k|^n p(f_k) \tag{2}$$

where n denotes the moment number, N is the variable number of histogram in horizontal direction. f_k is the k-order frequency of Fourier transform where $k = -N/2$, $\cdots, -1, 0, 1, \cdots, N/2$. $p(f_k)$ is the distribution of the amplitude of histogram after Fourier transform, i.e.

$$p(f_k) = \frac{|H(f_k)|}{\sum_{j=-N/2}^{N/2} |H(f_j)|} \tag{3}$$

where $|H(f_k)|$ is k-order amplitude of histogram $h(x_k)$ after Fourier transforming.

1.3 Image Feature Extraction

Given a JPEG image, operate DCT (Discrete Cosine Transform) on it and the obtained DCT coefficients are taken as the object to be processed.

Operate discrete wavelet transform on the picture. In this paper, 3 level wavelet decomposition is adopted. 12 sub-band images can be obtained. Calculate the histograms of 12 sub-band images and picture. Operate Fourier transform on the histograms and calculate the 3-order histogram moment according to Eq. (2). We can obtain 39 features in total.

2 Iterative MapReduce

There are many parallel algorithms with simple iterative structures. Most of them can be found in the domains such as data clustering, dimension reduction, link analysis, machine learning, and computer vision. These algorithms can be implemented with iterative MapReduce computation. Professor Fox developed the first iterative MapReduce computation model Twister. It has several components, i.e. MapReduce main job, Map job, Reduce job, and combine job. Twister's programming model can be described as in Fig. 1.

Twister is a distributed in-memory MapReduce runtime optimized for iterative MapReduce computations. It reads data from local disks of the worker nodes and handles the intermediate data in distributed memory of the worker nodes. All communication and data transfers are handled via a publish/subscribe messaging infrastructure. During configuration, the client assigns MapReduce methods to the job, prepares KeyValue pairs and prepares static data for MapReduce tasks through the partition file if required. Map daemons operate on computation nodes, loading the Map classes and starting them as Map workers. During initialization, Map workers load static data from the local disk according to records in the partition file and cache the data into memory. Most computation tasks defined by the users are executed in the Map workers. Reduce daemons operate on computation nodes. The reduce jobs depend on the computation results of Map jobs. The communication between daemons is through messages. Combine job is to collect MapReduce results.

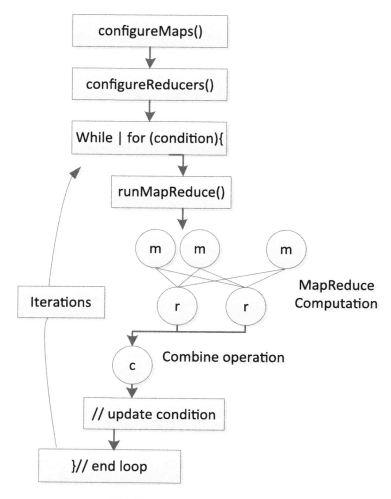

Fig. 1. Program model of Twister.

3 Parallel SVM Based on Iterative MapReduce

3.1 SVM

SVM first maps the input points into a high-dimensional feature space with a nonlinear mapping function Φ and then carries through linear classification or regression in the high-dimensional feature space. The linear regression in high-dimension feature space corresponds to the nonlinear classification or regression in low-dimensional input space. The general SVM can be described as follows.

Let l training samples be $T = \{(x_1, y_1), \ldots, (x_l, y_l)\}$, where $x_i \in R^n$, $y_i \in \{1, -1\}$ (classification) or $y_i \in R$ (regression), $i = 1, 2, \ldots, l$. Nonlinear mapping function is

$k(x_i, x_j) = \phi(x_i)\phi(x_j)$. Classification SVM can be implemented through solving the following equations.

$$\min_{w, \xi_i, b}\left\{\frac{1}{2}\|w\|^2 + C\sum_i \xi_i\right\}$$
$$s.t.\ y^i(\Phi^T(X_i)w + b) \geq 1 - \xi_i\ \forall i = 1, 2, \ldots, n \qquad (4)$$
$$\xi_i \geq 0\ \forall i = 1, 2, \ldots, n$$

By introducing Lagrangian multipliers, the optimization problem can be transformed into its dual problem.

$$\min_\alpha \sum_{i,j}\alpha_i\alpha_j y_i y_j k(x_i, x_j) - \sum_{i=1}^l \alpha_i$$
$$s.t.\ y^T\alpha = 0 \qquad (5)$$
$$0 \leq \alpha_i < C,\ i = 1, 2, \ldots, l$$

After obtaining optimum solution a^*, b^*, the following decision function is used to determine which class the sample belongs to.

$$f(x) = \mathrm{sgn}\left(\sum_{i=1}^l y_i\alpha_i^* K(x_i, x) + b^*\right) \qquad (6)$$

The classification precision of the SVM model can be calculated as

$$Accuracy = \frac{\#correctly\ predicted\ data}{\#total\ testing\ data} \times 100\%$$

3.2 Parallel SVM Based on Twister

The parallel SVM is based on the cascade SVM model. The SVM training is realized through partial SVMs. Each subSVM is used as filter. This makes it straightforward to drive partial solutions towards the global optimum, while alternative techniques may optimize criteria that are not directly relevant for finding the global solution. Through the parallel SVM model, large scale data optimization problems can be divided into independent, smaller optimizations. The support vectors of the former subSVM are used as the input of later subSVMs. The subSVM can be combined into one final SVM in hierarchical fashion. The parallel SVM training process can be described as in Fig. 2.

In the architecture, the sets of support vectors of two SVMs are combined into one set and to be input a new SVM. The process continues until only one set of vectors is left. In this architecture a single SVM never has to deal with the whole training set.

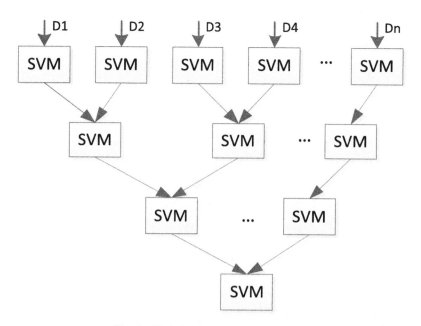

Fig. 2. Training flow of parallel SVM.

If the filters in the first few layers are efficient in extracting the support vectors then the largest optimization, the one of the last layer, has to handle only a few more vectors than the number of actual support vectors. Therefore, the training sets of each sub-problems are much smaller than that of the whole problem when the support vectors are a small subset of the training vectors.

4   Image Steganalysis

Large scale image steganalysis can be summarized as follows.

Step 1: Sample generate

Collect large scale raw images from available resources. Resize them into same size.

Step 2: Steganography

Write hide information into those resized images with different steganography methods, such as F5, Outguess and so on. Resized images and steganography images are combined together to generate sample images.

Step 3: Feature extraction

Operate feature extraction introduced as in Sect. 2 on the sample images. We get the features of all samples. They are divided into 2 parts. One part is taken as training samples and the others are taken as test samples.

Step 4: Classifier training

The training samples are input into the parallel SVM introduced as in Sect. 4. The test samples are used to test the performance of the trained classifier.

Step 5: Performance evaluation

The performance of the steganalysis classifier is evaluated with the 4 following indices.

(1) TPR (True Positive Rate)

$$TPR = TP/(TP + FN) \tag{7}$$

(2) TNR (True Negative Rate)

$$TNR = TN/(TN + FP) \tag{8}$$

(3) FPR (False Positive Rate)

$$FPR = FP/(FP + TN) \tag{9}$$

(4) FNR (False Negative Rate)

$$FNR = FN/(FN + TP) \tag{10}$$

where TP (True Positive) the number of positive samples are estimated positive, TN (True Negative) is the number of negative samples are estimated negative, FP (False Positive) is the number of negative samples are estimated positive, and FN (False Negative) is the number of positive samples are estimated negative.

5 Example

5.1 Data Source

We download Imagenet database from website http://image-net.org. There are 1.3 million images in total. They are all JPEG images. We select 50000 images randomly for study. They are resized into 256×256 firstly. Operate steganography on 25000 images with LSB method and on the other 25000 images with F5 method. The 100000 images are taken as the samples.

5.2 Feature Extraction

Operate feature extraction method on the 100000 images. We get 100000 vectors with 39 dimensions. 75000 vectors are selected as training samples and the others as taken as test sample.

5.3   Steganalysis Identify

The example is analyzed in Shandong Cloud platform. Twister0.9 software is deployed in each computational node. ActiveMQ is used as message broker. The configuration of each virtual machine is as follows. Each node is installed Ubuntu Linux OS. The processor is 3 GHz Intel Xeon with 4 GB RAM. 75000 training vectors are partitioned into 4, 2 and 1 sections respectively. They are trained with parallel SVM introduced as in Sect. 4 on 4, 2, and 1 computational nodes respectively. The 25000 test sample used to test the performance of different training model. The test results are listed in Table 1.

Table 1. Steganalysis results based on different partition.

Computational node number	Training time(s)	TPR	TNR
1	12032	0.981	0.972
2	72003	0.978	0.971
4	48098	0.978	0.972

5.4   Result Comparison

From above analysis results we can find that parallel classifier based on MapReduce can process large scale training samples. It will improve the classifier's performance with large scale training samples. Based on the same training samples, more computational nodes can improve the training speed and the performance are similar. It is efficient in dealing with large-scale image steganalysis problems.

6   Conclusions

Image steganalysis plays more and more important role in our information society. How to improve the steganalysis performance is the research focus of information security. The parallel image steganalysis based on MapReduce can take full use of large scale training sample's information to build more precise steganalysis classifier. The experience results show that the proposed method is efficient in dealing with large scale image steganalysis problems. Big data technology will be the develop trend to resolve steganalysis problems.

Acknowledgments. This work is supported by the national science foundation (No. 61472230), National Natural Science Foundation of China (Grant No. 61402271), the Natural Science Foundation of Shandong Province (Grant No. ZR2015JL023 and Grant No. ZR2015FL025), Shandong science and technology development plan (Grant No. J15LN54), Key R & D program in Shandong Province (Grant No. 2015GGX101012).

References

1. Chanu, Y.J., Tuithung, T., Manglem, S.K.: A short survey on image steganography and steganalysis techniques. In: 3rd National Conference on Emerging Trends and Applications in Computer Science, pp. 52–55 (2012)
2. Bilal, I., Roj, M.S., Kumar, R., Mishra, P.K.: Recent advancement in audio steganography. In: 2014 International Conference on Parallel, Distributed and Grid Computing, pp. 402–405 (2014)
3. Mazurczyk, W., Caviglione, L.: Steganography in modern smartphones and mitigation techniques. IEEE Commun. Surv. Tutorials **17**(1), 334–357 (2015)
4. Khosravirad, S.R., Eghlidos, T., Ghaemmaghami, S.: Closure of sets: a statistically hypersensitive system for steganalysis of least significant bit embedding. IET Signal Process. **5**(4), 379–389 (2011)
5. Tang, M.W., Fan, M.Y., Wang, G.W.: An extential steganalysis of information hiding for F5. In: 2nd International Workshop on Intelligent Systems and Applications, pp. 1–4 (2010)
6. Zhang, Z., Hu, D.H., Yang, Y., Su, B.: A universal digital image steganalysis method based on sparse representation. In: 9th International Conference on Computational Intelligence and Security, pp. 437–441 (2013)
7. Gul, G., Kurugollu, F.: SVD-based universal spatial domain image steganalysis. IEEE Trans. Inf. Forensics Secur. **5**(2), 349–353 (2010)
8. Davidson, J., Jalan, J.: Steganalysis using partially ordered markov models. In: Böhme, R., Fong, P.W.L., Safavi-Naini, R. (eds.) IH 2010. LNCS, vol. 6387, pp. 118–132. Springer, Heidelberg (2010)
9. Dai, Z.H., Xiong, Q., Peng, Y., Gao, H.H.: Research on the large scale image steganalysis technology based on cloud computing and BP neutral network. In: The 8th International Conference on Intelligent Information Hiding and Multimedia Signal Processing, pp. 415–419 (2012)
10. Fox, G.C., Bae, S.H.: Parallel data mining from multicore to cloudy grids. In: High Performance Computing and Grids workshop, pp. 311–340 (2008)
11. Ekanayake, J., Li, H.: Twister: a runtime for iterative MapReduce. In: The First International Workshop on MapReduce and Its Applications of ACM HPDC, pp. 810–818 (2010)
12. Islam, N.S., Wasi-ur-Rahman, M., Lu, X.Y., Shankar, D., Panda, D.K.: Performance characterization and acceleration of in-memory file systems for Hadoop and Spark applications on HPC clusters. In: IEEE International Conference on Big Data, pp. 243–252 (2015)
13. Huang, C., Gao, J.J., Xuan, G.R., Shi, Y.Q.: Steganalysis based on moments in the frequency domain of wavelet sub- band's histogram for JPEG images. Comput. Eng. Appl. **30**, 95–97 (2006)

Edge Detection Using Convolutional Neural Network

Ruohui Wang[✉]

Department of Information Engineering, The Chinese University of Hong Kong,
Hong Kong, China
wr013@ie.cuhk.edu.hk

Abstract. In this work, we propose a deep learning method to solve the edge detection problem in image processing area. Existing methods usually rely heavily on computing multiple image features, which makes the whole system complex and computationally expensive. We train Convolutional Neural Networks (CNN) that can make predictions for edges directly from image patches. By adopting such networks, our system is free from additional feature extraction procedures, simple and efficient without losing its detection performance. We also perform experiments on various networks structures, data combination, pre-processing and post-processing techniques, revealing their influence on performance.

Keywords: Deep learning · Convolutional neural network · Image processing · Computer vision · Edge detection

1 Introduction

Edge detection is the task of identifying object boundaries within a still image (see Fig. 1). As a fundamental technique, it has been widely used in image processing and computer vision areas [1–5]. Accurate, simple and fast edge detection algorithms can certainly increase both performance and efficiency of the whole image processing system. However, edges form in diverse ways. Finding a universally applicable detection rule is hence not easy.

Conventional edge detection algorithms rely heavily on gradient computing [1]. Pixels with large gradient magnitude are labeled as edges. Other techniques, such as non-maximum suppression [6], are usually combined to yield a better result. These methods are all based on the assumption that color or intensity changes sharply on the boundary between different objects while it remains unchanged within one object. Unfortunately, this is not always true. Large color gradient can appear on texture within one object while small color gradient can also appear on object boundaries.

Having realized the limitation of local gradient cues, recent works start to introduce learning techniques when designing edge detection algorithms [7–10]. Correspondence between object boundaries and image are learned from data instead of based on man-made assumptions. However, traditional learning methods are usually not powerful enough to learn a direct mapping from image

© Springer International Publishing Switzerland 2016
L. Cheng et al. (Eds.): ISNN 2016, LNCS 9719, pp. 12–20, 2016.
DOI: 10.1007/978-3-319-40663-3_2

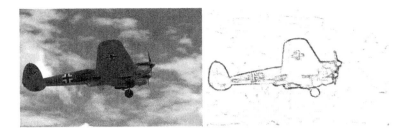

Fig. 1. An example of edge detection.

patches to edge predictions. People have to compute multiple color and gradient channels or extract self-similarity features in order to get a rich representation of the original image. As a result, such systems usually consist of multiple modules, which could be complex and inefficient. In the meantime, selecting proper features usually requires domain knowledge and can affect both efficiency and performance a lot.

Faced with such problems, we start to consider introducing powerful deep learning techniques, such as convolutional neural network (CNN) [11], into the edge detection problem. Unlike traditional shallow learning structures, deep neural networks can learn hierarchical feature representations in their multiple-layer structure. By adopting CNN, our edge detection system is free from extra feature extraction or multiple channel computation, thus becomes straightforward and efficient.

On the other hand, CNN tends to capture local patterns from images in its convolutional layers. This property also makes it a suitable tool for solving the edge detection problem, because edges are usually locally correlated and exhibit specific patterns, such as straight lines, corners, T-junctions and Y-junctions.

Motivated by these intuitions, we design a simple and efficient edge detection system with CNN being adopted as the central part. In order to further simplify the computation, we removed pooling layers in our networks. In order to select an optimal network structure, we performed a lot of experiments on the popular BSDS500 [1] dataset, comparing different network structures as well as data combination, preprocessing and post-processing techniques (see Sects. 2 and 3 for detail). The best performance is achieved on a simple three-layer network taking raw RGB color image patches as input without any preprocessing. By adding non-maximum suppression to the whole system, the performance can be further improved a little.

The rest of this paper is arranged as follow. In Sect. 2, we give an overview of our CNN edge detector as well as some preprocessing and post-processing techniques. In Sect. 3, we exploit different structures and data construction in our experiments, and obtain insightful observations. Section 4 concludes the paper.

2 Edge Detection System

Figure 2 provides an overview of our edge detection system. Given an image, we can first apply some preprocessing techniques [12] for noise removal. Then a convolutional neural network scans over the entire image, making edge prediction for every pixel based on the image patch centered on it. At last, non-maximal suppression [6] or morphological operations can be further applied as a post-processing step to thin the output edge map so as to increase localization accuracy.

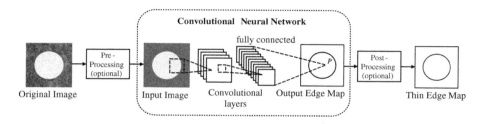

Fig. 2. An overview of our edge detection system.

2.1 Preprocessing

Natural images are sometimes degraded by noise. It is straightforward to apply some noise removal algorithm first. Here, we choose a slight and smart algorithm [12], which increases system complexity little.

2.2 Convolutional Neural Network

A Convolutional Neural Network works as the core component of our system. It takes image patches as input and makes predictions on whether their central pixels locate on an edge or not. Any network that fulfills this task can be adopted in our system, but choosing the best network structure is not straight-forward. We performed a thorough comparison on different network structures and the best performance was achieved on a three-layered network whose structure and parameters are summarized in Table 1.

Table 1. Network structure and parameters adopted in our system.

#	Input size	Layer type	Filter size	Stride	Nonlinearity	Output size
1	23*23*3	convolutional	5*5*3	3	ReLU	7*7*32
2	7*7*32	convolutional	3*3*32	2	ReLU	3*3*256
3	3*3*256	fully connected	N/A	N/A	Logistic	1

2.3   Post-Processing

As a common phenomenon, edges produced by detection algorithms usually cover several pixels and are considered to be inaccurate [6, 9, 10]. In this regard, non-maximal suppression [6] or morphological operation can be adopted to serve as a post-precessing step, rendering a thinner edge map in the final output.

3   Experiments and Results

3.1   Data Set and Evaluation Method

We selected the most popular BSDS500 data set [1, 13] for training and evaluating our edge detector. This dataset contains 500 natural images. For each image, several people were asked to draw a contour map separating different objects based on their own understanding. All 500 images are divided into 3 subsets, with 200 for training, 100 for validation and 200 for testing. We strictly followed the official guidelines [13] to train and tune our system exclusively on the train and validation subsets, and to evaluate our results on the entire test subset with provided benchmarking code [13].

3.2   Preparing Training Data

Before training our network, we need to prepare image patches and corresponding ground truth that can be acceptable by our neural networks. The procedure is summarized in Fig. 3.

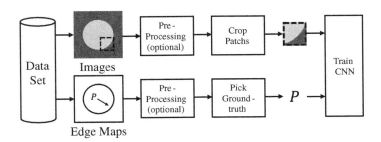

Fig. 3. Work flow on preparing training data.

First, we could apply some preprocessing techniques [12] to remove noise from the original images. For each image in BSDS500, there are multiple corresponding edge maps. In order to determine a single ground truth, we selected the most sparse one or averaging the most spare two or three. Then we cropped images into patches. For each image patch, we located it on the edge map and picked the value at its center as corresponding ground truth label. At last, these patch-label pairs were sent for training the neural network.

One key problem is that there are large numbers of negative samples, i.e. patches whose center do not belong to any edge, compared to positive samples (about 50:1). In order to balance the ratio between positive and negative samples, we selected all positive samples and a small random partition of negative samples for training the network, resulting the negative/positive ratio near 2:1.

3.3 Training Neural Networks

We used the cuda-convnet toolbox [14] to train our CNN on an NVIDIA Tesla K40c GPU. The training process would last about 40 min for a three-layer network, or nearly one day for a complex network. We always try to tune the network parameters, such as number of filters and filters' size, to avoid either underfitting or overfitting. We stopped the training process if the error curve converge or the error rate on the validation set achieve a minimum. Ordinary error curves during the training steps are provided in Fig. 4.

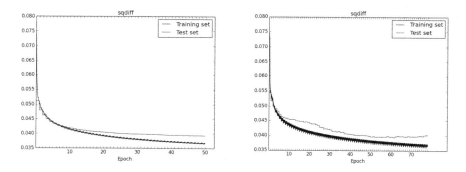

Fig. 4. Ordinary error curves on training and validation sets.

3.4 Result

We evaluated our CNN edge detector using the original benchmark code provided along with BSDS500 [13]. We tested on several networks with different structures and image color channel combinations. The best performance was achieved by a three-layered network taking raw RGB image patches as input. It is summarized in Fig. 5 and Table 2 with comparison to other mainstream methods, including the most widely used non-learning-based Canny edge detector [6] and several learning-based algorithms [1,8,9,15]. Figure 6 shows some sample results of our detector that of sketch tokens [9]. As shown in the comparison, our edge detector is fast while achieving comparable performance with state-of-the-art methods. Since our neural network contains only convolutional and fully connected layers, algorithms such as [16] can also be adopted in our system to further speed it up by several times or even tens of times.

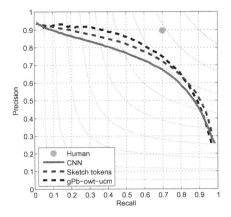

Fig. 5. Precession-recall curves.

Table 2. Performance measurements.

	ODS	OIS	AP	Speed
Human	.80	-	-	-
CNN (GPU)	.69	.71	.71	2 s
Canny [6]	.60	.64	.58	1/15 s
BEL [8]	.67	-	-	10 s
gPb [1]	.71	.74	.65	60 s
gPb-owt-ucm [1]	.73	.76	.73	240 s
SCG [15]	.74	.76	.77	280 s
Sketch tokens [9]	.73	.75	.78	1 s

Our network can also capture local edge structures. As shown in Fig. 7, filters in the first convolutional show different color changes, capturing step edges or textures. Both of them are local structures we want.

3.5   Comparison on Different Configurations

We have also tried different network structure, channel configuration, input size of image patches and preprocessing techniques in our experiments.

Fig. 6. Results on images with textures. The first and second column shows the original image and the ground truth respectively. The third and forth column shows the raw output of our CNN and sketch token [9] respectively.

Fig. 7. Learned filters in the first convolutional layer.

Table 3 compares the performance when using different network structures. We tried a complex network architecture consisting of three convolutional layers, three pooling layers and two fully connected layers. However, this complex structure can hardly improve the performance while it takes much longer time on training and overfits easily. Usually a network with two convolutional layers and one fully-connected layer is enough for this problem. So we chose using this structure in most our experiments.

Table 3. Performance comparison on using different network structures. L = locally connected layer, P = pooling layer, C = convolutional layer, F = fully connected layer.

Image patch	Network	ODS	OIS	AP
gray	LPLPLPFF	.64	.66	.57
L0 smooth	CCF	.67	.68	.68
17*17*1	CCF(bias=0)	.66	.68	.66
CIE+grad	LPLPLPFF	.64	.65	.64
17*17*8	CCF	.64	.65	.63
CIE+grad	CCF	.68	.70	.70
21*21*8	CCFF	.68	.69	.66

In our experiments, we tried four types of channel configuration of the input image patches, (1) single gray scale channel, (2) RGB channels, (3) multiple gray scale channels on different scales and (4) a combination of CIR-LUV, one gradient magnitude and four gradient orientation channels, which is similar with [9]. We found that gray level inputs worked a little bit worse than RGB inputs. Multi-scale input hardly contributed to the performance. Multiple feature channels performed almost the same as RGB while requiring much more time on training and consuming huge amount of memory. Based on these results, we can demonstrate that there is no need to design extra feature extraction steps in our method.

For patch size and preprocessing techniques, we found that the performance would decrease if patch size was less than 17*17 and would not increase a lot if patch size was beyond 21*21. L0 smooth [12] would usually increase the performance when using single-channel gray-scaled image patches as input while decrease the performance when using RGB. Table 4 is a conclusion of our

Table 4. Performance comparison when using different input image patches.

Network	Image Patch		ODS	OIS	AP
	Channel	Size			
CCFF	CIE+grad	21*21*8	.68	.70	.70
CCFF	RGB	35*35*3	.66	.68	.62
CF	RGB	35*35*3	.63	.65	.60
CCFF	RGB	25*25*3	.68	.70	.68
CCF	RGB	23*23*3	.68	.70	.70
CCF	RGB+L0	23*23*3	.66	.68	.62
CCF	RGB	21*21*3	.68	.69	.65
CCF	RGB	17*17*3	.67	.69	.63
CCF	gray+L0	31*31*1	.67	.69	.70
CCF	gray+L0	25*25*1	.67	.69	.69
CCF	gray+L0	17*17*1	.67	.68	.68
CCF	gray+L0	13*13*1	.65	.66	.60
CCF	gray 3 scale	21*21*1	.65	.67	.65
CCF	gray 4 scale	21*21*1	.64	.65	.60

experiment results. These results were achieved when applying morphological thin in the post-processing step.

4   Conclusion

In this work, we developed a deep learning method for solving the edge detection problem by using convolutional neural networks (CNN). Unlike previous work, our approach does not need extra feature extraction process and can be very simple and fast while achieving good result. It is also very easy for people to implement and integrate our simple algorithm into their own computer vision systems. Moreover, with deep learning becoming more and more popular, people try using CNN in every field of computer vision. It is probably that our network can be concatenated in front of the network used in other application. The whole network can then be jointly fine tuned. This is a particular advantage of CNN and have been studied in recent papers [17].

References

1. Arbelaez, P., Maire, M., Fowlkes, C., Malik, J.: Contour detection and hierarchical image segmentation. IEEE Trans. Pattern Anal. Mach. Intell. **33**, 898–916 (2011)
2. Ferrari, V., Fevrier, L., Jurie, F., Schmid, C.: Groups of adjacent contour segments for object detection. IEEE Trans. Pattern Anal. Mach. Intell. **30**, 36–51 (2008)

3. Yokoyama, M., Poggio, T.: A contour-based moving object detection and tracking. In: 2nd Joint IEEE International Workshop on Visual Surveillance and Performance Evaluation of Tracking and Surveillance 2005, pp. 271–276. IEEE (2005)

4. Yilmaz, A., Li, X., Shah, M.: Contour-based object tracking with occlusion handling in video acquired using mobile cameras. IEEE Trans. Pattern Anal. Mach. Intell. **26**, 1531–1536 (2004)

5. Vacchetti, L., Lepetit, V., Fua, P.: Combining edge and texture information for real-time accurate 3d camera tracking. In: Third IEEE and ACM International Symposium on Mixed and Augmented Reality, ISMAR 2004, pp. 48–56. IEEE (2004)

6. Canny, J.: A computational approach to edge detection. IEEE Trans. Pattern Anal. Mach. Intell. **8**, 679–698 (1986)

7. Zheng, S., Yuille, A., Tu, Z.: Detecting object boundaries using low-, mid-, and high-level information. Comput. Vis. Image Underst. **114**, 1055–1067 (2010)

8. Dollar, P., Tu, Z., Belongie, S.: Supervised learning of edges and object boundaries. In: 2006 IEEE Computer Society Conference on Computer Vision and Pattern Recognition, vol. 2, pp. 1964–1971 (2006)

9. Lim, J.J., Zitnick, C.L., Dollár, P.: Sketch tokens: a learned mid-level representation for contour and object detection. In: 2013 IEEE Conference on Computer Vision and Pattern Recognition (CVPR), pp. 3158–3165 (2013)

10. Dollár, P., Zitnick, C.L.: Structured forests for fast edge detection. In: 2013 IEEE International Conference on Computer Vision (ICCV), pp. 1841–1848 (2013)

11. LeCun, Y., Boser, B., Denker, J.S., Henderson, D., Howard, R.E., Hubbard, W., Jackel, L.D.: Backpropagation applied to handwritten zip code recognition. Neural Comput. **1**, 541–551 (1989)

12. Xu, L., Lu, C., Xu, Y., Jia, J.: Image smoothing via l 0 gradient minimization. ACM Trans. Graph. (TOG) **30**, 174 (2011)

13. Berkeley Segmentation Data Set and Benchmarks 500 (bsds500). (http://aiweb. techfak.uni-bielefeld.de/content/bworld-robot-control-software/)

14. Krizhevsky, A., Sutskever, I., Hinton, G.E.: Imagenet classification with deep convolutional neural networks. In: Advances in Neural Information Processing Systems, pp. 1097–1105 (2012)

15. Xiaofeng, R., Bo, L.: Discriminatively trained sparse code gradients for contour detection. In: Advances in Neural Information Processing Systems, pp. 584–592 (2012)

16. Li, H., Zhao, R., Wang, X.: Highly Efficient Forward and Backward Propagation of Convolutional Neural Networks for Pixelwise Classification (2014). arXiv preprint arXiv:1412.4526

17. Ouyang, W., Wang, X.: Joint deep learning for pedestrian detection. In: 2013 IEEE International Conference on Computer Vision (ICCV), pp. 2056–2063 (2013)

Spectral-spatial Classification of Hyperspectral Image Based on Locality Preserving Discriminant Analysis

Min Han[1(✉)], Chengkun Zhang[1], and Jun Wang[2]

[1] Faculty of Electronic Information and Electrical Engineering,
Dalian University of Technology, Dalian 116023, China
minhan@dlut.edu.cn
[2] Department of Computer Science, City University of Hong Kong,
Kowloon Tong, Hong Kong
jwang.cs@cityu.edu.hk

Abstract. In this paper, a spectral-spatial classification method for hyperspectral image based on spatial filtering and feature extraction is proposed. To extract the spatial information that contain spatially homogeneous property and distinct boundary, the original hyperspectral image is processed by an improved bilateral filter firstly. And then the proposed feature extraction algorithm called locality preserving discriminant analysis, which can explore the manifold structure and intrinsic characteristics of the hyperspectral dataset, is used to reduce the dimensionality of both the spectral and spatial features. Finally, a support vector machine with a composite kernel is used to examine the performance of the proposed methods. Experiments results on a hyperspectral dataset demonstrate the effectiveness of the proposed algorithm in the classification tasks.

Keywords: Hyperspectral · Spatial filtering · Feature extraction · Manifold structure · Support vector machine with a composite kernel

1 Introduction

Hyperspectral image provides hundreds of narrow and contiguous spectral bands spanning the visible to infrared spectrum. The wealth of spectral and spatial information of hyperspectral image promotes the development of many application fields, such as environment monitoring, target detection and mineral prospection [1]. Image classification, where pixels are assigned to certain classes to represent different kinds of ground truth, is a very important procedure in the analysis of hyperspectral image. However, the high dimensional bands make Hughes phenomenon easily happen during classification, which reduces the accuracy of recognition [2]. As a result, dimensionality reduction plays a vital important part in hyperspectral image classification, especially when the number of available labeled training samples is limited [3–5].

Generally, dimensionality reduction consists of two kinds of categories, feature extraction and band selection. In this work, only transform-based feature extraction is taken into discussion. Principal component analysis (PCA) and linear discriminant

© Springer International Publishing Switzerland 2016
L. Cheng et al. (Eds.): ISNN 2016, LNCS 9719, pp. 21–29, 2016.
DOI: 10.1007/978-3-319-40663-3_3

analysis (LDA) seem to be the most typical feature extraction algorithms [6]. However, recent research has shown that the hyperspectral dataset is likely to lie in a nonlinear submanifold, which leads to linear algorithms inefficiency [7–9]. To deal with the problem, manifold learning-based feature extraction methods have been developed, which can explicitly discover the nonlinear manifold structure concealed in original high dimensional dataset, such as locality preserving projections (LPP) and neighborhood preserving embedding (NPE) [10, 11]. Another manifold-based algorithm, Local fisher discriminant analysis (LFDA), inherits the discrimination of LDA and preserves the local structure of submanifolds, can discover the intrinsic characteristics of high-dimensional data, and perform well in classification tasks [12].

In order to explore the intrinsic characteristics of the hyperspectral dataset and enhance the discriminative ability of the subspace, a new feature extraction algorithm called locality preserving discriminant analysis (LPDA) is proposed. Two statistics are defined in this work. The within-class similarity scatter is defined to describe the similarity of the samples in the same class. The between-class diversity scatter pays attention to the diversity of the samples in different classes. Then LPDA finds an optimal subspace by maximizing the between-class diversity scatter and minimizing the within-class similarity scatter, in which the local structure of the similar samples are preserved and the margins between different classes are increased as well.

The participation of spatial information is essential to improve the classification accuracy of hyperspectral image [13]. So the performance of proposed feature extraction algorithm is evaluated by a spectral-spatial classification framework. Firstly, a bilateral filter is used to denoise each gray color image of the hyperspectral dataset. The achieved dataset can be regarded as spatial features with respect to the original spectral features. Secondly, feature extraction is operated on both spectral and spatial features respectively. Finally, we use a support vector machine (SVM) with a composite spectral-spatial kernel to evaluate the performance of the proposed LPDA [14].

The remainder of this paper is organized as follows: The proposed classification framework and proposed LPDA are introduced in Sect. 2. The experimental results are presented in Sect. 3. Finally, some remarks are given in Sect. 4.

2   Methodology

In this section, the modified spatial information extraction method and the solutions of the proposed algorithm, LPDA, will be discussed in detail. A spectral-spatial classification framework, seen in Fig. 1, is adopted.

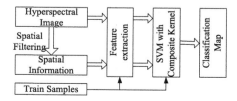

Fig. 1. Classification framework used in this research.

2.1   Spatial Information Extraction

Spatial filtering is an effective approach to extract spatial information. Bilateral filter takes spatial-domain and radiant-domain filtering into consideration, and can achieve smooth neighborhoods and distinct edges [15]. It can be represented mathematically:

$$h(z) = 1/k(z) \int_{\Omega} f(\omega) \cdot c(\omega, z) \cdot \varphi(f(\omega), f(z)) d\omega$$
$$k(z) = \int_{\Omega} c(\omega, z) \varphi(f(\omega), f(z)) d\omega$$

(1)

where z is a central pixel surrounded by neighboring pixels ω in a spatial domain Ω, $k(z)$ is the normalization factor. $f(z)$ denotes the radiant intensity of pixel z in input gray image, $h(z)$ is the corresponding output of bilaterally filtered.

The functions $c(\omega, z)$ and $\varphi(f(\omega), f(z))$ measure the spatial closeness and radiant similarity of neighborhood center z and a surrounding pixel ω respectively. Generally, $c(\omega, z) = \exp(-\|z - \omega\|^2 / 2\sigma_s)$ and $\varphi(f(\omega), f(z)) = \exp(-\|f(\omega) - f(z)\|^2 / 2\sigma_w)$, σ_s controls the spatial closeness, and σ_r defines how much the weight of a neighboring pixel decreases because of the radiant intensity difference.

Above all, a bilateral filter can realize edges-preserving and neighbors-smoothing, which is of great importance in remote sensing image classification. However, the hyperspectral image is in a data cube structure. The spectral vectors are the very features to distinguish different materials. The bilateral filter is designed for 2-D space, and suffers from limitation in hyper spectral denoise. Supposed $F(z)$ denotes the spectral feature of pixel z, it is obvious that the following function $\Phi(F(\omega), F(z))$ can measure the spectral similarity of pixels ω and z.

$$\Phi(F(\omega), F(z)) = \frac{\langle F(\omega), F(z) \rangle}{\|F(\omega)\| \cdot \|F(z)\|}$$

(2)

Then the modified bilateral filter can be summarized as:

$$h(z) = 1/k'(z) \int_{\Omega} f(\omega) c(\omega, z) \varphi(f(\omega), f(z)) \Phi(F(\omega), F(z)) d\omega$$
$$k'(z) = \int_{\Omega} c(\omega, z) \varphi(f(\omega), f(z)) \Phi(F(\omega), F(z)) d\omega$$

(3)

Taking advantage of the modified bilateral filter, we acquire spatial features which show smoother neighbors and sharper edges.

2.2   Locality Preserving Discriminant Analysis

Given a hyperspectral dataset $X = [x_1, x_2, \ldots, x_n] \in \mathbb{R}^{d \times n}$ with n pixels and c classes, $x_i \in \mathbb{R}^d (i = 1, 2, \ldots, n)$ is the vector pattern of the ith pixel. The purpose of LPDA is to find a transfer matrix $T \in \mathbb{R}^{d \times r}$ to mapping the dataset X into a low-dimensional but class-discriminative subspace $\mathbb{R}^r (r < d)$, in which the local structure of samples in the same class is preserved and the margin between different classes is enlarged. To achieve this, LPDA define within-class similarity scatter and between-class diversity scatter to describe the local structure of samples in the same class and the margin between different classes. By maximizing the between-class diversity scatter and minimizing the within-class similarity scatter, the desired subspace can be acquired.

Within-Class Similarity Scatter. Suppose $Z = [z_1, z_2, \ldots, z_n] \in \mathbb{R}^{r \times n}$ is the transformed dataset in a subspace with respect to the given hyperspectral dataset X by transformation $z_i = T^T x_i (i = 1, 2, \ldots, n)$. The vector $y = [y_1, y_2, \ldots, y_n]$ ($y_1 \in \{1, 2, \ldots, c\}$) is the corresponding class labels, $z_i \in \mathbb{R}^r$ is the r-dimensional mapping feature of the ith pixel. It is obviously that to preserve the local structure of the original dataset in a low dimensional subspace is beneficial to classification. The promoting feature extraction method LPP is designed to satisfy this purpose [10]. However, when it comes to classification tasks, LPP will be confronted with many limitations for its strictly unsupervised nature. The reasonable utilization of supervised information is beneficial to keep the local structure of the same class and make similar samples more compact. Construct cost function as follows:

$$
\begin{aligned}
J_w &= \min_T \frac{1}{2} \sum_{i=1}^n \sum_{j=1}^n \left\| z_i - z_j \right\|^2 W_{ij}^w \\
W_{i,j}^w &= \begin{cases} \exp\left(-\left\| x_i - x_j \right\|^2 / t\right) & \text{if } y_i = y_j \text{ and } x_i \in N_{k_1}(x_j) \cup x_j \in N_{k_1}(x_i) \\ 0 & \text{if } y_i \neq y_j \end{cases}
\end{aligned}
\tag{4}
$$

Where W^w is asymmetrical affinity matrix describing the neighborhood similarity of samples in the same class. $N_{k_1}(x_i)$ denotes the set made up by k_1 nearest neighbors of pixel x_i with same labels. t is decided by local scaling heuristic to make it adaptive [12]. $t = \sigma_i \sigma_j$, $\sigma_i = \left\| x_i - x_i^{knn} \right\|$, and x_i^{knn} is the kth nearest neighbors of pixel x_i. The cost function in Eq. (4) can be reformulated by $z_i = T^T x_i$:

$$
J_w = \min_T \text{tr} \left\{ T^T X (D^w - W^w) X^T T \right\} = \min_T \text{tr} \left(T^T X L^w X^T T \right)
\tag{5}
$$

where $D_{ii}^w = \sum_{j=1}^n W_{ij}^w$, $L^w = D^w - W^w$. Define $S^w = tr\left(T^T X L^w X^T T \right)$ as the within-class similar scatter to measure the similarity between the samples in the same class.

Between-Class Diversity Scatter. In terms of classification, to enlarge the separability between different classes is an effective approach to increase classification accuracy.

The between-class diversity scatter S^b is defined to measure the margin of different classes, shown as below:

$$S^b = \frac{1}{2} \sum_{i=1}^{n} \sum_{j=1}^{n} \left\| z_i - z_j \right\|^2 W_{ij}^b$$

$$W_{ij}^b = \begin{cases} 1 & if \left(x_i, x_j \right) \in P_{k_2}(y_i) \cup \left(x_i, x_j \right) \in P_{k_2}(y_j) \\ 0 & otherwise \end{cases} \tag{6}$$

where W^b is asymmetrical affinity matrix, $P_{k_2}(y)$ is a data pairs set with k_2 nearest pairs among the set $\left\{ (x_i, x_j), x_i \in y, x_j \notin y \right\}$. Then formulate Eq. (6) by $z_i = T^T x_i$:

$$S^b = \mathrm{tr}\left\{ T^T X (D^b - W^b) X^T T \right\} = \mathrm{tr}\left(T^T X L^b X^T T \right) \tag{7}$$

where $D_{ii}^b = \sum_{j=1}^{n} W_{ij}^b$, $L^b = D^b - W^b$. It is easily seen that, by maximizing S^b, the separability between different classes can be enlarged as far as possible in the mapping subspace, which improve the performance of dimensionality reduction.

Feature Extraction Criterion. From above all, minimizing the within-class similarity scatter and maximizing the between-class diversity scatter can preserve the local structure of samples in the same class, enlarge the separability of different classes as well, which is an important indicator in classification tasks. The LPDA transformation matrix $T \in \mathbb{R}^{d \times r}$ is defined as:

$$T = \arg\max_{T} \frac{\mathrm{tr}\left(T^T X L^b X^T T \right)}{\mathrm{tr}\left(T^T X L^w X^T T \right)} \tag{8}$$

in which d is the dimension of the hyper spectral dataset, r is the dimension of the desired subspace. Then, the optimal solutions of T are the eigenvectors corresponding to the r maximum eigenvalues of the following generalized eigenvalue problem:

$$X L^w X^T T = \lambda X L^b X^T T \tag{9}$$

It is easy to show that the matrices L^w and L^b are symmetric and positive semidefinite. We sort the eigenvalues of Eq. (9) in descending order and align the first r corresponding eigenvectors $w_0, w_1, \ldots, w_{r-1}$. $T = [w_0, w_1, \ldots, w_{r-1}]$ is the wanted matrix.

3   Experiment Results and Discussion

The Indian Pines hyper spectral dataset is used to test the performance of the proposed classification framework, as well as the manifold-based feature extraction method, LPDA. The lib SVM toolkit is used to build the SVM classifier with composite kernel (SVMCK) [14]. Parameters in SVMCK are determined based on cross validation.

The Indian Pines image has 220 bands and a spatial coverage of 145 × 145 pixels with a spatial resolution of 20 m per pixel. 20 water absorption bands are removed. The image contains 16 classes, Fig. 3(a) shows the color composite image and Fig. 3(b) shows the real materials. 10 % of all labeled samples are selected as training and the remaining as testing, seen in Fig. 3(c) and Table 1. To begin with, we calculate the spatial feature vectors x_i^w corresponding to hyper spectral pixel x_i^s by modified bilateral filter. The spatial domain is selected as 9 × 9 window. The experiments consist of three conditions, performing SVM on the original spectral feature x_i^s, performing SVMCK on the original spectral feature x_i^s and spatial feature x_i^w, performing SVMCK on dimensionality-reduced spectral and spatial features. SVM and SVMCK get accuracies of 82.95 % and 93.60 %, illustrating the effectiveness of spatial features. Here five kinds of dimensionality reduction algorithms are evaluated, including PCA, NWFE [16], LFDA [12], NPE [11], and the proposed LPDA.

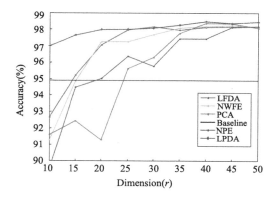

Fig. 2. The performance comparison of different methods. (Color figure online)

Table 1. Classification accuracy for the Indian pines image on the test set ($r = 20$).

No.	Train	Test	SVM	SVMCK	PCA	NWFE	LFDA	NPE	LPDA
1	6	48	77.27	**100.00**	79.54	93.18	**100.00**	97.73	**100.00**
2	144	1290	82.10	92.56	91.24	**96.97**	96.51	90.32	95.89
3	84	750	69.10	94.94	76.16	97.60	**98.80**	97.07	**98.80**
4	24	210	74.88	85.78	69.19	96.68	97.63	95.73	**98.58**
5	50	447	88.14	93.74	96.42	98.88	96.19	**99.11**	97.09
6	75	672	90.02	98.07	99.70	99.70	99.85	99.85	**100.00**
7	3	23	**100.00**	93.75	**100.00**	**100.00**	93.75	**100.00**	93.75
8	49	440	92.72	97.73	**100.00**	**100.00**	**100.00**	**100.00**	**100.00**
9	2	18	90.00	**100.00**	0.00	**100.00**	60.00	**100.00**	**100.00**
10	97	871	74.74	87.49	69.69	92.76	90.93	84.27	**96.33**
11	247	2221	84.96	94.60	96.35	96.21	97.79	94.73	**98.78**

(Continued)

Table 1. (*Continued*)

No.	Train	Test	SVM	SVMCK	PCA	NWFE	LFDA	NPE	LPDA
12	62	552	79.02	88.79	71.24	**98.19**	91.13	93.13	97.65
13	22	190	94.24	98.95	98.95	99.47	98.95	**100.00**	99.48
14	130	1164	95.45	97.08	99.82	98.88	99.65	**99.83**	99.14
15	38	342	52.92	84.80	88.01	97.36	97.66	**99.54**	96.78
16	10	85	84.70	**98.82**	**98.82**	96.47	**98.82**	**98.82**	**98.82**
OA(%)	1043	9323	82.95	93.60	89.86	97.24	97.06	95.04	**98.13**
AA(%)			83.14	94.19	83.45	97.65	94.86	96.82	**98.19**
κ			0.805	0.927	0.883	0.968	0.966	0.943	**0.979**

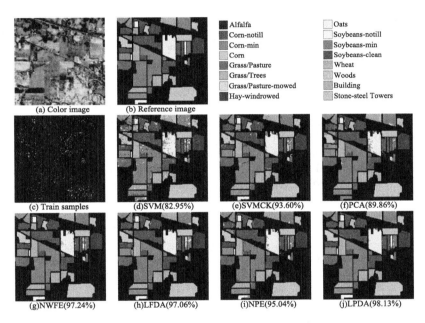

Fig. 3. Classification maps of Indian Pines image.

The effect of dimensionality, reduced by different algorithms, on the overall accuracy is shown in Fig. 2, in which the dimensionality of spectral and spatial features are reduced to r. SVMCK performing on the original dataset is used as a baseline. The results show that with increase of dimensionality, the accuracies of different algorithms show a growing trend. When the dimensionality satisfies $r \geq 25$, all experimented feature extraction algorithms have better accuracies than the baseline, proving the effectiveness of dimensionality reduction. When $d = 15$, LPDA gives an accuracy of 97.77 %, while the accuracies of the other algorithms are lower than the baseline.

From Fig. 2, $r = 20$ is chose for further discussed, which well represents the discriminative performance of each feature extraction algorithms. The results are tabulated in Table 1, including individual accuracy of each class, overall accuracy (OA), average accuracy (AA), kappa coefficient (κ). After dimensionality reduced by NWFE, LFDA, NPE and LPDA the classification results are better than the baseline. LPDA ($k_1 = 6$, $k_2 = 4500$) produces the best results than other algorithms with overall accuracy of 98.13 % and kappa coefficient of 0.979. In terms of the best individual accuracies, 9 of all 16 classes are acquired by LPDA. All of these denote that LPDA is significantly superior to other feature extraction algorithms. The classification maps obtained from different algorithms are shown in Fig. 3. It is obviously that by incorporating spatial features obtained from the modified bilateral filter, LPDA provides much smoother neighborhoods and distinct edges. It is an almost perfect classification map.

4 Conclusions

In this paper, a framework is proposed to improve the classification accuracy of hyper spectral image, including spatial information extraction by modified bilateral filter, dimensionality reduction by the proposed feature extraction algorithm LPDA, image classification by SVMCK. The experiment results have proven that the incorporation of spatial information can significantly improve the smoothness of the classification map. The experimented NWFE, LFDA, NPE and LPDA all produce better results than that without this preprocessing step. In particular, the proposed LPDA not only preserves the geometric structures of the original data, but also enlarges the separability of different classes. So LPDA is deserved to find the intrinsic characteristics of the hyper spectral data. For manifold learning-based algorithms, the graph structures mainly consist of k-nearest neighbor. However they are both sensitive to the data distribution and difficult to determine the parameters in complex structure dataset. So finding a new adaptive approach for nearest neighbor search will be the future work.

Acknowledgments. This work was supported by the National Natural Science Foundation of China (No. 61374154) and Special Fund for Basic Research on Scientific Instruments of the National Natural Science Foundation of China (No. 51327004).

References

1. He, K.S., Rocchini, D., Neteler, M., Nagendra, H.: Benefits of hyperspectral remote sensing for tracking plant invasions. Divers. Distrib. **17**(3), 381–392 (2011)
2. Pestov, V.: Is the k-NN classifier in high dimensions affected by the curse of dimensionality? Comput. Math Appl. **65**(10), 1427–1437 (2013)
3. Jia, X., Kuo, B.C., Crawford, M.M.: Feature mining for hyperspectral image classification. Proc. IEEE **101**(3), 676–697 (2013)
4. Feng, Z., Yang, S., Wang, S., Jiao, L.: Discriminative spectral-spatial margin-based semisupervised dimensionality reduction of hyperspectral data. IEEE Geosci. Remote Sens. Lett. **12**(2), 224–228 (2015)
5. Wu, S., Sun, M., Yang, J.: Stochastic neighbor projection on manifold for feature extraction. Neurocomputing **74**(17), 2780–2789 (2011)

6. Yan, S., Xu, D., Zhang, B., Zhang, H.J., Yang, Q., Lin, S.: Graph embedding and extensions: a general framework for dimensionality reduction. IEEE Trans. Pattern Anal. Mach. Intell. **29**(1), 40–51 (2007)
7. Bachmann, C.M., Ainsworth, T.L., Fusina, R.A.: Exploiting manifold geometry in hyperspectral imagery. IEEE Trans. Geosci. Remote Sens. **43**(3), 441–454 (2005)
8. Lunga, D., Prasad, S., Crawford, M.M., Ersoy, O.: Manifold-learning-based feature extraction for classification of hyperspectral data: a review of advances in manifold learning. IEEE Signal Process. Mag. **31**(1), 55–66 (2014)
9. Cui, Y., Fan, L.: A novel supervised dimensionality reduction algorithm: graph-based fisher analysis. Pattern Recogn. **45**(4), 1471–1481 (2012)
10. Niyogi, X.: Locality preserving projections. In: Neural Information Processing Systems, vol. 16, p. 153 (2004)
11. He, X., Cai, D., Yan, S., Zhang, H.J.: Neighborhood preserving embedding. In: IEEE International Conference on Computer Vision, pp. 1208–1213 (2005)
12. Sugiyama, M.: Dimensionality reduction of multimodal labeled data by local fisher discriminant analysis. J. Mach. Learn. Res. **8**, 1027–1061 (2007)
13. Fauvel, M., Tarabalka, Y., Benediktsson, J.A., Chanussot, J., Tilton, J.C.: Advances in spectral-spatial classification of hyperspectral images. Proc. IEEE **101**(3), 652–675 (2013)
14. Camps-Valls, G., Gomez-Chova, L., Muñoz-Marí, J., Vila-Francés, J., Calpe-Maravilla, J.: Composite kernels for hyperspectral image classification. IEEE Geosci. Remote Sens. Lett. **3**(1), 93–97 (2006)
15. Kang, X., Li, S., Benediktsson, J.A.: Spectral-spatial hyperspectral image classification with edge-preserving filtering. IEEE Trans. Geosci. Remote Sens. **52**(5), 2666–2677 (2014)
16. Kuo, B.C., Landgrebe, D.A.: Nonparametric weighted feature extraction for classification. IEEE Trans. Geosci. Remote Sens. **42**(5), 1096–1105 (2004)

Individual Independent Component Analysis on EEG: Event-Related Responses Vs. Difference Wave of Deviant and Standard Responses

Tiantian Yang[1,2], Fengyu Cong[1,2(✉)], Zheng Chang[2], Youyi Liu[3], Tapani Ristainiemi[2], and Hong Li[4]

[1] Department of Biomedical Engineering, Faculty of Electronic, Information and Electrical Engineering, Dalian University of Technology, Dalian, China
tiantian.yang1126@foxmail.com, cong@dlut.edu.cn
[2] Department of Mathematical Information Technology, University of Jyvaskyla, Jyvaskyla, Finland
{zheng.chang,tapani.ristaniemi}@jyu.fi
[3] State Key Laboratory of Cognitive Neuroscience and Learning, Beijing Normal University, Beijing, China
youyiliu@bnu.edu.cn
[4] College of Psychology and Sociology, Shenzhen University, Shenzhen, China
lihongszu@szu.edu.cn

Abstract. Independent component analysis (ICA) is often used to spatially filter event-related potentials (ERPs). When an oddball paradigm is applied to elicit ERPs, difference wave (DW, responses of deviant stimuli minus those of standard ones) is often used to remove the common responses between the deviant and the standard. Thus, DW can be produced first, and then ICA is used to decompose the DW. Or, ICA is performed on responses of the deviant and standard stimuli separately, and then DW is applied on the filtered responses. In this study, we compared the two approaches to analyzing mismatch negativity (MMN). We found that DW introduced noise in the time and space domains, resulting in more difficulty to obtain the spatial properties of MMN by ICA on DW. Thus, we suggest using ICA to spatially filter event-related responses of each stimulus; and then DW is produced by the filtered responses.

Keywords: Independent component analysis (ICA) · Event-related potential (ERP) · Difference wave (DW) · Deviant · Standard · Oddball

1 Introduction

Event-related potentials (ERPs) have been extensively used for studying cognitive functions [1]. Conventional data processing to obtain ERPs usually consists of filtering continuous EEG, segmentation/epoch, artifact detection and rejection, and averaging over trials [1]. It is assumed that data from each trial include the constant part of brain

© Springer International Publishing Switzerland 2016
L. Cheng et al. (Eds.): ISNN 2016, LNCS 9719, pp. 30–39, 2016.
DOI: 10.1007/978-3-319-40663-3_4

activity related to the experimental design, the randomly fluctuating part, and artifacts as well as noise [2]. By averaging over large numbers of single trials, the constant part is therefore enhanced and the randomly fluctuating part is attenuated. In some cases, artifacts cannot be completely rejected in the preprocessing, the randomly fluctuating part cannot be sufficiently attenuated by the averaging, and the constant part can include overlapped activity from several brain areas. This hinders the estimation of ERP peak parameters, particularly, the peak latency, since the peak of an ERP component cannot be well-shaped when interference and noise are present.

Therefore, advanced signal processing methods are further performed on the averaged EEG in order to remove additional artifacts and extract components of interest simultaneously, including the one-way analysis like the wavelet filtering [3], the two-way analysis, such as principal component analysis (PCA) [4] and independent component analysis (ICA) [5], and tensor decomposition for the multi-way data (called as tensor) analysis [6].

The oddball paradigm is often applied to elicit ERPs, particularly for mismatch negativity (MMN) and P300 [1]. For example, Fig. 1 shows the illustration of a simple oddball paradigm. Difference wave (DW, responses of deviant stimuli minus those of standard ones) is usually used to remove the common responses between the deviant and standard stimuli. Therefore, in this paradigm, ICA can be applied on the DW to extract the components of interest. Or, ICA can be performed to spatially filter the responses of the deviant and the standard stimuli, and then, DW is produced in terms of the spatially filtered responses of the deviant and the standard stimuli. Without comparison, it is difficult to determine which is superior given a participant's ERP dataset in the oddball paradigm. In this study, we propose the methodology to compare the two approaches in terms of the stability of ICA decomposition and the temporal and spatial properties of the ERP of interest.

S S S **D** S S S S S S **D** S S S S **D** S S S

Fig. 1. Illustration of a simple oddball paradigm. 'S' denotes the standard stimulus, and 'D' is for the deviant stimulus. Usually, there are at least two different deviants. (Color figure online)

2 Method

2.1 Data Description

The ERP data of an 8-year-old child were used to examine the proposed approach in this study. The one-standard and two-deviant passive oddball paradigm in the ratio 80/10/10 was used to elicit MMN. The standard stimulus was /ba/, and the two deviants were /ga/ and /da/. The 128-channel EGI equipment was used for EEG data collection. After the conventional preprocessing processing and averaging over about 100 single trials, the responses to the deviant and the standard stimuli were obtained.

The epoch started 100 ms before stimulus onset and stopped at 600 ms after the stimulus onset, and the sampling rate was 500 Hz. Thus, the averaged EEG trace of each channel included 350 temporal samples for one stimulus.

2.2   ICA Procedure to Spatially Filter ERP to Extract Components of Interest

ICA is based on the linear transformation model associating the EEG/ERP field data (x) along the scalp and the EEG/ERP sources (**s**). The model without noise reads as

$$x = \mathcal{A}s \tag{1}$$

where $x = [x_1, x_2, \cdots, x_I]^T$, $\mathbf{s} = [s_1, s_2, \ldots, s_J]^T$, $I > J$, I is the number of electrodes in the high-dense array (for example, 128 channels in this study), J is the number of electrical sources, and $\mathcal{A} \in \Re^{I \times J}$ with the full column rank is usually called as the mixing matrix in ICA. We name the mixing matrix as the mapping matrix containing coefficients to map the EEG sources to electrode locations along the scalp. For any source, its mapping can be illustrated as

$$x_r = a_r \cdot s_r, \tag{2}$$

where, $x_r = [x_{1,r}, x_{2,r}, \ldots, x_{I,r}]^T$, a_r is one column of $\mathcal{A}$, and $r \in [1, J]$. In this case, x_r is not the mixture like x in (1) any more, but is the sole information of one EEG source. Hence, one goal to apply ICA is to achieve the mapping of one source in (2) from the mixture in (1).

In this study, we assume the model (1) is over-determined since the number of sensors/electrodes is larger than the number of sources. By PCA and model order selection [7], the over-determined model is converted to the determined model by

$$\mathbf{x} = \mathbf{V}^T x = \mathbf{V}^T \mathcal{A}s = \mathbf{A}s \tag{3}$$

where, $\mathbf{A} = \mathbf{V}^T \mathcal{A}$, $\mathbf{A} \in \Re^{J \times J}$, $\mathbf{V} \in \Re^{I \times J}$ represents the first J eigenvectors of the covariance matrix of x whose eigenvalues are ranked decreasingly, and $\mathbf{x}$ is indeed of the leading components for further ICA decomposition. J is estimated by the model order selection method from the I eigenvalues [7]. Then, an unmixing matrix is learned by ICA [8], and then it transforms the mixture in (3) into independent components as

$$\mathbf{y} = \mathbf{W}\mathbf{x} \tag{4}$$

where $\mathbf{W} \in \Re^{J \times J}$, $\mathbf{y} = [y_1, y_2, \cdots, y_J]^T$, and $\mathbf{C} = \mathbf{W}\mathbf{A}$. Subsequently, any component of interest can be projected back to the electrode field with the expectation to correct the polarity and variance indeterminacy of the extracted component by

$$\mathbf{e}_k = \mathbf{V} \cdot \mathbf{b}_k \cdot y_k \tag{5}$$

where $\mathbf{e}_k = [e_{1,k}, e_{2,k}, \cdots, e_{I,k}]^T$, $\mathbf{b}_k$ is one column of $\mathbf{B}(= \mathbf{W}^{-1})$, y_k is one element of $\mathbf{y}$ and $k \in [1, J]$. Consequently, it is expected to obtain the determined magnitude of the electrical brain activity of interest in microvolt units [9]. If ICA decomposition is satisfactory in practice, y_k is dominated by one source, for example, $\mathbf{e}_k$ can be very similar to the mapping of the source, i.e., x_r in Eq. (2) [10]. In theory, it reads as the following under the global optimization [10]:

$$\mathbf{e}_k = \mathbf{V} \cdot \mathbf{b}_k \cdot c_{k,r} \cdot s_r \overset{\mathbf{BC=A}}{=} a_r \cdot s_r = x_r \tag{6}$$

Therefore, from Eqs. (1)–(6), the mixture in the electrode field is spatially filtered to approximate the mapping of individual sources in the electrode field in practice.

2.3   Proposed Data Processing Procedure on the Averaged EEG

Step 1: An appropriated wavelet filter was designed in terms of the properties of the MMN in this study [3]. The wavelet filter used the reversal biorthogonal wavelet with the order of 6.8. The ERP data were decomposed into nine levels with the coefficients from the sixth level to eighth level used to reconstruct the ERP of interest.

Step 2: PCA was applied on the wavelet-filtered 128-channel ERP data for the model order selection to determine the number of components to extract for each ERP dataset among the DW of deviant /da/, the DW of deviant /Ga/, responses of deviant /da/, responses of deviant /ga/, and responses of standard /ba/.

Step 3: ICA was applied on the DW of deviant /da/, the DW of deviant /Ga/, the responses of deviant /da/, the responses of deviant /ga/, and the responses of standard /ba/, separately. In details, Fast ICA [8] was performed on the selected PCA components for each of the five applications of ICA. ICASSO software was used to validate the robustness and stability of the extracted component by ICA [11]. One of schemes of ICASSO [11] includes: (1) initializing the unmixing matrix many times; (2) running one ICA algorithm with every initialized unmixing matrix; and (3) clustering all extracted components to seek common components. The default parameters for ICASSO were used here, and Fast ICA was run 100 times.

Step 4: In order to select the component of interest, the peak latency of each temporal component and the associated topography of the component were examined. When the temporal and the spatial properties of an ERP of interest were found, the temporal component and the associated topography were chosen to be projected back to the electrode for correcting their indeterminacy in the polarity and variance.

3   Result

In this section, we compare the results of ICA on DW of deviant /da/, ICA on DW of deviant /Ga/, ICA on responses of deviant /da/, ICA on responses of deviant /ga/, and ICA on responses of standard /ba/. Here, the 'responses' mean the averaged EEG data.

3.1 Waveform and Topography by Conventional Data Processing and Averaging

Figure 2a shows the waveforms of two deviant stimuli and one standard stimulus and their DW at the two representative electrode sites. Figure 2b shows the topography of the ERP components of interest in terms of the mean amplitude value between 230 ms and 330 ms. Obviously, the DW was much noisier in both the time and the spatial domain than the responses of two deviant stimuli and the standard stimulus.

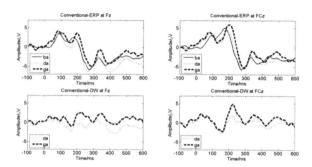

Fig. 2a. Waveforms of two deviant stimuli and the standard stimulus and DW.

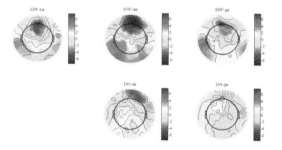

Fig. 2b. Topographies of two deviant stimuli and the standard stimulus and their DW in terms of the mean amplitude value between 230 ms and 330 ms.

3.2 Model Order Selection to Determine the Number of Extracted Components

Figure 3 shows the logarithms of the eigenvalues of each condition and the explained variance of accumulated eigenvalues. When 15 leading PCA components were selected, over 99 % of variance was explained for any condition. In terms of the gap method for model order selection [7], 21 components were suggested for the number of components to extract for each condition. With the trade-off among the explained variance, the estimated model order and the number of temporal samples within one epoch for ICA composition [9], we decided to extract 15 independent components from the one-epoch-long averaged EEG for each of five the applications.

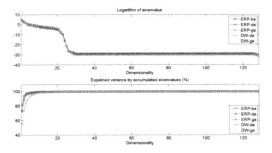

Fig. 3. Logarithms of eigenvalues and explained variance of accumulated eigenvalues.

3.3 Stability of ICA on DW and ICA on Responses

Using ICASSO, we investigated the stability of ICA decomposition. Here, at each time to run ICA, 15 components were extracted. Since ICA was run 100 times, 1500 components were produced for each application. Then, the 1500 components were clustered into 15 clusters. We found that no matter which application was (ICA on DW of any deviant or ICA on responses of any stimulus), the decomposition was stable. Figure 4 presents the clustered results for evaluating the stability of ICA on DW of deviant /ga/ and ICA on responses of deviant /ga/, showing no clear difference between the two applications regarding decomposition stability by visual inspection. Here, only the ICA on DW of deviant /ga/ and ICA on responses of deviant /ga/ among the five applications of ICA are shown due to the limitation of the paper lenth.

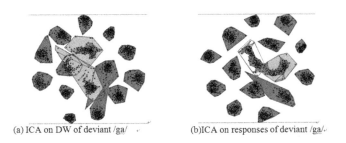

(a) ICA on DW of deviant /ga/   (b)ICA on responses of deviant /ga/

Fig. 4. Stability of ICA decomposition in terms of clustering 1500 components into 15 clusters (15 components in each run for ICA on 128-channel EEG, and 100 runs).

3.4 Temporal and Spatial Components and ICA on DW and ICA on Responses

Figure 5 shows the temporal components and the associated spatial maps for ICA on DW of deviant /ga/. Although the two temporal components (# 8&9) with the peak latencies within the interesting time window ranging from 230 ms to 330 ms read satisfactory, the associated spatial maps were scattered, indicating the unreasonable brain activities. This means that the extracted components were technical artifacts and

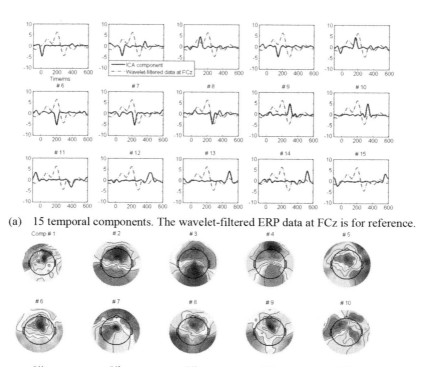

(a)   15 temporal components. The wavelet-filtered ERP data at FCz is for reference.

(b)   15 spatial components. Each is associated with the temporal one with the same
index

Fig. 5. ICA on DW of deviant /ga/. (Color figure online)

could not be accepted for the further analysis. For the ICA on DW of deviant /da/, the
similar temporal components with the scattered topographies were also extracted. Thus,
ICA on DW here was not reliable.

Figure 6 shows the temporal components and the associated spatial maps for ICA
on responses of deviant /ga/. Both the two temporal components (# 8&9) with the peak
latencies within the interesting time window and the associated spatial maps showing
dipolar patterns read satisfactory. This means that reasonable brain activities were
extracted out and acceptable for the further analysis. Then, they were selected and
projected back to the electrode field. The projected waveforms at FZ and FCz and the
topography are shown in Figs. 7a and 7b for ICA on the responses of deviant /ga/.

Regarding the applications of ICA on responses of deviant /da/ and standard /ba/
separately, the satisfactory temporal components of interest with the dipolar topographies were also extracted, and were selected to project back to the electrode field. The
extracted components are not shown due to the limited length of the paper.

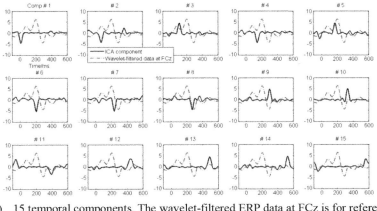

(a)   15 temporal components. The wavelet-filtered ERP data at FCz is for reference.

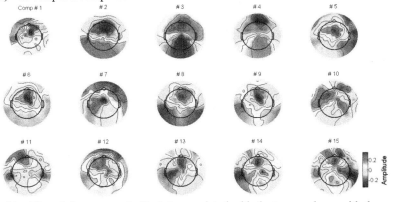

(b)   15 spatial components. Each is associated with the temporal one with the same index

Fig. 6. ICA on responses of deviant /ga/. (Color figure online)

3.5   Spatially Filtered Responses of Deviant and Standard Stimuli and Their DW

As mentioned previously, after ICA was performed on the responses of deviant /ga/ and /da/ and standard /ba/ separately, we all found the decompositions were reasonable and the reliable components were selected and projected back to the electrode field. Figures 7a and 7b shows the waveforms of the spatially filtered responses by ICA, their DW and the corresponding topographies. The dipolar topographies mean that the approach that ICA was applied on the responses of each stimulus and then the DW was produced was successful in this study.

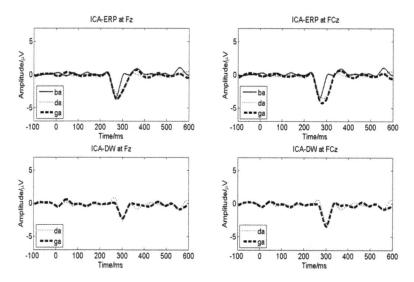

Fig. 7a. ICA on responses: waveforms of deviant stimuli and the standard stimulus and there DW of the spatially filtered responses.

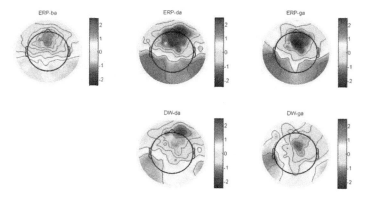

Fig. 7b. ICA on responses: topographies of two deviant stimuli and the standard stimulus and their DW in terms of the mean amplitude value between 230 ms and 330 ms.

4   Discussion

When ICA is used to decompose EEG/ERP data, unless the stability is stable and the temporal and spatial properties of the extracted components of interest match those of the underlying brain activity, the results are not accepted for the further analysis. In this study, for the ERP data of the child in the oddball paradigm including the deviant and the standard stimuli (see Fig. 1), we found that performing ICA to spatially filter the responses of the deviant and the standard separately and then producing the DW of the filtered responses by ICA was more useful to exact the MMN components than applying ICA on DW of responses of deviant and standard stimuli.

DW may introduce noise in the time and space domains, resulting in more difficulty for ICA to reveal ERP's spatial properties. Thus, we do not suggest performing ICA on DW directly when it is noisy. Demo of MATLAB codes and ERP data for this study are available via http://www.escience.cn/people/cong/index.html.

Acknowledgments. Authors thank the support from National Natural Science Foundation of China (Grant Nos. 81471742 & 81461130018). Cong thanks Dr. Nicole Landi in Haskins Laboratories in Yale University for providing their ERP data.

References

1. Luck, S.J.: An Introduction to the Event-Related Potential Technique. The MIT Press, Cambridge (2005)
2. Talsma, D.: Auto-adaptive averaging: detecting artifacts in event-related potential data using a fully automated procedure. Psychophysiology **45**(2), 216–228 (2008)
3. Cong, F., Leppanen, P.H., Astikainen, P., Hamalainen, J., Hietanen, J.K., Ristaniemi, T.: Dimension reduction: additional benefit of an optimal filter for independent component analysis to extract event-related potentials. J. Neurosci. Methods **201**(1), 269–280 (2011)
4. Dien, J.: Applying principal components analysis to event-related potentials: a tutorial. Dev. Neuropsychol. **37**(6), 497–517 (2012)
5. Kalyakin, I., Gonzalez, N., Karkkainen, T., Lyytinen, H.: Independent component analysis on the mismatch negativity in an uninterrupted sound paradigm. J. Neurosci. Methods **174**(2), 301–312 (2008)
6. Cong, F., Lin, Q.H., Kuang, L.D., Gong, X.F., Astikainen, P., Ristaniemi, T.: Tensor decomposition of EEG signals–a brief review. J. Neurosci. Methods **248**, 59–69 (2015)
7. Cong, F., Nandi, A.K., He, Z., Cichocki, A., Ristaniemi, T.: Fast and effective model order selection method to determine the number of sources in a linear transformation model. In: Proceeding of the 2012 European Signal Processing Conference (EUSIPCO-2012), pp. 1870–1874 (2012)
8. Hyvarinen, A.: Independent component analysis: recent advances. Proc. R. Soc. A Math. Phys. Eng. Sci. **371**, 1–19 (2013)
9. Makeig, S., Westerfield, M., Jung, T.P., Covington, J., Townsend, J., Sejnowski, T.J.: Functionally independent components of the late positive event-related potential during visual spatial attention. J. Neurosci. **19**(7), 2665–2680 (1999)
10. Cong, F., Kalyakin, I., Zheng, C., Ristaniemi, T.: Analysis on subtracting projection of extracted independent components from EEG recordings. Biomedizinische Technik/Biomed. Eng. **56**(4), 223–234 (2011)
11. Himberg, J., Hyvarinen, A., Esposito, F.: Validating the independent components of neuroimaging time series via clustering and visualization. NeuroImage **22**(3), 1214–1222 (2004)

Parallel Classification of Large Aerospace Images by the Multi-alternative Discrete Accumulation Method

Vladimir I. Vorobiev[✉], Elena L. Evnevich, and Dmitriy K. Levonevskiy

Institute for Informatics and Automation of Russian Academy of Science,
Saint Petersburg, Russia
{vvi,eva,dl}@iias.spb.su

Abstract. The paper deals with parallel large aerospace images processing. We considered a simple multi-alternative discrete accumulation method for reliable distinction of satellite imagery and implemented a parallel classification system to increase the algorithm efficiency. The process of development of the distinction algorithm and system architecture was described. The system prototype was successfully tested. The experiments allowed to draw conclusion about the system performance and to estimate the effect of using the parallel architecture. The considered approach could be used in complex neural networks processing.

Keywords: Parallel computing · Aerospace images · Satellite imagery · Neural networks · Image processing · Image classification · Geographic information systems · GIS

1 Introduction

Significant features of the contemporary stage of IT development include the increasingly rapid availability of the satellite data, wide possibilities of preliminary and on-topic data processing and creation of developed tools and appropriate web resources for advanced image analysis [1, 2]. Due to the increasing availability of large data amount the importance of knowledge extraction methods is growing too. Classification methods enable this process to be automated, and the classification systems may be used in cartography, forest fires detection, illegal deforesting and building prevention, flora structure analysis, oil spill detection [3]. Geographic information systems (Erdas Imagine, ArcGIS) on the basis of existing classification methods, principles and approaches (for example, kernel extreme learning machine approach [4], minimum description length principle [5]) successfully solve these and similar problems. However, these systems seldom support parallel classification algorithms, so the hardware resources are used inefficiently, and the scalability of methods is limited. This work is focused on creation of a system for the parallel classification of large data amount, and this implies development of a parallel classification algorithm using clusters and GRID systems possessing necessary hardware resources.

© Springer International Publishing Switzerland 2016
L. Cheng et al. (Eds.): ISNN 2016, LNCS 9719, pp. 40–48, 2016.
DOI: 10.1007/978-3-319-40663-3_5

2   Materials and Methods

The task is to design and develop a system for reliable automated distinction of objects on the large satellite images of the Earth surface on the basis of the multi-alternative discrete accumulation method [6, 7]. Classifying is the process of ranging image elements (pixels) into a finite number of classes according to the values of the element attributes. Comparison of classifying methods is shown in Table 1, where estimates vary between 5 (the highest rate) and 1 (the lowest one). We considered three groups of methods: unsupervised algorithms clustering data into different classes [8] (K-Means, ISODATA), supervised clustering algorithms that imply learning by examples [9] (in particular, flexible and fuzzy classifiers), and neural networks. Apparently there is no best classifying algorithm. Each algorithm possesses a certain set of features, and one should select the most relevant algorithm for a given problem. We have selected neural networks, taking into account higher classification accuracy of this approach. Learning can be performed using results of a supervised classifying process.

Table 1. Classifying approaches comparison

Approach	Accuracy	Sensitivity	Ease of knowledge application	Variability of the classes number	Mixed classes
Unsupervised classification	4	–	3	5	No
Supervised classification	3	5	5	3	Yes
Neural networks	5	3	2	1	Yes

The classification method [10] consists in overlapping etalon figures onto the analyzed ones, computing the product sum of corresponding brightness values and drawing a conclusion about the analyzed figure according to the values of the corresponding product sums. This implies preliminary computation of the virtual brightness value (which is a substitute for the black color brightness on a grayscale image) assuring faultless distinction of the samples. Thus, the optimization of the training set is carried out. The procedure described above is equivalent to counting the number of matching brightness of black while superimposing the etalon figures on the current one and assigning a number to it according to the majority of coincidences.

Here 7 images (channels) are used: RED + GREEN + BLUE (« gray » channel), separated RED, GREEN and BLUE channels and the half-sums of RED + GREEN, RED + BLUE and GREEN + BLUE. All channels are treated like a vector consisting of 7 numbers.

For each pixel there is defined the set of pixels falling into the pupil of the analyzing system focused on this pixel. Then the brightness vectors of this set are computed. The samples are superimposed on the selected set one by one, and the number of coincidences (votes) is computed for each sample. As soon as the coincidence numbers are known, the conclusion is that the image matches the class with the largest coincidence number. Then the analyzed pixel is colored by the conventional color for the selected class.

If the sum is zero, this pixel is treated as unclassified and colored white. The algorithm of producing the output image after the first voting may be represented with the diagram of an artificial inadaptive neuron (Fig. 1), where T is the number of pixels in the image, X_k is the set of the pixels falling into the pupil focused on the selected pixel, Y_i the set of pixels corresponding to the sample number i, $\sum$ means sum of the brightness coincidence numbers between X_k and Y_i, MAX_i is the number of the sample with the greatest number of coincidences in the sets X_k and Y_i.

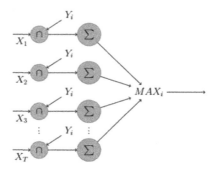

Fig. 1. Artificial inadaptive neuron

3   Implementation

Fast image classifying requires large computing resources and parallel data processing. The algorithm under consideration does not possess any computation branches or parts that can run in parallel. An obvious solution is to split the algorithm by the used data, i.e. different parts of the image are processed by different nodes.

Let us consider the parallel modification of the classification algorithm:

1. Load the image and the classification data.
2. Split the image into N parts, extend each part by the margins, each margin is of half-size of the pupil (Fig. 2).
3. Transmit the classification data to all processors.
4. Distribute the image parts among the processors.
5. Each processor classifies its image part.
6. Collect the classified parts.
7. Assemble the image throwing off the margins of each part.
8. Save the classified image.

The classifying system is divided into two subsystems (Fig. 3):

1. Control subsystem is constituted of the single process responsible for loading images and classification data, splitting images into parts, distributing the tasks, synchronizing, assembling and saving images. This subsystem is responsible for the general application logic, it should be aware of various image formats, color encoding modes, number of classifying processes, etc.

2. Classifying subsystem consists of multiple processes responsible for receiving images parts and classification data, classifying of the parts and sending the results back.

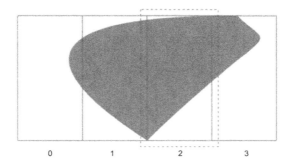

Fig. 2. Splitting the source image into parts, margins of part 2 are shown

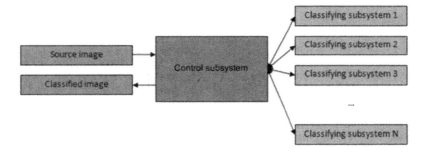

Fig. 3. Architecture of the parallel algorithm

Communication between parallel processes in systems with distributed memory is based on passing messages. This fact is reflected in the name of the MPI (Message Passing Interface) technology which is applied in this work. The MPI standard defines the interface that should be strictly implemented by programming systems on each platform and by developers creating their programs [11, 12].

During the implementation the system kernel was revised. On the application logic level the standard Java classes and patterns were applied as much as possible [13]. For instance, the wrapping class java.awt.image.DataBuffer was used to store images and etalon figures, java.awt.image.WritebleRaster to store results. The java.util.comparator interface was implemented to compare images.

4   Experiments

The experiments were performed on the computer cluster designed as shown below (Fig. 4).

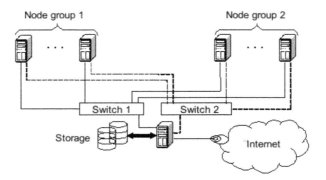

Fig. 4. Computer cluster architecture

The cluster consists of dual processor computers (nodes), each processor has 2, 4 or 8 cores (we used Intel Xeon E5-2650@2.0 Ghz Supermicro MBD-X9DRD-iF, Intel Xeon E5335@2.0 Ghz S5000VCL and Intel Xeon 5130@2.0 Ghz S5000VSA4 models). The nodes communicate through Gigabit Ethernet and use ISCSI Infortrend ES A12E-G2121 storage. The master node provides the user interface and connection between nodes, storage and WAN.

The system was tested on three images:

1. Geometrical figures: a simple test for faults in the algorithm implementation.
2. Satellite images of two islands: an attempt to distinguish a large island having marked the patterns on the small one.
3. Aerospace image of a part of the Mayskoe forestry terrain with the scale 1:25000. Almost a typical classification task. Two large classes are to be distinguished: hayfields and forests.

We shall perform the first test to determine whether the system works correctly. The source image and the result are shown on Fig. 5. The second test (Fig. 6) enables classifying of ocean (blue color), snow (red) and earth (green) on the map. The third test (Fig. 7) illustrates classifying of hayfield (red) and forest (green).

Fig. 5. Geometrical figures: the source image and the classified image (Color figure online)

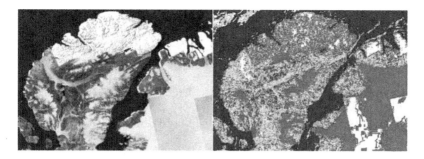

Fig. 6. Map classification (Color figure online)

Fig. 7. Mayskoe forestry classifying results (Color figure online)

Classification time in the first test is about 4 times less that in the third one (Table 2). That may be explained by small source image size, because the process communication overheads exceed the useful working time.

Table 2. Classification time

Test No.	Image size	Number of samples	Pupil size	Time, s
1	200 × 200	5	3	2.7
2	569 × 426	3	3	8.1
3	500 × 500	3	2	11.5

The results indicate the following drawback of the system: if 10 processes are available, the system tries to run all of them simultaneously regardless the input data. The more adequate behavior would be to define the minimum image part size, so that the smaller parts should not be split.

The performance measurement for more appropriate images while using different number of processes will be considered below.

5   Results

Consider the performance of the cluster-based system on an image of a size close to that of real tasks. In this case the time of transmitting images, partitioning and assembling becomes insignificant, so it will be possible to evaluate the system performance in comparison with a single-processor case and to determine the optimum image part size to be classified by a single process.

We consider again the satellite image of the Mayskoe forestry terrain (Fig. 8).

Fig. 8. The satellite photograph of the Mayskoe forestry terrain (4609 × 4872 pixels)

Two basic patterns were classified here, hayfield and forest. The size of the pupil was 3. The classification time for different numbers of processors is shown in Table 3. Thus, the optimum processor number for this case is 25.

Table 3. Performance of the cluster-based system

Number of processes	Maximum time of classification by one process, s	Total time, s	Part size, Mpix
2	102.32	217.70	17.846
3	48.32	136.56	8.921
5	27.34	101.98	4.460
10	12.26	80.84	1.982
15	8.04	75.63	1.273
20	7.36	74.07	0.937
25	4.85	71.93	0.743
30	8.73	76.06	0.612

6   Further Research

The achieved efficiency in the considered simple algorithm enables using parallel computing in more complicated neural networks. For example, it is possible to increase the distinction accuracy using multilayer perceptron where layers are RGB channels, images processed by Gaussian filters with different frequencies, brightness levels, positive and negative images. This approach also makes possible to reveal various objects on the satellite images (i.e. buildings, roads, vehicles). Building such networks is the subject of further research. It also implies multiple experiments on varied data sets in order to guarantee the required accuracy in various cases.

7   Conclusion

Survey and comparison of the existing classification algorithms presented in this paper has shown that the simple and efficient multi-alternative discrete accumulation method is appropriate for satellite image classifying. We implemented it firstly in a single-processor mode, but in order to process large images rapidly, a parallel MPI-based modification of the method was suggested. The system prototype was successfully implemented and tested on a multiprocessor system. The classification time was reduced 2.6 times in comparison to the single-processor version. Optimal number of processors to solve the real life tasks could be determined during the experiment. For the considered task the best efficiency was achieved using 25 processors.

References

1. Bartelev, S.A., Lupyan, E.A.: Research and works of ISR RAS on satellite monitoring of the vegetative cover. Mod. Probl. Earth Studying Space **10**, 197–214 (2013)
2. Terentyev, I.V.: Reliable object recognition on the aerospace images of the surface. Study Studying Earth Space **5**, 57–64 (1999)
3. Jensen, J.R.: Introductory Digital Image Processing, 3rd edn. Prentice Hall, Upper Saddle River (2005)
4. Yao, W., Zeng, Z., Lian, C., Tang, H.: A kernel ELM classifier for high-resolution remotely sensed imagery based on multiple features. In: Zeng, Z., Li, Y., King, I. (eds.) ISNN 2014. LNCS, vol. 8866, pp. 270–277. Springer, Heidelberg (2014)
5. Potapov, A.S.: Principle of representational minimum description length in image analysis and pattern recognition. Pattern Recogn. Image Anal. **22**, 82–91 (2012)
6. Jia, S., Liu, H., Sun, F.: Aerial scene classification with convolutional neural networks. In: Hu, X., Xia, Y., Zhang, Y., Zhao, D. (eds.) ISNN 2015. LNCS, vol. 9377, pp. 258–265. Springer, Heidelberg (2015). doi:10.1007/978-3-319-25393-0_29
7. Berezin, V.I., Vorobiev, V.I., Vasiliev, N.P., Morus, G.A., Terentyev, I.V., Shubina, M.A.: Reliable distinction of forestry objects. Earth Studying Space **4**, 55–62 (2003)
8. Shapiro, L., Stockman, G.: Computer Vision. Moscow, Binom (2006)
9. Zubkov, I.A., Skripachev, V.O.: Application of unsupervised classification algorithms to process the earth remote sounding data. In: The 4th Russian Open Conference "Modern Problems of the Earth Studying from Space" (2006)
10. Terentyev, I.V.: Distinction of separate hand-written symbols by means of optimization techniques. Comput. Model. **2005**, 407–408 (2005)
11. Gropp, W., Hoefler, T., Thakur, R., Lusk, E.: Modern Features of the Message-Passing Interface. The MIT Press, Cambridge (2014)

12. MPI: A Message-Passing Interface Standard. Message Passing Interface Forum. http://www.mpi-forum.org/docs/mpi-3.0/mpi30-report.pdf
13. Gamma, E., Helm, R., Johnson, R., Vlissides, J.: Elements of Reusable Object-Elemented Software. Design Patterns. St. Petersburg, Piter (2001)

Instantaneous Wavelet Correlation
of Spectral Integrals Related to Records
from Different *EEG* Channels

Sergey V. Bozhokin[1] and Irina B. Suslova[2(✉)]

[1] Department of Theoretical Physics, Peter the Great Saint-Petersburg
Polytechnic University, Saint-Petersburg, Russia
bsvjob@mail.ru
[2] Department of Theoretical Mechanics, Peter the Great Saint-Petersburg
Polytechnic University, Saint-Petersburg, Russia
ibsus@mail.ru

Abstract. The paper presents new algorithm to evaluate instantaneous correlation of nonstationary Electroencephalogram (*EEG*) signals related to different
brain leads. The method includes calculation of instantaneous values of spectral
integrals by Continuous Wavelet Transform (*CWT*). To adjust spectral and
temporal resolution of the signals we introduce the adaptive Morlet mother
wavelet function with a control parameter. To test the technique we composed
mathematical models of *EEG* signals based on real *EEG* record and special
short-time elementary signals. We discuss the application of the method in
different fields of physics.

Keywords: Adaptive mother wavelet · Instantaneous correlation · Spectral
integrals

1 Introduction

It is known that human EEG shows electrical activity of a great number of brain
neurons [1, 2]. *EEG* signals have time-varying properties as well as all the records of
processes occurring in living systems. We treat the signal as a nonstationary if its
spectral and statistic characteristics vary in time [1–4]. Considering *EEG* records, we
can observe transients both in the state of rest and during various functional tests (photo
stimulation, hyperventilation). We also can detect the appearance and disappearance of
correlations that occur between the measurements obtained from different areas of the
brain. To analyze the spectra of nonstationary signals (*NS*) we can apply Short Time
Fourier Transform (*STFT*) [3, 4]. In this case we divide the total time of measurement
T into the sequence of N windows with the duration $W = T/N$. Coupling in time of
two processes $Z_J(t)$ and $Z_K(t)$ obtained from different brain leads J и K can be
characterized by the value known as coherence [5]. Coherence function $\gamma_{JK}(f)$ depends
on frequency f and implies smoothing ($<\ldots>$) of Fourier components $\hat{Z}_J^{(n)}(f)$ и $\hat{Z}_K^{(n)}(f)$
corresponding to the signals over a large number of realizations ($n = 1, 2..N$).

© Springer International Publishing Switzerland 2016
L. Cheng et al. (Eds.): ISNN 2016, LNCS 9719, pp. 49–55, 2016.
DOI: 10.1007/978-3-319-40663-3_6

$$\gamma_{JK}^2(f) = \frac{\left|<\hat{Z}_J(f)\hat{Z}_K^*(f)>\right|^2}{<\left|\hat{Z}_J(f)\right|^2> <\left|\hat{Z}_K(f)\right|^2>}. \tag{1}$$

By computing the coherence value for different *EEG* leads, we can study normal brain activity as well as to detect synchronization in schizophrenia, dementia and Parkinson's disease [6–9].

Phase coherence is characterized by Phase Locking Value [10]

$$PLV_{JK}(f) = \sqrt{<\cos(\varphi_J(f) - \varphi_K(f))>^2 + <\sin(\varphi_J(f) - \varphi_K(f))>^2}. \tag{2}$$

We have $PLV_{JK}(f) \approx 0$ for two signals not synchronized in phase, and $PLV_{JK}(f) \approx 1$ in the opposite situation. Fourier analysis works successfully when we take as a window a stationary portion of the signal with unchanging set of frequencies. In [5] it was shown that the squared coherence of two signals depends on the way of averaging, the choice of the window width, the form of window function and on the size of the window shift. Therefore, we cannot consider (1) as sufficiently accurate and reliable measure of the correlation between two signals $Z_J(t)$ and $Z_K(t)$.

Wavelet transform [4] is a powerful tool in the studies of nonstationary *EEG*. The review of other methods can be found in [11, 12]. Cross Wavelet Spectrum

$$CWS_{JK}(v,t) = V_J(v,t)V_K^*(v,t), \tag{3}$$

where $V_J(v,t)$ and $V_K(v,t)$ are the continuous wavelet transforms (*CWT*) corresponding to signals $Z_J(t)$ and $Z_K(t)$, have been often used to detect interactions between oscillatory components in *EEG*. By averaging of $CWS_{JK}(v,t)$ over frequency v or time t intervals and normalizing it, we can obtain squared Wavelet Coherence Function $\Gamma_{JK}^2(v,t)$ [2, 13, 14]. We should emphasize that the values of $\Gamma_{JK}^2(v,t)$ also depend on the mother wavelet function and on the averaging procedure over v and t scales. The generalization of (2) to the case of wavelet phase coherence was carried out in [15, 16].

The purpose of this article is to find the dynamic behavior of correlation between spectral integrals calculated for the *EEG* signals from different brain leads. The wavelet transform involves the adaptive Morlet mother wavelet function with a control parameter, which allows us to modify the spectral and temporal resolution of *NS*. The advantage of the proposed method over other methods [2, 13, 14, 17] of assessing the coherence among different brain channels is in the fact that using of spectral integrals technique gives high temporal resolution precisely in the case when amplitude-frequency characteristics vary in time very rapidly.

2   Methods and Results

2.1   Continuous Wavelet Transform (*CWT*)

Continuous Wavelet Transform (*CWT*) maps nonstationary signal $Z(t)$ onto a plane with the coordinates of time t (s) and frequency v (Hz) [2, 18–22]

$$V(v,t) = v \int\limits_{-\infty}^{\infty} Z(t')\, \psi^*\big(v(t'-t)\big)\, dt', \qquad (4)$$

where $\psi(x)$ is the mother wavelet function, symbol * denotes complex conjugation. We propose the adaptive Morlet mother wavelet function (AMW) involving control parameter m:

$$\psi(x) = D_m \exp\left(-\frac{x^2}{2m^2}\right)\left[\exp(2\pi i x) - \exp\left(-\Omega_m^2\right)\right]. \qquad (5)$$

Herein $\Omega_m = m\pi\sqrt{2}$, and D_m is a constant determined by the condition $\|\psi(x)\|^2 = 1$. The length of AMW along x-axis is $\Delta_x \approx m/\sqrt{2}$, and along the axis of dimensionless frequencies is $\Delta_F \approx 1/(\sqrt{8}\pi m)$. The relation (5) shows that by changing m, in accordance with [4], we can make Δ_x and Δ_F shorter or longer, thus adapting the window function $\psi(x)$ to the form, as well as to the spectral composition of the signal under study. This is the way of controlling the time and frequency resolution of NS. If $m = 1$, then AMW (5) transforms into the standard Morlet mother wavelet function [2, 18–22]. For the signals $Z_K(t)$ and $Z_J(t)$ corresponding to different EEG brain leads we have the spectral integrals $E_\mu(K,t)$ and $E_\mu(J,t)$ calculated in the frequency range $\mu = \{\delta, \theta, \alpha, \beta\}$. Correlation of spectral integrals can be measured by Pearson correlation coefficient [11].

$$R_{KJ}(\mu,t) = \frac{<E_\mu(K,t)E_\mu(J,t)> \; - \; <E_\mu(K,t)> <E_\mu(J,t)>}{\sqrt{\Big[<E_\mu^2(K,t)> \; - \; <E_\mu(J,t)>^2\Big]\Big[\big[<E_\mu^2(K,t)> \; - \; <E_\mu(J,t)>^2\big]\Big]}}. \qquad (6)$$

2.2   Nonstationary EEG Simulation

Let us consider the example of evaluating the correlation of two signals. We take the real human EEG record $Z_K(t)$ (Fig. 1) as the first signal.

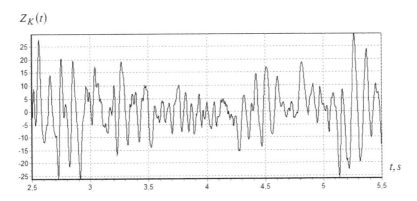

Fig. 1. Fragment of EEG signal $Z_K(t)$ within time interval t = [2.5; 5.5 s].

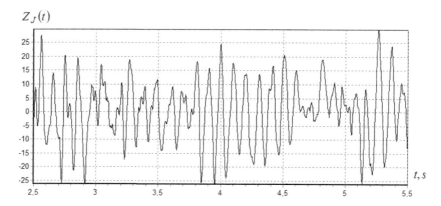

Fig. 2. Fragment of superposition $Z_J(t) = Z_K(t) + Z_L(t)$ within time interval t = [2.5; 5.5 s] at phase value $\beta_L = 0$.

CWT analysis of $Z_K(t)$ shows minimal activity at $t \approx 4$ s in α-frequency range, $\alpha = [7.5\text{--}14\text{ Hz}]$. To compose the second signal $Z_L(t)$ we add model signal

$$Z_L(t - t_L) = \frac{b_L}{2\tau_L\sqrt{\pi}}\exp\left(-\frac{(t - t_L)^2}{4\tau_L^2}\right)\cos(2\pi f_L(t - t_L) + \beta_L) \tag{7}$$

with the parameters $L = (b_L; f_L; t_L; \tau_L; \beta_L) = (15; 10; 4; 0.2; \beta_L)$ to the first *EEG* signal $Z_K(t)$. Subsequently, we will vary phase β_L of such an elementary model of *NS* (7). Signal $Z_L(t)$ shows maximal activity at frequency $f_L = 10$ Hz within time interval $[t_L - 4\tau_L; t_L + 4\tau_L]$, where $t_L = 4$ s, $\tau_L = 0.2$ s. Signal $Z_J(t) = Z_K(t) + Z_L(t)$ (Fig. 2) is a specially composed superposition, which will allow us to test our method. It differs from *EEG* record $Z_K(t)$ only within time interval $[t_L - 4\tau_L; t_L + 4\tau_L]$.

In Fig. 3 we represent time dependence of spectral integrals $E_\alpha(t)$ in α-range of *EEG* rhythm. The curve denoted in Fig. 3 by 0 (the thinnest one) shows spectral integral $E_\alpha(t)$ corresponding to $Z_K(t)$ (Fig. 1). The analysis of this curve demonstrates that spectral integral $E_\alpha(t)$ in α-range has minimal value at $t \approx t_L = 4$ s. The curve 1 in Fig. 3 shows time dependence of $E_\alpha(t)$ for superposition $Z_J(t) = Z_K(t) + Z_L(t)$ at the phase of the additional model signal $\beta_L = 0$. Signal $Z_J(t)$ in α-range demonstrates maximal activity near $t \approx t_L$. Hereinafter, we set index number $K = 0$ for the human *EEG* at rest (Fig. 1). For superposition $Z_J(t)$ at $\beta_L = 0$ (Fig. 2) and $\beta_L = \pi/4$ we use indices $J = 1$ and $J = 2$. A slight change in the phase (from $\beta_L = 0$ to $\beta_L = \pi/4$) of the additional signal (see curve 2 in Fig. 3) doesn't affect the behavior of spectral integral $E_\alpha(t)$ near $t \approx t_L$. However, at $t \approx t_L \pm 2\tau_L$, where $\tau_L = 0.2$ s, the curves 1 and 2 in Fig. 3 differ noticeably. Curve 3 (bold line) in Fig. 3 corresponds to the spectral integral of $Z_J(t)$ with phase $\beta_L = 3\pi/4$.

It is important to note that all four curves for $E_\alpha(t)$ in Fig. 3 coincide at $t > t_L + 4\tau_L, t < t_L - 4\tau_L$, because in this case the contribution of model signal $Z_L(t)$ (7) to the superposition signal $Z_J(t) = Z_K(t) + Z_L(t)$ is negligibly small. Figure 4 shows the graphs of Pearson correlation coefficients $R_{JK}(\alpha, t)$ (6) depending on time.

$E_\alpha(t)$

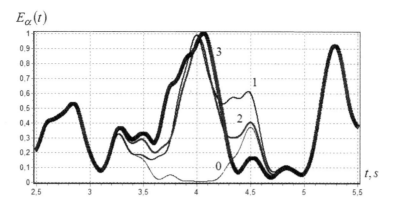

Fig. 3. Time dependence of spectral integrals $E_\alpha(t)$ computed in α-frequency range ($\alpha = [7.5; 14\ \text{Hz}]$) of EEG rhythm.

$R_{JK}(\alpha, t)$

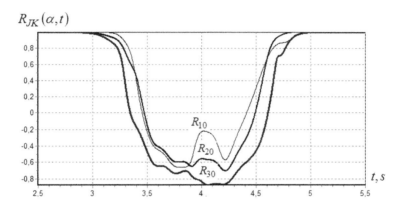

Fig. 4. Time dependence of Pearson correlation coefficient $R_{JK}(\alpha, t)$.

The upper curve R_{10} in Fig. 4 demonstrates the dependence in time of spectral integrals' correlation $R_{10}(\alpha, t)$ in α-frequency range between the curves 0 ($K = 0$) and 1 ($J = 1$) in Fig. 3. Time dependence of $R_{20}(\alpha, t)$ shows correlation in time between the curves 0 ($K = 0$) and 2($J = 2$, $\beta_L = \pi/4$) in Fig. 3. Similarly, time dependence $R_{30}(\alpha, t)$ corresponds to the correlation between curve 0 ($K = 0$) and 3 ($J = 3$, $\beta_L = 3\pi/4$). Figure 4 shows that at $t > t_L + 4\tau_L$, $t < t_L - 4\tau_L$ correlation coefficient $R_{JK}(\alpha, t) = 1$. Within the interval $t_L - 4\tau_L \leq t \leq t_L + 4\tau_L$ we observe increasing correlation with the lowest correlation value $R_{30}^{\min}(\alpha, t) \approx -0.83$ at $t \approx 4.15$ s. The fact can be explained by considering the spectral integral of the first signal (curve 0), which increases and the second one (curve 3), which decreases near $t \approx 4.15$ s. Thus, by evaluating the correlation by the technique of spectral integrals we obtain the expected (in the simulation) result.

3 Discussion and Conclusions

The basic idea of this work is to develop the technique for the evaluation of instantaneous correlation of two nonstationary signals. We composed mathematical models based on real human *EEG* record and elementary nonstationary signal in the form of Gaussian envelope multiplied by oscillating function with fixed frequency in the range of *EEG* alpha rhythm. Due to this type of simulation, we knew in advance the coherence behavior of our model signals. The peculiarity of this model is that we can change the phase of the oscillating function and study the changes in the degree of instantaneous correlation of two nonstationary signals. The use of adaptive Morlet mother wavelet significantly improves frequency and temporal resolution. The method based on calculating the spectral integrals' behavior in time proposed in this paper has some advantages over the traditional methods, when only long-term correlations allow accurate evaluation. The technique gives the opportunity to detect the characteristic times of the appearance and disappearance of correlations during such nonstationary functional tests as photic stimulation, hyperventilation, and solution of cognitive tasks. The proposed method can be applied successively to determine short-term changes in synchronous states in brain electrical activity [10], EEG bursts [11], and evoked potentials [23]. It can be useful in testing neural networks and in the fields of physics related to the studies of quickly changing (flare) processes.

Acknowledgement. The work have been carried out with the financial support of State Assignment N 3.1446.2014/K.

References

1. Gnezditskii, V.V.: A Reverse EEG Problem and Clinical Electroencephalography. MEDpress-inform, Moscow (2004). (in Russian)
2. Hramov, A.E., Koronovckii, A.A., Makarov, V.A., Pavlov, A.N., Sitnikova, E.: Wavelets in Neuroscience. Springer Series in Synergetics. Springer, Berlin (2015)
3. Kaplan, A.Y., Fingelkurts, A.A., Fingelkurts, A.A., Borisov, S.V., Darkhovsky, B.S.: Nonstationary nature of the brain activity as revealed by EEG/MEG: methodological, practical and conceptual challenges. Sig. Process. **85**, 2190–2212 (2005)
4. Mallat, S.: A Wavelet Tour of Signal Processing, 3rd edn. Academic Press, New York (2008)
5. Kulaichev, A.P.: The Informativeness of Coherence Analysis in EEG Studies. Neurosci. Behav. Physiol. **41**, 321–328 (2011)
6. Lachaux, J.-P., Lutz, A., Rudrauf, D., Cosmelli, D., Quyen, L.V., Martinerie, J., Varela, F.: Estimating the timecourse of coherence between single-trial brain signals: an introduction to wavelet coherence. Neurophysiol. Clin. **32**, 157–174 (2002)
7. Uhlhaas, P.J., Singer, W.: Abnormal neural oscillations and synchrony in schizophrenia. Neurosci. Nat. Rev. **11**, 100–113 (2010)
8. Roik, A.O., Ivanitskii, G.A., Ivanitskii, A.M.: The human cognitive space: coincidence of models constructed on the basis of analysis of brain rhythms and psychometric measurements. Neurosci. Behav. Physiol. **44**, 692–701 (2014)

9. Bazanova, O.: Comments for current interpretation EEG alpha activity. J. Behav. Brain Sci. **2**, 239–248 (2012)
10. Trofimov, A.G., Kolodkin, I.V., Ushakov, V.L., Velichkovski, B.M.: Agglomerative method for isolating microstates EEG related to the characteristics of the traveling wave. In: Proceedings of the XVII - Russian Scientific and Technical Conference "Neuroinformatics–2015", Moscow Engineering Physics Institute, 19 January–23 January 2015, Collection of Scientific Papers, pp. 66–77. In 3 parts, Part.1.M.: MEPhI. ISBN 978-5-7262-2043-7 (2015). (in Russian)
11. Bozhokin, S.V., Suslova, I.B.: Wavelet-based analysis of spectral rearrangements of EEG patterns and of non-stationary correlations. Phys. A **421**, 151–160 (2015)
12. Rosso, O.A., Martin, M.T., Figliola, A., Keller, K.: EEG analysis using wavelet-based information tools. J. Neurosci. Methods **153**, 163–182 (2006)
13. Klein, A., Sauer, T., Jedynak, A., Skrandies, W.: Conventional and wavelet coherence applied to sensory-evoked electrical brain activity. IEEE Trans. Biomed. Eng. **53**, 266–272 (2006)
14. Li, X., Yau, X., Fox, J., Jefferys, J.G.: Interactions dynamics of neuronal oscillations analysed using wavelet transform. J. Neurosci. Methods **160**, 178–185 (2007)
15. Bernjak, A., Stefanovska, A., McClintock, P.V.E., Owen-Lynch, P.J., Clarkson, P.B.M.: Coherence between fluctuations in blood flow and tissue oxygen saturation. Fluctuation Noise Lett. **11**, 1240013 (2012)
16. Sheppard, L.W., Stefanovska, A., McClintock, P.V.E.: Testing for time-localized coherence in bivariate data. Phys. Rev. E **85**, 046205 (2012)
17. Cohen, E.A.K., Walden, A.T.: Wavelet coherence for certain nonstationary bivariate processes. IEEE Trans. Sig. Process. **59**, 2522–2531 (2011)
18. Chui, C.K., Jiang, Q.: Applied Mathematics. Data Compression, Spectral Methods, Fourier Analysis, Wavelets and Applications. Mathematics Textbooks for Science and Engineering, vol. 2. Atlantis Press, Paris (2013)
19. Bozhokin, S.V., Suvorov, N.B.: Wavelet analysis of transients of an electroencephalogram at photostimulation. Biomed. Radioelektron N3, 21–25 (2008). (in Russian)
20. Bozhokin, S.V.: Wavelet analysis of learning and forgetting of photostimulation rhythms for a nonstationary electroencephalogram. Tech. Phys. **55**, 1248–1256 (2010)
21. Bozhokin, S.V.: Continuous wavelet transform and exactly solvable model of nonstationary signals. Tech. Phys. **57**(7), 900–906 (2012)
22. Bozhokin, S.V., Suslova, I.B.: Wavelet analysis of non-stationary signals in medical cyber-physical systems (*MCPS*). In: Balandin, S., Andreev, S., Koucheryavy, Y. (eds.) NEW2AN/ruSMART 2014. LNCS, vol. 8638, pp. 467–480. Springer, Heidelberg (2014)
23. Borodina, U.V., Aliev, R.R.: Wavelet spectra of visual evoked potentials: time course of delta, theta, alpha and beta bands. Neurocomputing **121**, 551–556 (2013)

The Theory of Information Images: Modeling of Communicative Interactions of Individuals

Alexandr Y. Petukhov[✉] and Sofia A. Polevaya

Lobachevsky Nizhny Novgorod State University, Nizhny Novgorod, Russia
Lectorr@yandex.ru

Abstract. The current work represents a formalized description of information and communicative interactions of individuals on the basis of theory of information images. It also demonstrates how important it is to choose the models type adequate to the systems under research.

It also introduces an explication of the possibility to create the model of information and communication interactions that can illustrate transmission of information between two and more individuals. Methods and approaches suggested in the current article allow us to compare different levels of the described processes depending on the chosen architecture of the model; they are also able to are able to simulate the processes of distortion and generation of information images while information and communication social interaction.

Expansion and addition of information images theory in terms of transmission of information among individuals enables us to speak about the space of individual information images. The existence of such a space and creation of correct formalized model help us to explain a number of characteristic phenomena of human thinking processes. As a result of the current research, the authors introduce an equation that describes the spatial and structural evolution of individual information images.

There is also an example of modeling on the basis of the current theory taking into account the results of the experiment (bilingual Stroop test).

Keywords: Communication field · Information image · The information images space · Virtual particles · Bilingual Stroop test

1 Introduction

Natural-science aspects of influence of the environmental information factors on the living systems are poorly studied. Many new electrophysiological methods have been created: foremost the registration of caused and event-related potentials, being registered in the experimental behavioral situations. These methods enabled us to come closer to studying physiological mechanisms of particular stages of process of information processing, such as sensory analysis, mobilization of attention, image formation, memory standards removing, decision making, etc. Studying of time parameters of electrophysiological responses to different types of stimuli and in different conditions made timekeeping possible for the first time ever, i.e. the assessment of duration of separate stages of information processing directly at the level of the cerebral substrate [3]. A lot of foundational

© Springer International Publishing Switzerland 2016
L. Cheng et al. (Eds.): ISNN 2016, LNCS 9719, pp. 56–65, 2016.
DOI: 10.1007/978-3-319-40663-3_7

facts were received by classical methods of psychology concerning the analysis and studying of the subjects behavior in different social situations. Along with the cognitive psychophysiology a relatively new section of neurobiology, which is neuroinformatics, has appeared. As well as cognitive psychophysiology, neuroinformatics actually represents. As well as cognitive psychophysiology, neuroinformatics is actually an application of computer metaphor for the analysis of mechanisms of creation and processing of information in brain of human-being and animals.

The problem of description of information and communication interactions between people is a fundamental one for the modern cognitive science. Not so long ago new unique natural-science models of information transmission from one individual to another have appeared [3, 7, 10], and some others in other fields of biology and cognitive psychology [4, 6, 8, 11, 16, 19]. However a lot of represented models are poorly scalable and do not allow fundamentally explain the processes of transmission of information images and their distortions that root in the interactions with the external communication environment. Along with that, a vitally important question for the modern cognitive science devoted to the living systems is the following: how generation and distortion of information images in the human mind are held, as well as which model is able to describe and correctly predict those processes (that can be verified through an experiment).

2   The Theory of Information Images

The basis of our model was the idea to describe the information interactions between people as a process of interaction between Information Images (hereinafter – II) created by them. Information images can be determined as the displays of objects and events, i.e. the sequence of operational actions in the situation context, in the set space of characteristics. We concern them as an elementary object of any virtual environment or as an elementary information quantum of human thinking (taking into account that images can have a complicated composition). A human being, thanks to direct individual experience and cognition, becomes an owner of individual information images space with a specific typology and boundary conditions. Thus, engrams of both real and unreal events and information image rooted in external resources can coexist equally in the subjective human world; these images of events can be incorporated by external resources or people that a human being had to trust due to certain consequences. In the case if the brain loses information about the nature of information sources of an image (real or virtual one), there is a phenomenon of «false memory» [1].

The material basis of the subjective II space (subjective virtual space) are the specific patterns of activity in the neural networks of a particular person. These neural codes are unique insofar that gene, life experience of each individual and the event context corresponding to the very moment of formation of the subjective II, are unique as well [12].

A human being cannot submit the image that exists in his/her mind or in his II space to another person directly. To do this, he/she uses different tools of communicative apparatus that were formed through the social communication superstructure or communication field (CF) [4]). CF if an information commonness of individual experience and

collective unconscious formed as a result of being of an individual in a society. Communicative apparatus contains such tools as speaking, visual instruments of information
transmission, tactile and symbolic ways, etc.

In this regard, II is transferred into some sort of code, which was created by the
internal human encoder (which is formed by religion, culture, language, etc. – CF) and
is transferred to another person (with a light, touch, sound, etc.). The brain of other
person sends this image to be decoded by his internal decoder (in accordance with the
data in his\her own CF) and the received result has practically no chances to match with
the initial II, however if the individuals share common linguistic, cultural, etc. basis, the
mismatch is relatively small and an information channel is created with distortions that
the interacting subjects consider to be acceptable or insignificant.

3   Generation of Information Images

It is also necessary to define the mechanism of generation of information image (hereafter – II). Simplified scheme of II generation in a human brain can be seen at Fig. 1.

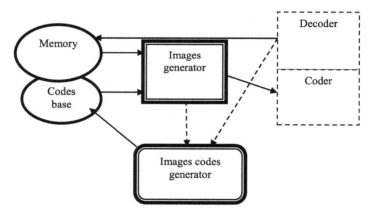

Fig. 1. The schematic diagram of the II generation in the human brain

There is a nominal images` generator that can be presented as an image consisting
from the elements of real and virtual memory (i.e. II can be made up from memories
about events that really did not happen, imaginary elements, the so-called «false
memory»). It is important to clarify that for a generator there is no fundamental difference whether real or virtual memory is used for an image, if the latter is available, it will
be perceived by a generator as an equivalent one.

The II sequence and shape are determined by its code which is generated by an image
code generator and stored in the memory. A code generator works most properly in the
childhood and adolescence, when the person makes own idea of the world and accu-
mulates initial knowledge. When a person meets something unknown in an elder age
he/she usually tries to cope with it using already familiar images, searches for analogy
and only in vase of absence or serious inconsistency between II and acquired experience,
a person commands to create a new code and a new image [8, 13].

Thus, the code of an image can be defined as a sequence or a set of values and a certain order of signs that determine the layout of initial elements of real and virtual memory.

Assembled image enters the II space, where it interacts with the other images. We will later discuss the way how II further act in the individual space.

Obviously, not all images are equally active and important while decision-making process. Part of the images define human lifestyle, their habits of communication and affection, part of which practically does not take part in real life and switches on only in very specific situations.

In this regard, we must define several types or levels of information images

1. Dominant. Dominant images define the so-called dominant [3] of human behavior and its primary information portrait. These are the images that are defined through the behavioral algorithms that have unquestionable preference among individuals and form their behavior in the very first turn; these algorithms are also partially influence on how the people interacting with this or that individual judge him\her (in accordance with his\her behavioral algorithms).
2. Active. Active images as are some of the key in the information portrait of human. However, their impact is somewhat less dominant and to reach them, more "dense" communication is required.
3. Passive. Passive images are usually hidden in human activities and are responsible for the virtual behavior - i.e., dreams, fantasies, displays in a virtual environment, for example, in social networks under false names, in online gaming, and so on.
4. Deferred. These images are almost no outward in a conscious activity of the individual and include, as a rule, only in certain cases, or have their effects without their awareness and do not become a need.

Apparently, the real basic information human activity is determined only by the first two types of images and partially by the third one, the main task of the 3rd and the 4th group of images is to determine the virtual activity.

It is important to note that while activity of daily living images and formation algorithms related to them cam transfer from one type to another in accordance with the type and strength of emotional estimation for particular events of the current activity.

4   Model Description of the II Space

Obviously, an information image cannot be transferred unaltered through communication between individuals. After all, each II is unique as for an individual and as a whole, because each individual has a unique experience, or as noted above - a unique social superstructure of communication - communication field (CF). Simplistically it can be represented as – Fig. 2.

It means that actually the following chain is used:

Image → communicative field → message-signal → disturbance in the communicative field of another individual → image → perception

CF features are formed by the internal space of individual information images.

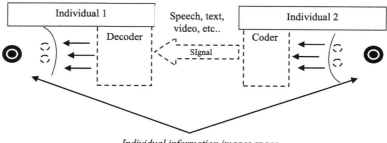

Individual information images space

Fig. 2. Schematic diagram of interaction between the communicative fields of individuals.

Description of model of inter-social interaction via communication field has been already proposed in some early works of the authors, but without the use of the apparatus of information images [15]. This model proved to be quite useful for the simulation of a number of special cases, information and psychological influence on public opinion in the social systems and not only.

Communication fields can accumulate on each other while interacting and impact on the social superstructure of individual and through it on the election of the algorithms used for the recognition and activity from the 6 priority levels mentioned above.

The way of interaction of II is shown graphically on Fig. 3.

However, schematic image of layering of the individuals CF has a specific informational and psychological meaning. The more layering is, the higher is the chance of inclusion of interaction of not only the dominant images that show the greatest activity information, but also lying deeper in the CF. Actually, this is due to the fact that the "denser" people communicate, the more they open up to others and open the diversity of their own II in the situation of interaction, as well as activate deeper II layers.

As a result of the interaction, the II can change themselves and other II both "native", and of other individuals, as well as change the parameters of the individual (by changing the cognitive systems and real and virtual memory).

There is a probabilistic characteristic of CF internal interactions, and as a result II change each other without any external influence, however, as a rule, external influence is a catalyst for change in the internal structure of the individual CF and its characteristics.

By parametrising information, it is possible to define several basic parameters that determine their essential qualities.

The first parameter, the introduction of which is absolutely logical given the already mentioned types of II on the basis of levels, is the energy of information image. II levels interpreted as energy ones that brings us back to an analogy with elementary particles.

Accordingly, by increasing or decreasing their energy due to internal or external influences II can transfer to another energy level and change its type (e.g., active for the dominant, or vice versa).

For the integration of the proposed model in the approaches that are able to describe the processes of public processes, you can determine the energy of II as social energy of the "socio-energy approach" [15]. In this case, the parameter acquires a concretized value.

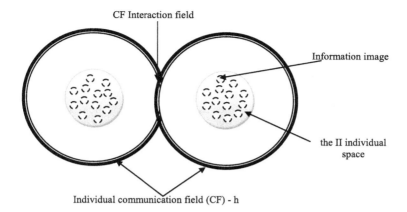

Fig. 3. The interaction of two individuals and their CF

5 Description of the II Movements in the Individual Space

By defining a certain space of information images of the human mind, one should speak about a potential field $\vec{U}$ in this space, which affects II.

This field structures and differentiates them according to the energy level and persistence (m - a measure of the II persistence).

Top layer - active II with low m-value, high energy.

The lower layers - a relatively passive II, high m values, low levels of energy.

Areas of high II density, especially close to the center of the space, lead to "blocking" overlay due to the acting communication field, as well as the interaction of images, overlapping and mutual displacement of parameters. Also, images can be organized into systems and subsystems in the form of multi-component structures of some of the images-supraimages with increased complexity that are another complex topic of research.

Description of the II movements is possible through the diffusion equation for Brownian motion:

$$\frac{\partial}{\partial t}\xi(x, t) = \sum_{i=1}^{n} f\big(o_i, o\big)\delta\big(x - x_i\big) + D_0 \Delta \xi_0(x, t) \tag{1}$$

Taking into account the field potential $\vec{U} = \nabla \Psi$

$$\vec{F}(\vec{r}, t) = -\nabla U(\vec{r}, t)$$

Thus, the II movement can be described as the following:

$$\frac{dx_i}{dt} = m_i \nabla_x \xi_0(x_i, t)\big|_{x_i} + \sqrt{2D_n}\varepsilon_i(t) - \int \nabla U(x_i, t)$$

Where $\varepsilon_i(t)$− stochastic force.

Information images interact between each other in a similar way that the particles of Brownian motion; the only difference is that the clash happens with the changes of not only on parameters of kinetic energy and the speed vector, but also in the specific II parameters.

Here is an example of modeling based on this approach. We used the software package MatLab 2009b.

We have led the computerized bilingual Stroop test that is fairly well known in psychophysiology [18]. The Stroop effect is defined as response delay while reading words, when the color of the words does not match the written words (for example, the word "red" written in blue). The appropriate test is used to identify characteristic patterns.

The test measures the time for decision-making and a certain number of errors in the sample of individuals of a special computer program in the 4 contexts:

the 1st context: words denoting color, written in black letters and presented on a light background; the objective function - the choice of a color sample in accordance with the meaning of the word;

the 2nd context - the color of the letters and the meaning of words are equivalent; the objective function - the choice of a color sample in accordance with the meaning of the word;

the 3rd context - the color of the letters does not match the meaning of the words; the objective function - the choice of a color sample in accordance with the meaning of the word;

the 4th context - the color of the letters does not match the meaning of the words; the objective function - to select the color of the sample in accordance with the color of the letters.

We used the Stroop test implemented on a multi-platform WEB-Apway.ru.

We will not clarify the core of the experiment in details, as in this case, the article is devoted to a greater extent of the model itself. Mainly interested in the simulation results, which are generally correctly predicted the results of the test under the specific conditions (test was conducted in two languages).

In fact, we are dealing with the influence of external information on the individual, where in cases 3 and 4 are two information streams are contrary to each other and cause cognitive dissonance. The process of choosing the color of the image corresponds to a specific activation (correct or incorrect), respectively, we can calculate the time of its activation.

The dependencies on the activisation of information images for this experiment were build on the basis of the developed model.

The first graph series refer to the display of characteristic dependence of activation (i.e., move to a higher level in the space IS) for a single II (Fig. 4). The y-axis demonstrates the total change in the spatial coordinates of the II. X is for the time frame.

In general, simulation and experimental results showed that the concentration of one of the information flow, without activating the other requires deliberate blocking of unconscious reactions to familiar stimuli (such as the choice of color in accordance with color), and the effect of cognitive dissonance itself is explained here by the mismatch activity of information images involved in mixed disturbance of the communication field, inconsistent with the unconscious adaptive responses.

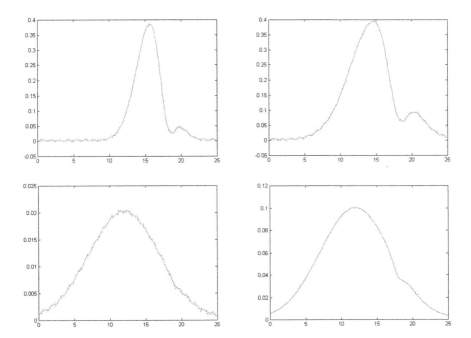

Fig. 4. Comparing the profile of activation for the cases 3 and 4 of the Stroop test for an individual for two different languages (upper - for foreign, lower - for the native)

The need of mind in building of new logical chains, by passing the existing and comparison of other codes of image with the codes of information perturbations of communication field leads to the information "failure", as a result the individuals involved in the experiment "stalled" over the choice of color, tugging the arrow on the screen from one to another. Only after the removal of the conflicting images of the structure, the final choice was made.

6   Conclusion

Thus, it is shown that the information image can be divided into 4 types according to their energy levels we can create the quantity description of dynamical processes connected with the transfer and understanding of information images between the interacting individuals, as well as estimate the degree of activity in the communication processes.

The most efficient in terms of authors' algorithm of submission of information and communication exchange via the communicative field is suggested.

A new way of describing the II space using the diffusion equation is revealed using the diffusion equation thereby creating the basis of a mathematical model capable of correctly simulating the processes of II generation and distortion while information and communication and social interaction.

The example of activity modeling of individual information images while external information influence (including inconsistent ones).

Based on the proposed model, the authors plan to further develop an approach to implement an experimental confirmation of the theory, to parameterize the main features and use these techniques to predict some of the social and political processes.

The theoretical part of the research was performed with support by the grant of the Russian Science Foundation (project №15-18-00047).

The experimental part of the research was performed with support by grants from the Board President of the Russian Federation (Project MK-7165.2015.6).

References

1. Aleksandrov, Y.: Laws actualization of individual experience, and the reorganization of its system structure: a comprehensive study. Proc. ISA RAS **T.61**(3), 3–25 (2011)
2. Ashby, F.G., Sebastien, H.A.: Tutorial on computational cognitive neuroscience: modeling the neurodynamics of cognition. J. Math. Psychol. **55**, 273–289 (2011)
3. Chernavskii, D.S.: Synergetics and Information. Dynamic Information Theory. URSS (2009)
4. Chuprakova, N.S., Petukhov, A.Y.: Threshold effects in the social and political processes. Social-Energy Approach. Sovremennye issledovaniya sotsialnykh problem [Modern Research of Social Problems] **8**, 69 (2013)
5. Chuprakova, N.S., Petukhov, A.Y.: Development of vulnerability to manipulations in the consciousness of adolescents with the help of modern virtual-communicative means. AYER **1**, 49–55 (2014)
6. Faugeras, O., Inglis, J.: Stochastic neural field equations: a rigorous footing. J. Math. Biol. **71**, 259–300 (2015)
7. Griffith, D., Greitzer, F.: Neo-symbiosis: the next stage in the evolution of human-information interaction. Cogn. Inf. Nat. Intell. **1**, 39–52 (2007)
8. Haazebroek, P., van Dantzig, S., Hommel, B.A.: Computational model of perception and action for cognitive robotics. Cogn. Process. **12**, 355–365 (2011)
9. Helsel, S.K., Roth, J.P. (eds.): Virtual Reality: Theory, Practice, and Promise. Meckler Corporation, Westport, CT (1991)
10. Lee, T.M., Liu, H.L., Chan, C.C. (eds.): Neural correlates of feigned memory impairment. Neuroimage **28**, 305–313 (2005)
11. Kooi, B.W.: Modelling the dynamics of traits involved in fighting-predators–prey system. J. Math. Biol. (2015)
12. Kryukov, V.: The model of attention and memory, based on the principle of dominant and comparator function of the hippocampus. J. High. Nerv. Activity **T.54**(1), 10–29 (2004)
13. Mustajoki, A.A.: Speaker-oriented multidimensional approach to risks and causes of miscommunication. Lang. Dialogue **4**, 216–243 (2011)
14. Scholtz, J.: Metrics for evaluating human information interaction systems. Interact. Comput. **18**, 507–527 (2006)
15. Petukhov, A.Y.: Branched chain reaction in complex social systems. Fractal Simul. Engl. **1**, 20–28 (2013)
16. Petukhov, A.Y.: Modeling of branched chain reactions in political and social processes. Glob. J. Pure Appl. Math. **11**, 3401–3408 (2015)
17. Petukhov, A.Y., Polevaya, S.A.: Modeling of information images dynamics through the communicative field method. Int. J. Biomath. (2016)

18. Stroop, J.R.: Studies of interference in serial verbal reactions. J. Exp. Psychol. **18**, 643–662 (1935)
19. Vandekerckhove, J.: A cognitive latent variable model for the simultaneous analysis of behavioral and personality data. J. Math. Psychol. **60**, 58–71 (2014)

A Two-Stage Channel Selection Model for Classifying EEG Activities of Young Adults with Internet Addiction

Wenjie Li[1,2], Ling Zou[1,2(✉)], Tiantong Zhou[1,2], Changming Wang[3,4], and Jiongru Zhou[1,2]

[1] School of Information Science and Engineering, Changzhou University, Changzhou 213164, China
zouling0@aliyun.com
[2] Changzhou Key Laboratory of Biomedical Information Technology, Changzhou 213164, China
[3] Beijing Anding Hospital, Capital Medical University, Beijing 100088, China
[4] Beijing Institute for Brain Disorders, Beijing 100088, China

Abstract. Full scalp electroencephalography (EEG) recording is generally used in brain computer interface (BCI) applications with multi-channel electrode cap. The data not only has comprehensive information about the application, but also has irrelevant information and noise which makes it difficult to reveal the patterns. This paper presents our preliminary research in selecting the optimal channels for the study of internet addiction with visual "Oddball" paradigm. A two-stage model was employed to select the most relevant channels about the task from the full set of 64 channels. First, channels were ranked according to power spectrum density (PSD) and Fisher ratio separately for each subject. Second, the occurrence rate of each channel among different subjects was computed. Channels whose occurrences was more than twice consisted the optimal combination. The optimal channels and other comparison combinations of channels (including the whole channels) were used to distinguish between the target and non-target stimuli with Fisher linear discriminant analysis method. Classification results showed that the channel selection method greatly reduced the abundant channels and guaranteed the classification accuracy, specificity and sensitivity. It can be concluded from the results that there is attention deficit on internet addicts.

Keywords: Channel selection · Electroencephalogram (EEG) · Internet addiction · Oddball · Power spectrum density · Fisher linear discriminant analysis

1 Introduction

Internet addiction (IA) has emerged as a potential problem in adolescents and young adults, especially in south-east Asian countries [1]. Excessive use of internet could affect learning and working performances of adolescents and young adults, worsen social relationships with other people and even cause mental and behavioral disorders.

© Springer International Publishing Switzerland 2016
L. Cheng et al. (Eds.): ISNN 2016, LNCS 9719, pp. 66–73, 2016.
DOI: 10.1007/978-3-319-40663-3_8

EEG and event-related potential (ERP) features are extensively investigated in neuromodulation and brain-computer interface studies. The P300 ERP signal is of much concern in the research of internet addition. It has been shown that internet addicts has decreased P300 amplitude compared to controls, this amplitude reflects attention allocation [2]. A lot of methods have been proposed to extract spatial information and select optimal channels. Common spatial patterns (CSP) is a powerful method for feature extraction and channel optimization in motor imagery BCI studies [3, 4]. Compared to commonly used unsupervised principal component analysis (PCA) and independent component analysis (ICA) methods [5, 6], supervised feature extraction method is more popular in EEG classification studies [7, 8]. One difficult issue for PCA is to determine the number of principal components. Genetic algorithm is also used to determine the optimal subset of channels [9, 10]. It has a drawback of complicate computation that is time consuming. This is not suitable for on-line applications.

In this paper, a two-step model that combines channel ranking and occurrence analysis is used in channel selection. This two-stage model may be used for the detection and treatment of internet addiction subjects.

2 Materials and Methods

2.1 Subjects and Experiment

The subjects, who are healthy, right handed and with normal vision, are adopted in all the sessions of experiment. The subjects are divided into two groups. Control group has ten subjects, five male and five female, with a mean age of 22.4 years and a standard deviation of 1.65 years. Internet addict group has seven subjects, all male, with a mean age of 21.4 years and a standard deviation of 0.53 years.

The EEG acquisition process is shown in Fig. 1. We indicate the procedures followed in the manuscript were in accordance with the ethical standards of the responsible committee on human experimentation (institutional and national) and with the Helsinki Declaration of 1975, as revised in 2004(14).

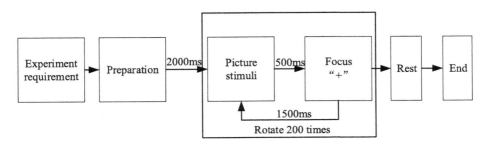

Fig. 1. Acquisition process of EEG with oddball paradigm

There are two kinds of picture stimulus, one is called target stimuli ("XXXXX"), and the other is non-target stimuli ("OOOOO"). They would be shown randomly in the experiment. In the whole experiment, target stimuli appears 42 times, and non-target stimuli appears 558 times.

2.2   Channel Selection and Feature Extraction

Channel Ranking with PSD Analysis. On macroscopic aspect, the channels were selected by analyzing the mean EEG signals of target stimuli and non-target stimuli in frequency domain. At first, the 558 non-target trials and 42 target trials were averaged, and then, power spectrum estimation was done. In order to determine which electrodes have more interested information, PSD analysis was done on each channel, and specific frequency band was analyzed. Channels with high power distribution on the frequency band may be important. The Welch method is a nonparametric method, makes no assumption about how the data were generated, and makes some improvements to the Bartlett method. Channels can be selected and ranked according to the average power or peak value on the interested frequency band.

Channel Ranking with Fisher Ratio. On microscopic aspect, the channels were selected by analyzing each single trial in time domain. The mean value of the interval that covers P300 was calculated for each single trial on each channel, it was taken as the P300 feature. Then, the averaged P300 features were calculated for 558 non-target trials and 42 target trials. The Fisher ratio was calculated on each channel. The 64 channels were ranked according to the Fisher ratio. Channels which ranked behind were abandoned. In our experiment the last 30 % were abandoned.

Feature Extraction. The channels selected with PSD analysis and Fisher ratio were compared, and the intersects were found out for each subject. The main difference between target stimuli and non-target stimuli is that the P300 amplitude of target stimuli is higher than the one of non-target stimuli, it can be used as the feature to distinguish between one from the other.

Classification with Fisher Linear Discriminant Analysis. Fisher linear discriminant analysis is generally used in BCI applications, has good and robust performance in revealing the patterns. Specially, in our experiment, Fisher ratio is used in channel selection, so fisher linear discriminant analysis is taken as granted, it is employed to classify target stimuli and non-target stimuli.

3   Results and Discussion

3.1   Observation of P300 EEG

The P300 EEG characteristic was observed in time field by averaging the signals of all trials and subjects. It can be seen on many probes that there was an event related potential at about 300 ms or later after the stimuli was presented. This phenomenon appeared on both target stimuli and non-target stimuli, but it was much stronger for target stimuli case. The ERP signals on one parietal probe are plotted in Figs. 2 and 3. It can be seen that the P300 event related potential on control subjects is stronger than that on internet addicts, no matter what stimuli type it is. This may indicate that internet addicts have attention deficit compared with control subjects.

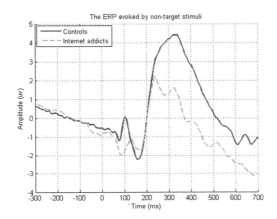

Fig. 2. Observation of the ERP signals evoked by non-target stimuli

In order to observe the spatial distributing of the P300 EEG signals, the topographical plot of averaged EEG signals is shown in Fig. 4, where X represents target stimuli and O represents non-target stimuli. Because the P300 ERP time intervals of target and non-target stimuli are different, which is shown in Figs. 2–3, the topographical plot time is selected at 300 ms and 400 ms separately.

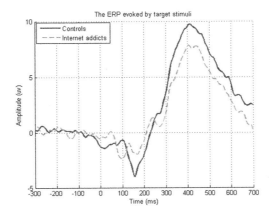

Fig. 3. Observation of the ERP signals evoked by target stimuli

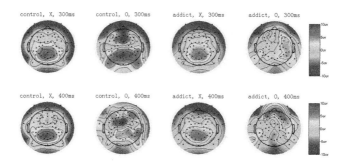

Fig. 4. Topographical plot of the mean EEG signal at 300 ms and 400 ms

3.2 Channel Selection Results

The statistics is shown in Table 1. Channels selected from control subjects are listed in plain text, channels selected from internet addicts are listed in underlined text.

Table 1. Illustration of selected EEG channels under different occurrence rate

Number of occurrences	Channels selected from 10 control subjects	Channels selected from 7 internet addicts
7	{34}	none
6	{33, 36, 47}	none
5	{5, 29, 30, 31, 32}	{34}
4	{1, 3, 8, 10, 35, 38, 40}	{31, 33, 36, 38, 47}
3	{6, 9, 11, 27, 28}	{1, 5, 29, 40}
2	{2, 7, 17, 21, 37, 41, 42, 43, 44, 45, 55, 60}	{3, 8, 17, 28, 30, 32, 41}
1	{4, 12, 13, 16, 20, 24, 26, 39, 52, 57, 58}	{2, 4, 6, 7, 9, 10, 11, 13, 16, 20, 21, 26, 27, 35, 42, 52, 55, 57, 60}

3.3 Cross-Validated Classification Results

Table 2 summarizes the performance of classifying target and non-target stimuli with different set of selected channels on control subjects. The results were got by averaging 10 times 3-fold cross validation results. It can be seen from Table 2 that, although accuracy and specificity increase following the number of channels, sensitivity increases at first, but decreases later. The percentage of correctly classified target on whole channels is not the highest, it is only 57.14 %. The highest sensitivity is 63.64 %, it was got on the optimal channels whose number of occurrences is more than 2 times.

Table 2. Classification rates comparison of different selected channels on control subjects

The least occurrences	Number of channels	Accuracy (%) Mean (%) ± SD	Sensitivity (%) Mean (%) ± SD	Specificity (%) Mean (%) ± SD
7	1	52.71 ± 1.02	53.21 ± 2.90	52.67 ± 1.20
6	4	66.82 ± 0.32	57.67 ± 1.53	67.51 ± 0.37
5	9	73.76 ± 0.30	60.40 ± 1.09	74.76 ± 0.33
4	16	77.15 ± 0.23	62.36 ± 1.33	78.26 ± 0.27
3	21	77.85 ± 0.36	61.36 ± 1.19	79.09 ± 0.39
2	33	81.35 ± 0.21	63.64 ± 1.46	82.68 ± 0.30
1	44	82.51 ± 0.37	62.38 ± 1.70	84.03 ± 0.44
0	64	84.00 ± 0.30	57.14 ± 1.64	86.02 ± 0.36

Table 3 summarizes the performance of classifying target and non-target stimuli with different set of selected channels on internet addicts. The trend is similar to the results of control subjects. The highest sensitivity is 53.81 %. Compare with the results in Table 2, it can be find that the performance on internet addicts is worse than on control subjects, the reason may be that there is deficit on attention of internet addicts.

Table 3. Classification rates comparison of different selected channels on internet addicts

The least occurrences	Number of channels	Accuracy (%) Mean (%) ± SD	Sensitivity (%) Mean (%) ± SD	Specificity (%) Mean (%) ± SD
5	1	55.18 ± 1.29	45.27 ± 2.37	55.93 ± 1.46
4	6	61.80 ± 0.50	51.60 ± 2.21	62.57 ± 0.58
3	10	65.62 ± 0.52	51.73 ± 1.15	66.67 ± 0.54
2	17	71.48 ± 0.48	53.81 ± 2.44	72.81 ± 0.58
1	36	75.56 ± 0.68	52.55 ± 2.15	77.29 ± 0.77
0	64	78.43 ± 0.55	48.23 ± 2.11	80.70 ± 0.59

The sensitivity generalization error in our experiment was calculated according to different number of selected channels on both control subjects and internet addicts, and is shown in Fig. 5.

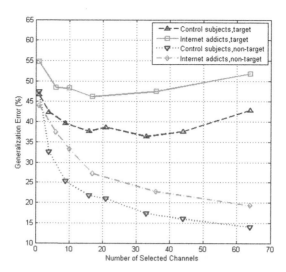

Fig. 5. Generalization error comparison of different number of selected channels

4 Conclusions

This paper focused on a two-stage channel selection model in internet addiction research. Our purpose was to reduce the number of channels for performance and convenient consideration. This method can greatly reduce the redundant channels.

Channel selection result agreed well with the observation. Classification of target and non-target stimuli was done with optimal channels and other comparison sets of channels. The result showed that the optimal channels yielded the best cross validation classification sensitivity. The trend of sensitivity generalization error curve was consistent with the ideal schematic curve. The classification rate of internet addicts is lower than that of control subjects. This indicates that the difference between targets and non-targets on internet addicts is not as obvious as on control subjects. The conclusion is that there is attention deficit on internet addicts.

Acknowledgements. This work has been partially supported by the National Natural Science Foundation of China (61201096, 81501155), Science and Technology Program of Changzhou City (CE20145055) and Qing Lan Project of Jiangsu Province.

References

1. Kuss, D.J., van Rooij, A.J., Shorter, G.W., Griffiths, M.D., van de Mheen, D.: Internet addiction in adolescents: prevalence and risk factors. Comput. Hum. Behav. **29**(5), 1987–1996 (2013)
2. Kuss, D.J., Griffiths, M.D.: Internet and gaming addiction: a systematic literature review of neuroimaging studies. Brain Sci. **2**(3), 347–374 (2012)
3. Wang, Y.J., Gao, S.K., Gao, X.R.: Common spatial pattern method for channel selection in motor imagery based brain-computer interface. In: Proceedings of IEEE Engineering in Medicine and Biology Society, pp. 5392–5395. IEEE Press, New York (2005)

4. Fattahi, D., Nasihatkon, B., Boostani, R.: A general framework to estimate spatial and spatio-spectral filters for EEG signal classification. Neurocomputing **119**(7), 165–174 (2013)
5. Zou, L., Pu, H., Sun, Q., Su, W.: Analysis of attention deficit hyperactivity disorder and control participants in EEG using ICA and PCA. In: Wang, J., Yen, G.G., Polycarpou, M.M. (eds.) ISNN 2012, Part I. LNCS, vol. 7367, pp. 403–410. Springer, Heidelberg (2012)
6. Zou, L., Xu, S.K., Ma, Z.H.: Automatic removal of artifacts from attention deficit hyperactivity disorder electroencephalograms based on independent component analysis. Cogn. Comput. **5**(2), 225–233 (2013)
7. Fan, J., Shao, C., Ouyang, Y., Wang, J., Li, S., Wang, Z.-C.: Automatic seizure detection based on support vector machines with genetic algorithms. In: Wang, T.-D., Li, X., Chen, S.-H., Wang, X., Abbass, H.A., Iba, H., Chen, G.-L., Yao, X. (eds.) SEAL 2006. LNCS, vol. 4247, pp. 845–852. Springer, Heidelberg (2006)
8. Schröder, M., Lal, T.N., Hinterberger, T., Bogdan, M., Hill, N.J., Birbaumer, N., Rosenstiel, W., Schölkopf, B.: Robust EEG channel selection across subjects for brain-computer interfaces. EURASIP J. Adv. Sig. Process. **2005**, 3103–3112 (2005)
9. He, L., Hu, Y.P., Li, Y.Q., Li, D.L.: Channel selection by Rayleigh coefficient maximization based genetic algorithm for classifying single-trial motor imagery EEG. Neurocomputing **121**(9), 422–433 (2013)
10. Yang, J.H., Singh, H., Hines, E.L., Schlaghecken, F., Iliescu, D.D., Leeson, M.S., Stocks, N. G.: Channel selection and classification of electroencephalogram signals: an artificial neural network and genetic algorithm-based approach. Artif. Intell. Med. **55**(2), 117–126 (2012)

Text-independent Speaker Recognition Using Radial Basis Function Network

Anton Yakovenko and Galina Malychina$^{(\boxtimes)}$

Institute of Computer Science and Technology, Peter the Great Saint-Petersburg Polytechnic University, Polytechnicheskaya 21, 194021 St. Petersburg, Russia
yakovenko_aa@spbstu.ru g_f_malychina@mail.ru

Abstract. Radial Basis Function Neural Network (RBFNN) is proposed as a solution to the text-independent speaker recognition problem. Recognition is based on estimation of sufficiently large set of acoustic features, construction of multidimensional histograms and approximation arbitrary distributions of the components of acoustic features with probability density functions (PDF), with possibility of wide shape variation. Proposed method allowed to reduce the probability of errors in the decision making.

Keywords: Speaker recognition · Text-independence · Feature extraction · RBFNN

1 Introduction

Biometric identification allows us to find the right connection to authorize any person in information system. In recent years, the interest in voice biometrics have increased. This is completely in demand in the areas of access permission organization for information systems, biometric solving search and forensic accounting, in banking systems, contact centers, etc. Identification based on speakers voice provides a unique opportunity to secure access to information, remote maintenance and examination to establish the identity.

General statement of speaker recognition problem includes two tasks: identification and verification [1]. In case of speaker verification, the goal is to verify whether the tested phonogram (voiceprint) matches to the voice of speaker, which has been registered in verification system: the object is to confirm a person's identity, using their voice. It is a true or false scenario because there are only two possible outcomes - it is either the clamed speaker or an impostor.

The objective of speaker identification is to identify unknown voice from a set of known voices. As number of speakers in the set increases, the likelihood of making a false identification also increases. For this reason speaker identification for a large population is generally considered more difficult task. Identification based on recognition of speaker's voice reduces to the problem of deciding which of the plurality of speakers most likely belongs to the tested phonogram.

© Springer International Publishing Switzerland 2016
L. Cheng et al. (Eds.): ISNN 2016, LNCS 9719, pp. 74–81, 2016.
DOI: 10.1007/978-3-319-40663-3_9

As well, identification and verification of speaker is divided into two main categories: a text-dependent and text-independent recognition, and can run using open or closed set of speakers. The problem is solved on a closed set of speakers, when the tested phonogram relate to a particular individual, which registered in recognition system. But if the phonogram does not belong to any candidate, then the problem is solved on an open set of speakers.

When biometric system trained in advance to recognize passphrase, that defined by speaker, then it is a text-dependent recognition task. Phonemic dictionary and phrase structure in this case requires less amount of training speech data. The necessity of pronouncing passphrase during training and operation of the system limits the practical range of its application.

Recognition based on text-independent approach does not contain information about the uttered phrase. This approach suggests training and then testing on arbitrary voice and speech data. The performance of such biometric systems is lower than based on the text-dependent technique, becouse it requires longer training and testing utterances to achieve good performance. But speaker recognition in this case more conveniently and has broader application, since knowledge of uttered phrase is optional.

Development of a recognition system occurs in three main steps. Thus, in general, system for speaker recognition includes stage, which comprise analysis of primary feature vectors of the speech signal, and the speaker's voice simulation stage, which are divided according to the tasks [1]. Becouse the human speech is regarded as an acoustic signal, the analysis of the signal takes place by means of digital processing. Since actual recordings made under conditions, there are many extraneous signals, various kinds of noise, impulse noise and congested areas of speech, preprocessing and noise removing stage can improve the efficiency further processing. Special preprocessing algorithms of the entire signal, perform the selection of speech segments, and feature extraction for each segment. Therefore, the operation of the automatic text-independent announcer recognition consists of the following stages: feature extraction, modeling of speakers, comparison of the speaker models and finally decision-making process is carried out. Given phonogram is mapped to the phonogram of reference speaker, by comparing the decision, whether a voice recording belong to this person or different people.

2   RBF Neural Network Approach

Radial Basis Functions (RBF) are variant of feed-forward artificial neural network, that consists of at least three layers of neurons: an input layer, hidden layer and an output layer, where each hidden unit implements a radial activated function [2]. The input into an RBF network is nonlinear while the output is linear. An RBF network with D inputs, M hidden units and K outputs is shown in Fig. 1. The output layer forms a linear combiner which calculates the weighted sum of the outputs of the hidden units.

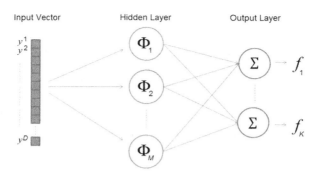

Fig. 1. Neural network structure

The k output of an RBF neural network has the form:

$$f_k(Y) = w_{0k} + \sum_{j=1}^{M} w_{jk}\Phi_j(Y)$$

$j = 1, ..., M$ and $k = 1, ..., K$, where w_{jk} - weights of the network.
For an RBF network the activation function is:

$$\Phi_j(Y) = \exp\{-\frac{1}{2.\sigma_j^2}\|Y - c_j\|^2\} \tag{1}$$

where $\|.\|$ denotes the Euclidean distance.

In (1) Φ_j is activation function, $Y = \{y_t, t = 1, ..., T\}$ is the input vector of length T and dimention D, c_j - function centers, σ_j - the function width.

By means of training, the neural network models the underlying function of a certain mapping. The hidden layer neurons represent a series of centres in the input data space. Each of these centres has an activation function, typically Gaussian. The activation depends on the distance between the presented input vector and the centre. The further the vector is from the centre, the lower is the activation and vice versa. The generation of the centres and their widths is done using an unsupervised k-means [3] clustering algorithm. The centres and widths created by this algorithm then form the weights and biases of the hidden layer, which remain unchanged once the clustering has been done. RBF networks has both a supervised and unsupervised component to its learning [4], but they are used mainly in supervised applications. Fully supervised training to find neuron centers, widths, and amplitude. In a supervised application, used a training set of data samples for which the corresponding network outputs are known. In this case the network parameters are found such that they minimize a cost function.

3 Feature Extraction

Feature extraction process inherently is not specific to the tasks of speaker identification, but rather is common to most areas of speech technology. For the analysis in the speech signal is assumed to use a set of features such as signal energy, linear prediction coefficients, coefficients of smoothed power spectrum, coefficients of real cepstrum, formant frequencies and pitch period for voiced phonemes. Present correlation between features can be reduced by applying principal component analysis to the vector features.

Energy of signal:

$$E(n) = \sum_{m=-N}^{N} x^2(m)w(n-m),$$

where $w(n-m)$ - window function, for example, a Hamming window:

$$h = [0.54 - 0.46 \cos 2\pi n/(N-1)], 0 \le n \le N-1.$$

Linear prediction coefficients, which are the result of solving a system of linear equations Yule - Walker:

$$\sum_{k=1}^{p} a_k R_n(i-k) = R(i), 1 < i \le p,$$

where p - prediction order, $R(i)$ - autocorrelation function, a_k - linear prediction coefficients $1 < k \le p$.

Hamming window reduces the prediction error, as the first p samples of a rectangular window with linearly unpredictable. Autocorrelation function calculated with a window:

$$R_n(k) = \frac{1}{N-1-k} \sum_{m=0}^{N-1-k} [x(m)h(n-m)x] \times [x(m+k)h(m-k-m)].$$

Formant frequency is determined by the smoothed power spectrum:

$$|H(z)|^2 = \left| \frac{G}{A(z)} \right|^2,$$

where $z = e^j w$, $H(z)$ - the transfer function of the vocal tract, $A(z)$ - z-transform of linear prediction coefficient sequences.

Cepstral coefficients:

$$c(n) = \frac{1}{N} \sum_{k=0}^{N-1} \log |X(k)| e^{j\frac{2\pi kn}{N}},$$

where $X(k) = \sum_{n=0}^{N-1} x(n)e^{j\frac{2\pi kn}{N}}$ - Fourier transform of signal frame. Pitch period is determined using the window $l(n)$ for cepstrum:

$$T(n) = c(n)l(n)$$

$$l(n) = \begin{cases} 0 \ |n| < n_0 \\ 1 \ |n| \geq n_0 \end{cases}$$

$$n_0 = \arg\max(T(n)) \neq 0,$$

where n_0 - pitch period. Obtained characteristics from the feature vectors X for each speech segment.

4   Modeling of Speaker

A set of multivariate Probability Density Functions (PDF) describe the hidden acoustic classes of feature vectors. PDF is suitable to approximate arbitrary distributions of the components of acoustic features, making PDF quite convenient for applications in text-independent speaker identification and verification.

Usually in the problem of text independent speaker recognition a Gaussian Mixture Models (GMM) are used. GMM is a speaker probabilistic model for multivariate probability density functions. This model has the obvious disadvantage is that the distribution of acoustic features of speech signals are non-Gaussian, distributions are more peaked. A family of PDF of various shapes, are characterized by three parameters: the expectation m_x, standard deviation σ_x^2 and shape parameter α.

$$f(x) = \frac{\alpha}{2\lambda \sum_x \cdot \Gamma(\frac{1}{\alpha})} \cdot \exp(-\left|\frac{x - m_x}{\lambda \sum_x}\right|^\alpha) \tag{2}$$

where $\Gamma(\alpha) \equiv \int\limits_0^\infty x^{\alpha-1} \exp(-x)dx$.

The distribution function has the form:

$$F(x) = \int\limits_{-\infty}^{(\frac{x-m_x}{\lambda\sigma_x})^\alpha} \frac{1}{2\widetilde{A}(\frac{1}{\alpha})} \exp^{-\zeta} \zeta^{\frac{1}{\alpha}-1}d\zeta.$$

Scale parameter $\beta = \lambda\sigma_x$ of distribution depends on the multiplier λ, which is expressed in terms of the shape parameter α according to the relationship:

$$\lambda = \sqrt{\frac{\widetilde{A}(\frac{1}{\alpha})}{\widetilde{A}(\frac{3}{\alpha})}}.$$

Centers of GMM are proposed to determine using RBF network. Centers of RBF and other parameters of network undergo a supervised learning process. The most convenient for RBF network learning is a gradient descent algorithm that represents a generalization of the Least Mean Square (LMS) algorithm.

The family of RBF networks is broad enough to uniformly approximate any continuous function on a compact set and consists of functions represented by:

$$F(\mathbf{x}) = \sum_{i=1}^{m} a_i \phi(\mathbf{w}_i^T \mathbf{x}) \tag{3}$$

where m - the number of neurons in the first layer, a_i, $\mathbf{w}_i$ - coefficients of neural network, $\phi(.)$ - activation function.

As the activation function $\phi(\mathbf{w}_i^T \mathbf{x})$ in the expression (3) a family of exponential distributions (2) with the shape parameter α is proposed.

Calculating the mean square error of approximation of the mixture of multidimensional sampling distributions:

$$\varepsilon = \frac{1}{2}\sum_{j=1}^{N} e_j^2,$$

where N is the size of the training sample.

Error signal defined by:

$$e_j = d_j - \sum_{i=1}^{M} w_i f(\mathbf{x_j} - \mathbf{m}_i),$$

where d_j - data.

The requirement is to find parameters w_i, $\mathbf{m}_i$, Σ, α_i. For better convergence of the algorithm initial values of parameters are selected. Clustering of the sample data is performed according to the method of k-means, which estimates initial value of the centers $\mathbf{m}_i$, the initial values of α_i are chosen close to the $\alpha_i = 2$, correlation matrix Σ is chosen close to diagonal, weights are initialized with random values.

Neural network training procedure is performed incrementally. Changing weights on the next step:

$$w_i(n+1) = w_i(n) - \eta_1 \frac{\partial \varepsilon(n)}{\partial w_i(n)}, i = 1, ..., m_i$$

$$\frac{\partial \varepsilon(n)}{\partial w_i(n)} = \sum_{j=1}^{N} e_j(n) f(\mathbf{x_j} - \mathbf{m}_i(n)).$$

Adjustment of the position of the centers:

$$t_i(n+1) = t_i(n) - \eta_2 \frac{\partial \varepsilon(n)}{\partial t_i(n)}, i = 1, ..., m_i$$

$$\frac{\partial \varepsilon(n)}{\partial t_i(n)} = \alpha w_i(n) \sum_{j=1}^{N} e_j(n) f'(x_j - m_i(n)) \times \Sigma^{-1}(x_j - t_i(n))^{\alpha-1}.$$

Adjustment of distribution width:

$$\Sigma_i^{-1}(n+1) = \Sigma_i^{-1}(n) - \eta_3 \frac{\partial \varepsilon(n)}{\partial \Sigma_i^{-1}(n)}, i = 1, ..., m_i$$

$$\frac{\partial \varepsilon(n)}{\partial \Sigma_i^{-1}(n)} = -\alpha w_i(n) \sum_{j=1}^{N} e_j(n) f'(\mathbf{x_j} - \mathbf{m}_i(n)) \mathbf{Q}_{ij}(n)$$

$$\mathbf{Q}_{ij}(n) = (\mathbf{x_j} - \mathbf{m}_i(n))^{\alpha-1}(\mathbf{x_j} - \mathbf{m}_i(n))^T.$$

Adjustment of the PDF shape parameter:

$$\alpha_i(n+1) = \alpha_i(n) - \eta_2 \frac{\partial \varepsilon(n)}{\partial \alpha_i(n)}, i = 1, ..., m_i$$

$$\frac{\partial \varepsilon(n)}{\partial \alpha_i(n)} = 2\omega_i(n) \sum_{j=1}^{N} e_j(n)f(x_j - m_i(n))\alpha^{-1} + f'(x_j - m_i(n)) \left(\alpha \left| \frac{x - m(n)}{\lambda\Sigma} \right|^{\alpha-1} \right).$$

5   Results of Experiment

Experimental results are based on a 15-speaker database recorded in the laboratory conditions and consists of two word test utterances, read in random order for every speaker, with normal spoken manner. As shown in [5] such conditions provide good performance for recognition. Analyzed 10 male and 5 female (both true talkers and imposters), for each phonogram obtained multidimensional histogram features. An example is shown in Fig. 2, initial values of distribution centers obtained by k-means clustering. Number of PDF ranged from 10 to 500.

The speaker recognition system has been evaluated in terms of both speaker verification and closed-set speaker identification tasks. The Identification Rate (IR) is used for evaluating the performance of a closed set speaker identification systems. The Equal Error Rate (EER) is used for evaluating the performance of the speaker verification. The EER is defined as the point at which two errors False Rejection Rate (FRR) and False Acceptance Rate (FAR) are equal.

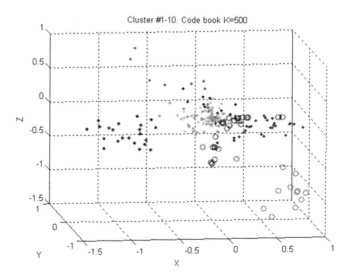

Fig. 2. Projection of histogram on the three main components

Estimation of errors of the first and second kind for different size of RBF neural network for different sample sizes and different speakers, in order to determine the optimal parameters for recording speaker recognition.

Each speaker to be recognized had their own RBF network model [6] which is trained to recognise spectral feature vectors representative of his speech. The best performance is obtained from an RBF network with 384 RBF nodes in the hidden layer, given an 8% true talker rejection rate for a fixed 1% imposter acceptance rate.

6   Conclusion

This paper presents mechanism of the speaker recognition system and several text-independent speaker identification and verification experiments based on RBF networks. The RBF neural network classifies the input feature space of the speakers into clusters and subsequently selects appropriate neural net for determining the speaker under a class. Main consideration in the speaker recognition problem has been given to the selection of features. Experimental results demonstrate that the signal energy, linear prediction coefficients, coefficients of smoothed power spectrum, real cepstrum, formant frequencies and pitch period in the speech signal together can identify the speaker to high accuracy. The results show that RBF networks offer fast learning speed and good generalization, and they are very robust in detecting impostors for clean speech.

References

1. Campbell, J.P.: Speaker recognition: a tutorial. Proc. IEEE **85**(9), 1437–1462 (1997)
2. Bors, A.G.: Introduction of the radial basis function (RBF) networks. Online Symp. Electron. Eng. **1**, 1–7 (2001)
3. Horne, B.G.: Progress in supervised neural networks. IEEE Signal Process. Mag. **10**(1), 8–39 (1993)
4. Balaska, N., Ahmida, Z., Goutas, A.: Speaker recognition using artificial neural networks: RBFNNs vs. EBFNNs. In: Proceedings of the 4th International Conference on Computer Integrated Manufacturing (2007)
5. Oglesby, J., Mason, J.S.: Radial basis function networks for speaker recognition. In: Proceedings of ICASSP-91, pp. 393–396 (1991)
6. Lian, H., Wang, Z., Wang, J., Zhang, L.-M.: Speaker identification using reduced RBF networks array. In: Yin, F.-L., Wang, J., Guo, C. (eds.) ISNN 2004. LNCS, vol. 3173, pp. 924–929. Springer, Heidelberg (2004)

Usage of DNN in Speaker Recognition: Advantages and Problems

Oleg Kudashev[1], Sergey Novoselov[1(✉)], Timur Pekhovsky[1,2],
Konstantin Simonchik[2], and Galina Lavrentyeva[1,2]

[1] Speech Technology Center, Krasutskogo str. 4, 196084 St. Peterburg, Russia
{kudashev,novoselov,tim,lavrentyeva}@speechpro.com
[2] ITMO University, St. Petersburg, Russia
simonchik@speechpro.com

Abstract. In this paper we consider different approaches of artificial neural networks application for speaker recognition task. We investigated the performance of DNN application at different levels of speaker recognition system: i-vector extraction level and model Back-End level. Results of our study perform high efficiency of the proposed neural network based approaches for solving this problem. It is shown that the use of DNN technology at different levels increases the reliability of speaker recognition system independently. However, there are some disadvantages of such systems, which are also described in this paper.

Keywords: DNN · Speaker recognition · PLDA

1 Introduction

State-of-the-art technology of text-independent speaker recognition is based on the i-vector extraction paradigm. Typically this framework can be decomposed into three stages: the collection of sufficient statistics, the extraction of i-vectors and a probabilistic linear discriminant analysis (PLDA) backend [1–4].

Sufficient statistics are collected by using a sequence of feature vectors (e.g., mel-frequency cepstral coefficients (MFCC)) which are usually represented by the Baum-Welch statistics obtained with respect to a GMM, refered to as universal background model (UBM). These statistics are converted into a single low-dimensional feature vector — an i-vector — that represents important information about the speaker and all other types of variability. After i-vectors are extracted, a PLDA model is used to produce verification scores by comparing i-vectors extracted from different speech segments.

Successful application of deep neural networks (DNN) [5, 6] in automatic speech recognition has provided a strong motivations to searching attempts of possible gains from applying DNN to speaker recognition task. For example, DNN posteriors instead of GMM posteriors have been used by Lei et al. [1], Kenny et al. [2] to derive sufficient statistics for alternative i-vectors calculation allowing to discriminate speakers at triphone level. According to recent results, this approach significantly outperforms a conventional UBM-TV-i-vectors scheme in speaker recognition on telephone speech.

© Springer International Publishing Switzerland 2016
L. Cheng et al. (Eds.): ISNN 2016, LNCS 9719, pp. 82–91, 2016.
DOI: 10.1007/978-3-319-40663-3_10

In the paper [7] authors also reported good achievement in DNN-based speaker identification (SID) performance on microphone speech. Two approaches of DNN-based SID were considered: one uses the DNN to extract features, and another uses the DNN for feature modeling. Modeling is performed using the i-vector framework, in which the traditional universal background model is replaced with a DNN.

Alternative way of succesfull applying DNN in SID task is extracting of bottleneck features from the DNN. Once bottleneck features are extracted, standard UBM based approach can be used to produce effective i-vectors [7].

In our work [4] we investigated the efficiency of neural networks (NN) implementation at the Back-End level. We proposed to use Denoising autoencoder (DAE) for compensation of the within-speaker variability on the model level. The training procedure of our DAE contain two stages: (I) supervized pre-training using Denoising Restricted Boltzmann Machines (RBM) and (II) fine tuning of DAE weights. Final DAE-system uses standard PLDA as back-end. Trained in the described way DAE-PLDA system demonstrated the significant improvement compared to the standard Baseline-PLDA scheme based on the traditional UBM – TV - i-vector approach. In paper [4], as well, we tested our DAE-system on the new i-vectors [1, 2], obtained by using posteriors from DNN Automatic Speech Recognition (ASR) system. DAE gives additional contribution in high performance of DNN- posteriors system.

In this paper we summarized the results of our research for DNN application at different system levels for text-independent speaker recognition task. We produced comprehensive investigation of new DNN-based approaches with the use of variable training and test data corpus. We also considered advantages and disadvantages of the proposed methods.

2   Extraction of i-Vectors

2.1   GMM-Based Approach

GMM based approach for i-vector extraction consists of mean values estimation of speaker voice features attributed to each component of Universal Background Model (UBM). The distribution of concatenated mean vectors (mean supervector) is assumed as:

$$s = m_0 + T\omega$$

where m_0 is mean supervector of UBM, T is the matrix defining the basis in the reduced feature space, ω is the i-vector in the reduced feature space, with prior $p(\omega) = \mathcal{N}(0, I)$.

T and m_0 are global parameters which can be estimated using Factor Analysis and Expectation-Maximization algorithm. The reader can refer to [13] for more details.

UBM is obtained by unsupervised training using huge amount of representative unlabeled speech data. Thus, UBM is used for posteriors calculation on each speech frame followed by Baum-Welch statistics accumulation [2].

2.2   ASR DNN - Based Approach

GMM and ASR DNN are both intended for initial clusterization of the processed speech signal acoustic features. However, despite the GMM, which is trained by the EM-algorithm in an unsupervised way, ASR DNN is trained on the base of preliminary labeled data in a supervised way in order to minimize loss function value:

$$\mathcal{L}(X) = \sum_{t=1}^{T} \ln p(x_t, s_t)$$

where $X = \{x_1, \ldots, x_T\}$ - ASR voice features set; s_t - target class, which vector x_t should be attended to, according to ground truth labels; $p(x_t, s_t)$- value of output soft-max neural network layer for s_t class.

In this paper for ASR DNN training back propagation method was used. As a target classes we use triphone sequences of russian and english languages, as well as, the set of non-speech classes for noise, silence, laugh etc. ASR DNN was trained on the spontaneous speech recordings of phone conversations. Beforehand we aligned the class labels by the standard ASR approach based on HMM.

During our experiments we used DNN with the structure demonstrated in Fig. 1.

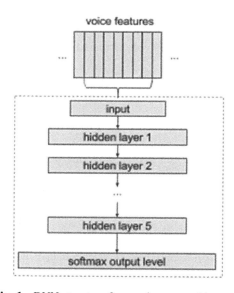

Fig. 1. DNN structure for speaker recognition task

This DNN consists of input layer, 5 inner layers and output soft-max layer. As the input data we used vector of concatenated consecutive ASR voice features, which were performed by 13 first MFCC features (including C_0). This approach is widely used in ASR systems and is called Long-Term voice features.

After that values from output soft-max DNN layer are used for Baum-Welsh statistics estimation for each class separately:

$$N_c = \sum_t p(x_t, c), F_c = \sum_t \hat{x}_t \cdot p(x_t, c),$$

where $\hat{x}_t$- speaker identification (SID) voice features, which can differ from ASR voice features in general case. In this research we chose 20 first MFCC features (including C_0) with its first and second order derivatives as SID voice features for our system.

Similar to GMM case, described statistics are further used for i-vectors extraction. But despite the GMM, DNN statistics also contain non-speech components in addition to voice components. These non-speech components should be preliminary removed. The removal of these non-speech components allows to produce "soft" removal of non-speech segments contribution to overall statistics, despite the "hard"-decision VAD. General scheme of the DNN-based i-vector extraction procedure is shown in Fig. 2.

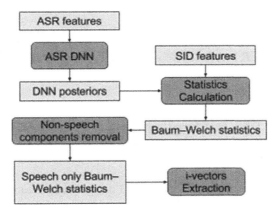

Fig. 2. Scheme of ASR DNN-based i-vectors extraction

3 Applying DNN as a Back-End

3.1 Baseline PLDA

Among state-of-the-art speaker verification systems, leading positions are occupied by PLDA-based systems [1–4, 8] working in the i-vector space.

$$i_{s,r} = \mu + V_y y_s + \epsilon_{s,h} \tag{1}$$

where $i_{s,h}$ is an f-dimensional i-vector from set $\{i_1, \ldots, i_H\}$, obtained from H utterances belonging to speaker s, and $y, \epsilon_{s,h}$ is hidden speaker factors and Gaussian noise, respectively. V_y- is an eigenvoices matrix.

In order to carry out a large number of experiments we did not use costly EM-algorithm [8] to estimate the model parameters. Instead of that, similar to our study [4], we estimated between-speaker and within-speaker covariances, respectively, according to formulas:

$$\Sigma_B = \frac{1}{S}\sum_s (i_s - \mu)(i_s - \mu)^T, \tag{2}$$

$$\Sigma_W = \frac{1}{S}\sum_s^S \frac{1}{H}\sum_h (i_{s,h} - i_s)(i_{s,h} - i_s)^T, \tag{3}$$

The vector $i_{s,h}$ is an i-vector extracted from the h-th session of the s-th speaker, the vector i_s is the average over all sessions of this speaker. i_s can be viewed as the maximum likelihood estimate in the Gaussian model of within-speaker variability. μ is the dataset mean and S is the number of speakers in the training set. Given a pair of i-vectors i_1 and i_2, assuming zero mean and skipping the scalar term, the commonly used PLDA verification score can be written as:

$$Score = i_1^T Q i_1 + i_2^T Q i_2 + 2i_1^T P i_2 \tag{4}$$

where square matrices P and Q can be expressed in terms of (2) and (3).

3.2 Denoising Auto-Encoders Based Back-End

An unconventional, DAE [11] based scheme is presented in Fig. 2. It aims is to minimize within-class variability of input i-vectors. In the scheme B denoting binary hidden layer, **V** and **W** are weight matrices. **G** in Fig. 2 reflects the fact that both parts of the input layer are real valued.

First, the hidden layer is pretrained as RBM using standard contrastive divergence method. The learning technique is similar to that described earlier [9], except that an average of all i-vectors referring to speaker s is employed instead of the speaker label whereas each i-vector from the speaker s together with its label was used in [9].

Following pretraining, discriminative fine-tuning of the model is performed with the same data (Fig. 2, right). Speaker averaged i-vectors are used as targets and individual session/speaker i-vectors as inputs for the model. The latter may be considered as a standard DAE [11] trained to produce the speaker i-vector while compensating for within-speaker variability (noise). MSE cost function is used at the fine-tuning stage. The resulting autoencoder is then applied to the training set, its outputs being used as inputs to train a classical PLDA model serving as back-end to provide final verification scores (Fig. 3).

3.3 Replacing Back-End Parameters for DAE

During our experiments we found out that whitening and i-vector length normalization (LN) [10] are critical for training RBM. Following the standard recipe, first, we estimate dataset mean vector and the whitening matrix A (see Fig. 3), then we project

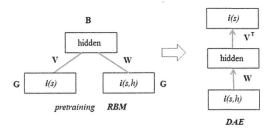

Fig. 3. Schematic diagram of DAE training.

whitened i-vector on the unit hypersphere [10] (hereinafter these three steps are referred to as normalization). According to the diagram in Fig. 3, we also use the same normalization block before DAE training. In Fig. 3 there are also Baseline PLDA scheme and RBM-PLDA-based verification system in the center.

We found out that we get the best performance for the DAE system if we use the set of parameters $\{P, Q, A, \mu\}$ estimated on the i-vectors passed through the RBM instead of DAE. This parameter transfer is depicted by arrows on the Fig. 3. At that time we could not find a convincing theoretical explanation why RBM transform provides better parameters for DAE then DAE itself. It is still an open question (Fig. 4).

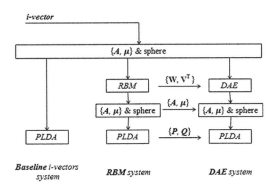

Fig. 4. Block diagram of speaker recognition systems based on DAE. Arrows denote transferring of parameters and dashed rectangle denotes discriminative training of the encircled part.

4   Experimental Settings and Results

4.1   Data Bases

In our research we used various spontaneous speech corpus in russian and english languages. Among the wide variety of speech data bases appropriate for investigation in speaker recognition field the most widespread are bases collected by the National Institute of Standards and Technology (USA) during the regular international competitions on voice biometrics. In our study we used telephone dialogs data bases in English from NIST SRE 1998-2008 (16618 sessions from 1763 male speakers) for extractors and back-end models training. Data bases from NIST 2010 (det-5 protocol) and NIST 2012 (C2 protocol) were used as test bases in English. In our experiments we used only male voices.

We also investigated described approaches on the corpus of russian spontaneous speech. For this purpose we used RusTelecom data base to train extractors and back-end models. RusTelecom - is the russian speech corpus of telephone data, collected by call-centers in Russia. In this case we also used only male voices during our experiments. Train part of RusTelecom data base consists of 33678 sessions from 6508 male speakers. For our experiments as a test part we used russian telephone corpus of russian dialog conversations, collected by the call-center of Speech Technology Center company (STC-Base-PHN). Evaluation part consists of 50 male speakers recordings. Evaluation protocol (single segments comparisons) contains 5200 target trials and 78645 imposter trials. In order to check our results in microphone channel we used STC-Base-MIC base, that contain the same speakers recordings as STC-Base-PHN, but recorded in microphone channel.

Our English ASR DNN (ENG-DNN) was trained with KALDI toolkit [12] on the SwitchBoard RC2 speech corpus and had 2720 outputs (senones). Our Russian ASR DNN (RUS-DNN) was also trained with KALDI on the russian telephone corpus of russian dialog conversations and had the same 2720 outputs. Accordingly, the UBM in GMM-based systems included 2720 components. The networks comprised 6 hidden layers 2048 nodes each with sigmoid activation function. The hidden layers were generatively pretrained as RBMs. Then the network has been discriminatively fine-tuned with stochastic gradient descent algorithm.

4.2   Results

Table 1 demonstrates the comparative results of investigation of different speaker recognition systems based on T-extractors, that used UBM-GMM and DNN for state-posterior calculation. These results were obtained with the use of Baseline PLDA as Back-end model for scoring.

By analyzing results performed in first and second lines of Table 1, it can be concluded that DNN-based english language system provides significantly better

Table 1. Experiment results of speaker verification systems with Baseline PLDA Back-End in telephone channel.

System configuration			NIST'2010	NIST'2012	Stc-Base-PHN
T-extractor	Extractor train base	Back-end train base	EER [%]/ minDCF	EER [%]/ minDCF	EER [%]/ minDCF
GMM-UBM	NIST's	NIST's	2.73/0.51	4.73/0.31	6.05/0.33
ENG-DNN	NIST's	NIST's	1.64/0.36	3.25/0.25	3.62/0.33
GMM-UBM	NIST's + RusTelecom	NIST's + RusTelecom	–	6.53/0.38	**4.91/0.32**
RUS-DNN	NIST's + RusTelecom	NIST's + RusTelecom	–	**5.38/0.35**	**2.64/0.20**

quality verification results at EER and minDCF points [14], achieving 1.64 % EER for NIST 2010 det 5. It can be mentioned that english language systems showed worse results on the russian language than on the english language test bases. Which proves that system based on DNN posteriors are sensitive to language mismatch conditions. Results for russian language systems (second and third lines in Table 1) also confirm the success of the DNN-based extractor and language-dependency of described speaker recognition systems. Notice that we managed to achieve relative reduction of 46 % in EER point for RUS-DNN based extractor on STC-Base-PHN, compared to the standard GMM-UBM approach.

Experiment results for microphone channel are presented In Table 2. DNN-based system shows degradation in this conditions, in comparisson to GMM-UBM extractor. Thus, system with DNN extractor tuned for telephone channel conditions proves to be ineffective in case of microphone channel. This fact indicates a low robustness of such systems in case of transition to unknown recording conditions.

Results in Table 3 were obtained for the same extractors as in Table 1. These results were obtained with the use of DAE with replaced Back-End parameters from RBM PLDA as Back-end model. They demonstrate success of NN technologies implementation in model level of speaker recognition systems. Applying of channel-compensating autoencoder in Back-End level allows to additionally improve recognition quality for all considered test bases. Regardless of the DNN technology implementation at i-vectors extraction level, DAE application reduces the verification errors rate for baseline UBM-GMM systems and DNN-based either. As it was mentioned before, results in Table 3 were obtained by using the whitening and PLDA parameters from RBM projector. By the way there is no theoretical explanation of the effect of system enhancement, that was obtained by additional substitution of upper level hyperparameters, which is disadvantage of this approach.

Table 2. Experiment results of speaker verification systems with Baseline PLDA Back-End in microphone channel.

System configuration			Stc-Base-MIC
T-extractor	Extractor train base	Back-end train base	EER [%]/minDCF
GMM-UBM	NIST's	NIST's	4.40/0.35
ENG-DNN	NIST's	NIST's	5.95/0.46

Table 3. Experiment results for speaker verification systems with DAE- PLDA Back-End.

System configuration			Nist'2010	NIST'2012	Stc-Base-PHN
T-extractor	Extractor train base	Back-end train base	EER [%]/ minDCF	EER [%]/ minDCF	EER [%]/ minDCF
GMM-UBM	NIST's	NIST's	**2.31/0.50**	**3.68/0.29**	**4.47/0.30**
ENG-DNN	NIST's	NIST's	1.44/0.29	2.5/0.22	**3.25/0.21**
GMM-UBM	NIST's + RusTelecom	NIST's + RusTelecom	–	5.8/0.38	**4.12/0.28**
RUS-DNN	NIST's + RusTelecom	NIST's + RusTelecom	–	**4.42/0.31**	**2.29/0.18**

5 Conclusions

In this paper we investigated the efficiency of DNN technology application for text-independent speaker recognition task. We considered DNN implementation at different levels of speaker recognition system: i-vector extraction level and model Back-End level. Results obtained during our research confirm high efficiency of the proposed neural network based approaches for SID. We managed to achieve relative reduction of 46 % in EER point for STC-Base-PHN date base by applying DNN based extractor. Using of DAE at the model level reduced EER by 13 % compared to Baseline PLDA system. It should be mentioned, that use of DNN technology at various levels improves the speaker recognition system quality independently. Nevertheless, the main disadvantages of the described systems are language dependency and low robustness in case of transition to unknown recording conditions. In order to enhance the system reliability in such cases it is necessary to tune the extractor to work in broader conditions. This can be done by expanding the training data base by different conditions of voice recording. The lack of theoretical explanation of some occurring effects is disadvantage of system with DAE model Back-End.

Acknowledgments. This work was partially financially supported by the Government of the Russian Federation, Grant 074-U01.

References

1. Lei, Y., Scheffer, N., Ferrer, L., McLaren, M.: A novel scheme for speaker recognition using a phonetically aware deep neural network. In: 2014 IEEE International Conference on Acoustics, Speech, Signal Process, pp. 1695–1699 (2014)
2. Kenny, P., Gupta, V., Stafylakis, T., Ouellet, P., Alam, J.: Deep Neural Networks for extracting Baum-Welch statistics for Speaker Recognition. The Speaker and Language Recognition Workshop (2014). http://cs.uef.fi/odyssey2014/program/pdfs/28.pdf
3. Stafylakis, T., Kenny, P., Senoussaoui, M., Dumouchel, P.: PLDA using gaussian restricted Boltzmann machines with application to speaker recognition. In: 3th Annual Conference of the International Speech Communication Association, Portland, OR, USA, pp. 1692–1696 (2012)
4. Novoselov, S., Pekhovsky, T., Kudashev, O., Mendelev, V., Prudnikov, A.: Non-linear PLDA for i-vector speaker verification. In: 16th Annual Conference of the International Speech Communication Association, Dresden, Germany, pp. 214–218 (2015)
5. Hinton, G.E., Osindero, S., Teh, Y.W.: A fast learning algorithm for deep belief nets. Neural Comput. **18**(7), 1527–1554 (2006)
6. Hinton, G., Deng, L., Yu, D., Dahl, G., Mohamed, A., Jaitly, N., Senior, A., Vanhoucke, V., Nguyen, P., Sainath, T., Kingsbury, B.: Deep neural networks for acoustic modeling in speech recognition. IEEE Sig. Process **29**(6), 82–97 (2012)
7. McLaren., M., Lei, Y., Ferrer, L.: Advances in deep neural network approaches to speaker recognition. In: International Conference on Acoustics, Speech and Signal Processing (ICASSP). IEEE Press (2015)

8. Prince, S.J.D., Elder, J.H.: Probabilistic linear discriminant analysis for inferences about identity. In: IEEE 11th International Conference on Computer Vision, Rio de Janeiro, Brazil, pp. 1–8 (2007)
9. Novoselov, S., Pekhovsky, T., Simonchik, K.: STC Speaker Recognition System for the NIST i-Vector Challenge. The Speaker and Language Recognition Workshop. http://cs.uef.fi/odyssey2014/program/pdfs/25.pdf
10. Daniel, G.R., Carol, Y.E.W.: Analysis of i-vector length normalization in speaker recognition systems. In: 12th Annual Conference of the International Speech Communication Association, Florence, Italy, pp. 249–252 (2011)
11. Vincent, P., Larochelle, H., Bengio, Y., Manzagol, P.: Extracting and composing robust Features with denoising autoencoders. In: 25th International Conference on Machine Learning, Helsinki, Finland (2008)
12. Povey, D., Ghoshal, A., Boulianne, G., Burget, L., Glembek, O., Goel, N., Hannemann, M., Motlıcek, P., Qian, Y., Schwarz, P., Silovsky, J., Stemmer, G., Vesely, K.: The Kaldi speech recognition toolkit. In: IEEE Automatic Speech Recognition and Understanding Workshop (2011)
13. Kenny, P.: A study of interspeaker variability in speaker verification. IEEE Trans. Audio Speech Lang. Process. **16**(5), 980–988 (2008)
14. The NIST Year 2012 Speaker Recognition Evaluation Plan. http://www.nist.gov/itl/iad/mig/upload/NIST_SRE12_evalplan-v17-r1.pdf

Boosted Inductive Matrix Completion for Image Tagging

Yuqing Hou[(✉)]

Key Laboratory of Machine Perception (MOE), School of EECS, Peking University,
Beijing 100871, China
houyuqing1988@gmail.com

Abstract. Search engines have traditionally used manual image tagging for indexing and retrieving image collections. Manual tagging is expensive and labor intensive, motivating the research on automatic tag completion. However, existing tag completion approaches suffer from deficient or inaccurate tags. In this study, we formulate the task in the boosted inductive matrix completion (BIMC) framework, which combines the power of the inductive matrix completion (IMC) model together with a standard matrix completion (MC) model. We incorporates visual-tag correlation and semantic-tag correlation properties into the model for better exploration of the latent connection between image features and tags. We exploit CNN features and word vectors to narrow the semantic gap. The proposed method achieves good performance on several benchmark datasets with missing and noisy tags.

Keywords: Image tag completion · Boosted inductive matrix completion · Visual-tag correlation · Semantic-tag correlation · CNN features · Word vectors

1 Introduction and Motivation

Many machine learning methods have been developed for the image tag completion task. However, most methods are usually region-based, depending heavily on the image segmentation accuracy. In recent years, matrix completion-based methods [2–4] stand out owing to their robustness and efficiency property since they avoid the image segmentation and features similarity calculation procedures.

Matrix completion-based methods usually operate on the tag matrix $\mathbf{O} \in \mathbb{R}^{N_{im} \times N_{tg}}$, where each row corresponds to one image, each column corresponds to one tag, and N_{im} and N_{tg} denote the number of images and tags respectively. $o_{ij} = 1$ only if image i is annotated with tag j and 0 otherwise. Thus one can get a completed tag matrix $\hat{\mathbf{O}}$ by completing the matrix $\mathbf{O}$ [2,5,6]. During the matrix completion procedure, if we can observe that o_{ij} is nonzero (zero) but $\hat{o}_{ij}$ becomes zero (nonzero), we say that the algorithm removes (adds) tag j from (to) image i.

L. Cheng et al. (Eds.): ISNN 2016, LNCS 9719, pp. 92–99, 2016.
DOI: 10.1007/978-3-319-40663-3_11

Semantically similar images usually have similar tags. However, the relationship between images and tags can hardly be characterized by linear models and traditional matrix completion-based methods can not take full advantage of side information such as user activity (e.g., like and reblog) and rich content (e.g., tags and images) [1]. In this work, we exploit the recently proposed BIMC [1] model, which combines the power of IMC [7] together with MC. BIMC have demonstrated its scalability and capability of exploiting side information in blog recommendation [1], where it effectively combines heterogeneous user and blog features from multiple sources for more accurate recommendations. To make the most of the side information, we model the visual-tag correlation and semantic-tag correlation properties in the BIMC model. Word vectors and CNN features are further utilized for extracting the tag and the visual features, respectively. These features have high level semantic meanings and can narrow the semantic gap effectively.

2   Boosted Inductive Matrix Completion

This section introduces the original BIMC model. Sections 2.1, 2.2 and 2.3 introduces the formulation for standard MC model, the IMC model and the BIMC model briefly.

2.1   Standard Matrix Completion

Low rank matrix completion (MC) recover the underlying low rank matrix by using the observed entries of $\mathbf{O}$, which is typically formulated as follows:

$$\min_{\mathbf{P},\mathbf{Q}} \|\mathbf{U} \odot (\mathbf{O} - (\mathbf{P}\mathbf{Q}^\top))\|_F^2 + \lambda(\mathrm{rank}(\mathbf{P}\mathbf{Q}^\top)) \tag{1}$$

where $\mathbf{P} \in \mathbb{R}^{N_{im}}$ and $\mathbf{Q} \in \mathbb{R}^{N_{tg}}$ with r being the dimension of the latent feature space; $\mathbf{U}$ is the 0/1 binary mask with the same size as $\mathbf{O}$. The entry value 0 means that the corresponding entry in $\mathbf{O}$ is not observed, and 1 otherwise. The operator $\odot$ is the Hadamard entry-wise product. λ is a regularization parameter. The low-rank constraint on $\mathbf{P}\mathbf{Q}^\top$ is NP-hard to solve. The standard relaxation of the rank constraint is the trace norm, which is equivalent to minimizing $\frac{1}{2}(\|\mathbf{P}\|_F^2 + \|\mathbf{Q}\|_F^2)$ [7]:

$$\min_{\mathbf{P},\mathbf{Q}} \|\mathbf{U} \odot (\mathbf{O} - (\mathbf{P}\mathbf{Q}^\top))\|_F^2 + \frac{\lambda}{2}(\|\mathbf{P}\|_F^2 + \|\mathbf{Q}\|_F^2) \tag{2}$$

Note that MC only utilizes the observed entries of $\mathbf{O}$.

2.2   Inductive Matrix Completion

Standard MC methods is restricted to their transductive setting, thus cannot predict tags for new images. Further more, MC suffers performance with extreme sparsity in the data [1]. IMC is proposed to alleviate data sparsity issues as well as enable predictions for new images and tags by incorporating side information.

Let $\mathbf{v}_i \in \mathbb{R}^{f_{im}}$ denote the feature vector of image i, and $\mathbf{t}_j \in \mathbb{R}^{f_{tg}}$ denote the feature vector of tag j. Let $\mathbf{V} \in \mathbb{R}^{N_{im} \times f_{im}}$ denote the feature matrix of N_{im} images, where the i-th row is the image feature vector $\mathbf{v}_i$, and $\mathbf{T} \in \mathbb{R}^{N_{tg} \times f_{tg}}$ denote the feature matrix of N_{tg} tags, where the i-th row is the tag feature $\mathbf{t}_i$.

IMC assume that the tag matrix is generated by applying feature vectors associated with its row as well as column entities to a underlying low-rank matrix $\mathbf{M} = \mathbf{W}\mathbf{H}^\top$, where $\mathbf{W} \in \mathbb{R}^{f_{im} \times r}, \mathbf{H} \in \mathbb{R}^{r \times f_{tg}}$ are of rank $r \ll N_{im}, N_{tg}$:

$$\min_{\mathbf{W},\mathbf{H}} \quad \text{loss}(\mathbf{O}_{i,j}, (\mathbf{V}\mathbf{W}\mathbf{H}^\top\mathbf{T}^\top)_{i,j}) + \frac{\lambda}{2}(\|\mathbf{W}\|_F^2 + \|\mathbf{H}\|_F^2) \qquad (3)$$

A common choice for the loss function is the squared loss:

$$\min_{\mathbf{W},\mathbf{H}} \|\mathbf{U} \odot (\mathbf{O} - \mathbf{V}\mathbf{W}\mathbf{H}^\top\mathbf{T}^\top)\|_F^2 + \frac{\lambda}{2}(\|\mathbf{W}\|_F^2 + \|\mathbf{H}\|_F^2) \qquad (4)$$

2.3   Boosted Inductive Matrix Completion

IMC is too rigid as it heavily depends on the image feature matrix $\mathbf{V}$ and tag feature matrix $\mathbf{T}$. BIMC tackle the problem by combine both standard MC and IMC, and thereby better utilize the power of both. BIMC combine the power of MC to reduce the noise level in the input data as well as the advantage of IMC to incorporate side information of users and items [1]. BIMC models $\mathbf{O}_{i,j}$ as

$$\mathbf{O}_{i,j} = (\mathbf{P}\mathbf{Q}^\top)_{i,j} + \alpha \mathbf{v_i}^\top \mathbf{M}\mathbf{t_j} \qquad (5)$$

where the parameter α adjusts the contribution of features in the final prediction.

BIMC first learn the latent factor matrices $\mathbf{P}$ and $\mathbf{Q}$ of the MC model as in (2). The resulting approximation error or residual matrix $\mathbf{R} = \mathbf{O} - \mathbf{P}\mathbf{Q}^\top$ can then be modeled with IMC as:

$$\mathbf{R}_{i,j} = \mathbf{O}_{i,j} - (\mathbf{P}\mathbf{Q}^\top)_{i,j} = \mathbf{v_i}^\top \mathbf{M}\mathbf{t_j} \qquad (6)$$

Thus, choosing the squared loss, the object function of BIMC is

$$\min_{\mathbf{W},\mathbf{H}} \quad \|\mathbf{U} \odot (\mathbf{O} - \mathbf{P}\mathbf{Q}^\top - \mathbf{V}\mathbf{W}\mathbf{H}^\top\mathbf{T}^\top)\|_F^2 + \frac{\lambda}{2}(\|\mathbf{W}\|_F^2 + \|\mathbf{H}\|_F^2) \qquad (7)$$

We will introduce the incorporation of visual-tag correlation and semantic-tag correlation in Sect. 3.

3   Incorporating Visual-Tag Correlation and Semantic-Tag Correlation

To make the most of side information, we incorporate the visual-tag correlation and semantic-tag correlation in the BIMC model.

Let the ith row of the residual matrix $\mathbf{R}$ as $\mathbf{R}_i$, corresponding to the residual tag vector of image i. Thus we can measure the correlation between image i and image j in two ways: (1) similarity between image features $\mathbf{v}_i$ and $\mathbf{v}_j$, (2) similarity between residual tag vectors $\mathbf{R}_i$ and $\mathbf{R}_j$. Since semantically similar images usually have similar tags, these two kinds of similarities should be correlated.

Similarly, since each column of the the residual matrix $\mathbf{R}$ represents the feature of a tag, we can measure the correlation between tag i and tag j in two ways: (1) similarity between their corresponding word vectors $\mathbf{t}_i$ and $\mathbf{t}_j$, (2) similarity between $\mathbf{R}^i$ and $\mathbf{R}^j$. These two kinds of similarities should be correlated, too.

We define $g_{ij} = \cos(\mathbf{v}_i, \mathbf{v}_j)$ and $h_{ij} = \cos(\mathbf{t}_i, \mathbf{t}_j)$ to measures the similarity between $\mathbf{v}_i, \mathbf{v}_j$ and $\mathbf{t}_i, \mathbf{t}_j$, respectively. Similar to [8], we model the two kinds of correlation using Graph Laplacian technique [9]:

$$\min_{\mathbf{W},\mathbf{H}} \mathrm{Tr}(\mathbf{R}\mathbf{L_v}\mathbf{R}^\top + \mathbf{R}^\top\mathbf{L_s}\mathbf{R}) = \tag{8}$$

$$\min_{\mathbf{W},\mathbf{H}}[\mathrm{Tr}(\mathbf{VWH}^\top\mathbf{T}^\top\mathbf{L_v}\mathbf{THW}^\top\mathbf{V}^\top) + \mathrm{Tr}(\mathbf{THW}^\top\mathbf{V}^\top\mathbf{L_s}\mathbf{VWH}^\top\mathbf{T}^\top)]$$

where $\mathbf{L_v} = \mathrm{diag}(\mathbf{G1}) - \mathbf{G}$ is the Graph Laplacian matrix of visual similarity matrix $\mathbf{G}$, and $\mathbf{L_s} = \mathrm{diag}(\mathbf{H1}) - \mathbf{H}$ is the Graph Laplacian matrix of semantic similarity matrix $\mathbf{H}$.

Thus we can incorporate the two kinds of correlation into IMC:

$$\min_{\mathbf{W},\mathbf{H}} \quad \|\mathbf{U} \odot (\mathbf{O} - \mathbf{PQ}^\top - \mathbf{VWH}^\top\mathbf{T}^\top)\|_F^2 + \frac{\lambda_1}{2}(\|\mathbf{W}\|_F^2 + \|\mathbf{H}\|_F^2) + \tag{9}$$

$$\lambda_2[\mathrm{Tr}(\mathbf{VWH}^\top\mathbf{T}^\top\mathbf{L_v}\mathbf{THW}^\top\mathbf{V}^\top) + \mathrm{Tr}(\mathbf{THW}^\top\mathbf{V}^\top\mathbf{L_s}\mathbf{VWH}^\top\mathbf{T}^\top)]$$

We set a same weight parameter λ_2 for both visual-tag correlation and semantic-tag correlation for optimization efficiency.

4   Optimization

The objective function is non-convex. We utilize the same MC and IMC solvers adopted in [1] to solve our formulation. First we solve MC subproblem using [10], then we use LELM [11] method, which naturally fits for the large-scale multi-label learning with missing labels task, to solve the improved IMC subproblem. The solver uses alternating minimization (fix $\mathbf{W}$ and solve for $\mathbf{H}$ and vice versa) to optimize the function. When $\mathbf{W}$ or $\mathbf{H}$ is fixed, the resulting problem in one variable ($\mathbf{H}$ or $\mathbf{W}$) is solved using the Conjugate Gradient iterative procedure. For example, fixing $\mathbf{H}$, the gradient of the above objective in matrix form is given as:

$$2\mathbf{U} \odot \mathbf{V}^\top[\mathbf{U} \odot (\mathbf{O} - \mathbf{PQ}^\top - \mathbf{VWH}^\top\mathbf{T}^\top)]\mathbf{TH} + \lambda_1\mathbf{W} + \tag{10}$$

$$2\lambda_2(\mathbf{V}^\top\mathbf{VWHT}^\top\mathbf{L_v}\mathbf{TH} + \mathbf{V}^\top\mathbf{L_s}\mathbf{VWHT}^\top\mathbf{TH})$$

5   CNN Features and Semantic Vectors

We utilize DeCAF$_6$ [12] to extract visual features, which have high level semantic meanings thus are more representative than low level visual features. And we adopt pre-trained word2vec [13] to calculate the word vectors for each tag, which could keep their semantic meanings precisely.

6   Experimental Evaluation

The proposed model is denoted as BITMC (Boosted Inductive Tag Matrix Completion). We follow the same experimental settings in [8] and evaluate BITMC on three benchmark datasets: Corel5K, Labelme [14] and MIRFlickr-25K [15].

6.1   Datasets and Experimental Setup

LabelMe dataset is collected through an online tagging project. MIRFlickr-25K is collected from Flickr. Compared to Corel5K and Labelme, tags in MIRFlickr-25K are much more noisy. Hence, a pre-processing procedure is performed. We match each tag with entries in a Wikipedia thesaurus and only retain the tags in accordance with Wikipedia. We extract tag vectors and visual features for all the datasets.

To study the tag completion performance, multiple models are employed as the baselines, including matrix completion-based models (LRES [5], TCMR [3], RKML [4], 4 Priors* [8] and ITMC), search-based models (JEC [16], TagProp [17] and TagRelevance [18]), mixture models (CMRM [19] and MBRM [20]) and CCA based model FastTag [21]. Note that we denote our model without the MC part as ITMC. 4 Priors is based on IMC, incorporating the same visual-tag correlation, semantic-tag correlation and inhomogeneous errors properties [8]. For the sake of fair comparison, we remove the inhomogeneous errors term to get the 4 Priors*.

We tuned λ_1, λ_2 using cross validation, and the parameters of adopted baselines are also carefully tuned on the validation set of the three datasets using the same strategies as in [6].

We adopt the same evaluation metrics used in [8]. All models are evaluated in terms of *average precision@N* (i.e. *AP@N*), *average recall@N* (i.e. *AR@N*) and *coverage@N* (i.e. *C@N*). In the top N completed tags, *precision@N* is to measure the ratio of correct tags in the top N competed tags and *recall@N* is to measure the ratio of missing ground-truth tags, both averaged over all test images. *Coverage@N* is to measure the ratio of test images with at least one correctly completed tag.

6.2   Evaluation and Observation

Tables 1, 2 and 3 show performance comparisons on the three datasets. Top 3 in each measure is shown in bold.

Table 1. Performance comparison on Corel5K

	Corel5K											
	N = 2			N = 3			N = 5			N = 10		
	AP	AR	C	AP	AR	C	AP	AR	C	AP	AR	C
BITMC	**0.58**	**0.43**	**0.52**	**0.49**	**0.48**	**0.64**	**0.44**	0.56	0.65	**0.38**	0.62	0.89
ITMC	0.56	**0.42**	0.49	0.46	**0.49**	0.57	0.41	**0.55**	0.63	0.35	**0.64**	0.88
4 Priors* [8]	**0.58**	0.41	**0.50**	0.48	**0.49**	0.62	0.42	0.58	0.65	0.37	0.62	**0.91**
LRES [5]	**0.58**	0.39	0.47	**0.48**	**0.48**	0.57	0.41	0.53	0.62	**0.37**	0.62	0.85
TCMR [3]	0.57	0.39	0.49	**0.48**	0.47	**0.58**	**0.44**	**0.55**	**0.66**	**0.38**	0.61	0.88
RKML [4]	0.29	0.21	0.24	0.25	0.24	0.29	0.23	0.25	0.34	0.19	0.29	0.67
JEC [16]	0.36	0.34	0.39	0.31	0.40	0.47	0.27	0.32	0.59	0.20	0.33	0.76
TagProp [17]	0.46	0.40	**0.50**	0.38	**0.48**	0.57	0.33	0.51	0.63	0.26	0.54	0.86
TagRel [18]	0.43	**0.41**	0.48	0.37	0.47	0.57	0.31	0.50	0.60	0.26	0.53	0.90
CMRM [19]	0.29	0.20	0.23	0.24	0.24	0.27	0.21	0.25	0.35	0.16	0.27	0.63
MBRM [20]	0.35	0.29	0.35	0.28	0.34	0.42	0.24	0.24	0.39	0.17	0.28	0.70
FastTag [21]	0.54	0.31	0.45	0.46	0.44	0.51	0.40	0.52	0.63	0.36	**0.62**	0.82

Table 2. Performance comparison on Labelme

	Labelme											
	N = 2			N = 3			N = 5			N = 10		
	AP	AR	C	AP	AR	C	AP	AR	C	AP	AR	C
BITMC	0.47	0.34	0.41	**0.48**	0.37	0.51	**0.45**	0.47	0.60	0.34	0.60	0.77
ITMC	0.43	**0.35**	0.39	0.43	0.35	0.48	0.36	**0.46**	0.58	**0.32**	0.57	0.77
4 Priors* [8]	**0.50**	**0.35**	**0.42**	0.47	**0.39**	**0.52**	0.42	**0.47**	**0.61**	**0.32**	0.60	0.74
LRES [5]	0.42	0.32	0.39	0.40	0.36	0.50	0.35	0.45	0.55	0.27	0.56	0.69
TCMR [3]	**0.44**	0.32	**0.42**	0.41	0.36	**0.51**	**0.37**	0.45	**0.60**	0.29	0.55	**0.75**
RKML [4]	0.21	0.14	0.20	0.20	0.16	0.21	0.19	0.20	0.23	0.14	0.22	0.28
JEC [16]	0.33	0.29	0.31	0.30	0.32	0.37	0.27	0.38	0.45	0.20	0.48	0.58
TagProp [17]	0.39	0.31	0.36	0.35	**0.37**	0.45	0.33	0.45	0.52	0.25	0.56	0.64
TagRel [18]	0.43	0.32	0.36	0.37	0.35	0.44	0.34	0.45	0.51	0.27	0.55	0.62
CMRM [19]	0.20	0.14	0.18	0.18	0.15	0.20	0.18	0.19	0.25	0.12	0.22	0.29
MBRM [20]	0.23	0.14	0.18	0.21	0.16	0.21	0.18	0.20	0.25	0.12	0.27	0.37
FastTag [21]	0.43	**0.34**	0.40	**0.48**	0.36	0.44	**0.37**	0.44	0.53	0.28	**0.57**	0.70

We can observe that methods achieve better performance on Corel5K and Labelme than MIRFlickr-25K, since tags in MIRFlickr-25K are much more noisy. Matrix completion-based semi-supervised methods, such as BITMC, LRES, TCMR, ITMC and 4 Priors* usually achieve the best performances owing to the advantage of exploiting both labeled (few) and large number of unlabeled information. In all cases, BITMC, ITMC and 4 Priors* achieve satisfactory performances. BITMC combines the power of ITMC and standard MC, which is

Table 3. Performance comparison on MIRFlickr-25K

| | MIRFlickr-25K | | | | | | | | | | | |
| | N = 2 | | | N = 3 | | | N = 5 | | | N = 10 | | |
	AP	AR	C	AP	AR	C	AP	AR	C	AP	AR	C
BITMC	**0.50**	**0.36**	**0.44**	**0.48**	**0.43**	**0.54**	**0.38**	**0.44**	**0.60**	**0.32**	**0.61**	**0.81**
ITMC	0.45	**0.36**	0.43	0.44	0.41	**0.53**	0.37	**0.44**	0.56	**0.29**	0.58	0.78
4 Priors* [8]	**0.52**	0.35	0.41	**0.47**	0.40	0.50	**0.38**	0.43	0.57	**0.29**	0.56	0.74
LRES [5]	0.43	0.35	0.40	0.40	0.39	**0.53**	0.32	0.40	0.57	0.26	0.45	0.73
TCMR [3]	0.45	0.35	0.44	0.43	0.38	**0.54**	0.35	0.41	**0.60**	0.28	0.48	0.77
RKML [4]	0.21	0.15	0.15	0.23	0.22	0.25	0.13	0.23	0.31	0.13	0.22	0.55
JEC [16]	0.33	0.30	0.32	0.31	0.38	0.45	0.25	0.34	0.55	0.19	0.35	0.66
TagProp [17]	0.39	0.35	0.39	0.36	0.42	0.51	0.28	0.37	**0.59**	0.20	0.41	0.73
TagRel [18]	0.42	0.34	0.37	0.37	**0.43**	0.52	0.30	0.37	0.57	0.20	0.40	**0.78**
CMRM [19]	0.20	0.15	0.16	0.18	0.21	0.24	0.13	0.18	0.30	0.11	0.20	0.50
MBRM [20]	0.22	0.16	0.18	0.17	0.30	0.35	0.13	0.18	0.33	0.10	0.22	0.55
FastTag [21]	0.43	0.35	0.38	0.39	**0.43**	0.51	0.30	0.41	0.57	0.27	0.42	0.75

verified on the performance comparisons between BITMC and ITMC. Note that as the dataset become more noisy, their difference becomes larger. The reason for this phenomenon is that as the data becomes noisier, the benefit of side information (IMC) becomes relatively small comparing to the benefit of low-rankness (MC). Note that BITMC is more efficient than 4 Priors* because BIMC do not have to explicitly form $\mathbf{O} - \mathbf{PQ}^\top$ [1] and the optimization procedure has closed-form solutions.

7   Conclusion

We have improved the powerful BIMC model and proposed an effective model BITMC for tag completion, which takes low-rankness, visual-tag correlation, semantic-tag correlation into consideration. We utilize word vectors to calculate semantic-tag correlation and CNN features to measure tag-visual correlation. BITMC outperforms several state-of-the-art methods on benchmark datasets.

References

1. Shin, D., Cetintas, S., Lee, K., Dhillon, I.: Tumblr blog recommendation with boosted inductive matrix completion. In: Proceedings of the 24th ACM International on Conference on Information and Knowledge Management. ACM (2015)
2. Goldberg, A., Recht, B., Xu, J., Nowak, R., Zhu, X.: Transduction with matrix completion: three birds with one stone. In: Advances in Neural Information Processing Systems (2010)
3. Feng, Z., Feng, S., Jin, R., Jain, A.K.: Image tag completion by noisy matrix recovery. In: Fleet, D., Pajdla, T., Schiele, B., Tuytelaars, T. (eds.) ECCV 2014, Part VII. LNCS, vol. 8695, pp. 424–438. Springer, Heidelberg (2014)

4. Feng, Z., Jin, R., Jain, A.: Large-scale image annotation by efficient and robust kernel metric learning. In: Proceedings of the IEEE International Conference on Computer Vision (2013)

5. Zhu, G., Yan, S., Ma, Y.: Image tag refinement towards low-rank, content-tag prior and error sparsity. In: Proceedings of the International Conference on Multimedia. ACM (2010)

6. Wu, L., Jin, R., Jain, A.: Tag completion for image retrieval. IEEE Trans. Pattern Anal. Mach. Intell. **35**(3), 716–727 (2013)

7. Jain, P., Dhillon, I.: Provable inductive matrix completion (2013). arXiv preprint arXiv:1306.0626

8. Hou, Y.: Image annotation incorporating low-rankness, tag and visual correlation and inhomogeneous errors. In: Jiang, J., et al. (eds.) ISVC 2015. LNCS, vol. 9474, pp. 71–81. Springer, Heidelberg (2015). doi:10.1007/978-3-319-27857-5_7

9. Chung, F.: Spectral Graph Theory. American Mathematical Society, Providence (1997)

10. Jain, P., Netrapalli, P., Sanghavi, S.: Low-rank matrix completion using alternating minimization. In: Proceedings of the Forty-Fifth Annual ACM Symposium on Theory of Computing. ACM (2013)

11. Yu, H., Jain, P., Kar, P., Dhillon, I.: Large-scale multi-label learning with missing labels. In: Proceedings of The 31st International Conference on Machine Learning (2014)

12. Donahue, J., Jia, Y., Vinyals, O., Hoffman, J., Zhang, N., Tzeng, E., Darrell, T.: DeCAF: a deep convolutional activation feature for generic visual recognition (2013). arXiv preprint arXiv:1310.1531

13. Mikolov, T., Chen, K., Corrado, G., Dean, J.: Efficient estimation of word representations in vector space (2013). arXiv preprint arXiv:1301.3781

14. Russell, B., Torralba, A., Murphy, K., Freeman, W.: Labelme: a database and web-based tool for image annotation. Int. J. Comput. Vis. **77**(1), 157–173 (2008)

15. Huiskes, M., Lew, M.: The MIR flickr retrieval evaluation. In: Proceedings of the 1st ACM International Conference on Multimedia Information Retrieval. ACM (2008)

16. Makadia, A., Pavlovic, V., Kumar, S.: A new baseline for image annotation. In: Forsyth, D., Torr, P., Zisserman, A. (eds.) ECCV 2008, Part III. LNCS, vol. 5304, pp. 316–329. Springer, Heidelberg (2008)

17. Guillaumin, M., Mensink, T., Verbeek, J., Schmid, C.: TagProp: discriminative metric learning in nearest neighbor models for image auto-annotation. In: Proceedings of the IEEE 12th International Conference on Computer Vision, pp. 309–316 (2009)

18. Li, X., Snoek, C., Worring, M.: Learning social tag relevance by neighbor voting. IEEE Trans. Multimedia **11**(7), 1310–1322 (2009)

19. Jeon, J., Lavrenko, V., Manmatha, R.: Automatic image annotation and retrieval using cross-media relevance models. In: Proceedings of the 26th Annual International ACM SIGIR Conference on Research and Development in Informaion Retrieval. ACM (2003)

20. Feng, S., Manmatha, R., Lavrenko, V.: Multiple Bernoulli relevance models for image and video annotation In: Proceedings of the 2004 IEEE Computer Society Conference on Computer Vision and Pattern Recognition (2004)

21. Chen, M., Zheng, A., Weinberger, K.: Fast image tagging. In: Proceedings of the 30th International Conference on Machine Learning (2013)

Neurological Classifier Committee Based on Artificial Neural Networks and Support Vector Machine for Single-Trial EEG Signal Decoding

Konstantin Sonkin[1(✉)], Lev Stankevich[1], Yulia Khomenko[2], Zhanna Nagornova[3], Natalia Shemyakina[3], Alexandra Koval[1], and Dmitry Perets[1]

[1] St. Petersburg State Polytechnic University, 29 Polytechnicheskaya st., 195251 St. Petersburg, Russia
sonkink@gmail.com
[2] N.P. Bechtereva Institute of Human Brain, Russian Academy of Sciences, 9 Akademika Pavlova st., 197376 St. Petersburg, Russia
[3] I.M. Sechenov Institute of Evolutionary Physiology and Biochemistry, Russian Academy of Sciences, 44 Toreza pr., 194223 St. Petersburg, Russia

Abstract. This study aimed to finding effective approaches for electroencephalographic (EEG) multiclass classification of imaginary movements. The combined classifier of EEG signals based on artificial neural network (ANN) and support vector machine (SVM) algorithms was applied. Effectiveness of the classifier was shown in 4-class imaginary finger movement classification. Nine right-handed subjects participated in the study. The mean decoding accuracy using combined heterogeneous classifier committee was -60 ± 10 %, max: 77 ± 5 %, while application of homogeneous classifier based on committee of ANNs -52 ± 9 % and 65 ± 5 % correspondingly. This work supports the feasibility of the approach, which is presumed suitable for imaginary movements decoding of four fingers of one hand. These results could be used for development of effective non-invasive BCI with enlarged amount of degrees of freedom.

Keywords: Electroencephalography · Classifier committee · Artificial neural network · Support vector machine · Imaginary finger movements

1 Introduction

Classification of electroencephalographic (EEG) signals is important for the solution of a number of biological and medical tasks, for example, for decoding of mental states, diagnosing of epileptic seizures etc. One of the most important applications of the EEG signal classification – implementation in non-invasive brain-computer interfaces (BCI), which allow people to communicate with the environment without using neuromuscular pathways. This is extremely important for the rehabilitation of immobilized patients because BCI may serve for communication, rehabilitation and restoring of motor functions [1–3]. Consequently, development of non-invasive, mobile, high-precision BCI with negligible time delays and capable to use the large number of mental commands is one of the main tasks of fundamental and applied science.

© Springer International Publishing Switzerland 2016
L. Cheng et al. (Eds.): ISNN 2016, LNCS 9719, pp. 100–107, 2016.
DOI: 10.1007/978-3-319-40663-3_12

One of the promising approaches (in BCI) is the use of different types of imaginary movements as the control commands. For this type BCI realization the reliable EEG pattern classifier with enlarged amount of degrees of freedom is required. Classification of imaginary fine movements might be an important task for the increasing of degrees of freedom. Few research groups in the world work at the similar tasks. In the study [4] achieved average decoding accuracy of 5 finger movements was of 39,7 %. Quandt et al. reported about decoding of 4 fingers movements with the average accuracy 43 %, maximal accuracies for several subjects were 46 % and 56 % for 4- and 5-class classification [5].

Support vector machine (SVM) [6] and artificial neural network (ANN) [7] are known as the most applicable methods for classification of EEG patterns. The approach of the committee of homogeneous ANNs for the imaginary movements classification was reported [8]. In previous research we used SVM and ANN classifiers separately and found out that they were sensitive to different EEG signal features [6, 9]. We also developed committee of ANNs that allowed increasing the decoding accuracy of imaginary movements [9], but it was insufficient for the practical implementation in BCI. New approaches are proposed in the current study: the neurological committee of heterogeneous classifiers for motor imagery pattern decoding and the preliminary mapping of the most informative channel pairs for individual channel selection.

Thus, the study aimed to development means of classification for single-trial decoding of EEG patterns of imaginary movements. In the work we combined different types of classification methods in the heterogeneous classifier committee and compared its efficiency in contrast with the homogeneous ANN committee for problem of imaginary finger movement decoding.

2   System of Imaginary Movements EEG-Pattern Classification

Classification of EEG patterns of imaginary movements was proposed to carry out by two types of the classifier committees: the neurological committee of heterogeneous classifiers (SVMs and ANNs) and the homogeneous committee of ANNs.

The committee of ANNs consists of two ANNs at the fist level and the generalizing ANN at the second level. Each ANN at the first level analyzes only one of two feature spaces. The results of the first level ANN classification is generalized by the second level ANN, which makes the final decision on the assigning of the trial to one of the classes.

The committee of ANNs is used for the joint analysis of the two types of features. The advantage of the committee is the decision making approach based on the learning of the top-level network, which allows to provide selection of optimal solutions of lower level networks. In addition, such a committee is scalable, that is, when adding a new feature space a committee can be expanded by adding new lower level ANN for new feature space analysis.

The committee is based on ANNs with a backward propagation of errors, one input layer, two hidden layers and one output layer. As activation function for the hidden layer neurons the sigmoid (hyperbolic tangent) function was used, and for the neurons in the output layer the linear function was used. The ANN training process was carried out once for each type of movements. The learning process was performed until the assigned classification accuracy was achieved for the learning sample which comprised 70 % of the trials. The testing sample included the subsequent 30 %. The samples did not overlap.

The neurological committee of heterogeneous classifiers based on artificial neural networks and support vector machines was developed and applied as a mean of classification. The selected approaches are effective means of EEG pattern classification, including EEG patterns of imaginary movements (for review – [7]). It was supposed that the committee of heterogeneous classifiers will combine the advantages of ANN and SVM. Two types of features were used simultaneously (length and square under the curve). The structure of the heterogeneous committee is presented at Fig. 1.

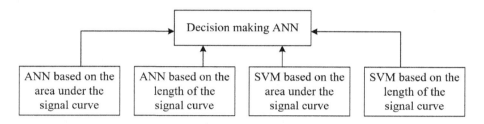

Fig. 1. The structure of the heterogeneous committee of classifiers

The committee used ANNs of the same type as in the committee of ANNs described earlier.

SVM is a method of classification that constructs a separating hyperplane or a set of hyperplanes in a high- or infinite-dimensional space [9]. Based on the results of the studies on the selection of the preferred type of SVM [6, 10, 11] a Gaussian radial basis function was used as the kernel $K(x_i, x_j) = \exp(-\gamma \|x_i - x_j\|^2)$, for $\gamma > 0$. The SVM classifier was implemented using MATLAB LIBSVM package [11] "one-vs-one" scheme was applied to solve a multiclass problem.

The results of the first-level classifiers formed a feature vector for the second level generalizing ANN. Two types of features were used simultaneously (length and square under the curve). After receiving responses of the four first-level classifiers, the second-level generalizing ANN made the final decision on the assigning of the trial to one of the classes.

3   Application of the Classification System

The developed means of classification were applied for decoding of imaginary movements of fingers. The decoding accuracies of applied homogeneous and heterogeneous committees of classier were compared.

3.1 Methods

Subjects. A total of 9 healthy, right-handed volunteers (3 male, 6 female, mean age: 29 ± 5[SD] years) participated in the study. All procedures were carried out in accordance with the Helsinki declaration (1974).

Tasks in EEG Investigation. In order to increase speed and efficiency of classification of imaginary finger movements were used sequences of short trials (600 ms) with fine movement imagination in the previously suggested rhythm. Trials used for decoding didn't have external stimuli in them, but we were able to predict time of imagination because of previously suggested rhythm of imagination. Description of the full paradigm of rhythmically repeated trials with movement imagination for multiclass decoding is performed in [12, 13]. In the conditions of practical realization of BCI short time intervals for realization of executive commands are preferable, so as the short time necessary for classification.

EEG Registration and Data Filtering. Electroencephalogram (EEG) was recorded using a Mitsar 32 channel EEG system (Mitsar, Ltd. St. Petersburg, http://www.mitsar-medical.com) from 19 sites - Fp1, Fp2, F7, F3, Fz, F4, F8, T3, C3, Cz, C4, T4, T5, P3, Pz, P4, T6, O1, O2 (10–20 %) with ears linked reference, filtered between 0.53 and 30 Hz, 50 Hz notch filter, sample rate −500 Hz. The ground electrode was placed on the forehead. The electrode impedance was kept at less than 5 kΩ. For EEG analysis was used WinEEG software (Ponomarev, V.A., Kropotov, Ju.D., registration no. 2001610516 at 08.05.2001).

The raw EEG data was transformed to a weighted average reference montage (WAR montage). In comparison to common averaged reference montage, the WAR montage greatly reduces the topographic displacement of a widespread potential, such as alpha rhythms [14]. Extra high (20–30 Hz, over 35 μV) and low frequency (0–2 Hz, over 50 μV) activities were automatically marked as artifacts and the trials containing artifacts were excluded from further analysis.

Feature Extraction. In the present work we analyze single-trial EEG signal in time domain. This approach is also used by several researchers [5, 15] for classification of finger movements. As we used ERP paradigm and short trials with some synchronized endogenous events, it is important to take into consideration time localization of features.

Preprocessed EEG-signals were used for evaluation of two types of features – the area under the curve of a signal and the length of the curve – calculated in sliding time windows of analysis (window length – 100 ms, shift – 50 ms). Two types of features are used for joint feature set generation for further classification. Both high and low frequency characteristics of signals may contribute to the efficiency of classification, so we used and the area under the signal curve (low frequency signal characteristic) and the length of the curve (high frequency signal characteristic). Besides that the use of these features is small time-consuming due to the relatively high speed of computation.

3.2   Individual Best Channels Choice (Mapping)

A separate EEG channels' accuracy "mapping" of the imaginary movements (one type – tapping) of four fingers showing the best results for each individual subject was applied for classification. The choice of individual parameters, such as informative intervals and EEG signal topographies, is an actual approach for decoding accuracy increase in classification of EEG of patterns, that is realized by various methods [16, 17]. The special attention is paid for the choice of individual states (with higher decoding accuracy), the tasks for classification, and also for the problem of individual choice of features and channels for classification [18–20].

3.3   Results

Repeated measures ANOVA within subjects design was applied for statistical data analysis.

Significant differences were revealed between decoding accuracies of imagery movements of fingers with application of combined classifier committee based on ANNs and SVM and committee of neural networks classifier: $F(1,8) = 19.9$, $p < 0.003$.

The mean decoding accuracy using combined heterogeneous classifier and choice of best classified channels was -60 ± 10 %, max: 77 ± 5 %, while application of homogeneous classifier based on committee of ANNs -52 ± 9 % and 65 ± 5 % correspondingly.

As could be seen in the Table 1 the most number of the subjects had higher or comparable decoding accuracy of 4-class classification with application of committee of heterogeneous classifiers based on ANNs and SVM in comparison with the application of homogeneous classifier based on ANNs for the same pairs of the electrodes. Committee of heterogeneous classifiers has demonstrated higher decoding accuracies of EEG signals.

Table 1. Decoding accuracy for 4-class imaginary fingers movements classification

Subjects	Sites	Accuracy % using combined committee of classifiers (ANNs and SVM)	Accuracy % using committee of neural networks classifier (ANNs)
Subj.1	F7-F8	72 ± 7	65 ± 4
Subj.2	Fp2-T5	53 ± 5	46 ± 6
Subj.3	F7-C4	49 ± 6	41 ± 6
Subj.4	Fp2-F3	64 ± 7	47 ± 5
Subj.5	F3-F4	61 ± 5	53 ± 7
Subj.6	Fz-P3	56 ± 6	57 ± 5
Subj.7	O1-O2	63 ± 4	51 ± 5
Subj.8	T3-O1	77 ± 5	65 ± 5
Subj.9	Fp1-Cz	45 ± 5	43 ± 5
Average		$\mathbf{60 \pm 10}$	$\mathbf{52 \pm 9}$

4   Discussion

The average decoding accuracy of 4-class classification for the committee of heterogeneous classifiers was 60 ± 10 %, maximal -77 ± 5 % (guessing level for the 4-class problem is 25 %). The obtained results were higher for the most of subjects (Subj. 1–5,7,8) in comparison with the decoding accuracies showed by the homogeneous committee of ANNs. Thus, the introduction of new types of classifiers into the committee of classifiers did not reduce the decoding accuracy; conversely, it increased by 8 % on average and by 12–17 % in several subjects, which was significant for multiclass classification. Consequently, higher decoding accuracy was shown by the committee of heterogeneous classifiers in comparison with the committee of artificial neural networks.

In previous research it was supposed that different types of classifiers showed different sensitivity to different EEG signal characteristics. Thus, the decoding accuracy of the SVM classifier was higher for the accumulated EEG signal, while classifier on the basis of ANN was more efficient for the single-trial decoding [6, 9].

In the present work method of individual informative channels selection was implemented. As it was shown before, individual channels selection allowed to increase the decoding accuracy in average by 10 % in 4-class classification [12].

Different scientific groups solved the problem of the individual informative channels selection in different ways. Thus, [21] applied the common spatial pattern method, [22] implicated modified regression algorithm, [4] used coefficient of determination for the evaluation of spectral features topographies. The mapping method realized in this work allowed obtaining decoding accuracies for each channel that makes the selection of informative channels numerically validated but requires computational resources. However, individual adjustment of the BCI systems is usually performed before the user exploitation, so the computing costs could be rational in case of the significant increase of the decoding accuracy.

The obtained decoding accuracy of single-trial EEG signals of imaginary movements of fingers are comparable with the results of the studies reported in the literature (average accuracy 45.2 % for 5 finger movements in time series analysis [15]; 43 % - for EEG and 57 % for MEG for 4 class classifications [5]) and in several cases even exceeds them.

We suggested an approach of the neurological committee of classifiers implementation for motor imagery classification. According to the approach we tested two designs of the classifier committee - heterogeneous and homogeneous. Implementation of the homogeneous committee of ANNs for EEG-signal classification of 4-class imaginary movements of large body parts (hands, legs, tongue) was reported by [8] and the decoding accuracies were in the range 48 to 74 % [8]. In our study the homogeneous classifier committee of fine motor imagery reached 52 ± 9 % in average in the range from 41 to 65 %. While the heterogeneous committee of classifiers demonstrated higher accuracy -60 ± 10 % in average in the range from 45 to 77 %.

5   Conclusions

The combined committee of heterogeneous classifiers based on artificial neural networks and support vector machines was developed and the method of the informative EEG channels selection was applied. The implemented approaches allowed achieving the average 4-class decoding accuracy $60 \pm 10\%$, maximal — $77 \pm 5\%$ (theoretical guessing level — 25 %) that exceeded results obtained with the homogeneous artificial neural networks committee. Thus, application of the combined classifier for analysis of EEG signal characteristics could be more promising than the use of different separate classifiers. This work supports the feasibility of the approach, which is presumed suitable for imaginary movements decoding of four fingers of one hand. These results could be applied in the elaboration of multiclass BCI systems.

Acknowledgements. The study was supported by the RFBR foundation grant № 13-01-12059 ofi-m.

References

1. Chai, R., Ling, S.H., Hunter, G.P., Tran, Y., Nguyen, H.T.: Brain-computer interface classifier for wheelchair commands using neural network with fuzzy particle swarm optimization. IEEE J. Biomed. Health. Inform. **18**(5), 1614 (2014)
2. Brunner, C., Birbaumer, N., Blankertz, B., Guger, C., Kübler, A., Mattia, D., Millan, J.R., Miralles, F., Nijholt, A., Opisso, E., Ramsey, N., Salomon, P., Müller-Putz, G.R.: BNCI Horizon 2020: towards a roadmap for the BCI community. Brain-Comput. Interfaces **2**(1), 1–10 (2015)
3. Wolpaw, J.R., Wolpaw, E.W.: Brain Computer Interfaces: Principles and Practice. Oxford Univ. Press, New York (2012)
4. Xiao, R., Ding, L.: EEG resolutions in detecting and decoding finger movements from spectral analysis. Front Neurosci. **9**, 308 (2015)
5. Quandt, F., Reichert, C., Hinrichs, H., Heinze, H.J., Knight, R.T., Rieger, J.W.: Single trial discrimination of individual finger movements on one hand: a combined MEG and EEG study. NeuroImage **59**, 3316–3324 (2012)
6. Sonkin, K.M., Stankevich, L.A., Khomenko, Y., Nagornova, Z., Shemyakina, N.V.: Classification of electroencephalographic patterns of imagined and real movements by one hand fingers using the support vectors method. Pac. Med. J. **2**, 30–35 (2014)
7. Lotte, F., Congedo, M., Lecuyer, A., Lamarche, F., Arnaldi, B.: Review of classification algorithms for EEG-based brain-computer interfaces. J. Neural Eng. **4**, 1–24 (2007)
8. Lazurenko, D.M., Shepelev, I.E., Kiroy, V.N., Aslanyan, E.V., Bakhtin, O.M., Minyaeva, N.R.: Ideomotor EEG patterns in the profile of brain-computer interface. In: Ivanova, G.E. (ed.) Selected Topics of Neurorehabilitation: Proceedings of VII International Congress Neurorehabilitation-2015, pp. 246–249 (2015). (Russian)
9. Sonkin, K.M., Stankevich, L.A., Khomenko, J., Nagornova, Z., Shemyakina, N.V.: Development of electroencephalographic pattern classifiers for real and imaginary thumb and index finger movements of one hand. Artif. Intell. Med. **63**(2), 107–117 (2015)
10. Shawe-Taylor, J., Cristianini, N.: Kernell Methods for Pattern Analysis. Cambridge Univesity Press, New York (2004)
11. Chang, C.C., Lin, C.J.: LIBSVM: a library for support vector machines. ACM Trans. Intell. Syst. Technol. **2**, 1–27 (2011)
12. Stankevich, L.A., Sonkin, K.M., Shemyakina, N.V., Nagornova, Z., Khomenko, Y., Perets, D.S., Koval, A.V.: Pattern decoding of rhythmic individual finger imaginary movements of one hand. Hum. Physiol. **42**(1), 32–42 (2016)

13. Stankevich, L.A., Sonkin, K.M., Nagornova, Z.V., Khomenko, J.G., Shemyakina, N.V: Classification of electroencephalographic patterns of imaginary one-hand finger movements for brain-computer interface development. SPIIRAS Proc. **3**(40), 163–182 (2015)

14. Lemos, M.S., Fisch, B.J.: The weighted average reference montage. Electroencephalogr. Clin. Neurophysiol. **79**(5), 361 (1991)

15. Xiao, R., Ding, L.: Evaluation of EEG features in decoding individual finger movements from one hand. Comput. Math. Methods Med., 243 (2013)

16. Sotnikov, P.I.: Optimal EEG signal frequency ranges selection in brain-computer interface. Sci. Educ., 217–234 (2015)

17. Basterrech, S., Bobrov, P., Frolov, A., Húsek, D.: Nature-inspired algorithms for selecting EEG sources for motor imagery based BCI. In: Rutkowski, L., Korytkowski, M., Scherer, R., Tadeusiewicz, R., Zadeh, L.A., Zurada, J.M. (eds.) ICAISC 2015. LNCS, vol. 9120, pp. 79–90. Springer, Heidelberg (2015)

18. Friedrich, E.V.C., Neuper, C., Scherer, R.: Whatever works: a systematic user-centered training protocol to optimize brain-computer interfacing individually. PLoS ONE **8**(9), e76214 (2013)

19. Daly, I., Billinger, M., Laparra-Hernandez, J., Aloise, F., Garcia, M.L., Faller, J., Scherer, R., Muller-Putz, G.: On the control of brain-computer interfaces by users with cerebral palsy. Clin. Neurophysiol. **124**, 1787–1797 (2013)

20. Liao, K., Xiao, R., Gonzalez, J., Ding, L.: Decoding individual finger movements from one hand using human EEG signals. PLoS ONE **9**(1), e85192 (2014)

21. Asensio-Cubero, J., Gan, J.Q., Palaniappan, R.: Multiresolution analysis over graphs for a motor imagery based online BCI game. Comput. Biol. Med. **68**(1), 21–26 (2016)

22. Shan, H., Xu, H., Zhu, S., He, B.: A novel channel selection method for optimal classification in different motor imagery BCI paradigms. Biomed. Eng. Online **14**, 93 (2015)

Calculation of Analogs for the Largest Lyapunov Exponents for Acoustic Data by Means of Artificial Neural Networks

German A. Chernykh[1], Yuri A. Kuperin[1], Ludmila A. Dmitrieva[1(✉)], and Angelina A. Navleva[2]

[1] Saint Petersburg State University, Universitetskaya Nab. 7/9, 199034 Saint Petersburg, Russia
{g.chernykh,y.kuperin,l.dmitrieva}@spbu.ru
[2] Saint-Petersburg State Polytechnic University, Grazhdanskii Pros., 28, 195220 Saint Petersburg, Russia
lina.navleva@mail.ru

Abstract. A method for calculating the largest Lyapunov exponents analogs for the numerical series obtained from acoustic experimental data is proposed. It is based on the use of artificial neural networks for constructing special additional series which are necessary in the process of calculating the Lyapunov exponents. The musical compositions have been used as acoustic data. It turned out that the error of the largest Lyapunov exponent computations within a single musical composition is sufficiently small. On the other hand for the compositions with different acoustic content there were obtained various numerical values Lyapunov exponents. This enables to make conclusion that the proposed procedure for calculating the Lyapunov exponents is adequate. It also allows to use the obtained results as an additional macroscopic characteristics of acoustic data for comparative analysis.

Keywords: Largest Lyapunov exponent · Acoustic characters · Time series · Artificial neural network · Mel-frequency cepstral coefficients

1   Introduction

The set of Lyapunov exponents, i.e. Lyapunov exponent spectrum components is well-known characteristic of the dynamical systems states [1]. According to the nu-merical values of the Lyapunov exponents is possible to determine the key properties of attractors of dynamical systems. In the case of deterministic chaos in the phase space of the dynamical system there appears a strange attractor. If the largest Lyapunov exponent is positive the letter serves as the main quantitative criterion of instability of trajectories on the strange attractor. Or, in other words, the largest Lyapunov exponent is a measure of state of chaos in the system. Usually the so-called global and local Lyapunov exponents are introduced. Global Lyapunov exponents characterize the attractor in whole and

© Springer International Publishing Switzerland 2016
L. Cheng et al. (Eds.): ISNN 2016, LNCS 9719, pp. 108–114, 2016.
DOI: 10.1007/978-3-319-40663-3_13

local Lyapunov exponents reflect the instability of the system in local areas of the attractor. In certain sense, the global Lyapunov exponents are calculated as the average of local Lyapunov exponents. Now there are quite a large number of methods of numerical calculations of largest Lyapunov exponents for various situations [1–3].

However, in any case these calculations require the presence of set of paths on the attractor, which at certain times are localized in sufficiently small phase volumes. But when the analytic representation of a dynamical system is unknown, we can deal only with experimental numerical time series. Among the numerical algorithms for calculating the largest Lyapunov exponent directly from time series let us mention the Wolf algorithm [4], Sano and Sawada [5] and Eckmann [6] algorithms. Separately stand the Rosenstein, Collins and Luca algorithm [7] and Kantz algorithm [8,9]. Let us also mention the paper [16]. There exists also the neural networks approach to estimation of the largest Lyapunov exponent [10]. In this case, the set of additional paths on the reconstructed attractor can be obtained by using the artificial neural networks. Neural networks are trained on the available time series, and then are applied to the perturbed time series. It also is necessary to implement the attractor recovery procedure on the base of observed data [8]. In the simplest case of one-dimensional time series the attractor reconstruction process requires the determination of the so-called lag and the dimension of the embedding space. In the present study investigated objects are temporal sequence of vectors constructed from acoustic data. In such a situation it is quite problematic to talk about attractors with specific properties. Therefore, our goal was not to build an attractor and determine its properties. The aim of this study was to develop a new macroscopic characteristics for acoustic time series based on modification of neural network method for calculating largest Lyapunov exponent mentioned above. That is, we consider analog of largest Lyapunov exponent applied to the acoustic data. The criterion for the adequacy of the obtained results has been the invariance of desired characteristics with respect to varying the parameters of the computational algorithm within a given error. Numerical experiments were performed on acoustic data representing audio files with musical compositions. Stable results were obtained with up to three significant figures in the mantissa.

2   The Procedure for Calculating Lyapunov Exponent

An algorithm for calculating largest Lyapunov exponent is described by the following sequence of steps:

1. Finding the-MFCC-features from acoustic data.
2. EMD-filtering for each component of the vector-signs.
3. Building neural networks and their training on the filtered data.
4. The use of trained neural networks for obtaining alternative types of vector-signs.
5. The calculation of the largest Lyapunov exponent.

Below we give a detailed description of each of the four steps of the algorithm.

MFCC-features or Mel-frequency cepstral coefficients are now widely used in artificial intelligence systems for processing various sound signals: speech recognition, the identification of the human voice and the identification of other sound sources, the definition of tone, etc. (See., e.g. [11–14]). In particular, MFCC-decomposition are widely used to transmit compressed acoustic information in mobile communication systems. Moreover, MFCC-signs are usually used at the lowest level of primary audio processing. Algorithm of the MFCC-features extraction is well described in the literature and is one of the standard methods in the above areas. However, it contains some number of parameters, which are selected empirically for the most effective solution to this or that application. The main parameter of MFCC-decomposition is dimension of the obtained coefficients. In this paper, this parameter is selected to be equal to three, based on the criterion of reducing the computational complexity. Moreover, we exclude the first component, which is the energy component. Thus, after applying MFCC-expansion to the sound file we have a sequence of three-dimensional vectors. Vectors of this sequence can be interpreted as a trajectory in three-dimensional space.

EMD-decomposition (Empirical mode decomposition) [15] it is one of the well known methods of representation of the time series as a sum of simpler than the original series components or modes. In particular, EMD-decomposition can be used for filtering and smoothing the signals, which contain noise. Like other

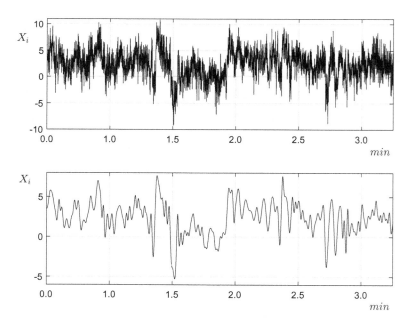

Fig. 1. The time dependence of one of the components of MFCC-expansion before and after EMD-filtration (upper and lower graphs respectively).

smoothing and filtering algorithms, EMD-expansion has undesirable edge effects. But in our case, the presence of edge effects is not critical, because we can always delete the initial and final interval time series without a substantial effect on the result of the Lyapunov exponents calculations. We use the EMD-decomposition (Empirical Mode Decomposition) [15] to remove rapidly oscillating modes from the signal obtained after MFCC-decomposition. It should be noted that the EMD-smoothing time dependences of MFCC-signs applies to every component of the vectors separately. In the process of the Lyapunov exponent calculations we have shown that it is necessary to remove the first four or five modes of EMD-decomposition in order to obtain a stable result. The stability of the result is ultimately determined by good statistics of neural network training. It is clear that by removing of such amount of EMD-modes we can talk about a significant loss of information that was originally contained in the test audio data. But here it is necessary to take into account the following. Based on the physical meaning of the MFCC-decomposition, elimination of the first few EMD-modes leads to the smoothing of the amplitude modulation of some bands in the power spectrum of the original audio signal. Thus, high-frequency modulations are filtered separately for different spectral intervals. Thus, the calculated Lyapunov exponent will contain information not only on the slower changes in the acoustic data, but also information about the changes including both high and low frequency harmonics. Example of influence of the EMD-filtering on MFCC-features is represented in the graphs (see Fig. 1) showing the time dependence of one component of MFCC-expansion before and after EMD-filtration.

Neural network approach for the calculation of Lyapunov exponents, where neural networks are used for alternative beam path is described in detail in [11]. As neural networks to obtain additional data there were taken feed forward neural networks (perceptrons) consisting of two hidden layers with sigmoid activation functions. The dimension of the input vectors was equal to 3 and has been a multiple of dimension MFCC features used for the construction of the training pattern. As the target pattern we always have chosen one MFCC-vector, which directly followed the last vectors in a particular training pattern. For example, for a set of six three-dimensional MFCC features there has been built one learning pair in which the first 5 vectors formed training pattern and the sixth vector was a target. In this case neural network had 25 inputs and 3 outputs. Along with learning sample the validating and testing samples have been used.

The peculiarity of our approach for calculating the analog of Lyapunov exponent consists in the procedure for obtaining additional paths segments on the basis of the original sequence of MFCC-features. We show this on the following example. Suppose we have an interval from N pieces of MFCC features. In this case N vectors are necessary for the prediction of $N + 1$-th vector. Then, around the point defined by the coordinates of the last vector from N available vectors, the sphere of sufficiently small radius is constructed. On this sphere M points are randomly selected. Then M additional sequences consisting of N points in each are constructed. Thus the first $N - 1$ points in each sequence are the points from the initial interval and the last point is one of points distributed on the sphere.

With the help of trained neural network, which uses the obtained patterns, M forecasts are constructed. Thus, the neural network is used to monitor the local transformation of the phase space, which is bounded by the original sphere. In view of the linearity of the phase volume transformations, the sphere in the phase space on the small time interval is transformed into an ellipsoid. It is sufficient to use about one thousand points randomly distributed on the initial sphere to obtain appropriate numerical results for local Lyapunov exponent. From the value of the logarithm of the ratio of the major axis of the ellipsoid to the radius of the sphere, one can estimate the value of the largest local Lyapunov exponent. Global largest Lyapunov exponent is calculated by averaging the local Lyapunov exponents.

3   Results

To assess the accuracy of calculating the Lyapunov exponent, as well as to prove the invariance of the result with respect to calculation parameters, the following quantities were varied: number of MFCC-features, which form the training patterns, the radius of the sphere (see Sect. 2), the number of neurons in the layers of the neural network, the neural network initialization parameters (the matrix elements, weights) and so on. We have shown that the error of the Lyapunov exponent calculation caused by the initialization of the neural network completely included errors of the variation of other algorithm parameters. Note that calculation of Lyapunov exponent and evaluation of error of calculations were carried out using neural networks committee, including neural networks with the same structure, but with different initialization. Lyapunov exponents were calculated on the basis of the audio files with duration in a few minutes. For comparative analysis the musical compositions in the presence of a vocal component and without it have been taken. Table 1 shows some of the results obtained for the Lyapunov exponents.

Table 1. Lyapunov exponents with the errors calculation for six musical compositions.

Instrumental compositions	Vocal compositions
1.154 ± 0.002	1.22 ± 0.08
1.140 ± 0.004	1.240 ± 0.002
1.19 ± 0.003	1.219 ± 0.002

The reliability of calculations for various compositions correspond to two or three significant digits of the mantissa, which is pretty good accuracy in the calculation of Lyapunov exponents for standard situations. In particular, it was shown that for the vocal musical compositions largest Lyapunov exponent appreciably higher than in the case of the purely instrumental compositions. It turned out that the error of computations within a single musical composition is sufficiently small. On the other hand for the compositions with different acoustic content various numerical values Lyapunov exponents have been obtained. This enables to make conclusion that the proposed procedure for calculating the Lyapunov exponents is adequate. It also allows to use the results as an additional macroscopic characteristics of acoustic data for comparative analysis. It should be noted that in this study the influence of various parameters of MFCC-decomposition on the results was not carried out. In addition, one can specify some other parameters of the algorithm, which probably would affect the numerical values of obtained Lyapunov exponents. Consequently, the value of the results consists in possibility of the comparative analysis of acoustic data, based on processing by means of artificial neural networks.

References

1. Benettin, G., Galgani, L., Giorgilli, A., Strelcin, J.M.: Lyapunov characteristic exponents for smooth dynamical systems and for hamiltonian systems: a method for computing all of them. pt. I: theory. pt. II: numerical applications. Meccanica **15**, 9–30 (1980)
2. Geist, K., Parlitz, U., Lauterborn, W.: Comparison of different methods for computing Lyapunov exponents. Progr. Theor. Phys. **83**, 875–893 (1990)
3. Ershov, S.V., Potapov, A.B.: On the concept of stationary Lyapunov basis. Phys. D **118**, 167–198 (1998)
4. Wolf, A., Swift, J., Swinney, H., Vastano, J.: Determining Lyapunov exponents from a time series. Phys. D **16**, 285–301 (1985)
5. Sano, M., Sawada, Y.: Measurements of the Lyapunov spectrum from a chaotic time series. Phys. Rev. Lett. **55**, 1082–1085 (1985)
6. Eckman, J.P., Kamphorst, S.O., Ruelle, D., Ciliberto, S.: Lyapunov exponents from time series. Phys. Rev. A **34**, 4971–4979 (1986)
7. Rosenstein, M.T., Collins, J.J., De Luca, C.J.: A practical method for calculating largest Lyapunov exponents from small data sets. Phys. D **65**, 117–134 (1993)
8. Kantz, H.: A robust method to estimate the maximal Lyapunov exponent of a time series. Phys. Lett. A **185**, 77–87 (1994)
9. Kantz, H., Schreiber, T.: Nonlinear Time Series Analysis. Cambridge University Press, Cambridge (1997)
10. Golovko, M.A.: Neural network techniques for chaotic signal processing. scientific session of the moscow engineering physics institute. In: VII All-Russian Scientific and Technical Conference "Neuroinformatics-2005". Lectures on Neuroinformatics, pp. 43–91. Moscow Engineering Physics Institute, Moscow (2005). (in Russian)
11. Jeong, J., Moir, T.J.: Kepstrum approach to real-time speech-enhancement methods using two microphones. Res. Lett. Inf. Math. Sci. **7**(1), 135–145 (2005)
12. Jun, Z., et al.: Using mel-frequency cepstral coefficients in missing data technique. EURASIP J. Appl. Sig, Process. **3**(1), 340–346 (2004)
13. Logan, B., et al.: Mel frequency cepstral coefficients for music modeling. In: ISMIR (2000)

14. Sung, B.K., Chung, M.B., Ko, I.J.: A feature based music content recognition method using simplified MFCC. Int. J. Princ. Appl. Inf. Sci. **2**(1), 13–23 (2008)
15. Huang, N.E., et al.: The empirical mode decomposition and the hilbert spectrum for nonlinear and non-stationary time series analysis. Proc. Roy. Soc. Lond. A: Math. Phys. Eng. Sci. **454**, 903–995 (1998)
16. Brown, R., Bryant, P., Abarbanel, H.: Computing the Lyapunov spectrum of a dynamical system from an observed time series. Phys. Rev. A **43**, 2787–2806 (1991)

Robust Acoustic Emotion Recognition Based on Cascaded Normalization and Extreme Learning Machines

Heysem Kaya[1(✉)], Alexey A. Karpov[2,3], and Albert Ali Salah[4]

[1] Department of Computer Engineering, Çorlu Faculty of Engineering,
Namik Kemal University, Çorlu, Tekirdağ, Turkey
hkaya@nku.edu.tr
[2] St. Petersburg Institute for Informatics and Automation
of Russian Academy of Sciences, St. Petersburg, Russia
karpov@iias.spb.su
[3] ITMO University, St. Petersburg, Russia
[4] Department of Computer Engineering, Boğaziçi University,
Bebek, Istanbul, Turkey
salah@boun.edu.tr

Abstract. One of the challenges in speech emotion recognition is robust and speaker-independent emotion recognition. In this paper, we take a cascaded normalization approach, combining linear speaker level, non-linear value level and feature vector level normalization to minimize speaker-related effects and to maximize class separability with linear kernel classifiers. We use extreme learning machine classifiers on a four class (i.e. joy, anger, sadness, neutral) problem. We show the efficacy of our proposed method on the recently collected Turkish Emotional Speech Database.

Keywords: Acoustic emotion recognition · Speech emotion recognition · Cascaded normalization · Extreme learning machines · ELM

1 Introduction

Automatic emotion recognition from audio is a popular research branch, with a range of applications in human-computer interaction and automatic analysis of human-human interactions. The topic is studied under several disciplines, such as multi-modal affect recognition, computational paralinguistics and speaker state and trait recognition.

The particular focus on emotion stems from the fact that emotion is intrinsically related to other *states* (such as mood, depression) and traits (such as personality) [1]. From a machine learning perspective, features and classifiers that work well in solving the emotion recognition problem can be expected to succeed in other related tasks.

As the search for the "optimal" set of features and classifiers that handle the problem robustly and efficiently is ongoing, speaker variability remains a

© Springer International Publishing Switzerland 2016
L. Cheng et al. (Eds.): ISNN 2016, LNCS 9719, pp. 115–123, 2016.
DOI: 10.1007/978-3-319-40663-3_14

major issue. The speech signal carries information about multiple aspects of the speaker in addition to emotion. The extracted acoustic features reflect these properties, and speaker identity and gender play a more prominent role compared to the affective content. To minimize the effect of these factors, the most common approaches taken in the literature are speaker normalization [16] and variability compensation via i-Vector modeling [10,20]. While the first approach is simple and effective, it does not fully compensate for the speaker variability. The i-Vector approach, which is originally proposed by Dehak *et al.* [2] for speaker recognition, suffers from high computational cost and is limited to linear transformations through factor analysis.

In the computer vision domain, problems with high-dimensional feature vectors and very high number of instances paved the way for the development of alternative normalization methods. One idea is to exploit low cost linear kernels with the capability of nonlinear discrimination. In this vein, Perronin *et al.* have shown that power normalization followed by instance level L_2 normalization gives a boost in recognition performance [12].

Considering the issues in paralinguistic speech processing and the efficient approaches in the computer vision domain, we have recently proposed a method that benefits from the cascaded normalization strategy [8]. Motivated by the good results obtained on the INTERSPEECH 2015 computational paralinguistic challenge, we propose in this paper to apply the cascaded normalization strategy to the problem of acoustic emotion recognition. The proposed approach combines linear speaker level normalization with nonlinear value level and feature vector level L_2 normalization to minimize speaker-related effects and to maximize class separability with linear kernel classifiers. To further promote efficient learning with high generalization capability, we use Extreme Learning Machine [6] as classifier and compare its performance with the popular Support Vector Machine (SVM).

The layout of the paper is as follows. In the next section, we give details of the proposed method. In Sect. 3, we introduce the database and the protocol used in the study. In Sect. 4 we provide the experimental results. Section 5 gives an overview of our findings and indicates some future directions.

2   Proposed Method

Our proposed method has three main components, namely speech feature extraction, cascaded normalization and classification. For the first step we use a freely available acoustic feature extractor called openSMILE [3]. Our contribution is in the second step, where we apply a refined and extended version of cascaded normalization in comparison to the one proposed in [8]. We then employ ELM for classification, which is recently successfully applied to multi-modal emotion recognition in the wild [7]. We compare the performance of ELM with SVM and make extensive tests with alternative normalization combinations to show the advantage of our approach.

The overall pipeline is summarized in Fig. 1 and the details of each step are explained in the following subsections.

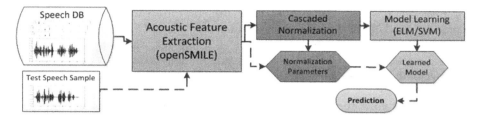

Fig. 1. The overall pipeline of the proposed method. The dashed lines illustrate the processing steps for a new test sample.

2.1   Speech Feature Extraction

The open-source[1] openSMILE tool [3] is popularly used to extract acoustic features of international paralinguistic and multi-modal challenges. The idea is to obtain a large pool of potentially relevant features by passing an extensive set of summarizing functionals on the low level descriptor (LLD) contours. Despite the high dimensionality, this simple approach gives state-of-the-art results in a wide range of paralinguistic tasks such as emotion, autism and depression detection [15,19].

We use the openSMILE toolbox with a standard feature configuration that served as the INTERSPEECH 2010 Paralinguistic challenge baseline set [14]. The 1582 dimensional feature set covers a range of popular LLDs such as Fundamental Frequency (F0), Mel Frequency Cepstral Coefficients (MFCC [0–14]), and Line Spectral Pairs Frequency [0–7], mapped to a fixed-length feature vector by means of functionals such as arithmetic mean and extrema.

2.2   Cascaded Normalization

Perronnin *et al.* propose a simple but efficient nonlinear normalization scheme to be used in linear classifiers (e.g. Linear Kernel Support Vector Machines) with power normalization, followed by instance level L_2 normalization [12]. The authors argue that power normalization helps "unsparsify" the distribution of feature values, consequently improving discrimination:

$$f(x) = sign(x)|x|^\alpha, \tag{1}$$

where $0 \leq \alpha \leq 1$ is a parameter to optimize. In [12] the authors empirically choose $\alpha = 0.5$.

The flowchart of the normalization steps we applied on the standard openSMILE features is given in Fig. 2. We use the combination of speaker, feature, value and instance level normalization strategies. Note that without any feature level normalization, the performance is poor for the ELM or SVM classifier on openSMILE features.

---

[1] The tool is available at http://www.openaudio.eu/.

In *speaker level normalization*, the parameters are estimated and applied separately on data of each speaker. We refer to *feature normalization* where parameters are estimated from the whole training set, then applied on both the training and test sets. In both "speaker level" and "feature" normalization, the processing is linear (e.g. z-normalization) and applied to each feature of the dataset separately.

Value level normalization is nonlinear (e.g. power normalization) and it is applied to each value of the data matrix separately. The final step is *instance level*, where each instance (feature vector) x is separately L_2 normalized giving $\hat{x}$ with $||\hat{x}^T \hat{x}||_2 = 1$.

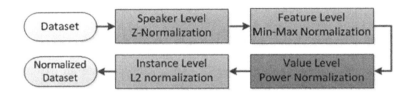

Fig. 2. Proposed cascaded normalization pipeline.

2.3   Model Learning

To learn a classification model, we use the Kernel ELM approach due to its fast and accurate learning capability [4]. Initially, ELM is proposed as a fast learning method for Single Hidden Layer Feedforward Networks: an alternative to back-propagation [5]. Hence, the architecture of basic ELM resembles a multilayer perceptron with a single hidden layer. ELM proposes the random generation of the hidden node output matrix $\mathbf{H} \in \mathbb{R}^{N \times h}$, where N and h denote the number of instances and the hidden neurons, respectively. The actual learning takes place in the second layer between $\mathbf{H}$ and the label matrix $\mathbf{T} \in \mathbb{R}^{N \times L}$, where L is the number of classes. $\mathbf{T}$ is composed of continuous annotations in case of regression, and therefore, is a vector. In the case of L-class classification, $\mathbf{T}$ is represented with one vs. all coding:

$$\mathbf{T}_{t,l} = \begin{cases} +1 \text{ if } y^t = l, \\ -1 \text{ if } y^t \neq l. \end{cases} \tag{2}$$

The second level weights $\beta \in \mathbb{R}^{h \times L}$ are learned by least squares solution to a set of linear equations $\mathbf{H}\beta = \mathbf{T}$. The output weights can be learned via:

$$\beta = \mathbf{H}^\dagger \mathbf{T}, \tag{3}$$

where $\mathbf{H}^\dagger$ is the Moore-Penrose generalized inverse [13] that gives the minimum L_2 norm solution to $||\mathbf{H}\beta - \mathbf{T}||$, simultaneously minimizing the norm of $||\beta||$.

This extreme learning rule is generalized to use any kernel $\mathbf{K}$ with a regularization parameter C, without generating $\mathbf{H}$ [4], relating ELM to Least Square SVM [18]:

$$\beta = (\frac{\mathbf{I}}{C} + \mathbf{K})^{-1}\mathbf{T}, \tag{4}$$

where $\mathbf{I}$ is the $N \times N$ identity matrix. In our experiments, we use Kernel ELM learning rule given in Eq. (4).

3 The Database and Protocol

The Boğaziçi University emotional database (BUEMODB) used in this study was collected by Meral *et al.* in 2003 [11]. However, it is only recently introduced for automatic emotion classification [9]. The corpus is portrayed by 11 amateur theater actors/actresses using the *Stanislavski effect* for generating emotional sentences, where the actor imagines the conditions that trigger an emotion [17]. The sentences are portrayed in four emotion classes that are evenly distributed over activation/valence axes: joy $(+/+)$, neutral $(-/+)$, anger $(+/-)$, and sadness $(-/-)$. The sentences used in the database are affirmative-informative and have different syntactic structures [11]:

1. Dışarıda kar yağıyor. (It is snowing outside.)
2. Sınavdan yetmiş aldım. (I got seventy from the exam.)
3. Hoca bana yetmiş verdi. (The instuctor gave me seventy.)
4. Galatasaray maçı iki-sıfır kazandı. (Galatasaray won the game two-nil.)
5. Beni çok şaşırttın. (You surprised me much.)
6. Telefonum çalıyor. (My phone is ringing.)
7. Kurs yarın bitiyor. (The course finishes tomorrow.)
8. Yarın kar yağacakmış. (Tomorrow, it will snow.)
9. Dersler iki hafta ertelendi. (The courses are postponed two weeks.)
10. Kapı açık kalmış. (The door had been left open.)
11. Dersi sadece iki kişi geçemedi. (Only two people failed the course.)

In total, 4 emotions $\times$ 11 actors/actresses $\times$ 11 sentences $= 484$ samples are collected. Out of 11 subjects, 7 are female. Here we follow the experimental protocol introduced by Kaya *et al.* [9] for comparability. The speaker independent partitioning of the data is given in Table 1.

In our experiments, we use the training and validation sets to learn a model and to optimize the hyper parameters. Then, we use the parameters of the top performing systems to re-train on the combination of training and validation sets, and cast our predictions on the test set. The defined protocol ensures both speaker independence and avoids over-estimation. The baseline four-way emotion classification accuracies are 57.6 % and 64.2 % for validation and test sets, respectively [9].

Table 1. Speaker independent data partitioning. F: Female, M:Male

	Train	Validation	Test
Gender distribution	2F+2M	2F+1M	3F+1M
Number of samples for joy	44	33	44
Number of samples for anger	44	33	44
Number of samples for neutral	44	33	44
Number of samples for sadness	44	33	44
Total number of samples	176	132	176

4   Experimental Results

In our experimental setting, we assume that speaker identities are known for the training and test sets. From an application perspective, this is a realistic assumption for ethical and legal reasons. First, it is possible to record people's speech, however these data should not be used for emotional analysis, unless there is informed consent from the subject. Second, it may be difficult to drive people into emotional states like anger. We postulate that the system will be used in personalization of affective systems, and therefore, the scope of the application will involve a closed set of people. In this case, one can apply speaker normalization after some initial data are collected, for instance sufficiently many samples to calculate a realistic standard deviation for the used features.

While the database is portrayed in four classes (three basic emotions plus the neutral state), we can map these classes into binary arousal and valence axes. Arousal can be defined as the state of vitality versus sleepiness, whereas valence is the dimension of positiveness of emotion. Therefore, we have three classification tasks; one for the original four-class problem and two for emotion primitives, respectively.

After the normalization steps, we kernelize the data using linear kernel prior to classification. As mentioned in the previous section, we use the training set to train a classifier and validation set to optimize the hyper-parameters such as the complexity parameters of Kernel ELM and SVM. For all normalization alternatives, the complexity parameter of these linear kernel classifiers are searched in the set $2^{\{-6,-5,...,5\}}$. Training ELM classifiers and testing on the validation set with the given set of hyper-parameters take an average of 6.9 milliseconds (ms) with a standard deviation of 3.6 ms. On the other hand, the SVM counterpart of these statistics are 254.8 ms $\mp$ 16.3 ms. Empirically, we observe that ELM training is 40 times faster compared to SVM.

In addition to simple z-normalization (i.e. standardization to zero mean and unit variance) and min-max normalization into [0,1] range, we applied six different combinations. Comparative results of eight normalization schemes and two classifiers over three affective classification tasks are given in Table 2. The table is sorted from proposed to prototypical normalization methods.

Table 2. Best validation set accuracies (%) of three affective classification tasks with alternative normalization schemes. zn: z-norm, spzn: speaker level z-norm, mmn: min-max norm, pn: power norm, l2n:L_2 norm. Best results per kernel-classifier combination are shown in **bold**.

Normalization	Four-class emotion		Binary arousal		Binary valence	
	ELM	SVM	ELM	SVM	ELM	SVM
spzn+mmn+pn+l2n	**65.2**	**67.4**	**91.7**	**90.2**	**62.1**	**66.7**
spzn+pn+l2n	**65.9**	67.4	**91.7**	**90.2**	**62.1**	62.9
spzn+mmn	60.6	62.1	87.1	87.1	**62.1**	64.4
spzn	59.1	63.6	86.4	86.4	60.6	65.2
mmn+pn+l2n	58.3	59.1	87.1	84.8	57.6	57.6
zn+pn+l2n	58.3	57.6	85.6	83.3	61.4	61.4
mmn	53.8	55.3	88.6	84.8	58.3	53.8
zn	50.8	56.8	85.6	84.1	56.8	59.1

Table 3. Validation and test set accuracies (%) with the proposed normalization approach.

Task	Validation		Test	
	ELM	**SVM**	**ELM**	**SVM**
Four-class emotion	65.2	67.4	79.0	77.3
Binary arousal	91.7	90.2	94.3	94.3
Binary valence	62.1	66.7	73.9	73.9

We see that (i) in all three classification tasks, there is a dramatic jump of performance between prototypical and proposed approach; (ii) recognition of valence is the poorest, a typical situation when only speech acoustics is used for emotion recognition. When the best overall performances are considered, we see that while ELM is better in arousal classification, SVM is better in the other two tasks. However, as can be seen in Table 3, ELM performance is either better or on-par-with SVM on the sequestered test set. Note also that SVM training is an order of magnitude slower than that of ELM.

Combining simple and computationally efficient normalization schemes, we also attain a dramatic improvement over the best test set accuracy (four-way emotion - 64.2 %) reported in [9].

5   Conclusion

In this work, we propose the application of cascaded normalization for robust speaker independent acoustic emotion recognition. We empirically show the breakdown of performance improvement due to speaker level and nonlinear normalization. The proposed approach outperforms typically employed linear feature normalization schemes including speaker normalization. The proposed approach can be efficiently and effectively applied to real-life scenarios, where speaker identity is known. When this information is missing with experimental data, it is possible to employ speaker clustering methods. Application of the proposed method on other emotional speech corpora with the challenging cross-corpus recognition setting constitutes the nearest future work.

Acknowledgments. This research is partially supported by the Council for Grants of the President of the Russian Federation (Project № MD-3035.2015.8) and by the Government of the Russian Federation (Grant № 074-U01).

References

1. Cowie, R., Sussman, N., Ben-Ze'ev, A.: Emotion-Oriented Systems: The Humaine Handbook, pp. 9–32. Springer, Heidelberg (2011)
2. Dehak, N., Kenny, P., Dehak, R., Dumouchel, P., Ouellet, P.: Front-end factor analysis for speaker verification. IEEE Trans. Audio Speech Lang. Process. **19**(4), 788–798 (2011)
3. Eyben, F., Wöllmer, M., Schuller, B.: Opensmile: the Munich versatile and fast open-source audio feature extractor. In: Proceedings of the International Conference on Multimedia, pp. 1459–1462. ACM (2010)
4. Huang, G.B., Zhou, H., Ding, X., Zhang, R.: Extreme learning machine for regression and multiclass classification. IEEE Trans. Syst. Man Cybern. Part B Cybern. **42**(2), 513–529 (2012)
5. Huang, G.B., Zhu, Q.Y., Siew, C.K.: Extreme learning machine: a new learning scheme of feedforward neural networks. Proc. IEEE Int. Joint Conf. Neural Netw. **2**, 985–990 (2004)
6. Huang, G.B., Zhu, Q.Y., Siew, C.K.: Extreme learning machine: theory and applications. Neurocomputing **70**(1), 489–501 (2006)
7. Kaya, H., Gürpinar, F., Afshar, S., Salah, A.A.: Contrasting and combining least squares based learners for emotion recognition in the wild. In: Proceedings of the 2015 ACM International Conference on Multimodal Interaction, pp. 459–466. ACM (2015)
8. Kaya, H., Karpov, A.A., Salah, A.A.: Fisher vectors with cascaded normalization for paralinguistic analysis. In: INTERSPEECH, pp. 909–913 (2015)
9. Kaya, H., Salah, A.A., Gurgen, S.F., Ekenel, H.: Protocol and baseline for experiments on Bogazici University Turkish emotional speech corpus. In: Proceedings of the 22nd IEEE Signal Processing and Communications Applications Conference (SIU), pp. 1698–1701 (2014)
10. Kua, J.M.K., Sethu, V., Le, P., Ambikairajah, E.: The UNSW submission to INTERSPEECH 2014 compare cognitive load challenge. In: INTERSPEECH, pp. 746–750 (2014)
11. Meral, H.M., Ekenel, H.K., Ozsoy, A.: Analysis of emotion in Turkish. In: XVII National Conference on Turkish Linguistics (2003)
12. Perronnin, F., Sánchez, J., Mensink, T.: Improving the Fisher Kernel for large-scale image classification. In: Proceedings of the 11th European Conference on Computer Vision, pp. 143–156 (2010)

13. Rao, C.R., Mitra, S.K.: Generalized Inverse of Matrices and its Applications. Wiley, New York (1971)
14. Schuller, B., Steidl, S., Batliner, A., Burkhardt, F., Devillers, L., Müller, C.A., Narayanan, S.S.: The INTERSPEECH 2010 paralinguistic challenge. In: INTER-SPEECH, pp. 2794–2797 (2010)
15. Schuller, B., Steidl, S., Batliner, A., Vinciarelli, A., Scherer, K., Ringeval, F., Chetouani, M., Weninger, F., Eyben, F., Marchi, E., Mortillaro, M., Salamin, H., Polychroniou, A., Valente, F., Kim, S.: The INTERSPEECH 2013 computational paralinguistics challenge: social signals, conflict, emotion, autism. In: INTER-SPEECH, pp. 148–152 (2013)
16. Schuller, B., Vlasenko, B., Eyben, F., Wollmer, M., Stuhlsatz, A., Wendemuth, A., Rigoll, G.: Cross-corpus acoustic emotion recognition: variances and strategies. IEEE Trans. Affect. Comput. **1**(2), 119–131 (2010)
17. Stanislavski, C.: An Actor Prepares. Routledge, London (1989)
18. Suykens, J.A., Vandewalle, J.: Least squares support vector machine classifiers. Neural Process. Lett. **9**(3), 293–300 (1999)
19. Valstar, M., Schuller, B., Smith, K., Almaev, T., Eyben, F., Krajewski, J., Cowie, R., Pantic, M.: AVEC 2014–3D dimensional affect and depression recognition challenge. In: Proceedings of the 4th ACM International Workshop on Audio/Visual Emotion Challenge, AVEC 2014 (2014)
20. Van Segbroeck, M., Travadi, R., Vaz, C., Kim, J., Black, M.P., Potamianos, A., Narayanan, S.S.: Classification of cognitive load from speech using an i-vector framework. In: INTERSPEECH, pp. 751–755 (2014)

Dynamical Behaviors of Recurrent Neural Networks

Matrix-Valued Hopfield Neural Networks

Călin-Adrian Popa[✉]

Department of Computer and Software Engineering, Polytechnic University
Timişoara, Blvd. V. Pârvan, No. 2, 300223 Timişoara, Romania
calin.popa@cs.upt.ro

Abstract. In this paper, we introduce matrix-valued Hopfield neural networks, for which the states, outputs, weights and thresholds are all square matrices. Matrix-valued neural networks represent a generalization of the complex-, hyperbolic-, quaternion- and Clifford-valued neural networks that have been intensively studied over the last few years. The dynamics of these networks is studied by giving an expression for the energy function, and proving that it is indeed an energy function for the proposed network.

Keywords: Clifford-valued neural networks · Hopfield neural networks · Energy function · Matrix-valued neural networks

1 Introduction

In the last few years, there has been an increasing interest in the study of neural networks with values in multidimensional domains. The most popular form of multidimensional neural networks are complex-valued neural networks, which were first introduced in the 1970's (see, for example, [23]), but have received more attention in the 1990's and in the past decade, because of their numerous applications, starting from those in complex-valued signal processing and continuing with applications in telecommunications and image processing (see, for example, [5, 14]).

Another type of neural networks, defined on the 2-dimensional algebra of hyperbolic numbers, are hyperbolic-valued neural networks, see [3, 12, 16]. It has been shown that their decision boundary, consisting of two hypersurfaces, can have any angle, unlike the complex-valued neural networks, which always have orthogonal decision boundaries.

Neural networks defined on the 4-dimensional quaternion algebra gained more interest in the last few years. Quaternion-valued neural networks were also first introduced in the 1990's, in the beginning as a generalization of the complex-valued neural networks, see [1, 2, 15]. Later, quaternion-valued neural networks were applied to chaotic time series prediction, the 4-bit parity problem, and, recently, to quaternion-valued signal processing. Another emerging application field for these networks is 3-dimensional and color image processing, because three dimensional objects and color pixels can be represented using quaternions.

© Springer International Publishing Switzerland 2016
L. Cheng et al. (Eds.): ISNN 2016, LNCS 9719, pp. 127–134, 2016.
DOI: 10.1007/978-3-319-40663-3_15

The complex, hyperbolic and quaternion algebras are all special cases of Clifford algebras, which have dimension 2^n, $n \geq 1$. Also called geometric algebras, they have numerous applications in physics and engineering, which made them appealing for use in the field of neural networks, also. Clifford-valued neural networks were defined in [17,18], and later discussed, for example, in [4,10,12]. Because of the close relation between Clifford algebras and geometry, they can offer, in the future, a way to solve many problems arising in the design of intelligent systems, allowing them to process different geometric objects and apply different geometric models to data.

Complex, hyperbolic, quaternion and Clifford numbers can all be written in matrix form. For example, a complex number $a + ib$, $i = \sqrt{-1}$, can be written in the form

$$\begin{pmatrix} a & -b \\ b & a \end{pmatrix},$$

a hyperbolic number $a + ub$, $u^2 = 1$, $u \neq \pm 1$ in the form

$$\begin{pmatrix} a & -b \\ b & a \end{pmatrix},$$

and a quaternion $a + ib + jc + kd$, $i^2 = j^2 = k^2 = ijk = -1$ in the form

$$\begin{pmatrix} a & b & c & d \\ -b & a & -d & c \\ -c & d & a & -b \\ -d & -c & b & a \end{pmatrix}.$$

Thus, each of these algebras can be seen as a subalgebra of the algebra of square matrices, with the natural addition and multiplication of the matrices. This led to the natural idea of defining a generalization of all the above neural networks with matrix inputs, outputs, weights and biases, first in the form of feedforward networks, see [19], and now in the form of Hopfield networks. Because of their degree of generality, these neural networks are bound to have many applications in the future at solving problems at which traditional neural networks have failed or performed poorly.

Hopfield first proposed the idea of introducing an energy function in order to study the dynamics of fully connected recurrent neural networks at the beginning of the 1980's, see [6–8,20]. He showed that combinatorial problems can be solved by using this type of network. Since then, Hopfield neural networks have been applied to the synthesis of associative memories, image processing, speech processing, control, signal processing, pattern matching, etc.

Because of the fact that complex-valued Hopfield networks were introduced in [11,13], hyperbolic-valued Hopfield networks in [9,12], quaternion-valued Hopfield networks in [10,21], and Clifford-valued Hopfield networks in [10,22], we considered an interesting idea to introduce matrix-valued Hopfield neural networks. These networks can be applied to the synthesis of matrix-valued associative memories, and also to image processing and pattern matching, where the data can be treated in matrix form.

Thus, the remainder of this paper is organized as follows: Sect. 2 gives the definition of matrix-valued Hopfield neural networks, and an expression for the energy function, showing that the given function is indeed an energy function for the proposed network. Section 3 is dedicated to presenting the conclusions of the study.

2 Matrix-Valued Hopfield Neural Networks

Consider the algebra $\mathcal{M}_n$ of square matrices of order n with real entries.

In what follows, we will define Hopfield neural networks for which the states, outputs, weights and thresholds are all from $\mathcal{M}_n$, which means that they are square matrices. The network is described by the set of differential equations

$$\tau_i \frac{dV_i(t)}{dt} = -V_i(t) + \sum_{j=1}^{N} W_{ij} f(V_j(t)) + B_i, \ i \in \{1, \ldots, N\}, \tag{1}$$

where $\tau_i \in \mathbb{R}$, $\tau_i > 0$ is the time constant of neuron i, $V_i(t) \in \mathcal{M}_n$ is the state of neuron i at time t, $W_{ij} \in \mathcal{M}_n$ is the weight connecting neuron j to neuron i, $f : \mathcal{M}_n \to \mathcal{M}_n$ is the nonlinear matrix-valued activation function, and B_i is the threshold of neuron i, $\forall i \in \{1, \ldots, N\}$. The derivative is taken to be the matrix formed by the derivatives of each element $[V_i(t)]_{ab}$ of the matrix $V_i(t)$ with respect to t:

$$\frac{dV_i(t)}{dt} := \left(\frac{d([V_i(t)]_{ab})}{dt} \right)_{1 \le a,b \le n}.$$

If we denote by $X_j(t) := f(V_j(t))$ the output of neuron j, the above set of differential equations can be written as

$$\tau_i \frac{dV_i(t)}{dt} = -V_i(t) + \sum_{j=1}^{N} W_{ij} X_j(t) + B_i, \ i \in \{1, \ldots, N\}.$$

The activation function f is formed of n^2 functions $f^{ab} : \mathcal{M}_n \to \mathbb{R}$, $1 \le a, b \le n$:

$$f(V) = \left(f^{ab}(V) \right)_{1 \le a,b \le n}.$$

In order to study the stability of the above defined network, we need to make a series of assumptions about the activation function.

The first assumption is that the functions f^{ab} are continuously differentiable with respect to each $[V]_{cd}$, $\forall 1 \le c, d \le n$, $\forall 1 \le a, b \le n$, and the function f is bounded: $\exists M > 0$, $||f(V)|| \le M$, $\forall V \in \mathcal{M}_n$, where $||X||$ is the Frobenius norm of matrix X, defined by $||X|| = \sqrt{\mathrm{Tr}(XX^T)}$, and $\mathrm{Tr}(X)$ represents the trace of matrix X. In this setting, the $n^2 \times n^2$ Jacobian matrix of the function f can be defined as

$$\mathbf{Jac}_f(V) = \left(\frac{\partial f^{ab}(V)}{\partial [V]_{cd}}\right)_{\substack{1\leq a,b\leq n \\ 1\leq c,d\leq n}}.$$

The second assumption that we have to make is that f is injective and $\mathbf{Jac}_f(V)$ is symmetric and positive definite, $\forall V \in \mathcal{M}_n$. This, together with the above assumption, assures the existence of the inverse function of f, $g : \mathcal{M}_n \to \mathcal{M}_n$, $g = f^{-1}$. We can thus write $g(X_i(t)) = V_i(t)$, $\forall i \in \{1, \ldots, N\}$. Now, we can define a function $G : \mathcal{M}_n \to \mathbb{R}$,

$$G(X) = \sum_{a,b=1}^n \int_0^{[X]_{ab}} g^{ab}(Y^{ab})dy,$$

where $g^{ab} : \mathcal{M}_n \to \mathbb{R}$ are the component functions of g and the matrices Y^{ab} have the following form

$$[Y^{ab}]_{cd} = \begin{cases} y, & (c,d) = (a,b) \\ [X]_{cd}, & \text{else} \end{cases}, \ \forall 1 \leq a,b \leq n.$$

For example, for 2×2 matrices, we have that

$$G(X) = \int_0^{[X]_{11}} g^{11}\left(\begin{pmatrix} y & [X]_{12} \\ [X]_{21} & [X]_{22} \end{pmatrix}\right)dy + \int_0^{[X]_{12}} g^{12}\left(\begin{pmatrix} [X]_{11} & y \\ [X]_{21} & [X]_{22} \end{pmatrix}\right)dy$$

$$+ \int_0^{[X]_{21}} g^{21}\left(\begin{pmatrix} [X]_{11} & [X]_{12} \\ y & [X]_{22} \end{pmatrix}\right)dy + \int_0^{[X]_{22}} g^{22}\left(\begin{pmatrix} [X]_{11} & [X]_{12} \\ [X]_{21} & y \end{pmatrix}\right)dy.$$

This function satisfies

$$\frac{\partial G(X)}{\partial [X]_{ab}} = g^{ab}(X), \ \forall 1 \leq a,b \leq n.$$

The above condition can also be written in matrix form as

$$\frac{\partial G(X)}{\partial X} = g(X). \tag{2}$$

The last assumption concerns the weights of the network, which must satisfy:

$$W_{ji} = W_{ij}^T, \ \forall i,j \in \{1, \ldots, N\}.$$

Having made all the above assumptions, we can define the energy function $E : \mathcal{M}_n^N \to \mathbb{R}$ of the Hopfield network (1) as:

$$E(\mathbf{X}(t)) = -\frac{1}{2}\sum_{i=1}^N \sum_{j=1}^N \text{Tr}(X_i(t)^T W_{ij} X_j(t)) + \sum_{i=1}^N G(X_i(t)) - \sum_{i=1}^N \text{Tr}(B_i^T X_i(t)).$$

$$\tag{3}$$

A function E is an energy function for the Hopfield network (1) if the derivative of E along the trajectories of network, denoted by $\frac{dE(\mathbf{X}(t))}{dt}$, satisfies the condition $\frac{dE(\mathbf{X}(t))}{dt} \leq 0$ and $\frac{dE(\mathbf{X}(t))}{dt} = 0 \Leftrightarrow \frac{dX_i(t)}{dt} = 0$, $\forall i \in \{1, \dots, N\}$. We will show that the function E defined in (3) is indeed an energy function for the network (1).

For this, we start by applying the chain rule:

$$\frac{dE(\mathbf{X}(t))}{dt} = \sum_{i=1}^{N} \sum_{a,b=1}^{n} \frac{\partial E(\mathbf{X}(t))}{\partial [X_i(t)]_{ab}} \frac{d[X_i(t)]_{ab}}{dt}$$

$$= \sum_{i=1}^{N} \text{Tr} \left(\left(\frac{\partial E(\mathbf{X}(t))}{\partial X_i(t)} \right)^{T} \frac{dX_i(t)}{dt} \right), \tag{4}$$

where by $\frac{\partial E(\mathbf{X}(t))}{\partial [X_i(t)]_{ab}}$ we denoted the partial derivative of the function E with respect to each element $[X_i(t)]_{ab}$ of the matrix $X_i(t)$, $\forall 1 \leq a, b \leq n$, $\forall i \in \{1, \dots, N\}$. Taking (3) into account, using the fact that

$$\frac{d\text{Tr}(X^T A)}{dX} = \frac{d\text{Tr}(A^T X)}{dX} = A,$$

relation (2), the assumption $W_{ji} = W_{ij}^T$, and also the set of equations given by (1), the expression of the partial derivative $\frac{\partial E(\mathbf{X}(t))}{\partial X_i(t)} = \left(\frac{\partial E(\mathbf{X}(t))}{\partial [X_i(t)]_{ab}} \right)_{1 \leq a,b \leq n}$ is computed as:

$$\frac{\partial E(\mathbf{X}(t))}{\partial X_i(t)} = -\sum_{j=1}^{N} W_{ij} X_j(t) + g(X_i(t)) - B_i$$

$$= -\left(\sum_{j=1}^{N} W_{ij} X_j(t) - V_i(t) + B_i \right)$$

$$= -\tau_i \frac{dV_i(t)}{dt}, \ \forall i \in \{1, \dots, N\}.$$

If we denote by $\text{vec}(X)$ the vectorization of matrix X, and use the identity $\text{Tr}(A^T B) = \text{vec}(A)^T \text{vec}(B)$, $\forall A, B \in \mathcal{M}_n$, we can now write equation (4) as:

$$\frac{dE(\mathbf{X}(t))}{dt} = \sum_{i=1}^{N} \text{Tr} \left(\left(-\tau_i \frac{dV_i(t)}{dt} \right)^{T} \frac{dX_i(t)}{dt} \right)$$

$$= -\sum_{i=1}^{N} \tau_i \left[\text{vec} \left(\frac{dV_i(t)}{dt} \right) \right]^{T} \text{vec} \left(\frac{dX_i(t)}{dt} \right)$$

$$= -\sum_{i=1}^{N} \tau_i \left[\text{vec} \left(\frac{dX_i(t)}{dt} \right) \right]^{T} [\mathbf{Jac}_g(X_i(t))]^{T} \text{vec} \left(\frac{dX_i(t)}{dt} \right)$$

$$\leq 0, \tag{5}$$

where, from $g(X_i(t)) = V_i(t)$, we obtained that

$$\text{vec}\left(\frac{dg(X_i(t))}{dt}\right) = \mathbf{Jac}_g(X_i(t))\text{vec}\left(\frac{dX_i(t)}{dt}\right), \ \forall i \in \{1, \dots, N\}.$$

Because $\mathbf{Jac}_f(V)$ is symmetric and positive definite, we deduce that $\mathbf{Jac}_g(X)$ is also symmetric and positive definite, and thus

$$\left[\text{vec}\left(\frac{dX_i(t)}{dt}\right)\right]^T [\mathbf{Jac}_g(X_i(t))]^T \text{vec}\left(\frac{dX_i(t)}{dt}\right) \geq 0, \ \forall i \in \{1, \dots, N\},$$

which allowed us to write the last inequality in relation (5). Equality is attained when

$$\frac{dE(\mathbf{X}(t))}{dt} = 0 \Leftrightarrow \text{vec}\left(\frac{dX_i(t)}{dt}\right) = 0 \Leftrightarrow \frac{dX_i(t)}{dt} = 0, \ \forall i \in \{1, \dots, N\},$$

thus ending the proof that E is indeed an energy function for the network (1).

We now give two examples of activation functions that satisfy the above assumptions, inspired by the ones used in real-valued and complex-valued neural networks:

$$f(V) = \frac{V}{1 + ||V||}, \ \forall V \in \mathcal{M}_n,$$

$$f\left(([V]_{ab})_{1 \leq a,b \leq n}\right) = (\tanh[V]_{ab})_{1 \leq a,b \leq n}, \ \forall V \in \mathcal{M}_n.$$

3   Conclusions

The definition of matrix-valued Hopfield neural networks was given. Matrix-valued neural networks represent a generalization of the complex-, hyperbolic-, quaternion- and Clifford-valued neural networks, because complex, hyperbolic, quaternion and Clifford algebras can be seen as matrix subalgebras of the square matrix algebra with the natural operations of addition and multiplication of the matrices.

The existence conditions for the energy function of the proposed network were detailed. Then, the expression of the energy function was given, showing that the proposed function is indeed an energy function for the defined matrix-valued Hopfield neural network.

It is natural to think that the future will bring even more applications for the complex- and quaternion-valued neural networks, and also for their direct generalization, namely the Clifford-valued neural networks. Taking into account the fact that Clifford algebras have dimension 2^n, $n \geq 1$, it is possible that the applications do not need such a large dimension for the values of the input data. This is where matrix-valued neural networks might come into play, because their dimension is only n^2, and thus it allows for more efficient memory use than in the case of Clifford-valued neural networks.

The present work represents only a first step done towards a more general framework for neural networks, which could benefit not only from increasing the number of hidden layers and making the architecture ever more complicated, but also from increasing the dimensionality of the data that is being handled by the network.

References

1. Arena, P., Fortuna, L., Muscato, G., Xibilia, M.: Multilayer perceptrons to approximate quaternion valued functions. Neural Netw. **10**(2), 335–342 (1997)
2. Arena, P., Fortuna, L., Occhipinti, L., Xibilia, M.: Neural networks for quaternion-valued function approximation. In: International Symposium on Circuits and Systems (ISCAS), vol. 6, pp. 307–310. IEEE (1994)
3. Buchholz, S., Sommer, G.: A hyperbolic multilayer perceptron. In: International Joint Conference on Neural Networks (IJCNN), vol. 2, pp. 129–133. IEEE (2000)
4. Buchholz, S., Sommer, G.: On Clifford neurons and Clifford multi-layer perceptrons. Neural Netw. **21**(7), 925–935 (2008)
5. Hirose, A.: Complex-Valued Neural Networks, Studies in Computational Intelligence, vol. 400. Springer, Heidelberg (2012)
6. Hopfield, J.: Neural networks and physical systems with eemergent collective computational abilities. Proc. Natl. Acad. Sci. U.S.A. **79**(8), 2554–2558 (1982)
7. Hopfield, J.: Neurons with graded response have collective computational properties like those of two-state neurons. Proc. Natl. Acad. Sci. U.S.A. **81**(10), 3088–3092 (1984)
8. Hopfield, J., Tank, D.: "Neural" computation of decisions in optimization problems. Biol. Cybern. **52**(3), 141–152 (1985)
9. Kobayashi, M.: Hyperbolic hopfield neural networks. IEEE Trans. Neural Netw. Learn. Syst. **24**(2), 335–341 (2013)
10. Kuroe, Y.: Models of cClifford recurrent neural networks and their dynamics. In: International Joint Conference on Neural Networks (IJCNN), pp. 1035–1041. IEEE (2011)
11. Kuroe, Y., Hashimoto, N., Mori, T.: On energy function for complex-valued neural nnetworks and its applications. In: International Conference on Neural Information Processing (ICONIP), vol. 3, pp. 1079–1083. IEEE (2002)
12. Kuroe, Y., Tanigawa, S., Iima, H.: Models of hopfield-type Clifford neural networks and their energy functions - hyperbolic and dual valued networks. In: Lu, B.-L., Zhang, L., Kwok, J. (eds.) ICONIP 2011, Part I. LNCS, vol. 7062, pp. 560–569. Springer, Heidelberg (2011)
13. Kuroe, Y., Yoshida, M., Mori, T.: On activation functions for complex-valued neural networks - existence of energy functions. Artificial Neural Networks and Neural Information Processing - ICANN/ICONIP. LNCS, pp. 985–992. Springer, Heidelberg (2003)
14. Mandic, D., Goh, S.: Complex Valued Nonlinear Adaptive Filters Noncircularity, Widely Linear and Neural Models. Wiley, New York (2009)
15. Nitta, T.: A quaternary version of the back-propagation algorithm. In: International Conference on Neural Networks, pp. 2753–2756, vol. 5. IEEE (1995)
16. Nitta, T., Buchholz, S.: On the decision boundaries of hyperbolic neurons. In: International Joint Conference on Neural Networks (IJCNN), pp. 2974–2980. IEEE (2008)

17. Pearson, J., Bisset, D.: Back propagation in a Clifford algebra. In: International Conference on Artificial Neural Networks, vol. 2, pp. 413–416 (1992)
18. Pearson, J., Bisset, D.: Neural networks in the Clifford domain. In: International Conference on Neural Networks, vol. 3, pp. 1465–1469. IEEE (1994)
19. Popa, C.A.: Matrix-Valued neural networks. In: Matoušek, R. (ed.) International Conference on Soft Computing (MENDEL). Advances in Intelligent Systems and Computing, vol. 378, pp. 245–255. Springer International Publishing, Heidelberg (2015)
20. Tank, D., Hopfield, J.: Simple "neural" optimization networks: an a/d converter, signal decision circuit, and a linear programming circuit. IEEE Trans. Circ. Syst. **33**(5), 533–541 (1986)
21. Valle, M.: A novel continuous-valued quaternionic hopfield neural network. In: Brazilian Conference on Intelligent Systems (BRACIS), pp. 97–102. IEEE (2014)
22. Vallejo, J., Bayro-Corrochano, E.: Clifford hopfield neural networks. In: International Joint Conference on Neural Networks (IJCNN), pp. 3609–3612. IEEE, June 2008
23. Widrow, B., McCool, J., Ball, M.: The Complex Lms Algorithm. Proc. IEEE **63**(4), 719–720 (1975)

Synchronization of Coupled Neural Networks with Nodes of Different Dimensions

Manchun Tan$^{(\boxtimes)}$ and Desheng Xu

College of Information Science, Jinan University, Guangzhou 510632, China
`tanmc@jnu.edu.cn`

Abstract. A class of coupled neural networks with nodes of different coupling time-delays and different state dimensions is investigated in this paper. Based on Lyapunov stability theory, some sufficient conditions for synchronization of coupled neural networks are derived. The coupling configuration matrix is not necessary to be symmetric or irreducible, and the inner coupling matrix need not be symmetric. Finally, numerical examples are presented to demonstrate the effectiveness of the designed method.

Keywords: Coupled neural networks · Synchronization · Nodes of different dimensions · Different coupling time delays

1 Introduction

Over the last decades, the dynamical behaviors of neural networks (NNs) have been extensively investigated (see [1–5] and references therein). Particularly, the combination of a set of neural networks could achieve higher level of information processing, and the dynamical behaviors of coupled networks are more complex than those of any subsystem. The stabilization and synchronization in an array of coupled neural networks have attracted increasing attention of many researchers in the past few years [6–15]. Most of studies (for instance, [9–11]) on synchronization of coupled NNs are based on Lyapunov functional approach integrated with LMI technique. In [9], neural networks with hybrid coupling and interval time-varying delay are analyzed, by constructing the augmented functional with multiple Kronecker product operations. Instead of using Kroneker product method, which often treats the coupled complex networks as a whole dynamical system, in [10], the authors deal with the isolated neural networks directly. In [11], the author transforms both of the problems of global synchronization of dynamics and convergence of dynamics into solving a corresponding homogeneous system of linear algebraic equations. In [12], the authors analyze the finite-time synchronization in an array of coupled neural networks with discontinuous activation functions. In [13], the synchronization of linearly coupled dynamical systems with discontinuous identical nodes and time delay is studied. In [14], the authors investigate the stability and synchronization of memristor-based coupling neural networks with time-varying delays via intermittent control.

© Springer International Publishing Switzerland 2016
L. Cheng et al. (Eds.): ISNN 2016, LNCS 9719, pp. 135–142, 2016.
DOI: 10.1007/978-3-319-40663-3_16

Most of the aforementioned works have assumed that the state vectors have the same dimensions [16–18]. Recently, there are more and more studies on complex networks with non-identical nodes, because most dynamical networks in engineering have different nodes [19–24]. If a network is constructed by nodes with different state dimension, the network will exhibit different dynamical behaviors and the previous methods of stabilization will be invalid [25–29]. In reality, complex dynamical networks are more likely to have different time-delay coupling for different nodes. It is necessary to investigate the dynamics of neural networks with different coupling time-delays.

Motivated by the above discussion, in this paper, we consider the synchronization of coupled neural networks with non-identical nodes, which may possess different coupling time-delays and different state dimensions. Some sufficient conditions guaranteeing the synchronization of NNs are developed. Numerical simulations will be given to illustrate the validity of the theoretical results.

2   Model Description and Preliminaries

In this section, the function projective synchronization is considered between two coupled neural networks. The drive neural network is given by

$$\dot{x}_i(t) = A_i x_i(t) + f_i(x_i(t)) + \sum_{j=1}^{N} c_{ij} H_{ij} x_j(t - \tau_j), \tag{1}$$

where $x_i = (x_{i1}, x_{i2}, \cdots, x_{in_i})^T \in R^{n_i}$ is the state vector of the i-th node; $A_i \in R^{n_i \times n_i}$ is a constant matrix; $f_i(\cdot) \in R^{n_i}$ is a continuous vector function; $\tau_i > 0$ is the known time delay; $H_{ij} \in R^{n_i \times n_j}$ is the inner coupling matrices, $i, j = 1, 2, \cdots, N$; $C = (c_{ij})_{N \times N}$ is the outer coupling matrix representing the coupling strength and the topological structure of the network. The matrix C is defined as follows: if there is a connection from node j to node i $(i \neq j)$, then $c_{ij} \neq 0$; otherwise $c_{ij} = 0$.

The response network is given by

$$\dot{y}_i(t) = B_i y_i(t) + g_i(y_i(t)) + \sum_{j=1}^{N} d_{ij} G_{ij} y_j(t - \tau_j) + u_i, \quad i = 1, 2, \cdots, N, \tag{2}$$

where $y_i = (y_{i1}, y_{i2}, \cdots, y_{im_i})^T \in R^{m_i}$ is the state vector of the i-th node of the response network; $B_i \in R^{m_i \times m_i}$ is a constant matrix; $g_i(\cdot) \in R^{m_i}$ is a continuous vector function; $\tau > 0$ is the known time delay; $G_{ij} \in R^{m_i \times m_j}$ is the inner coupling matrices, $i, j = 1, 2, \cdots, N$; $D = (d_{ij})_{N \times N}$ is the outer coupling matrix representing the coupling strength and the topological structure of the network. The matrix D is defined as follows: if there is a connection from node j to node i $(i \neq j)$, then $d_{ij} \neq 0$; otherwise $d_{ij} = 0$.

Define the function projective synchronization errors as

$$e_i(t) = y_i(t) - \alpha_i x_i(t), \qquad i = 1, 2, \cdots, N, \tag{3}$$

where $\alpha_i \in R^{m_i \times n_i}$ is a constant matrix.

The time derivative of $e_i(t)$ is $\dot{e}_i(t) = \dot{y}_i(t) - \alpha_i \dot{x}_i(t)$. From Eqs. (2) and (3), we obtain the error dynamical system:

$$\dot{e}_i(t) = B_i y_i(t) + g_i(y_i(t)) + \sum_{j=1}^{N} d_{ij} G_{ij} y_j(t - \tau_j) - \alpha_i \dot{x}_i(t) + u_i. \tag{4}$$

The controller u_i is designed as

$$u_i = \alpha_i \dot{x}_i(t) - B_i \alpha_i x_i(t) - g_i(\alpha_i x_i(t)) - \sum_{j=1}^{N} d_{ij} G_{ij} \alpha_j x_j(t - \tau_j) - k_i e_i(t), \tag{5}$$

where $k_i (i = 1, 2, \cdots, N)$ are positive numbers to be determined.

Remark 1. The nodes in the neural networks (1) and (2) could have different dimensions. Model (1) generalizes some of models in literature. For example, if $\tau_i = 0$ and $f_i = 0$ hold, Eq. (1) reduces to the model studied in [25]. If $f_i = 0$ holds, Eq. (1) reduces to the model studied in [26]. If $\tau_i = \tau$ holds, Eq. (1) reduces to the model studied in [29]. Although neural networks with nonidentical nodes and different dimensions are considered in [18,27], the nodes have the same state dimensions in the drive and response systems in these literatures, respectively. If $n_i = n$ and $m_i = m$ hold, the drive-response systems (1) and (2) reduce to the models studied in [18].

Assumption 1 (H1): There exists a positive definite diagonal matrix δ_i such that

$$(\xi_1 - \xi_2)^T (g_i(\xi_1) - g_i(\xi_2)) \leq \delta_i (\xi_1 - \xi_2)^T (\xi_1 - \xi_2)$$

holds for all $\xi_1, \xi_2 \in R^{m_i}, \xi_1 \neq \xi_2, i = 1, 2, ..., N$.

3   Main Result

Theorem 1. Under the assumption (H1), the drive network (1) and response network (2) can achieve synchronization by the controller (5), if there exist positive definite symmetric matrices $Q_i \in R^{m_i \times m_i}$, and positive numbers ε and $k_i (i = 1, \cdots, N)$, such that the following condition is satisfied

$$\Xi = \begin{bmatrix} \Xi_{11} & \Xi_{12} \\ \Xi_{12}^T & \Xi_{22} \end{bmatrix} < 0, \tag{6}$$

where

$$
\begin{aligned}
\Xi_{11} = \operatorname{diag}\{ &\frac{1}{2}(B_1 + B_1^T) + (\delta_1 - k_1)I_{m_1} + \varepsilon Q_1, \\
&\frac{1}{2}(B_2 + B_2^T) + (\delta_2 - k_2)I_{m_2} + \varepsilon Q_2, \cdots, \\
&\frac{1}{2}(B_N + B_N^T) + (\delta_N - k_N)I_{m_N} + \varepsilon Q_N\},
\end{aligned}
$$

$$
\Xi_{12} = \frac{1}{2}
\begin{bmatrix}
d_{11}G_{11} & d_{12}G_{12} & \cdots & d_{1N}G_{1N} \\
d_{21}G_{21} & d_{22}G_{22} & \cdots & d_{2N}G_{2N} \\
\cdots & \cdots & \cdots & \cdots \\
d_{N1}G_{N1} & d_{N2}G_{N2} & \cdots & d_{NN}G_{NN}
\end{bmatrix},
$$

$$
\Xi_{22} = -\varepsilon \cdot \operatorname{diag}\{Q_1, Q_2, \cdots, Q_N\}. \tag{7}
$$

Proof. From (4) and (5), we obtain

$$
\dot{e}_i(t) = (B_i - k_i I_{m_i})e_i(t) + \bar{g}_i(e_i(t)) + \sum_{j=1}^{N} d_{ij}G_{ij}e_j(t - \tau_j), \tag{8}
$$

where $\bar{g}_i(e_i(t)) = g_i(y_i(t)) - g_i(\alpha_i x_i(t))$.

Select a Lyapunov-Krasovskii functional as

$$
V(t) = \frac{1}{2}\sum_{i=1}^{N} e_i^T(t)e_i(t) + \varepsilon \sum_{i=1}^{N} \int_{t-\tau_i}^{t} e_i^T(s)Q_i e_i(s)ds.
$$

Calculating the derivative of $V(t)$ along the trajectories of Eq. (8), we get

$$
\begin{aligned}
\dot{V}(t) = &\sum_{i=1}^{N} e_i^T(t)\{(B_i - k_i I_{m_i})e_i(t) + \bar{g}_i(e_i(t)) + \sum_{j=1}^{N} d_{ij}G_{ij}e_j(t - \tau_j)\} \\
&+\varepsilon \sum_{i=1}^{N} e_i^T(t)Q_i e_i(t) - \varepsilon \sum_{i=1}^{N} e_i^T(t - \tau_i)Q_i e_i(t - \tau_i).
\end{aligned}
$$

From assumption (H1), we obtain

$$
e_i^T(t)\bar{g}_i(e_i(t)) \leq \delta_i e_i^T(t)e_i(t).
$$

Hence, we have

$$
\begin{aligned}
\dot{V} \leq &\sum_{i=1}^{N} e_i^T(t)(B_i + \delta_i I_{m_i} - k_i I_{m_i})e_i(t) + \sum_{i=1}^{N} e_i^T(t)\sum_{j=1}^{N} d_{ij}G_{ij}e_j(t - \tau_j) \\
&+\varepsilon \sum_{i=1}^{N} e_i^T(t)Q_i e_i(t) - \varepsilon \sum_{i=1}^{N} e_i^T(t - \tau_i)Q_i e_i(t - \tau_i) \\
= &\, Y^T(t)\Xi Y(t),
\end{aligned}
$$

where $Y(t) = (e_1^T(t), e_2^T(t), \cdots, e_N^T(t), e_1^T(t - \tau_1), e_2^T(t - \tau_2), \cdots, e_N^T(t - \tau_N))^T$.

It follows from the condition (6) that $\dot{V}(t)$ is negative definite. By using the Lyapunov stability theories, we conclude that the trajectories $e_i(t)$ of Eq. (4) will converge to 0. This means that the function projective synchronization between networks (1) and (2) can be achieved. The proof is completed.

4   Numerical Example

Example 1: In this example, two different neural networks are constructed to show the effectiveness of Theorem 1. In the simulations, the drive network (1) is described by the following Lü chaotic systems:

$$\begin{pmatrix} \dot{x}_{i1} \\ \dot{x}_{i2} \\ \dot{x}_{i3} \end{pmatrix} = \begin{pmatrix} -36 & 36 & 0 \\ 0 & 20 & 0 \\ 0 & 0 & -3 \end{pmatrix} \begin{pmatrix} x_{i1} \\ x_{i2} \\ x_{i3} \end{pmatrix} + \begin{pmatrix} 0 \\ -x_{i1}x_{i3} \\ x_{i1}x_{i2} \end{pmatrix}, \quad i = 1, 2.$$

For the response network (2), the first node is described by the following hyperchaotic Lorenz system:

$$\begin{pmatrix} \dot{y}_{11} \\ \dot{y}_{12} \\ \dot{y}_{13} \\ \dot{y}_{14} \end{pmatrix} = \begin{pmatrix} -10 & 10 & 0 & 1 \\ 28 & -1 & 0 & 0 \\ 0 & 0 & -8/3 & 0 \\ 0 & 0 & 0 & 1.3 \end{pmatrix} \begin{pmatrix} y_{11} \\ y_{12} \\ y_{13} \\ y_{14} \end{pmatrix} + \begin{pmatrix} 0 \\ -y_{11}y_{13} \\ y_{11}y_{12} \\ -y_{11}y_{13} \end{pmatrix}.$$

Choose the second node dynamics as the following Chen chaotic system:

$$\begin{pmatrix} \dot{y}_{21} \\ \dot{y}_{22} \\ \dot{y}_{23} \end{pmatrix} = \begin{pmatrix} -35 & 35 & 0 \\ -7 & 28 & 0 \\ 0 & 0 & -3 \end{pmatrix} \begin{pmatrix} y_{21} \\ y_{22} \\ y_{23} \end{pmatrix} + \begin{pmatrix} 0 \\ -y_{21}y_{23} \\ y_{21}y_{22} \end{pmatrix}.$$

According to Refs. [30,31], there exist some constants $M_{11} = 25$, $M_{12} = 25$, $M_{13} = 45$, $M_{14} = 180$, $M_{21} = 23$, $M_{22} = 32$, $M_{23} = 61$, such that Chaotic attractors of two systems satisfied: $|s_{11}| \leq M_{11}$, $|s_{12}| \leq M_{12}$, $|s_{13}| \leq M_{13}$, $|s_{14}| \leq M_{14}$, $|s_{21}| \leq M_{21}$, $|s_{22}| \leq M_{22}$, $|s_{23}| \leq M_{23}$. It yields

$$\|g_1(x_1) - g_1(s_1)\| \leq \sqrt{3M_{11}^2 + 2M_{13}^2 + M_{12}^2}\,\|x_1 - s_1\|,$$

$$\|g_2(x_2) - g_2(s_2)\| \leq \sqrt{2M_{21}^2 + M_{22}^2 + M_{23}^2}\,\|x_2 - s_2\|.$$

Choose $\delta_1 = 71.764$, $\delta_2 = 76.177$, then assumption (H1) holds. The outer coupling matrices are taken arbitrarily as $C = \begin{pmatrix} 2 & 0.5 \\ 3 & 2 \end{pmatrix}$, $D = \begin{pmatrix} -1 & 3 \\ 1 & 2 \end{pmatrix}$, and the inner coupling matrices are given respectively as follows:

$$H_{11} = \begin{pmatrix} 3 & 4 & 1 \\ 2 & 1 & 3 \\ 1 & 3 & 5 \end{pmatrix}, \quad H_{12} = \begin{pmatrix} 0 & 1 & 3 \\ 1 & 3 & 0.5 \\ 1 & 0 & 1 \end{pmatrix}, \quad H_{21} = \begin{pmatrix} 3 & 0 & 1 \\ -1 & 2 & 0 \\ 0 & 0 & -1 \end{pmatrix},$$

$$H_{22} = \begin{pmatrix} 3 & 0 & 1 \\ 0 & 2 & -1 \\ -2 & 1 & 2 \end{pmatrix}, \quad G_{11} = \begin{pmatrix} 3 & 4 & 0 & 1 \\ 5 & -2 & 1 & 0 \\ 2 & 0 & 0 & 3 \\ 0 & 2 & 4 & -1 \end{pmatrix}, \quad G_{12} = \begin{pmatrix} 0.1 & 2 & 1 \\ 3 & 4 & 0 \\ 0 & -2 & 0.1 \\ 0.5 & 0.6 & 4 \end{pmatrix},$$

$$G_{21} = \begin{pmatrix} -1 & 0.3 & 0 & 0.5 \\ 2 & 4 & -0.2 & 6 \\ 1 & 0 & 0 & 5 \end{pmatrix}, \quad G_{22} = \begin{pmatrix} 0.1 & 0 & 0 \\ 0 & 2 & 0 \\ 0 & 0 & 4 \end{pmatrix}.$$

We take the scaling matrices

$$\alpha_1 = \begin{pmatrix} 2 & 0 & 0 \\ 0 & -1 & 0 \\ 1 & 0 & 0 \\ 0 & 0.1 & 1 \end{pmatrix}, \quad \alpha_2 = diag(1.5, 0.5, -2)$$

and time delay $\tau_1 = 0.3$, $\tau_2 = 1$. Let $\varepsilon = 1, k_1 = 110$ and $k_2 = 115$. It's easily verified that the condition (6) in Theorem 1 is satisfied. The initial values of the state vectors $x_i(0), y_i(0)$ are chosen randomly in $(-4, 4)$. According to Theorem 1, function projective synchronization between networks (1) and (2) can be achieved as shown in Fig. 1.

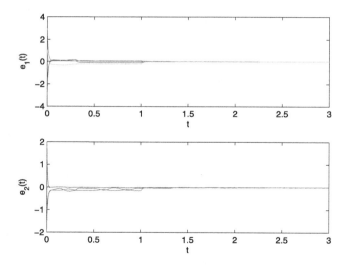

Fig. 1. The error curves of states of the drive-response networks.

5   Conclusion

In this paper, we study the synchronization of coupled neural networks (NNs) forced by nonlinear feedback control. The coupled NNs are more general than that exist in the literature. By constructing suitable Lyapunov functionals, a set of synchronization criteria have been established to ensure that the proposed NNs are synchronized. Illustrative simulations including some chaotic NNs have been provided to verify the correctness and effectiveness of the obtained results.

Acknowledgments. The research is supported by grants from the National Natural Science Foundation of China (No.11471083 and No. 61572233), the Natural Science Foundation of Guangdong Province in China (No.9151001003000005), and the Fundamental Research Funds for the Central Universities (No. 21612443).

References

1. Lu, W.L., Chen, T.P.: Synchronization of coupled connected neural networks with delays. IEEE Trans. Circ. Syst. **51**, 2491–2503 (2004)
2. Cao, J.D., Chen, G.R., Li, P.: Global synchronization in an array of delayed neural networks with hybrid coupling. IEEE Trans. Syst. Man. Cybern. B **38**(2), 488–498 (2008)
3. Xiao, J., Zhong, S., Li, Y.: Improved passivity criteria for memristive neural networks with interval multiple time-varying delays. Neurocomputing **171**, 1414–1430 (2016)
4. Zheng, C.D., Shan, Q.H., Zhang, H.G.: On stabilization of stochastic Cohen-Grossberg neural networks with mode-dependent mixed time-delays and Markovian switching. IEEE Trans. Neural Netw. Learn. Syst. **24**, 800–811 (2013)
5. Zeng, Z.G., Zheng, W.X.: Multistability of two kinds of recurrent neural networks with activation functions symmetrical about the origin on the phase plane. IEEE Trans. Neural Netw. Learn. Syst. **24**, 1749–1762 (2013)
6. Cao, J.D., Li, P., Wang, W.W.: Global synchronization in arrays of delayed neural networks with constant and delayed coupling. Phys. Lett. A **353**, 318–325 (2006)
7. Song, Q.K.: Synchronization analysis of coupled connected neural networks with mixed time delays. Neurocomputing **72**, 3907–3914 (2009)
8. Wang, G., Yin, Q., Shen, Y.: Exponential synchronization of coupled fuzzy neural networks with disturbances and mixed time-delays. Neurocomputing **106**, 77–85 (2013)
9. Zhang, H.G., Gong, D.W., Chen, B., Liu, Z.W.: Synchronization for coupled neural networks with interval delay: a novel augmented Lyapunov-Krasovskii functional method. IEEE Trans. Neural Netw. Learn. Syst. **24**, 58–70 (2013)
10. Wang, Z.S., Zhang, H.G.: Synchronization stability in complex interconnected neural networks with nonsymmetric coupling. Neurocomputing **108**, 84–92 (2013)
11. Tseng, J.P.: Global asymptotic dynamics of a class of nonlinearly coupled neural networks with delays. Discrete Contin. Dyn. Syst. **33**, 4693–4729 (2013)
12. Yang, X.S., Song, Q., Liang, J.L., He, B.: Finite-time synchronization of coupled discontinuous neural networks with mixed delays and nonidentical perturbations. J. Frankl. Inst. **352**, 4382–4406 (2015)
13. Wang, Z., Huang, L.H.: Synchronization analysis of linearly coupled delayed neuralnetworks with discontinuous activations. Appl. Math. Model. **39**, 7427–7441 (2015)
14. Zhang, W., Li, C.D., Huang, T.W., Huang, J.J.: Stability and synchronization of mmemristor-bbased ccoupling neural networks with time-varying delays via intermittentcontrol. Neurocomputing **173**, 1066–1072 (2016)

15. Tan, M.C.: Stabilization of coupled time-delay neural networks with nodes of different dimensions. Neural Process. Lett. **43**, 255–268 (2016)
16. Chen, G.R.: Pinning control and synchronization on complex dynamical networks. Int. J. Control Autom. Syst. **12**, 221–230 (2014)
17. Huang, T., Yang, Z., Li, C.: Theory and Applications of Complex Networks. Math. Probl. Eng. 315059 (2014)
18. Sun, Y., Li, W., Ruan, J.: Finite-time generalized outer synchronization between two different complex networks. Commun. Theor. Phys. **58**, 697–703 (2012)
19. Cai, S., He, Q., Hao, J., Liu, Z.: Exponential synchronization of complex networks with nonidentical time-delayed dynamical nodes. Phys. Lett. A **374**, 2539–2550 (2010)
20. Wu, Z., Fu, X.: Cluster projective synchronization between community networks with nonidentical nodes. Phys. A **391**, 6190–6198 (2012)
21. Wu, X., Lu, H.: Generalized function projective (lag, anticipated and complete) synchronization between two different coupled complex with nonidentical nodes. Commun. Nonlinear Sci. Numer. Simul. **17**, 3005–3021 (2012)
22. Yang, X., Cao, J., Lu, J.: Stochastic synchronization of complex networks with nonidentical nodes via hybrid adaptive and iimpulsive control. IEEE Trans. Circ. Syst. **I**(59), 371–384 (2012)
23. Yang, X., Cao, J., Lu, J.: Synchronization of Markovian coupled neural networks with nnonidentical node-delays and random coupling strengths. IEEE Trans. Neural Netw. Learn. Syst. **23**, 60–71 (2012)
24. Du, H.: Function projective synchronization in drive-response dynamical networks with nonidentical nodes. Chaos Solitons Fractals **44**, 510–514 (2011)
25. Wang, Y., Fan, Y., Wang, Q., Zhang, Y.: Stabilization and synchronization of complex dynamical networks with different dynamics of nodes via decentralized controllers. IEEE Trans. Circ. Syst. I **59**(8), 1786–1795 (2012)
26. Fan, Y., Wang, Y., Zhang, Y., Wang, Q.: The synchronization of complex dynamical networks with similar nodes and coupling time-delay. Appl. Math. Comput. **219**, 6719–6728 (2013)
27. Dai, H., Jia, L., Zhang, Y.: Adaptive generalized mmatrix projective lag synchronization between two different complex networks with nonidentical nodes and different dimensions. Chin. Phys. B. **21**, 120508 (2012)
28. Dai, H., Si, G., Zhang, Y.: Adaptive generalized function mmatrix projective lag synchronization of uncertain complex dynamical networks with different dimensions. Nonlinear Dyn. **74**, 629–648 (2013)
29. Tan, M.C., Tian, W.X.: Finite-time stabilization and synchronization of complex dynamical networks with nonidentical nodes of different dimensions. Nonlinear Dyn. **79**(1), 731–741 (2015)
30. Wu, Z.Y.: Cluster synchronization in colored community network with different order node dynamics. Commun. Nonlinear Sci. Numer. Simul. **19**, 1079–1087 (2014)
31. Tan, M.C., Pan, Q., Zhou, X.: Adaptive stabilization and synchronization of nondiffusively coupled complex networks with nonidentical nodes of different dimensions. Nonlinear Dyn. 1–14 (2016). doi:10.1007/s11071-016-2686-4

Asymptotic Behaviors for Non-autonomous Difference Neural Networks with Impulses and Delays

Shujun Long[1] and Bing Li[2(✉)]

[1] College of Mathematics and Information Science, Leshan Normal University,
Leshan 614004, China
[2] College of Mathematics and Statistics, Chongqing Jiaotong University,
Chongqing 400074, China
libingcnjy@163.com

Abstract. In this paper, the asymptotic behaviors of non-autonomous difference neural network with impulses and distributed delays are studied by using difference inequality and properties of $\mathcal{M}$-matrix. Some new results on attracting set and periodic solution of networks are obtained.

Keywords: Invariant set · Attractor · Periodic solution · Exponential stability · Neural networks

1 Introduction

In the past decades, neural networks have attracted more and more attentions because of their important applications in many fields. With the development of computer techniques, when implementing the continuous-time neural networks for computer simulation, experiment and computation, it is essential to formulate difference neural networks which is an analogue of the continuous-time types. Recently, there has been much literature on the dynamical behaviors of difference neural networks with delays [1–13].

Impulses are encountered in many fields such as bursting rhythm models in pathology, optimal control models, etc. There have been many works on the dynamical behaviors of difference equations with impulses and delays [14–22]. The authors in [19] established an inequality to study the stability of impulsive difference equations with distributed delays, but the inequality is not effective for the attracting and invariant sets of non-autonomous difference system. The authors in [21] derived out an inequality to study the asymptotic behaviors of non-autonomous impulsive difference equation with delays, but they required the common factor of the variable coefficients and did not consider the periodic solution.

Motivated by previous works, the main purpose of this paper is to establish a new difference inequality such that it is effective for non-autonomous difference

© Springer International Publishing Switzerland 2016
L. Cheng et al. (Eds.): ISNN 2016, LNCS 9719, pp. 143–151, 2016.
DOI: 10.1007/978-3-319-40663-3_17

neural networks with impulses and distributed delays. Based on the new inequality, we investigate the global attracting, invariant sets and the periodic solution of the considered network. Our main results improve some related results in previous literature. One example is given to illustrate the effectiveness of our conclusion.

2   Model Description and Preliminaries

Let Z be the integer set and $Z^+ = \{k \in Z | k \geq 0\}$, $\mathcal{N} = \{1, 2, \cdots, n\}$. R^n and $R^{n \times n}$ denote by the set of n-dimensional real column vectors and $n \times n$ real matrices, respectively. E denotes an $n \times n$ unit matrix. For $A, B \in R^{n \times n}$ and $x \in R^n$, $A \geq B$ means that each pair of corresponding elements of A and B satisfies the inequality "$\geq$". Especially, $A \in R^{n \times n}$ is called a nonnegative matrix if $A \geq 0$. $[A]^+ = (|a_{ij}|)_{n \times n}$, $[x]^+ = (|x_1|, \cdots, |x_n|)^T$. $\rho(A)$ is an eigenvalue of A and its eigenspace is denoted by $\Omega_\rho(A) \triangleq \{x \in R^n | Ax = \rho(A)x\}$. For $\omega \in Z$ and $0 < \omega \leq +\infty$, let $Z_\omega = \{k \in Z | - \omega \leq k \leq 0\}$. $\mathcal{C}_\omega$ denotes the set of all bounded functions $\phi : Z_\omega \to R^n$. For any $\phi(k) \in \mathcal{C}_\omega$, $[\varphi(k)]^+_\omega = ([\varphi_1(k)]_\omega, \ldots, [\varphi_n(k)]_\omega)^T$, where $[\varphi_i(k)]_\omega = \sup_{s \in Z_\omega} \{|\varphi_i(k + s)|\}$, $i \in \mathcal{N}$. $\|A\| = \max_{i \in \mathcal{N}} \sum_{j=1}^n |a_{ij}|$, $\|x\| = \max_{i \in \mathcal{N}} \{|x_i|\}$, $\|\varphi\| = \max_{i \in \mathcal{N}} \{[\varphi_i(k)]_\omega\}$.

Consider difference neural networks with impulses and delays

$$\begin{cases} x(k+1) = A(k)x(k) + B(k)f(x(k)) + (C(k) \circ \sum_{r=1}^{\omega} S(r))g(x(k - \tau_r(k))) \\ \qquad + \tilde{J}(k), k \in Z^+, k \neq k_v, \\ x(k+1) = I_v(x(k)), k = k_v, \\ x(k) = \varphi(k), \quad k \in Z_\omega, \end{cases} \tag{1}$$

where the neurons $x(k) = (x_1(k), \ldots, x_n(k))^T$, the activity functions $f(x) = (f_1(x_1), \ldots, f_n(x_n))^T$, $g(x) = (g_1(x_1), \ldots, g_n(x_n))^T$, the external input $\tilde{J}(k) = (\tilde{J}_1(k), \ldots, \tilde{J}_n(k))^T$, the impulsive jumps $I_v(x(k)) = (I_{1v}(x(k)), \ldots, I_{nv}(x(k)))^T$. $A(k) = \text{diag}\{a_1(k), \ldots, a_n(k)\}$, $B(k) = (b_{ij}(k))_{n \times n}$, $C(k) = (c_{ij}(k))_{n \times n}$, $S(r) = (s_{ij}(r))_{n \times n}$, $C(k) \circ \sum_{r=1}^{\omega} S(r) = (c_{ij}(k) \sum_{r=1}^{\omega} s_{ij}(r))_{n \times n}$, $\varphi \in \mathcal{C}_\omega$, $I_v(x) : R^n \to R^n$. The delays $0 < \tau_r(k) \leq \tau_r \in Z^+$. The impulsive moments $k_v \in Z^+ (v \in \mathbb{N})$ satisfy $0 < k_1 < k_2 < \cdots$ and $\lim_{v \to \infty} k_v = \infty$. The constants $s_{ij}(r) (i, j \in \mathcal{N})$ satisfy the following convergence conditions

$$(H) : \sum_{r=1}^{\infty} e^{\gamma_0 r} |s_{ij}(r)| < \infty, \ i, j \in \mathcal{N} \text{ for } \omega = +\infty, \tag{2}$$

where γ_0 is a positive constant.

The following definitions will be used in later discussion.

Definition 1. [21] The set $S \subset C_\omega$ is called a positive invariant set of (1), if for any initial value $\varphi \in S$, we have the solution $x(k, \varphi) \in S$ for $k \geq k_0$.

Definition 2. [21] The set $S \subset C_\omega$ is called a global attracting set of (1), if for any initial value $\varphi \in C_\omega$, the solution $x(k, \varphi)$ converges to S as $k \to +\infty$. That is,

$$\text{dist}(x(k, \varphi), S) \to 0 \text{ as } k \in Z^+, \ k \to +\infty,$$

where $\text{dist}(\varphi, S) = \inf_{\phi \in S} \text{dist}(\varphi, \phi)$, $\text{dist}(\varphi, \phi) = \sup_{s \in Z_\omega} |\varphi(k + s) - \phi(k + s)|$ for $\varphi \in C_\omega$.

Definition 3. [21] The zero solution of (1) is said to be globally exponentially stable if system (1) has a zero solution and for any solution $x(k, \varphi)$, there exist constants $\gamma > 0$ and $M \geq 1$ such that

$$\|x(k, \varphi)\| \leq M\|\varphi\|e^{-\gamma k}, \ k \in Z^+.$$

For a nonsingular $\mathcal{M}$-matrix D, let $\Omega_\mathcal{M}(D) \overset{\Delta}{=} \{z \in R^n | Dz > 0, \ z > 0\}$.

3   Main Results

For network (1), we give the following assumptions.

(H_1) There are nonnegative matrices $L = \text{diag}\{l_1, \ldots, l_n\}$, $M = \text{diag}\{m_1, \ldots, m_n\}$ such that for any $x, y \in R^n$,

$$[f(x) - f(y)]^+ \leq L[x - y]^+, \ [g(x) - g(y)]^+ \leq M[x - y]^+.$$

(H_2) Suppose that $\widehat{\Pi} = -(\widehat{P} - E + \widehat{Q})$ is an $\mathcal{M}$−matrix, where $\widehat{P}(k) = [A(k)]^+ + [B(k)]^+ L \leq \widehat{P}$, $\widehat{Q}(k, r) = ([C(k)]^+ \circ [S(r)]^+)M \leq (\widetilde{C} \circ [S(r)]^+)M = \widehat{Q}(r)$. Denote $\widehat{Q} = \sum\limits_{r=1}^{\omega} \widehat{Q}(r)$, $\hat{J}(k) = [B(k)]^+[f(0)]^+ + ([C(k)]^+ \circ \sum\limits_{r=1}^{\omega} [S(r)]^+)[g(0)]^+ + [\tilde{J}(k)]^+ \leq \hat{J}$.

(H_3) For any $x \in R^n$, there exist nonnegative matrices $R_v = (r_{ij}^v)_{n \times n}$ such that

$$[I_v(x)]^+ \leq R_v[x]^+, \ v \in \mathbb{N}.$$

(H_4) The set $\Omega = \bigcap\limits_{v=1}^{\infty} \Omega_\rho(R_v) \bigcap \Omega_\mathcal{M}(\widehat{\Pi})$ is nonempty (i.e., $\Omega \neq \emptyset$) and for a given $z \in \Omega$, the scalar $\gamma \in (0, \gamma_0)$ satisfies

$$((\gamma - 1)E + \widehat{P} + \sum_{r=1}^{\omega} \widehat{Q}(r)e^{\gamma \tau_r})z \leq 0. \tag{3}$$

(H_5) There exists a constant μ such that

$$\frac{\ln\mu_v}{k_v - k_{v-1}} \leq \mu < \gamma, \text{ and } \delta = \sum_{v=1}^{\infty} \ln\delta_v < \infty, \tag{4}$$

where $\mu_v \geq \max\{1, e^\gamma \rho(R_v)\}$, $\delta_v \geq 1$ satisfy $R_v \widehat{\Pi}^{-1}\hat{J} \leq \delta_v \widehat{\Pi}^{-1}\hat{J}$, $v \in \mathbb{N}$.

(H_6) $a_i(k)$, $b_{ij}(k)$, $c_{ij}(k)$, $\tilde{J}_i(k)$, $\tau_r(k)$ are T-periodic functions, where T is a positive integer, $i, j \in \mathcal{N}$, $r = 1, \ldots, \omega$.

(H_7) There exists a positive integer q such that $k_{v+q} = k_v + T$, $I_{i,v+q}(x) = I_{iv}(x)$ for $v = 1, 2, \ldots$, $i \in \mathcal{N}$.

First, we propose a difference inequality as follows.

Theorem 1. Let $k_0 \in Z^+$ and $u(k) = (u_1(k), \ldots, u_n(k))^T \in R_+^n$ be a solution of the following difference inequality

$$\begin{cases} u(k+1) \leq P(k)u(k) + \sum_{r=1}^{S} Q(k,r)u(k - \tau_r(k)) + J(k), \ k \geq k_0, \\ u(k_0 + k) \in \mathcal{C}_{S-k_0}, \ k \in Z_S, \end{cases} \tag{5}$$

where $1 \leq S \leq +\infty$, $J(k) \in R_+^n$, $P(k) \in R_+^{n\times n}$, $Q(k,r) = (q_{ij}(k,r))_{n\times n} \geq 0$, $0 < \tau_r(k) \leq \tau_r \in Z^+$ with $P(k) \leq P$, $Q(k,r) \leq Q(r) \overset{\Delta}{=} (q_{ij}(r))_{n\times n} \in R_+^{n\times n}$, $J(k) \leq J$ for all $k \geq k_0$. If $S = +\infty$, then there exists a positive constant λ_0 such that $\sum_{r=1}^{\infty} e^{\lambda_0 \tau_r} q_{ij}(r) < \infty$ $(i, j \in \mathcal{N})$. Denote $Q = (q_{ij})_{n\times n} \overset{\Delta}{=} (\sum_{r=1}^{S} q_{ij}(r))_{n\times n} \in R_+^{n\times n}$. Suppose that $\Pi = -(P - E + Q)$ is an $\mathcal{M}$−matrix, then we have

$$u(k) \leq \nu z e^{-\lambda(k-k_0)} - (P - E + Q)^{-1}J, \ k \geq k_0, \tag{6}$$

provided that the initial condition satisfies

$$u(s) \leq \nu z e^{-\lambda(s-k_0)} - (P - E + Q)^{-1}J, \ k_0 - S \leq s \leq k_0, \tag{7}$$

where $\nu \geq 0$ is a constant, $\lambda \in (0, \lambda_0)$ and $z = (z_1, \ldots, z_n)^T \in \Omega_{\mathcal{M}}(\Pi)$ satisfy

$$\left((\lambda - 1)E + P + \sum_{r=1}^{S} Q(r)e^{\lambda\tau_r}\right) z \leq 0. \tag{8}$$

Proof. The proof is similar to those of [19, 21], so omitted.

Remark. In Theorem 1, we do not require the common factor in time-varying coefficients $P(k)$ and $Q(k,r)$, which implies it is an improved version of [19, 21].

Theorem 2. Assume that $(H_1) - (H_5)$ hold. Then $S = \{\varphi \in \mathcal{C}_\omega | [\varphi(0)]_\omega^+ \leq e^\delta \widehat{\Pi}^{-1}\hat{J}\}$ is a global attracting set of (1). Especially, $S = \{\varphi \in \mathcal{C}_\omega | [\varphi(0)]_\omega^+ \leq \widehat{\Pi}^{-1}\hat{J}\}$ is a positive invariant set of (1) if $\delta_v = 1$ for $v \in \mathbb{N}$.

Proof. From (H_1), (H_2) and (1), we get for $k \neq k_v$

$$[x(k+1)]^+ \leq [A(k)]^+[x(k)]^+ + [B(k)]^+[f(x(k)) - f(0)]^+$$

$$+ ([C(k)]^+ \circ \sum_{r=1}^{\omega}[S(r)]^+)[g(x(k - \tau_r(k))) - g(0)]^+$$

$$+ [B(k)]^+[f(0)]^+ + ([C(k)]^+ \circ \sum_{r=1}^{\omega}[S(r)]^+)[g(0)]^+ + [\tilde{J}(k)]^+$$

$$\leq \widehat{P}(k)[x(k)]^+ + \sum_{r=1}^{\omega} \widehat{Q}(k,r)[x(k - \tau_r(k))]^+ + \hat{J}(k). \tag{9}$$

Since $\widehat{\Pi}$ is an $\mathcal{M}-$matrix and Ω is nonempty, we know there exist constant $\gamma \in (0, \gamma_0)$ and $z \in \Omega$ such that (3) holds. In addition, $\widehat{\Pi}^{-1} \geq 0$, and so $\widehat{\Pi}^{-1}\hat{J} \geq 0$. For the initial conditions $x(k) = \varphi(k) \in \mathcal{C}_\omega$, $k \in Z_\omega$, we have

$$[x(k)]^+ \leq \nu_0 z, \ \nu_0 = \frac{\|\varphi\|}{\min_{i \in \mathcal{N}}\{z_i\}}, \ k \in Z_\omega, \tag{10}$$

and so

$$[x(k)]^+ \leq \nu_0 z e^{-\gamma k} + \widehat{\Pi}^{-1}\hat{J}, \ k \in Z_\omega. \tag{11}$$

From (9)–(11) and Theorem 1, we can get

$$[x(k)]^+ \leq \nu_0 z e^{-\gamma k} + \widehat{\Pi}^{-1}\hat{J}, \ k \in [0, k_1], \ k \in Z^+. \tag{12}$$

Suppose that for all $u = 1, 2, \ldots, v$, $k_{u-1} + 1 \leq k \leq k_u$, the inequalities

$$[x(k)]^+ \leq \mu_0 \mu_1 \cdots \mu_{u-1} \nu_0 z e^{-\gamma k} + \delta_0 \delta_1 \cdots \delta_{u-1} \widehat{\Pi}^{-1}\hat{J}, \tag{13}$$

hold, where $\mu_0 = \delta_0 = 1$. Then, from (H_3), (H_5) and (13), we have

$$[x(k_v + 1)]^+ \leq R_v[x(k_v)]^+$$

$$\leq \mu_0 \mu_1 \cdots \mu_{v-1} \mu_v \nu_0 z e^{-\gamma(k_v+1)} + \delta_0 \delta_1 \cdots \delta_{v-1} \delta_v \widehat{\Pi}^{-1}\hat{J} \tag{14}$$

Noting $\mu_v \geq 1$, $\delta_v \geq 1$ and letting $\tilde{z} = \mu_0 \mu_1 \cdots \mu_{v-1} \mu_v \nu_0 e^{-\gamma(k_v+1)} z$, it follows from the property of $\mathcal{M}-$cone, we get for $-\omega \leq k \leq k_v + 1$

$$[x(k)]^+ \leq \tilde{z} e^{-\gamma(k-(k_v+1))} + \delta_0 \delta_1 \cdots \delta_{v-1} \delta_v \widehat{\Pi}^{-1}\hat{J}. \tag{15}$$

In addition, from (9), we have for $k_v + 1 \leq k < k_{v+1}$

$$[x(k+1)]^+ \leq \widehat{P}(k)[x(k)]^+ + \sum_{r=1}^{\omega} \widehat{Q}(k,r)[x(k-\tau_r(k))]^+ + \delta_0 \delta_1 \cdots \delta_{v-1} \delta_v \hat{J}(k). \tag{16}$$

It follows from (15), (16) and Theorem 1 that

$$[x(k)]^+ \leq \tilde{z} e^{-\gamma(k-(k_v+1))} + \delta_0 \delta_1 \cdots \delta_{v-1} \delta_v \widehat{\Pi}^{-1}\hat{J}$$

$$= \mu_0 \mu_1 \cdots \mu_{v-1} \mu_v \nu_0 z e^{-\gamma k} + \delta_0 \delta_1 \cdots \delta_{v-1} \delta_v \widehat{\Pi}^{-1}\hat{J}. \tag{17}$$

By the mathematical induction, we know for $k_{v-1} + 1 \leq k \leq k_v$, $v \in \mathbb{N}$

$$[x(k)]^+ \leq \mu_0 \mu_1 \cdots \mu_{v-1} \nu_0 z e^{-\gamma k} + \delta_0 \delta_1 \cdots \delta_{v-1} \widehat{\Pi}^{-1} \hat{J}. \tag{18}$$

From (H_5), we know

$$\mu_v \leq e^{\mu(k_v - k_{v-1})}, \ \delta_0 \delta_1 \cdots \delta_{v-1} \leq e^{\delta}. \tag{19}$$

Combining (18) with (19), we can get

$$\begin{aligned}
[x(k)]^+ &\leq e^{\mu k_1} \cdots e^{\mu(k_{v-1} - k_{v-2})} \nu_0 z e^{-\gamma k} + \delta_0 \delta_1 \cdots \delta_{v-1} \widehat{\Pi}^{-1} \hat{J} \\
&\leq \nu_0 z e^{\mu k} e^{-\gamma k} + e^{\delta} \widehat{\Pi}^{-1} \hat{J} \\
&= \nu_0 z e^{-(\gamma - \mu)k} + e^{\delta} \widehat{\Pi}^{-1} \hat{J}, \ 0 \leq k \leq k_v, \ v \in \mathbb{N}, \ k \in Z.
\end{aligned} \tag{20}$$

Furthermore, the proof implies that $S = \{\varphi \in \mathcal{C}_\omega | [\varphi(0)]_\omega^+ \leq \widehat{\Pi}^{-1} \hat{J}\}$ is a positive invariant set provided $\delta_v = 1$ for $v \in \mathbb{N}$. The proof is complete.

Theorem 3. Assume that $(H_1) - (H_7)$ with $\delta_v = 1, v \in \mathbb{N}$ hold. Then the neural network (1) with $\omega < +\infty$ has a unique periodic solution which is globally exponentially stable.

Proof. Since $\omega < +\infty$, then $\mathcal{C}_\omega$ is a Banach space with norm $\|\varphi\| = \max_{i \in \mathcal{N}} \{[\varphi_i(k)]_\omega\}$, and $S = \{\varphi \in \mathcal{C}_\omega | [\varphi(0)]_\omega^+ \leq \widehat{\Pi}^{-1} \hat{J}\}$ is a non-empty closed subset of $\mathcal{C}_\omega$. From Theorem 2, we know $S \subset \mathcal{C}_\omega$ is a positive invariant set of (1).

For any solutions $x(k, \phi)$, $x(k, \psi)$ of neural network (1) with initial conditions $\phi, \psi \in S$, by similar proof to Theorem 1, we can get

$$[x(k, \phi) - x(k, \psi)]^+ \leq \nu_0 z e^{-(\gamma - \mu)k}, \ k \geq 0. \tag{21}$$

From (21), we know there must be a positive integer p such that

$$\|x(k, \phi) - x(k, \psi)\| \leq \kappa \|\phi - \psi\| \text{ for all } k \geq pT + \omega, \tag{22}$$

where $\kappa \in (0, 1)$ is a given constant.

We define $F : S \rightarrow S$ by

$$F\phi = x(k + T, \phi) \text{ for } k \in Z_\omega.$$

Now $x(k + T, \phi)$ is a solution for $k \geq 0$ with initial function $F\phi$. It is easy to get

$$x(k + T, \phi) = x(t, F\phi) \tag{23}$$

by uniqueness. Next, $F^2 \phi = x(k + T, F\phi)$ for $k \in Z_\omega$ and $x(k + T, F\phi)$ is a solution with initial function $F^2\phi$. Hence,

$$x(k + T, F\phi) = x(t, F^2\phi) \tag{24}$$

by uniqueness. Let k be replaced by $k+T$ in (23). Combining with (24), we can get

$$x(k+2T,\phi) = x(k+T,F\phi) = x(k,F^2\phi).$$

In general, we have $x(k+uT,\phi) = x(k,F^u\phi) = F^u\phi$.

From (22), we have

$$|F^p\phi - F^p\psi| = \kappa\|\phi - \psi\|,\ \kappa \in (0,1). \tag{25}$$

According to the Banach fixed point theorem, we know operator F has a fixed point φ^* in S, that is,

$$x(k+T,\varphi^*) = F\varphi^* = \varphi^* \text{ for } k \in Z_\omega. \tag{26}$$

That is to say $x(k,\varphi^*)$ and $x(k+T,\varphi^*)$ are both solutions with the same initial condition. Thus, we get $x(k,\varphi^*) = x(k+T,\varphi^*)$ by uniqueness. This implies that $x(k,\varphi^*) \in S$ is an T-periodic solution which is exponentially stable by (21).

4   Numerical Example

Example. Consider model

$$\begin{cases}
x_1(k+1) = 0.5x_1(k) + 0.02\sin(\frac{k\pi}{2})f_1(x_1(k)) - 0.01\cos(\frac{k\pi}{2})f_2(x_2(k)) \\
\qquad +0.03\cos(k\pi)g_1(x_1(k - \sin^2(\frac{(2k+1)\pi}{2}))) \\
\qquad -0.03\sin(k\pi)g_2(x_2(k - \cos^2(k\pi))) + 4\cos(\frac{k\pi}{4}), \\
x_2(k+1) = 0.4x_1(k) - 0.01\cos(\frac{k\pi}{2})f_1(x_1(k)) \\
\qquad +0.02\sin(k\pi)f_2(x_2(k)) - 0.04\sin(\frac{k\pi}{2})g_1(x_1(k - \cos^2(k\pi))) \\
\qquad +0.03\cos(\frac{k\pi}{2})g_2(x_2(k - \sin^2(\frac{(2k+1)\pi}{2}))) + 3\sin(\frac{k\pi}{4}), k \neq k_v, \\
x_1(k_v + 1) = \beta_v x_1(k_v),\ x_2(k_v + 1) = \beta_v x_2(k_v),\ k = k_v = 2v + 2,\ v \in Z^+,
\end{cases}$$

where $f_1(s) = g_1(s) = s$, $f_2(s) = g_2(s) = \sin s$.

We compute $\widehat{P} = \begin{pmatrix} 0.52 & 0.01 \\ 0.01 & 0.42 \end{pmatrix}$, $\widehat{Q} = \begin{pmatrix} 0.03 & 0.03 \\ 0.04 & 0.03 \end{pmatrix}$, $R_v = \begin{pmatrix} \beta_v & 0 \\ 0 & \beta_v \end{pmatrix}$.

$\widehat{\Pi} = \begin{pmatrix} 0.45 & -0.04 \\ -0.05 & 0.55 \end{pmatrix}$ is an $\mathcal{M}$-matrix and $\Omega_{\mathcal{M}}(\widehat{\Pi}) = \{(z_1,z_2)^T > 0 | \frac{4}{45}z_2 < z_1 < 11z_2\}$. Choosing $z = (1,1)^T \in \bigcap_{v=1}^{\infty} \Omega_\rho(R_v) \bigcap \Omega_{\mathcal{M}}(\widehat{\Pi})$ and $\gamma = 0.38$, we

have $((\gamma - 1)E + \widehat{P} + \widehat{Q}e^{\gamma\tau})z = (-0.0023, -0.0876)^T < (0,0)^T$, where $\tau = 1$.

If $\beta_v = e^{\frac{1}{5v}}$, $v = 1,2,\cdots$, then $\mu_v = e^{\gamma + \frac{1}{5v}}$, $\delta_v = e^{\frac{1}{5v}}$. Thus, $\frac{\ln\mu_v}{k_v - k_{v-1}} < 0.39$ and $\delta = \frac{1}{4}$. Then $S = \{\varphi \in C_\omega | [\varphi(0)]_\omega^+ \leq e^\delta \widehat{\Pi}^{-1}\hat{J} = (12.1342, 8.1069)^T\}$ is a global attracting set.

If $\beta_v = 0.7$, $v = 1,2,\cdots$, then this is an 8-periodic system. We can obtain $\mu = 0.0166 < 0.39 = \gamma$ and $\delta = 0$. There exists $q = 4$ such that $k_{v+q} = k_v + T$ and $I_{i,v+q}(x) = I_{iv}(x)$ for $v = 1,2,\cdots, i \in \mathcal{N}$. Therefore, the system has a unique exponentially stable periodic solution. However, those results in [8,9,16,17] are invalid for this example.

Acknowledgments. This work is supported by National Natural Science Foundation of China (11271270, 11501065), Scientific Research Fund of Sichuan Provincial Education Department (16TD0029), the Natural Science Foundation of Chongqing (cstc2015jcyjA00033), Project of Leshan Normal University (Z1324) and the Doctoral Foundation of Chongqing Jiaotong University (2014kjc-II-019).

References

1. Song, Q.K., Wang, Z.D.: A delay-dependent LMI approach to dynamics analysis of discrete-time recurrent neural networks with time-varying delays. Phys. Lett. A **368**, 134–145 (2007)
2. Liu, Y.R., Wang, Z.D., Serrano, A., Liu, X.H.: Discrete-time recurrent neural networks with time-varying delays: exponential stability analysis. Phys. Lett. A **362**, 480–488 (2007)
3. Zhang, B.Y., Xu, S.Y., Zou, Y.: Improved delay-dependent exponential stability criteria for discrete-time recurrent neural networks with time-varying delays. Neurocomputing **72**, 321–330 (2008)
4. Wu, Z.G., Su, H.Y., Chu, J., Zhou, W.N.: New results on robust exponential stability for discrete recurrent neural networks with time-varying delays. Neurocomputing **72**, 3337–3342 (2009)
5. Udpin, S., Niamsup, P.: New discrete type inequalities and global stability of nonlinear difference equations. Appl. Math. Lett. **22**, 856–859 (2009)
6. Yu, J.J., Zhang, K.J., Fei, S.M.: Exponential stability criteria for discrete-time recurrent neural networks with time-varying delay. Nonlinear Anal. Real World Appl. **11**, 207–216 (2010)
7. Baker, C.T.H.: Development and application of halanay-type theory: evolutionary differential and difference equations with time lag. J. Comput. Appl. Math. **234**, 2663–2682 (2010)
8. Song, Y.F., Shen, Y., Yin, Q.: New discrete halanay-type inequalities and applications. Appl. Math. Lett. **26**, 258–263 (2013)
9. Hien, L.V.: A novel approach to exponential stability of nonlinear non-autonomous difference equations with variable delays. Appl. Math. Lett. **38**, 7–13 (2014)
10. Yang, R., Wu, B., Liu, Y.: A halanay-type inequality approach to the stability analysis of discrete-time neural networks with delays. Appl. Math. Comput. **265**, 696–707 (2015)
11. Feng, Z., Zheng, W.X.: On extended dissipativity of discrete-time neural networks with time delay. IEEE Trans. Neural Netw. Learn. Syst. **26**, 3293–3300 (2015)
12. Banu, L.J., Balasubramaniam, P.: Robust stability analysis for discrete-time neural networks with time-varying leakage delays and random parameter uncertainties. Neurocomputing **179**, 126–134 (2016)
13. Singh, J., Barabanov, N.: Stability of discrete time recurrent neural networks and nonlinear optimization problems. Neural Netw. **74**, 58–72 (2016)
14. Zhu, W., Xu, D.Y., Yang, Z.C.: Global exponential stability of impulsive delay difference equation. Appl. Math. Comput. **181**, 65–72 (2006)
15. Yang, Z.G., Xu, D.Y.: Mean square exponential stability of impulsive stochastic difference equations. Appl. Math. Lett. **20**, 938–945 (2007)
16. Zhang, H., Chen, L.S.: Asymptotic behavior of discrete solutions to delayed neural networks with impulses. Neurocomputing **71**, 1032–1038 (2008)
17. Song, Q.K., Cao, J.D.: Dynamical behaviors of discrete-time fuzzy cellular neural networks with variable delays and impulses. J. Franklin Inst. **345**, 39–59 (2008)

18. Zhu, W.: Invariant and attracting sets of impulsive delay difference equations with continuous variables. Comput. Math. Appl. **55**, 2732–2739 (2008)
19. Li, D.S., Long, S.J., Wang, X.H.: Difference inequality for stability of impulsive difference equations with distributed delays. J. Inequalities Appl. **8**, 1–9 (2011)
20. Li, D.S., Long, S.J.: Attracting and quasi-invariant sets for A class of impulsive stochastic difference equations. Adv. Differ. Equ. **3**, 1–9 (2011)
21. Li, B., Song, Q.K.: Asymptotic behaviors of non-autonomous impulsive difference equation with delays. Appl. Math. Model. **35**, 3423–3433 (2011)
22. Zhang, Y.: Exponential stability analysis for discrete-time impulsive delay neural networks with and without uncertainty. J. Franklin Inst. **350**, 737–756 (2013)

Optimal Real-Time Price in Smart Grid via Recurrent Neural Network

Haisha Niu, Zhanshan Wang$^{(\boxtimes)}$, Zhenwei Liu, and Yingwei Zhang

School of Information Science and Engineering, Northeastern University,
Shenyang 110819, China
Zhanshan_wang@163.com

Abstract. In this paper, we consider a smart power infrastructure which contains a single energy provider, several load subscribers and a regulatory authority. Considering the importance of energy pricing as an essential tool to develop efficient demand side management strategies, we propose a neural network modeled by a differential inclusion to solve real-time pricing problem in smart grid based on optimization theory. Compared with the existing algorithms, our model has fewer variables, our model is not only parallel computational model, but also can be implemented with the schematic block diagram. Moreover, the solution of proposed network converges to the feasible region in finite time and to the particular element in the optimal solution set with the smallest norm, which indicates that the proposed neural network is globally attractive. Finally, simulation results confirm the effectiveness and performance of the proposed network.

Keywords: Real-time price · Optimization · Neural network · Differential inclusion · Smart grid

1 Introduction

Recently, the smart grid has received much attention, it is through information and communication technologies to collect suppliers and consumer behavior, in order to improve the efficiency of electricity production and distribution, reliability, economy and sustainability. Given the importance of smart grid, more and more researchers from different backgrounds begin to study this issue. Demand-side management (DSM) is a key element of the smart grid.

There is a wide range of DSM techniques such as voluntary load management programs [2] and direct load control [3]. However, smart pricing is known as one of the most common tools that can encourage users to consume wisely and more efficiently. Demand side management (DSM) has been an attempt to reduce peak power consumption [4, 5]. In recent studies on various mechanisms designed for DSM, Pedram, Samadi et al. [6] proposed a dynamic policy, which is an effective method to reduce peak demand, and proposed a distributed sub-gradient algorithm to solve the model. Conejo et al. [18] proposed a simple linear programming based algorithm capable of forecasting energy consumption levels of a particular consumer. Gatsis and Giannakis [7] considered cooperative multi-residence demand scheduling approach. The distributed algorithm

© Springer International Publishing Switzerland 2016
L. Cheng et al. (Eds.): ISNN 2016, LNCS 9719, pp. 152–159, 2016.
DOI: 10.1007/978-3-319-40663-3_18

based on Lagrangian multiplier is used to solve these optimization problems. However, on the one hand, this method converged slowly or didn't converge. On the other hand, in applications of the demand-side management optimization problem in smart grid, real-time solutions are often required. He et al. [20] considered the neural network to solve the real-time price. But in the proof of the finite time convergence to the feasible region, the interior of the feasible region is nonempty, and the neural network is effective relying on the exact penalty parameters.

In order to overcome these difficulties, we apply a new neural network to solve real-time price in smart grid. Different from the existing algorithms, the proposed network has a Tikhonov regularization item ε, which let the proposed network solve the optimal problem effectively without any evaluation on the exact penalty parameters and the solution of proposed network is convergent to the optimal solution set of optimization problem. By another item $\varepsilon^2(t)y(t)$, the proposed network is globally attractive.

This paper is organized as follows. In the next section, the system model is given. In Sect. 3, the neural network model formulation and the performance of the proposed neural network are analyzed. In Sect. 4, simulation results are reported. Conclusions are drawn in Sect. 5.

2   System Model

Denote $\mathfrak{R}$ as the set of customers or users requiring electricity, $R \triangleq |\mathfrak{R}|$ and $i \in \mathfrak{R}$ is the each user. The intended time cycle for the operation of the customers is divided into K time slots, where $K \triangleq |\Psi|$.

Let x_i^k be the power consumption demand at time slot k of customer i. The power consumption is bounded, $x_i^k \in [m_i^k, M_i^k]$. Let L_k be the regulatory capacity in each time slot $k \in \Psi$. We model the behavior of different users through their different choices of utility functions [11]. For all users, let $U(x_i^k, \omega_i^k)$ represent the corresponding utility function. Let $C_k(L_k)$ indicate the cost of providing L_k units of energy offered by the energy provider.

The price optimal model is formulated as follows:

$$\underset{i \in \mathfrak{R}, k \in \Psi}{\max \text{imize}} \quad \sum_{k \in \Psi} \sum_{i \in \mathfrak{R}} U(x_i^k, \omega_i^k) - \sum_{k \in \Psi} C_k(L_k) \tag{1}$$

$$s.t. \quad m_i^k \leq x_i^k \leq M_i^k$$

We assume that utility function and cost function fulfill the following properties:

Property I: [20] Utility function is non-decreasing, namely

$$\frac{\partial U(x_i^k, \omega_i^k)}{\partial x_i^k} \geq 0 ; \tag{2}$$

Property II: The utility function is concave.

$$\frac{\partial^2 U(x_i^k, \omega_i^k)}{\partial (x_i^k)^2} \leq 0 ; \tag{3}$$

Property III: [6] We assume, for a fixed consumption level x, a larger ω implies a larger $U(x, \omega)$, which can be expressed as

$$\frac{\partial U(x_i^k, \omega_i^k)}{\partial \omega_i^k} > 0 ; \tag{4}$$

Property IV: We assume the general expectation that no power consumption brings no benefit, so we have

$$U(0, \omega_i^k) = 0, \quad \forall \omega > 0 ; \tag{5}$$

Many different utility functions can be used as long as the functions can satisfy Properties I–IV. In this paper, we think about the following utility functions.

$$U(x, \omega) = \begin{cases} \omega x - \frac{\alpha}{2} x^2 , & 0 \leq x \leq \frac{\omega}{\alpha}, \\ \frac{3}{20} \left(\frac{\omega}{\alpha}\right)^2, & x > \frac{\omega}{\alpha}, \end{cases} \tag{6}$$

$$U(x, \omega) = \begin{cases} \omega x - \frac{\alpha}{4} x^4 , & 0 \leq x \leq \sqrt[3]{\frac{\omega}{\alpha}}, \\ \frac{3}{4} \left(\frac{\omega}{\alpha}\right)^{\frac{4}{3}}, & x > \sqrt[3]{\frac{\omega}{\alpha}}, \end{cases} \tag{7}$$

Property V: The cost functions are increasing,

$$C_k(\tilde{L}_k) \geq C_k(\bar{L}_k), \quad \forall \tilde{L}_k \geq \bar{L}_k ; \tag{8}$$

Property VI: The cost functions are strictly convex [13],

$$C_k(\varepsilon \bar{L}_k + (1 - \varepsilon)\tilde{L}_k) \leq \varepsilon C_k(\bar{L}_k) + (1 - \varepsilon)C_k(\tilde{L}_k), \ \forall \ 0 \leq \varepsilon \leq 1, k \in \Psi, \tilde{L}_k, \bar{L}_k \geq 0 ; \tag{9}$$

In this paper, we consider the (10) and (11) that satisfy Property V-VI:

$$C_k(L_k) = a_k L_k^2 + b_k L_k + c_k , \tag{10}$$

$$C_k(L_k) = \begin{cases} a_k L_k + b_k, & L_k \leq \varphi , \\ \widehat{a}_k L_k + \widehat{b}_k, & L_k > \varphi , \end{cases} \tag{11}$$

Where $a_k, b_k, c_k, \widehat{a}_k, \widehat{b}_k, \varphi$ are positive parameters.

To develop the neural network for solving problem (1) and simplify our discussion, we introduce the following notations: $x = \left(x_1^1, x_2^1, \ldots, x_R^1, x_1^2, x_2^2, \ldots, x_R^2, \ldots, x_1^\Psi, \ldots, x_R^\Psi\right)^T; L = (L_1, L_2, \ldots, L_\Psi)^T$ and $y = \left(x^T, L^T\right)^T$.

Then problem (1) can be rewritten as follows:

$$
\begin{aligned}
\min \quad & f(y) \\
\text{s.t} \quad & y_i^k - m_i^k \geq 0 \\
& M_i^k - y_i^k \geq 0 \\
& y_{R+1}^k - L_k^{\min} \geq 0 \\
& L_k^{\max} - y_{R+1}^k \geq 0 \\
& y_{R+1}^k - \sum_{i \in R} y_i^k \geq 0
\end{aligned}
\tag{12}
$$

Where $f(y) = \sum_{k \in \Psi} L_k(C_k) - \sum_{k \in \Psi} \sum_{i \in R} U\left(x_i^k, \omega_i^k\right)$.

3   Neural Network Description

3.1   The Neural Network Model

To solve problem (12), based on generalized nonlinear programming circuit, the neural network proposed in [15] is described as the following nonautonomous differential inclusion:

$$
\dot{y}(t) \in -\partial P(y(t)) - \varepsilon(t)\partial f(y(t)) - \varepsilon^2(t)y(t).
\tag{13}
$$

The penalty functions is $P(y) = \sum_{i=1}^{R} \max\{0, G_i(y)\}$, where $G_i(y)$ are the constraints functions, $\{y : P(y) \leq 0\} = S$. $\varepsilon : [0, \infty) \to (0, \infty)$ is defined by $\varepsilon(t) = \varepsilon_0 / \sqrt[\beta]{t + a}$, $\varepsilon_0 > 0$, $a > 0$ and $\beta \geq 2$.

3.2   Theoretical Analysis

Lyapunov method is employed to study the properties of solutions of network (13). We employ the following Lyapunov energy function throughout this paper.

$$
W(y, t) = P(y) + \varepsilon(t)f(y) + \frac{1}{2}\varepsilon^2(t)\|y\|^2,
\tag{14}
$$

and we denote

$$
\partial W(y, t) = \partial P(y) + \varepsilon(t)\partial f(y) + \varepsilon^2(t)y.
\tag{15}
$$

Theorem 3.1. For $y_0 \in R^n$, there is a unique solution of (13) defined on.

Proof. Denote $z : [0, T) \to R^n$ a solution of (13) with initial point z_0. Differentiating $\frac{1}{2}\|y(t) - z(t)\|^2$ along the two solutions of (13), there exist $\bar{\xi}^y \in \partial P(y(t))$, $\bar{\eta}^y \in \partial f(y(t))$ and $\bar{\xi}^z \in \partial P(z(t))$, $\bar{\eta}^z \in \partial f(z(t))$ such that $\frac{d}{dt}\frac{1}{2}\|y(t) - z(t)\|^2 = \langle y(t) - z(t), \dot{y}(t) - \dot{z}(t)\rangle = \langle y(t) - z(t), -\bar{\xi}^y - \varepsilon(t)\bar{\eta}^y - \varepsilon^2(t)y(t) + \bar{\xi}^z + \varepsilon(t)\bar{\eta}^z + \varepsilon^2(t)z(t)\rangle.$

Combining the monotonicity of $\partial f(y)$ and the above inequality, we have

$$\|y(t) - z(t)\| \leq \|y_0 - z_0\|, \forall t \in [0, T). \tag{16}$$

Choose $h \in (0, T)$, $z_0 = y(h)$, $\lim_{t \to T}\|y(t)\| \leq \|y(T - h)\| + \|y(0) - y(h)\|$. Differentiating $W(y(t), t)$ along the solution of (13), there is

$$\frac{d}{dt}W(y(t), t) = \langle m(\partial W(y(t), t)), \dot{y}(t)\rangle + \dot{\varepsilon}(t)f(y(t)) + \varepsilon(t)\dot{\varepsilon}(t)\|y(t)\|^2. \tag{17}$$

From (13) and (17), it gives $\frac{d}{dt}W(y(t), t) = -\|\dot{y}(t)\|^2 + \dot{\varepsilon}(t)f(y(t)) + \varepsilon(t)\dot{\varepsilon}(t)\|y(t)\|^2$.

Combining (16) and (17), it derives $\langle m(\partial W(y(t), t)), \dot{y}(t)\rangle = -\|\dot{y}(t)\|^2$, which implies that $\|\dot{y}(t)\| \leq \|m(-\partial W(y(t), t))\|$, $a.e.t \in [0, +\infty)$. Integrating (16) from 0 to t, we obtain $\int_0^t \|\dot{y}(s)\|^2 ds \leq W(y(0), 0) - W(y(t), t)$. Combining the above result with the nonincreasing properties of $\|\dot{y}(t)\|$ on $[0, +\infty)$, we obtain $\lim_{t \to \infty}\|\dot{y}(t)\| = 0$.

Theorem 3.2. When $int(S) \neq \emptyset$, the solution of (13) with initial point y_0 converges to feasible region S in finite time $T_S \geq 0$.

Theorem 3.3. The solution of (13) converges to the optimal solution of (12) with the smallest 2-norm $\lim_{t \to +\infty}\|y(t) - y^*\| = 0$, which indicates that the proposed network (13) is globally attractive.

Remark 3.1. Using nonsmooth analysis, the details of the proof in Theorem 3.2, 3.3 are similar with that of reference [15, 17, 19]. From Theorems 3.2, it can be seen that the solution of the proposed neural network is convergent to the optimal solution in finite time. From Theorem 3.3, it can be seen that the proposed network is globally attractive. These properties are very important in engineering applications.

4   Numerical Simulation

Example 1 [20]: We assume there are $R = 3$ subscribers. The entire time cycle is divided into 3 time slots representing the peak hours, mid-peak hours and off-peak hours. $C_k(L_k)$ and $U(x, \omega)$ are defined in (6) and (10). We also assume the ω parameter of each user is selected randomly from the interval $[1, 4]$. For the interactions between the users and the energy provider, the price optimal model of smart grid in real time is formulated as follows:

$$\min \quad \sum_{k \in 3} \sum_{i \in 3} U\left(x_i^k, \omega_i^k\right) - \sum_{k \in 3} C_k(L_k)$$

$$s.t \quad 1 \le y_1^1 \le 36, \qquad 1 \le y_3^1 \le 40$$
$$5 \le y_1^2 \le 60, \qquad 5 \le y_3^2 \le 60$$
$$10 \le y_1^3 \le 80, \quad 10 \le y_3^3 \le 76$$
$$1 \le y_2^1 \le 40, \qquad 1 \le y_4^1 \le 36$$
$$5 \le y_2^2 \le 64, \qquad 5 \le y_4^2 \le 60$$
$$10 \le y_2^3 \le 72, \quad 10 \le y_4^3 \le 80$$
$$\sum_{i \in 3, k \in 3} y_i^k \le y_4^k$$

where we choose $\varepsilon(t) = 1/\sqrt[3]{t+1}$.

From the simulation Fig. 1, we can see that the neural network (13) has better convergence performance than [20] in off-peak hours. From the Fig. 2, we confirm that the proposed neural network (13) is globally attractive in solving (18).

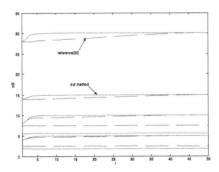

Fig. 1. Convergent property of (13) with any four initial points.

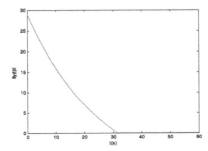

Fig. 2. Covergent property of y with different random initial point

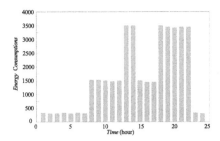

Fig. 3. Power consumption for the proposed neural network.

Example 2: We assume there are $R = 200$ subscribers. The entire time cycle is divided into 24 time slots representing the 24 h of the day. Utility function $U(x, \omega)$ and cost function $C_k(L_k)$ are given in (7) and (12). Figure 3 shows the optimal energy consumption of all users in each slot.

5   Conclusions

In this paper, based on optimization theory, we apply a neural network modeled by a differential inclusion to solve real-time price based on utility maximization in smart grid. It can be implemented in a neural network manner to maximize the aggregate utility of all users and minimize the cost imposed to the energy provider while keeping the total power consumption below the generating capacity. The proposed network has a Tikhonov regularization item ε, which let the proposed network solve (12) effectively without any evaluation on the exact penalty parameters. By another item $\varepsilon^2(t)y(t)$, the proposed network is globally attractive and the solution of it converges to the optimal solution of (12) with the smallest 2-norm. Simulation results confirmed that by using the proposed optimization-based real-time pricing model are used to demonstrate the theoretical results. The ideal developed in this paper can be extended in several directions for other DSM optimization problems in smart grid.

Acknowledgments. This work was supported by the National Natural Science Foundation of China (Grant Nos. 61473070, 61433004), the Fundamental Research Funds for the Central Universities (Grant No. N130104001), and SAPI Fundamental Research Funds (Grant No. 2013 ZCX01).

References

1. Tsoukalas, L.H., Gao, R.: From smart grids to an energy internet: assumptions, architectures, and requirements. In: Proceeding of Third International Conference on Electric Utility Deregulation and Restructuring and Power Technologies, Nanjing, China, 6–9 April, pp. 94–98 (2008)
2. Faranda, R., Pievatolo, A., Tironi, E.: Load shedding: a new proposal. IEEE Trans. Power Syst. **22**, 2086–2093 (2007)

3. Ruiz, N., Cobelo, I., Oyarzabal, J.: A direct load control model for virtual power plant management. IEEE Trans. Power Syst. **24**, 959–966 (2009)
4. U.S. Department of Energy: The Smart Grid: An introduction (2008)
5. Gellings, C.: The concept of demand-side management for electric utilities. Proc. IEEE **73** (10), 1468–1470 (1985)
6. Samadi, P., Mohsenian-Rad, A.H., et al.: Optimal real-time pricing algorithm based on utility maximization for smart grid. In: Proceedings of the First IEEE International Conference on Smart Grid Communications, pp. 415–420 (2010)
7. Gatsis, N., Giannakis, G.: Cooperative multi-residence demand response scheduling. In: Proceedings of the 45th Annual Conference on Information Sciences and Systems (CISS), pp. 1–6 (2011)
8. Tarasak, P.: Optimal real-time pricing under load uncertainty based on utility maximization for smart grid. In: Proceedings of the Second IEEE International Conference on Smart Grid Communication, pp. 321–326 (2011)
9. McArthur, S.D.J., et al.: Multi-agent systems for power engineering applications, part i:concepts, approaches, and technical challenges. IEEE Trans. Power Syst. **22**, 1743–1752 (2007)
10. Liu, Q., Wang, J.: A one-layer projection neural network for nonsmooth optimization subject to linear equalities and bound constraints. IEEE Trans. Neural Netw. Learn. Syst. **24**, 812–824 (2013)
11. Gellings, C.: The concept of demand-side management for electric utilities. Proc. IEEE **73**, 1468–1470 (1985)
12. Mohsenian-Rad, A.H.V., Wong, W., Jatskevich, S.J., Schober, R.: Optimal and autonomous incentive-based energy consumption scheduling algorithm for smart grid. In: Proceeding of IEEE PES Conference on Innovative Smart Grid Technologies, Gaithersburg, MD, January, pp. 1–6 (2010)
13. Fahrioglu, M., Alvarado, F.: Using utility information to calibrate customer demand management behavior models. IEEE Trans. Power Syst. **16**, 317–322 (2001)
14. Fahrioglu, M., Fern, M., Alvarado, F.: Designing cost effective demand management contracts using game theory. In: Proceeding of IEEE Power Engineering Society 1999 Winter Meeting, New York, 1 January, pp. 427–432 (1999)
15. Boyd, S., Vandenberghe, L.: Convex Optimization. Cambridge University Press, Cambridge (2004)
16. Cichocki, A., Unbehauen, R.: Neural Networks for Optimization and Signal Processing. Wiley, New York (1993)
17. Bian, W., Xue, X.: Neural network for solving constrained convex optimization problems with global attractivity. IEEE Trans. Circuits Syst. I Reg. Pap. **60**, 710–723 (2013)
18. Conejo, A.J., Morales, J.M., Baringo, L.: Real- time demand response model. IEEE Trans. Smart Grid **1**, 236–242 (2010)
19. Yu, R., Yang, W., Rahardja, S.: Optimal real-time price based on a statistical demand elasticity model of electricity. In: 2011 First IEEE International Workshop on Smart Grid Modeling and Simulation (SGMS), pp. 90–95 (2011)
20. He, X., Huang, T., Li, C., Che, H., Dong, Z.: A recurrent neural network for optimal real-time price in smart grid. Neurocomputing **149**, 608–612 (2015)

Exponential Stability of Anti-periodic Solution of Cohen-Grossberg Neural Networks with Mixed Delays

Sitian Qin[1(✉)], Yongyi Tan[1], and Fuqiang Wang[2]

[1] Department of Mathematics, Harbin Institute of Technology, Weihai 264209,
People's Republic of China
qinsitian@163.com
[2] School of Automobile Engineering, Harbin Institute of Technology, Weihai 264209,
People's Republic of China

Abstract. In this paper, we study the global exponential stability of anti-periodic solution of Cohen-Grossberg neural networks with mixed delays and distributed delays. Based on Lyapunov function and contraction mapping theorem, we introduce some sufficient conditions to ensure the existence and exponential stability of anti-periodic solution of Cohen-Grossberg neural networks. Finally, some numerical examples are provided to show the effectiveness of the obtained results.

Keywords: Cohen-Grossberg neural networks · Anti-periodic solution · Exponential stability · Contraction mapping

1 Introduction

Since Cohen and Grossberg raised Cohen-Grossberg neural network models in 1983 (see [5]), this topic has been comprehensively studied in theory and successfully applied in practice. The applications of Cohen-Grossberg neural networks rely on their dynamical properties, including the stability, periodicity and chaos and so on. On the other hand, delays can more precisely reflect the perturbations that unavoidably exist in real physics world. Therefore, it's valuable to study the dynamical properties of Cohen-Grossberg neural networks with different delays.

Recently, the dynamical properties of Cohen-Grossberg neural networks have received considerable attention, and massive important results have been obtained (see [9,12,14]). Among those, the existence and stability of the periodic solution is one of the most attractive problem (see [2,4,8,13]). For instance, based on the continuation theorem of coincidence degree theory, authors in [2]

W. Fuqiang—This research is supported by the national science fund of grant (61403101), and Weihai Science and technology Development Plan Project (2014DXGJ07).

© Springer International Publishing Switzerland 2016
L. Cheng et al. (Eds.): ISNN 2016, LNCS 9719, pp. 160–167, 2016.
DOI: 10.1007/978-3-319-40663-3_19

investigated the existence and stability of periodic solutions for Cohen-Grossberg neural networks with delays and impulses. In [10], Li studied the existence and global exponential stability of periodic solution for impulsive Cohen-Grossberg-type bidirectional associative memory neural networks with continuously distributed delays. Meanwhile, more and more scholars observe that anti-periodicity frequently appears in practical world, such as the anti-periodic trigonometric polynomials in interpolation problems (see [6]), anti-periodic wavelets (see [3]) and so on. Some researchers have studied the anti-periodic solution of Cohen-Grossberg neural networks. For example, Li in [11]) studied a class of Cohen-Grossberg neural networks with bounded or unbounded delays. Gongcite in [7] considered the Cohen-Grossberg neural networks model with time-varying delays and continuously distributed delays. And Abdurahman et al. in [1] investigated the Cohen-Grossberg neural networks with impulsive effects. These works gave out a number of results of the existence and exponential stability of the anti-periodic solutions. But as far as we concerned, there are few results of the global exponential stability of anti-periodic solution of Cohen-Grossberg neural networks.

Inspired by above researches, we will study the existence and stability of the anti-periodic solution of Cohen-Grossberg neural networks in this paper. The rest of the paper organized as follows. In Sect. 2, the proposed model is introduced and several related preliminaries are given. Some sufficient conditions for the existence and exponential stability of the anti-periodic solution are presented in Sect. 3. And some numerical examples are provided to show the effectiveness of the obtained results in Sect. 4.

2 Neural Networks Model and Preliminaries

In this paper, we study the following Cohen-Grossberg neural network with mixed time-varying delays:

$$
\begin{aligned}
\dot{x}_i(t) = c_i(x_i(t))[&-a_i(x_i(t)) + \sum_{j=1}^{n} b_{ij}(t)f_j(x_j(t)) + \sum_{j=1}^{n} d_{ij}(t)g_j(x_j(t - \tau_{ij}(t))) \\
&+ q_{ij}(t) \int_{-h_{ij}(t)}^{0} K_{ij}(s)p_j(x_j(t+s))ds + I_i(t)]
\end{aligned}
$$

$$(1)$$

where n is the number of neurons, $x_i(t)$ is the state variable of the ith neuron at time t. $c_i(x_i(t))$ is amplification function. $a_i(x_i(t))$ is self-inhibition terms. $f_i(x)$, $g_i(x)$ and $p_i(x)$ are active functions. $b_{ij}(t)$, $d_{ij}(t)$ and $q_{ij}(t)$ are connection weights. $K_{ij}(s)$ are the kernel of distributed state. $I_i(t)$ are external inputs. $\tau_{ij}(t)$ and $h_{ij}(t)$ are transmission time-varying delays. We assume $c_i(\cdot)$, $b_{ij}(\cdot)$, $d_{ij}(\cdot)$, $q_{ij}(\cdot)$, $f_i(\cdot)$, $g_i(\cdot)$, $p_i(\cdot)$, $K_{ij}(\cdot)$, $I_i(\cdot)$, $\tau_{ij}(\cdot)$ and $h_{ij}(\cdot)$ are all continuous.

Definition 1. *A continuous function $v : \mathbb{R} \to \mathbb{R}$ is said to be T anti-periodic on $\mathbb{R}$ if,*

$$v(t+T) = -v(t), \text{ for all } t \in \mathbb{R} \qquad (2)$$

In this paper, for any $i, j \in \{1, ..., n\}$, we introduce the following assumptions.

Assumption 1. For any $t \in \mathbb{R}$, the following holds,

$$
\begin{array}{lll}
a_i(t) = -a_i(-t) & b_{ij}(t+T) = -b_{ij}(t) & d_{ij}(t+T) = -d_{ij}(t) \\
q_{ij}(t+T) = -q_{ij}(t) & f_j(t) = f_j(-t) & g_j(t) = g_j(-t) \\
\tau_{ij}(t+T) = \tau_{ij}(t) & c_i(-t) = c_i(t) & p_i(t) = p_i(-t)
\end{array}
\tag{3}
$$

Assumption 2. There exist two constants $\overline{\tau}_{ij}$ and $\overline{h}_{ij}$, such that for any $t > 0$, we have $0 \le \tau_{ij}(t) \le \overline{\tau}_{ij}$, $0 \le h_{ij}(t) \le \overline{h}_{ij}$, $\dot{\tau}_{ij}(u(t)) < 1$, where $u(t)$ is the inverse function of the function $t - \tau_{ij}(t)$.

Assumption 3. There exist $\underline{c}_i$ and $\overline{c}_i$ such that for any $t \in \mathbb{R}$, we have $0 < \underline{c}_i \le c_i(t) \le \overline{c}_i$.

Assumption 4. There exists a constant μ_i such that

$$
\frac{a_i(s) - a_i(t)}{s - t} \ge \mu_i > 0
$$

for all $s, t \in \mathbb{R}$ with $s \ne t$.

Assumption 5. The neuron activation function f_i, g_i, p_i, satisfy

$$
L_{i1}^- \le \frac{f_i(x) - f_i(y)}{x - y} \le L_{i1}^+, \ L_{i2}^- \le \frac{g_i(x) - g_i(y)}{x - y} \le L_{i2}^+, \ L_{i3}^- \le \frac{p_i(x) - p_i(y)}{x - y} \le L_{i3}^+
\tag{4}
$$

for all $x, y \in \mathbb{R}$ with $x \ne y$. Here, L_{i1}^-, L_{i1}^+, L_{i2}^-, L_{i2}^+, L_{i3}^- and L_{i3}^+ are all positive constants.

Assumption 6. There exist $\lambda_i > 0$, $\alpha_{ij}, \beta_{ij} \in \mathbb{R}$ and $\gamma > 0$, such that

$$
h(t) \triangleq \min_{0 \le i \le n} \left\{ -2\underline{c}\mu_i + \sum_{j=1}^{n} \overline{c}[\frac{\lambda_j}{\lambda_i} |b_{ji}(t)| L_{i1}^{2(1-\alpha_{ji})} + |b_{ij}(t)| L_{j2}^{2\alpha_{ij}} + |d_{ij}(t)| L_{j2}^{2\beta_{ij}} \right.
$$
$$
\left. + \frac{d_{ji}^+ L_i^{2(1-\beta_{ji})} \lambda_j}{(1 - \dot{\tau}_{ij}(u(t)))\lambda_i}] + \frac{\lambda_j}{\lambda_i} (\sum_{i=1}^{n} \overline{q}_{ji} \overline{K}_{ji} \overline{L}_{i3} \overline{c}_i \overline{h}_{ji})^2 + 1 \right\} \le -\gamma
$$

$$
\tag{5}
$$

where $d_{ij}^+ = \max_{0 \le t \le T} |d_{ij}(t)|$.

3   Main Results

In this section, we present some sufficient conditions for the existence and exponential-stability of the anti-periodic solution of the Cohen-Grossberg neural network (1).

Theorem 1. *Under Assumptions 1–6, the Cohen-Grossberg neural network (1) has a unique $T-$anti-periodic solution, which is globally exponentially stable.*

Proof. For ϕ and $\varphi \in C([-\tau, 0]; \mathbb{R}^n)$, let $x(t) = (x_1(t), x_2(t), ..., x_n(t))^T$ and $y(t) = (y_1(t), y_2(t), ..., y_n(t))^T$ be two solutions of the Cohen-Grossberg neural network (1) through ϕ and φ. Let $z_i(t) = \int_{y_i(t)}^{x_i(t)} \frac{ds}{c_i(s)}$. By Assumption 3, we have $\frac{|x_i(t)-y_i(t)|}{\overline{c}_i} \le |z_i(t)| \le \frac{|x_i(t)-y_i(t)|}{\underline{c}_i}$. After simple calculation, we get

$$
\begin{aligned}
\dot{z}_i(t) &= \frac{\dot{x}_i(t)}{c_i(x_i(t))} - \frac{\dot{y}_i(t)}{c_i(y_i(t))} \\
&= a_i(x_i(t)) + a_i(y_i(t)) + \sum_{j=1}^{n} b_{ij}(t)[f_j(x_j(t)) - f_j(y_j(t))] \\
&\quad + \sum_{j=1}^{n} d_{ij}(t)[g_j(x_j(t - \tau_{ij}(t))) - f_j(y_j(t - t - \tau_{ij}(t)))] \\
&\quad + \sum_{j=1}^{n} q_{ij}(t) \int_{-h_{ij}(t)}^{0} K_{ij}(s)(p_j(x_j(t+s)) - p_j(y_j(t+s)))ds \\
&\le -\underline{c}_i \mu_i(t)|z_i(t)| + \sum_{j=1}^{n} |b_{ij}(t)|L_{j1}\overline{c}_j||z_j(t)| + \sum_{j=1}^{n} |d_{ij}(t)|L_{j2}\overline{c}_j||z_j(t - \tau_{ij}(t))| \\
&\quad + \sum_{j=1}^{n} \overline{q}_{ij}\overline{K}_{ij}L_{j3}\overline{c}_j \int_{-\overline{h}_{ij}}^{0} |z_j(t+s)|ds
\end{aligned}
$$
(6)

Next we construct the following Lyapunov function,

$$
V(t) = \sum_{i=1}^{n} \lambda_i z_i^2(t) + \sum_{i,j=1}^{n} \lambda_i \int_{t-\tau_{ij}}^{t} \frac{\overline{c}d_{ij}^{+}L_{j2}^{2(1-\beta_{ij})}}{1-\dot{\tau}_{ij}(u(\theta))} z_j^2(\theta)d\theta + \sum_{i=1}^{n} \lambda_i \eta_i \int_{-\overline{h}_{ij}}^{0} \int_{t+s}^{t} z_j^2(\rho)d\rho ds,
$$
(7)

where $\eta_i \triangleq (\sum_{j=1}^{n} \overline{q}_{ij}\overline{K}_{ij}\overline{p}_j\overline{L}_{j3}\overline{c}_j\overline{h}_{ij}^{\frac{1}{2}})^2$.

Calculating the derivative of V, we can get

$$
\begin{aligned}
\frac{dV}{dt} &= \sum_{i=1}^{n} \lambda_i [2z_i(t)\dot{z}_i(t) + \sum_{j=1}^{n} \frac{\overline{c}d_{ij}^{+}L_{j1}^{2(1-\beta_{ij})}}{(1-\dot{\tau}_{ij}(u(t)))} z_j^2(t) - \sum_{j=1}^{n} \overline{c}d_{ij}^{+}L_{j2}^{r(1-\beta_{ij})} z_j^2(t - \tau_{ij}(t)) \\
&\quad + \eta_i \int_{-\overline{h}_{ij}}^{0} z_j^2(t)ds - \eta_i \int_{-\overline{h}_{ij}}^{0} z_j^2(t+s)ds]
\end{aligned}
$$
(8)

From (6), we have

$$
\begin{aligned}
2z_i(t)\dot{z}_i(t) &\le -2\underline{c}_i\mu_i(t)|z_i(t)| + \sum_{j=1}^{n} 2\overline{c}_j[|b_{ij}(t)|L_{j1}|z_j(t)||z_i(t)| \\
&\quad + |d_{ij}(t)|L_{j2}|z_j(t - \tau_{ij}(t))||z_i(t)|] + \sum_{j=1}^{n} 2|z_i(t)|\overline{q}_{ij}\overline{K}_{ij}L_{j3}\overline{c}_j \int_{-\overline{h}_{ij}}^{0} |z_j(t+s)|ds \\
&\le -2\underline{c}_i\mu_i(t)|z_i(t)| + \sum_{j=1}^{n} 2\overline{c}_j[|b_{ij}(t)|L_{j1}^{1-\alpha_{ij}}|z_j(t)|L_{j1}^{\alpha_{ij}}|z_i(t)| \\
&\quad + |d_{ij}(t)|L_{j2}^{1-\beta_{ij}}|z_j(t - \tau_{ij}(t))|L_{j2}^{\beta_{ij}}|z_i(t)|] + \sum_{j=1}^{n} 2|z_i(t)|\overline{q}_{ij}\overline{K}_{ij}L_{j3}\overline{c}_j \int_{-\overline{h}_{ij}}^{0} |z_j(t+s)|ds
\end{aligned}
$$
(9)

By Young inequality,

(i) $L_{j1}^{1-\alpha_{ij}}|z_j(t)|L_{j1}^{\alpha_{ij}}|z_i(t)| \le \frac{1}{2}L_{j1}^{1-\alpha_{ij}}|z_j(t)|^2 + \frac{1}{2}L_{j1}^{2\alpha_{ij}}|z_i(t)|^2$

(ii) $L_{j2}^{1-\beta_{ij}}|z_j(t - \tau_{ij}(t))|L_{j2}^{\beta_{ij}}|z_i(t)| \le \frac{1}{2}L_{j2}^{1-\beta_{ij}}|z_j(t - \tau_{ij}(t))|^2 + \frac{1}{2}L_{j2}^{2\beta_{ij}}|z_i(t)|^2$

(10)

Then,

$$2|z_i(t)| \sum_{j=1}^{n} \overline{q}_{ij} \overline{K}_{ij} L_{j3} \overline{c}_j \int_{-\overline{h}_{ij}}^{0} |z_j(t+s)| ds \leq |z_i(t)|^2 + (\sum_{j=1}^{n} \overline{q}_{ij} \overline{K}_{ij} L_{j3} \overline{c}_j \int_{-\overline{h}_{ij}}^{0} |z_j(t+s)| ds)^2$$

$$(11)$$

Using Hölder inequality, it is obtained that

$$\int_{-\overline{h}_{ij}}^{0} |z_j(t+s)| ds \leq (\int_{-\overline{h}_{ij}}^{0} |z_j(t+s)|^2 ds)^{\frac{1}{2}} (\int_{-\overline{h}_{ij}}^{0} 1^2 ds)^{\frac{1}{2}} = (\overline{h}_{ij} \int_{-\overline{h}_{ij}}^{0} |z_j(t+s)|^2 ds)^{\frac{1}{2}}$$

$$(12)$$

So, by (11) and (12), we have

$$2|z_i(t)| \sum_{j=1}^{n} \overline{q}_{ij} \overline{K}_{ij} L_{j3} \overline{c}_j \int_{-\overline{h}_{ij}}^{0} |z_j(t+s)| ds$$

$$\leq |z_i(t)|^2 + (\sum_{j=1}^{n} \overline{q}_{ij} \overline{K}_{ij} L_{j3} \overline{c}_j)^2 \overline{h}_{ij} \int_{-\overline{h}_{ij}}^{0} |z_j(t+s)|^2 ds$$

$$(13)$$

Meanwhile, from (8), (10) and (13), the following holds

$$\frac{dV}{dt} \leq \sum_{i=1}^{n} \lambda_i \{ -2\underline{c}_i \mu_i(t)|z_i(t)| + \sum_{j=1}^{n} 2\overline{c}_j [|b_{ij}(t)| \frac{1}{2} L_{j1}^{1-\alpha_{ij}} |z_j(t)|^2$$

$$+ \frac{1}{2} L_{j1}^{2\alpha_{ij}} |z_i(t)|^2 + |d_{ij}(t)| \frac{1}{2} L_{j2}^{1-\beta_{ij}} |z_j(t)|^2 + \frac{1}{2} L_{j2}^{2\beta_{ij}} |z_i(t)|^2] + |z_i(t)|^2$$

$$+ (\sum_{j=1}^{n} \overline{q}_{ij} \overline{K}_{ij} L_{j3} \overline{c}_j)^2 \overline{h}_{ij} \int_{-\overline{h}_{ij}}^{0} |z_j(t+s)|^2 ds - \sum_{j=1}^{n} \overline{c} d_{ij}^{+} L_{j2}^{2(1-\beta_{ij})} |z_j(t-\tau_{ij}(t))|^2$$

$$+ \sum_{j=1}^{n} \overline{q}_{ij} \overline{K}_{ij} \overline{p}_j \overline{L}_{j3} \overline{c}_j \overline{h}_{ij}^{\frac{1}{2}^2} \int_{-\overline{h}_{ij}}^{0} |z_j(t)|^2 ds - \sum_{j=1}^{n} \overline{q}_{ij} \overline{K}_{ij} \overline{p}_j \overline{L}_{j3} \overline{c}_j \overline{h}_{ij}^{\frac{1}{2}^2} \int_{-\overline{h}_{ij}}^{0} |z_j(t+s)|^2 ds \}$$

$$\leq \sum_{i=1}^{n} \lambda_i \{ -2\underline{c} \mu_i(t) + \sum_{j=1}^{n} \overline{c} [\frac{\lambda_j}{\lambda_i} |b_{ji}| L_{i1}^{2(1-\alpha^{ji})} + |b_{ij}(t)| L_{j2}^{2\alpha_{ij}} + |d_{ij}| L_j^{2\beta_{ij}} + \frac{d_{ji}^{+} L_i^{2(1-\beta_{ji})} \lambda_j}{(1-\tau_{ij}(u(t)))\lambda_i}]$$

$$+ \frac{\lambda_j}{\lambda_i} (\sum_{i=1}^{n} \overline{q}_{ji} \overline{K}_{ji} \overline{L}_{i3} \overline{c}_i \overline{h}_{ji})^2 + 1 \} |z_i(t)|^2 \leq -h(t) V(t)$$

By Assumption 4, we have

$$\frac{dV}{dt} \leq -\gamma V(t), \forall t \geq 0$$

which means $V(t) \leq V(0) e^{-\gamma t}$, $\forall t \geq 0$. So, for all $t \geq 0$, $\lambda_0^{\frac{1}{2}} |x(t) - y(t)| \leq V(t)^{\frac{1}{2}} \leq V(0) e^{-\frac{\gamma}{2} t}$, where $\lambda_0 = \min_{1 \leq i \leq n} \lambda_i$. Then, for all $t \geq 0$,

$$\|x(t) - y(t)\| \leq \frac{M}{(\min_{1 \leq i \leq n} \lambda_i)^{\frac{1}{2}}} \|\phi - \varphi\| e^{-\frac{\gamma}{2} t},$$

$$(14)$$

where $M^2 = \max_{1 \leq i \leq n} \{ \lambda_i + \lambda_i \sum_{j=1}^{n} (\frac{\overline{c} d_{ij}^{+} L_{j1}^{2(1-\beta_j i)}}{1-\overline{\tau}} + \overline{q}_{ji} \overline{K}_{ji} \overline{p}_{ji} \overline{L}_{ji} \overline{c}_{ji} \overline{h} \overline{s}) \}$.

Next, we will prove the existence of the anti-periodic solution of (1) using contraction mapping theory. Construct the following map

$$F : C([-\tau, 0] : \mathbb{R}^n) \longrightarrow C([-\tau, 0] : \mathbb{R}^n)$$
$$\varphi \longmapsto -x_T(\varphi)$$

$$(15)$$

where $x_T(\varphi) = x(t+T)$ and $x(t+T)$ is the solution of (1) through initial values φ. For all $\phi, \varphi \in C$, by (14), we get

$$
\begin{aligned}
||F\phi - F\varphi|| &= \max_{-\tau \le t \le 0} x_T\phi(t) - x_T\varphi(t) \le \max_{-\tau \le t \le 0} \frac{M}{(\min_{1 \le i \le n} \lambda_i)^{\frac{1}{2}}} ||\phi - \varphi|| e^{-\frac{\gamma}{2}t} \\
&\le \frac{M}{(\min_{1 \le i \le n} \lambda_i)^{\frac{1}{2}}} ||\phi - \varphi|| e^{-\frac{\gamma}{2}(T+t)} \le \frac{M}{(\min_{1 \le i \le n} \lambda_i)^{\frac{1}{2}}} e^{-\frac{\gamma}{2}(T-\tau)} ||\phi - \varphi||
\end{aligned}
$$

(16)

For all $m \in \mathbb{N}$, after simple calculation,

$$
||F^m \phi - F^m \varphi|| \le \frac{M}{(\min_{1 \le i \le n} \lambda_i)^{\frac{1}{\tau}}} e^{-\frac{\gamma}{2}(mT-\tau)} ||\phi - \varphi||
$$

(17)

Here, $F^m = F_1 \circ F_2 \circ ... F_m$ with $F_i = F$. It is clear that there must be a large enough positive integer m, such that $0 < Me^{-\frac{\gamma}{2}(mT-\tau)}/(\min_{1 \le i \le n} \lambda_i)^{\frac{1}{\tau}} < 1$. Hence, $F^m : C([-\tau, 0], \mathbb{R}^n) \longrightarrow C([-\tau, 0], \mathbb{R}^n)$ is a contraction map. So by the contraction mapping theory, F has a unique fixed point $\psi \in C([-\tau, 0], \mathbb{R}^n)$. That is, $x_T(\psi) = -\psi$.

Let $x(t)$ be the solution of (1) from the initial ψ, and we construct the following function,

$$
X(t) = \begin{cases}
\psi(t), & t \in [-\tau, 0] \\
x(t), & t \in [0, T] \\
-x(t-T), & t \in [T, 2T] \\
... \\
(-1)^k x(t-kT), & t \in [kT, (k+1)T]
\end{cases}
$$

(18)

It is obvious that $X(t)$ is a $T-$anti-periodic function. By Assumption 1, $a_i(\cdot)$ is odd function, $b_{ij}(\cdot), d_{ij}(\cdot), q_{ij}(\cdot)$ are all $T-$antiperiodic functions and $c_i(\cdot)$, $f_{ij}(\cdot), g_{ij}(\cdot), p_{ij}(\cdot)$ are all even functions, for all $i, j \in \{1, ..., n\}$. It is easy to see that $X(t)$ is a solution of Cohen-Grossberg neural network (1) through the initial ψ. Due to the uniqueness of the solution, we get that

$$
X(t) = x(t), \text{ for all } t \ge 0
$$

Hence, $x(t)$ is a $T-$anti-periodic solution of the Cohen-Grossberg neural network (1). By the above statement (14), the Cohen-Grossberg neural network (1) has a unique anti-periodic solution $x(t)$ which is globally exponentially stable.

4   Numerical Examples

Example 1. Consider a Cohen-Grossberg neural network (1) with the following network parameters,

$$
(a_i(t)) = \begin{pmatrix} 30t \\ 10t \end{pmatrix}, (b_{ij}(t)) = \begin{pmatrix} \sin 10t, \ 2\cos 10t \\ \cos 10t, \ \cos 10t \end{pmatrix}, (d_{ij}(t)) = \begin{pmatrix} \cos 10t, \ 0 \\ 3\sin 10t, \ -\sin 10t \end{pmatrix}
$$

$$
(q_{ij}(t)) = \begin{pmatrix} \sin 10t, \ \cos 10t \\ \cos 10t, \ 0 \end{pmatrix}, (f_i(t)) = (g_i(t)) = (p_i(t)) = \frac{1}{t^2+1} \begin{pmatrix} 1 \\ 1 \end{pmatrix}
$$

$$
c_i(t) = 1, \tau_{ij}(t) = 1, (I_i(t)) = \begin{pmatrix} \sin t \\ \cos t \end{pmatrix}
$$

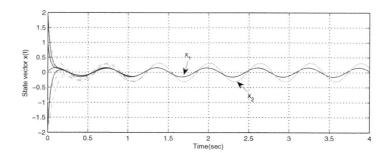

Fig. 1. The state trajectories of Cohen-Grossberg neural network (1) in Example 1.

We can verify that the Assumptions 1–6 hold. Hence, from Theorem 1, Cohen-Grossberg neural network (1) defined above has a unique anti-periodic solution which is globally exponentially stable. Figure 1 displays the state trajectory of the neural network with random initial values.

Example 2. Consider a Cohen-Grossberg neural network (1) with the following network parameters

$$(a_i(t)) = \begin{pmatrix} 15t \\ 50t \end{pmatrix}, (b_{ij}(t)) = \begin{pmatrix} \frac{1}{3}\cos 15t, & \frac{1}{4}\cos 15t \\ \cos 15t, & \cos 15t \end{pmatrix}, (d_{ij}(t)) = \begin{pmatrix} \cos 15t, & 0 \\ 3\sin 15t, & -\sin 15t \end{pmatrix},$$

$$(q_{ij}(t)) = \begin{pmatrix} \frac{1}{6}\sin 15t, & \cos 15t \\ \cos 15t, & 0 \end{pmatrix}, (f_i(t)) = \begin{pmatrix} e^{-t^2} \\ e^{-t^2} \end{pmatrix}, (g_i(t)) = \begin{pmatrix} \frac{1}{t^2+1} \\ \frac{1}{t^2+1} \end{pmatrix},$$

$$(p_i(t)) = \begin{pmatrix} |t| \\ |t| \end{pmatrix}, c_i(t) = 1, \tau_{ij}(t) = 1, (I_i(t)) = \begin{pmatrix} \sin t \\ \cos t \end{pmatrix}$$

Similarly, Cohen-Grossberg neural network (1) defined in this example satisfies the Assumptions 1–6, and from the following figure, we can see that the solution is anti-periodic and exponentially stable (Fig. 2).

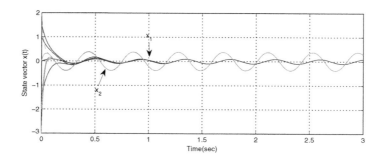

Fig. 2. The state trajectories of Cohen-Grossberg neural network (1) in Example 2.

5   Conclusions

In this paper, we present some new sufficient criteria for global exponential stability of anti-periodic solution of Cohen-Grossberg neural networks with mixed delays and distributed delays. By Lyapunov function and contraction mapping theorem, we obtain the related results. We also give some numerical examples to illustrate the obtained result.

References

1. Abdurahman, A., Jiang, H.: The existence and stability of the anti-periodic solution for delayed Cohen-Grossberg neural networks with impulsive effects. Neurocomputing **149**, 22–28 (2015)
2. Bai, C.: Global exponential stability and existence of periodic solution of Cohen-Grossberg type neural networks with delays and impulses. Nonlinear Anal. Real World Appl. **9**, 747–761 (2008)
3. Chen, H.: Antiperiodic wavelets. J. Comput. Math. Int. Ed. **14**, 32–39 (1996)
4. Chen, X., Song, Q.: Global exponential stability of the periodic solution of delayed Cohen-Grossberg neural networks with discontinuous activations. Neurocomputing **73**, 3097–3104 (2010)
5. Cohen, M., Grossberg, S., et al.: Absolute stability of global pattern formation and parallel memory storage by competitive neural networks. IEEE Trans. Syst. Man Cybern. **5**, 815–826 (1983)
6. Delvos, F.J., Knoche, L.: Lacunary interpolation by antiperiodic trigonometric polynomials. BIT Numer. Math. **39**, 439–450 (1999)
7. Gong, S.: Anti-periodic solutions for a class of Cohen-Grossberg neural networks. Comput. Math. Appl. **58**, 341–347 (2009)
8. Li, B., Xu, D.: Existence and exponential stability of periodic solution for impulsive Cohen-Grossberg neural networks with time-varying delays. Appl. Math. Comput. **219**, 2506–2520 (2012)
9. Li, G., Sun, C.: Global stability of Cohen-Grossberg neural network with time-varying delays via nonlinear measure. J. Comput. Inf. Syst. **9**, 1389–1398 (2013)
10. Li, X.: Existence and global exponential stability of periodic solution for impulsive Cohen-Grossberg-type bam neural networks with continuously distributed delays. Appl. Math. Comput. **215**, 292–307 (2009)
11. Li, Y., Yang, L.: Anti-periodic solutions for Cohen-Grossberg neural networks with bounded and unbounded delays. Commun. Nonlinear Sci. Numer. Simul. **14**, 3134–3140 (2009)
12. Song, Q., Zhang, J.: Global exponential stability of impulsive Cohen-Grossberg neural network with time-varying delays. Nonlinear Anal. Real World Appl. **9**, 500–510 (2008)
13. Wang, D., Huang, L.: Periodicity and global exponential stability of generalized Cohen-Grossberg neural networks with discontinuous activations and mixed delays. Neural Netw. **51**, 80–95 (2014)
14. Zheng, C.D., Shan, Q.H., Wang, Z.: Novel stability criteria of Cohen-Grossberg neural networks with time-varying delays. Int. J. Circuit Theory Appl. **40**, 221–235 (2012)

Stability of Complex-Valued Cohen-Grossberg Neural Networks with Time-Varying Delays

Zhenjiang Zhao[1]([✉]) and Qiankun Song[2]

[1] Department of Mathematics, Huzhou University, Huzhou 313000, China
zhaozjcn@163.com
[2] Department of Mathematics, Chongqing Jiaotong University,
Chongqing 400074, China
qiankunsong@163.com

Abstract. In this paper, the complex-valued Cohen-Grossberg neural networks model with time-varying delays is considered. By employing the idea of vector Lyapunov function, M-matrix theory and inequality technique, a new sufficient condition is obtained to ensure the existence, uniqueness and global exponential stability of equilibrium point for the considered neural networks. The provided result generalizes a few previous known ones. An example with simulations is given to show the effectiveness of the obtained results.

Keywords: Complex-valued Cohen-Grossberg neural networks · Stability · Time-varying delays · Equilibrium point

1 Introduction

The Cohen-Grossberg neural network models, first proposed and studied by Cohen and Grossberg [1], have been widely applied within various engineering and scientific fields such as neurobiology, population biology, and computing technology. In such applications, it is of prime importance to ensure that the designed neural networks be stable [2]. In implementation of neural networks, however, time delays are unavoidably encountered owing to the finite switching speed of the amplifiers and communication time [3]. It has been found that the existence of time delays often causes undesirable dynamic behaviors such as performance degradation, oscillation, or even instability of the systems [4]. Therefore, stability analysis of Cohen-Grossberg neural networks with time delays has received much attention.

As an extension of real-valued neural networks, complex-valued neural networks with complex-valued state, output, connection weight, and activation functions become strongly desired because of their practical applications in physical systems dealing with electromagnetic, light, ultrasonic, and quantum waves [5]. In fact, complex-valued neural networks make it possible to solve some problems which cannot be solved with their real-valued counterparts [6]. Recently, there have been some researches on the stability of various complex-valued neural

© Springer International Publishing Switzerland 2016
L. Cheng et al. (Eds.): ISNN 2016, LNCS 9719, pp. 168–176, 2016.
DOI: 10.1007/978-3-319-40663-3_20

networks, for example, see [6–11] and references therein. In [7–9], the method used to analyze stability of complex-valued neural networks was to separate the problem into real parts and imaginary parts, and then recast it into equivalent real-valued neural networks. But this method encounters two problems. One is that the dimension of the real-valued neural networks is double that of complex-valued neural networks, which leads to difficulties on the analysis. The other is that this method needs an explicit separation of complex-valued activation function into its real part and imaginary part, however, this separation is not always expressible in an analytical form. In [10,11], when the considered complex-valued neural networks are not separated into their real and imaginary parts, several sufficient conditions for checking the stability of complex-valued neural networks with time-varying delays were obtained. Very recently, authors considered the complex-valued Cohen-Grossberg neural networks model with constant delay, and investigated the stability problem by separating its real and imaginary parts [12].

In this paper, the complex-valued Cohen-Grossberg neural networks model with time-varying delays is considered, a sufficient criterion to ensure the existence, uniqueness and global exponential stability of equilibrium points is established without separating the real and imaginary parts of model. The provided result has generalized a few previous researches.

Notations: The notations are quite standard. Throughout this paper, i shows the imaginary unit, i.e., $i = \sqrt{-1}$. For complex number $z = x + iy$, the notation $|z| = \sqrt{x^2 + y^2}$ stands for the module of z. E represents the unitary matrix with appropriate dimensions. $\mathbb{C}$, $\mathbb{C}^n$ and $\mathbb{C}^{n \times m}$ denote, respectively, the set of all complex numbers, the set of all n-dimensional complex-valued vectors and the set of all $n \times m$ complex-valued matrices. $\overline{A}$ and A^* show the conjugate and conjugate transpose of complex-valued matrix A, respectively. For a complex-valued vector $u = (u_1, u_2, \cdots, u_n)^T \in \mathbb{C}^n$, $|u|$ denotes the module vector given by $|u| = (|u_1|, |u_2|, \cdots, |u_n|)^T$, while the notation $\|u\|$ is the Euclidean norm of u. For a complex-valued matrix $A = (a_{ij})_{n \times n} \in \mathbb{C}^{n \times n}$, $|A|$ denotes the module matrix given by $|A| = (|a_{ij}|)_{n \times n}$, while $\|A\|$ denotes a matrix norm defined by $\|A\| = \sqrt{A^* A}$. $\rho(A)$ denotes the spectral radius of matrix A.

2   Problem Formulation and Preliminaries

In this paper, we consider the following complex-valued Cohen-Grossberg neural networks with time-varying delays

$$\dot{z}_i(t) = a_i(z_i(t))\{-c_i z_i(t) + \sum_{j=1}^{n} a_{ij} f_j(z_j(t)) + \sum_{j=1}^{n} b_{ij} f_j(z_j(t - \tau_{ij}(t))) + J_i\} \,(1)$$

for $t \geq 0$, $i = 1, 2, \cdots, n$, where $z(t) = (z_1(t), z_2(t), \cdots, z_n(t))^T \in \mathbb{C}^n$, $z_i(t)$ is the state of the ith neuron at time t; $a_i(z_i(t))$ represents an amplification function at time t; $f(z(t)) = (f_1(z_1(t)), f_2(z_2(t)), \cdots, f_n(z_n(t)))^T \in \mathbb{C}^n$, and $f(z(t - \tau(t))) = (f_1(z_1(t - \tau(t))), f_2(z_2(t - \tau(t))), \cdots, f_n(z_n(t - \tau(t))))^T \in \mathbb{C}^n$, are the vector-valued activation functions without and with time delays whose elements consist of complex-valued nonlinear functions; $\tau_{ij}(t)$ corresponds to the transmission delay along the axon of the jth unit from the ith unit and satisfies $0 \leq \tau_{ij}(t) \leq \tau_{ij}$ (τ_{ij} is a constant); $C = \mathrm{diag}\{c_1, c_2, \cdots, c_n\} \in \mathbb{R}^{n \times n}$ is the self-feedback connection weight matrix, where $c_i > 0$; $A = (a_{ij})_{n \times n} \in \mathbb{C}^{n \times n}$ and $B = (b_{ij})_{n \times n} \in \mathbb{C}^{n \times n}$ are the connection weight matrices; $J = (J_1, J_2, \cdots, J_n)^T \in \mathbb{C}^n$ is the input vector.

The initial conditions of model (1) are of the form $z_i(s) = \phi_i(s)$, $s \in (-\infty, 0]$, where ϕ_i is bounded and continuous on $[-\tau, 0]$, and $\tau = \max\limits_{1 \leq i,j \leq n} \{\tau_{ij}\}$.

Throughout this paper, we make the following assumptions:

Assumption 1. Each function $a_i(u)$ is continuous and $0 < \underline{a}_i \leq a_i(u)$ for all $u \in R$, where $\underline{a}_i$ is a positive constant, $i = 1, 2, \cdots, n$.

Assumption 2. For any $i \in \{1, 2, \cdots, n\}$, there exists a positive diagonal matrix $L = \mathrm{diag}\{l_1, l_2, \cdots, l_n\}$ such that

$$|f_i(\alpha_1) - f_i(\alpha_2)| \leq l_i |\alpha_1 - \alpha_2|$$

for all $\alpha_1, \alpha_2 \in \mathbb{C}$.

Definition 1. The equilibrium point $\widetilde{z} = (\widetilde{z}_1, \widetilde{z}_2, \cdots, \widetilde{z}_n)^T$ of model (1) is said to be globally exponentially stable, if there exist constants $\varepsilon > 0$ and $M > 0$ such that

$$\|z(t) - \widetilde{z}\| \leq M \|\phi - \widetilde{z}\| e^{-\varepsilon(t - t_0)}$$

for all $t > 0$, where $z(t) = (z_1(t), z_2(t), \cdots, z_n(t))^T$ is any solution of model (1), and $\|\phi - \widetilde{z}\| = \sup\limits_{s \in [-\tau, 0]} \left(\sum\limits_{i=1}^{n} |\phi_i(s) - \widetilde{z}_i|^2 \right)^{\frac{1}{2}}$.

3   Main Result

Theorem 1. Under Assumption 1 and 2, model (1) has a unique equilibrium point, which is globally exponentially stable if

$$W = C - (|A| + |B|)L$$

is a nonsingular M-matrix.

Proof. Let $\tilde{z} = (\tilde{z}_1, \tilde{z}_2, \cdots, \tilde{z}_n)^T$ be equilibrium point of model (1), then we have from Assumption 1 that

$$-c_i\tilde{z}_i + \sum_{j=1}^{n}(a_{ij} + b_{ij})f_j(\tilde{z}_j) + J_i = 0, \quad i = 1, 2, \cdots, n.$$

Let $H(u) = (H_1(u), H_2(u), \cdots, H_n(u))^T$, where

$$H_i(u) = -c_iu_i + \sum_{j=1}^{n}(a_{ij} + b_{ij})f_j(u_j) + J_i, \quad i = 1, 2, \cdots, n.$$

In the following, we shall prove that $H(u)$ is a homeomorphism of $\mathbb{C}^n$ onto itself.

First, we prove that $H(u)$ is an injective map on $\mathbb{C}^n$.

In fact, if there exist $u = (u_1, u_2, \cdots, u_n)^T$ and $v = (v_1, v_2, \cdots, v_n)^T \in \mathbb{C}^n$ and $u \neq v$ such that $H(u) = H(v)$, then

$$c_i(u_i - v_i) = \sum_{j=1}^{n}(a_{ij} + b_{ij})(f_j(u_j) - f_j(v_j)), \quad i = 1, 2, \cdots, n.$$

From Assumption 2, we get that

$$c_i|u_i - v_i| \leq \sum_{j=1}^{n}(|a_{ij}| + |b_{ij}|)l_j|u_j - v_j|, \quad i = 1, 2, \cdots, n.$$

That is,

$$(C - (|A| + |B|)L)(|u_1 - v_1|, |u_2 - v_2|, \cdots, |u_n - v_n|)^T \leq 0.$$

From $W = C - (|A| + |B|)L$ is a nonsingular M-matrix, we can get

$$u_i = v_i, \quad i = 1, 2, \cdots, n,$$

which is a contradiction. Hence $H(u)$ is an injective on $\mathbb{C}^n$.

Second, we prove that $\|H(u)\| \to +\infty$ as $\|u\| \to +\infty$.

Since W is a nonsingular M-matrix, there exists a positive diagonal matrix $P = diag(p_1, p_2, \cdots, p_n)$ such that $PW + W^T P$ is a positive definite matrix. Let $\psi(u) = (\psi_1(u), \psi_2(u), \cdots, \psi_n(u))^T$, $\lambda_{min}(PW + W^T P)$ denotes the minimum eigenvalue of matrix $PW + W^T P$, where

$$\psi_i(u) = -c_iu_i + \sum_{j=1}^{n}(a_{ij} + b_{ij})(f_j(u_j) - f_j(0)), \quad i = 1, 2, \cdots, n.$$

Calculating

$$u^* P \psi(u) + \psi^*(u) P u$$

$$= \sum_{i=1}^{n} (\overline{u}_i p_i \psi_i(u) + \overline{\psi_i(u)} p_i u_i)$$

$$= 2 \sum_{i=1}^{n} \operatorname{Re}(\overline{u}_i p_i \psi_i(u))$$

$$= \sum_{i=1}^{n} \left[-2 p_i c_i |u_i|^2 + 2 \sum_{j=1}^{n} \operatorname{Re}\left(\overline{u}_i p_i (a_{ij} + b_{ij})(f_j(u_j) - f_j(0)) \right) \right]$$

$$\leq \sum_{i=1}^{n} \left[-2 p_i c_i |u_i|^2 + 2 \sum_{j=1}^{n} |\overline{u}_i p_i (a_{ij} + b_{ij})(f_j(u_j) - f_j(0))| \right]$$

$$\leq \sum_{i=1}^{n} \left[-2 p_i c_i |u_i|^2 + 2 \sum_{j=1}^{n} |u_i| p_i (|a_{ij}| + |b_{ij}|) l_j |u_j| \right]$$

$$= -2 \Big(|u_1|, |u_2|, \cdots, |u_n| \Big) P W \Big(|u_1|, |u_2|, \cdots, |u_n| \Big)^T$$

$$= - \Big(|u_1|, |u_2|, \cdots, |u_n| \Big) (PW + W^T P) \Big(|u_1|, |u_2|, \cdots, |u_n| \Big)^T$$

$$\leq -\lambda_{min}(PW + W^T P) \sum_{i=1}^{n} |u_i|^2$$

$$= -\lambda_{min}(PW + W^T P) \|u\|^2.$$

Hence

$$\lambda_{min}(PW + W^T P) \|u\|^2 \leq 2\|u\| \cdot \|P\| \cdot \|\psi(u)\|.$$

When $\|u\| \neq 0$, we have $\|\psi(u)\| \geq \frac{1}{2} \lambda_{min}(PW + W^T P) \frac{\|u\|}{\|P\|}$. Therefore $\|\psi(u)\| \to +\infty$ as $\|u\| \to +\infty$, which implies $\|H(u)\| \to +\infty$ as $\|u\| \to +\infty$.

Thus, we know that $H(u)$ is a homeomorphism of R^n. So model (1) has a unique equilibrium point.

Let $\widetilde{z} = (\widetilde{z}_1, \widetilde{z}_2, \cdots, \widetilde{z}_n)^T$ is the equilibrium point of model (1). Denote

$$u_i(t) = z_i(t) - \widetilde{z}_i, \quad \alpha_i(u_i(t)) = a_i(z_i(t) + \widetilde{z}_i), \quad g_j(u_j(t)) = f_j(u_j(t) + \widetilde{z}_j) - f_j(\widetilde{z}_j).$$

Then model (1) can be rewritten as

$$\dot{u}_i(t) = \alpha_i(u_i(t)) \left[-d_i u_i(t) + \sum_{j=1}^{n} a_{ij} g_j(u_j(t)) + \sum_{j=1}^{n} b_{ij} g_j(u_j(t - \tau_{ij}(t))) \right]. \quad (2)$$

From M-matrix theory, since W is a nonsingular M-matrix, there exists $\xi = (\xi_1, \xi_2, \cdots, \xi_n)^T \in \mathbb{R}^n$ such that

$$-\xi_i c_i + \sum_{j=1}^{n} \xi_j (|a_{ij}| + |b_{ij}|) l_j < 0.$$

Furthermore, there exists $\varepsilon > 0$ such that

$$\xi_i(-c_i + \frac{\varepsilon}{a_i}) + \sum_{j=1}^{n} \xi_j |a_{ij}| l_j + e^{\varepsilon\tau} \sum_{j=1}^{n} \xi_j |b_{ij}| l_j < 0, \quad i = 1, 2, \cdots, n. \tag{3}$$

Let

$$v_i(t) = e^{\varepsilon t} |u_i(t)| = e^{\varepsilon t} \sqrt{u_i(t)\overline{u_i(t)}}, \quad i = 1, 2, \cdots, n.$$

Calculating the derivative of $v_i(t)$ along the solutions of (2), we get from Assumption 1 and Assumption 2 that

$$\dot{v}_i(t) = \varepsilon e^{\varepsilon t} |u_i(t)| + \frac{e^{\varepsilon t}}{|u_i(t)|} \mathrm{Re}\left(\dot{u}_i(t)\overline{u_i(t)}\right)$$

$$= \varepsilon e^{\varepsilon t} |u_i(t)| + \frac{e^{\varepsilon t}}{|u_i(t)|} \alpha_i(u_i(t)) \left[-c_i |u_i(t)|^2 + \sum_{j=1}^{n} \mathrm{Re}\left(a_{ij} g_j(u_j(t))\overline{u_i(t)}\right) \right.$$

$$\left. + \sum_{j=1}^{n} \mathrm{Re}\left(b_{ij} g_j(u_j(t - \tau_{ij}(t)))\overline{u_i(t)}\right) \right]$$

$$\leq \alpha_i(u_i(t)) e^{\varepsilon t} \left[(\frac{\varepsilon}{a_i} - c_i)|u_i(t)| + \frac{1}{|u_i(t)|} \sum_{j=1}^{n} |a_{ij} g_j(u_j(t))\overline{u_i(t)}| \right.$$

$$\left. + \frac{1}{|u_i(t)|} \sum_{j=1}^{n} |b_{ij} g_j(u_j(t - \tau_{ij}(t)))\overline{u_i(t)}| \right]$$

$$\leq \alpha_i(u_i(t)) e^{\varepsilon t} \left[(\frac{\varepsilon}{a_i} - c_i)|u_i(t)| + \sum_{j=1}^{n} |a_{ij}| l_j |u_j(t)| + \sum_{j=1}^{n} |b_{ij}| l_j |u_j(t - \tau_{ij}(t))| \right]$$

$$\leq \alpha_i(u_i(t)) \left[(\frac{\varepsilon}{a_i} - c_i)v_i(t) + \sum_{j=1}^{n} |a_{ij}| l_j v_j(t) + e^{\varepsilon\tau} \sum_{j=1}^{n} |b_{ij}| l_j v_j(t - \tau_{ij}(t)). \tag{4}$$

Let $\gamma = \frac{(1+\delta)\|\phi - \tilde{z}\|}{\min\limits_{1 \leq i \leq n} \{\xi_i\}}$ (δ is a positive constant), then

$$v_i(s) = e^{\varepsilon s} |u_i(s)| \leq |u_i(s)| = |\phi_i(s) - \tilde{z}_i| \leq \|\phi - \tilde{z}\| < \xi_i \gamma, \quad s \in [-\tau, 0].$$

In the following, we will prove that

$$v_i(t) < \xi_i \gamma \tag{5}$$

hold for $t \geq 0$. In fact, if (5) is not true, then there exist some i_0 and $\bar{t} \geq 0$ such that

$$v_{i_0}(\bar{t}) = \xi_{i_0}\gamma, \quad \dot{v}_{i_0}(\bar{t}) \geq 0, \quad \text{and} \quad v_j(t) \leq \xi_j \gamma, \quad -\tau \leq t \leq \bar{t}. \tag{6}$$

It follow from (3), (4) and (6) that

$$\dot{v}_{i_0}(\bar{t}) \leq \alpha_{i_0}(u_{i_0}(\bar{t}))\Big[(\frac{\varepsilon}{a_{i_0}} - c_{i_0})v_{i_0}(\bar{t}) + \sum_{j=1}^{n}|a_{i_0j}|l_jv_j(\bar{t})$$

$$+ e^{\varepsilon\tau}\sum_{j=1}^{n}|b_{i_0j}|l_jv_j(t - \tau_{i_0j}(\bar{t}))\Big]$$

$$\leq \alpha_{i_0}(u_{i_0}(\bar{t}))\Big[(\frac{\varepsilon}{a_{i_0}} - c_{i_0})\xi_{i_0}\gamma + \sum_{j=1}^{n}|a_{i_0j}|l_j\xi_j\gamma + e^{\varepsilon\tau}\sum_{j=1}^{n}|b_{i_0j}|l_j\xi_j\gamma\Big]$$

$$< 0,$$

this is a contradiction. Hence inequality (5) holds. That is

$$|u_i(t)| \leq \xi_i\gamma e^{-\varepsilon t}, \quad t \geq 0, \quad i = 1, 2, \cdots, n. \tag{7}$$

So,

$$\|z(t) - \tilde{z}\| \leq M\|\phi - u^*\|e^{-\varepsilon t}$$

for $t \geq 0$, where $M = \sqrt{\sum_{i=1}^{n}\Big(\frac{(1+\delta)\xi_i}{\min\limits_{1\leq i\leq n}\{\xi_i\}}\Big)^2} \geq 1$. This means that the equilibrium

point of model (1) is globally exponentially stable. The proof is completed.

4   Example

We consider consider a two-neuron CVNN (1), where $c_1 = 3$, $a_{11} = b_{11} = -2 + i$, $a_{12} = b_{12} = 2 - i$, $J_1 = 0$, $c_2 = 4$, $a_{21} = b_{21} = -1 + 2i$, $a_{22} = b_{22} = -1 - -2i$, $J_2 = 0$, $\tau_{11}(t) = \tau_{12}(t) = \tau_{21}(t) = \tau_{22}(t) = |\sin(3t)|$, $a_1(z_1(t)) = a_2(z_2(t)) = 1$, $f_j(z_j) = \frac{1}{5}(|x_j| + i|y_j|)$, $z_j = x_j + iy_j$, $j = 1, 2$.

It is easy to check that $(0, 0)^T$ is an equilibrium point of this neural network, and Assumptions 1 and 2 are satisfied with $F = \frac{1}{5}\begin{pmatrix} 1 & 0 \\ 0 & 1 \end{pmatrix}$. It is also easy to compute that

$$W = C - (|A| + |B|)F = \begin{pmatrix} 3 - \frac{2}{\sqrt{5}} & -\frac{2}{\sqrt{5}} \\ -\frac{2}{\sqrt{5}} & 4 - \frac{2}{\sqrt{5}} \end{pmatrix}$$

is an M-matrix. From Theorem 1, we know that this model has a unique equilibrium point which is globally exponentially stable. Figure 1 shows the real parts and imaginary parts of time responses of this model with initial value $z_1(s) = 0.3\cos(0.6s) + 2\sin(0.9s)i$, $z_2(s) = -0.4\cos(6s) + 7\sin(0.7s)i$ for $s \in [-2, 0]$. It confirms that the equilibrium point of this model is globally exponentially stable.

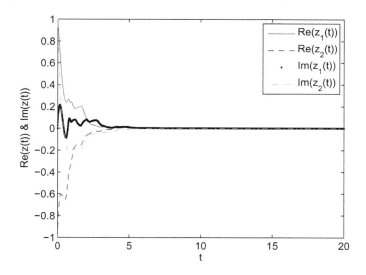

Fig. 1. Real part of state trajectories. (Color figure online)

5   Conclusions

In this paper, the complex-valued Cohen-Grossberg neural networks model with time-varying delays has been investigated. By employing the idea of vector Lyapunov function, M-matrix theory and inequality technique, a new sufficient condition has been obtained to ensure the existence, uniqueness and global exponential stability of equilibrium point for the considered neural networks. The provided result generalizes a few previous known ones. An example with simulations has been provided to show the effectiveness of the obtained results.

Acknowledgments. This work was supported by the National Natural Science Foundation of China under Grants 61273021, 61473332 and 61403051.

References

1. Cohen, M., Grossberg, S.: Absolute stability of global pattern formation and parallel memory storage by competitive neural networks. IEEE Trans. Syst. Man Cybern. **13**, 815–826 (1983)
2. Arik, S., Orman, Z.: Global stability analysis of Cohen-Grossberg neural networks with time varying delays. Phys. Lett. A **341**, 410–421 (2005)
3. Cao, J., Song, Q.: Stability in Cohen-Grossberg type BAM neural networks with time-varying delays. Nonlinearity **19**, 1601–1617 (2006)
4. Song, Q., Cao, J.: Impulsive effects on stability of fuzzy Cohen-Grossberg neural networks with time-varying delays. IEEE Trans. Syst. Man Cybern. **37**, 733–741 (2007)

5. Hirose, A.: Dynamics of fully complex-valued neural networks. Electron. Lett. **28**, 1492–1494 (1992)
6. Lee, D.: Relaxation of the stability condition of the complex-valued neural networks. IEEE Trans. Neural Netw. **12**, 1260–1262 (2001)
7. Zhou, B., Song, Q.: Boundedness and complete stability of complex-valued neural networks with time delay. IEEE Trans. Neural Netw. Learn. Syst. **24**, 1227–1238 (2013)
8. Zhang, Z., Lin, C., Chen, B.: Global stability criterion for delayed complex-valued recurrent neural networks. IEEE Trans. Neural Netw. Learn. Syst. **25**, 1704–1708 (2014)
9. Chen, X., Song, Q.: Global stability of complex-valued neural networks with both leakage time delay and discrete time delay on time scales. Neurocomputing **121**, 254–264 (2013)
10. Song, Q., Zhao, Z., Liu, Y.: Stability analysis of complex-valued neural networks with probabilistic time-varying delays. Neurocomputing **159**, 96–104 (2015)
11. Song, Q., Zhao, Z.: Stability criterion of complex-valued neural networks with both leakage delay and time-varying delays on time scales. Neurocomputing **171**, 179–184 (2016)
12. Zhang, Z., Yu, S.: Global asymptotic stability for a class of complex-valued cohen-grossberg neural networks with time delays. Neurocomputing **171**, 1158–1166 (2016)

Space-Time Structures of Recurrent Neural Networks with Controlled Synapses

Vasiliy Osipov[✉]

St. Petersburg Institute for Informatics
and Automation of the Russian Academy of Sciences,
39, 14 Liniya, St. Petersburg 199178, Russia
osipov_vasiliy@mail.ru

Abstract. A model of recurrent neural network with controlled synapses is considered. Issues of improvement of space-time structures of neural networks are discussed. Possible impacts on associative interactions between signals in recurrent neural networks are studied by means changing their spatial parameters depending on the current states of network layers. A model of the recurrent neural network with deep associative and spatial signal processing in real-time is proposed. Aspects that ensure sustainability of such neural networks are discussed. Simulation results of associative and spatial signal processing in recurrent neural networks with controlled synapses are given. Recommendations on development of advanced neural networks are formulated.

Keywords: Neural network structure · Synapses · Signals · Associative and spatial processing

1 Introduction

Search for the ways of improving artificial recurrent neural networks (RNN) is of great scientific and practical interest. Based on RNN, scientists expect to create thinking and learning machines. The main feature of RNN is its associative memorizing and extraction of signals from memory. However, associatively itself is not enough to endow these networks with thinking. None of them is able to solve a wide range of creative tasks in real time. Capabilities of the known artificial RNNs do not go beyond pattern recognition and classification [1, 2]. The depth of information processing in RNN remains fairly low. There exists a suppression of reverse recognition results by input signals; as well as high redundancy of information storing; certain difficulties at associative retrieval of previously stored information. Moreover, problems persist in ensuring the RNN sustainability. It's not possible to exercise control over the return of information processing results to effectors. The above issues are caused, to a certain extent, by an inadequate structure and methods of associative and spatial signal processing in RNNs.

© Springer International Publishing Switzerland 2016
L. Cheng et al. (Eds.): ISNN 2016, LNCS 9719, pp. 177–184, 2016.
DOI: 10.1007/978-3-319-40663-3_21

Two types of artificial RNN could be singled out based on the degree of real-time requirements satisfaction. The first type is formed by real-time RNNs and incorporates, first of all, neural networks based on perceptrons [3, 4]. Such RNNs can process signals in real time, but not deeply enough. The second type includes RNNs intended for deep and detailed information processing. These RNNs include associative memory devices [3–6] and self-organizing networks [3, 4, 7]. Actually, for solving the existing problems RNNs need to combine properties of these two types of networks.

Classic recurrent neural networks are characterized by purely associative information processing. Spatial processing in these RNNs is barely manifested. For a long time spatial processing was given a consideration exclusively in radial neural networks of direct distribution [3, 4, 8].

Lately, associative-spatial processing of signals in RNN received significantly increased attention. New results [9–15] extended RNN capabilities for intelligent information processing. In [13] it was proposed to carry out spatial shifts of signals in RNNs when transferring them from one layer to another, and thus to form appropriate logical network structures. In [14, 15] it was recommended to control associative interaction between signals depending on the current states of neurons in the interacting layers. Nevertheless, a number of aspects related to deep associative and spatial signal processing in real time is still somewhat neglected. It is necessary to solve problems of finding the effective space-time structures of RNNs.

The paper analyzes potential space and time structures of recurrent neural networks with controlled synapses. An approach to control signals' associative interactions in such RNNs was developed. The structure of a recurrent neural network with deep associative and spatial signal processing in real time is disclosed and requirements for other RNN prospective structures are formulated.

2 Model of Recurrent Neural Network with Controlled Synapses

Let us consider the structure of a recurrent neural network with controlled synapses. This structure at the level of separate neurons and synapses is shown in Fig. 1, where 1.1,…,1.n, 2.1,…,2.n are neurons of the first and the second layers, respectively; and n is a number of neurons in every layer; SD is a single delay. Synapses are shown as grey ovals. In the general case, each neuron of one layer is related to all the neurons of another layer by means of its synapses. The farther neurons are from each other, the greater is attenuation at synapses. Neurons from the same layer are not connected with each other. Three states are specific for neurons of this network: standby, excitation and refractoriness. Any neuron transfers to the excitation state as soon as the total input potential in the standby state exceeds the excitation threshold. As a result, the neuron generates a single pulse. Excitation is followed by refractoriness, and each neuron stays in this state for a period greater than the delay time of single pulses in the created network circuits.

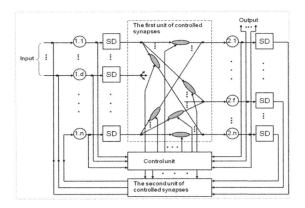

Fig. 1. Recurrent neural network with controlled synapses.

Such network is provided with decomposed signals (e.g., acoustic and/or optic ones) in the input layer consistent basis. Whereas every component, before entering the network, is converted into a sequence of single pulses (SP) at the recurrence rate as a preliminary set function of its amplitude. Then successive SP populations (SPPs) containing all data about input signals are sent to the network input. Information about frequency and spatial components of the signal is assigned to the numbers of created sequences of single pulses. Phase characteristics are undoubtedly related to delays of single pulses. One-to-one correspondence between components of input signals and components of output signals is ensured due to the priority of short connections between neurons over long connections.

When the current population of single pulses enters the first layer of an RNN with regard to recognition reverse results, after excitation of relevant neurons, a new population is created at the layer's output. This new population contains data about input signals, as well as stored signals and signals circulating in the network. Then each single pulse from the received population enters the matching population of synapses after a delay. In the general case, synapses connect each excited neuron with all the rest neurons of the second layer. Divergent SPs are created due to such branching; the received SPs are stored at synapses, and data about previous impacts are read off. Depending on the control actions arriving from the control unit, spatial shifts of the SP populations are realized along the layers. Similarly, the SP populations are processed while being transferred from the second layer to the first layer. Results of signal processing in an RNN are expressed as successive populations of single pulses taken from the output layer and converted into signals similar to the input ones.

Depending on the implemented signal spatial transformations, the neural network could be endowed with a variety of space-time structures with their own capabilities for associative and spatial information processing. These capabilities still remain unstudied to a considerable degree.

3   Linear and Spiral Structure of RNN

Different linear and spiral space-time structures of RNNs with controlled synapses have been investigated; the examined objects include neural networks with up to nine thousand neurons (N) in each layer and up to several million synapses. Characteristics (L × W/C × R) of investigated RNNs are shown in Table 1, where L, W are the length and the width of the layer, respectively; C, R are the number of columns and rows in the layer, respectively. Examples of two types of spiral structures are shown in Figs. 2 and 3.

Table 1. Characteristics of the investigated RNN structures.

Name of structure	The number (N) of neurons in each layer										
	360	540	720	900	1080	1260	1440	1620	1800	2016	9000
«Linear»	60×6 / 10×1	90×6 / 15×1	12×0×6 / 20×1	15×0×6 / 25×1	18×0×6 / 30×1	21×0×6 / 35×1	24×0×6 / 40×1	27×0×6 / 45×1	30×0×6 / 50×1	33×6×6 / 40×1	1500×6 / 25×0×1
«Spiral-12»	12×30 / 2×5	—	12×60 / 2×10	—	12×90 / 2×15	—	12×120 / 2×20	—	12×150 / 2×25	—	—
«Spiral-30»	30×12 / 5×2	30×18 / 5×3	30×24 / 5×4	30×30 / 5×5	30×36 / 5×6	30×42 / 5×7	30×48 / 5×8	30×54 / 5×9	30×60 / 5×10	—	—
«Spiral-60»	60×6 / 10×1	—	60×12 / 10×2	—	60×18 / 10×3	—	60×24 / 10×4	—	60×30 / 10×5	—	—
«Spiral-84»	—	—	—	—	—	—	—	—	—	96×21 / 14×4	—
«Spiral-120»	—	—	12×0×6 / 20×1	—	—	—	120×12 / 20×2	—	—	—	—

Linear structures represent a particular case of spiral schemes, when they represent a half of spiral turn. Capabilities of RNNs with these structures were estimated based on two criteria: the average time T(N) of active interaction between single pulses in the plane of the network layers and the increase of the total synapses' weight ΔW(t) at each step (t) of the network with respect to the previous step. It was proved that the spiral structure with a spiral variable diameter (Fig. 3) has essential advantages over linear structures by the first criterion.

The spiral structure due to signal propagation in opposite directions allows for their strong associative interaction. In case of a linear structure, same space-time signal relationships are stored repeatedly in different synapses. This significantly limits the possibilities of synaptic memory due to excessive storing of the same signals. So, the following conclusion could be drawn: linear structures are good for signal transmission, rather than for their storing at processing in real time. Thereby, in some cases connections inside spiral beams could be neglected; however, this is not applicable to connections between the spiral turns. The primary role at storing information in such RNNs belongs to synaptic connections between spiral turns. Thus, the number of synaptic connections in artificial RNNs could be significantly reduced. RNN single- and double-spiral structures may be estimated by the second criterion: the increase of the total synapse weight at each step of the network functioning with respect to the previous step. It was proved that recurrent neural networks with the layer structure in form of a double spiral have lower redundancy as well as lower delay at the start of signal pronounced associative interaction compared with single spiral structures. Therefore, they have higher preference than RNNs with a simple spiral structure of layers. Recurrent neural networks with such space-time structures can be regarded as single signal systems. Despite these advantages, their ability to perform deep information processing in real time is significantly limited. In such RNNs reciprocal results of recognition are somewhat suppressed by input signals.

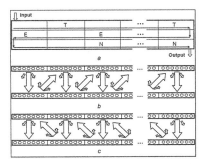

Fig. 2. The spiral structure of a recurrent neural network: a – top view on the first layer; b, c-cross-sections of two layers along the first and second lines.

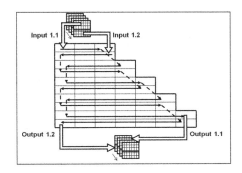

Fig. 3. Example of the longitudinal structure of a recurrent neural network in form of a double spiral.

4   Recurrent Neural Network with Three Signaling Systems

A new space-time structure of recurrent neural network is proposed, that is devoid of the disadvantages noted in Sect. 3. Under certain conditions a base RNN with controlled synapses can be endowed with three signaling systems. The above could be achieved by adding to RNN the ability to control spatial parameters of divergent beams of single pulses being transmitted from layer to layer. Such control is feasible depending on the current state of the interacting neurons. Examples of smooth cross-sections of such divergent beams can be different, and include one- and multi-lobes cross-sections. In particular, the cross-sections may be elliptical. So, different signaling systems can be possibly linked through control over forming copies of processed signals. An example of an RNN structure with three signaling systems is shown in Fig. 4. If the RNN structure were considered at the level of neural network channels of forwarding single pulses sets along the layers, then it could be of a view represented in Fig. 5.

   In accordance with this structure, the RNN first signaling system may be responsible for forming conditional reflex connections and reflexes based on the results of the stimuli impact on the receptors. The second signaling system is able to process information independently of direct perception of the reality. Presence of the second signaling system can significantly increase the depth of information processing an RNN. Also, by controlling spatial parameters of divergent beams of single pulses in the

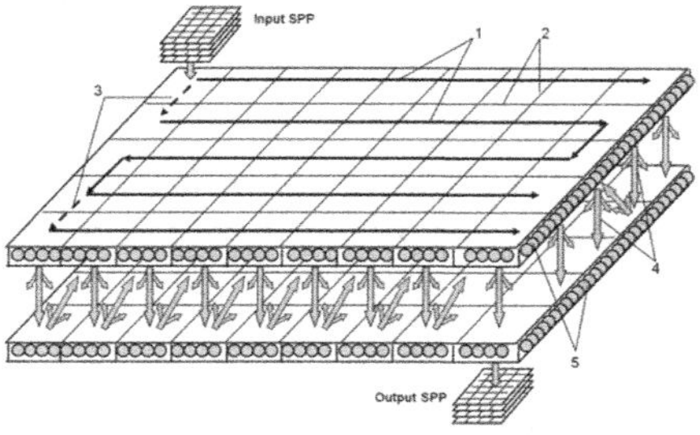

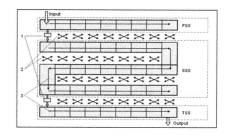

Fig. 4. Example of an RNN three signaling systems: 1, 3, 4-directions of forwarding these SPPs along the layers and between them; 2-lines, breaking down the layers into logical fields due to SPP shifts along the layers; 5-neurons.

Fig. 5. The structure of a recurrent neural network with three signaling systems at the level of neural network channels: FSS, SSS, TSS-the first, second, and third signaling systems; 1-auxiliary channels; 2-main associative interactions; 3-directions of forwarding these SSPs along the layers.

first, second and third signaling systems, it becomes feasible to process signal cycles and switch over one task to another. Presence of the third signaling system an RNN allows the second system to "comprehend" them, and then give out results to effectors. The third signaling system may also have a number of functions to form a stable signal chain to implement common actions through effectors. In modeling RNN with three signaling systems a possibility of these systems controlled binding can be confirmed. If, for instance, at the initial step the first and second signaling systems receive the same additional signals (secondary noise) for a long time, then unambiguous associative connections would be set between these systems. After that, the first signaling system can form signal copies to be sent to the second signaling system. Besides, unambiguous associations between the second and third signaling systems could be set. In principle, this compatibility between signaling systems can be implemented at the phase of their construction. Modeling also confirmed the ability of RNN to carry out stable deep processing of signals from the input flow (Fig. 6). The above is ensured by cyclic associative call signals from the memory of the second signaling system. Cyclic call of information from the RNN associative memory is feasible at the orientation of diverging beams single pulses to neurons capable of generating the following matching signals. This RNN is furnished with a flexible associative and spatial addressable memory. RNN stable functioning is feasible due to eliminating the overloads of the memory different types, like, memory for synapses and neurons themselves.

Elimination of the synaptic memory overload is possible due to partial erasing of outdated information. To prevent the RNN hyper excitability it is possible to control the threshold of neuron excitation. The number of simultaneously excited neurons in RNN should not exceed a half of the network neurons.

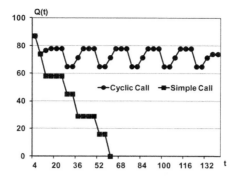

Fig. 6. The number Q(t) of excited neurons in SSS depending on the time (t) under cyclic and simple call signals from the memory.

5 Conclusions

The reported study of different space-time structures of recurrent neural networks with controlled synapses demonstrated high dependency between the network capacity and implemented methods of associative and spatial signal processing. Additional spatial processing of signals in RNN significantly increases their state space and adds new properties. The new possibilities are: control of associative interaction of signals in RNN; exclusion of recognition reciprocal result suppression by the input signal flow; switching from one task to another; solving of various creative tasks by the same neural network. Significant reduction in synaptic memory redundancy and, consequently, extension of the associative memory are attainable. Associative cyclic call of information from the RNN memory is facilitated. It is worth noting that due to the priority of short connections of the long ones, such RNNs have clear logical structures. Implementation of the studied RNNs is possible in a number of ways. To create a small RNN (tens and hundreds of thousands of neurons in a layer) conventional computing technologies could be applied. Implementation of large and extra large RNN is also feasible through creation of analog chips based on controlled memristor matrices. The results of the study can be applied to development of advanced recurrent neural networks for cognitive processing of diverse information.

References

1. Haikonen, P.O.A.: The role of associative processing in cognitive computing. Cogn. Comput. **1**(1), 42–49 (2009)
2. Palm, G.: Neural associative memories and sparse coding. Neural Netw. **37**(1), 165–171 (2013)
3. Haykin, S.: Neural Networks and Learning Machines, 3rd edn. Prentice Hall, New York (2008)

4. Galushkin, A.I.: Neural Networks Theory. Springer Science & Business Media, Berlin (2007)
5. Hopfield, J.J.: Neural networks and physicals systems with emergent collective computational abilities. Proc. Nat. Acad. Sci. USA **79**(8), 2554–2558 (1982)
6. Kosko, B.: Bidirectional associative memories. IEEE Trans. Syst. Man Cybern. **18**(1), 49–60 (1988)
7. Kohonen, T.: Essentials of the self-organizing map. Neural Netw. **37**, 52–65 (2013)
8. Cover, T.M.: Geometrical and statistical properties of systems of linear inequalities with applications in pattern recognition. IEEE Trans. Electron. Comput. **3**, 326–334 (1965)
9. Madl, T., Chen, K., Montaldi, D., Trappl, R.: Computational cognitive models of spatial memory in navigation space: a review. Neural Netw. **65**, 18–43 (2015)
10. Chrtier, S., Giguere, G., Langlois, D.: A new bidirectional heteroassociative memory encompassing correlational. Competitive Topological Prop. Neural Netw. **22**(5), 568–578 (2009)
11. Jeong, S., Lee, M.: Adaptive object recognition model using incremental feature representation and hierarchical classification. Neural Netw. **25**, 130–140 (2012)
12. Montazer, G.A., Giveki, D.: An improved radial basis function neural network for object image retrieval. Neurocomputing **168**, 221–233 (2015)
13. Osipov, V., Osipova, M.: Method and device of intellectual processing of information in neural network. RU Patent No. 2413304 (2011)
14. Osipov, V.: Method for intelligent information processing in neural network. RU Patent No. 2427914 (2011), RU Patent No. 2502133 (2013)
15. Osipov, V.: Associative and spatial addressing to memory of recurrent neural networks. Informacionnye Technologii **21**(8), 631–637 (2015)

A Practical Simulator of Associative Intellectual Machine

Sergey Baranov[(⊠)]

SPIIRAS, ITMO University, St. Petersburg, Russia
SNBaranov@iias.spb.su

Abstract. A software program AIM simulating the behavior of an associative intellectual machine (AIM) in various configurations and paths of information transmission among neuron layers is described. The aim is to study the impact of various AIM parameters on the effectiveness and efficiency of information processing in various configurations and modifications of AIMs. The program is written in the interpretative programming language Forth and provides a wide range of options for studying the AIM behavior. It has an open user interface which allows to easily extend or modify the existing toolkit and to conjugate it with other tools for data analysis and visualization.

Keywords: Simulation · Forth · UI · Software engineering · Visualization

1 Introduction

Associative Intellectual Machines (AIM) allow to substantially extend the capabilities of automated means of information processing [1] due to their architecture based on recurrent neural networks (RNN) [2]. This gives rise to a great variety of strategies for transmitting information among the network layers and ways of information transformation and processing. After a certain boom in instrument creation a number of software tools currently exist [3, 4] which help the researchers to study various aspects of RNN-based solutions, Matlab [5] probably being among mostly widespread ones. However, most of these tools look like "dinosaurs" – they are huge and inflexible for running sophisticated experiments with carefully carved parameters and features.

An alternative approach based on the "small is beautiful" paradigm [6] was successfully used to overcome some of these hurdles, tools based on the Python language [7, 8] being quite successful and thus encouraging to try other options.

The AIM program described in this paper was written in Forth of the ANS Forth 200× standard [9] to run on top of the VFX Forth for Windows IA32 by MPE Ltd. (UK) [10] under MS Windows. A freeware option for the platform is gForth [11]. AIM allows to specify various cases and combinations of "experiment parameters", to simulate a particular AIM behavior, and to visualize the obtained results as well as the very process of their development and to objectively estimate and compare their characteristics. Forth was selected as the implementation language due to the flexibility it provides for implementing programming solutions and its interoperability with other software tools. It allows to use only fixed-point arithmetic in calculations and thus

© Springer International Publishing Switzerland 2016
L. Cheng et al. (Eds.): ISNN 2016, LNCS 9719, pp. 185–195, 2016.
DOI: 10.1007/978-3-319-40663-3_22

avoids floating point with related issues and trade-offs. The program size is around 2 KLOC in Forth. The simulator employs a simple model of a multi-layer RNN and allows to easily modify the employed user interface and to use other advanced tools for further analysis and visualization of simulation results.

2   Internal Representation of an RNN

The core of the AIM program consists of the module which simulates the work of a multi-layer RNN of a specified structure under a particular scenario of input signals incoming. The program maintains a discrete counter of system time, and all simulated events are attached to the system time axis with a respective time-stamp.

Each neuron v of the network may be in one of two states: *excited* or *unexcited*, and is characterized by the value of its potential U_v: $0 \le U_v \le U_{max}$. The neurons are set in K *layers* of the same size. Each layer forms a matrix of $M \times N$ *fields*, each field being itself a matrix of $m \times n$ neurons, $1 \le K,M,N,m,n \le 32$. Thus, the total number of neurons in an RNN equals to $K \times M \times N \times m \times n$.

Let's consider a two-layer ($K = 2$) RNN with layers L_0 (the upper layer) and L_1 (the lower layer) numbered by 0 and 1. Fields within a layer number l ($0 \le l \le K-1$) are numbered as elements of the matrix $F_l[i,j]$, $0 \le i \le M-1$, $0 \le j \le N-1$ (in Forth it's more convenient to start any numbering with 0 rather than 1); neurons inside each field are numbered as elements of the matrix $Neu[i,j]$, $0 \le i \le m-1$, $0 \le j \le n-1$. Thus, the full address of a neuron consists of its layer number, its field indices in this layer, and the neuron indices in this field. Fields reside in their layer matrix elements (see Fig. 1).

$F_0[0,0]$	$F_0[0,1]$	$F_0[0,2]$	$F_0[0,3]$	$F_0[0,4]$	$F_0[0,5]$	$F_1[0,0]$	$F_1[0,1]$	$F_1[0,2]$	$F_1[0,3]$	$F_1[0,4]$	$F_1[0,5]$
$F_0[1,0]$	$F_0[1,1]$	$F_0[1,2]$	$F_0[1,3]$	$F_0[1,4]$	$F_0[1,5]$	$F_1[1,0]$	$F_1[1,1]$	$F_1[1,2]$	$F_1[1,3]$	$F_1[1,4]$	$F_1[1,5]$
$F_0[2,0]$	$F_0[2,1]$	$F_0[2,2]$	$F_0[2,3]$	$F_0[2,4]$	$F_0[2,5]$	$F_1[2,0]$	$F_1[2,1]$	$F_1[2,2]$	$F_1[2,3]$	$F_1[2,4]$	$F_1[2,5]$
$F_0[3,0]$	$F_0[3,1]$	$F_0[3,2]$	$F_0[3,3]$	$F_0[3,4]$	$F_0[3,5]$	$F_1[3,0]$	$F_1[3,1]$	$F_1[3,2]$	$F_1[3,3]$	$F_1[3,4]$	$F_1[3,5]$
$F_0[4,0]$	$F_0[4,1]$	$F_0[4,2]$	$F_0[4,3]$	$F_0[4,4]$	$F_0[4,5]$	$F_1[4,0]$	$F_1[4,1]$	$F_1[4,2]$	$F_1[4,3]$	$F_1[4,4]$	$F_1[4,5]$
Upper neuron layer						*Lower neuron layer*					

Fig. 1. Representation of a two-layer RNN of 30 fields as two matrices of 5×6 elements

Any two neurons from adjacent layers may be linked via *synapses*, while neurons from the same layer are not linked at all. The synapse $v \rightarrow \eta$ connecting a neuron $v = Neu[x_v,y_v]$ from the field $F = F_k[x_F,y_F]$ of the layer L_k and a neuron $\eta = Neu[x_\eta,y_\eta]$ from a field $F* = F_{k\pm1}[x_{F*}, y_{F*}]$ of an adjacent layer $L_{k\pm1}$ is characterized by its weight $w_{v\eta}$ which may change in the process of RNN functioning, and the distance $d_{v\eta}$ between these neurons as points in the 3D space and in general case is calculated as

$$d_{v\eta}^2 = D_z^2 + D_{xy}^2 \times \left(\left(|x_v-x_\eta| + |x_F-x_{F*}| \times n \right)^2 + \left(|y_v-y_\eta| + |y_F-y_{F*}| \times m \right)^2 \right),$$

$D_z > 0$ being the distance between adjacent layers, and $D_{xy} > 0$ being the distance between two adjacent neurons in a layer; these values are assumed to be the same for all neurons. Only neurons with the distance d_{vn} not exceeding the value D_{max} are linked via synapses, the "scaling factors" D_{max}, D_z, and D_{xy} being specified while configuring the AIM program. A particular case is formed by the so called "neighboring" neurons, located in adjacent fields of signal propagation paths specified by the user. In this case the distance between neurons with w.r.t. axis Z equals to D_z independently of the shift between the fields relatively to each other with respect to axes X and Y.

An RNN signal propagation path is specified through nonrecurring enumeration of RNN fields, each two adjacent elements in this list belonging to adjacent layers (in case of a two-layer RNN this means alternating). The first element of this list is its entry – it may accept signals from the RNN environment in form of an input unitary image (IUI) which is a matrix of $m \times n$ binary values; each bit being mapped to a neuron in the entry field with the same matrix indices. If this neuron is unexcited, then it *accepts* the respective binary signal which may result in a change of the neuron potential while the neuron itself becomes excited for a period of time equal to some Δt_{excite}; during that period the neuron accepts no other signals.

In the time period $\Delta t_{receive} \leq \Delta t_{excite}$ the excited neuron passes the accepted signal to all unexcited neurons it is linked to by synapses, and if the recipients become excited they propagate the accepted signal further in the similar way in the time period $\Delta t_{pass} \leq \Delta t_{excite}$. Thus in a number of steps the accepted signal may reach the last field in the path which is its exit – when the signal reaches its neurons, they excite and form the output unitary image (OUI) in a similar way to IUI, this image being sent to the RNN environment as the output from the RNN in the time period $\Delta t_{send} \leq \Delta t_{excite}$.

Signal propagation may "die" when it doesn't reach the output field in the path for various reasons; thus the signal becomes lost.

The fields of each layer of the RNN in Fig. 2 form a matrix of $M \times N$ ($M = 5$, $N = 6$) with $m \times n$ neurons ($m = 6$, $n = 7$). A signal propagation path is specified by the sequence:

$\{F_0[0,0], F_1[0,0], F_0[0,1], F_1[0,1], F_0[0,2], F_1[0,2], F_0[0,3], F_1[0,3], F_0[0,4], F_1[0,4], F_0[0,5], F_1[0,5],$
$F_0[1,5], F_1[1,5], F_0[1,4], F_1[1,4], F_0[1,3], F_1[1,3], F_0[1,2], F_1[1,2], F_0[1,1], F_1[1,1], F_0[1,0], F_1[1,0],$
$F_0[2,0], F_1[2,0], F_0[2,1], F_1[2,1], F_0[2,2], F_1[2,2], F_0[2,3], F_1[2,3], F_0[2,4], F_1[2,4], F_0[2,5], F_1[2,5],$
$F_0[3,5], F_1[3,5], F_0[3,4], F_1[3,4], F_0[3,3], F_1[3,3], F_0[3,2], F_1[3,2], F_0[3,1], F_1[3,1], F_0[3,0], F_1[3,0],$
$F_0[4,0], F_1[4,0], F_0[4,1], F_1[4,1], F_0[4,2], F_1[4,2], F_0[4,3], F_1[4,3], F_0[4,4], F_1[4,4], F_0[4,5], F_1[4,5]\},$

its projection on the upper layer is marked with a dotted line. The path input field $F_0[0,0]$ resides in the upper left corner of the layer L_0, and its exit field $F_1[4,5]$ is in the bottom right corner of the lower layer L_1. This propagation path tours through all fields of the RNN, alternating from the upper layer to the lower layer and vice versa. Several such paths may be specified, each two of them having no fields in common.

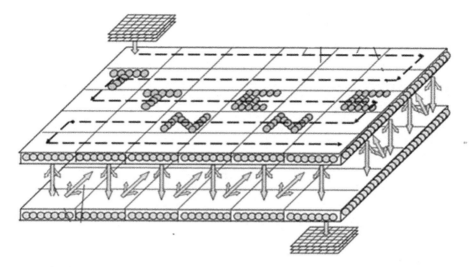

Fig. 2. A two-layer RNN of 5 × 6 = 30 fields with 6 × 7 = 42 neurons each

3   Event Types and Signal Propagation

When simulation starts, all neurons are unexcited, their potentials are equal to U_0, and the weights of all synapses are all equal to some initial value w_0. Signals propagate through RNN in accordance with the event list which is modified as a result of event processing through eliminating the processed events and adding new events created at this processing. Event processing consists of the following actions performed as reactions to events with the time stamp t (the current system time).

1. Receive – receive/accept the next IUI. In accordance with the specified scenario of incoming IUIs, at the moment t of system time an IUI appears at the entry field of the specified propagation path in form of a binary matrix of $n \times m$ bits.

   If this is not the last IUI in the given scenario, then a new event is added to the event list – receive the next IUI in its time $t' \geq t$. The neurons of the entry field which are unexcited at the moment t form the set of "entry" neurons. They all become excited and a new event is added to the event list for the time moment $t + \Delta t_{excite}$ – all neurons of this field become unexcited. The accepted IUI changes the potential $U(v)$ of each entry neuron v to U_0 or U_1, depending on what particular bit (0 or 1) resides in the IUI matrix position of this neuron. A new event is added for the time moment $t + \Delta t_{receive}$ – pass the received signal from the entry field neurons to all other unexcited neurons linked to them via synapses.

2. Pass – propagate accepted signals from all excited neurons of the given field F_k to all neurons of the field $F_{k\pm1}$ from the adjacent layer linked to them via synapses; the recipients simultaneously become excited, their potentials and the weights of the connecting synapses through which signals pass to them change accordingly.

The sets $Neu_{excite} \subseteq F_k$ and $Neu'_{excite} \subseteq F_{k\pm1}$ of excited neurons of the fields F_k and $F_{k\pm1}$ are considered. For each neuron $v \in Neu_{excite}$ with the potential U_v it is determined whether the signal may be passed from it to each neuron $\eta \in (F_{k\pm1} \setminus Neu'_{excite})$ with the potential U_η through the connecting synapse $v \to \eta$ with the weight $w_{v\eta}$. This propagation occurs only if $|U_v - U_\eta| \times w_{v\eta} \geq U_{min}$ for some predefined threshold value U_{min}. If this condition is satisfied by no neuron from Neu_{excite}, then no signal propagation takes place (propagation dies). Otherwise, if this condition is satisfied for some neuron set $\tilde{N} \subseteq Neu_{excite}$, then for each $v \in \tilde{N}$ a non-empty set $A(v) \subseteq (F_{k\pm1} \setminus \tilde{N})$ of neurons η from the field $F_{k\pm1}$ which receive the signal from the neuron $v \in \tilde{N}$ is defined. Each such neuron η in its turn becomes excited, and a new event is added to the event list for the time moment $t + \Delta t_{excite}$ – all excited neurons of the field $F_{k\pm1}$ become unexcited. For each neuron $\eta \in A(v)$ the set $S(\eta)$ of neurons connected with v by incoming synapses from neurons of the set $\tilde{N}$ is determined and a new potential U'_η of η is computed as the sum of the charges (taking into account the sign of the potential difference) transmitted to the neuron η from all neurons σ form the set $S(\eta)$ through the connecting synapses $\sigma \to \eta$ as $U'_\eta = \sum_{\{\sigma \in S(\eta)\}} (U_\sigma - U_\eta) \times w_{\sigma\eta}$. Depending on its value, U'_η is converted into U_0 or U_1, and the weight $w_{\sigma\eta}$ of each acting synapse changes at the value of the current which passed through it, taking into account the sign: $w'_{\sigma\eta} = w_{\sigma\eta} + \left(U_\sigma - U'_\eta\right) \times w_{\sigma\eta}$.

A new event is added to the event list: if $F_{k\pm1}$ is not the exit field of some path, then pass the signal to other neurons connected with the given one via synapses at the time moment $t + \Delta t_{pass}$; otherwise, form an OUI from potentials of the neurons of this field and send it as output to the environment at the time moment $t + \Delta t_{send}$.

3. Unexcite – change the state of excited neurons of the given field to "unexcited" and make their potentials U_0. No new events are added to the event list.
4. Send – send an OUI to the external environment. The potential U_η of each excited neuron of the given exit field is converted into a binary value $I_\eta \in \{0, 1\}$ depending on its value: $I_\eta = $ if $U_\eta > U_{min}$ then 1 else 0 fi; an OUI is formed from these bits as a binary matrix of $m \times n$ elements matching the neuron indices and this OUI is sent to the external environment. No new events are added to the event list.

When an unexcited neuron receives a binary signal either from an IUI, or from another neuron from the adjacent layer linked to the given neuron through a synapse, this neuron becomes excited in its turn for a period of time Δt_{excite}, specified while configuring the AIM program.

4 The Simulation Process

The overall workflow of the AIM program is presented in Fig. 3. The process starts with configuring the RNN structure, defining a scenario for incoming IUIs and other necessary parameters; after that simulation of the RNN behavior with the specified AIM parameters begins.

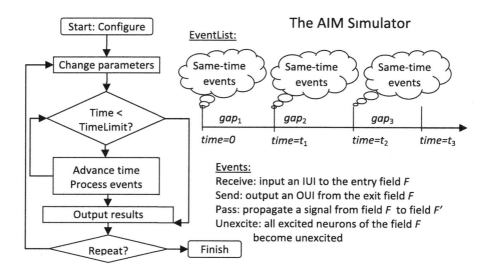

Fig. 3. The overall structure of the AIM simulator

The outer loop starts every time with a new set of AIM parameters (they are changed through assigning new values to the respective variables). The inner loop executes with the specified AIM parameters until the pre-set time limit is exhausted or the scenario of incoming IUIs terminates.

In case of incorrect specifications of AIM parameters or the IUI scenario, a respective error message is sent to the output stream and the work terminates abnormally.

The simulation process is controlled by a list of events EventList ordered w.r.t. their timestamps: 0, t_1, t_2, t_3,... and assembled into same-time event groups. The main simulation loop consists in advancing the system time counter to the nearest time stamp of the events in this list and processing this event which may produce new events in this list with the same or later time stamp.

The main loop reiterates until it becomes exhausted or the overall time limit for system time is reached.

5 System Start-up and Configuring

The AIM constraints are specified as values of particular variables like Max#Events (the maximal number of scheduled future events at any moment of the system time) or MaxFileName (the maximal length of the file name for the IUIs or OUIs). Their change assumes recompiling of the source code of the AIM program. In case of violating any of these constraints AIM terminates its work with a respective error message. After a successful compilation of the source code in the selected Forth environment, an invitation "ok" appears for setting the AIM parameters.

AIM parameter setting is performed through the operator Include parameterized with the name of a file which is a regular Forth text with assignments to certain variables and specification of the RNN structure and the scenario for IUI input. E.g., the following input is provided for a file with the name "c:\mpe\SanityAIM_2.txt": Include c:\mpe \SanityAIM_2.txt for which the AIM simulator responds with the message "Including c: \mpe\SanityAIM_2.txt ok" in case of a successful opening this file.

Figure 4 provides an example of AIM configuring and parameter setting. This Forth text assigns values to respective variables with the assign operator ("!"), creates an RNN of the specified size with the operator CreateRNN, defines signal propagation paths between the RNN layers, and specifies the names of files with incoming and outgoing unitary images though the operators InputFile" and OutputFile". The assign operator has two parameters at its left side: the value to be assigned and the name of the variable which this value is assigned to. The text to the right of a backslash ("\") up to the end of the line is a comment. Spaces, tabulations, and end-of-lines serve as operator delimiters.

```
8 Texcite !  \ Time interval for an excited neuron to become unexcited
5 Tpass !  \ Time interval for passing the signal to other neurons linked through synapses
3 Umin ! \ Minimal potential difference for a signal to pass between two neurons
200 D2max ! \ The squared maximal distance between two neurons for a synapse to exist
2 5 6 6 7 CreateRNN \ Create a two layer RNN of 5x6 fields, each field of 6x7 neurons
0 <Path> \ Specify a single propagation path with number 0
<F 0 0 0 F> <F 1 0 0 F> <F 0 0 1 F> <F 1 0 1 F> <F 0 0 2 F> <F 1 0 2 F> <F 0 0 3 F> <F 1 0 3 F>
<F 0 0 4 F> <F 1 0 4 F> <F 0 0 5 F> <F 1 0 5 F> <F 0 1 5 F> <F 1 1 5 F> <F 0 1 4 F> <F 1 1 4 F>
<F 0 1 3 F> <F 1 1 3 F> <F 0 0 2 F> <F 1 1 2 F> <F 0 1 1 F> <F 1 1 1 F> <F 0 1 0 F> <F 1 1 0 F>
<F 0 2 0 F> <F 1 2 0 F> <F 0 2 1 F> <F 1 2 1 F> <F 0 2 2 F> <F 1 2 2 F> <F 0 2 3 F> <F 1 2 3 F>
<F 0 2 4 F> <F 1 2 4 F> <F 0 2 5 F> <F 1 2 5 F> <F 0 3 5 F> <F 1 3 5 F> <F 0 3 4 F> <F 1 3 4 F>
<F 0 3 3 F> <F 1 3 3 F> <F 0 3 2 F> <F 1 3 2 F> <F 0 3 1 F> <F 1 3 1 F> <F 0 3 0 F> <F 1 3 0 F>
<F 0 4 0 F> <F 1 4 0 F> <F 0 4 1 F> <F 1 4 1 F> <F 0 4 2 F> <F 1 4 2 F> <F 0 4 3 F> <F 1 4 3 F>
<F 0 4 4 F> <F 1 4 4 F> <F 0 4 5 F> <F 1 4 5 F> </Path>
InputFile" c:\mpe\AIM_02.txt" \ The name of a file with incoming IUIs
OutputFile" c:\mpe\AIM_02_.txt" \ The name of a file with outgoing OUIs
```

Fig. 4. Example of a set file c:\mpe\SanityAIM_2.txt

A signal propagation path (there may be several of them) is specified in an XML-like technique, which turned out to be quite convenient for this purpose. The pair of Forth words <Path> and </Path> frame the whole path, while the pair <F and F> frame the coordinates of each field in this path in the respective order. Field coordinates are specified as triples of numbers: the layer number (from 0 to $K-1$), the row number, (from 0 to $M-1$), and the column number (from 0 to $N-1$) for the field as an element of a matrix of $M \times N$ fields.

Incoming IUIs are specified in a similar way in a file, whose name stands to the right of the operator InputFile" (see Fig. 5):

The word pair <Images> and </Images> frames the whole series of IUIs, while another pair <I and I> frames each IUI inside this series. The word <I is preceded by two numbers – the moment of the system time, when this image comes to the RNN and

```
<Images>
1     0  <I 0001100 0010110 0100011 1111111 1000011 1000011 I>
72    0  <I 1111000 1100100 1111100 1100110 1100011 1111110 I>
148   0  <I 1111111 1001100 0001100 0001100 0001100 0001100 I>
232   0  <I 0111110 1100011 1100011 1100011 1100011 0111110 I>
321   0  <I 1100011 0010110 1001011 1001011 1000011 1000011 I>
414   0  <I 0001100 0010110 0100011 1111111 1000011 1000011 I>
515   0  <I 1111111 1001100 0001100 0001100 0001100 0001100 I>
625   0  <I 1111110 1100011 1100011 1111110 1100000 1100000 I>
745   0  <I 1100011 1100011 1100011 1111111 1100011 1100011 I>
868   0  <I 0111110 1100011 1100000 1100000 1100011 0111110 I>
</Images>
```

Fig. 5. An example of a file with incoming IUIs

the number of a propagation path which the signal is expected to propagate through. Binary representations of the m rows of each unitary image:

$$\begin{bmatrix} 0001100 \\ 0010110 \\ 0100011 \\ 1111111 \\ 1000011 \\ 1000011 \end{bmatrix}, \begin{bmatrix} 1111000 \\ 1100100 \\ 1111100 \\ 1100110 \\ 1100011 \\ 1111110 \end{bmatrix}, \begin{bmatrix} 1111111 \\ 1001100 \\ 0001100 \\ 0001100 \\ 0001100 \\ 0001100 \end{bmatrix}, \ldots, \begin{bmatrix} 0111110 \\ 1100011 \\ 1100000 \\ 1100000 \\ 1100011 \\ 0111110 \end{bmatrix}$$

reside between the words <I and I> (leading zeros may be omitted). The specified ten IUIs arrive to the entry field $F_0[0,0]$ of the only path of this example with the number 0 which starts in the upper left corner of the upper layer L_0 in successive moments of the system time 1, 72, 148, 232, 321, 414, 515, 625, 745 and 868 in the specified order. These timings should form a non-decreasing series. Operator Simulate starts running a simulation session with the specified parameters and constraints.

6   Experimental Results, Their Processing and Presentation

Upon starting simulation with the operator Simulate as each inner AIM loop (Fig. 3) is performed, the incoming IUIs are read from the input file and respective events are created for them – receiving the next IUI by the entry field of the respective propagation path at the specified moment of the system time. The generated OUIs are accumulated in the specified output file and the protocol of AIM work with its messages is displayed on the console. Figure 6 presents an example of the protocol.

Successful termination of a simulation session is indicated by respective message suited by information of the elapsed time, the achieved system time, the number of formed and processed events, and the numbers of input and output unitary images. The OUIs have the same format as the IUIs. In this example ten equal OUIs were formed at the exit field F_1 [4, 5] of the specified propagation path number 0 in the successive moments 301, 372, 448, 532, 621, 714, 815, 925, 1045, and 1168 of the system time, and stored in the output file:

$$\begin{bmatrix} 1111111 \\ 1111111 \\ 1111111 \\ 1111111 \\ 1111111 \\ 1111111 \end{bmatrix}, \begin{bmatrix} 1111111 \\ 1111111 \\ 1111111 \\ 1111111 \\ 1111111 \\ 1111111 \end{bmatrix}, \begin{bmatrix} 1111111 \\ 1111111 \\ 1111111 \\ 1111111 \\ 1111111 \\ 1111111 \end{bmatrix}, \cdots, \begin{bmatrix} 1111111 \\ 1111111 \\ 1111111 \\ 1111111 \\ 1111111 \\ 1111111 \end{bmatrix}.$$

In case of an abnormal termination, the AIM program outputs a respective informative message.

As the output from the AIM simulator is just plain text, it is quite easy to use it as input for other sophisticated tools for data analysis and visualization; e.g., such as MS Excel or Matlab.

```
include c:\mpe\sanityAIM_2.txt
Including c:\mpe\sanityAIM_2.txt ok
Simulate
Simulator version: C:\MPE\AIM_206   Session on 15.01.2016 at 12:56:37
TimeLimit=1000000 ImageSize=6x7 LayerSize=5x6
NumberOfNeurons=2520 NumberOfSynapses=104076 MaxSynLengthSquared=170
Input file: c:\mpe\AIM_02.txt
Output file name: c:\mpe\AIM_02_.txt
SIMULATION SUCCESSFULLY COMPLETED
Time elapsed=16 ms SystemTime=1176 NumberOfEvents=1210
NumberOfInputImages=10 NumberOfOutputImages=10  ok
```

Fig. 6. Example of a protocol of a simulation session

7   System Log

The AIM program allows to register processed events within the specified interval of the system time in a system log for further analysis. The interval is specified through assigning respective values to the variables LogStartTime and LogStopTime when setting up the AIM parameters.

If the current moment of system time when processing a group of same-time events turns out to be within the specified interval, then respective data is sent to the output stream in form of Time = *moment* followed by a list of events which occurred at this time. An event is denoted with a letter R, S, P, L or U followed by one or two parameters of this event; e.g., the denotations of neuron fields (see Fig. 7 for an example).

Other event data, like neuron potentials, synapse weights, and unitary images may be easily added through respective user defined Forth operators.

Time=1 R I0 F0[0,0]	Time=72 R I1 F0[0,0]	Time=109 U F0[1,1]
Time=6 P F0[0,0] F1[0,0]	Time=74 U F1[1,5]	Time=110 U F0[0,3]
Time=11 P F1[0,0] F0[0,1]	Time=76 P F0[1,4] F1[1,4]	Time=111 P F1[1,1] F0[1,0]
Time=14 U F1[0,0]	Time=77 P F0[0,0] F1[0,0]	Time=112 P F1[0,3] F0[0,4]
Time=16 P F0[0,1] F1[0,1]	Time=79 U F0[1,4]	Time=114 U F1[1,1]
Time=19 U F0[0,1]	Time=81 P F1[1,4] F0[1,3]	....
Time=21 P F1[0,1] F0[0,2]	Time=82 P F1[0,0] F0[0,1]	Time=1123 P F0[4,1] F1[4,1]
Time=24 U F1[0,1]	Time=84 U F1[1,4]	Time=1126 U F0[4,1]
Time=26 P F0[0,2] F1[0,2]	Time=85 U F1[0,0]	Time=1128 P F1[4,1] F0[4,2]
Time=29 U F0[0,2]	Time=86 P F0[1,3] F1[1,3]	Time=1131 U F1[4,1]
Time=31 P F1[0,2] F0[0,3]	Time=87 P F0[0,1] F1[0,1]	Time=1133 P F0[4,2] F1[4,2]
Time=34 U F1[0,2]	Time=89 U F0[1,3]	Time=1136 U F0[4,2]
Time=36 P F0[0,3] F1[0,3]	Time=90 U F0[0,1]	Time=1138 P F1[4,2] F0[4,3]
Time=39 U F0[0,3]	Time=91 P F1[1,3] F0[1,2]	Time=1141 U F1[4,2]
Time=41 P F1[0,3] F0[0,4]	Time=92 P F1[0,1] F0[0,2]	Time=1143 P F0[4,3] F1[4,3]
Time=44 U F1[0,3]	Time=94 U F1[1,3]	Time=1146 U F0[4,3]
Time=46 P F0[0,4] F1[0,4]	Time=95 U F1[0,1]	Time=1148 P F1[4,3] F0[4,4]
Time=49 U F0[0,4]	Time=96 P F0[1,2] F1[1,2]	Time=1151 U F1[4,3]
Time=51 P F1[0,4] F0[0,5]	Time=97 P F0[0,2] F1[0,2]	Time=1153 P F0[4,4] F1[4,4]
Time=54 U F1[0,4]	Time=99 U F0[1,2]	Time=1156 U F0[4,4]
Time=56 P F0[0,5] F1[0,5]	Time=100 U F0[0,2]	Time=1158 P F1[4,4] F0[4,5]
Time=59 U F0[0,5]	Time=101 P F1[1,2] F0[1,1]	Time=1161 U F1[4,4]
Time=61 P F1[0,5] F0[1,5]	Time=102 P F1[0,2] F0[0,3]	Time=1163 P F0[4,5] F1[4,5]
Time=64 U F1[0,5]	Time=104 U F1[1,2]	Time=1166 U F0[4,5]
Time=66 P F1[1,5] F1[1,5]	Time=105 U F1[0,2]	Time=1168 S I9 F1[4,5]
Time=69 U F0[1,5]	Time=106 P F0[1,1] F1[1,1]	Time=1171 U F1[4,5]
Time=71 P F1[1,5] F0[1,4]	Time=107 P F0[0,3] F1[0,3]	Time=1176 U F1[4,5]

Fig. 7. Fragments of a system log with processed events

8 Conclusions

The described simulator is a relatively simple but powerful tool for studying various RNN structures and various combinations of signal propagation paths and RNN parameters under various circumstances. Forth allows the user to easily specify the respective user interfaces and interoperate with other powerful tools for representing and visualization of experimental results.

The described programming solution based on the event list structure which controls the simulation process turned out to be both effective and efficient, so it worth for reuse in other applications or subject domains. The described simple system log allows for relatively easy detecting violations and errors in the simulation process and helps in debugging the simulator and its input data.

The described simulator demonstrated acceptable performance on a regular laptop with relatively small RNNs of up to one million of synapses. However, its performance can be further improved with the assembler option offered by most Forth systems, which allows for direct programming performance critical words in assembler, thus ensuring the most efficient realization of such critical data structures and respective processing means.

The future work will be focused on developing a variety of user interfaces and typical solutions, as well as accumulating and analyzing the results of experiments with various RNN structures and signal data.

This work was partially financially supported by the Government of the Russian Federation, Grant 074-U01.

References

1. Osipov, V.Yu.: Associative intellectual machine with three signaling systems. Inf. Control Syst. **5**, 12–17 (2014). (in Russian)
2. Haykin, S.S., et al.: Neural Networks and Learning Machines, 3rd edn. Pearson Education, Upper Saddle River (2009)
3. Brette, R., et al.: Simulation of networks of spiking neurons: a review of tools and strategies. J. Comput. Neurosci. **23**(3), 349–398 (2007)
4. Zell, A., et al.: SNNS (stuttgart neural network simulator). In: Skrzypek, J. (ed.) Neural Network Simulation Environments. The Kluwer International Series in Engineering and Computer Science, vol. 254, pp. 165–186. Springer, New York (1994)
5. Demuth, H., Beale, M.: Neural Network Toolbox for Use with MATLAB. The MathWorks, Natick (1998)
6. Elman, J.L.: Learning and development in neural networks: the importance of starting small. Cognition **48**, 71–99 (1993)
7. Goodman, D., Brette, R.: Brian: a simulator for spiking neural networks in python. Front. Neuroinform. 2, article 5 (2008). http://www.frontiersin.org
8. Davison, A., et al.: PyNN: a common interface for neuronal network simulators. Front. Neuroinform. 2, article 11 (2009). http://www.frontiersin.org
9. Forth 200x (2016). http://www.forth200x.org/forth200x.html
10. VFX Forth for Windows. User Manual. Manual Revision 4.70, MPE Ltd., Southampton (2014). http://www.mpeforth.com/
11. gForth. Free Software Foundation, Inc. (2016). https://www.gnu.org/software/gforth/

Hopfield Network with Interneuronal Connections Based on Memristor Bridges

Mikhail S. Tarkov[(⊠)]

A.V. Rzhanov Institute of Semiconductor Physics SB RAS,
Novosibirsk, Russia
tarkov@isp.nsc.ru

Abstract. A scheme for the Hopfield associative memory hardware implementation with interneuronal connections through bridges using memristors is proposed. The Hopfield associative memory is realized as a network of coupled phase oscillators. It is shown how to use the CMOS transistor switches to control the memristance (memristor resistance) value.

Keywords: Memristor · Memristance · Bridge · Hopfield network · Weight matrix · LTSPICE

1 Introduction

Memristor was predicted theoretically in 1971 by Chua [1]. First physical memristor implementation in 2008 was demonstrated by a laboratory from Hewlett Packard as a thin film structure TiO_2 [2]. In Russia the first TiO_2 memristor was obtained in 2012 by the Tyumen State University [3]. The memristor has many advantages such as non-volatile storage media, low power consumption, high density integration and excellent scalability. The unique ability to retain traces of the device excitation makes it an ideal candidate for the implementation of electronic synapses in neural networks [4].

The memristor-based Hopfield networks are intensively investigated now [5, 6]. In this article we propose a new version of the memristor-based Hopfield associative memory implemented as a network of coupled phase oscillators [7]. Such network has only error-free retrieval states.

2 Memristor

A memristor behaves like a synapse: it "remembers" the total electric charge passed through it [8]. The memristor-based memory can reach its very high integration degree of 100 Gbits/cm^2, several times higher than that based on the flash memory technology [9]. These unique properties make the memristor a promising device for creating massively parallel neuromorphic systems [10–12].

© Springer International Publishing Switzerland 2016
L. Cheng et al. (Eds.): ISNN 2016, LNCS 9719, pp. 196–203, 2016.
DOI: 10.1007/978-3-319-40663-3_23

A memristance (the memristor resistance) (Fig. 1) can be represented [13] as

$$M(p) = p \cdot R_{on} + (1-p) \cdot R_{off}, \qquad (1)$$

where $0 \leq p \leq 1$ is the doping front position relative to the total film thickness h of TiO_2, R_{on} is the memristor minimum resistance, R_{off} is the memristor maximum resistance. Setting the memristor to the desired level of memristance M_d depends on the ratio between M_d and the initial memristance value M_0. For model (1), the memristance adjustment is made by applying a constant voltage $V > V_{th}$ for $M_0 > M_d$ or a voltage $V < - V_{th}$ for $M_0 < M_d$ to the memristor due some time [14]

$$\tau = \begin{cases} \frac{M_0^2 - M_d^2}{2k(V - V_{th})}, & V > V_{th}, \\[2mm] \frac{M_0^2 - M_d^2}{2k(V + V_{th})}, & V < - V_{th}, \end{cases} \qquad (2)$$

where $k = \mu_v \frac{R_{off}}{h^2}(R_{off} - R_{on})$.

Here μ_v is the average ion mobility, h is the total memristor film thickness, $V(t)$ is the current voltage value on the memristor, V_{th} is the threshold voltage.

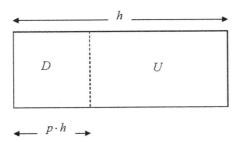

Fig. 1. The memristor structure: D – the low resistance region, U – the high resistance region, $p \in [0,1]$ is the doping front position relative to the total film thickness h of TiO_2

3   Weighting Input Signals by Bridge Circuits

Both positive and negative weighing coefficients values can be implemented on the base of the bridge resistor circuits [15]. Consider a bridge circuit (Fig. 2). The relationship between input and output voltages of this circuit is described by

$$U_{out} = w \cdot U_{in},$$
$$w = \frac{M2}{M1 + M2} - \frac{M4}{M3 + M4}, \qquad (3)$$

where w is the bridge circuit weight. Let $M2 = M3 = M4 = m$. Then, from (3), we have

$$w = \frac{1}{2} \cdot \frac{m - M1}{m + M1}. \tag{4}$$

From (4)

$$-\frac{1}{2} < w < \frac{1}{2}, \tag{5}$$

i.e. the bridge circuit can be used as the synapse whose weight may be a positive value (for $m > M1$), a negative value (for $m < M1$), and zero value (for $m = M1$).

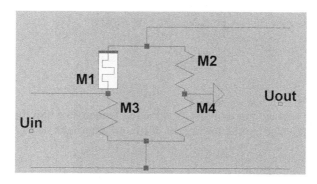

Fig. 2. Bridge circuit using memristor

By changing the memristor M1 resistance we can manage the weight value of synapse w.

From (4) and (5)

$$M1 = \frac{1 - 2w}{1 + 2w} \cdot m, \ w \in \left(-\frac{1}{2}, \frac{1}{2}\right). \tag{6}$$

An example of the boolean function AND implementation by an electronic neuron model, in which the weighing coefficients are realized by bridge circuits, and the step activation function is realized by the operational amplifier operating in the comparator

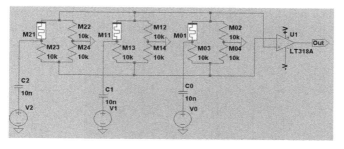

Fig. 3. The boolean function AND implementation by an electronic neuron model

mode, is shown in Fig. 3. V1 and V2 inputs, and threshold V0 are realized as bipolar pulses (Fig. 4a, and b), which allows keeping the memristor resistance unchanged. The corresponding weights are $w_1 = w_2 = 0.01$, $w_0 = -0.01$. According to (6), we get M11 = M21 = 9.6 kohm, M01 = 10.6 kohm.

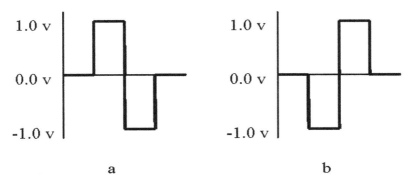

a   b

Fig. 4. Input signals 1 (a) и −1 (б)

4 Hopfield Network with Weights on the Base of Bridge Circuits

The character images L, T, and X with dimension 3 × 3 pixels are shown in Fig. 5. These images can be described by the vectors (by-line image scanning), respectively:

$$L = (1, -1, -1, 1, -1, -1, 1, 1, 1),$$
$$T = (1, 1, 1, -1, 1, -1, -1, 1, -1),$$
$$X = (1, -1, 1, -1, 1, -1, 1, -1, 1),$$

where 1 corresponds to the white pixel, and −1 corresponds to the black one.

  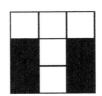 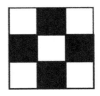

Fig. 5. The character images L, T, and X with dimension 3 × 3 pixels

Let us calculate (by the Hebb's rule) the matrix

$$W' = L^T \cdot L + T^T \cdot T + X^T \cdot X.$$

Zeroing the matrix main diagonal we obtain the Hopfield network weight matrix:

$$
W = \begin{bmatrix}
0 & -1 & 1 & -1 & 1 & -3 & 1 & 1 & 1 \\
-1 & 0 & 1 & -1 & 1 & 1 & -3 & 1 & -3 \\
1 & 1 & 0 & -3 & 3 & -1 & -1 & -1 & -1 \\
-1 & -1 & -3 & 0 & -3 & 1 & 1 & 1 & 1 \\
1 & 1 & 3 & -3 & 0 & -1 & -1 & -1 & -1 \\
-3 & 1 & -1 & 1 & -1 & 0 & -1 & -1 & -1 \\
1 & -3 & -1 & 1 & -1 & -1 & 0 & -1 & 3 \\
1 & 1 & -1 & 1 & -1 & -1 & -1 & 0 & -1 \\
1 & -3 & -1 & 1 & -1 & -1 & 3 & -1 & 0
\end{bmatrix}.
$$

According to (6), the matrix W is multiplied by 0.01. Table 1 specifies the correspondence between weight w and the memristance M1 value when the other resistances values are equal to $m = 10$ kohm.

Table 1. Correspondence between the weight and the memristance M1 value

w	−0.03	−0.01	0	0.01	0.03
M1 (kohm)	11.3	10.4	10	9.6	8.9

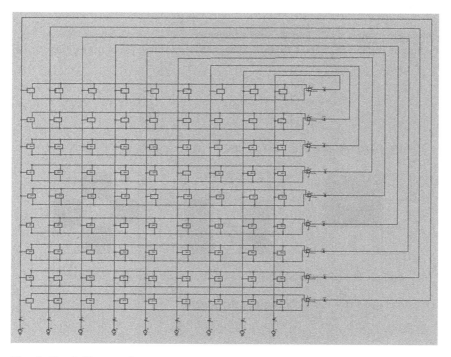

Fig. 6. Hopfield network as an autoassociative memory for the Fig. 5 character images

An example of the Hopfield network which weights (represented by rectangles) are implemented on the base of the bridge circuits with memristors is shown in Fig. 6. The step bipolar neuron activation function is implemented here on the base of the operational amplifier working in a comparator mode. This network can be considered as a modification of the multivibrator (oscillator) shown in Fig. 7. The multivibrator bridge circuit sets a positive feedback. The multivibrator input pulse sets the oscillation phase, i.e. the change of the input pulse sign leads to the multivibrator output signal inversion.

When applying the input, the Hopfield network converges to one of the Fig. 5 images and continue to make the transitions from this image to its inversion and back indefinitely long (see Fig. 8 example). The alternating of positive and negative pulses at the Hopfield network neurons outputs allows us to keep the memristors resistances unchanged.

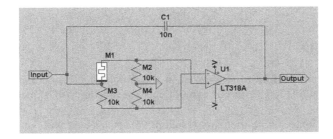

Fig. 7. The multivibrator using operational amplifier as a comparator

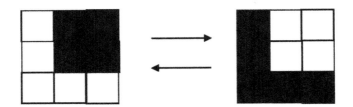

Fig. 8. Example of the Hopfield network output oscillations

5 Setting the Hopfield Network Weights

To change the memristance, the additional circuits based on CMOS transistors are connected to the bridge input and control the memristor current (Fig. 9). The second pole of the memristor is grounded by nmos-transistor M7.

If it is necessary to reduce the memristance, a positive pulse with a required duration is supplied to the drain In + of the nmos-transistor M5 opened by voltage +V.

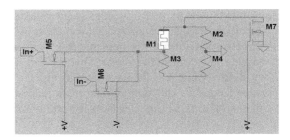

Fig. 9. Circuits for the memristance M1 setting

If it is necessary to increase the memristance, a negative pulse with a required duration is supplied to the drain In − of the nmos-transistor M6 opened by the voltage −V.

In the Hopfield network operation mode, transistor M5 is open continuously and the input In + is used to feed the signal to the bridge.

6   Experiments

The Hopfield network (Fig. 6) is realized in the simulator LTspice IV [16] with the MATLAB interface [17, 18]. In our experiments we use the SPICE memristor model from [19, 20].

We used the images with the pixels random values selected with equal probability from the set {−1, 1} as a test. We carried out 100 experiments. In all experiments, the network in Fig. 6 initially converges to one of the images of Fig. 5 or its inversion, i.e. the image, which is obtained by the image pixels inversion. We have no error states in all experiments. It is a property of the oscillatory associative memory [7].

7   Conclusion

Memristor (memory resistor) is a perspective element for hardware synapses implementation. In this paper we propose an approach to the implementation of electronic associative memory based on the Hopfield network with tunable weights based on the resistor bridges containing memristors. The Hopfield associative memory is implemented as a network of coupled phase oscillators.

It is shown how to use the CMOS transistors switches to control the memristance value. The experiments using LTSPICE-Hopfield network model show that for the reference binary images with size 3 × 3 the network converges to the reference images (and, accordingly, to their inversion) with a random uniform distribution of binary pixel values of the input images. We have no error states in all experiments. It is a property of the oscillatory associative memory.

References

1. Chua, L.: Memristor – the missing circuit element. IEEE Trans. Circuit Theory **18**, 507–519 (1971)
2. Strukov, D.B., Snider, G.S., Stewart, D.R., Williams, R.S.: The missing memristor found. Nature **453**, 80–83 (2008)
3. University of Tyumen. http://www.utmn.ru/presse/teleradiokanal-evrazion/videonovosti-tyumgu/89986/
4. Pershin, Y., Di Ventra, M.: Experimental demonstration of associative memory with memristive neural networks. Neural Netw. **23**, 881–886 (2010)
5. Wu, A., Zhang, J., Zeng, Z.: Dynamic behaviors of a class of memristor-based Hopfield networks. Phys. Lett. A **375**, 1661–1665 (2011)
6. Liu, B., Chen, Y., Wysocki, B., Huang, T.: Reconfigurable neuromorphic computing system with memristor-based synapse design. Neural Process. Lett. **41**, 159–167 (2015)
7. Nishikawa, T., Hoppensteadt, F.C., Lai, Y.-C.: Oscillatory associative memory network with perfect retrieval. Physica D **197**, 134–148 (2004)
8. Chua, L.: Resistance switching memories are memristors. Appl. Phys. A Mater. Sci. Process. **102**, 765–783 (2011)
9. Ho, Y., Huang, G.M., Li, P.: Nonvolatile memristor memory: device characteristics and design applications. In: Proceedings of IEEE/ACM International Conference on Computer-Aided Design (ICCAD), pp. 485–490 (2009)
10. Jo, S.H., Chang, T., Ebong, I., Bhadviya, B.B., Mazumder, P., Lu, W.: Nanoscale memristor device as synapse in neuromorphic systems. Nanoletters **10**, 1297–1301 (2010)
11. Kavehei, O.: Memristive devices and circuits for computing, memory, and neuromorphic applications. Ph.D. thesis, The University of Adelaida, Australia (2011)
12. Lehtonen, E.: Memristive Computing. University of Turku, Finland (2012)
13. Wu, Q., Liu, B., Chen, Y., Li, H., Chen, Q., Qiu, Q.: Bio-inspired computing with resistive memories – models, architectures and applications. In: Proceedings of 2014 IEEE International Symposium on Circuits and Systems (ISCAS), pp. 834–837 (2014)
14. Tarkov, M.S.: Mapping weight matrix of a neural network's layer onto memristor crossbar. Opt. Mem. Neural Netw. (Inf. Opt.) **24**, 109–115 (2015)
15. Kim, H., Sah, M.P., Yang, C., Roska, T., Chua, L.O.: Memristor bridge-based artificial neural weighting circuit. In: Adamatzky, A., Chua, L.O. (eds.) Memristor Networks, pp. 249–266. Springer, Switzerland (2014)
16. LTSPICE IV User Manual. http://ecee.colorado.edu/ ~ mathys/ecen1400/pdf/scad3.pdf
17. Wagner, P.: The LTspice2Matlab Function. http://www.mathworks.com/matlabcentral/fileexchange/23394-fast-import-of-compressed-binary-raw-files-created-with-ltspice-circuit-simulator
18. Dorran, D.: The RunLTspice Function. https://dadorran.wordpress.com/tag/run-ltspice-matlab/
19. Biolek, Z., Biolek, D., Biolkova, V.: SPICE model of memristor with nonlinear dopant drift. Radioengineering **18**, 210–214 (2009)
20. Falatic, M.: Memristor Simulation with LTspice. http://www.falatic.com/index.php/69

Two-Dimensional Fast Orthogonal Neural Networks

A. Yu. Dorogov$^{(\boxtimes)}$

Saint Petersburg Electrotechnical University "LETI", Saint Petersburg, Russia
vaksa2006@yandex.ru

Abstract. A new method of learning fast two-dimensional orthogonal transformations is considered. Tunable orthogonal transformations are regarded as special neural networks. The learning takes a finite number of steps. The learning algorithm does not have the error feedback and is absolutely stable. The method is based on fractal filtering of signals and images. Linguistic models are used to determine the topology and structure of fast transformations. Examples are given.

Keywords: Orthogonal fast transform · Learning algorithm · Fractal filtering · System model

1 Introduction

Signal recognition and classification often require preprocessing procedures that help remove redundancy and select informative features. Use of orthogonal transformations for this purpose allows input information to be presented as independent spectral components. It is known that the greatest reduction of data redundancy is ensured by using Karhunen-Loeve orthogonal transformation formed by the signal covariance matrix eigenvectors. However, using this transformation involves a lot of computations. For this reason Karhunen-Loeve method is not used in processing of data of large dimensionality. However, if only one, most important eigenvector (the principal component) is used in Karhunen-Loeve transformation, the computation load decreases significantly. The orthogonal transformation tuned to one principal component belongs to the class of adapted transformations.

In 1970s Andrews and Caspari [1] were the first who suggested the idea of generalized orthogonal transformation. First learning algorithms for this kind of transformations were developed by Solodovnikov and his colleagues [2]. This class of transformations was also known as tunable fast transformations at that time. Introducing activation functions and offsets turns a fast transformation into a fast neural network [3] which can be trained by gradient algorithms like for example the error back propagation algorithm. These algorithms can also be used to train linear tunable fast transformations (given bounding nonlinearity in the learning circuit). The drawback of gradient algorithms consists in potentially unstable learning procedures and the presence of dead ends and hanging up at local minima. Structural properties of fast transformations allow specific learning methods free from the above-mentioned

© Springer International Publishing Switzerland 2016
L. Cheng et al. (Eds.): ISNN 2016, LNCS 9719, pp. 204–210, 2016.
DOI: 10.1007/978-3-319-40663-3_24

drawbacks. Considered below, the method of training adapted transformations is based on the possibility of multiplicative decomposition of fast-transformation matrix elements proved by Good [4] (1958). Methods of fractal finite-interval filtration of signals [5, 6] are used for multiplicative decomposition.

2   System Models of Two-Dimensional Fast Transformations

Orthogonal transformations with fast execution algorithm are usually used in image processing. The aim of the processing usually involves filtration and compression of an image.

Let us denote an $N_y \times N_x$ image matrix as $F(U_y, U_x)$. When we subject an image to linear transformation $H(U_y, U_x; V_y, V_x)$, we get an array of $M_y \times M_x$ coefficients. The two-dimensional transformation complies with the following rule:

$$S(V_y, V_x) = \sum_{Uy=0}^{Ny-1} \sum_{Ux=0}^{Nx-1} F(U_y, U_x) H(U_y, U_x; V_y, V_x). \tag{1}$$

A two-dimensional transformation is called an orthogonal (unitary, to be exact) transformation if the following condition is met:

$$\sum_{Uy=0}^{Ny-1} \sum_{Ux=0}^{Nx-1} H(U_y, U_x; V_y, V_x) \bar{H}(U_y, U_x; V_y', V_x') = \begin{cases} 1 \ \ if \ \ V_y = V_y' \ and \ V_x = V_x' \\ 0 \ if \ V_y \neq V_y' \ or \ V_x \neq V_x' \end{cases}. \tag{2}$$

where the over line sign marks the complex-conjugate transformation. For an orthogonal transformation we have $N_y = M_y$, $N_x = M_x$. The necessary condition of existence of a fast algorithm is the possibility of multiplicative decomposition of the values of both dimensions of an image into the same number of multiplicands:

$$N_y = p_0^y p_1^y \cdots p_{n-1}^y,$$
$$N_x = p_0^x p_1^x \cdots p_{n-1}^x.$$

Here indices x, y indicate to which coordinate axis the original image belongs to. The above condition is not a very strict limitation because some multiplicands can have unit values. Nevertheless, the greater the number of non-unit multiplicands, the higher the computation efficiency of a fast algorithm. Using multiplicands N_y and N_x let us express the coordinates of image points in a positional numeration of compound bases:

$$U_y = \langle u_{n-1}^y u_{n-2}^y \cdots u_1^y u_0^y \rangle,$$
$$U_x = \langle u_{n-1}^x u_{n-2}^x \cdots u_1^x u_0^x \rangle, \tag{3}$$

where weight of the m-th digit is determined as $p_{m-1}^* p_{m-2}^* \cdots p_1^* p_0^*$, and u_m^* is a digit variable taking values $[0, p_m^* - 1]$ (the asterisk replaces indices x and y here). Similarly we can express the coordinates of spectral coefficients in plane $[V_y, V_x]$:

$$V_y = \left\langle v_{n-1}^y v_{n-2}^y \cdots v_1^y v_0^y \right\rangle,$$
$$V_x = \left\langle v_{n-1}^x v_{n-2}^x \cdots v_1^x v_0^x \right\rangle.$$

The algorithm of a fast transformation is usually represented as a graph of different topologies. In the case of Tukey-Cooley topology with time down sampling, the graph can be described as a linguistic sentence [6]:

$$\left[\begin{array}{l} \left\langle u_{n-1}^* u_{n-2}^* \cdots u_1^* u_0^* \right\rangle \left\langle u_{n-1}^* u_{n-2}^* \cdots u_1^* v_0^* \right\rangle \cdots \\ \cdots \left\langle u_{n-1}^* u_{n-2}^* \cdots u_{m+1}^* u_m^* v_{m-1}^* v_{m-2}^* \cdots v_0^* \right\rangle \cdots \left\langle v_{n-1}^* v_{n-2}^* \cdots v_1^* v_0^* \right\rangle \end{array} \right].$$

The first and the last words of the sentence correspond to the coordinates of values in the spatial and spectral areas. The number of words in the sentence is $n+1$. The words in the middle determine the coordinates U_y^m, U_x^m and V_y^m, V_x^m of points of an input image in the inner layers of the fast algorithm. If the algorithm has a regular compact topology, the following condition is true:

$$U_y^{m+1} = V_y^m, \qquad U_x^{m+1} = V_x^m. \tag{4}$$

In the general case the topologies for x and y axes can differ. The graph of a fast algorithm in layer m holds base operations $W_{i_x^m, i_y^m}^m \left(u_m^y u_m^x; v_m^y v_m^x \right)$ which are four-dimensional matrices of dimensionality $\left[p_m^y, p_m^x; p_m^y, p_m^x \right]$ (transformation kernels). The relation between base operations is determined by the fast transformation structural model. For the given topology the graph of the structural model is described by the linguistic sentence:

$$\left[\begin{array}{l} \left\langle u_{n-1}^* u_{n-2}^* \cdots u_1^* \right\rangle \left\langle u_{n-1}^* u_{n-2}^* \cdots u_2^* v_0^* \right\rangle \cdots \\ \cdots \left\langle u_{n-1}^* u_{n-2}^* \cdots u_{m+1}^* v_{m-1}^* v_{m-2}^* \cdots v_0^* \right\rangle \cdots \left\langle v_{n-2}^* v_{n-3}^* \cdots v_1^* v_0^* \right\rangle \end{array} \right].$$

Each word in the sentence defines the number of base operation i_*^m in layer m. The number of words in the sentences is n.

Figure 1 shows the structural model of a two-dimensional transformation for an 8×8 image. The input image goes to the bottom layer, while spectral coefficients are produced in the top layer. Base operations relate to the model nods. A kernel in layer m makes the two-dimensional transformation of a $p_m^y \times p_m^x$ space block:

$$S^m \left(V_y^m, V_x^m \right) = \sum_{u_m^y} \sum_{u_m^x} F^m \left(U_y^m, U_x^m \right) W_{i_x^m, i_y^m}^m \left(u_m^y u_m^x; v_m^y v_m^x \right). \tag{5}$$

Correspondences $U_*^m \leftrightarrow \left(i_*^m, u_*^m \right)$ are determined one-one by the linguistic sentences of the topological and structural models.

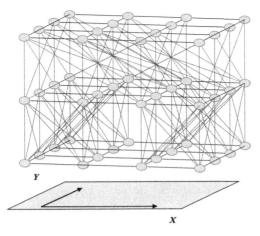

Fig. 1. The structural model of a two-dimensional fast transformation

3 Multiplicative Decompositions of Matrix Elements of the Two-Dimensional Fast Transformation

Setting specific values for all digit variables u_m^*, v_m^* (where m takes $0, 1, \ldots, n-1$ successively) defines a particular path in the topological graph between a couple of elements from the start and finish layers. It follows from the single-valuedness of the digit representation of numbers that this path is unique for each pair combination of space points of the input and output layers. This fact allows a convenient analytic expression connecting the fast transformation matrix elements with kernel elements. From (1) it follows:

$$H\left(U_y, U_x; V_y, V_x\right) = \frac{\partial S\left(V_y, V_x\right)}{\partial F\left(U_y, U_x\right)}. \tag{6}$$

Differentiating (6) as a compound function, we get:

$$H\left(U_y, U_x; V_y, V_x\right) = \frac{\partial S^{n-1}}{\partial F^{n-1}} \frac{\partial F^{n-1}}{\partial S^{n-2}} \frac{\partial S^{n-2}}{\partial F^{n-2}} \cdots \frac{\partial F^1}{\partial S^0} \frac{\partial S^0}{\partial F^0}.$$

From condition (4) it follows that for all m we have $\frac{\partial F^m}{\partial S^{m-1}} = 1$, and from (5) it follows that $\frac{\partial S^m}{\partial F^m} = W_{i_x^m, i_y^m}^m \left(u_m^y u_m^x; v_m^y v_m^x\right)$. So we find that each element of the four-dimensional transformation matrix H is expressed with the help of kernel elements as the product:

$$H\left(U_y, U_x; V_y, V_x\right) = W_{i_x^{n-1}, i_y^{n-1}}^{n-1} \left(u_{n-1}^y u_{n-1}^x; v_{n-1}^y v_{n-1}^x\right) \cdot$$
$$W_{i_x^{n-2}, i_y^{n-2}}^{n-2} \left(u_{n-2}^y u_{n-2}^x; v_{n-2}^y v_{n-2}^x\right) \ldots W_{i_x^0, i_y^0}^0 \left(u_0^y u_0^x; v_0^y v_0^x\right), \tag{7}$$

where the digit expressions of kernels indices of layer m for the given topology has the form:

$$
\begin{aligned}
i_x^m &= \left\langle u_{n-1}^x u_{n-2}^x \cdots u_{m+1}^x v_{m-1}^x v_{m-2}^x \cdots v_0^x \right\rangle, \\
i_y^m &= \left\langle u_{n-1}^y u_{n-2}^y \cdots u_{m+1}^y v_{m-1}^y v_{m-2}^y \cdots v_0^y \right\rangle.
\end{aligned}
\tag{8}
$$

By putting expression (7) in the orthogonality condition (2), we find that it will be fulfilled when all kernels are orthogonal, i.e. for any m, i_x^m, i_y^m the following takes place:

$$
\sum_{u_m^y} \sum_{u_m^x} W_{i_x^m, i_y^m}^m \left(u_m^y u_m^x; v_m^y v_m^x \right) \bar{W}_{i_x^m, i_y^m}^m \left(u_m^y u_m^x; {}^\backprime v_m^y {}^\backprime v_m^x \right) =
\begin{cases}
1 \ \ if \ \ v_m^y = {}^\backprime v_m^y \ \ and \ \ v_m^x = {}^\backprime v_m^x \\
0 \ \ if \ \ v_m^y \neq {}^\backprime v_m^y \ \ or \ \ v_m^y \neq {}^\backprime v_m^x
\end{cases} .
$$

4   Fractal Learning of Two-Dimensional Fast Transformations

The learning algorithm for two-dimensional fast transformations uses the ideas of fractal filtration which were discussed in [6]. In the case of two dimensions the fractal filtration is multiple-scale image processing which successively compresses an image down to a single point. The flowchart of the fractal filtration can be pictured as a pyramid shown in Fig. 2. The pyramid base is an original image whose $F(U_y, U_x)$ has arguments U_y, U_x are expressed in a positional numeration (see (3)). Let us fix all digits except for two low order digits u_0^y and u_0^x in this positional representation.

If we make these digits take all possible values, we get a two-dimensional sampling of size $p_0^y \times p_0^x$. We regard the fractal filter as an arbitrary functional Φ defined over this sampling. Formally it can be written in the form:

$$
F_1\left(\left\langle u_{n-1}^y u_{n-2}^y \cdots u_1^y \right\rangle, \left\langle u_{n-1}^x u_{n-2}^x \cdots u_1^x \right\rangle \right) = \underset{\left(u_0^y, u_0^x \right)}{\Phi} \left[F\left(\left\langle u_{n-1}^y u_{n-2}^y \cdots u_1^y u_0^y \right\rangle, \left\langle u_{n-1}^x u_{n-2}^x \cdots u_1^x u_0^x \right\rangle \right) \right].
$$

It is obvious that image F_1 will be scaled down comparing the size of the original image. The functional may be, for example, the rule of computing the mean of the sampling or its median. The original image can be formally represented as the product:

$$
\begin{aligned}
F\left(\left\langle u_{n-1}^y u_{n-2}^y \cdots u_1^y u_0^y \right\rangle, \left\langle u_{n-1}^x u_{n-2}^x \cdots u_1^x u_0^x \right\rangle \right) = \\
= F_1\left(\left\langle u_{n-1}^y u_{n-2}^y \cdots u_1^y \right\rangle, \left\langle u_{n-1}^x u_{n-2}^x \cdots u_1^x \right\rangle \right) f_{j_0^y j_0^x}\left(u_0^y, u_0^x \right),
\end{aligned}
$$

where $f_{j_0^y j_0^x}\left(u_0^y, u_0^x \right)$ is a set of two-dimensional function-multiplicands which depend on digit variables u_0^y and u_0^x, and indices j_0^y, j_0^x selects the two-dimensional function from the set. The values of the indices are set equal to the values of the arguments of image F_1, i.e. $j_0^y = \left\langle u_{n-1}^y u_{n-2}^y \cdots u_1^y \right\rangle$ and $j_0^x = \left\langle u_{n-1}^x u_{n-2}^x \cdots u_1^x \right\rangle$. In order to get function-multiplicands, it is enough to scalarly divide image F by image F_1, varying all digit

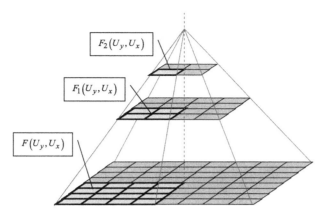

Fig. 2. The diagram of fractal filtration of an image

variables in the process. Image F_1 in turn can also be represented as the product of image F_2 and multiplicands from set $f_{j_1^y j_1^x}(u_1^y, u_1^x)$. Repeating the operations of fractal filtration and decomposition over and over again, we will reach the peak of the image pyramid and get multiplicative decomposition of images:

$$F\left(\langle u_{n-1}^y u_{n-2}^y \cdots u_1^y u_0^y \rangle, \langle u_{n-1}^x u_{n-2}^x \cdots u_1^x u_0^x \rangle\right) = f_{j_{n-1}^y j_{n-1}^x}(u_{n-1}^y, u_{n-1}^x) f_{j_{n-2}^y j_{n-2}^x}(u_{n-2}^y, u_{n-2}^x) \cdots$$
$$\cdots f_{j_1^y j_1^x}(u_1^y, u_1^x) f_{j_0^y j_0^x}(u_0^y, u_0^x), \tag{9}$$

where $j_m^y = \langle u_{n-1}^y u_{n-2}^y \cdots u_{m+1}^y \rangle$ and $j_0^x = \langle u_{n-1}^x u_{n-2}^x \cdots u_{m+1}^x \rangle$. If we compare the decomposition of an image with the fast-transformation decomposition (7), we will see that they are similar. Moreover, the set of indices of kernels covers the set of indices of function-multiplicands. This fact allows the conclusions that the fast transformation will be accommodated to the image when the transformation kernels are accommodated to decomposition functions (9). In fact, for each two-dimensional function-multiplicand it is necessary to complete orthogonal functions to the full set of basic functions of the orthogonal kernel. It is simple to do by using Gram-Schmidt orthoganalization procedure. The sets of function-multiplicands with deficiency of kernels can be supplemented by themselves. The choice of complementary functions does not have effect on the accommodation condition of the fast transformation, but change the form of other basic functions of the transformation.

5 Conclusion

We have considered the algorithms of learning orthogonal neural networks that have no closed loop of error-based weight correction. Due to this property the algorithms are always stable and stop running in a finite number of steps, the number of learning steps is proportional to the logarithm of the transformation dimensionality. The precision of learning is fundamentally determined by the processor precision. The learning algorithms for non-orthogonal transformations do not change radically except that there is no Gram-Schmidt orthoganalization procedure.

Acknowledgements. The paper was prepared in SPbETU and is supported by the Contract № 02.G25.31.0149 dated 01.12.2015 (Board of Education of Russia).

References

1. Andrews, H.C., Caspari, K.L.: A general techniques for spectral analysis. IEEE. Tr. Comput. **C19**, 16–25 (1970)
2. Solodovnikov, A.I., Spivakovskii, A.M.: Osnovy Teorii I Metody Spektralnoi Obrabotki Informatsii. Leningrad (1986). (in Russian)
3. Dorogov, A.Y.: Bystrye Neironnye Seti: Proektirovanie, Nastroika, Prilojenie. http://bookfi. org/book/805420. (in Russian)
4. Good, I.J.: The interaction algorithm and practical fourier analysis. J. R. Statistic. Soc. Ser. B **20**, 361–372 (1958)
5. Dorogov, A.Y., Shestopalov, M.Y.: Neirosetevoe Modelirovanie Regulyarnyh Fraktalov, in Neirocompyutery: razrabotka I primenenie, no. 6 (2007). (in Russian)
6. Dorogov, A.Y.: Fractal learning of fast orthogonal neural networks. J. Opt. Mem. Neural Netw. (Information Optics) **21**, 105–118 (2012)

Existence of Periodic Solutions to Non-autonomous Delay Cohen-Grossberg Neural Networks with Impulses on Time Scales

Zhouhong Li[✉]

Department of Mathematics, Yuxi Normal University, Yuxi, Yunnan 653100, China
zhouhli@yeah.net

Abstract. In this paper, without assuming the boundedness of activation functions, by applying continuous theorem of coincidence degree theory and the theory of calculus on time scales, we obtain some criteria for the existence exponential stability of periodic solutions to impulses Cohen-Grossberg neural networks with delay on time scales. Finally, an example is given to illustrate our results.

Keywords: Periodic solution · Delay · Cohen-Grossberg neural networks · Impulse · Time scales

1 Introduction

As we well known, Cohen-Grossberg neural networks(CGNNs) was first proposed in 1983 [1]. In the past two decades, the dynamical behaviours of the Cohen-Grossberg neural networks have been widely investigated, such as fruitfully applied in signal and image processing, pattern recognition, optimization and so on. Many important results on the existence and uniqueness of equilibrium point, global asymptotic stability, anti-periodic solutions, periodic solutions, almost periodic solutions and almost automorphic solutions have been established, see e.g. [2–11] and the references therein.

Recently, the existence of periodic solutions to neural network on time scales is proposed in [12]. However, both continuous and discrete systems are very important in implementing and applications. It is well known that the theory of time scales has received a lot of attention which was introduced by Stefan Hilger in order to unify continuous and discrete analysis. Therefore, it is meaningful to study dynamic systems on time scales which can unify differential and difference systems [4,13,15]. Moreover, in [16], the authors investigated the global stability of complex-valued recurrent neural networks.

Motivated by the above works, in this paper, we consider the following delayed Cohen-Grossberg with impulsive on time scales

© Springer International Publishing Switzerland 2016
L. Cheng et al. (Eds.): ISNN 2016, LNCS 9719, pp. 211–220, 2016.
DOI: 10.1007/978-3-319-40663-3_25

$$
\begin{cases}
x_i^{\Delta}(t) = -a_i(x_i(t))\Big[b_i(x_i(t)) - \sum_{j=1}^{n} c_{ij}(t)f_j\left(x_j(t)\right) - \sum_{j=1}^{n} d_{ij}(t)g_j(x_j(t \\
\qquad\qquad -\tau_{ij}(t))) - e_i(t)\Big], \qquad\qquad t \in \mathbb{T}^+,\ t \neq t_k,\ k \in \mathbb{N}, \\
\Delta x_i(t_k) = x_i(t_k^+) - x_i(t_k^-) = I_{ik}(x_i(t_k)), \quad i = 1, 2, \dots, n,
\end{cases} \tag{1}
$$

where $\mathbb{T}$ is an ω-periodic time scale which has the subspace topology inherited from the standard topology on $\mathbb{R}$, and for each interval $L_{\mathbb{T}} = L \cap \mathbb{T}$, $L^+ = L \cap [0, \infty)$, $\Delta x_i(t_k) = x_i(t_k^+) - x_i(t_k^-)$ are the impulses at moments t_k and $t_1 < t_2 < \dots$ is a strictly increasing sequence such that $\lim_{k \to \infty} t_k = +\infty$.

The initial conditions for the system (1) are chosen as:

$$
x_i(s) = \varphi_i(s), \quad s \in [-\tau, 0]_{\mathbb{T}}, \quad \tau = \max\{\tau_{ij}^+\}, 1 \le i, j \le n, \tag{2}
$$

where $\tau_{ij}^+ = \max_{t \in [0,\omega]} \tau_{ij}(t)$, $\varphi_i(t)(i = 1, 2, \dots, n)$ are continuous ω-periodic functions defined on $[-\tau, 0]_{\mathbb{T}}$. To the best of our knowledge, there is no paper applying the method of coincidence degree to investigate the existence of periodic solutions to impulsive Cohen-Grossberg neural networks with delays on time scales. The main aim of this paper is to study the existence and global exponential stability of the periodic solutions of system (1) by using the method of coincidence degree and Lyapunov method.

We have made the following assumptions:

(H_1) $\tau_{ij}(t) \ge 0$, $c_{ij}, d_{ij}, e_i \in C(\mathbb{T}, (0, \infty))$ are positive continuous functions with period ω and $t - \tau_{ij}(t) \in \mathbb{T}, i, j = 1, 2, \dots, n$.

(H_2) Functions $f_j, g_j \in C(\mathbb{R}, \mathbb{R})(j = 1, 2, \dots, n.)$ are neuron functions and there exist positive constants F_j and G_j such that

$$
|f_j(u) - f_j(v)| \le F_j|u - v| \quad \text{and} \quad |g_j(u) - g_j(v)| \le G_j|u - v|, \forall u, v \in \mathbb{R}.
$$

(H_3) $I_{ik} \in C(\mathbb{R}, \mathbb{R})$ and there exist positive constants G_{ik} such that

$$
|I_{ik}(u) - I_{ik}(v)| \le G_{ik}|u - v|, \forall u, v \in \mathbb{R}, k \in \mathbb{N}, i = 1, 2, \dots, n.
$$

For the sake of convenience, we introduce the following notations:

$$
h^m = \min_{t \in [0,\omega]_{\mathbb{T}}} |h(t)|, \quad h^M = \max_{t \in [0,\omega]_{\mathbb{T}}} |h(t)|, \quad \|h\|_2 = \left(\int_0^{\omega} |h(t)|^2 \Delta t\right)^{1/2},
$$

where $h(t)$ is an ω-periodic function.

2   Preliminaries

Definition 1 [10]. *A time scale $\mathbb{T}$ is an arbitrary nonempty closed subset of the real set $\mathbb{R}$ with the topology and ordering inherited from $\mathbb{R}$. The forward and backward jump operators σ, $\rho : \mathbb{T} \to \mathbb{T}$ and the graininess $\mu : \mathbb{T} \to \mathbb{R}^+$ are defined, respectively, by*

$$
\sigma(t) := \inf\{s \in \mathbb{T} : s > t\}, \quad \rho(t) := \sup\{s \in \mathbb{T} : s < t\}, \quad \mu(t) := \sigma(t) - t.
$$

The point $t \in \mathbb{T}$ is called left-dense, left-scattered, right-dense or right-scattered if $\rho(t) = t$, $\rho(t) < t$, $\sigma(t) = t$ or $\sigma(t) > t$, respectively. Points that are right-dense and left-dense at the same time are called dense. If $\mathbb{T}$ has a left-scattered maximum m, defined $\mathbb{T}^k = \mathbb{T} - \{m\}$; otherwise, set $\mathbb{T}^k = \mathbb{T}$.

Definition 2 [13]. *If $F^{\Delta}(t) = f(t)$, then we define the delta integral by*

$$\int_a^t f(s)\Delta s = F(t) - F(a).$$

Definition 3 [13]. *For each $t \in \mathbb{T}$, let N be a neighborhood of t. Then we defined the generalized derivative(or Dini derivative), $D^+u^{\Delta}(t)$ to mean that, given $\epsilon > 0$, there exists a right neighborhood $N(\epsilon) \subset N$ of t such that*

$$\frac{u(\sigma(t)) - u(s)}{\sigma(t) - s} < D^+u^{\Delta}(t) + \epsilon$$

for each $s \in N(\epsilon)$, $s > t$.
In case t is right-scattered and $u(t)$ is continuous at t, this reduce to

$$D^+u^{\Delta}(t) = \frac{u(\sigma(t)) - u(t)}{\sigma(t) - t}.$$

Next, we shall give the definition of periodic function on a time scale as following:

Definition 4 [10]. *We say that a time scale $\mathbb{T}$ is periodic if there exists $p > 0$ such that if $t \in \mathbb{T}$, then $tp \in \mathbb{T}$. For $\mathbb{T} \neq \mathbb{R}$, the smallest positive p is called the period of the time scale. Let $\mathbb{T} \neq \mathbb{R}$ be a periodic time scale with period p. We say that the function $f : \mathbb{T} \to \mathbb{R}$ is ω-periodic if there exists a natural number n such that $\omega = np$, $f(t + \omega) = f(t)$ for all $t \in \mathbb{T}$ and ω is the smallest number such that $f(t + \omega) = f(t)$. If $\mathbb{T} = \mathbb{R}$, we say that f is ω-periodic if ω is the smallest positive number such that $f(t + \omega) = f(t)$ for all $t \in \mathbb{T}$.*

Lemma 1 [13]. *Assume that f, $g : \mathbb{T} \to \mathbb{R}$ are delta differentiable at $t \in \mathbb{T}^k$. Then*

$$(fg)^{\Delta}(t) = f^{\Delta}(t)g(t) + f(\sigma(t))g^{\Delta}(t) = f(t)g^{\Delta}(t) + f^{\Delta}(t)g(\sigma(t)).$$

Lemma 2 [13]. *Let t_1, $t_2 \in [0, \omega]_{\mathbb{T}}$. If $x : \mathbb{T} \to \mathbb{R}$ is ω-periodic, then*

$$x(t) \leq x(t_1) + \int_0^{\omega} |x^{\Delta}(s)|\Delta s \quad \text{and} \quad x(t) \geq x(t_2) - \int_0^{\omega} |x^{\Delta}(s)|\Delta s.$$

Lemma 3 [13]. *Let a, $b \in \mathbb{T}$. For rd-continuous functions f, $g : [a,b]_{\mathbb{T}} \to \mathbb{R}$ we have*

$$\int_a^b |f(t)||g(t)|\Delta t \leq \left(\int_a^b |f(t)|^2 \Delta t\right)^{1/2} \left(\int_a^b |g(t)|^2 \Delta t\right)^{1/2}.$$

Lemma 4 [13]. *Let $f \in C(\mathbb{T}, \mathbb{R})$ is Δ-differentiable at t. Then*

$$\Delta^r |f(t)| \leq \text{sign}(f^\sigma * (t)) f^\Delta(t) \text{ where } f^\sigma(t) = f(\sigma(t)).$$

Definition 5. *The periodic solution $x^*(t) = (x_1^*(t), x_2^*(t), \ldots, x_n^*(t))^T$ of system (1) is said to be globally exponentially stable if there exist positive constants λ and $M = M(\lambda) \geq 1$, for any solution $x(t) = (x_1(t), x_2(t), \ldots, x_n(t))^T$ of system (1) with the initial value $\varphi(t) = (\varphi_1(t), \varphi_2(t), \ldots, \varphi_n(t))^T \in C([-\tau, 0]_\mathbb{T}, \mathbb{R}^n)$, such that*

$$\sum_{i=1}^n |x_i(t) - x_i^*(t)| \leq M(\lambda) e_{\ominus \lambda}(t, \alpha) \|\varphi - x^*\|,$$

where $\|\varphi - x^*\| = \sum_{i=1}^n \sup_{s \in [-\tau, 0]_\mathbb{T}} |\varphi_i(s) - x_i^*(s)|, \quad s \in [-\tau, 0]_\mathbb{T}$.

The following fixed point theorem of coincidence degree is crucial in the arguments of our main results, we can refer to [17], here we omit it.

Lemma 5 [17]. *Let $\mathbb{X}$, $\mathbb{Y}$ be two Banach spaces, $\Omega \subset \mathbb{X}$ be open bounded and symmetric with $0 \in \Omega$. Suppose that $L : D(L) \subset \mathbb{X} \to \mathbb{Y}$ is a linear Fredholm operator of index zero with $D(L) \cap \bar{\Omega} \neq \emptyset$ and $N : \bar{\Omega} \to \mathbb{Y}$ is L-compact. Further, we also assume that*

(H) $Lx - Nx \neq \lambda(-Lx - N(-x))$ for all $D(L) \cap \partial\Omega$, $\lambda \in (0, 1]$.

Then equation $Lx = Nx$ has at least one solution on $D(L) \cap \bar{\Omega}$.

3   Main Results

Theorem 1. *Assume that (H_1)–(H_3) hold. Suppose further that (H_4) $\Pi = (h_{ij})_{n \times n}$ is a nonsingular M matrix, where*

$$h_{ij} = a_i^m \left(\omega - a_i^M \omega^2 \sum_{k=1}^q G_{ik} \right) - \sum_{k=1}^q G_{ik} - a_i^M \omega(1 + a_i^m \omega) \sum_{j=1}^n (c_{ij}^M F_j + d_{ij}^M G_j), \quad i = j,$$

$$h_{ij} = -a_i^M \omega(1 + a_i^m \omega) \sum_{j=1}^n (c_{ij}^M F_j, +d_{ij}^M G_j), i \neq j.$$

Then system (1) has at least one ω-periodic solution.

Proof. Let $C^k[0, \omega; t_1, t_2, \ldots, t_q]_\mathbb{T} = \{x : [0, \omega]_\mathbb{T} \to \mathbb{R} | x^r(t)$ is a piecewise continuous map with first-class discontinuous points in $[0, \omega]_\mathbb{T} \cap \{t_k : k \in \mathbb{N}\}$ and at each discontinuous point it is continuous on the left$\}$, $r = 0, 1$. Take

$$\mathbb{X} = \{x \in C[0, \omega; t_1, t_2, \ldots, t_q]_\mathbb{T} : x(t + \omega) = x(t), \forall t \in [0, \omega]_\mathbb{T}, \mathbb{Y} = \mathbb{X} \times \mathbb{R}^{n \times q}$$

be two Banach spaces with the norms

$$\|x\|_{\mathbb{X}} = \sum_{i=1}^{n} |x_i|_0 \quad \text{and} \quad \|z\|_{\mathbb{Y}} = \|x\|_{\mathbb{X}} + \|y\| \quad \forall x \in \mathbb{X}, y \in \mathbb{R}^{n \times q},$$

in which $|x_i|_0 = \max\limits_{t \in [0,\omega]_{\mathbb{T}}} |x_i(t)|$, $i = 1, 2, \ldots, n$, $\| \cdot \|$ is any norm of $\mathbb{R}^{n \times q}$. It is clear that Ω satisfies all the requirement in Lemma 5 and the condition (H_4) is satisfied. In view of all the discussions above, we conclude from Lemma 5 that system (1) has at least one ω-periodic solution. This completes the proof.

Remark 1. The full proof of Theorem 1 is very similar to that of Theorem 3.1 in [12]. Here we omit the details here.

Next, we will construct some suitable Lyapunov functions to study the global exponential stability of this positive periodic solution.

Theorem 2. *Assume that* (H_1)-(H_4) *hold. Suppose further that*

(H_5) *There exist positive constants* f_j^M, g_j^M *such that* $|f_j(u)| \leq f_j^M$ *and* $|g_j(u)| \leq g_j^M$ *for all* $u \in \mathbb{R}$, $j = 1, 2, \ldots, n$.
(H_6) *There exist positive constants* A_i *such that*

$$|a_i(u) - a_i(v)| \leq A_i|u - v| \quad \text{for all } u, v \in \mathbb{R}, i = 1, 2, \ldots, n.$$

(H_7) *There exist positive constants* l_i *such that*

$$(a_i(u)b_i(u) - a_i(v)b_i(v))(u - v) \geq 0 \quad \text{and} \quad |a_i(u)b_i(u) - a_i(v)b_i(v)| \geq l_i|u - v|$$

for all $u, v \in \mathbb{R}$, $i = 1, 2, \ldots, n$.
(H_8) *There exists a positive constant* ϵ *such that* $\max_{t \in [0,\omega]_{\mathbb{T}}} \Theta_i(\epsilon, t) < 0$, *where*

$$\Theta_i(\epsilon, t) = \epsilon - K_i(1 + \epsilon\mu(t)) + (1 + \epsilon\mu(t)) \sum_{j=1}^{n} a_j^M c_{ji}^L L_i$$

$$+ \sum_{j=1}^{n} (1 + \epsilon\mu(t + \tau_{ji})) e_\epsilon(t + \tau_{ji}, t) a_j^M d_{ji}^L P_i, \, i = 1, 2, \ldots, n,$$

$$K_i = l_i - \nu_i e_i^M - \nu_i \sum_{j=1}^{n} c_{ij}^L f_j^M - \nu_i \sum_{j=1}^{n} d_{ij}^L g_j^M, \, i = 1, 2, \ldots, n.$$

(H_9) *The impulsive operators* $I_{ik}(x_i(t))$ *satisfy*

$$I_{ik}(x_i(t_k)) = -\gamma_{ik} x_i(t_k), \, 0 \leq \gamma_{ik} \leq 2, \, i = 1, 2, \ldots, n, \, k \in \mathbb{N}.$$

Then the system (1) *is globally exponentially stable.*

Proof. According to Theorem 1, we known that system (1) has an ω-periodic solution $x^*(t) = (x_1^*(t), x_2^*(t), \ldots, x_n^*(t))^T$, suppose that $x(t) = (x_1(t), x_2(t), \ldots, x_n(t))^T$ is an arbitrary solution of system (1). Then it follows from system (1) that

$$
\begin{cases}
(x_i(t) - x_i^*(t))^\Delta = -\big[a_i(x_i(t))b_i(x_i(t)) - a_i(x_i^*(t))b_i(x_i^*(t))\big] \\
+a_i(x_i(t))\Big[\sum\limits_{j=1}^{n} c_{ij}(t)f_j(x_j(t)) + \sum\limits_{j=1}^{n} d_{ij}(t)g_j(x_j(t - \tau_{ij})) + e_i(t)\Big] \\
-a_i(x_i^*(t))\Big[\sum\limits_{j=1}^{n} c_{ij}(t)f_j(x_j^*(t)) + \sum\limits_{j=1}^{n} d_{ij}(t)g_j(x_j^*(t - \tau_{ij})) + e_i(t)\Big], \\
t \in \mathbb{T}^+, \ t \neq t_k, \ k \in \mathbb{N}, \\
\Delta(x_i(t_k) - x_i^*(t_k)) = -\gamma_{ik}(x_i(t_k) - x_i^*(t_k)), \quad i = 1, 2, \ldots, n.
\end{cases}
\tag{3}
$$

Set $x_i(t) - x_i^*(t) = u_i(t), i = 1, 2, \ldots, n$. In view of system (3), for $t \in \mathbb{T}^+, t \neq t_k$, $k \in \mathbb{N}, i = 1, 2, \ldots, n$. Hence we can obtain from (H_5)–(H_7) that

$$
D^+|u_i(t)|^\Delta \leq -K_i|u_i(t) + \mu(t)u_i^\Delta(t)| + \mu(t)K_i|u_i^\Delta(t)| + a_i^M \sum_{j=1}^{n} c_{ij}^M F_j|u_j(t)|
$$

$$
+\mu(t)u_j^\Delta(t)| - a_i^M \sum_{j=1}^{n} c_{ij}^M F_j|u_j^\Delta(t)| + a_i^M \sum_{j=1}^{n} d_{ij}^M G_j|u_j(t - \tau_{ij})|
$$

for $i = 1, 2, \ldots, n$. And we have from (H_9) that

$$
|u_i(t_k^+)| = |1 - \gamma_{ik}||x_i(t_k) - x_i^*(t_k)| \leq |x_i(t_k) - x_i^*(t_k)| = |u_i(t_k)|.
$$

For any $\alpha \in [-\tau, 0]_\mathbb{T}$, we consider the Lyapunov functional:

$$
V(t) = V_1(t) + V_2(t),
$$

$$
V_1(t) = \sum_{i=1}^{n} e_\epsilon(t, \alpha)|u_i(t)|,
$$

$$
V_2(t) = \sum_{i=1}^{n}\sum_{j=1}^{n} \int_{t-\tau_{ij}}^{t} (1 + \epsilon\mu(s + \tau_{ij}))e_\epsilon(s + \tau_{ij}, \alpha)a_i^M d_{ij}^L P_j|x_j(s) - x_j^*(s)|\Delta s.
$$

For $t \in \mathbb{T}^+, t \neq t_k, k \in \mathbb{N}$, calculating the delta derivative $D^+V(t)^\Delta$ of $V(t)$ along system (3), we can get

$$
D^+V_1(t)^\Delta \leq \sum_{i=1}^{n} \Big(\epsilon - \Big[l_i - \nu_i\Big(e_i^L + \sum_{j=1}^{n} c_{ij}^L f_j^M + \sum_{j=1}^{n} d_{ij}^L g_j^M\Big)\Big](1 + \epsilon\mu(t))\Big)e_\epsilon(t, \alpha)
$$

$$
\times|x_i(t) - x_i^*(t)| + (1 + \epsilon\mu(t))e_\epsilon(t, \alpha)\sum_{i=1}^{n}\sum_{j=1}^{n} a_i^M c_{ij}^L L_j|x_j(t) - x_j^*(t)|
$$

$$
+(1 + \epsilon\mu(t))e_\epsilon(t, \alpha)\sum_{i=1}^{n}\sum_{j=1}^{n} a_i^M d_{ij}^L P_j|x_j(t - \tau_{ij}) - x_j^*(t - \tau_{ij})|
$$

and

$$D^+V_2(t)^\Delta \le \sum_{i=1}^n \sum_{j=1}^n (1 + \epsilon\mu(t + \tau_{ij}))e_\epsilon(t + \tau_{ij}, \alpha)a_i^M d_{ij}^L P_j |x_j(t) - x_j^*(t)|$$

$$-(1 + \epsilon\mu(t))e_\epsilon(t, \alpha) \sum_{i=1}^n \sum_{j=1}^n a_i^M d_{ij}^L P_j |x_j(t - \tau_{ij}) - x_j^*(t - \tau_{ij})|.$$

From the assumption (H_8), it concludes that

$$D^+(V(t))^\Delta = D^+(V_1(t) + V_2(t))^\Delta$$

$$\le \sum_{i=1}^n \left\{ \epsilon - \left[l_i - \nu_i \left(e_i^L + \sum_{j=1}^n c_{ij}^L f_j^M + \sum_{j=1}^n d_{ij}^L g_j^M \right) \right] (1 + \epsilon\mu(t)) \right.$$

$$+(1 + \epsilon\mu(t)) \sum_{j=1}^n a_j^M c_{ji}^L L_i + \sum_{j=1}^n (1 + \epsilon\mu(t + \tau_{ji}))e_\epsilon(t + \tau_{ji}, t)a_j^M$$

$$\left. \times d_{ji}^L P_i \right\} e_\epsilon(t, \alpha)|x_i(t) - x_i^*(t)| \le 0, \quad t \in \mathbb{T}^+, \ t \ne t_k, \ k \in \mathbb{N}.$$

Similar,

$$V(t_k^+) = V_1(t_k^+) + V_2(t_k^+)$$

$$\le \sum_{i=1}^n e_\epsilon(t_k, \alpha)|x_i(t_k) - x_i^*(t_k)| + \sum_{i=1}^n \sum_{j=1}^n \int_{t_k - \tau_{ij}}^{t_k} (1 + \epsilon\mu(s + \tau_{ij}))$$

$$\times e_\epsilon(s + \tau_{ij}, \alpha)a_i^M d_{ij}^L P_j |x_j(s) - x_j^*(s)|\Delta s = V(t_k), \quad k \in \mathbb{N}.$$

It follows that $V(t) \le V(0)$ for all $t \in \mathbb{T}^+$.

On the other hand, we have

$$V(0) = V_1(0) + V_2(0) \le \Gamma(\epsilon) \sum_{i=1}^n \sup_{s \in [-\tau, 0]_\mathbb{T}} |\varphi_i(s) - x_i^*(s)|,$$

where

$$\Gamma(\epsilon) = \max_{1 \le i \le n} \left\{ \sup_{\alpha \in [-\tau, 0]_\mathbb{T}} \left\{ e_\epsilon(0, \alpha) + \sum_{j=1}^n \int_{-\tau_{ji}}^0 (1 + \epsilon\mu(s + \tau_{ji}))e_\epsilon(s + \tau_{ji}, \alpha) \times a_j^M d_{ji}^L P_i \Delta s \right\} \right\}.$$

It is obviously that

$$\sum_{i=1}^n e_\epsilon(t, \alpha)|x_i(t) - x_i^*(t)| \le V(t) \le V(0) \le \Gamma(\epsilon) \sum_{i=1}^n \sup_{s \in [-\tau, 0]_\mathbb{T}} |\varphi_i(s) - x_i^*(s)|.$$

So we can finally get

$$\sum_{i=1}^n |x_i(t) - x_i^*(t)| \le \Gamma(\epsilon)e_{\ominus\epsilon}(t, \alpha)\|\varphi - x^*\|.$$

Since $\Gamma(\epsilon) \ge 1$, from Definition 5, positive periodic solution of system (1) is globally exponential stable. This completes the proof.

4   An Example

Example 1. Consider the system (1), let

$$
(a_i(x_i))_{2\times 1} = \begin{pmatrix} 10 + \frac{2}{3}\arctan|x_1| \\ 11 + \frac{2}{\pi}\arctan|x_2| \end{pmatrix}, \quad (b_i(x_i))_{2\times 1} = \frac{1}{220\pi}\begin{pmatrix} x_1 \\ x_2 \end{pmatrix},
$$

$$
(f_j(x_j))_{2\times 1} = (g_j(x_j))_{2\times 1} = \frac{1}{2}\begin{pmatrix} \sin x_1 \\ \sin x_2 \end{pmatrix}, \quad (c_{ij})_{2\times 2} = \frac{1}{2640\pi}\begin{pmatrix} \sin t & \cos t \\ \cos t & \sin t \end{pmatrix},
$$

$$
(d_{ij})_{2\times 2} = \frac{1}{2640\pi}\begin{pmatrix} \sin^2 t & \cos^2 t \\ \cos^2 t & \sin^2 t \end{pmatrix}, \quad (e_i)_{2\times 1} = \frac{\pi}{2}\begin{pmatrix} \sin t \\ \cos t \end{pmatrix},
$$

$$
(\tau_{ij})_{2\times 2} = \begin{pmatrix} 1 & 1 \\ 1 & 1 \end{pmatrix}, \quad I_{ik} = \frac{1}{240},
$$

$$
\omega = 2\pi, \quad [0, 2\pi]_{\mathbb{T}} \cap \{t_k : k \in \mathbb{N}\} = \{t_1, t_2\}.
$$

When $\mathbb{T} = \mathbb{R}$ or $\mathbb{Z}$, system (1) has at least one exponentially stable 2π-periodic solution.

Proof. By calculation, we have $a_1^m = 10$, $a_2^m = 11$, $a_1^M = 11$, $a_2^M = 12$, $\nu_1 = \nu_2 = \frac{2}{\pi}$, $l_1 = 9 - \frac{1}{\pi}$, $l_2 = 10 - \frac{1}{\pi}$, $\rho_1 = \rho_2 = \frac{1}{220\pi}$, $\delta_1 = \delta_2 = \frac{1}{220\pi}$, $L_1 = L_2 = \frac{1}{2}$, $P_1 = P_2 = \frac{1}{2}$, $f_1^M = f_2^M = \frac{1}{2}$, $g_1^M = g_2^M = \frac{1}{2}$, $c_{11}^L = c_{12}^L = c_{21}^L = c_{22}^L = \frac{1}{2640\pi}$, $d_{11}^L = d_{12}^L = d_{21}^L = d_{22}^L = \frac{1}{2640\pi}$, $e_1^L = e_2^L = \frac{\pi}{2}$, $G_{11} = G_{12} = G_{21} = G_{22} = \frac{1}{240}$. It is obvious that (H_1)–(H_3), (H_5)–(H_9) are satisfied. Furthermore, we can easily calculate that

$$
\Pi = (h_{ij})_{2\times 2} \approx \begin{pmatrix} 7 & -1 \\ -1.1 & 8 \end{pmatrix}
$$

is a nonsingular M matrix, thus (H_4) is satisfied.

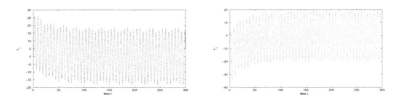

Fig. 1. Trajectory of x(t) of the CGNNs in the example with impulses.

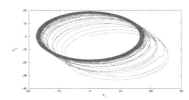

Fig. 2. Phase portrait of the CGNNs in the example with impulses.

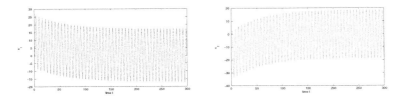

Fig. 3. Trajectory of x(t) of the CGNNs in the example without impulses.

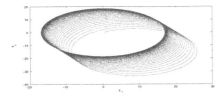

Fig. 4. Phase portrait of the CGNNs in the example without impulses.

Case 1. When $\mathbb{T} = \mathbb{R}$, $\mu(t) = 0$. Take $\epsilon = 1$, we have that

$$\Theta_1(\epsilon, t) = \Theta_1(1, t) \approx -6.67468 < 0 \quad \text{and} \quad \Theta_2(\epsilon, t) = \Theta_2(1, t) \approx -7.67468 < 0.$$

Hence (H_8) holds. By Theorems 1 and 2, system (1) has at least one exponentially stable 2π-periodic solution.

Case 2. When $\mathbb{T} = \mathbb{Z}$, $\mu(t) = 1$. Take $\epsilon = 1$, we have that

$$\Theta_1(\epsilon, t) = \Theta_1(1, t) \approx -15.3514 < 0 \quad \text{and} \quad \Theta_2(\epsilon, t) = \Theta_2(1, t) \approx -17.3514 < 0.$$

Hence (H_8) holds. By Theorems 1 and 2, system (1) has at least one exponentially stable 2π-periodic solution.

In fact, by simulations, in view of Theorems 1 and 2, the unique 2π-periodic solution $x^*(t)$ is globally exponentially stable. Figures 1, 2, 3 and 4 demonstrate the validity of our results (initial values are $[x_1(0), x_2(0)]^T \equiv [0.2, -0.1]^T$). For their corresponding asymptotic 2π-periodic sequences.

Acknowledgments. This work is supported by the National Natural Sciences Foundation of Peoples Republic of China under Grant 11561070, and the Natural Scientific Research Fund Project of Yunnan Province (No. 2014FD049), and the Young Teacher Program of Yuxi Normal University.

References

1. Cohen, M., Grossberg, S.: Absolute stability and global pattern formation and parallel memory storage by competitive neural networks. IEEE Trans. Syst. Man Cybern. **13**, 815–816 (1983)

2. Cao, J., Song, Q.: Stability in cohen-grossberg-type bidirectional associative memory neural networks with time-varying delays. Nonlinearity **19**, 1601–1617 (2006)
3. Chen, Z., Ruan, J.: Global stability analysis of impulsive cohen-grossberg neural networks with delay. Phys. Lett. A **345**, 101–111 (2005)
4. Xiang, H., Cao, J.: Almost periodic solution of cohen-grossberg neural networks with bounded and unbounded delays. Nonlinear Anal. Real World Appl. **10**(4), 2407–2419 (2009)
5. Li, Y.: Existence and Stability of Periodic Solutions for Cohen-Grossberg Neural Networks with Multiple Delays. Chaos Solitons Fractals **20**, 459–466 (2004)
6. Li, C., Li, Y., Ye, Y.: Exponential stability of fuzzy cohen-grossberg neural networks with time delays and impulsive effects. Commun. Nonlinear Sci. Numer. Simul. **15**(11), 3599–3606 (2010)
7. Li, Y.: Almost automorphic solution for neutral type high-order hopfield neural networks with delays inleakage terms on time scale. Appl. Math. Comput. **242**, 679–693 (2014)
8. Li, Y., Chen, X., Zhao, L.: Stability and existence of periodic solutions to delayed cohen-grossberg BAM neural networks with impulses on time scales. Neurocomputing **72**, 1621–1630 (2009)
9. Li, Y., Wang, C.: Pseudo almost periodic functions and pseudo almost periodic solutions to dynamic equations on time scales. Adv. Differ. Eqn. 77 (2012)
10. Bohner, M., Peterson, A.: Dynamic Equations on Time Scales: An Introduction with Applications. Birkhauser, Boston (2001)
11. Yang, Z., Xu, D.: Impulsive effects on stability of cohen-grossberg neural networks with variable delays. Appl. Math. Comput. **177**(1), 63–78 (2006)
12. Li, Y.: Periodic solutions of non-autonomous cellular neural networks with impulses and delays on time scales. IMA J. Math. Control Inf. **13**, 273–291 (2014)
13. Lakshmikantham, V., Vatsala, A.S.: Hybird systems on time scales. J. Comput. Appl. Math. **141**, 227–235 (2002)
14. Kaufmann, E., Raffoul, Y.: Periodic solutions for a neutral nonlinear dynamical equation on a time scale. J. Math. Anal. Appl. **319**(1), 315–325 (2006)
15. Agarwal, R., Bohner, M., Peterson, A.: Inequalities on times scales: a survey. Math. Inequalities Appl. **4**(4), 535–557 (2001)
16. Gong, W., Liang, J., Cao, J.: Matrix measure method for global exponential stability of complex-valued recurrent neural networks with time-varying delays. Neural Netw. **70**, 81–89 (2015)
17. Cho, Y.J., Chen, Y.Q.: Topological Degree Theory and Application. Taylor & Francis Group, Boca Raton, London, New York (2006)

Intelligent Control

Improved Direct Finite-control-set Model Predictive Control Strategy with Delay Compensation and Simplified Computational Approach for Active Front-end Rectifiers

Xing Liu, Dan Wang[✉], and Zhouhua Peng

School of Marine Engineering, Dalian Maritime University, 116026 Dalian,
People's Republic of China
liuxing@dlmu.edu.cn, dwangdl@gmail.com, zhouhuapeng@gmail.com

Abstract. In this paper, an improved direct finite-control-set model predictive control with delay compensation and simplified computational approach is proposed for active front-end rectifiers. Specifically, an active voltage vector which only requires one exploration is directly selected and applied to avoid the exhaustive exploration for testing all feasible voltage vectors during one switching period. Meanwhile, a delay compensation method is presented. The control delay caused by the calculation effort can be compensated. The performance of the proposed method can be improved. Simulation results are presented to demonstrate the efficacy of the proposed method.

Keywords: Direct finite-control-set model predictive control · Active front-end rectifiers · Delay compensation

1 Introduction

In various application areas, such as electrical drives [1] and distributed generation units (DGs) [2], active front-end rectifiers (AFEs) are often employed to connect the electrical system to the grid. To obtain high power quality and fast dynamic response, advanced control techniques are essential. Recently, various control schemes have been proposed for AFEs. Direct power control (DPC) is one of the most popular control methods because of its predominant transient performance [3,4]. This strategy is based on the evaluation of the active and reactive instantaneous power errors values without any internal current control loops and no pulse-width modulation (PWM) block. However, large power ripple and variable switching frequency are two major drawbacks of this approach. To overcome these issues, the authors of literature [5] present a predictive DPC (P-DPC) technique using deadbeat control principle. This method eliminates the classic linear controller by a predictive model of the system. This model is used to calculate the required reference voltage. Then, the required voltage vector is applied by using a pulse width modulator. Constant switching frequency can be thus achieved, and less current harmonics can be obtained.

© Springer International Publishing Switzerland 2016
L. Cheng et al. (Eds.): ISNN 2016, LNCS 9719, pp. 223–232, 2016.
DOI: 10.1007/978-3-319-40663-3_26

An alternative control strategy, based on finite-control-set model predictive control (FCS-MPC), is presented in [6]. Compared with the method in [5], the proposed strategy utilizes the inherent discrete nature of the power converter without a modulator to solve the optimization problem by using a simple iterative algorithm. Besides, the FCS-MPC exhibits highly dynamic response and excellent performance [7–9]. However, the disadvantage of the approach is the large amount of calculation efforts, which will result in the time delay of control command. The time delay can deteriorate the performance of the system [10]. Therefore, the control delay caused by the calculation effort is a very important factor that needs to be considered in the real-time implementation of FCS-MPC algorithm.

An improved direct finite-control-set model predictive control (DFCS-MPC) with delay compensation and simplified computational approach is proposed for AFEs in this paper. It considers the computational efforts and the time delay compensation simultaneously. In order to reduce the high amount of calculation burdens, an active voltage vector which only requires one exploration is selected and applied by using a sector distribution method. The exhaustive exploration can be avoided, the process of selection of desired voltage vector can be optimized. Meanwhile, a delay compensation method is presented. The control delay caused by the calculation effort can be compensated. The performance of the proposed method can be improved. Simulation results are presented to demonstrate the efficacy of the proposed method.

This paper is organized as follows: Section 2 introduces FCS-MPC method of the AFEs. In Sect. 3, the simplified DFCS-MPC with delay compensation for AFEs is described. Section 4 provides the simulation results to illustrate the proposed control scheme. Section 5 concludes this article.

2   FCS-MPC Method of AFEs

The common topology of the AFEs is shown in Fig. 1. The AFEs with six switches is connected to the three-phase supply voltages through the input coupling inductances L_g and resistances R_g at its input side.

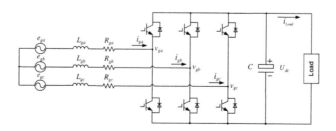

Fig. 1. The common topology of the AFEs.

From Fig. 1, the three-phase input voltages and input currents can be expressed in the $\alpha - \beta$ reference frame as

$$e_g = \frac{2}{3}(e_{ga} + \alpha e_{gb} + \alpha^2 e_{gc}),\tag{1}$$

$$i_g = \frac{2}{3}(i_{ga} + \alpha i_{gb} + \alpha^2 i_{gc}),\tag{2}$$

where $\alpha = e^{j(2\pi/3)}$. The rectifier voltage vector is calculated from the switching state and DC-link voltage, and can be expressed as follows:

$$v_g = S_g U_{dc},\tag{3}$$

where U_{dc} is the DC-link voltage, and S_g is the switching state vector of the rectifier defined as

$$S_g = \frac{2}{3}(S_{ga} + \alpha S_{gb} + \alpha^2 S_{gc}),\tag{4}$$

where S_{ga}, S_{gb}, and S_{gc} are the switching states of each rectifier leg, as shown in Fig. 1, and represent the value of 0 if S_{gx} is OFF, or 1 if S_{gx} is ON (x = a, b, c).

To obtain a dynamic model for the grid input currents, the currents on the grid sides are described

$$L_g \frac{di_g}{dt} = e_g - v_g - R_g i_g,\tag{5}$$

The derivative of the input currents in the continuous-time model are then approximated on the basis of the forward Euler approximation with one switching period T_s as

$$\frac{di_g}{dt} \approx \frac{i_g(k+1) - i_g(k)}{T_s},\tag{6}$$

The predicted input currents are calculated using the discrete-time equation

$$i_g(k+1) = (1 - \frac{R_g T_s}{L_g})i_g(k) + \frac{T_s}{L_g}\left[e_g(k) - v_g(k)\right].\tag{7}$$

The proposed predictive control algorithm block is shown in Fig. 2. As could be seen in Fig. 2, under the assumption of balanced three-phase system, amounts of supply voltage and input current of the rectifier are measured and transformed to the stationary reference frame $\alpha - \beta$, then active and reactive powers are calculated as [11]:

$$\begin{aligned}P_g(k+1) &= \mathrm{Re}\left\{e_g(k+1)\bar{i}_g(k+1)\right\}\\ &= (e_{g\alpha}i_{g\alpha} + e_{g\beta}i_{g\beta}),\end{aligned}\tag{8}$$

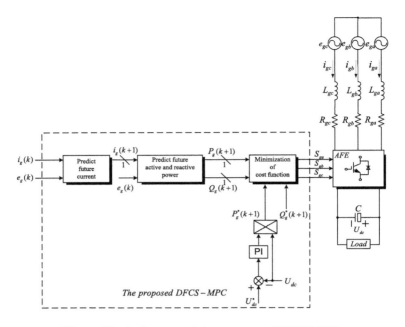

Fig. 2. Block diagram of the proposed DFCS-MPC.

$$Q_g(k+1) = \text{Im}\left\{e_g(k+1)\bar{i}_g(k+1)\right\}$$
$$= (e_{g\beta}i_{g\alpha} - e_{g\alpha}i_{g\beta}), \tag{9}$$

where $e_g(k+1)$ and $i_g(k+1)$ are the three-phase grid predicted input voltages and input currents, respectively.

Then, a cost function is applied as a criterion to select the optimal states

$$f_{AFE} = \left|P_g^* - P_g(k+1)\right|^2 + \left|Q_g^* - Q_g(k+1)\right|^2, \tag{10}$$

where $P_g(k+1)$ and $Q_g(k+1)$ are the predicted active and reactive power, P_g^* and Q_g^* are the reference active and reactive power.

If the sampling period is assumed to be small in comparison with the period of the power-source voltage

$$e_g(k+1) \approx e_g(k). \tag{11}$$

3   The Simplified DFCS-MPC Method with Delay Compensation for AFEs

One of the major disadvantages of conventional FCS-MPC method has the large amount of computational burdens, which will result in the time delay of control command. This will hinder the implementation of the FCS-MPC strategy. To resolve this issue, a simplified computational approach is proposed in this section.

3.1 Calculating the Required Rectifier Voltage Vector with DB-PDPC Solution

From Fig. 1, the relationship between the voltages and currents can be expressed in the $\alpha - \beta$ reference frame as

$$L_g \frac{d}{dt} \begin{bmatrix} i_{g\alpha}(t) \\ i_{g\beta}(t) \end{bmatrix} = \begin{bmatrix} e_{g\alpha}(t) \\ e_{g\beta}(t) \end{bmatrix} - \begin{bmatrix} v_{g\alpha}(t) \\ v_{g\beta}(t) \end{bmatrix} - R_g \begin{bmatrix} i_{g\alpha}(t) \\ i_{g\beta}(t) \end{bmatrix}. \tag{12}$$

The values of input resistances are very small. Hence neglecting the effect of input resistances and using the approximate Euler formula, the variation of input current vector is obtained as follows:

$$\begin{bmatrix} i_{g\alpha}(k+1) - i_{g\alpha}(k) \\ i_{g\beta}(k+1) - i_{g\beta}(k) \end{bmatrix} = \frac{T_s}{L_g} \left(\begin{bmatrix} e_{g\alpha}(k) \\ e_{g\beta}(k) \end{bmatrix} - \begin{bmatrix} v_{g\alpha}(k) \\ v_{g\beta}(k) \end{bmatrix} \right). \tag{13}$$

The variation of active and reactive powers during one switching period T_s is calculated as follows:

$$\begin{bmatrix} P_g(k+1) - P_g(k) \\ Q_g(k+1) - Q_g(k) \end{bmatrix} = \frac{T_s}{L_g} \begin{bmatrix} e_{g\alpha}(k) & e_{g\beta}(k) \\ e_{g\beta}(k) & -e_{g\alpha}(k) \end{bmatrix} \tag{14}$$

$$\times \left(\begin{bmatrix} e_{g\alpha}(k) \\ e_{g\beta}(k) \end{bmatrix} - \begin{bmatrix} v_{g\alpha}(k) \\ v_{g\beta}(k) \end{bmatrix} \right).$$

The objective of DB-PDPC for the following fixed period T_s is to control the instantaneous active and reactive

$$\begin{bmatrix} \Delta P_g(k+1) \\ \Delta Q_g(k+1) \end{bmatrix} = \begin{bmatrix} P_g^*(k+1) - P_g(k+1) \\ Q_g^*(k+1) - Q_g(k+1) \end{bmatrix} = \begin{bmatrix} 0 \\ 0 \end{bmatrix}, \tag{15}$$

where $P_g^*(k+1)$ and $Q_g^*(k+1)$ are the predicted reference active and reactive power.

Therefore, the required rectifier voltage vector is expressed as follows:

$$\begin{bmatrix} v_{g\alpha}^*(k) \\ v_{g\beta}^*(k) \end{bmatrix} = \begin{bmatrix} e_{g\alpha}(k) \\ e_{g\beta}(k) \end{bmatrix} - \frac{L_g}{T_s(e_{g\alpha}^2(k) + e_{g\beta}^2(k))} \tag{16}$$

$$\times \begin{bmatrix} e_{g\alpha}(k) & e_{g\beta}(k) \\ e_{g\beta}(k) & -e_{g\alpha}(k) \end{bmatrix} \times \begin{bmatrix} P_g^*(k+1) - P_g(k) \\ Q_g^*(k+1) - Q_g(k) \end{bmatrix}.$$

If the tracking error of DC-link voltage is assumed constant over two successive sampling periods

$$\begin{bmatrix} P_g^*(k+1) \\ Q_g^*(k+1) \end{bmatrix} = \begin{bmatrix} 2P_g^*(k) - P_g^*(k-1) \\ Q_g^*(k) \end{bmatrix}. \tag{17}$$

As a consequence, the required rectifier voltage vector to be applied during each switching period is given by the following equation:

$$\begin{bmatrix} v_{g\alpha}^*(k) \\ v_{g\beta}^*(k) \end{bmatrix} = \begin{bmatrix} e_{g\alpha}(k) \\ e_{g\beta}(k) \end{bmatrix} - \frac{L_g}{T_s(e_{g\alpha}^2(k) + e_{g\beta}^2(k))} \tag{18}$$

$$\times \begin{bmatrix} e_{g\alpha}(k) & e_{g\beta}(k) \\ e_{g\beta}(k) & -e_{g\alpha}(k) \end{bmatrix} \times \begin{bmatrix} 2P_g^*(k) - P_g^*(k-1) - P_g(k) \\ Q_g^*(k) - Q_g(k) \end{bmatrix}.$$

3.2   The Optimized Exploration of DFCS-MPC with Space Vector Modulation (SVM)

A space-vector diagram containing the eight output vectors and the corresponding switching states of the rectifier can be generated, and is shown in Fig. 3.

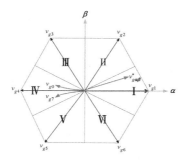

Fig. 3. Sector information distribution of voltage vector space.

The corresponding sector information of the $v_{g\alpha\beta}^*$ are obtained by using space vector modulation technique. If the $v_{g\alpha\beta}^*$ falls into sectors I~VI, the v_{gn} located in the corresponding sector will be closest to $v_{g\alpha\beta}^*$. Therefore, only one rectifier voltage vector (nonzero voltage vector) is, in the proposed DFCS-MPC method, directly utilized for power predictions, not only to perform DFCS-MPC but also to reduce the calculation burden.

3.3   Delay Compensation Method for AFEs

The DFCS-MPC algorithm needs the big calculation efforts, as compared with a classical control scheme. When the DFCS-MPC is experimentally implemented, the time delay of control command is inevitable. The time delay can cause some problems without proper compensation. In this section, a delay compensation method is presented by using the reorganized cost function for AFEs. The proposed algorithm is shown in Fig. 4. The red line in Fig. 4 represents the reference power P_g^*. The broken blue lines represent reference predictive value P_g^p, while the solid ones represent the actual power. The ideal arrangement of sampling and updating points depicted in Fig. 4(a) cannot be implemented in practical systems due to the time boundedness for voltage/current samplings and the algorithm calculations.

In ideal case, the power value is calculated at time t_k, and the optimal switching state is instantly finished. The switching state that minimizes the error at time t_k is selected and applied at time t_{k-1}. Then, the power reaches the predicted value at t_k. When implementing the predictive controller in a real system, a high amount of calculation will result in the time delay of control command, as shown in Fig. 4(b). The operation of the predictive control with a compensation delay is shown in Fig. 4(c). The previous values of the power reference

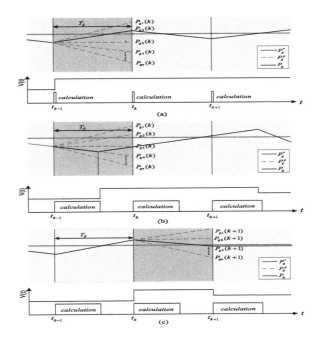

Fig. 4. Predictive power control algorithm. (a) Without delay. (b) With delay and without compensation. (c) With delay and compensation. (Color figure online)

are used to predict the power reference value at time t_{k+1}. Thus, the switching state that minimizes the error at time t_{k+1} is selected and applied at time t_k. Therefore, the time delay of control command can be compensated.

To achieve this, the modified cost function is defined as

$$f_{AFE} = \left| P_g^*(k+1) - P_g(k+1) \right|^2 \tag{19}$$
$$+ \left| Q_g^*(k+1) - Q_g(k+1) \right|^2,$$

where the reference reactive power $Q_g^*(k+1) = Q_g^*(k)$ is set to zero for unity power factor operation.

However, the future reference power value required by (19) is unknown. Therefore, it has to be predicted from the previous values of the power reference using a four-order extrapolation given by

$$P_g^*(k+1) = 10 P_g^*(k-1) - 20 P_g^*(k-2) + 15 P_g^*(k-3) \tag{20}$$
$$- 4 P_g^*(k-4).$$

4   Simulation Results

Simulation results are presented, illustrating the effectiveness of the proposed DFCS-MPC method in this section. The system simulations are carried out

230   X. Liu et al.

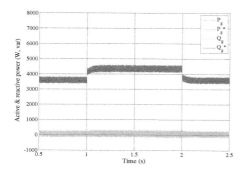
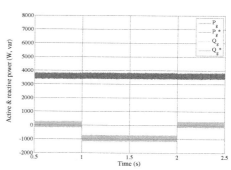

Fig. 5. Active power tracking with proposed DFCS-MPC. (Color figure online)

Fig. 6. Reactive power tracking with proposed DFCS-MPC. (Color figure online)

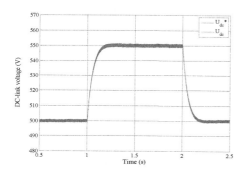
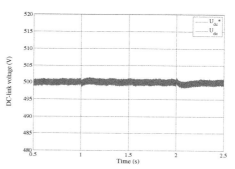

Fig. 7. DC-link voltage with a variation in P_g^*. (Color figure online)

Fig. 8. DC-link voltage with a variation in Q_g^*. (Color figure online)

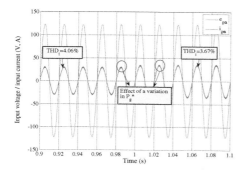
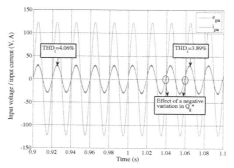

Fig. 9. Input voltage/current with a variation in P_g^*. (Color figure online)

Fig. 10. Input voltage/current with a variation in Q_g^*. (Color figure online)

Table 1. System and control parameters.

line resistance	R_g	$0.01\,\Omega$
line inductance	L_g	$10\,\text{mH}$
DC-link capacitor	C	$1100\,\mu\text{F}$
load resistance	R_L	$70\,\Omega$
phase voltage peak	e_g	$100\,\text{V}$
grid frequency	f	$50\,\text{Hz}$
DC-link reference voltage	U_{dc}	$500\,\text{V}$
sampling and control period	T_s	$25\mu\,\text{s}$

in MATLAB/Simulink environment. The system and control parameters are presented in Table 1.

The simulation results are given in Figs. 5, 6, 7, 8, 9 and 10. The dynamic behaviors of AFEs under step changes in DC-link voltage reference U_{dc}^* and reactive power reference Q_g^* are shown in Figs. 5 and 6. They show that the active/reactive power of the AFEs are able to track the reference value accurately. Thus, the power waveform seems smooth and stays constant both in transient and steady-state, indicating excellent effects of DFCS-MPC. Figure 7 shows transient responses when the DC-link voltage reference changes from 500 V to 550 V at $t = 1\,\text{s}$ and back to 500 V at $t = 2\,\text{s}$. The results show that fast and accurate tracking of dynamic DC-link voltage reference can be achieved. Figure 8 shows that the DC-link voltage is able to track the reference value accurately. The satisfactory steady-state performances can be obtained. Figure 9 shows the quasi-sinusoidal supply current with a total harmonic distribution (THD) value of 4.06 % before the step $t = 1\,\text{s}$ to 3.67 % after the step in active power reference. Since in an active front end the exchange of power is between the AC side and DC side, increasing the DC-link voltage reference essentially means that the supply current is increased as well. Figure 10 shows the input current with a THD value of 4.06 % before the step $t = 1\,\text{s}$ to 3.89 % after the step in reactive power reference.

5   Conclusions

An improved DFCS-MPC method for AFEs is proposed in this paper. It takes into account the computational effort and delay compensation method simultaneously. Firstly, the DFCS-MPC and space vector modulation technique are brought together to reduce the high amount of calculation efforts. Compared with the conventional FCS-MPC method, the proposed strategy avoids the exhaustive exploration for testing all feasible voltage vectors, which reduces the execution time for the prediction process. Secondly, the proposed method can mitigate performance degradation caused by the time delay of control command. The results confirm that the efficacy of the proposed method.

Acknowledgement. This work was in part supported by the National Nature Science Foundation of China under Grants 61273137, 51209026, 51579023, 51579022, and in part by the China Postdoctoral Science Foundation under Grant 2015M570247, and in part by the Fundamental Research Funds for the Central Universities under Grants 3132016201, 3132014321.

References

1. Davari, S.A., Khaburi, D.A., Kennel, R.: An improved FCS-MPC algorithm for an induction motor with an imposed optimized weighting factor. IEEE Trans. Power Electron. **27**(3), 1540–1551 (2012)
2. Dragicevic, T., Lu, X., Vasquez, J.C., Guerrero, J.M.: DC microgrids-Part II: a review of power architectures, applications, and standardization issues. IEEE Trans. Power Electron. **31**(5), 3528–3549 (2016)
3. Bouafia, A., Gaubert, J.P, Krim F.: Analysis and design of new switching table for direct power control of three-phase PWM rectifier. In: International Power Electronics and Motion Control Conference, pp. 703–709 (2008)
4. Martínez, J.A., García, J.E., Martín, D.S., Arnaltes, S.: A new variable-frequency optimal direct power control algorithm. IEEE Trans. Ind. Electron. **60**(4), 1442–1451 (2013)
5. Zhang, Y.C., Xie, W., Zhang, Y.C.: Deadbeat direct power control of three-phase pulse-width modulation rectifiers. IET Power Electron. **7**(6), 1340–1346 (2014)
6. Rodríguez, J., Pontt, J., Silva, C.A., Correa, P., Lezana, P., Cortés, P., Ammann, U.: Predictive current control of a voltage source inverter. IEEE Trans. Power Electron. **54**(1), 495–503 (2007)
7. Cho, Y., Lee, K.B.: Virtual-flux-based predictive direct power control of three-phase PWM rectifiers with fast dynamic response. IEEE Trans. Power Electron. **31**(4), 3348–3359 (2016)
8. Tarisciotti, L., Zanchetta, P., Watson, A., Clare, J.C., Degano, M., Bifaretti, S.: Modulated model predictive control for a three-phase active rectifier. IEEE Trans. Ind. Applications **51**(2), 1610–1620 (2015)
9. Cheng, L., Liu, W.C., Hou, Z.G., Yu, J.Z., Tan, M.: Neural-network-based nonlinear model predictive control for piezoelectric actuators. IEEE Trans. Ind. Appl. **62**(12), 7717–7727 (2015)
10. Hu, J.B.: Improved dead-beat predictive DPC strategy of grid-connected DC-AC converters with switching loss minimization and delay compensations. IEEE Trans. Ind. Inf. **9**(2), 728–738 (2013)
11. Peng, F.Z., Lai, J.S.: Generalized instantaneous reactive power theory for three-phase power systems. IEEE Trans. Instru. Meas. **45**(1), 293–297 (1996)

Distributed Tracking Control of Uncertain Multiple Manipulators Under Switching Topologies Using Neural Networks

Long Cheng[1]([✉]), Ming Cheng[2], Hongnian Yu[3], Lu Deng[4], and Zeng-Guang Hou[1]

[1] State Key Laboratory of Management and Control for Complex Systems, Institute of Automation, Chinese Academy of Sciences, Beijing 100190, China
long.cheng@ia.ac.cn
[2] Department of Petroleum Engineering, Harbin Institute of Petroleum, Harbin 150028, China
[3] Department of Computing, Bournemouth University, Poole BH12 5BB, UK
[4] School of Statistics and Mathematics, Central University of Finance and Economics, Beijing 100081, China

Abstract. The distributed tracking control of a group of manipulators under switching directed topologies is studied. Each manipulator is modeled by the Euler-Lagrange dynamics which includes uncertainties and external disturbances. The proposed controller has the neural network approximation unit for compensating uncertainties and the robust term for counteracting external disturbances. It can be proved that when the communication topology switches among a set of graphes which have a spanning tree and have no loop structure, the final tracking error can be reduced as small as possible.

Keywords: Leader-following problem · Manipulator · Uncertainty · Neural networks · Switching topologies

1 Introduction

The distributed coordination control of a group of manipulators has been studied extensively due to its potential applications in the industrial assembly and manufacturing. In the literature, many research papers have been published to investigate this topic with two control objectives: the leaderless consensus problem [1–9] and the leader-following/tracking problem [10–17]. For the leaderless consensus problem, all manipulators are required to achieve a consensus on certain value in the joint space [2–8] or the task space [1,9]. However, the leaderless consensus can only be regarded as the "self-organization behavior". To make the whole group have some predesigned global motions, the leader-following problem should be considered, which is relatively more difficult to be solved compared to the consensus problem. And the leader-following problem becomes more challenging if the leader's state is set as a time-varying trajectory rather than a fixed value.

© Springer International Publishing Switzerland 2016
L. Cheng et al. (Eds.): ISNN 2016, LNCS 9719, pp. 233–241, 2016.
DOI: 10.1007/978-3-319-40663-3_27

From the communication constraint aspect, most results in the literature assume that the communication topologies among manipulators are time-invariant. Because of the link failure and the packet drop, this assumption cannot always hold in practice. To address this issue, few attempts have been made to study the coordination control of multiple-manipulator systems under switching topologies. In [7], the consensus of a group of manipulators with actuator faults has been considered and the communication topology can switch among a family of connected graphs. In [5,9], the consensus problem of networked manipulators is investigated and it is proved that the consensus can be achieved in the joint space [5] and in the task space [9] if the switching topologies are jointly connected. However, these algorithms only work when the manipulator has the uncertain parameters satisfying the "parameter-in-linearity" condition. In [13], the leader-following problem of multiple-manipulator systems is taken into account where the leader's trajectory is generated by an exosystem. The switching topologies are only required to jointly have a spanning tree. However, every follower needs to know the state matrix of the exosystem which belongs to the global information. By the above analysis, it can be seen that there are still rooms left for further improvements.

This paper studies the distributed leader-following/tracking control of a group of manipulators whose dynamics includes uncertainties not satisfying the "parameter-in-linearity" condition. Neural networks are employed to approximate these uncertainties, and the approximation error and external disturbances are counteracted by the robust term. The communication topology among manipulators is directed and time-varying. It has been proved that when the communication topology switches among a set of directed graphs which have the spanning tree and have no loop structure, the steady-state tracking error can be reduced as small as possible if parameters in the proposed controller are appropriately chosen. This paper is a continuation of the previous work [12]. The main improvements are that: the controlled object is the robotic manipulator rather than the one with the "relative degree of one" dynamics; and the switching topology is studied.

The following notations will be used throughout this paper: $1_n = (1, 1, \cdots, 1) \in \mathbb{R}^n$; $0_n = (0, 0, \cdots, 0) \in \mathbb{R}^n$; I_n denotes the $n \times n$ dimensional identity matrix; $\otimes$ denotes the Kronecker operator; For a given matrix X, $\|X\|$ denotes its Euclidean norm; $\|X\|_F$ denotes its Frobenius norm.

2   Preliminary Results and Problem Formulation

Consider a group of N manipulators whose topological connections are modeled by a graph $\mathcal{G} = (\mathcal{V}_\mathcal{G}, \mathcal{E}_\mathcal{G}, \mathcal{A}_\mathcal{G})$. Here $\mathcal{V}_\mathcal{G} = \{v_1, \cdots, v_N\}$ denotes the node set and each node represents one manipulator; $\mathcal{E}_\mathcal{G} = \{e_{ij} \triangleq (v_j, v_i)\}$ denotes the edge set and $e_{ij} \in \mathcal{E}_\mathcal{G}$ implies that there exists a directed information link from the jth manipulator to the ith manipulator; $\mathcal{A}_\mathcal{G} = [a_{ij} \geq 0]_{N \times N}$ denotes the weight matrix, and $a_{ij} > 0$ if $e_{ij} \in \mathcal{E}_\mathcal{G}$, $a_{ij} = 0$ if $e_{ij} \notin \mathcal{E}_\mathcal{G}$. It is assumed that there is no self-loop in the system, i.e., $e_{ii} \notin \mathcal{E}_\mathcal{G}$. A sequence of directed edges $e_{i_2,i_1}, e_{i_3,i_2}, \cdots, e_{i_k,i_{k-1}}$

is called a directed path from the node v_{i_1} to the node v_{i_k}. Define the Laplacian matrix of $\mathcal{G}$ as $L_{\mathcal{G}} = \text{diag}\{\sum_{i=1}^{N} a_{1i}, \cdots, \sum_{i=1}^{N} a_{Ni}\} - \mathcal{A}_{\mathcal{G}}$.

The dynamics of each manipulator is described by the following Euler Lagrange model

$$M_i(q_i(t))\ddot{q}_i(t) + C_i(q_i(t), \dot{q}_i(t))\dot{q}_i(t) + G_i(q_i(t)) = \tau_i(t) + \tau_{id}(t), \quad i = 1, \cdots, N, \tag{1}$$

where $q_i(t) \in \mathbb{R}^n$, $\dot{q}_i(t) \in \mathbb{R}^n$, $\ddot{q}_i(t) \in \mathbb{R}^n$ denote the joint position, the joint velocity and the joint acceleration of the ith manipulator, respectively; $M_i(q_i(t)) \in \mathbb{R}^{n \times n}$, $C_i(q_i(t), \dot{q}_i(t)) \in \mathbb{R}^{n \times n}$ and $G_i(q_i(t)) \in \mathbb{R}^n$ denote the inertia matrix, the centripetal-Coriolis matrix and the gravity vector of the ith manipulator, respectively; $\tau_i(t) \in \mathbb{R}^n$ and $\tau_{id}(t) \in \mathbb{R}^n$ denote the joint torque and the external disturbance of the ith manipulator, respectively. To simplify the notions, $q_i(t)$, $\dot{q}_i(t)$, $\ddot{q}_i(t)$, $\tau_i(t)$ and $\tau_{id}(t)$ are written as q_i, $\dot{q}_i$, $\ddot{q}_i$, τ_i and τ_{id}, respectively, if no confusion is caused.

Regarding the dynamics defined by (1), there are two properties which are useful for the study conducted in this paper.

Property 1 ([18]). *The inertia matrix $M_i(q_i)$ is symmetric and positive definite, and satisfies the following inequalities: $\alpha_1\|y\|^2 \leq y^T M_i(q_i)y \leq \alpha_2\|y\|^2$, $\forall y \in \mathbb{R}^n$, $i = 1, \cdots, N$, where α_1 and α_2 are known positive constants.*

Property 2 ([18]). *The time derivative of the inertia matrix and the centripetal-Coriolis matrix satisfy the skew symmetric relation; that is, $y^T(\dot{M}_i(q_i) - 2C(q_i, \dot{q}_i))y = 0$, $\forall y \in \mathbb{R}^n$.*

The following assumption on the external disturbance should also be made.

Assumption 1. *The external disturbance τ_{id} in (1) is bounded by a given constant Δ_i: $\|\tau_{id}\| \leq \Delta_i$.*

The desired joint position trajectory is given by $q_0(t)$. The following assumption on $q_0(t)$ should be made, which always holds in practice.

Assumption 2. *The desired joint position q_0 and its derivatives up to the third order are all bounded. That is: there exists a constant M_0 such that $\|q_0\| + \|\dot{q}_0\| + \|\ddot{q}_0\| + \|\dddot{q}_0\| < M_0$.*

This desired trajectory is only accessible to a part of manipulators. Define a vector $b_{\mathcal{G}} = (b_1, \cdots, b_N)^T \in \mathbb{R}^N$. If the ith manipulator has an access to the desired joint information, then $b_i > 0$; otherwise $b_i = 0$. If the desired joint information is considered as a new node v_0, the augmented graph $\bar{\mathcal{G}} = (\mathcal{V}_{\bar{\mathcal{G}}}, \mathcal{E}_{\bar{\mathcal{G}}}, \mathcal{A}_{\bar{\mathcal{G}}})$ can be obtained, where $\mathcal{V}_{\bar{\mathcal{G}}} \triangleq \{v_0\} \bigcup \mathcal{V}_{\mathcal{G}}$, $\mathcal{E}_{\bar{\mathcal{G}}} \triangleq \{e_{i0}|b_i > 0\} \bigcup \mathcal{E}_{\mathcal{G}}$, and $\mathcal{A}_{\bar{\mathcal{G}}} = \begin{pmatrix} 0, b_{\mathcal{G}}^T \\ 0_N, \mathcal{A}_{\mathcal{G}} \end{pmatrix}$. If there is a directed path from v_0 to any other node v_i in $\bar{\mathcal{G}}$, then $\bar{\mathcal{G}}$ is called to have a spanning tree.

Lemma 1. *If $\bar{\mathcal{G}}$ has a spanning tree, then all eigenvalues of the matrix $(L_{\mathcal{G}}+B_{\mathcal{G}})$ have the positive real parts ($B_{\mathcal{G}} \in \mathbb{R}^{N \times N}$ is the diagonal matrix whose diagonal entries are $b_1, \cdots, b_N$).*

The control objective of this paper is to solve the distributed tracking problem. That is: to design the joint torque τ_i for the ith manipulator such that $\|q_i(t) - q_0(t)\|$ can be reduced as small as possible as time goes to infinity. In addition, when designing the ith manipulator's joint torque, only the information of connected manipulators $\{v_j | e_{ij} \in \mathcal{E}_{\bar{\mathcal{G}}}\}$ can be used.

Finally, some basic knowledge on the approximation ability of the radial basis function neural network (RBFNN) is introduced. A continuous function $h(z) : \mathbb{R}^m \to \mathbb{R}^m$ can be approximated by RBFNN to be a given accuracy ε_N over a compact set $\Omega_z \subset \mathbb{R}^m$. That is: there exist an ideal weight matrix $W^* \in \mathbb{R}^{p \times m}$ and the number of neurons p such that $h(z) = W^{*T} S(z) + \varepsilon$, where the input vector $z \in \Omega \subset \mathbb{R}^m$ and $S(z) = [s_1(z), \cdots, s_p(z)]^T$ is the activation function, and $s_i(z) = \exp[-(z - \mu_i)^T (z - \mu_i)/\sigma_i^2]$, $i = 1, \cdots, p$, $\mu_i = [\mu_{i1}, \mu_{i2}, \cdots, \mu_{im}]^T$ is the center of the receptive field and σ_i is the width of the Gaussian function. ε is the bounded function approximation error satisfying $|\varepsilon| < \varepsilon_N$ in Ω_z.

It is worth emphasizing that the ideal matrix W^* is only the quantity required for analytical purpose. For real applications, its estimation $\hat{W}$ is used for the practical function approximation. The estimation of $h(z)$ can be given by $\hat{h}(z) = \hat{W}^T S(z)$. $\hat{W}$ will be updated on-line by the designed adaptive law.

The following lemma will be used to analyze the closed-loop system.

Lemma 2. *Let $V(t) \geq 0$ be a bounded continuous function, and $\dot{V}(t) \leq -\gamma V(t) + \kappa$, where γ and κ are positive constants, then $V(t) \leq V(0)e^{-\gamma t} + \frac{\kappa}{\gamma}(1 - e^{-\gamma t})$.*

3   Distributed Tracking Controller Design Under Switching Topologies

Due to some unavoidable reasons such as the link failure and the packet drop, the communication topology among manipulators is usually time-varying. As an attempt to study the switching topology, it is assumed that the communication topology switches at the time instances $\{t_0 = 0, t_1, t_2, \cdots\}$. And the dwelling time $\Delta^* = \inf\{t_{k+1} - t_k | k = 0, 1, \cdots\} > 0$ is assumed to be lower bounded. During each time interval $[t_k, t_{k+1})$, the communication topology keeps invariant. In the following paper, if a variable regarding the communication topology is appended by (t_k), it means the value of this variable in the interval $[t_k, t_{k+1})$.

In each time interval $[t_k, t_{k+1})$, the tracking controller is designed by the back-stepping approach.

Step 1: Inspired by the controller proposed in [12], the auxiliary joint velocity $\dot{q}_{di}$ of the ith manipulator can be designed as follows

$$v_{di} = \rho_i s_i + \dot{\bar{q}}_i, \quad \bar{q}_i = \frac{\left(\sum_{j=1}^N a_{ij}(t_k)q_j + b_i(t_k)q_0\right)}{\sum_{j=1}^N a_{ij}(t_k) + b_i(t_k)}, \quad i = 1, 2, \cdots, N, \quad (2)$$

where $\rho_i > 0$ is the constant control gain, $a_{ij}(t_k)$ is the (i,j)-entry of $\mathcal{A}_{\mathcal{G}(t_k)}$, $b_i(t_k)$ is the ith entry of $b(t_k)$, $s_i = \bar{q}_i - q_i$.

Then the dynamics of error $r_i = v_{di} - \dot{q}_i$ becomes:

$$M_i(q_i)\dot{r}_i = M_i(q_i)\dot{v}_{di} - M_i(q_i)\ddot{q}_i = M_i(q_i)\dot{v}_{di} + C_i(q_i, \dot{q}_i)\dot{q}_i + G_i(q_i) - \tau_i - \tau_{id}. \quad (3)$$

Step 2: The distributed tracking controller can therefore be designed as follows.

$$\tau_i = -\eta_i r_i - \hat{W}_i^T S_i(\dot{v}_{di}, q_i, \dot{q}_i) - \delta_{M_i} \tanh\left(\frac{n k_u \delta_{M_i} r_i}{\epsilon_i}\right) + \dot{v}_{di} + s_i. \quad (4)$$

where $\eta_i > 0$ is the control gain, the RBFNN $\hat{W}_i^T S_i(\dot{v}_{di}, q_i, \dot{q}_i)$ is used to approximate $M_i(q_i)\dot{v}_{di} + C_i(q_i, \dot{q}_i)\dot{q}_i + G_i(q_i)$. By the knowledge in Sect. 2, there exists an optimal weight matrix W_i^* such that over a compact set Ω_i, $\|(W_i^*)^T S_i(\dot{v}_{di}, q_i, \dot{q}_i) - (M_i(q_i)\dot{v}_{di} + C_i(q_i, \dot{q}_i)\dot{q}_i + G_i(q_i))\| \le \varepsilon_i$ where $\varepsilon_i > 0$ is the optimal approximation error. $\delta_{M_i} \tanh\left(\frac{n k_u \delta_{M_i} r_i}{\epsilon_i}\right)$ is the robust term where $k_u = 0.2785$, $\delta_{M_i} > \varepsilon_i + \Delta_i$ is the gain of the robust term, and $\epsilon_i > 0$ determines the slope of the robust term.

The updating law for the weight matrix $\hat{W}_i$ of RBFNN is given as follows

- if $\mathrm{Tr}\left(\hat{W}_i^T \hat{W}_i\right) < W_{\max i}$ holds, then $\dot{\hat{W}}_i = \chi_i S_i(\dot{v}_{di}, q_i, \dot{q}_i) r_i^T$;
- if $\mathrm{Tr}\left(\hat{W}_i^T \hat{W}_i\right) = W_{\max i}$ and $r_i^T \hat{W}_i^T S_i(\dot{v}_{di}, q_i, \dot{q}_i) < 0$ hold, then $\dot{\hat{W}}_i = \chi_i S_i(\dot{v}_{di}, q_i, \dot{q}_i) r_i^T$;
- if $\mathrm{Tr}\left(\hat{W}_i^T \hat{W}_i\right) = W_{\max i}$ and $r_i^T \hat{W}_i^T S_i(\dot{v}_{di}, q_i, \dot{q}_i) \ge 0$ hold, then $\dot{\hat{W}}_i = \chi_i S_i(\dot{v}_{di}, q_i, \dot{q}_i) r_i^T - \chi_i \frac{r_i^T \hat{W}_i^T S_i(\dot{v}_{di}, q_i, \dot{q}_i)}{\mathrm{Tr}(\hat{W}_i^T \hat{W}_i)} \hat{W}_i$;

where $\chi_i > 0$ is the adaption gain, $W_{\max i} > 0$ is used to limit the range of $W_i(t)$. $W_{\max i}$ needs to satisfy the following two conditions: $\mathrm{Tr}((W_i^*)^T W_i^*) < W_{\max i}$ and $\mathrm{Tr}(\hat{W}_i^T(0)\hat{W}_i(0)) < W_{\max i}$.

4   Performance Analysis

Theorem 1. *Assume that during each time interval $[t_k, t_{k+1})$, $\bar{\mathcal{G}}(t_k)$ has a spanning tree and there is no loop in $\bar{\mathcal{G}}(t_k)$. Then under the proposed controller defined by (4), the steady-state tracking error can be reduced as small as possible if parameters in the controller are chosen appropriately.*

PROOF: First, similar to the one given in [4], it can be proved that $\mathrm{Tr}(\hat{W}_i^T(t)\hat{W}_i(t)) < W_{\max i}$ and $\|\tilde{W}_i\|_F \le 2\sqrt{W_{\max i}}$. Then the closed-loop performance can be analyzed by the Lyapunov-like approach. Construct the following energy function

$$E_i = \frac{1}{2} s_i^T s_i + \frac{1}{2} r_i^T M_i(q_i) r_i + \mathrm{Tr}\left(\frac{1}{2\chi_i} \tilde{W}_i^T \tilde{W}_i\right). \quad (5)$$

If the control gain satisfies $\eta_i > \alpha_2\rho_i$, the time derivative of E_i is

$$
\begin{aligned}
\dot{E}_i =\,& s_i^T \dot{s}_i + r_i^T M_i(q_i)\dot{r}_i + \text{Tr}(\tilde{W}_i^T \dot{\tilde{W}}_i/\chi_i) \\
=\,& s_i^T \dot{s}_i + r_i^T(M_i(q_i)\dot{v}_{di} + C_i(q_i,\dot{q}_i)\dot{q}_i + G_i(q_i) - \tau_i - \tau_{id}) + \text{Tr}(\tilde{W}_i^T \dot{\tilde{W}}_i/\chi_i) \\
=\,& s_i^T(r_i - \rho_i s_i) + r_i^T(-\eta_i r_i + \tilde{W}_i^T S_i(\dot{v}_{di},q_i,\dot{q}_i) - s_i - \tau_{id} - \varepsilon_i + \delta_{M_i}\tanh(nk_u\delta_{M_i}r_i/\epsilon_i)) \\
& + \text{Tr}(\tilde{W}_i^T \dot{\tilde{W}}_i/\chi_i) \\
\leq\,& -\rho_i s_i^T s_i - \eta_i r_i^T r_i + \epsilon_i \\
\leq\,& -\rho_i s_i^T s_i - \rho_i r_i^T M_i(q_i)r_i - \rho_i\text{Tr}(\tilde{W}_i^T \tilde{W}_i/\chi_i) + \rho_i\text{Tr}(\tilde{W}_i^T \tilde{W}_i/\chi_i) + \epsilon_i \\
\leq\,& -2\rho_i E_i(t) + 4\rho_i W_{\max i}/\chi_i + \epsilon_i,
\end{aligned}
\tag{6}
$$

where the fact that $\text{Tr}(\tilde{W}_i^T(\dot{\tilde{W}}_i/\chi_i + S_i(\dot{v}_{di},q_i,\dot{q}_i)s_i^T)) \geq 0$ is used (see [4] for proof).

Then, by Lemma 2, it can be obtained that

$$
E_i(t_{k+1}) \leq E_i(t_k)e^{-2\rho_i(t_{k+1}-t_k)} + \delta_i(1 - e^{-2\rho_i(t_{k+1}-t_k)}) \leq E_i(t_k)e^{-2\rho_i(t_{k+1}-t_k)} + \delta_i,
\tag{7}
$$

where $\delta_i = 4\rho_i W_{\max i}/\chi_i + \epsilon_i$. By the definition of E_i, it follows that

$$
\|s_i(t_{k+1})\|^2 + \|r_i(t_{k+1}))\|^2 \leq (\|s_i(t_k)\|^2 + \|r_i(t_k)\|^2)e^{-2\rho_i\Delta^*} + \bar{\delta}_i,
\tag{8}
$$

where $\bar{\delta}_i = 2\delta_i + 2W_{\max i}/\chi_i$.

By the definition of s_i, it can be obtained that $s = (s_1^T, \cdots, s_N^T)^T$ satisfies that

$$
s = ((D_{\mathcal{G}(t_k)}^{-1}(L_{\mathcal{G}(t_k)} + B_{\mathcal{G}(t_k)})) \otimes I_n)(q - 1_N \otimes q_0),
\tag{9}
$$

where $q = (q_1^T, \cdots, q_N^T)^T$ and $D_{\mathcal{G}(t_k)} = \text{diag}(\sum_{j=1}^N a_{1j}(t_k) + b_1(t_k), \cdots, \sum_{j=1}^N a_{Nj}(t_k) + b_N(t_k))$. Here $D_{\mathcal{G}(t_k)}$ is invertible because $\bar{\mathcal{G}}(t_k)$ has a spanning tree.

By (9), it follows that

$$
\lambda_{\min}^2(t_k)\|q - 1_N \otimes q_0\|^2 \leq \sum_{i=1}^N \|s_i\|^2, \quad \lambda_{\max}^2(t_k)\|q - 1_N \otimes q_0\|^2 \geq \sum_{i=1}^N \|s_i\|^2,
\tag{10}
$$

where $\lambda_{\min}(t_k)$ and $\lambda_{\max}(t_k)$ are the minimal and maximal eigenvalues of $((L_{\mathcal{G}(t_k)}^T + B_{\mathcal{G}(t_k)}^T)D_{\mathcal{G}(t_k)}^{-2}(L_{\mathcal{G}(t_k)} + B_{\mathcal{G}(t_k)})) \otimes I_n$, respectively. Since $\bar{\mathcal{G}}(t_k)$ has a spanning tree, by Lemma 1, $\lambda_{\min}^2(t_k) > 0$ and $\lambda_{\max}^2(t_k) > 0$.

By (8) and (10), it can be obtained that

$$
\begin{aligned}
\lambda_{\min}^2(t_k)\|q(t_{k+1}) - 1_N \otimes q_0(t_{k+1})\|^2 + \sum_{i=1}^N \|r_i(t_{k+1}))\|^2 \leq\,& (\lambda_{\max}^2(t_k)\|q(t_k) - 1_N \otimes q_0(t_k)\|^2 \\
& + \sum_{i=1}^N \|r_i(t_k))\|^2)e^{-2\rho_{\min}\Delta^*} + \sum_{i=1}^N \bar{\delta}_i,
\end{aligned}
\tag{11}
$$

where $\rho_{\min} = \min\{\rho_1, \cdots, \rho_N\}$.

Since the number of possible topology graphs $\bar{\mathcal{G}}(t_k)$ is finite, let $c_{\min}^* = \min\{1, \lambda_{\min}^2(0), \lambda_{\min}^2(t_1), \lambda_{\min}^2(t_2), \cdots\}$ and $c_{\max}^* = \max\{1, \lambda_{\max}^2(0), \lambda_{\max}^2(t_1), \lambda_{\max}^2(t_2), \cdots\}$.

Then, (11) can be further relaxed to $c^*_{\min}(\|q(t_{k+1}) - 1_N \otimes q_0(t_{k+1})\|^2 + \sum_{i=1}^{N} \|r_i(t_{k+1}))\|^2) \le c^*_{\max}(\|q(t_k) - 1_N \otimes q_0(t_k)\|^2 + \sum_{i=1}^{N} \|r_i(t_k))\|^2)e^{-2\rho_{\min}\Delta^*} + \sum_{i=1}^{N} \bar{\delta}_i$.

Therefore,

$$\|q(t_{k+1}) - 1_N \otimes q_0(t_{k+1})\|^2 \le (\|q(0) - 1_N \otimes q_0(0)\|^2 + \sum_{i=1}^{N} \|r_i(0))\|^2)\gamma^k$$
$$+ \sum_{i=1}^{N} \bar{\delta}_i (1 - \gamma^{k+1})/c^*_{\min}(1 - \gamma), \quad (12)$$

where $\gamma = e^{-2\rho_{\min}\Delta^*} c^*_{\max}/c^*_{\min}$.

By (12), it can be proved that if ρ_i, χ_i and ϵ_i ($i = 1, \cdots, N$) are appropriately chosen, the steady-state tracking error can be reduced as small as desired as time goes to infinity. Finally, by the similar approach in [12], it can be proved that there exists a compact set Ω_i such that the input $(\dot{v}_{di}^T, q_i^T, \dot{q}_i^T)^T$ of RBFNN is always in this compact set. □

5   A Numerical Example

Consider a group of five manipulators. The parameters in the manipulator's dynamics are same to the ones in [11]. The reference signal is $q_0(t) = (\pi/6 + \sin(t), \sin(t), -\pi/6 + \sin(t))^T$. Three possible communication topology graphs

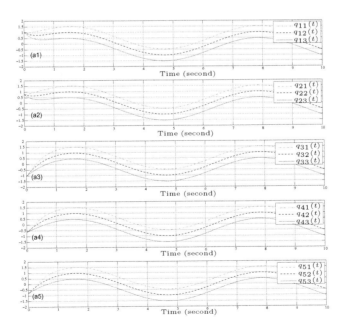

Fig. 1. The joint trajectory profiles of all manipulators: (a1) $q_1(t)$; (a2) $q_2(t)$; (a3) $q_3(t)$; (a4) $q_4(t)$; (a5) $q_5(t)$. (Color figure online)

are set as follows. $\mathcal{G}_1$: a_{ij} in $\mathcal{A}_{\mathcal{G}_1}$ satisfies that $a_{31} = 0.8$, $a_{41} = 0.3$, $a_{42} = 0.5$, $a_{43} = 0.9$, $a_{53} = 0.2$, $a_{54} = 1.2$, and all other a_{ij} are zero; b_i in $b_{\mathcal{G}_1}$ satisfies that $b_1 = 1.5$, $b_2 = 0.5$ and other elements in $b_{\mathcal{G}_1}$ are zero. $\mathcal{G}_2$: a_{ij} in $\mathcal{A}_{\mathcal{G}_2}$ satisfies that $a_{21} = 0.5$, $a_{31} = 0.8$, $a_{41} = 0.3$, $a_{42} = 0.5$, $a_{43} = 0.9$, $a_{53} = 0.2$, $a_{54} = 1.2$, and all other a_{ij} are zero; b_i in $b_{\mathcal{G}_2}$ satisfies that $b_1 = 1.5$ and other elements in $b_{\mathcal{G}_2}$ are zero. $\mathcal{G}_3$: a_{ij} in $\mathcal{A}_{\mathcal{G}_3}$ satisfies that $a_{21} = 0.5$, $a_{31} = 0.8$, $a_{41} = 0.3$, $a_{42} = 0.5$, $a_{43} = 0.9$, $a_{53} = 0.2$, and all other a_{ij} are zero; b_i in $b_{\mathcal{G}_3}$ satisfies that $b_1 = 1.5$, $b_2 = 0.5$, and other elements in $b_{\mathcal{G}_3}$ are zero. The topology switching signal is defined as $[t_k, t_{k+1}) = [k, k+1)$, $k = 0, 1, \cdots$, and the communication topology during the interval $[t_k, t_{k+1})$ is $\mathcal{G}_{\mathrm{mod}(k,3)}$. The simulation results are given in Fig. 1, from which it can be seen that all manipulators can track the desired reference well even under switching topologies.

6   Conclusions

This paper studies the distributed tracking problem of networked manipulator systems with uncertainties. The communication topology among manipulators is directed and time-variant. A tracking algorithm is proposed which can guarantee the steady-state tracking error to be as small as possible if the communication topology switches among a family of graphes which have a spanning tree and have no loop structure. In the future, more effort is to be made on the communication noises between manipulators. By incorporating the noise attenuation technique proposed in [19,20], this topic will be very challenging.

Acknowledgement. This work was supported in part by the National Natural Science Foundation of China (Grants 61422310, 61370032, 71401189) and the Beijing Natural Science Foundation (Grant 4162066).

References

1. Cheng, L., Hou, Z.-G., Tan, M., Liu, D., Zou, A.: Multi-agent based adaptive consensus control for multiple manipulators with kinematic uncertainties. In: 2008 IEEE Internationial Symposium on Intelligent Control, pp. 189–194. San Antonio (2008)
2. Cheng, L., Hou, Z.-G., Tan, M.: Decentralized Adaptive Consensus Control for Multi-Manipulator System with Uncertain Dynamics. In: 2008 IEEE International Conference on Systems, Man, and Cybernetics, pp. 2712–2717. Singapore (2008)
3. Ren, W.: Distributed leaderless consensus algorithms for networked euler-lagrange systems. int. j. control. **82**, 2137–2149 (2009)
4. Hou, Z.-G., Cheng, L., Tan, M.: Decentralized robust adaptive control for the multiagent system consensus problem using neural networks. IEEE Trans. Syst. Man Cybern. Part B: Cybern. **39**, 636–647 (2009)
5. Liu, Y., Min, H., Wang, S., Liu, Z., Liao, S.: Distributed adaptive consensus for multiple mechanical systems with switching topologies and time-varying delay. Syst. Control Lett. **64**, 119–126 (2014)

6. Wang, H., Cheng, L.: Second-order consensus of networked mechanical systems with communication delays. In: The 11th World Congress on Intelligent Control and Automation, pp. 2126–2131. Shenyang (2014)
7. Mehrabian, A., Khorasani, K., Tafazoli, S.: Reconfigurable synchronization control of networked euler-lagrange systems with switching communication topologies. asian j. control. **16**, 830–844 (2014)
8. Zhao, X., Ma, C., Xing, X., Zheng, X.: A stochastic sampling consensus protocol of networked euler-lagrange systems with application to two-link manipulator. IEEE Trans. Ind. Inf. **11**, 907–914 (2015)
9. Liu, Y.-C.: Distributed synchronization for heterogeneous robots with uncertain kinematics and dyanmics under switching topologies. J. Franklin Inst. **352**, 3808–3826 (2015)
10. Cheng, L., Hou, Z.-G., Tan, M.: Decentralized adaptive leader-follower control of multi-manipulator system with uncertain dynamics. In: The 34th Annual Conference of the IEEE Industrial Electronics Society, pp. 1608–1613. Orlando (2008)
11. Hou, Z.-G., Cheng, L., Tan, M., Wang, X.: Distributed adaptive coordinated control of multi-manipulator systems using neural networks. In: Liu, H., Gu, D., Howlett, R., Liu, Y. (eds.) Robot Intelligence: An Advanced Knowledge Processing Approach, pp. 49–69. Springer-Verlag, London (2010)
12. Cheng, L., Hou, Z.-G., Tan, M., Lin, Y., Zhang, W.J.: Neural-network-based adaptive leader-following control for multi-agent systems with uncertainties. IEEE Trans. Neural Netw. **21**, 1351–1358 (2010)
13. Cai, H., Huang, J.: Leader-following consensus of multiple uncertain euler-lagrange systems under switching network topology. Int. J. Gen. Syst. **43**, 294–304 (2014)
14. Wang, Y., Cheng, L., Hou, Z.-G., Tan, M., Bian, G.: Polynomial trajectory tracking of networked Euler-Lagrange Systems. In: The 33rd Chinese Control Conference, pp. 1568–1573. Nanjing (2014)
15. Cheng, L., Wang, Y., Ren, W., Hou, Z.-G., Tan, M.: Containment Control of Multi-Agent Systems with Dynamic Leaders based on a PI^n-type Approach. IEEE Trans. Cybern (2016). doi:10.1109/TCYB.2015.2494738
16. Mei, J., Ren, W., Li, B., Ma, G.: Distributed containment control for multiple unknown second-order nonlinear systems with application to networked lagrangian systems. IEEE Trans. Neural Netw. Learn. Syst. **26**, 1885–1899 (2015)
17. Yang, Z., Shibuya, Y., Qin, P.: Distributed robust control for synchronised tracking of networked euler-lagrange systems. Int. J. Syst. Sci. **46**, 720–732 (2015)
18. Lewis, F.L., Abdallah, C.T., Dawson, D.W.: Control of Robot Manipulators. Macmillan, New York (1993)
19. Cheng, L., Wang, P., Ren, W., Hou, Z.-G., Tan, M.: On convergence rate of leader-following consensus of linear multi-agent systems with communication noises. IEEE Trans. Autom. Control (2017). doi:10.1109/TAC.2016.2522647
20. Wang, P., Cheng, L., Ren, W., Hou, Z.-G., Tan, M.: Seeking consensus in networks of linear agents: communication noises and Markovian switching topologies. IEEE Trans. Autom. Control. **60**, 1374–1379 (2015)

A Novel Emergency Braking Method with Payload Swing Suppression for Overhead Crane Systems

He Chen, Yongchun Fang$^{(\boxtimes)}$, and Ning Sun

Institute of Robotics and Automatic Information System (IRAIS),
Tianjin Key Laboratory of Intelligent Robotics (tjKLIR), Nankai University,
Tianjin 300353, China
chenh@mail.nankai.edu.cn, {fangyc,sunn}@nankai.edu.cn

Abstract. In practice, to deal with emergency situations, emergency braking of overhead cranes plays an important role to ensure safety. However, a sudden braking of the trolley may cause uncontrollable swing of the payload, which is very dangerous and can probably lead to collision and even accidents. Therefore a proper emergency braking method with the consideration of payload swing suppression is of great importance. In this paper, we propose a novel method to achieve the emergency braking objective of overhead crane systems. In particular, after deep analysis, the control objective is divided into two parts. Then two kinds of control methods are proposed to achieve the corresponding objective. After that, we combine these control methods together and propose a novel emergency braking control method, which can ensure trolley braking, as well as payload swing suppression simultaneously. At last, simulation results are included to illustrate the superior control performance of the proposed method.

Keywords: Emergency braking · Overhead cranes · Underactuated systems · Swing suppression

1 Introduction

. In industry, to transport heavy payloads to desired positions, cranes systems, including overhead cranes, tower cranes, boom cranes, and offshore cranes [1], are widely used. To simplify the mechanical structure and reduce the cost, the payload is usually linked to the trolley or the boom by a rope, which leads to the fact that the payload cannot be controlled directly. This kind of design usually results in the fact that the number of control inputs of crane systems is less than to-be-controlled degrees of freedom. Systems with this behavior are known as underactuated systems [2], which are more difficult to be controlled properly compared with full actuated systems, due to the unactuated property. As a typical underactuated system, the overhead crane system is always operated by experienced workers. However, long time working may cause fatigue and

L. Cheng et al. (Eds.): ISNN 2016, LNCS 9719, pp. 242–249, 2016.
DOI: 10.1007/978-3-319-40663-3_28

operation errors, which is very dangerous. Therefore, automatic control design for overhead crane systems is very important.

So far, the control problem of overhead crane systems has attracted attentions of researchers with a series of control methods presented. For example, Singhose et al. [3–5] propose a series of open loop methods based on the idea of input shaping, which can suppress the payload swing with few sensors. By deeply analyzing the system energy, Sun et al. present some passivity based control methods which can achieve asymptotically stable results in [6,7]. To deal with unknown disturbance/uncertainties, in [8–10], some sliding mode control methods are proposed, which show great robustness. Considering that system parameters cannot be obtained accurately, researchers propose some adaptive methods, which can obtain satisfactory performance w.r.t. parameter uncertainties [11,12]. Trajectory planning methods are also used to control overhead crane systems, which can treat various of constraints conveniently [13,14].

In recent years, intelligent control methods are developing quickly and a lot of intelligence based control methods, such as fuzzy control methods [16], genetic algorithm (GA)-based methods [17], and neural network-based methods [18–21], have been proposed for overhead crane systems. For example, in [18], a novel control method is proposed by combining the principles of neural networks and variable structure systems, which can drive the cart smoothly, rapidly and with limited payload swing. [20] proposes a recurrent neural network control method with a hybrid evolutionary algorithm to achieve the control objective of a three dimension tower crane system. An anti-sway position control method is design for an automated transfer crane in [21], which uses the neural network method to tune the PID parameters.

These methods stated above can achieve the transportation objective of overhead crane systems. However, in practice, the working environment of an overhead crane may be very complex and some emergency situations may occur. To deal with this, overhead cranes may need to brake emergently to avoid accidents like collision. As far as we are concerned, the common used braking method in industry is directly set the actual force to zero. It should be noted that there exists high coupling between the trolley motion and the payload swing and irrelevant operation to the trolley can cause large payload swing which is very dangerous. Due to the lack of consideration for the payload swing angle, the crane braking method in industry can only ensure the stop of the trolley while the objective of swing suppression is ignored. To deal with the emergency braking problem, only few methods have been presented. In [22], Ma et al. propose a switching based braking method to deal with this problem, however, the trolley braking and the swing suppression still cannot be dealt with simultaneously.

To achieve the objectives of fast trolley braking and payload swing suppression at the same time, we propose a novel emergency braking method in this paper. In particular, by deep analysis, we first divide the control objective into two parts. Then corresponding controllers are designed for both parts to ensure the sub-objectives separately. After that, a novel emergency braking method is proposed by combing these two controllers together. At last, the performance of our method is verified by simulation tests. Different from the existing method, the proposed method can ensure the trolley being within safe domain (never exceeding the safe limit) and the payload swing suppression simultaneously, which can make sure safety as much as possible.

The rest of this paper is organized as follows. The crane dynamics, as well as the control objectives of emergency braking, are described in Sect. 2. In Sect. 3, we show the main design process of the proposed method. Section 4 exhibits some numerical simulation results to illustrate the effectiveness of this method. The main work is finally concluded and summarized in Sect. 5.

2   Problem Statement

In this paper, we focus on the emergency braking problem of overhead crane systems, whose dynamics are shown as follows:

$$(M + m)\ddot{x} + ml\ddot{\theta}\cos\theta - ml\dot{\theta}^2\sin\theta = F, \tag{1}$$

$$ml^2\ddot{\theta} + ml\cos\theta\ddot{x} + mgl\sin\theta = 0, \tag{2}$$

where M and m denote the trolley mass and the payload mass, respectively; $x(t)$ is the trolley displacement while $\theta(t)$ represents the payload swing angle; l is the length of the massless rope; g is the gravity acceleration constant. From the system dynamics, it is seen that there exist two system states, $x(t)$ and $\theta(t)$, and only one control input $F(t)$, which leads to the fact that the overhead crane system is a typical underactuated system. We can also find that strong coupling exists between the trolley motion and the payload swing. In summary, it is difficult to control overhead crane systems properly.

The control objective of the emergency braking problem is to develop a proper control strategy to regulate the trolley velocity, together with the payload swing angle and swing angular velocity, to zero. Thus we have

$$\dot{x}(t) \to 0, \ \theta(t) \to 0, \ \dot{\theta}(t) \to 0, \tag{3}$$

At the same time, during the entire braking process, the trolley displacement should not exceed the safe limit, in the sense that,

$$x(t) < \lambda, \tag{4}$$

wherein λ represents the safe limit.

Considering the emergency braking problem, the trolley should stop as soon as possible to avoid collision while the payload swing should be as small as possible to ensure safety. Based on this, the control objective of emergency braking can be divided into two parts, known as fast trolley braking and payload swing suppression. Due to the coupling behavior, sudden braking of the trolley may probably cause large swing of the payload which is very dangerous. Therefore, it is difficult to achieve these two parts simultaneously, and it is very important and useful to design a proper emergency braking control method for overhead cranes. Subsequently, we will design a suitable control method for each part and then combine them together to propose a combined method to achieve the entire control objective.

3   Controller Design and Analysis

In this section, a novel emergency braking controller is proposed for overhead crane systems to achieve fast trolley braking, as well as payload swing suppression. Specifically, control objectives of trolley braking and payload swing suppression are considered separately with suitable control methods being designed. Then we combine the designed controllers together to obtain a novel emergency braking controller.

3.1   Braking Control for the Trolley

To regulate the trolley velocity to zero, we can design the following controller directly:

$$F_{1a} = -k_1\dot{x}, \tag{5}$$

where $k_1 \in \mathbf{R}$ represents a positive control gain. Though controller (5) can regulate the trolley velocity to zero successfully, it cannot ensure the constraint in (4). Based on this fact, a potential-function-like item is added into the controller with the following expression:

$$F_{1p} = -\frac{k_2\dot{x}}{(\lambda - x_b)^2 - (x - x_b)^2}, \tag{6}$$

where λ is the safe position and x_b denotes the trolley position when emergency braking begins. $k_2 \in \mathbf{R}$ is a positive control gain. From the structure of (6), it is seen that if the trolley gets too close to the safe limit, this part can provide a large braking force to make the trolley decelerate fast.

Then we can design the trolley braking controller as follows:

$$\begin{aligned} F_1 &= F_{1a} + F_{1_p} \\ &= -k_1\dot{x} - \frac{k_2\dot{x}}{(\lambda - x_b)^2 - (x - x_b)^2}. \end{aligned} \tag{7}$$

Using (7), we can ensure that the trolley can brake fast and will not reach the safe limit.

3.2   Swing Suppression Control for the Payload

Though the controller in (7) can achieve the objective of trolley braking, it may also cause large payload swing which can be very dangerous and may lead to accidents. Therefore, the objective of payload swing suppression needs also to be taken into consideration. Then, the following controller is proposed to suppress the payload swing:

$$F_2 = k_3\theta + k_4\dot{\theta}, \tag{8}$$

where k_3, $k_4 \in \mathbf{R}$ represent positive control gains. Utilizing this controller, the payload swing angle, as well as the swing angular velocity, will converge to zero.

3.3   Emergency Braking Control with Swing Suppression

Using (7) and (8), the objective of trolley braking and swing suppression can be achieved respectively. To achieve these objectives simultaneously, (7) and (8) are combined to obtain a novel emergency braking controller as follows:

$$F = F_1 + F_2 = -k_1\dot{x} - \frac{k_2\dot{x}}{(\lambda - x_b)^2 - (x - x_b)^2} + k_3\theta + k_4\dot{\theta}. \tag{9}$$

Utilizing (9), the objective of emergency braking, shown in (3) and (4), can be achieved.

4   Simulation Results

To illustrate the effectiveness of the proposed method, some simulation tests are carried out in the environment of MATLAB/Simulink. The system parameters are set the same as the actual self-built crane test-bed:

$$m = 1 \text{ kg}, \ M = 6.5 \text{ kg}, \ l = 0.75 \text{ m}.$$

Without loss of generality, the initial trolley position and the initial payload swing angle are both set as zero, in the sense that,

$$x(0) = 0, \theta(0) = 0.$$

To test the proposed emergency braking method, the original target position of the trolley is selected as $x_d = 0.8$ m and we choose an eleven-order polynomial based trajectory planning method as the original crane control method. The expression of the planned trajectory is shown as follows:

$$x_r(t) = x_d \left(x_p + \frac{l}{g}\ddot{x}_p \right), \tag{10}$$

where g is the gravity acceleration constant which is chosen as $g = 9.8$ m/s^2. x_p represents an auxiliary trajectory with the following expression:

$$x_p = -252\tau^{11} + 1386\tau^{10} - 3080\tau^9 + 3456\tau^8 - 1980\tau^7 + 462\tau^6, \tag{11}$$

wherein $\tau = \frac{t}{T}$ and T denotes the designed transportation time which is chosen as $T = 6$ s.

Emergency braking begins at 3 s and the controller is switched to the proposed emergency braking controller. The trolley position at this time is $x_b = 0.4$ m. To verify the performance, in the simulation test, we choose two safe limits as $\lambda = 0.6$ m and $\lambda = 0.5$ m. The control gains are chosen as $k_1 = 1$, $k_2 = 1$, $k_3 = 20$, $k_4 = 10$ when $\lambda = 0.6$ m, and $k_1 = 1$, $k_2 = 2$, $k_3 = 200$, $k_4 = 50$ when $\lambda = 0.5$ m.

The simulation results are shown in Figs. 1 and 2. From these figures, it is seen that using the proposed emergency braking controller, we can make the

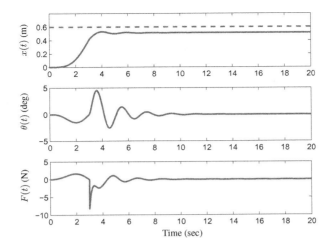

Fig. 1. Simulation results for emergency braking control. Solid line: proposed method; Dashed line: safe limit $\lambda = 0.6$ m.

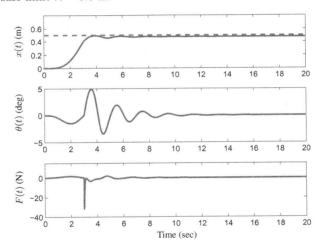

Fig. 2. Simulation results for emergency braking control. Solid line: proposed method; Dashed line: safe limit $\lambda = 0.5$ m.

trolley brake quickly and during the entire process, the trolley position never exceeds the given safe limits. At the same time, the controller can suppress the payload swing obviously, which can avoid large payload swing and make the braking process safer. In summary, it is concluded that the proposed controller can achieve the objectives of fast trolley braking and payload swing suppression.

5 Summary and Conclusion

In this paper, a novel emergency braking control strategy, which can ensure trolley braking and payload swing suppression simultaneously, is proposed for overhead crane systems. In particular, by deeply analyzing the control objective, it is divided into two parts which are considered separately. For each part, a proper controller is designed, and then a novel emergency braking control method is obtained by combined these two controllers together. Different from the common used braking method that directly makes the trolley stop, the proposed method can also achieve the objective of payload swing suppression which ensures safety. Simulation results are included to illustrate the satisfactory performance of the proposed controller in the sense of fast trolley braking and payload swing suppression. In our further work, we will concentrate on design intelligent control methods, for example, neural network-based control method, to finish the emergency braking objective. Also, we will try to use the neural network method to optimize the control gains of this proposed method to obtain better performance.

Acknowledgements. This work is supported by the National Natural Science Foundation of China under Grant 11372144 and 61503200 and by the Natural Science Foundation of Tianjin under Grant 15JCQNJC03800.

References

1. Fang, Y., Wang, P., Sun, N., Zhang, Y.: Dynamics analysis and nonlinear control of an offshore boom crane. IEEE Trans. Industr. Electron. **61**, 414–427 (2014)
2. Liu, Y., Yu, H.: A survey of underactuated mechanical systems. IET Control Theory and Appl. **7**, 921–935 (2013)
3. Singhose, W., Kim, D., Kenison, M.: Input shaping control of double-pendulum bridge crane oscillations. ASME J. Dyn. Syst. Meas. Control **130**, 1–7 (2008)
4. Blackburn, D., Singhose, W., Kitchen, J., Patrangenaru, V., Lawrence, J., Kamoi, T., Taura, A.: Command shaping for nonlinear crane dynamics. J. Vib. Control **16**, 477–501 (2010)
5. Vaughan, J., Kim, D., Singhose, W.: Control of tower cranes with double-pendulum payload dynamics. IEEE Trans. Control Syst. Technol. **18**, 1345–1358 (2010)
6. Sun, N., Fang, Y., Zhang, X.: Energy coupling output feedback control of 4-DOF uunderactuated cranes with saturated inputs. Automatica **49**, 1318–1325 (2013)
7. Sun, N., Fang, Y.: New energy analytical results for the regulation of underactuated overhead cranes: and end-effector motion-based approach. IEEE Trans. Industr. Electron. **59**, 4723–4734 (2012)
8. Almutairi, N.B., Zribi, M.: Sliding mode control of a three-dimensional overhead crane. J. Vib. Control **15**, 1679–1730 (2009)
9. Xi, Z., Hesketh, T.: Discrete time integral sliding mode control for overhead crane with uncertainties. IET Control Theory Appl. **4**, 2071–2081 (2010)
10. Sun, N., Fang, Y., Chen, H.: A new antiswing control method for underactuated cranes with unmodeled uncertainties: theoretical design and hardware eexperiments. IEEE Trans. Industr. Electron. **62**, 453–465 (2015)

11. Yang, J., Yang, K.: Adaptive coupling control for overhead crane systems. Mechatronics **17**, 143–152 (2007)

12. Yang, J., Shen, S.: Novel approach for adaptive tracking control of a 3-D overhead crane system. J. Intell. Rob. Syst. **62**, 59–80 (2011)

13. Lee, H.: Motion planning for three-dimensional overhead cranes with high-speed lload hoisting. Int. J. Control **78**, 875–886 (2005)

14. Uchiyama, N., Ouyang, H., Sano, S.: Simple rotary crane dynamics modeling and open-loop control for residual load sway suppression by only horizontal boom motion. Mechatronics **23**, 1223–1236 (2013)

15. Wu, Z., Xia, X.: Optimal motion planning for overhead cranes. IET Control Theory Appl. **8**, 1833–1842 (2014)

16. Zhao, Y., Gao, H.: Fuzzy-model-based control of an overhead crane with input delay and actuator saturation. IEEE Trans. Fuzzy Syst. **20**, 181–186 (2012)

17. Nakazono, K., Ohnishi, K., Kinjo, H., Yamamoto, T.: Load swing suppression for rotary crane system using direct gradient descent controller optimized by genetic algorithm. Trans. Inst. Syst. Control Inf. Eng. **22**, 303–310 (2011)

18. Lee, L.-H., Huang, P.-H., Shih, Y.-C., Chiang, T.-C., Chang, C.-Y.: Parallel neural network combined with sliding mode control in overhead crane control system. J. Vib. Control **20**, 749–760 (2012)

19. Nakazono, K., Ohnishi, K., Kinjo, H., Yamamoto, T.: Vibration control of load for rotary crane system using neural network with GA-based training. Artif. Life Robot. **13**, 98–101 (2008)

20. Duonga, S.C., Uezatob, E., Kinjob, H., Yamamotoc, T.: A hybrid evolutionary algorithm for recurrent neural network control of a three-dimensional tower crane. Autom. Constr. **23**, 55–63 (2012)

21. Suh, J.-H., Lee, J.-W., Lee, Y.-J., Lee, K.-S.: Anti-sway position control of an automated ttransfer crane based on neural network predictive PID controller. J. Mech. Sci. Technol. **19**, 505–519 (2005)

22. Ma, B., Fang, Y., Zhang, Y.: Switching-based emergency braking control for an overhead canre system. IET Control Theory Appl. **4**, 1739–1747 (2010)

Neural Network Approximation Based Multi-dimensional Active Control of Regenerative Chatter in Micro-milling

Xiaoli Liu[1], Chun-Yi Su[1,2($\boxtimes$)], and Zhijun Li[1]

[1] College of Automation Science and Engineering,
South China University of Technology, Guangzhou 510641, China
[2] Department of Mechanical and Industrial Engineering, Concordia University,
Montréal, QC H3B 1R6, Canada
`chunyi.su@gmail.com`

Abstract. In this paper, an active control approach with the employments of two piezoelectric actuators, the Neural Networks (NNs) as approximators and the Lyapunov-Krasovskii functional which is used to deal with the time delayed tool vibrations is investigated for suppressing the 2-dof regenerative chatter in micro-milling. A dynamic model of micro-milling process and corresponding controlled system are established. Simulations are presented to validate the control performances of developed control approach.

Keywords: Micro-milling · Regenerative chatter · Adaptive neural control · Lyapunov-Krasovskii functional

1 Introduction

Regenerative chatter of micro-milling can lead to significant wear and even devastating breakage of cutting tool, devoting itself to manufacture workpiece with poor surface finish [1]. Therefore, suppressing the negative impacts of regenerative chatter is urgent.

One of the challenges of controlling chatter in micro-milling is considering the actually unknown dynamic uncertainties of cutting which are responsible for contributing to chatter instability, for instance the gyroscopic effects of spindle system, the deformations of bearings and spindle shafts of the spindle system and the cutting coefficients [2–6]. In fact, there are a lot of uncertainties during micro-milling including external disturbances. Another concern of suppressing chatter is the time delay effect of tool vibrations, which plays an important role in forming the phase differences between two successive cuts and in the presence of regenerative chatter [7].

There are few research works directly involving with chatter control of micro-milling. Specially, a time-frequency method to decrease chatter vibration of micro-milling has been proposed by [8]. Nevertheless, the actuators utilized to realize control instructions were not presented. The mechanism of regenerative

© Springer International Publishing Switzerland 2016
L. Cheng et al. (Eds.): ISNN 2016, LNCS 9719, pp. 250–259, 2016.
DOI: 10.1007/978-3-319-40663-3_29

chatter of micro-milling is similar to that of traditional milling except that the micro-milling is different from conventional milling mainly due to the miniature size of the cutting tool [9–12]. According to the principles of reducing chatter instability, the control schemes can probably be classified into two groups. The first group refers to the passive strategy, composed of designing passive absorbers similar to [13] and regulating cutting parameters on the basis of their influences on chatter like altering spindle velocities in [14]. However, this control pattern may restrict the production efficiency. The active control approaches with the applications of active elements [6,15–17] represent the second group. Actually, piezoactuators are characterized by nanoresolution, high speed of positioning and wide range of forces. Hence, it is prospective to suppress chatter of micro-milling via employing piezoactuators as control actuators and designing controller with good performances.

The aim of this paper is to decrease chatter instability. First, a dynamic model of micro-milling with consideration of the process damping effect and corresponding controlled system are established. Then, an active control strategy is developed by utilizing NN (properly designed NN is capable of approximating system uncertainties [18–20]) to approximate the unknown dynamics with uncertainties which are assumed to be finally reflected in the dynamic model of micro-milling, by the usage of Lyapunov-Krasovskii functional to address the time delay effect of tool vibrations and by employing two piezoactuators (PZTAs) as control elements. At the end, simulations are carried out to validate the effectiveness of developed controller.

2   Modeling

As depicted in Fig. 1, the dynamics of micro-milling can be expressed in two coupled equations [8,21] as a lumped mass-spring-damper system, where the tool motion in Z direction is presumed to be neglected.

$$
\left.\begin{aligned}
m_X \ddot{X}(t) + c_X \dot{X}(t) + k_X X(t) &= F_X(t) \\
m_Y \ddot{Y}(t) + c_Y \dot{Y}(t) + k_Y Y(t) &= F_Y(t) \\
X(t) &= x_p(t) + x(t) \\
Y(t) &= y_p(t) + y(t)
\end{aligned}\right\}, \tag{1}
$$

where $m_X, m_Y,\ c_X, c_Y,\ k_X, k_Y,\ F_X, F_Y,\ X, Y,\ x_p, y_p$ and x, y are the modal mass, effective damping coefficients, relative stiffness, cutting forces, tool displacements, chatter free tool motions and tool vibrations in feed and normal directions of the tool-workpiece system. And [22] $F_X(t) = \sum_{n=1}^{N}[-F_{tn}(\phi_n)\cos\phi_n - F_{rn}(\phi_n)\sin\phi_n]$, $F_Y(t) = \sum_{n=1}^{N}[+F_{tn}(\phi_n)\sin\phi_n - F_{rn}(\phi_n)\cos\phi_n]$ with $F_{tn}(\phi_n) = K_t a h_n + C_t a \dot{v}/V_c$ as the tangential force and $F_{rn}(\phi_n) = K_r a h_n + C_r a \dot{v}/V_c$ as the radial force applied on the nth cutter with the cutting velocity V_c, axial depth of cut a and chip thickness $h_n = c\sin\phi_n + v(t) - v(t - T_p)$ (K_t, K_r, C_t, C_r are the cutting force coefficients and the damping coefficients in X and Y directions separately, $v(t), v(t - T_p)$,

$\dot{v} = \dot{x}(t)\sin\phi_n + \dot{y}(t)\cos\phi_n$ are the tool vibrations at time t, that at time $t - T_p$ with T_p as the tooth period and velocity of tool vibration in the radial direction, $\dot{x}$ and $\dot{y}$ on which the process damping forces depend are velocities of tool vibrations that are correlated with process damping effect induced by the contact between the miniature edge of micro-mill and wavy surface of machined workpiece, ϕ_n is the immersion angle of the nth tooth, Ω is the spindle speed and $\phi_p = 2\pi/N$ is the pitch angle of the micro-mill with a teeth number of N). Figures 2 and 3 show the schematic illustration and the diagram used in this paper to suppress chatter, where two piezoactuators are implanted on the extension near to the tip of cutting tool in X and Y directions respectively, similar to the passive pattern in [13]. The regenerative chatter is going to be controlled by transmitting tool vibrations and control forces generated by PZTAs via the radial bearing. Adding the control forces $u = \{u_x \quad u_y\}^T$ provided in X and Y directions to the right side of (1) yields

$$\left.\begin{array}{l} m_X \ddot{X}(t) + c_X \dot{X}(t) + k_X X(t) = F_X(t) + u_x \\ m_Y \ddot{Y}(t) + c_Y \dot{Y}(t) + k_Y Y(t) = F_Y(t) + u_y \end{array}\right\}. \tag{2}$$

Considering only the dynamic parts of the cutting forces that contribute to the regenerative chatter and the linear relationship of $u(t) = \alpha V_{pz}$ ($V_{pz} = \{V_{pzx} \quad V_{pzy}\}^T$ are the input voltages to PZTAs in X and Y directions, α is the proportionality coefficient) [23], (2) can be reformed as

$$\left.\begin{array}{l} \dot{x}_{j,1} = x_{j,2} \\ \dot{x}_{j,2} = f_{j,2}(x_{1,1}, x_{2,1}, x_{1,2}, x_{2,2}) + g_{j,2}V_{pz,j} + h_{j,2}\left(x_{\tau_{1,1}}, x_{\tau_{2,1}}\right) \\ y_j = x_{j,1} \end{array}\right\}. \tag{3}$$

where $x_{1,1} = x$, $x_{2,1} = y$, $x_{1,2} = \dot{x}$, $x_{2,2} = \dot{y}$, $g_{j,2} = \alpha/m_j$, $m_1 = m_X$, $m_2 = m_Y$, $V_{pz,1} = V_{pzx}$, $V_{pz,2} = V_{pzy}$, $x_{\tau_{j,1}} = x_{j,1}(t - \tau_{j,1})$ ($x_{\tau_{j,1}}(0) = 0$), $\tau_{j,1} = T_p$ for $j = 1, 2$, and

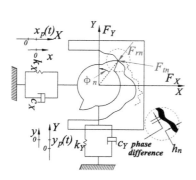

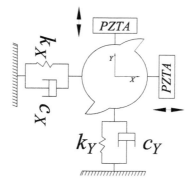

Fig. 1. The schematic mechanical model of micro-milling system.

Fig. 2. The schematic illustration of active control system.

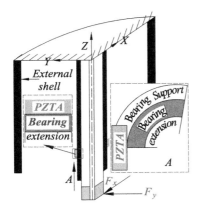

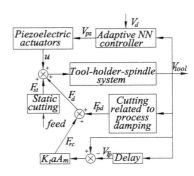

Fig. 3. Schematic diagram of the extension of tool-holder-spindle.

Fig. 4. The control block of active control strategy with adaptive NN controller ($V_{tool} = \{x(t) \quad y(t)\}^T, V_{T_p} = \{x(t-T_p) \quad y(t-T_p)\}^T$).

$$f_{1,2} = -\frac{C_r a}{m_X V_c} \sum_{n=1}^{N} [(c_t sin\phi_n cos\phi_n + sin^2\phi_n) x_{1,2} + (c_t cos^2\phi_n + sin\phi_n cos\phi_n) x_{2,2}]$$
$$- \frac{K_t a}{m_X} \sum_{n=1}^{N} [(sin\phi_n cos\phi_n + k_r sin^2\phi_n) x_{1,1} + (cos^2\phi_n + k_r sin\phi_n cos\phi_n) x_{2,1}]$$
$$- \frac{1}{m_X} (k_X x_{1,1} + c_X x_{1,2}).$$

$$h_{1,2} = \frac{K_t a}{m_X} \sum_{n=1}^{N} [(sin\phi_n cos\phi_n + k_r sin^2\phi_n) x_{\tau_1,1} + (cos^2\phi_n + k_r sin\phi_n cos\phi_n) x_{\tau_2,1}].$$

$$f_{2,2} = \frac{C_r a}{m_Y V_c} \sum_{n=1}^{N} [(c_t sin^2\phi_n - sin\phi_n cos\phi_n) x_{1,2} + (c_t sin\phi_n cos\phi_n - cos^2\phi_n) x_{2,2}]$$
$$+ \frac{K_t a}{m_Y} \sum_{n=1}^{N} [(sin^2\phi_n - k_r sin\phi_n cos\phi_n) x_{1,1} + (sin\phi_n cos\phi_n - k_r cos^2\phi_n) x_{2,1}]$$
$$- \frac{1}{m_Y} (k_Y x_{2,1} + c_Y x_{2,2}).$$

$$h_{2,2} = -\frac{K_t a}{m_Y} \sum_{n=1}^{N} [(sin^2\phi_n - k_r sin\phi_n cos\phi_n) x_{\tau_1,1} + (sin\phi_n cos\phi_n - k_r cos^2\phi_n) x_{\tau_2,1}].$$

3 Active Control Strategy

3.1 Adaptive NN Control Design

Three lemmas from [24,25] are demanded during the process of exploring the adaptive NN controller and analyzing the chatter stability.

Lemma 1. *For any continuous function* $h(\xi_1, ..., \xi_k) : R^{m_1} \times \cdots R^{m_k} \to R$ *satisfying* $h(0, ..., 0) = 0$, *where* $\xi_i \in R^{m_i}$ $(i = 1, 2, ..., k, m_i > 0)$, *there exist positive smooth functions* $\varrho_i(\xi_i) : R^{m_i} \to R$ $(i = 1, 2, ..., k)$ *satisfying* $\varrho_i(0) = 0$ *such that* $|h(\xi_1, ..., \xi_k)| \leq \sum_{i=1}^{k} \varrho_i(\xi_i)$.

Lemma 2. *For any constant* $\eta > 0$ *and variable* $z \in R$, $\lim_{z \to 0} \frac{tanh^2(z/\eta)}{z} = 0$.

Lemma 3. *For any variable* $z_{i,2}$ *out of the set* $\Omega_{c_{z_{i,2}}} := \{z_{i,2} | |z_{i,2}| < 0.8814\eta_{i,2}\}$, $i = 1, 2$, *it holds that* $[1 - 2tanh^2(\frac{z_{i,2}}{\eta_{i,2}})] \leq 0$.

The control block with adaptive NN controller to be developed is described in Fig. 4, where $V_d = \begin{bmatrix} y_{d1} & y_{d2} \end{bmatrix}^T$ (y_{d1} and y_{d2} are desired tool vibrations in X and Y directions respectively).

Two main steps based on (3) are required to derive the adaptive NN controller, where the Lyapunov function candidates and Lyapunov-Krasovskii functionals from [24] are developed with embedded backstepping approach.

Step $j, 1$ for $j = 1, 2$

The candidate Lyapunov function is established for the step $j, 1$ as

$$V_{j,1} = \frac{1}{2} z_{j,1}^2. \tag{4}$$

where $z_{j,1} = x_{j,1} - y_{dj}$ is the intermediate error defined for step $j, 1$ with $y_{dj} = 0$ when there are no tool vibrations. And the intermediate error for step $j, 2$ is expressed as $z_{j,2} = x_{j,2} - \alpha_{j,1}$ with the intermediate control law $\alpha_{j,1} = -\kappa_{j,1} z_{j,1}$. The derivative of $V_{j,1}$ becomes

$$\dot{V}_{j,1} \leq z_{j,1} z_{j,2} - (\kappa_{j,1} - \frac{1}{4\lambda}) z_{j,1}^2. \tag{5}$$

where $\lambda > 0$ and $\kappa_{j,1} > 0$ are design parameters. The term $z_{j,1} z_{j,2}$ shall be canceled in next step.

Step $j, 2$ for $j = 1, 2$

The candidate Lyapunov function defined for step $j, 2$ is described as

$$V_{z_{j,2}} = \frac{1}{2} z_{j,2}^2. \tag{6}$$

With the consideration of $z_{j,2}$, $\alpha_{j,1}$, (3) and Lemma 1, $\dot{V}_{z_{j,2}}$ can be written as

$$\dot{V}_{z_{j,2}} \leq z_{j,2} \left(f_{j,2} + g_{j,2} V_{pz,j} - \frac{\partial \alpha_{j,1}}{\partial x_{j,1}} x_{j,2} \right) + z_{j,2}^2 + \frac{1}{2} \sum_{i=1}^{2} [\varrho_{i,1}^{j,2}(x_{\tau_{i,1}})]^2. \tag{7}$$

where $\varrho_{i,1}^{j,2}(x_{\tau_{i,1}})$ means the boundary function composed of $x_{\tau_{i,1}}$ of $h_{j,2}$ with the consideration of Lemma 1. To cope with the last term of (7), the Lyapunov-Krasovskii functional is defined as $V_{U_{j,2}} = \frac{1}{2} \sum_{i=1}^{2} \int_{t-\tau_{i,1}}^{t} [\varrho_{i,1}^{j,2}(x_{i,1}(\tau))]^2 d\tau$ and $\dot{V}_{U_{j,2}}$ is expressed as

$$\dot{V}_{U_{j,2}} = \frac{1}{2} \sum_{i=1}^{2} \left[\left(\varrho_{i,1}^{j,2}(x_{i,1}) \right)^2 - \left(\varrho_{i,1}^{j,2}(x_{\tau_{i,1}}) \right)^2 \right]. \tag{8}$$

Combing (7) and (8) by taking $z_{j,2}$ and Lemma 2 into account results in the disappear of the delayed term of (7) and

$$\dot{V}_{z_{j,2}} + \dot{V}_{U_{j,2}} \leq z_{j,2} \left(F_{j,2}^*(Z_{j,2}) + g_{j,2} V_{pz,j} \right) + (1 - 2 tanh^2(\frac{z_{j,2}}{\eta_{j,2}})) U_{j,2}. \tag{9}$$

where $Z_{j,2} = [x_{1,1}, x_{1,2}, x_{2,1}, x_{2,2}]^T$, $U_{j,2} = 1/2 \sum_{i=1}^{2} [\varrho_{i,1}^{j,2}(x_{i,1})]^2$, $F_{j,2}^*(Z_{j,2}) = f_{j,2} + z_{j,2} - (\partial \alpha_{j,1}/\partial x_{j,1}) x_{j,2} + (2/z_{j,2}) tanh^2(z_{j,2}/\eta_{j,2}) U_{j,2}$ contains the effects of $f_{j,2}$ and $h_{j,2}$ of (3) which represent the unknown dynamics with uncertainties of cutting and will be approximated by RBFNNs (RBFNNs can be replaced by other approximation techniques, such as fuzzy systems [26,27])

$$F_{j,2}^* (Z_{j,2}) = W_{j,2}^{*T} S (Z_{j,2}) + \varepsilon_{j,2} (Z_{j,2}). \tag{10}$$

where $W_{j,2}^* \in R^l$ is the vector of weights with l as the number of NN nodes, $S(Z_{j,2}) = [s_1(Z_{j,2}), s_2(Z_{j,2}), ..., s_l(Z_{j,2})]^T \in R^l$ is the vector of Gaussian functions with $s_i(Z_{j,2}) = \exp\left(- (Z_{j,2} - \mu_i)^T (Z_{j,2} - \mu_i) / \zeta_i^2\right)$ ($\mu_i = [\mu_{i1}, \mu_{i2}, \mu_{i3}, \mu_{i4}]^T$ is the center of receptive field and ζ_i is the width of Gaussian function and $\varepsilon (Z_{j,2})$ is the approximation error satisfying $|\varepsilon (Z_{j,2})| \leq \bar{\varepsilon}$ with $\bar{\varepsilon}$ as an unknown constant).

Considering the error of weight estimation $\tilde{W}_{j,2} = \hat{W}_{j,2} - W_{j,2}^*$ with $\hat{W}_{j,2}$ as the estimation of NN weight, (5) and (9), the candidate Lyapunov function (6) is redefined for step $j, 2$ as

$$V_{j,2} = V_{j-1,2} + V_{j,1} + V_{z_{j,2}} + V_{U_{j,2}} + \frac{1}{2}\tilde{W}_{j,2}\Gamma_{j,2}^{-1}\tilde{W}_{j,2}. \tag{11}$$

where $\Gamma_{j,2} > 0$ is a design parameter and $V_{0,2} = 0$.

Choosing the adaptation laws of $V_{pz,j}$ and $\hat{W}_{j,2}$ to erase the term $z_{j,1}z_{j,2}$ in step $j, 1$ and to achieve the stability of (11) as

$$\left.\begin{array}{l} V_{pz,j} = -\frac{1}{g_{j,2}}[z_{j,1} + \kappa_{j,2}z_{j,2} + \hat{W}_{j,2}^T S(Z_{j,2})] \\ \dot{\hat{W}}_{j,2} = \Gamma_{j,2}[S(Z_{j,2})z_{j,2} - \sigma_{j,2}(\hat{W}_{j,2} - W_{j,2}^0)] \end{array}\right\}. \tag{12}$$

where $\sigma_{j,2} > 0$ and $W_{j,2}^0$ are constant parameters to be designed and combing (5), (10), (12) and the derivative of (11), we obtain

$$\dot{V}_{j,2} \leq - \sum_{i=1}^{j}[\sum_{k=1}^{2}(\kappa_{i,k} - \frac{1}{4\lambda})z_{i,k}^2 + \frac{1}{2}\sigma_{i,2}\|\tilde{W}_{i,2}\|^2 + (1 - 2tanh^2(\frac{z_{i,2}}{\eta_{i,2}}))U_{i,2}]$$

$$+ \sum_{i=1}^{j}[\lambda\bar{\varepsilon}_{i,2}^2 + \frac{1}{2}\sigma_{i,2}\|W_{i,2}^* - W_{i,2}^0\|^2]. \tag{13}$$

where $\kappa_{i,k} > 1/(4\lambda)$ is selected.

3.2 Stability Proof

Based on the above analyses, the Lyapunov function designed for the whole system can be formed as $V = V_{2,2}$ and the following expression can be achieved

$$\dot{V} \leq - \sum_{i=1}^{2}[\sum_{k=1}^{2}(\kappa_{i,k} - \frac{1}{4\lambda})z_{i,k}^2 + \frac{1}{2}\sigma_{i,2}\|\tilde{W}_{i,2}\|^2] + C$$

$$+ \sum_{i=1}^{2}(1 - 2tanh^2(\frac{z_{i,2}}{\eta_{i,2}}))U_{i,2}. \tag{14}$$

where $C = \sum_{i=1}^{2}[\lambda\bar{\varepsilon}_{i,2}^2 + \frac{1}{2}\sigma_{i,2}\|W_{i,2}^* - W_{i,2}^0\|^2]$ is a nonnegative constant.

Proof. The first summation part of (14) is negative, which suggests the boundedness of $\tilde{W}_{i,2}$ for $i = 1, 2$, the second part is constant and the sign of last summation part is on the basis of $z_{i,2}$ whose size can be divided into 3 cases to discuss.

Case 1: $z_{i,2} \in \Omega_{c_{z_{i,2}}}$, $\forall i = 1, 2$. As $z_{i,2}$ is bounded, we know from (14) that $z_{i,1}$ is bounded. That $y_{di} = \dot{y}_{di} = \ddot{y}_{di} = 0$ indicates the boundedness of $x_{i,1}$. Consequently, that $\alpha_{i,1}$ and V_{pzi} are bounded can be concluded. As such, all signals of closed-loop and V are bounded.

Case 2: $z_{i,2} \notin \Omega_{c_{z_{i,2}}}$, $\forall i = 1, 2$. From the definition of $U_{i,2}$ and Lemma 3, it implies that the last summation term of (14) is negative or 0. Then we can reform (14) as the following

$$\dot{V} \leq - \sum_{i=1}^{2} [\sum_{k=1}^{2} (\kappa_{i,k} - \frac{1}{4\lambda}) z_{i,k}^2 + \frac{1}{2} \sigma_{i,2} \|\tilde{W}_{i,2}\|^2] + C. \tag{15}$$

The boundedness of $z_{i,k}$ and $x_{i,1}$ can be found from (15) and $y_{di} = \dot{y}_{di} = \ddot{y}_{di} = 0$. Then the boundedness of all signals of the closed-loop are revealed.

Case 3: $\Sigma_I := \{z_{i,2}|z_{i,2} \in \Omega_{c_{z_{i,2}}}\}$ and $\Sigma_J := \{z_{j,2}|z_{j,2} \notin \Omega_{c_{z_{j,2}}}\}$ for $i \neq j$ and $(i, j) \in \{1, 2\}$. This case could be divided into two cases which are analysed in **Case 1** and **Case 2**. Using the same reasonings in the above analyses, it can be concluded that all signals of the whole system are bounded.

This completes the proof.   □

4   Simulation

A simulation study is carried out to verify the effectiveness of developed control strategy, where the dynamic model of micro-milling with process damping effect, the cutting parameters (where chatter emerges: the spindle speed $\Omega = 45450$ rev/min, the depth of cut a $= 30\,\mu m$ and the constant feed rate c $= 5\ \mu m$/tooth), the workpiece material (AISI 1045 steel), the tool parameters and cutting conditions (where the teeth number $= 2$, diameter of micromill $= 600\ \mu m$, rake angle $= 0°$, clearance angle $= 7°$, edge radius $= 4\ \mu m$, tool wear length $= 80\,\mu m$), the model parameters of micro-mill supposed to be equal in X and Y directions of the highest flexibility (where $\omega_n = 3.8\,kHz$, $\xi = 3.62\,\%$, $k = 0.89N/\mu m$), and the cutting coefficients (where $K_t = 4042MPa$, $K_r = 2814MPa$, $C_t = 18.72N/mm$, $C_r = 21.36N/mm$), are adopted from [22]. The initial condition $x(0) = y(0) = 0.1\mu m$, $\dot{x}(0) = \dot{y}(0) = 0.1$ mm/s is utilized for the simulations realized by MATLAB/SIMULINK. And the proportional coefficient $\alpha = 7.19N/V$ from [28] is employed.

The design parameters of adaptive NN controller are selected as: $l = 9$ for $S(Z_{j,2})$, the centers of Gaussian functions evenly spaced in $[-1, 1]$, and $\kappa_{j,1} = 1000$, $\kappa_{j,2} = 4500$, $\Gamma_{j,2} = 500$, $\sigma_{j,2} = 200$ for $j = 1, 2$ and $W_{i,2}^0 = \hat{W}_{i,2}(0) = 0$ for $i = 1, 2$.

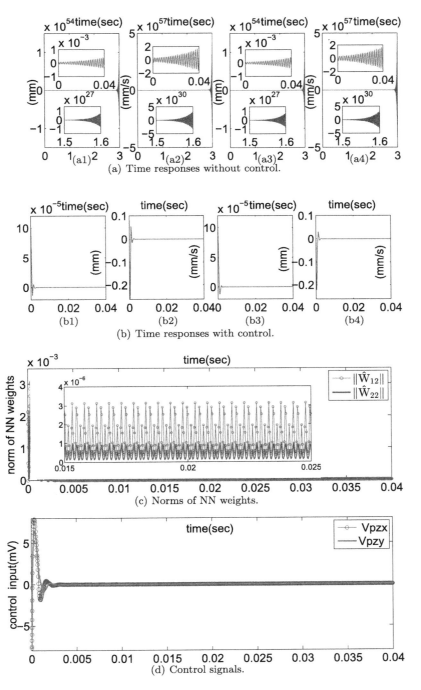

Fig. 5. Time responses. (a1), (b1) tool vibration in x-direction (mm); (a2), (b2) velocity of tool vibration in x-direction (mm/s); (a3), (b3) tool vibration in y-direction (mm); (a4), (b4) velocity of tool vibration in y-direction (mm/s). (Color figure online)

The time responses of the cutting system without control and that with control are described in Fig. 5(a) and (b) which suggest the unstable responses are suppressed to the neighborhood of 0 in less than 4 ms. The norms of NN weights and the control voltages to PZTAs are depicted in Fig. 5(c) and (d) and are found to be bounded.

Furthermore, larger $\kappa_{j,2}$ for $j = 1, 2$ or bigger numbers of NN nodes can offer better performance.

5   Conclusion

The paper develops an active control approach to suppress regenerative chatter of micro-milling with process damping effect through the usages of two piezoactuators as control actuators, the developed adaptive NN controller based on RBFNN approximation and the Lyapunov-Krasovskii functional. It can be seen that a fairly good performance of the developed control strategy is obtained via choosing appropriate design parameters. In our future work, we are trying to apply the developed theoretical work into practice.

References

1. Rahnama, R., Sajjadi, M., Park, S.S.: Chatter suppression in micro end milling with process damping. J. Mater. Process. Technol. **209**(17), 5766–5776 (2009)
2. Graham, E., Mehrpouya, M., Nagamune, R., Park, S.S.: Robust prediction of chatter stability in micro milling comparing edge theorem and LMI. CIRP J. Manuf. Sci. Technol. **7**(1), 29–39 (2014)
3. Graham, E., Mehrpouya, M., Park, S.S.: Robust prediction of chatter stability in milling based on the analytical chatter stability. J. Manuf. Process. **15**(4), 508–517 (2013)
4. Movahhedy, M.R., Mosaddegh, P.: Prediction of chatter in high speed milling including gyroscopic effects. Int. J. Mach. Tools Manuf. **46**(9), 996–1001 (2006)
5. Mayr, J., Jedrzejewski, J., Uhlmann, E., Alkan Donmez, M., Knapp, W., Härtig, F., Wendt, K., Moriwaki, T., Shore, P., Schmitt, R., Brecher, C., Würz, T., Wegener, K.: Thermal issues in machine tools. CIRP Ann. Manuf. Technol. **61**(2), 771–779 (2012)
6. Abele, E., Altintas, Y., Brecher, C.: Machine tool spindle units. CIRP Ann. Manuf. Technol. **59**(2), 781–802 (2010)
7. Liao, Y.S., Young, Y.C.: A new on-line spindle speed regulation strategy for chatter control. Int. J. Mach. Tools Manuf. **36**(5), 651–660 (1996)
8. Liu, M.-K., Halfmann, E.B., Suh, C.S.: Multi-dimensional time-frequency control of micro-milling instability. J. Vib. Control **20**(5), 643–660 (2014)
9. Chae, J., Park, S.S., Freiheit, T.: Investigation of micro-cutting operations. Int. J. Mach. Tools Manuf. **46**(3–4), 313–332 (2006)
10. Mian, A.J., Driver, N., Mativenga, P.T.: Identification of factors that dominate size effect in micro-machining. Int. J. Mach. Tools Manuf. **51**(5), 383–394 (2011)
11. Bissacco, G., Hansen, H.N., Slunsky, J.: Modelling the cutting edge radius size effect for force prediction in micro milling. CIRP Ann. Manuf. Technol. **57**(1), 113–116 (2008)

12. Malekian, M., Park, S.S., Jun, M.B.G.: Modeling of dynamic micro-milling cutting forces. Int. J. Mach. Tools Manuf. **49**(7–8), 586–598 (2009)
13. Saadabad, N.A., Moradi, H., Vossoughi, G.: Global optimization and design of dynamic absorbers for chatter suppression in milling process with tool wear and process damping. Procedia CIRP **21**, 360–366 (2014)
14. Kalinski, K.J., Galewski, M.A.: Chatter vibration surveillance by the optimal-linear spindle speed control. Mech. Syst. Signal Process. **25**(1), 383–399 (2011)
15. Pan, G., Xu, H., Kwan, C.M., Liang, C., Haynes, L., Geng, Z.: Modeling and intelligent chatter control strategies for a lathe machine. Control Eng. Pract. **4**(12), 1647–1658 (1996)
16. Dohner, J.L., Lauffer, J.P., Hinnerichs, T.D., Shankar, N., Regelbrugge, M., Kwan, C.-M., Xu, R., Winterbauer, B., Bridger, K.: Mitigation of chatter instabilities in milling by active structural control. J. Sound Vib. **269**(1–2), 197–211 (2004)
17. Aggogeri, F., Al-Bender, F., Brunner, B., Elsaid, M., Mazzola, M., Merlo, A., Ricciardi, D., de la O Rodriguez, M., Salvi, E.: Design of piezo-based AVC system for machine tool applications. Mech. Syst. Signal Process. **36**(1), 53–65 (2013)
18. He, W., Chen, Y., Yin, Z.: Adaptive neural network ccontrol of an uncertain robot with full-state constraints. IEEE Trans. Cybern. **46**(3), 620–629 (2016)
19. He, W., Dong, Y., Sun, C.: Adaptive neural impedance control of a robotic manipulator with input saturation. IEEE Trans. Syst. Man Cybern. **46**(3), 334–344 (2016)
20. Neural network Control of a Robotic Manipulator with Input Dead-zone and Output Constraint. http://ieeexplore.ieee.org/stamp/stamp.jsp?tp=& arnumber=7222457&isnumber=6376248
21. Insperger, T., Gradišek, J., Kalveram, M., Stépán, G., Winert, K., Govekar, E.: Machine tool chatter and surface location error in milling processes. J. Manuf. Sci. Eng. Trans. ASME **128**(4), 913–920 (2006)
22. Jin, X., Altintas, Y.: Chatter stability model of micro-milling with process damping. J. Manuf. Sci. Eng. Trans. ASME **135**(3), 031011-1–031011-9 (2013)
23. Liaw, H.C., Oetomo, D., Shirinzadeh, B., Alici, G.: Robust motion tracking control of piezoelectric actuation systems. In: 2006 IEEE International Conference on Robotics and Automation, ICRA 2006, pp. 1414–1419. IEEE Press, New York (2006)
24. Ge, S.S., Tee, K.P.: Approximation-based control of nonlinear MIMO time-delay systems. Automatica **43**(1), 31–43 (2007)
25. Lin, W., Qian, C.: Adaptive control of nonlinearly parameterized systems: the smooth feedback case. IEEE Trans. Autom. Control **47**(8), 1249–1266 (2002)
26. Li, Z., Su, C.-Y., Wang, L., Chen, Z., Chai, T.: Nonlinear disturbance observer-based control design for a robotic exoskeleton incorporating fuzzy approximation. IEEE Trans. Ind. Electron. **62**(9), 5763–5775 (2015)
27. Li, Z., Yang, C., Su, C.-Y., Deng, S., Sun, F., Zhang, W.: Decentralized fuzzy control of multiple cooperating robotic manipulators with impedance interaction. IEEE Trans. Fuzzy Syst. **23**(4), 1044–1056 (2015)
28. Badel, A., Qiu, J., Nakano, T.: Self-sensing force control of a piezoelectric actuator. IEEE Trans. Ultrason. Ferroelectr. Freq. Control **55**(12), 2571–2581 (2008)

A Distributed Delay Consensus of Multi-Agent Systems with Nonlinear Dynamics in Directed Networks

Li Qiu, Liuxiao Guo$^{(\boxtimes)}$, Jia Liu, and Yongqing Yang

School of Science, Jiangnan University, Wuxi 214122, People's Republic of China
guo_liuxiao@126.com, guoliuxiao@jiangnan.edu.cn

Abstract. This paper deals with the consensus problem of dynamical networks of multi-agents with communication delays. The communication topology is assumed to be directed and connected. The intrinsic nonlinear dynamics are introduced to reflect more realistic dynamical behaviors of the agent systems. We propose a complete consensus protocols that take into account the weighted sum historical information exchange over a time interval $[t - \tau, t]$. Using tools from differential equations, together with results from matrix theory and algebraic graph theory, sufficient conditions are derived to make all agents asymptotically achieve the consistency. Simulation results illustrate the theoretical results.

Keywords: Distributed delay protocol · Multi-agent system · Nonlinear dynamic · Directed network

1 Introduction

In recent years, the problem of consensus of multi-agent systems has attracted compelling attention from various scientific communities due to its extensive applications in real-world distributed computation, wireless sensor networks, satellite formation [1], the direction of fish or birds [2,3], distributed sensor filtering value [4], cooperative surveillance and so on. Indeed, some topics highly relevant to consensus are synchronization of complex networks [5–7].

In multi-agent systems, much progress has been recently achieved in investigating consensus in networks of identical linear systems, where consensus can be reached if the linear system satisfies some assumptions and the network is connected [8]. Note that typical multi-agent systems did not take full advantage of the powerful nonlinear dynamics. Recently some authors have considered the consensus problem with nonlinear agent dynamics [9–11]. Some typical nonlinear dynamics are the Euler-Lagrange [12,13] and classical chaotic systems. And most of real world networks are directed networks such as World Wide Web and mobile communication networks. Due to the algebraic graph theory has been well developed, very recently, the consensus problem without time delays in directed networks with non-linear dynamics has been discussed [14].

© Springer International Publishing Switzerland 2016
L. Cheng et al. (Eds.): ISNN 2016, LNCS 9719, pp. 260–268, 2016.
DOI: 10.1007/978-3-319-40663-3_30

Since the multi-agent systems cannot achieve consensus by itself, one critical issue arising from multi-agent systems is to develop the control strategies based only on local relative information that can guarantee the whole system to evolve into a coordinated behavior. To achieve the aim, suitable neighbor-based rules are usually adopted to interconnect the networked agents, such as impulsive control, adaptive control, distributed control [15–20]. One important challenge is the influence of time delays in the inter-agent information flows. In real systems, time delay always exists due to finite communication speed. Much works has recently been done on consensus control with time-delay. Olfati-Saber and Murray [21] studied the average consensus of first-order multiagent systems with constant and uniform communication time delays under fixed topology. Bliman and Ferrari Trecate [22] generalized the results of Reference in considering uniform and nonuniform time-varying time-delays. The consensus problem was investigated for delayed-input approach by introducing a nonlinear term describing the intrinsic dynamics of each agent [23], which discussed consensus problem with sampled-data information, and the protocol only uses a point information from the range $[t_k, t_{k+1}]$.

This paper propose the type of neighbor-based distributed delayed protocol of nonlinear continuous time in directed networks. The protocol makes use of a decaying weighted sum of historical information exchange over a time interval $[t - \tau, t]$. The weighted function maybe an exponentially decaying weighted function, maybe a piecewise function. The motivation is that outdated state information is within any control system and deserves consideration, and the memory is very cheap. Precisely, the designed protocol can guarantee that each individual state converges, whose expectation is right the weighted average of the states of the whole system. The form of the protocol are practical and general. The consensus protocols in Refs. [16,23] are the special cases of our new agreement. And our theoretical analysis does not require the Laplace transform, by properly selecting Lyapunov functions, we convert the convergence analysis of matrix products into that of scalar sequences, then the system achieves consensus. Finally, some numerical simulations are presented to demonstrate the effectiveness of the theoretical results.

The rest of the paper is organized as follows. In Sect. 2, some preliminaries and the model description of novel protocols are given. The main results of distributed consensus are discussed in Sect. 3. In Sect. 4, some numerical examples are given to illustrate the theoretical results. Conclusions are finally drawn in Sect. 5.

2   Preliminaries: Algebraic Directed Graph Theory

2.1   Network Topology

A directed graph $G = (V, \varepsilon, A)$ consists of a set of vertices $V = \{v_1, v_2, \ldots, v_N\}$, and a set of directed edges $\varepsilon \in V \times V$, and a weighted adjacency matrix $A = [a_{ij}]_{n \times n}$ having non-negative entries. An edge e_{ij} in graph G is denoted by the ordered pair of vertices (v_j, v_i), where v_j and v_i are called the parent and child vertices, respectively, and $e_{ij} \in \varepsilon$ if and only if $a_{ij} > 0$. For simplicity, denote

$G = (V, \varepsilon, A)$ by $G(A)$. A path between nodes v_i and v_j in G is a sequence of edges $(v_i, v_{i1}), (v_{i1}, v_{i2}), \ldots (v_{il}, v_j)$. A directed graph is called strongly connected if and only if there is a directed path between any pair of distinct vertices. The graph Laplacian L of the network is defined by $l_{ii} = -\Sigma_{j=1, j\neq i}^{N} a_{ij}$ and $l_{ij} = -a_{ij}$. As the connected graph is directed, the coupling topology matrix L is not necessarily symmetrical in our model networks. For more details, one can see Refs. [2, 23].

2.2   Consensus Protocols

Consider each node of the graph to be a dynamic agent with dynamics

$$\dot{x}_i(t) = f(x_i(t), t) + u_i(t) \ (i = 0, 1, \ldots, N) \tag{1}$$

where $f(x_i(t), t) \in R^n$ describes the intrinsic nonlinear dynamics of the ith agent. $u_i(t) \in R^n$ is a control input to be designed, i.e. only the information of v_i itself and its neighbors are available in forming the state feedback for the node v_i.

We propose the following neighbor-based protocol:

$$u_i = \sum_{v_j \in N_i} a_{ij} [\int_{t-\tau}^{t} K(t-s)x_j(s)ds - \int_{t-\tau}^{t} K(t-s)x_i(s))ds]; (i = 0, 1, \ldots, N) \tag{2}$$

where $K(t) = diag(k_1(t), \ldots, k_n(t))$, and the delay kernel $k_j(t)$ is a real-valued nonnegative continuous function defined on $[0, \infty)$. Here, suppose there exist two constants $v > 0$, $c > 0$, and matrix $\bar{K}(v) = diag(\bar{k}_1(v), \ldots, \bar{k}_n(v)) > 0$ such that

$$\int_0^\tau k_j(\theta)d\theta \leq c, \quad \int_0^\tau k_j(\theta)e^{v\theta}d\theta = \bar{k}_j(v) < \infty \tag{3}$$

Substituting (2) into (1) gives

$$\dot{x}_i(t) = f(x_i(t), t) + \sum_{v_j \in N_i} a_{ij}[\int_{t-\tau}^{t} K(t-s)(x_j(s) - x_i(s))ds] \ (i = 0, 1, \ldots, N) \tag{4}$$

Remark 1. Delay consensus protocol for continuous type of multi-agents system has been investigated by several authors [23, 24]. Here, we proposed the delay consensus criterion for nonlinear dynamics via the previous information, which uses a decaying weighted sum of historical information exchange over a time interval $[t-\tau, t]$. Compared with previous ones, this consensus criterion is a more general form of protocol. And as the weighted functions $k_i(s) = 1, when \ s = t - \tau; k_i(s) = 0, when \ s \in (t - \tau, t]$, $i = 1, 2, \ldots, n$, which are circumstances of Chen [9]. As $k_i(s) = 1, when \ s = t - \tau; k_i(s) = 0, when \ s \in (t - \tau, t); k_i(s) = 1, when \ s = t$, $i = 1, 2, \ldots, n$, are the two states of protocol in paper [16]. Certainly, an exponentially decaying weighted function $k_i(t) = exp(-t)$ is also a good choice to describe practical problem.

2.3   Network Dynamics

The collective dynamics of system (4) can be written in a compact form of the delay differential equation(DDE) as

$$\dot{x}(t) = f(x(t)) - L \int_{t-\tau}^{t} K(t-s)x(s)ds \tag{5}$$

where L is the graph Laplacian of the network, and $\Theta \in R^{n \times n^2}$ is a constant matrix defined by $\Theta = diag(\Theta_1, \dots \Theta_n)$, where Θ_i is an n-dimensional row vector given by $\Theta_i = [\sigma_{1i}, \sigma_{2i}, \dots, \sigma_{ni}]$.

Assumption 1. For any x, $y \in R^n$, there exist constants $\alpha > 0$ and $\beta > 0$, such that the nonlinearities $f(.)$ satisfy:

$$(x-y)^T[f(x) - f(y) - \alpha(x-y)] \le -\beta(x-y)^T(x-y), \quad \forall x, y \in R^n; \; \forall t \ge 0.$$

Remark 2. The assumption has been widely employed in the existing synchronization literature, which also can be proved to hold for some well-known neural networks, chaotic systems, and so on [25,26].

3   Main Results

In this section, we analyze the consensus properties of the dynamics of system (5).

Let $\delta_i(t) = x_i(t) - x_0(t)$ represent the position vector of the ith agent relative to the weighted average position of all the agents in system (5), where $x_0(t) = \Sigma_{j=1}^{N} \xi_j x_j(t)$, and $\xi = (\xi_1, \xi_2, \dots, \xi_N)^T$ is the positive left eigenvector of Laplacian matrix L associated with its zero eigenvalue, satisfying $\xi^T 1_N = 1$. Then, the error dynamical system can be written as

$$\dot{\delta}(t) = [(I - 1_N \xi^T) \otimes I_n][f(x(t)) - 1_N \otimes f(x_0(t))] - L \int_{t-\tau}^{t} K(t-s)\delta(s)ds)]dt \tag{6}$$

Theorem 1. *Suppose that the network $G(A)$ is directed and connected and Assumption 1 holds. Then, the consensus in system (5) is achieved if there exist some positive values, $c > 0$, $\alpha > 0$, $\beta > 0$, and a positive definite matrices $H = diag(h_1, \dots, h_n) \ge 0$, $\bar{K}(v) = diag(\bar{k}_1(v), \dots, \bar{k}_n(v)) > 0$, and the following LMI holds:*

$$\Gamma \doteq \begin{pmatrix} 2(\alpha - \beta)(I - 1_N \xi^T) \otimes I_N + H\bar{K}(v) & -L \\ -L & -\frac{1}{c}H \end{pmatrix} < 0 \tag{7}$$

Proof. The consensus analysis relies on the LyapunovCKrasovskii function candidate

$$V(\delta(t), t) = V_1(\delta(t), t) + V_2(\delta(t), t) \tag{8}$$

where

$$V_1(\delta(t), t) = \delta^T(t)\delta(t).$$

$$V_2(\delta(t), t) = \sum_{i=1}^{n} h_i \int_0^\tau k_i(\theta) \int_{t-\theta}^t \delta_i^2(s) ds d\theta. \tag{9}$$

Now, considering the derivative of V along the solution of system (8) with respect to t, we obtain

$$\dot{V}_1(\delta(t), t) = 2\delta^T(t)[((I - 1_N \xi^T) \otimes I_N)(f(x(t) - f(x_0(t)) - L \int_{t-\tau}^t K(t-s)\delta(s) ds]$$

$$= 2(\alpha - \beta)\delta^T(t)((I - 1_N \xi^T) \otimes I_N)\delta(t) - 2\delta(t)L \int_{t-\tau}^t K(t-s)\delta(s) ds$$

$$\tag{10}$$

$$\dot{V}_2(\delta(t), t) = \sum_{i=1}^{n} h_i \int_0^\tau k_i(\theta)[(\delta_i(t))^2 - \delta_i^2(t-\theta))] d\theta$$

$$\leq \sum_{j=1}^{N} h_i \bar{k}_i(v)\delta_i^2(t) - \frac{1}{c}\sum_{j=1}^{n} h_i[\int_0^\tau k_i(\theta)\delta_i(t-\theta) d\theta]^2$$

$$= \delta^T(t)H\bar{K}(v)\delta(t) - \frac{1}{c}[\int_{t-\tau}^t K(t-s)\delta(s) ds]^T H[\int_{t-\tau}^t K(t-s)\delta(s)) ds]$$

$$\tag{11}$$

Next, it follows from the conditions (7) and Assumption 1 that

$$\dot{V}(\delta(t), t) \leq \eta^T(t)\Gamma\eta(t) \tag{12}$$

where $\eta(t) = (\delta^T(t), (\int_{t-\tau}^t K(t-s)\delta(s) ds)^T)^T$

It can now be concluded from Lyapunov stability theory that the error system (6) is robustly, globally, asymptotically stable, which ensures the achievement of consensus in the multi-agent system (5). This completes the proof.

4   Simulations

Now, we perform some numerical simulations to illustrate our analysis by using MATLAB(7.0) programming.

Example: Consider three-dimensional multi-agent system [23] with the topology $G(A_1)$ as in Fig. 1, Let $f(x_i(t), t) = [a(-x_{i1}(t) + x_{i2} - \eta(x_{i1}(t))), x_{i1}(t) - x_{i2}(t) + x_{i3}(t), -cx_{i2}(t))]^T \in R^3$, where $\eta(x_{i1}(t)) = dx_{i2}(t) + 0.5(e - d)(|x_{i1}(t) + 1| - |x_{i1} - 1|)$, $i = 1, \ldots, N$. In this case, the isolated system is chaotic when $a = 10$, $c = 18$, $e = -\frac{4}{3}$ and $d = -\frac{3}{4}$, as shown in Fig. 2. $x_i(t) = (x_{i1}, x_{i2}, x_{i3})^T \in R^3$, $i = 1, \ldots, N$. Take the initial values randomly, let the initial conditions, $x_1(0) = (3, 0.05, 1)^T$, $x_2(0) = (0, 0.175, 1)^T$, $x_3(0) = (-3, 0, 1)^T$, $x_4(0) = (6, -0.75, 1)^T$, $x_5(0) = (9, -0.65, 1)^T$. Figure 3 shows the convergence is achieved. where we choose $\tau = 0.2$, $K(t) = diag(e^{-t}, 2e^{-2t}, e^{-t})$, $e_j = \sum_{i=1}^{N} e_{ij}^2(t)$, $j = 1, 2, 3$.

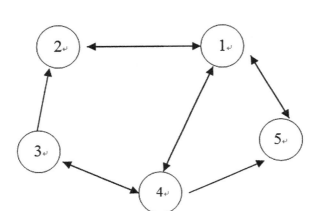

Fig. 1. Communication topology $G(A_1)$.

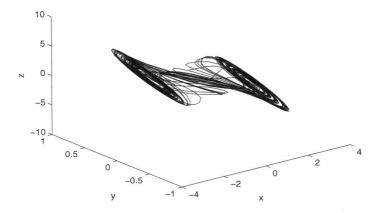

Fig. 2. The state trajectories of a single agent, for three-dimensional.

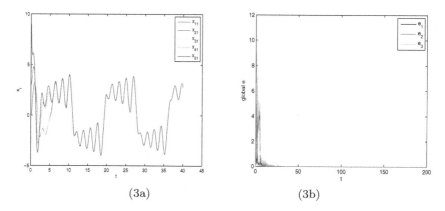

(3a)   (3b)

Fig. 3. (3a)Consensus of the states x_{i1}, (3b)Evolution of the errors $e_i(x)$. (Color figure online)

In simulations, we randomly choose the initial conditions. Under the control protocol (2), the state trajectories and the errors are shown in Fig. 3(3a) and 3(3b), respectively. It can be seen that all the states of the five agents achieve the consistency. The simulation results are in good agreements with the theoretical analysis.

5   Conclusion

In this paper, we have investigated the consensus problem of directed networks of multiple agents with intrinsic nonlinear dynamics and sampled-data information. A protocol is proposed firstly for nonlinear multi-agent systems with communication delays, which are used to model the effects of noise environment in reality. Distributed delay in control input considering a decaying weighted sum of historical information exchange over a time interval $[t - \tau, t]$ is put forward. Then the consensus problem for the proposed model is investigated based on graph theory, linear matrix inequality technique, stability theory and Lyapunov approach. A sufficient condition for the stability of the error dynamics is established in terms of a set of linear matrix inequalities. Our protocol contain several existing results as special cases. Finally, simulation results are given to verify the effectiveness of the proposed approach. Our future works will focus on the stochastic consensus behaviors of more practical models, such as multi-agent systems with nonlinear dynamics and second-order multi-agent systems. Continued research would be desirable.

Acknowledgments. This work was jointly supported by the National Natural Science Foundation of China under Grant 11202084, (Jiangnan University), Ministry of Education of China.

References

1. Beard, R.W., Lawton, J., Hadaegh, F.Y.: A coordination architecture for spacecraft formation control. IEEE Trans. Control Syst. Technol. **9**, 777–790 (2001)
2. Olfati-Saber, R.: Flocking for multi-agent dynamic systems: algorithms and theory. IEEE Trans. Autom. Control **51**, 401–420 (2006)
3. Tanner, H., Jadbabaie, A., Pappas, G.J.: Flocking in fixed and switching networks. IEEE Trans. Autom. Control **52**, 863–868 (2007)
4. Ren, W.: Multi-vehicle consensus with a time-varying reference state. Syst. Control Lett. **56**, 474–483 (2007)
5. Li, Z., Duan, Z., Chen, G.: Consensus of multiagent systems and synchronization of complex networks: a unified viewpoint. Ieee trans. circuits syst. i regul. pap. **57**, 213–224 (2010)
6. Hu, M., Yang, Y., Xu, Z.: Projective synchronization in drive-response dynamical networks. Phys. A Stat. Mech. Appl. **381**, 457–466 (2007)
7. Papachristodoulou, A., Jadbabaie, A., Munz, U.: Effects of delay in multi-agent consensus and oscillator synchronization. IEEE Trans. Autom. Control **55**, 1471–1477 (2010)
8. Dzhunusov, I.A., Fradkov, A.L.: Synchronization in networks of linear agents with output feedbacks. Autom. Remote Control **72**, 1615–1626 (2011)
9. Yu, W., Chen, G., Cao, M.: Second-order consensus for multiagent systems with directed topologies and nonlinear dynamics. IEEE Trans. Syst. Man Cybern. Part B Cybern. **40**, 881–891 (2010)
10. Song, Q., Cao, J., Yu, W.: Second-order leader-following consensus of nonlinear multi-agent systems via pinning control. Syst. Control Lett. **59**, 553–562 (2010)
11. Yu, W., Ren, W., Zheng, W.X.: Distributed control gains design for consensus in multi-agent systems with second-order nonlinear dynamics. Automatica **49**, 2107–2115 (2013)
12. Wang, H.: Consensus of networked mechanical systems with communication delays: a unified framework. IEEE Trans. Autom. Control **59**, 1571–1576 (2014)
13. Nuno, E., Ortega, R., Basanez, L., Hill, D.: Synchronization of networks of non-identical euler-lagrange systems with uncertain parameters and communication delays. IEEE Trans. Autom. Control **56**, 935–941 (2011)
14. Yu, W., Chen, G.R., Cao, M.: Consensus in directed networks of agents with nonlinear dynamics. IEEE Trans. Autom. Control **56**, 1436–1441 (2011)
15. Wang, B.C., Zhang, J.F.: Distributed control of multi-agent systems with random parameters and a major agent. Automatica **48**, 2093–2106 (2012)
16. Cao, Y., Ren, W.: Multi-agent consensus using both current and outdated states with fixed and undirected interaction. J. Intell. Robot. Syst. **58**(1), 95–106 (2010)
17. Kuang, Y.: Delay Differential Equations: with Applications in Population Dynamics. Academic Press, Boston (1993)
18. Liu, Y., Wang, Z., Liang, J.: Synchronization and state estimation for discrete-time complex networks with distributed delays. IEEE Trans. Syst. Man Cybern. Part B Cybern. **38**, 1314–1325 (2008)
19. Wan, X., Fang, H., Fu, S.: Observer-based fault detection for networked discrete-time infinite-distributed delay systems with packet dropouts. Appl. Math. Model. **36**, 270–278 (2008)
20. Wu, Z.G., Shi, P., Su, H.: Reliable control for discrete-time fuzzy systems with infinite-distributed delay. IEEE Trans. Fuzzy Syst. **20**, 22–31 (2012)

21. Olfati-Saber, R., Murray, R.M.: Consensus problems in networks of agents with switching topology and time-delays. IEEE Trans. Autom. Control **49**, 1520–1533 (2004)
22. Bliman, P.A., Ferrari-Trecate, G.: Average consensus problems in networks of agents with delayed communications. Automatica **44**, 1985–1995 (2008)
23. Wen, G., Duan, Z., Yu, W.: Consensus of multi-agent systems with nonlinear dynamics and sampled-data information. Int. J. Robust Nonlinear Control **23**, 602–619 (2013)
24. Liu, J., Liu, X., Xie, W.C.: Stochastic consensus seeking with communication delays. Automatica **47**, 2689–2696 (2011)
25. Liu, X., Chen, T.: Synchronization analysis for nonlinearly-coupled complex networks with an asymmetrical coupling matrix. Phys. A Stat. Mech. Appl. **387**, 4429–4439 (2008)
26. Wang, Y., Cao, J.: Pinning synchronization of delayed neural networks with nonlinear inner-coupling. Discrete Dynamics in Nature and Society (2011)

Discrete-Time Two-Player Zero-Sum Games for Nonlinear Systems Using Iterative Adaptive Dynamic Programming

Qinglai Wei[1(✉)] and Derong Liu[2]

[1] The State Key Laboratory of Management and Control for Complex Systems, Institute of Automation, Chinese Academy of Sciences, Beijing 100190, China
qinglai.wei@ia.ac.cn
[2] The School of Automation and Electrical Engineering, University of Science and Technology Beijing, Beijing 100083, China

Abstract. This paper is concerned with a discrete-time two-player zero-sum game of nonlinear systems, which is solved by a new iterative adaptive dynamic programming (ADP) method. In the present iterative ADP algorithm, two iteration procedures, which are upper and lower iterations, are implemented to obtain the upper and lower performance index functions, respectively. Initialized by an arbitrary positive semi-definite function, it is shown that the iterative value functions converge to the optimal performance index function if the optimal performance index function of the two-player zero-sum game exists. Finally, simulation results are given to illustrate the performance of the developed method.

Keywords: Adaptive critic designs · Adaptive dynamic programming · Approximate dynamic programming · Neuro-dynamic programming · Zero-sum game · Optimal control

1 Introduction

Dynamic programming is an effective method for optimal control problems [3]. However, it has to face the "curse of dimensionality" [2], which makes it often computationally untenable to run real dynamic programming to obtain the optimal solution. Adaptive dynamic programming (ADP), proposed by Werbos [20,21], overcomes the curse of dimensionality problem in dynamic programming by approximating the performance index function forward-in-time and becomes an important brain-like intelligent method of approximate optimal control for nonlinear systems [4,5,7,11–15,17,19]. In [8], for the first time a complex-valued ADP algorithm was discussed, where the optimal control problem of complex-valued nonlinear systems was successfully solved by ADP. In [9], based on

Q. Wei—This work was supported in part by the National Natural Science Foundation of China under Grants 61533017, 61374105, 61233001, 61304086, 61503379, and 61273140.

© Springer International Publishing Switzerland 2016
L. Cheng et al. (Eds.): ISNN 2016, LNCS 9719, pp. 269–276, 2016.
DOI: 10.1007/978-3-319-40663-3_31

neurocognitive psychology, a novel controller based on multiple actor-critic structures was developed for unknown systems and the proposed controller traded off fast actions based on stored behavior patterns with real-time exploration using current input-output data. In [10], an effective off-policy learning based integral reinforcement learning (IRL) algorithm was presented, which successfully solved the optimal control problem for completely unknown continuous-time systems with unknown disturbances.

In this paper, a new discrete-time iterative ADP algorithm is developed to solve infinite horizon optimal control problems for discrete-time two-player zero-sum games of nonlinear systems. It will be shown that the upper and lower iterative value functions converge to the upper and lower optimums, respectively. If the saddle-point equilibrium of the zero-sum game exists, the upper and lower iterative value functions will converge to the optimal solution of the zero-sum game, where the existence criteria of the saddle-point equilibrium in the traditional zero-sum ADP algorithms are not required. Finally, simulation results are given to illustrate the performance of the presented method.

2   Problem Formulations

In this paper, we will study the following discrete-time nonlinear system

$$x_{k+1} = F(x_k, u_k, w_k), \ k = 0, 1, 2, \ldots, \tag{1}$$

where $x_k \in \mathbb{R}^n$ is the state vector, $u_k \in \mathbb{R}^m$ and $w_k \in \mathbb{R}^l$ are the control vectors of Players I and II, respectively. Let x_0 be the initial state and let $F(x_k, u_k, w_k)$ be the system function. Let $\mathcal{U}$ and $\mathcal{W}$ denote policy spaces of Players I and II, respectively. Let $u \in \mathcal{U}$ and $w \in \mathcal{W}$ be the control laws of Players I and II, respectively. Then, the infinite-horizon performance index function $J : \mathcal{U} \times \mathcal{W} \to \mathbb{R}$ for state x_0 can be defined as

$$J(x_0, u, w) = \sum_{k=0}^{\infty} U(x_k, u_k, w_k), \tag{2}$$

where we let the utility function $U(x_k, u_k, w_k)$ be positive definite for x_k and u_k, and negative definite for w_k. The triple $\{J; \mathcal{U}, \mathcal{W}\}$ constitutes the *normal form* of the zero-sum game, in the context of which we can introduce the notion of a saddle-point equilibrium.

Definition 1. *Given a zero-sum dynamic game $\{J; \mathcal{U}, \mathcal{W}\}$ in a normal form, a pair of control laws $(u^*, w^*) \in \mathcal{U} \times \mathcal{W}$ constitutes a saddle-point solution if, for all $(u, w) \in \mathcal{U} \times \mathcal{W}$,*

$$J(x_k, u^*, w) \leq J^*(x_k) := J(x_k, u^*, w^*) \leq J(x_k, u, w^*). \tag{3}$$

The quantity $J^*(x_k)$ is the optimal performance index function of the game.

Given a zero-sum game $\{J; \mathcal{U}, \mathcal{W}\}$ in a normal form, we define the upper optimal performance index function as

$$\overline{J}^*(x_k) = \min_{u \in \mathcal{U}} \max_{w \in \mathcal{W}} J(x_k, u, w), \tag{4}$$

and the lower optimal performance index function can be defined as

$$\underline{J}^*(x_k) = \max_{w \in \mathcal{W}} \min_{u \in \mathcal{U}} J(x_k, u, w). \tag{5}$$

If the upper and lower optimal performance index functions are equal, i.e.,

$$\overline{J}^*(x_k) = \underline{J}^*(x_k) = J^*(x_k), \tag{6}$$

then optimal performance index function $J^*(x_k)$ for the zero-sum game is defined.

According to the principle of optimality [1], the upper optimal performance index function $\overline{J}^*(x_k)$ satisfies the following discrete-time Isaacs equation

$$\overline{J}^*(x_k) = \min_{u_k} \max_{w_k} \left\{ U(x_k, u_k, w_k) + \overline{J}^*(F(x_k, u_k, w_k)) \right\}. \tag{7}$$

The lower optimal performance index function $\underline{J}^*(x_k)$ satisfies the following discrete-time Isaacs equation

$$\underline{J}^*(x_k) = \max_{w_k} \min_{u_k} \left\{ U(x_k, u_k, w_k) + \underline{J}^*(F(x_k, u_k, w_k)) \right\}. \tag{8}$$

3 Iterative ADP Algorithm for Discrete-Time Zero-Sum Games

3.1 Derivations

In the developed iterative ADP algorithm, the value function and control law are updated at every iteration, with the iteration index i increasing from 0 to infinity. For $x_k \in \mathbb{R}^n$, let the initial function $\overline{\Psi}(x_k) \geq 0$ be an arbitrary positive semi-definite function. Then, let the upper initial value function be expressed as

$$\overline{V}_0(x_k) = \overline{\Psi}(x_k). \tag{9}$$

For $i = 0, 1, \ldots$, the iterative control law $\overline{\omega}_i(x_k, u_k)$ for the upper iterative value function can be computed as

$$\overline{\omega}_i(x_k, u_k) = \arg \max_{w_k} \left\{ U(x_k, u_k, w_k) + \overline{V}_i(x_{k+1}) \right\}$$
$$= \arg \max_{w_k} \left\{ U(x_k, u_k, w_k) + \overline{V}_i(F(x_k, u_k, w_k)) \right\} \tag{10}$$

where $\overline{V}_0(x_{k+1}) = \overline{\Psi}(x_{k+1})$. Then, the iterative control law $\overline{v}_i(x_k)$ can be obtained by

$$\overline{v}_i(x_k) = \arg \min_{u_k} \{ U(x_k, u_k, \overline{\omega}_i(x_k, u_k))$$
$$+ V_i(F(x_k, u_k, \overline{\omega}_i(x_k, u_k))) \}. \tag{11}$$

Letting $\overline{\omega}_i(x_k) = \overline{\omega}_i(x_k, \overline{v}_i(x_k))$, the upper iterative control pair $(\overline{v}_i(x_k), \overline{\omega}_i(x_k))$ can be expressed as

$$(\overline{v}_i(x_k), \overline{\omega}_i(x_k)) = \arg(\min_{u_k} \max_{w_k})\{U(x_k, u_k, w_k) + \overline{V}_i(x_{k+1})\}. \tag{12}$$

The upper iterative value function can be updated as

$$\begin{aligned}
\overline{V}_{i+1}(x_k) &= \min_{u_k} \max_{w_k}\{U(x_k, u_k, w_k) + \overline{V}_i(F(x_k, u_k, w_k))\} \\
&= U(x_k, \overline{v}_i(x_k), \overline{\omega}_i(x_k)) + \overline{V}_i(F(x_k, \overline{v}_i(x_k), \overline{\omega}_i(x_k))).
\end{aligned} \tag{13}$$

For $x_k \in \mathbb{R}^n$, let the initial function $\underline{\Psi}(x_k) \geq 0$ be an arbitrary positive semi-definite function. Let the lower initial value function be expressed as

$$\underline{V}_0(x_k) = \underline{\Psi}(x_k). \tag{14}$$

For $i = 0, 1, \ldots$, the iterative control law $\underline{v}_i(x_k, w_k)$ for lower iterative value function can be computed as

$$\begin{aligned}
\underline{v}_i(x_k, w_k) &= \arg \min_{u_k} \{U(x_k, u_k, w_k) + \underline{V}_i(x_{k+1})\} \\
&= \arg \min_{u_k} \{U(x_k, u_k, w_k) + \underline{V}_i(F(x_k, u_k, w_k))\}
\end{aligned} \tag{15}$$

where $\underline{V}_0(x_{k+1}) = \underline{\Psi}(x_{k+1})$. Then, the iterative control law $\underline{\omega}_i(x_k)$ can be obtained by

$$\begin{aligned}
\underline{\omega}_i(x_k) = \arg \max_{u_k} \{ &U(x_k, \underline{v}_i(x_k, w_k), w_k) \\
&+ V_i(F(x_k, \underline{v}_i(x_k, w_k), w_k))\}.
\end{aligned} \tag{16}$$

Letting $\underline{v}_i(x_k) = \underline{v}_i(x_k, \underline{\omega}_i(x_k))$, the lower iterative control pair $(\underline{v}_i(x_k), \underline{\omega}_i(x_k))$ can be expressed as

$$(\underline{v}_i(x_k), \underline{\omega}_i(x_k)) = \arg(\max_{w_k} \min_{u_k})\{U(x_k, u_k, w_k) + \underline{V}_i(x_{k+1})\}. \tag{17}$$

The lower iterative value function can be updated as

$$\begin{aligned}
\underline{V}_{i+1}(x_k) &= \max_{w_k} \min_{u_k}\{U(x_k, u_k, w_k) + \underline{V}_i(F(x_k, u_k, w_k))\} \\
&= U(x_k, \underline{v}_i(x_k), \underline{\omega}_i(x_k)) + \underline{V}_i(F(x_k, \underline{v}_i(x_k), \underline{\omega}_i(x_k))).
\end{aligned} \tag{18}$$

3.2   Properties

In [6,16], a "functional bound" method was proposed for the iterative ADP algorithm. Inspired by [6,16], new convergence analysis methods for the value iteration algorithm are developed in this subsection.

Theorem 1. *For $i = 0, 1, \ldots$, let $\overline{V}_i(x_k)$ and $(\overline{v}_i(x_k), \overline{\omega}_i(x_k))$ be obtained by the upper iteration (9)–(13). Then, if the optimal upper performance index function $\overline{J}^*(x_k)$ can be defined for all $x_k \in \Omega_x$, then the upper iterative value function $\overline{V}_i(x_k)$ converges to the upper optimal performance index function $\overline{J}^*(x_k)$ in (7) for all $x_k \in \Omega_x$, as $i \to \infty$.*

Proof. For any $x_k \in \Omega_x$, if $U(x_k, u_k, w_k) \neq 0$, then we assume that there exist positive constants α, β, which satisfy

$$\alpha \overline{J}^*(x_k) \leq \overline{\Psi}(x_k) \leq \beta \overline{J}^*(x_k), \tag{19}$$

where $0 \leq \alpha \leq 1 \leq \beta < \infty$. If the utility function $U(x_k, u_k, w_k) > 0$, then there exists a $\gamma > 0$ that satisfies

$$\overline{J}^*(F(x_k, u_k, w_k)) \leq \gamma U(x_k, u_k, w_k). \tag{20}$$

For $i = 0, 1, \ldots$, we can prove the following inequality

$$\left(1 + \frac{\alpha - 1}{(1 + \gamma^{-1})^i}\right) \overline{J}^*(x_k) \leq \overline{V}_i(x_k) \leq \left(1 + \frac{\beta - 1}{(1 + \gamma^{-1})^i}\right) \overline{J}^*(x_k) \tag{21}$$

holds for all $x_k \in \Omega_x \backslash \Omega_\epsilon$. Let $i \rightarrow \infty$ we can obtain the conclusion.

According to Theorem 1, we can derive that the lower iterative value function $\underline{V}_i(x_k)$ will also converge to the lower optimal performance index function. Thus, if the optimal performance index function of the two-player zero-sum exits, then it can derive the iterative upper and lower value function will converge to the optimum.

4   Simulation

We now consider the following system

$$\begin{pmatrix} x_{1(k+1)} \\ x_{2(k+1)} \end{pmatrix} = \begin{pmatrix} x_{1k} + \Delta T x_{2k} \\ -\Delta T x_{1k} + (1 + \Delta T) x_{2k} - \Delta T x_{1k}^2 x_{2k} \end{pmatrix} + \Delta T B u_k + \Delta T C w_k. \tag{22}$$

Let the performance index function be expressed by (2). The utility function is the quadratic form $U_3(x_k, u_k, w_k) = x_k^\mathsf{T} Q_3 x_k + u_k^\mathsf{T} R_3 u_k + w_k^\mathsf{T} S_3 w_k$, where $Q_3 = I_4$, $R_3 = I_5$, $S_3 = -5I_6$ and I_4, I_5, I_6 denote the identity matrices with suitable dimensions. To illustrate the effectiveness of the algorithm, four different initial value functions are considered. Let the upper initial value function be the quadratic form which are expressed by $\overline{\Phi}^j(x_k) = x_k^\mathsf{T} \overline{P}_j x_k$, $j = 0, 1$. Let $\overline{P}_0 = \begin{bmatrix} 7.98 & -1 \\ -1 & 25.97 \end{bmatrix}$ and $\overline{P}_1 = \begin{bmatrix} 8.98 & 2 \\ 2 & 30 \end{bmatrix}$. Let the lower initial value function be the quadratic form which are expressed by $\underline{\Phi}^j(x_k) = x_k^\mathsf{T} \underline{P}_j x_k$, $j = 0, 1$. Let $\underline{P}_0 = \begin{bmatrix} 24.98 & -0.5 \\ -0.5 & 9 \end{bmatrix}$ and $\underline{P}_1 = 0$. Implement the iterative zero-sum ADP algorithm for 25 iterations to reach the computation precision $\varepsilon = 0.01$. The convergence plots of the upper and lower iterative value functions, i.e., $\overline{V}_i(x_k)$ and $\underline{V}_i(x_k)$, which are initialized by $\overline{\Phi}^0(x_k)$ and $\underline{\Phi}^0(x_k)$, respectively, are shown in Figs. 1(a) and (b), respectively. The differences between the upper and lower

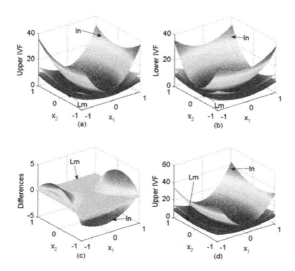

Fig. 1. Convergence plots of the iterative value functions. (a) $\overline{V}_i(x_k)$ with $\overline{\Phi}^0(x_k)$. (b) $\underline{V}_i(x_k)$ with $\underline{\Phi}^0(x_k)$. (c) Plots of $\overline{V}_i(x_k) - \underline{V}_i(x_k)$. (d) $\overline{V}_i(x_k)$ with $\overline{\Phi}^1(x_k)$.

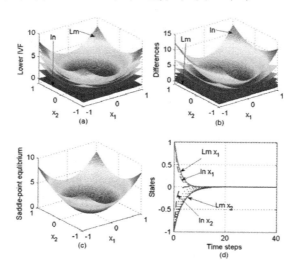

Fig. 2. Convergence plots of the iterative value functions and states. (a) $\underline{V}_i(x_k)$ with $\underline{\Phi}^1(x_k)$. (b) Plots of $\overline{V}_i(x_k) - \underline{V}_i(x_k)$. (c) Saddle-point equilibrium. (d) States by the upper iteration with $\overline{\Phi}^0(x_k)$.

iterative value functions are shown in Fig. 1(c). Initialized by $\overline{\Phi}^1(x_k)$ and $\underline{\Phi}^1(x_k)$, the convergence plots of the upper and lower iterative value functions are shown in Figs. 1(d) and 2(a), respectively. The differences between the upper and lower

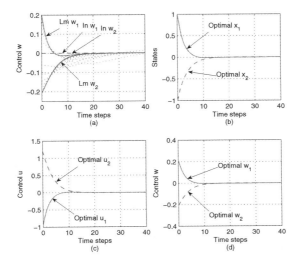

Fig. 3. Iterative and optimal trajectories. (a) Control w by the lower iteration with $\underline{\Phi}^1(x_k)$. (b) Optimal states. (c) Optimal control u. (d) Optimal control w.

iterative value functions are shown in Fig. 2(b). Hence, we can say that the saddle-point equilibrium of the zero-sum game exists, which is shown in Fig. 2(c). The state and control trajectories are shown in Figs. 2(d) and 3(a), respectively. The optimal states and controls are shown in Figs. 3(b), (c) and (d), respectively.

5   Conclusions

In this paper, a new iterative zero-sum ADP algorithm is developed to solve infinite horizon optimal control problems for zero-sum games of discrete-time nonlinear systems. In the developed algorithm, the upper and lower iterations permit arbitrary positive semi-definite functions to initialize the iterations. It is proven that the upper and lower iterative value functions can converge to the upper and lower optimal performance index functions, which satisfy the upper and lower Isaacs equations, respectively. Finally, a simulation example is given to illustrate the performance of the presented method.

References

1. Basar, T., Bernhard, P.: H_∞-Optimal Control and Related Minimax Design Problems: A Dynamic Game Approach, 2nd edn. Birkhauser, Boston (1995)
2. Bellman, R.E.: Dynamic Programming. Princeton University Press, New Jersey (1957)
3. Berkel, K., Jager, B., Hofman, T., Steinbuch, M.: Implementation of dynamic programming for optimal control problems with continuous states. IEEE Trans. Control Syst. Technol. **23**, 1172–1179 (2015)

4. Jiang, Y., Jiang, Z.P.: Robust adaptive dynamic programming and feedback stabilization of nonlinear systems. IEEE Trans. Neural Netw. Learn. Syst. **25**, 882–893 (2014)
5. Kiumarsi, B., Lewis, F.L.: Actor-critic based optimal tracking for partially unknown nonlinear discrete-time systems. IEEE Trans. Neural Netw. Learn. syst. **26**, 140–151 (2014)
6. Lincoln, B., Rantzer, A.: Relaxing dynamic programming. IEEE Trans. Autom. Control **51**, 1249–1260 (2006)
7. Ni, Z., He, H., Zhong, X., Prokhorov, D.V.: Model-free dual heuristic dynamic programming. IEEE Trans. Neural Netw. Learn. Syst. **26**, 1834–1839 (2015)
8. Song, R., Xiao, W., Zhang, H., Sun, C.: Adaptive dynamic programming for a class of complex-valued nonlinear systems. IEEE Trans. Neural Netw. Learn. Syst. **25**, 1733–1739 (2014)
9. Song, R., Lewis, F.L., Wei, Q., Zhang, H., Jiang, Z.P., Levine, D.: Multiple actor-critic structures for continuous-time optimal control using input-output data. IEEE Trans. Neural Netw. Learn. Syst. **26**, 851–865 (2015)
10. Song, R., Lewis, F. L., Wei, Q., & Zhang, H. Off-Policy Actor-Critic Structure for Optimal Control of Unknown Systems With Disturbances. IEEE Trans. Cybern. (2015) Article inpress. doi:10.1109/TCYB.2015.2421338
11. Wei, Q., Liu, D., Yang, X.: Infinite horizon self-learning optimal control of nonaffine discrete-time nonlinear systems. IEEE Trans. Neural Netw. Learn. Syst. **26**, 866–879 (2015)
12. Wei, Q., Wang, F., Liu, D., Yang, X.: Finite-approximation-error based discrete-time iterative adaptive dynamic programming. IEEE Trans. Cybern. **44**, 2820–2833 (2014)
13. Wei, Q., Liu, D.: Data-driven neuro-optimal temperature control of water gas shift reaction using stable iterative adaptive dynamic programming. IEEE Trans. Ind. Electron. **61**, 6399–6408 (2014)
14. Wei, Q., Liu, D., Shi, G., Liu, Y.: Optimal multi-battery coordination control for home energy management systems via distributed iterative adaptive dynamic programming. IEEE Trans. Ind. Electron. **42**, 4203–4214 (2015)
15. Wei, Q., Song, R., Yan, P.: Data-driven zero-sum neuro-optimal control for a class of continuous-time unknown nonlinear systems with disturbance using ADP. IEEE Trans. Neural Netw. Learn. Syst. **27**, 444–458 (2016)
16. Wei, Q., Liu, D., Lin, H.: Value iteration adaptive dynamic programming for optimal control of discrete-time nonlinear systems. IEEE Trans. Cybern. **46**, 840–853 (2016)
17. Wei, Q., Liu, D.: A novel iterative θ-adaptive dynamic programming for discrete-time nonlinear systems. IEEE Trans. Autom. Sci. Eng. **11**, 1176–1190 (2014)
18. Wei, Q., Liu, D.: Adaptive dynamic programming for optimal tracking control of unknown nonlinear systems with application to coal gasification. IEEE Trans. Autom. Sci. Eng. **11**, 1020–1036 (2014)
19. Wei, Q., Liu, D., Shi, G.: A novel dual iterative Q-learning method for optimal battery management in smart residential environments. IEEE Trans. Ind. Electron. **62**, 2509–2518 (2015)
20. Werbos, P.J.: Advanced forecasting methods for global crisis warning and models of intelligence. Gen. Syst. Yearb. **22**, 25–38 (1977)
21. Werbos, P.J.: A menu of designs for reinforcement learning over time. In: Miller, W.T., Sutton, R.S., Werbos, P.J. (eds.) Neural Networks for Control, pp. 67–95. MIT Press, Cambridge (1991)

Neural Network Technique in Boundary Value Problems for Ordinary Differential Equations

Elena M. Budkina[1], Evgenii B. Kuznetsov[1], Tatiana V. Lazovskaya[2], Sergey S. Leonov[1], Dmitriy A. Tarkhov[2], and Alexander N. Vasilyev[2(✉)]

[1] Moscow Aviation Institute, 4 Volokolamskoye Shosse, 125993 Moscow, Russia
emb0909@rambler.ru, kuznetsov@mai.ru, powerandglory@yandex.ru
[2] Peter the Great St. Petersburg Polytechnical University,
29 Politechnicheskaya Street, 195251 Saint-petersburg, Russia
tatianala@list.ru, dtarkhov@gmail.com, a.n.vasilyev@gmail.com

Abstract. This paper deals with two problems. The first of them is a boundary value problem for a nonlinear singular perturbed system of differential-algebraic equations having several solutions that can be applied to the simulation of the creep process of metal constructions. The second one is a practical problem of defining steady state stress in rotating solid disks at a constant temperature. The technique of neural network modeling is applied to solving these problems. The neural network approximate solutions agree well with the results of other authors obtained by the traditional methods.

Keywords: Modeling · Artificial neural network · Training · Global optimization · Error functional · Boundary value problem · Ordinary differential equation · Differential-algebraic equation

1 Introduction

Many theoretical and applied problems are reduced to solving a boundary value problem (BVP) for a system of ordinary differential (ODEs) or differential-algebraic equations (DAEs) [1,2]. But often the BVPs considered can be ill-posed and a process of their solution is followed by problem-solving difficulties. In some instances, these complexities are insurmountable without the use of the implicit techniques based on traditional methods in combination with special transformations, e.g. the best parametrization [3,4], or additional data about the behavior of solution, e.g. experimental data or another information about the physical process described [5]. However, in most cases, some additional data are not available or difficult to obtain, and each individual approach to solving ill-posed problems is essentially specialized.

We suggest the general artificial neural network (ANN) modeling method developed in the monograph [6] as a unified approach to the solution of the problems studied in the works [3,5]. This approach is independent of a particular task and allows obtaining a robust approximate solution (regularization) for

© Springer International Publishing Switzerland 2016
L. Cheng et al. (Eds.): ISNN 2016, LNCS 9719, pp. 277–283, 2016.
DOI: 10.1007/978-3-319-40663-3_32

these class problems. The ANN-technique can be considered as some generalization of Galerkin's methods [7] in the case of adjustable functional basis.

In present work the ANN-method [6] is tested on solving two different BVPs.
The first task is related to solving a system of nonlinear singular perturbed
DAEs, having several solutions, applied to the simulation of the creep process of
constructions under high temperature regimes and aggressive environments [3].
The second one is a practical problem of defining steady state stress in rotating
solid disks at a constant temperature [5]. The last of the problems is described by
a nonlinear system of ODEs with sub-definite boundary conditions. It is shown
that the ANN-method provides the solutions of these problems avoiding the use
of specific approaches described [3,5].

2   Scheme of Neural Network Approach

Using the results of the monograph [6], we consider the construction of an ANN-
approximation to a solution of the BVP for the system of ODEs with additional
algebraic relations named a system of differential-algebraic equations [2]

$$\begin{cases} \mathbf{F}(t, \mathbf{y}(t), \dfrac{d\mathbf{y}}{dt}, \mathbf{z}(t)) = 0, \\ \mathbf{G}(t, \mathbf{y}(t), \mathbf{z}(t)) = 0, \end{cases} \quad t \in [0, T], \tag{1}$$

and with boundary conditions

$$\mathbf{H}(\mathbf{y}(0), \mathbf{z}(0), \mathbf{y}(T), \mathbf{z}(T)) = 0, \tag{2}$$

where
$\mathbf{F} : R^1 \times R^n \times R^n \times R^m \to R^n$, $\mathbf{G} : R^1 \times R^n \times R^m \to R^m$, $\mathbf{H} : R^{2n+2m} \to R^n$
are vector-functions with components $F_i(t, \mathbf{y}(t), \dfrac{d\mathbf{y}}{dt}, \mathbf{z}(t))$, $G_j(t, \mathbf{y}(t), \mathbf{z}(t))$ and
$H_k(\mathbf{y}(0), \mathbf{z}(0), \mathbf{y}(T), \mathbf{z}(T))$ accordingly, i, $k = 1 \ldots n$, $j = 1 \ldots m$;
$\mathbf{y} : R^1 \to R^n$ and $\mathbf{z} : R^1 \to R^m$ – are some sought-for vector-functions
(solutions) with components $y_i(t)$ and $z_k(t)$ correspondingly.

We use the ANN-approach to solve this problem. The components $y_i(t)$ and
$z_k(t)$ of the sought-for vector-functions are approximated by outlets of ANNs
$\hat{y}_i(t)$ and $\hat{z}_k(t)$. In other words, we represent them in the form

$$\hat{y}_i(t) = \sum_{j=1}^{N_i} c_{ij} v(t, \mathbf{a}_{ij}); \ \hat{z}_k(t) = \sum_{j=1}^{\tilde{N}_k} d_{kj} w(t, \mathbf{b}_{kj}), \ i = \overline{1, n}, \ k = \overline{1, m}; \tag{3}$$

where $\mathbf{c} = (c_{ij})$, $\mathbf{d} = (d_{kj})$ are linear parameters, and $\mathbf{A} = (\mathbf{a}_{ij})$, $\mathbf{B} = (\mathbf{b}_{kj})$
are nonlinear input network weights; N_i and $\tilde{N}_k$ are numbers of components in
ANN-approximations (3); v, w are ANN basis elements.

Neural networks with radial basis elements are usually more effective in problems of interpolation and finding solutions in cases when we have reason to assume
their quite high smoothness. Sigmoidal basis functions are recommended to be

used in interpolation problems and for finding solutions with sharp fluctuations. For linear problems it is reasonable to use fundamental solutions as a functional basis [9]. In some tasks, heterogeneous neural networks are most effective. In particular, when constructing parametric solutions of differential equations we often use basis elements in which the dependence of the variables is Gaussian, and the dependence of the parameters of the problem is the sigmoid [8]. Some different forms of these elements are presented, e.g. in the book [6].

Using the Eq. (1) and the conditions (2), we obtain an error functional in the discrete form

$$J(\mathbf{A}, \mathbf{c}, \mathbf{B}, \mathbf{d}) = \sum_{i=1}^{M} \left(\sum_{j=1}^{n} \delta_j \left(F_j \left(t_i, \hat{\mathbf{y}}(t_i), \frac{d\hat{\mathbf{y}}}{dt}(t_i), \hat{\mathbf{z}}(t_i) \right) \right)^2 + \right.$$

$$\left. \sum_{j=1}^{n} \gamma_j \left(G_j \left(t_i, \hat{\mathbf{y}}(t_i), \hat{\mathbf{z}}(t_i) \right) \right)^2 \right) + \sum_{j=1}^{n} \beta_j \left(H_j \left(\hat{\mathbf{y}}(0), \hat{\mathbf{z}}(0), \hat{\mathbf{y}}(T), \hat{\mathbf{z}}(T) \right) \right)^2, \tag{4}$$

where $\delta_j, \gamma_j, \beta_j$ are positive penalty coefficients, $\{t_i\}_{i=1}^{M}$ is an ensemble of test points; the random components of this set are supposed to be uniformly distributed on the segment $[0, T]$.

The network weights are determined by the minimization of the error functional (4).

To avoid interruption of the minimization process at local minima, we make periodically some regeneration of the test points $\{t_i\}_{i=1}^{M}$ after several iterations of the minimization algorithm.

3   Differential-Algebraic Problem

Let us consider the BVP for the system of the singular perturbed DAEs [3]:

$$\begin{cases} y_1' = y_2, \\ \varepsilon y_2' = y_1 - z^2, \\ z - y_1^2 = 0. \end{cases} \tag{5}$$

with the boundary conditions

$$y_1(0) + y_2(0) = 0, \ y_1(1) = 1/2. \tag{6}$$

Using the shooting method and the best argument parametrization combination, authors showed that depending on the parameter ε task has different numbers of solutions and requires some customized approach [3].

We note that the term $y_2(t)$ can be eliminated from the set of equations if we pass on to ODE of the second order. Therefore, we consider the component $y_1(t)$ of the problem (5) and (6) solution.

We solved the problem (5) and (6) at various given values of ε using a general approach from the previous section. The ANN approximate solution $\hat{y}_1(t) = y(t)$ is sought in the form (3), and the basis functions used are Gaussians

$$v(t, \mathbf{a}) = \exp(-a_1^2(t - a_2)^2), \ \mathbf{a} = (a_1, a_2). \tag{7}$$

Sets of ANN approximate solutions for a given value of ε can be obtained by the restart method, i.e. the error functional minimization begins with various random starting weights. The ANN approximate solutions and their numbers are in satisfactory agreements with results from the work [3], where solutions were constructed using shooting method and method of solution continuation with respect to the best parameter [4]. The graphs of four ANN approximate solutions (for $\varepsilon = 0.04$) are shown in Fig. 1.

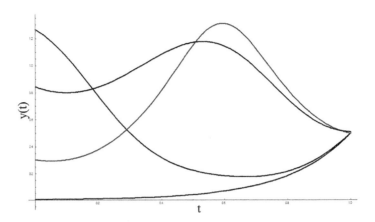

Fig. 1. Four ANN approximate solutions $y(t) = y_1(t)$ of the system (5), at $\varepsilon = 0.04$, obtained by restart method.

Another option is the introduction of additional parameter p arising from the shooting method [3]. In this case, the conditions (6) become the form

$$y_1(0) = p, \ y_2(0) = -p, \ z(0) = p^2, \ y_1(1) = 1/2. \qquad (8)$$

So, we choose the neural network function in the form [8]

$$v(t, \mathbf{a}, p) = \exp(-a_1^2(t - a_2)^2)\mathrm{th}(-a_3(p - a_4)), \ \mathbf{a} = (a_1, a_2, a_3, a_4). \qquad (9)$$

Likewise, the ANN approximate solution and according error functional depend on p additionally, and we change the ensemble of test points $\{t_i\}_{i=1}^M$ to the set $\{t_i, p_i\}_{i=1}^M$; the random components of the set are supposed to be uniformly distributed on the direct product $[0, T] \times [0, 1]$. This approach permits us to determine the number of task solutions and corresponding parameter p values at fixed value of ε. For every local minimum at Fig. 2 there is a standalone solution. This result is in agreement with the paper [3] as well.

4   Creep in Rotating Solid Disks

Let us consider the BVP of defining steady state creep of a uniformly heated rotating titan alloy aviation disk of constant thickness under creep conditions

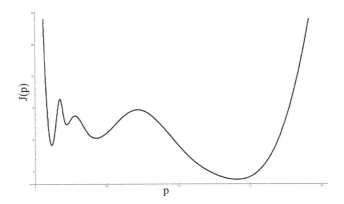

Fig. 2. Graph of the error functional of ANN-approach to solving the problem (5), at $\varepsilon = 0.04$, based on the parametrization (8) as a function of parameter p.

that described by the energy variant of the creep theory equations in the form [5, 10]

$$
\begin{cases}
\dfrac{d\sigma_\phi}{dx} = \dfrac{1}{H}\left\{ \dfrac{d\sigma_r}{dx}\left[1 + \dfrac{(\beta\sigma_i - 2)\left(2\sigma_i^2 - 3\sigma_\phi\sigma_r\right)}{2\sigma_i^2}\right] - \dfrac{3(\sigma_\phi - \sigma_r)}{x}\right\}, \\[4mm]
\dfrac{d\sigma_r}{dx} - \dfrac{\sigma_r - \sigma_\phi}{x} + Gx = 0,
\end{cases}
\tag{10}
$$

with the boundary conditions

$$
\sigma_r(0) = \sigma_\phi(0), \ \sigma_r(1) = 0,
\tag{11}
$$

where σ_r, σ_ϕ are radial and tangential stress components, $\sigma_i = \sqrt{\sigma_r^2 - \sigma_\phi\sigma_r + \sigma_\phi^2}$ is stress intensity, $H = 2 + \left(2\sigma_i^2\right)^{-1}(\beta\sigma_i - 2)(2\sigma_\phi - \sigma_r)^2$, x is dimensionless radius, and G is a level of stress applied.

The boundary condition on the rim of the disk is given only for the radial stress component. And the boundary condition in the center of the disk is given by the equality of the two stress components, but their values are unknown. In this case, traditional methods of solving boundary value problems are ineffective, because the boundary conditions are sub-definite. It is necessary to have additional data for solving similar problems. In the paper [5] this problem is solved by the shooting method. The elastic solution for rotating disk problem [11]

$$
\sigma_r(x) = G \cdot \frac{3+\nu}{8} \cdot \left(1 - x^2\right), \ \sigma_\phi(x) = G \cdot \left(\frac{3+\nu}{8} - \frac{1+3\nu}{8} \cdot x^2\right)
\tag{12}
$$

is used as an initial approximation for a creep problem to determine the boundary condition in the center of the disk. In relations (12) $\nu = 0, 4$ is the Poisson ratio. But for certain structures, solving the elastic problem itself may be very complicated. So, to solve the problem (10) and (11), we use our general method based on neural network modeling technique.

The ANN-approximate solutions $\hat{\sigma}_\phi(x)$, $\hat{\sigma}_r(x)$ are sought in the form (3). To solve the problem (10) and (11), we use different forms of the basis ANN-elements, e.g. Gaussians, hyperbolic tangent, the Cauchy functions and combinations thereof. But the best results were obtained with the use of the Cauchy functions

$$v(x, \mathbf{a}) = \left(1 + (a_1 x + a_2)^2\right)^{-1}, \quad \mathbf{a} = (a_1, a_2). \tag{13}$$

The error functional corresponds in the structure to the representation (4). The problem statement (10) and (11) is free of algebraic relations. The minimization problem was solved by the conjugate gradient method [12].

We find the BVP (10) and (11) solution for the OT-4 solid disk at constant temperature $T = 450\,^\circ\mathrm{C}$ and constant stress $G = 490.33$ MPa. The graphs of the stress components with respect to dimensionless radius are shown in Fig. 3. The ANN approximate solutions agree well with the results of the paper [5].

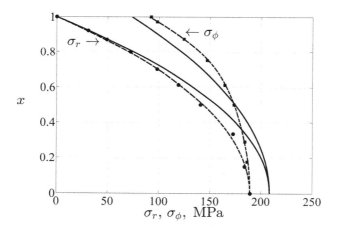

Fig. 3. ANN-approximation for stress components where dots and squares denote the solution obtained in [5] for radial and tangential stress components with the use of the shooting method, solid line is the elastic solution (12), and dashed line is the ANN approximate solution for the problem (10) and (11).

5   Conclusion

We considered two different BVPs and showed that they can be successfully solved using our unified ANN-approach. The ANN approximate solutions agree well with the results of an application of the traditional methods. Among other things, this general approach gives the following advantages:

- at each ε-parameter value of the BVP for the system of DAEs (5), the solution algorithm remains invariant while some traditional method application requires a customized approach;
- there is no need to find the elastic solution [5] for the rotating solid disk problem. The differential problem (10) and (11) can be transformed to the discrete form. The use of an ANN-approach modification in this statement increases the result accuracy;
- the obtained ANN-approximations for the solutions of both problems are the continuous functions of their arguments.

The neural network methodology presented in the book [6] offers the possibility for natural generalization of the approach introduced in the present paper to systems of higher order ODEs, DAEs, and partial differential equations.

Achnowledgements. The work was supported by the Russian Foundation for Basic Research, project numbers 16-08-00943, 14-01-00660, and 14-01-00733.

References

1. Hairer, E., Norsett, S.P., Wanner, G.: Solving Ordinary Differential Equations I: Nonstiff Problems. Springer, Heidelberg (1987)
2. Hairer, E., Wanner, G.: Solving Ordinary Differential Equations. II: Stiff and Differential-Algebraic Problems. Springer, Heidelberg (1996)
3. Budkina, E.M., Kuznetsov, E.B.: Solving of boundary value problem for differential-algebraic equations. In: Proceedings of the XIX International conference on computational mechanics and modern applied software systems, pp. 44–46. MAI Publishing House, Moscow (2015)
4. Shalashilin, V.I., Kuznetsov, E.B.: Parametric Continuation and Optimal Parametrization in Applied Mathematics and Mechanics. Kluwer Academic Publishers, Dordrecht/Boston/London (2003)
5. Sosnin, O.V., Gorev, B.V.: Energy Variant of the Theory of Creep and Long-term Strength. Report 3. Creep and Long-Term Strength of Rotating Disks. Problems of Strength, vol. 3, pp. 3–7 (in Russian) (1974)
6. Vasilyev, A.N., Tarkhov, D.A.: Neural Network Modeling, Principles, Algorithms, Applications. SPbSPU Publishing House, Saint-Petersburg (2009). (in Russian)
7. Fletcher, C.A.J.: Computational Galerkin Methods. Springer, New York (1984)
8. Lazovskaya, T.V., Tarkhov, D.A.: Fresh approaches to the construction of parameterized neural network solutions of a stiff differential equation. St. Petersburg Polytechnic Univ. J. Physics and Mathematics **1**, 192–198 (2015)
9. Kainov, N.U., Tarkhov, D.A., Shemyakina, T.A.: Application of neural network modeling to identification and prediction problems in ecology data analysis for metallurgy and welding industry. Nonlin. Phenom. Complex Syst. **17**, 57–63 (2014)
10. Sosnin, O.V., Nikitenko, A.F., Gorev, B.V.: Justification of the energy variant of the theory of creep and long-term strength of metals. J. Appl. Mech. Tech. Phys. **51**, 608–614 (2010)
11. Timoshenko, S.P., Goodier, J.N.: Theor. Elast. McGraw-Hill, New York (1970)
12. Polak, E.: Computational Methods in Optimization: A Unified Approach. Academic Press, New York (1971)

Transmission Synchronization Control of Multiple Non-identical Coupled Chaotic Systems

Xiangyong Chen[1,2], Jinde Cao[1], Jianlong Qiu[2(✉)], and Chengdong Yang[1,3]

[1] Department of Mathematics, Southeast University, Nanjing 210096, China
cxy8305@163.com, jdcao@seu.edu.cn
[2] School of Sciences, Linyi University, Linyi 276005, China
qiujianlong@lyu.edu.cn
[3] School of Informatics, Linyi University, Linyi 276005, China
yangchengdong@lyu.edu.cn

Abstract. In this paper, we investigate the transmission projective synchronization control problem for multiple, non-identical, coupled chaotic systems. By considering the influence of the occurrence of a fault between a driving system and a responding system, we define our new transmission synchronization scheme. After that, control laws are designed to achieve transmission projective synchronization and a simple stability criteria is obtained for reaching the transmission synchronization among multi-systems. A numerical example is used to verify the effectiveness of the synchronization within a desired scaling factor.

Keywords: Multiple coupled chaotic systems · Transmission projective synchronization control · Stability analysis

1 Introduction

Increasing interest has been devoted to the study of chaos synchronization [1]. New synchronization schemes for multiple chaotic systems have been reported in the literature, such as combination synchronization [2,3], hybrid Synchronization [4], targeting engineering synchronization [5], compound synchronization [6], and so on. These new synchronization schemes have advantages over conventional synchronization techniques for secure communication, information science, etc. Thus, the design of more effective synchronization schemes for an array of chaotic systems, has become a problem to be solved urgently.

Recently, a new synchronization phenomenon called transmission synchronization [7] has been observed in multiple chaotic systems. With conventional mode, multiple response systems only synchronize to one drive system, however for transmission synchronization, every system is not only a drive system, but a response system as well. Transmission synchronization is completed among multiple systems according to a step by step transmission method. In a real system, the occurrence of a synchronization fault between two of those systems

© Springer International Publishing Switzerland 2016
L. Cheng et al. (Eds.): ISNN 2016, LNCS 9719, pp. 284–291, 2016.
DOI: 10.1007/978-3-319-40663-3_33

is inevitable. With previous synchronization models, the occurrence of a fault would disrupt synchronization. In our model, synchronization can be achieved through the remaining systems to overcome the trouble without affecting their synchronization and performance. This increased reliability of achieving and maintaining transmission projective synchronization among multiple chaotic systems in the presence of faults is the motivation of the present study. For these reasons, it is highly desirable to develop the effective controller that realizes the transmission projective synchronization synchronization for multiple chaotic systems.

In recent years, the synchronization of an array of chaotic systems has became a hot topic in nonlinear research, especially for multiple coupled chaotic systems. Potential applications include multilateral communications, secret signaling and many other engineering areas. Much effort has been reported in the literature. In [4,8–10], the authors investigated some synchronization control problems in an array of coupled chaotic systems, such as complete synchronization [8], anti-synchronization [9], projective synchronization [10], hybrid synchronization [4]. Yu studied the global synchronization of three coupled chaotic systems with a ring connection [11]. Lv and Liu proposed the synchronization of N different coupled chaotic systems of with ring and chain connection [12]. Tang and Fang studied the synchronization of N-coupled fractional-order chaotic systems of with ring connection in Ref. [13]. Yang studied the synchronization of three identical systems and its application to secure communication with noise perturbations [14]. However, there has been very little effort on the transmission projective synchronization problem of an array of coupled chaotic systems.

In [15], Cheng et al. only discussed the transmission projective synchronization problem of identical coupled chaotic systems. In [7], Sun et al. studied the transmission projective synchronization of multiple systems with non-delayed and delayed coupling via impulsive control. Our article extends the work reported in [7,15], and provides a proper transmission projective synchronization criteria for multiple non-identical coupled chaotic systems into a nonlinear system with a special antisymmetric structure. In addition, we derive sufficient conditions to guarantee that the error systems asymptotically stabilize at the origin. Simulation results show the effectiveness of our presented synchronization control strategy.

2   Preliminaries and Problem Statement

In the section, we formulate a model of multiple coupled chaotic systems described as follows,

$$
\begin{cases}
\dot{x}_1 = A_1 x_1 + g_1(x_1) + D_1(x_N - x_1), \\
\dot{x}_2 = A_2 x_2 + g_2(x_2) + D_2(x_1 - x_2), \\
\vdots \\
\dot{x}_N = A_N x_N + g_n(x_N) + D_N(x_{N-1} - x_N),
\end{cases}
\tag{1}
$$

where $x_1, x_2, \cdots, x_N$ are the state vectors of the chaotic systems; $g_i(x_i)(i = 1, \cdots, N)$ is the continuous nonlinear function; $A_1, A_2, \cdots, A_N$ are constant matrices; $D_i = diag(d_{i1}, \cdots, d_{iN})$, and $d_{ij} \geq 0$ are the diagonal matrices which represent the coupled parameters. If the coefficient matrices $A_i \neq A_j (i, j = 1, \cdots N, i \neq j)$ and the functions $g_i(\cdot) \neq g_j(\cdot)$ then the system (1) is an array of non-identical chaotic systems.

Now the above simple coupling form is applied to investigate the transmission projective synchronization among multiple systems. The control terms are of the following form:

$$\begin{cases} \dot{x}_1 = A_1 x_1 + g_1(x_1) + D_1(x_N - x_1) + u_1 \\ \dot{x}_2 = A_2 x_2 + g_2(x_2) + D_2(x_1 - x_2) + u_2, \\ \vdots \\ \dot{x}_N = A_N x_N + g_n(x_N) + D_N(x_{N-1} - x_N) + u_N. \end{cases} \quad (2)$$

Before showing the main results of this paper, we first define transmission projective synchronization in an array of non-identical coupled chaotic systems.

Definition 1. For N non-identical coupled chaotic systems as described by (2), we say that they are in transmission projective synchronization if there exist controllers $u_1(t), \cdots, u_N(t)$ such that all trajectories $x_1(t), \cdots, x_N(t)$ in (2) with any initial condition $(x_1(0), \cdots, x_N(0))$ satisfy the following conditions.

$$\lim_{t \to \infty} \|e_i(t)\| = \lim_{t \to \infty} \|x_{i+1}(t) - \lambda_i x_i(t)\| = 0, i = 1, \cdots, N - 1, \quad (3)$$

$$\lim_{t \to \infty} \|e_N(t)\| = \lim_{t \to \infty} \|x_1(t) - \lambda_N x_N(t)\| = 0, \quad (4)$$

and the desired scaling factor $\lambda_i (i = 1, \cdots, N)$ satisfy

$$\lambda_1 \lambda_2 \cdots \lambda_N = 1. \quad (5)$$

Next, we introduce a lemma which is needed in the proof of the main theorem.

Lemma 1 [16,17]. Consider the systems with the state dependent coefficient:

$$\dot{y} = L(y)y, \quad (6)$$

where $y = [y_1, \cdots, y_n]^T$ is the state variable, $L(y)$ is the coefficient matrix. If $L(y) = L_1(y) + L_2$ with $L_1^T(y) = -L_1(y)$ and $L_2 = diag(l_1, \cdots, l_n), l_i < 0, (i = 1, \cdots, n)$, then the system (6) is asymptotically stable.

3   Transmission Synchronization of Multiple Non-identical Coupled Chaotic Systems

In this section, our objective is to design the appropriate controllers $u_i(t), (i = 1, \cdots, N)$ such that the state errors $e_1(t), \cdots, e_N(t)$ convergence to 0 as time t approaches to infinity, which implies the transmission projective synchronization of (2) is attained starting with arbitrary initial conditions.

Firstly, the following errors dynamic systems are obtained as,

$$
\dot{e} = \Gamma \begin{bmatrix} e_1 \\ e_2 \\ e_3 \\ e_4 \\ \vdots \\ e_N \end{bmatrix} + \begin{bmatrix} \left(\begin{pmatrix} (\lambda_1 - \lambda_1/\lambda_N)D_1 + (1 - \lambda_1)D_2 + \\ +\lambda_1(A_2 - A_1) \end{pmatrix} x_1 + \\ +g_2(x_2) - \lambda_1 g_1(x_1) + u_2 - \lambda_1 u_1 \right) \\ \left(\begin{pmatrix} (\lambda_2 - \lambda_2/\lambda_1)D_2 - (\lambda_2 - 1)D_3 + \\ +\lambda_2(A_3 - A_2) \end{pmatrix} x_2 + \\ +g_3(x_3) - \lambda_2 g_2(x_2) + u_3 - \lambda_2 u_2 \right) \\ \vdots \\ \left(\begin{bmatrix} (\lambda_{N-1} - \lambda_{N-1}/\lambda_{N-2})D_{N-1} - \\ -(\lambda_{N-1} - 1)D_N + \lambda_{N-1}(A_N - A_{N-1}) \end{bmatrix} x_{N-1} + \\ +g_N(x_N) - \lambda_{N-1}g_{N-1}(x_{N-1}) + u_N - \lambda_{N-1}u_{N-1} \right) \\ \left(\begin{bmatrix} \lambda_N(A_1 - A_N) + (1 - \lambda_N)D_1 + \\ +(\lambda_N - \lambda_N/\lambda_{N-1})D_N\lambda_N \end{bmatrix} x_N + \\ +g_1(x_1) - \lambda_N g_N(x_N) + u_1 - \lambda_N u_N \right) \end{bmatrix} \tag{7}
$$

where

$$
\Gamma = \begin{bmatrix} A_2 - D_2 - \frac{1}{\lambda_N}D_1 & -\frac{1}{\lambda_N\lambda_2}D_1 & -\frac{1}{\lambda_N\lambda_2\lambda_3}D_1 & \cdots & -\lambda_1 D_1 & 0 \\ (\lambda_2/\lambda_1)D_2 & A_3 - D_3 & 0 & \cdots & 0 & 0 \\ 0 & (\lambda_3/\lambda_2)D_3 & A_4 - D_4 & \cdots & 0 & 0 \\ \vdots & \vdots & \vdots & \cdots & \vdots & 0 \\ 0 & 0 & 0 & \cdots & (\lambda_N/\lambda_{N-1})D_N & A_1 - D_1 \end{bmatrix}
$$

And then, we choose the control inputs u_i to eliminate all known items that cannot be shown in the form of the error system e_i. The controller u_i can be given by

$$
\begin{cases} u_2 = v_1 - \begin{bmatrix} (\lambda_1 - \lambda_1/\lambda_N)D_1 + \\ +(1 - \lambda_1)D_2 + \lambda_1(A_2 - A_1) \end{bmatrix} x_1 - g_2(x_2) + \lambda_1 g_1(x_1) + \lambda_1 u_1 \\ u_3 = v_2 - \begin{bmatrix} (\lambda_2 - \lambda_2/\lambda_1)D_2 - \\ -(\lambda_2 - 1)D_3 + \lambda_2(A_3 - A_2) \end{bmatrix} x_2 - g_3(x_3) + \lambda_2 g_2(x_2) + \lambda_2 u_2 \\ \vdots \\ u_N = v_{N-1} - \left(\begin{bmatrix} (\lambda_{N-1} - \lambda_{N-1}/\lambda_{N-2})D_{N-1} - (\lambda_{N-1} - 1)D_N + \\ +\lambda_{N-1}(A_N - A_{N-1}) \end{bmatrix} x_{N-1} - \\ -g_N(x_N) + \lambda_{N-1}g_{N-1}(x_{N-1}) - \lambda_{N-1}u_{N-1} \right) \\ u_1 = v_N - \left(\begin{bmatrix} \lambda_N(A_1 - A_N) + (1 - \lambda_N)D_1 + (\lambda_N - \lambda_N/\lambda_{N-1})D_N\lambda_N \end{bmatrix} x_N + \\ -g_1(x_1) + \lambda_N g_N(x_N) - \lambda_N u_N \right) \end{cases} \tag{8}
$$

where $\begin{bmatrix} v_1 & v_2 & v_3 & \cdots & v_N \end{bmatrix}^{\mathrm{T}} = H \begin{bmatrix} e_1 & e_2 & e_3 & \cdots & e_N \end{bmatrix}^{\mathrm{T}}$, H is a coefficient matrix.
Then the error systems (7) with the controllers u_i can be rewritten by

$$
\dot{e} = L(e)e, \tag{9}
$$

where $L(e) = \Gamma + H$.

According to Lemma 1, we should choose the proper H to transform the error systems (9) into a nonlinear stable system, which conforms to the structure of system (6). The concrete form for the structure and the main results are given as follows.

Theorem 1. Consider the error dynamic system (9) with the state dependent coefficient $L(e) = L_1(e) + L_2$, if $L_1(e)$ and L_2 satisfy the following assumptions:

$$L_1^T(e) = -L_1(e), ;$$
$$L_2 = diag(l_1, \cdots, l_n), l_i < 0, (i = 1, \cdots, n)$$

then the system (9) is asymptotically stable, which means that N coupled chaotic systems (7) achieves the transmission projective synchronization.

Proof: Choose the Lyapunov function to be

$$V = \frac{1}{2} e^T e.$$

The derivative of V is

$$\dot{V} = \frac{1}{2} (\dot{e}^T e + e^T \dot{e}) = \frac{1}{2} e^T \left(L(e)^T + L(e) \right) e,$$

where $L_1^T(e) = -L_1(e)$ and $L_2 = diag(l_1, \cdots, l_n), l_i < 0, (i = 1, \cdots, n)$. Then we get that

$$\dot{V} = e^T L_2 e < 0.$$

From Lyapunov stability theory, the equilibrium $x = 0$ of the system (9) is global asymptotically stable. Then the transmission projective synchronization of N chaotic systems (7) is achieved.

According to Theorem 1, we can design the controllers by choosing the proper coefficient matrix H, which can guarantee $L(e)$ to be a special antisymmetric structure. We can find that there are many possible choices for H as long as it guarantees the error dynamic system (9) to be a stable system with a special antisymmetric structure. However, the selection of the coefficient matrix H is an important and difficult problem, because antisymmetric structure is related to the coefficient matrix H and the states of the original system. In the next section, we will demonstrate the proposed approaches for the special structure through a numerical examples.

4   Numerical Example and Simulation

In order to observe the chaos synchronization behavior for an array of non-identical coupled chaotic systems using the synchronous scheme in this paper, we consider the Chen system, Lü system and Lorenz system as drive system and response systems.

$$\begin{cases} \dot{x}_{11} = -35x_{11} + 35x_{12} + d_{11}(x_{31} - x_{11}) + u_{11}, \\ \dot{x}_{12} = -7x_{11} + 28x_{12} - x_{11}x_{13} + d_{12}(x_{32} - x_{12}) + u_{12}, \\ \dot{x}_{13} = -3x_{13} + x_{11}x_{12} + d_{13}(x_{33} - x_{13}) + u_{13}. \end{cases} \quad (10)$$

$$\begin{cases} \dot{x}_{21} = -36x_{21} + 36x_{22} + d_{21}(x_{11} - x_{21}) + u_{21}, \\ \dot{x}_{22} = 20x_{22} - x_{21}x_{23} + d_{22}(x_{12} - x_{22}) + u_{22}, \\ \dot{x}_{23} = -3x_{23} + x_{21}x_{22} + d_{23}(x_{13} - x_{23}) + u_{23}. \end{cases} \quad (11)$$

$$\begin{cases} \dot{x}_{31} = -10x_{31} + 10x_{32} + d_{31}(x_{21} - x_{31}) + u_{31}, \\ \dot{x}_{32} = 28x_{31} - x_{32} - x_{31}x_{33} + d_{32}(x_{22} - x_{32}) + u_{32}, \\ \dot{x}_{33} = -\frac{8}{3}x_{33} + x_{31}x_{32} + d_{33}(x_{23} - x_{33}) + u_{33}. \end{cases} \quad (12)$$

where

$$A_1 = \begin{bmatrix} -35 & 35 & 0 \\ -7 & 28 & 0 \\ 0 & 0 & -3 \end{bmatrix}, A_2 = \begin{bmatrix} -36 & 36 & 0 \\ 0 & 20 & 0 \\ 0 & 0 & -3 \end{bmatrix}, A_3 = \begin{bmatrix} -10 & 10 & 0 \\ 28 & -1 & 0 \\ 0 & 0 & -\frac{8}{3} \end{bmatrix},$$

$$g_1(x_1) = \begin{bmatrix} 0 \\ -x_{11}x_{13} \\ x_{11}x_{12} \end{bmatrix}, g_2(x_2) = \begin{bmatrix} 0 \\ -x_{21}x_{23} \\ x_{21}x_{22} \end{bmatrix}, g_3(x_3) = \begin{bmatrix} 0 \\ -x_{31}x_{33} \\ x_{31}x_{32} \end{bmatrix}$$

and $D_1 = diag(d_{11}, d_{12}, d_{13}), D_2 = diag(d_{21}, d_{22}, d_{23})$ and $D_3 = diag(d_{31}, d_{32}, d_{33})$ are the coupled matrices, $u_1 = [u_{11}, u_{12}, u_{13}]^T$ and $u_2 = [u_{21}, u_{22}, u_{23}]^T$ are the control inputs. Here we choose the scaling factor $\lambda_1 = \lambda_2 = 3$ and $\lambda_3 = 1/9$. Then, the synchronization error state be $\dot{e}_i = \dot{x}_{i+1} - 3\dot{x}_i, (i = 1, 2)$, and $\dot{e}_3 = \dot{x}_1 - 1/9\dot{x}_3$, the errors dynamic systems are written as,

$$\dot{e} = \begin{bmatrix} A_2 - D_2 - 9D_1 & -3D_1 & 0 \\ D_2 & A_3 - D_3 & 0 \\ 0 & (1/27)D_3 & A_1 - D_1 \end{bmatrix} \begin{bmatrix} e_1 \\ e_2 \\ e_3 \end{bmatrix} +$$

$$+ \begin{bmatrix} (-24D_1 - 2D_2 + 3(A_2 - A_1))x_1 + g_2(x_2) - 3g_1(x_1) + u_2 - 3u_1 \\ [2D_2 - 2D_3 + 3(A_3 - A_2)]x_2 + g_3(x_3) - 3g_2(x_2) + u_3 - 3u_2 \\ \begin{pmatrix} [(1/9)(A_1 - A_3) + (8/9)D_1 + (2/243)D_3]x_3 + \\ + g_1(x_1) - (1/9)g_3(x_3) + u_1 - (1/9)u_3 \end{pmatrix} \end{bmatrix} \quad (13)$$

We consider that there exists a fault between the system (10) and (12), then it is easy to get that $u_1 = 0$, and we can write the controllers as follows,

$$\begin{cases} u_2 - 3u_1 = v_1 + (24D_1 + 2D_2 - 3(A_2 - A_1))x_1 - g_2(x_2) + 3g_1(x_1) \\ u_3 - 3u_2 = v_2 - [2D_2 - 2D_3 + 3(A_3 - A_2)]x_2 - g_3(x_3) + 3g_2(x_2) \\ u_1 - \frac{1}{9}u_3 = v_3 - (\frac{1}{9}(A_1 - A_3) + \frac{8}{9}D_1 + \frac{2}{243}D_3)x_3 - g_1(x_1) + \frac{1}{9}g_3(x_3) \end{cases} \quad (14)$$

Design v_1, v_2 and v_3 to be

$$v_1 = \begin{bmatrix} 0 & 0 & 0 & 0 & 0 & 0 & 0 & 0 & 0 \\ -36 & 0 & 0 & 0 & 0 & 0 & 0 & 0 & 0 \\ 0 & 0 & 0 & 0 & 0 & 0 & 0 & 0 & 0 \end{bmatrix} e, v_3 = \begin{bmatrix} 0 & 0 & 0 & -\frac{1}{27}d_{31} & 0 & 0 & 0 & 0 & 0 \\ 0 & 0 & 0 & 0 & -\frac{1}{27}d_{32} & 0 & -28 & 0 & 0 \\ 0 & 0 & 0 & 0 & 0 & -\frac{1}{27}d_{33} & 0 & 0 & 0 \end{bmatrix} e.$$

$$v_2 = \begin{bmatrix} 3d_{11} - d_{21} & 0 & 0 & 0 & 0 & 0 & 0 & 0 & 0 \\ 0 & 3d_{12} - d_{22} & 0 & 0 & -38 & 0 & 0 & 0 & 0 \\ 0 & 0 & 3d_{13} - d_{23} & 0 & 0 & 0 & 0 & 0 & 0 \end{bmatrix} e,$$

From Theorem 1, we get the conditions

$$-36 - d_{21} - 9d_{11} < 0, 20 - d_{22} - 9d_{12} < 0, -3 - d_{23} - 9d_{13} < 0, -10 - d_{31} < 0,$$
$$-1 - d_{32} < 0, -\tfrac{8}{3} - d_{33} < 0, -35 - d_{11} < 0, 28 - d_{12} < 0, -3 - d_{13} < 0$$

and then, let the initial conditions of the drive system and the response systems be $(x_{11}(0), x_{12}(0), x_{13}(0)) = (4, 5, -3)$, $(x_{21}(0), x_{22}(0), x_{23}(0)) = (5, 2, -5)$ and $(x_{31}(0), x_{32}(0), x_{33}(0)) = (11, 15, 10)$ respectively. The initial value of the error states are $(e_{11}(0), e_{12}(0), e_{13}(0)) = (-7, -13, 4)$, $(e_{21}(0), e_{22}(0), e_{23}(0)) = (-4, 9, 25)$ and $(e_{31}(0), e_{32}(0), e_{33}(0)) = (25/9, 10/3, -37/9)$. In order to reduce the control cost, we choose that $d_{12} = 30$, $d_{11} = d_{22} = d_{32} = d_{13} = d_{23} = d_{33} = 0$, $d_{21} = 2$ and $d_{31} = 2$. The error state trajectories of the error dynamic systems are shown in Fig. 1. Figure 1 show that the error state trajectories have asymptotically converged to zero under the controllers (13). This implies that the transmission projective synchronization is realized.

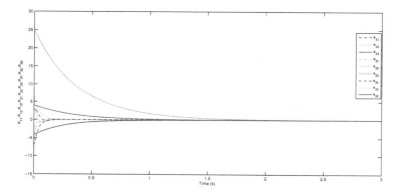

Fig. 1. Time behavior for the error state variables $e_{11}, e_{12}, e_{13}, e_{21}$, $e_{22}, e_{23}, e_{31}, e_{32}$ and e_{33} with the control strategies u_i. (Color figure online)

5   Conclusions

In this paper, we attained the transmission projective synchronization between multiple non-identical coupled chaotic systems. We realized the transmission projective synchronization between the different coupled chaotic systems by designing the controllers, and a new synchronization criteria is given for the synchronization of the chaotic systems. This technology will possess better theory and application value in engineering practice. Furthermore, our synchronization control strategy can ensure the strict synchronization of such chaotic systems.

Acknowledgement. This work was supported in part by the Applied Mathematics Enhancement Program (AMEP) of Linyi University and the National Natural Science Foundation of China (No.61403179, 61273012), by a Project of the Postdoctoral Sustentation Fund of Jiangsu Province under Grant 1402042B.

References

1. Pecora, L.M., Carroll, T.L.: Synchronization in chaotic systems. Phys. Rev. Lett. **64**, 821–824 (1990)
2. Sun, J., Shen, Y., Zhang, G.D., et al.: Combination-combination synchronization among four identical or different chaotic systems. Nonlinear Dyn. **73**, 1211–1222 (2013)
3. Jiang, C.M., Liu, S.T.: Generalized combination complex synchronization of new hyperchaotic complex Lü-like systems. Adv. Differ. Equ. **214**, 1–17 (2015)
4. Chen, X.Y., Qiu, J.L., Cao, J.D., He, H.B.: Hybrid synchronization behavior in an array of coupled chaotic systems with ring connection. Neurocomputing **173**, 1299–1309 (2016)
5. Bhowmick, S.K., Ghosh, D.: Targeting engineering synchronization in chaotic systems. Int. J. Mod. Phys. C 27, 1650006 (2016)
6. Sun, J., Shen, Y., Yin, Q., Xu, C.J.: Compound synchronization of four memristor chaotic oscillator systems and secure communication. Chaos 23, 013140 (2013)
7. Sun, J., Shen, Y., Zhang, G.D.: Transmission projective synchronization of multisystems with non-delayed and delayed coupling via impulsive control. Chaos 22, 043107 (2012)
8. Chen, X.Y., Qiu, J.L., Song, Q., Zhang, A.C.: Synchronization of N coupled chaotic systems with ring connection based on special antisymmetric structure. Abstr. Appl. Anal. 2013, 680604 (2013)
9. Chen, X.Y., Wang, C.Y., Qiu, J.L.: Synchronization and anti-synchronization of N different coupled chaotic systems with ring connection. Int. J. Mod. Phys. C **25**, 1–12 (2014)
10. Gong, H.C., Chen, X.Y., Qiu, J.L., et al.: Generalized projective synchronization of an array of non-identical coupled chaotic systems. In: The 26th Chinese Control and Decision Conference, pp. 119–123. IEEE Press, New York (2014)
11. Yu, Y., Zhang, S.: Global synchronization of three coupled chaotic systems with ring connection. Chaos Solitions Fractals **24**, 1233–1242 (2005)
12. Liu, Y., Lü, L.: Synchronization of N different coupled chaotic systems with ring and chain connections. Appl. Math. Mech. **29**, 1181–1190 (2008)
13. Tang, Y., Fang, J.A.: Synchronization of N-coupled Fractional-order chaotic systems with ring connection. Commun. Nonlinear Sci. Numer. Simul. **15**, 401–412 (2010)
14. Yang, L., Zhang, J.: Synchronization of three identical systems and its application for secure communication with noise perturbation. In: International Conference on Information Engineering and Computer Science, pp. 1–4. IEEE Press, New York (2009)
15. Cheng, L.Y., Chen, X.Y., Qiu, J.L., et al.: Transmission projective synchronization in an array of identical coupled chaotic systems. In: The 33th Chinese Control Conference, pp. 2789–2793. IEEE Press, New York (2014)
16. Cai, N., Jing, Y.W., Zhang, S.Y.: Generalized projective synchronization of different chaotic systems based on antisymmetric structure. Chaos Solitons Fractals **42**, 1190–1196 (2009)
17. Liu, B., Zhang, K.: Stability of nonlinear systems with tridiagonal structure and its applications. Acta Autom. Sin. **33**, 442–446 (2007)

Pneumatic Manipulator with Neural Network Control

Anton Aliseychik[1,2], Igor Orlov[1,2(✉)], Vladimir Pavlovsky[1],
Alexey Podoprosvetov[3], Marina Shishova[4], and Vladimir Smolin[1]

[1] Keldysh Institute of Applied Mathematics of RAS, Moscow, Russia
{aliseychik,i.orlov}@keldysh.ru,
{vlpavl,smolin}@keldysh.ru
[2] Department of Mechanics and Control of Machines,
Institute for Machine Science Named After A.A.Blagonravov of RAS,
Moscow, Russia
{aliseychik,i.orlov}@imash.ru
[3] Lomonosov Moscow State University, Moscow, Russia
llecxis@gmail.com
[4] Russian State University for Humanities, Moscow, Russia
to-be-e-e-gin@ya.ru

Abstract. Can the results of task solution be improved by means of neural net φ^{-1} implementation? Traditional methods gives ideal solution for ideal arm link control model. Real arm link and it's control can slightly differ from ideal model. During traditional control installation and tuning only few control parameters could be adjusted, while the function shape and direct proportionality control actions to pulse-width modulation (PWM) parameters are unalterable. Neural nets are representing transformation function in tabular form and are capable for functions form fine tuning during adjustments to real control conditions.

The reported study was funded by RFBR, according to the research project No. 16-38-60201 mol_a_dk.

Keywords: Manipulator · Control system · Neural network

1 Introduction

Manipulator ManGo with SCARA-like kinematics with pneumatic actuators considered in this paper. The original concept of position control based on neural network is proposed as a pneumatic actuator control system. Dynamic model and control system implemented in Matlab Simulink. The low-level control system is implemented on a microcontroller STM32F4 Discovery. Games "Go" and "Gomoku" selected as experimental tasks. Machine vision system for the recognition of the board and the game situation are implemented on the Android OS using OpenCV [1]. Task of recognising go board and stones is not new, as are methods which were used in process of solving it. Most of mathematical algorithms as FLANN and Hit-or-Miss algorithms are well known and commonly used, especially in computer vision. The main goal was to

© Springer International Publishing Switzerland 2016
L. Cheng et al. (Eds.): ISNN 2016, LNCS 9719, pp. 292–301, 2016.
DOI: 10.1007/978-3-319-40663-3_34

achieve a higher level of accuracy of recognition than of available solutions to the task. The robot was piloting, allowing to draw conclusions about the effectiveness of created software and hardware solutions to control manipulators with pneumatic actuators.

2   Manipulator ManGo

ManGo manipulator (Fig. 1) has a SCARA-like kinematic [2], that mostly suits object manipulation tasks on a plane, including desk games. The first steps in kinematic analysis were carried out during the creation of robots design in CAD soft complex, the pneumatics of Italian company Pneumax was used as the executing motor.

Fig. 1. ManGo robot

Optimal lengths of partsand attachment points of pneumatics were calculated to cover workspace of 500×500 cm size, which is enough to work with almost every knowledge-based logical desk game.

2.1   Control System Design

Due to simple two-section kinematic scheme the solution of inverse kinematic problem for robot control is trivial. Therefore, the most reliable in this case trajectory kinematic control with position feedback. The usage of pneumatics involves both advantages (cheapness, higher speeds of motion, great efforts, etc.) and drawbacks, the main of which is the difficulty of accurate control. In order to solve this problem the riveted PWM-control of the cylinders was realized. The control system of the manipulator ManGo for the "Go" and "Gomoky" games was realized on the pair microcontroller Stm32F4 – PC with Windows OS, and the control system – on the pair microcontroller Stm32F4 – smartphone with Android OS.

Matlab Simulink used to create the robot control system ManGo, which is divided into two components: the program decision-making and control of low-level system (firmware Stm32f4). These two components are connected by blocks of transmitting and receiving data: UART Rx and UART Tx respectively. Work firmware is built on iteration: each time a control program data arrive, the firmware performs only one operating cycle, sending the results back.

Let us consider the controller firmware. At each iteration, the firmware takes three integer signal: two represent the desired coordinates of the manipulator on the playing field, and the third signal A is determined that the program would do in any given moment.

A specific function is executed according to the signal value. Three types of functions exist: "go to the point", "take the stone", "put the stone".

Block-command "go to the point" consists of 3 components: performing, control valves and initiating exit from the function. Execution block is made up of two regulators – one for each joint. Each regulator compares the desired and actual value of the potentiometer readings and calculates the correction signals that are sent to the unit, which controls the valves. Execution block sends the new value to the block that initiates the output when the actual and desired values are the same. In this case, the signal A goes to the next state. Block-command "take the stone", and "put the stone" sent in a predetermined sequence of signals in the block, which controls the valves.

2.2   Android User Interface

An application for Android is a user interface, where it is possible to set rules of the game and start it. Once play has begun, it automatically tries to connect to robot via Bluetooth, after succeeding, if you choose to start a game, it enables you to "calibrate" the desk, recognize it with internal camera of the device. The internal tools of the application include GNU Go engine (Free Software Foundation http://www.gnu.org/), desk and stones recognition algorithms, and Bluetooth connection between ManGo robot and the device. GNU Go is a free Go playing program. It doesn't have a graphical interface, but it supports two protocols to "communicate" To other programs: Go Modem Protocol and Go Text Protocol (GTP). In this application GTP is used. The program is realized on C++ programming language. Image processing is realized on C++ language with usage of OpenCV (Open Computer Vision) library. The application interface, Bluetooth connection and camera treatment is realized on Java, with usage of Android features for developers (Android SDK), which enables us to control Android API.

3   Experiments

For acceptable control system maximum time to reach the 1 mm neighborhood of a chosen operating point should be 1–2 s.

Experiments on real prototype (Figs. 2 and 3) demonstrated, that for dynamically tuned relay control small but imminent environmental and settings changes like temperature, working time and throttling coefficient can cause unacceptable increase of orientation time in some areas of desk. Figure 2 shows that location of such areas may vary with variation of settings.

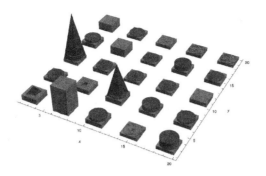

Fig. 2. Time to reach the point with different robot settings.

And essential control system for well-working robot should resist small environmental changes. Unfortunately, proportional position and PI-control do not significantly improve robust of control. Thus today we are working on neural network approach.

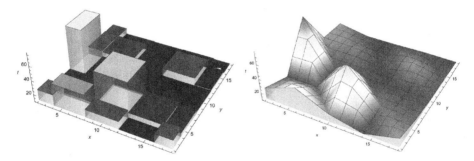

Fig. 3. Estimated time to achieve the position (on the left), average time to reach the point with relay control (on the right)

4 Neural Network Approach

All functions of manipulator control can be realized by means of neural network models. But before replacing control elements it is worth considering advantages and shortcomings of neural network models in contrast with traditional approaches.

The main convenience of neural network models is the capability of self-organization. Primary idea is just to remember some basic interrelations between input and output vector signals $X_i \rightarrow Y_i$ and make some kind of interpolations of Y for $X \neq X_i \in \{X_i\}$. Neural network shouldn't remember all input signals, but the set of remembered $\{X_i\}$ ought to represent correspondence $\{X\} \rightarrow \{Y\}$.

The conventional structure of formal neuron "j" is shown on Fig. 4. X, Y, Mj and Lj are vectors. X and Y are representing activity of input and output neurons and Mj and Lj are weights of input and output connections. The formal neuron "j" activity aj is calculated like some monotone increasing function of $X * M_j$ dot product.

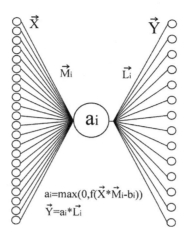

Fig. 4. Formal neuron.

Model of formal neuron can be considered as a memory cell. Weights of connections are responsible for data storage. Input weights Mj are the "address" of the cell. When X is close enough to Mj to make $aj = f(X * Mj) > 0$, neuron "j" plays back the data, stored by weights Lj. The contribution of neuron "j" to output layer activity Y is proportional to aj.

As a rule several neurons aj can add up to output neurons activity vector Y, as shown in Fig. 5. All formal neurons and connections in Fig. 5 are consimilar. But functions of layers X, A and Y are different. Layer X represents input signal, A is mapping the input signal space of states and Y should represent Y in correspondence $\{X\} \rightarrow \{Y\}$. Also the rules for M and L connections weights learning must be dissimilar. Interconnected layers pairs XA and AY only look alike. Pair XA has to remember the states of input layer X, while pair AY must memorize desirable states of output layer Y. Neuron "j" from layer A should remember the mean activity X while $a_j > 0$ and neuron "k" from layer Y – vary weights Lk to get suitable activity y_k for different X_i (and corresponding Ai).

Learning rules for vectors M_j and L_k should be like:

$$\Delta M_j = \eta_1 (X - M_j) a_j \Delta t \tag{1}$$

$$\Delta L_k = \eta_2 (y_k^s - y_k) A \Delta t \tag{2}$$

where y_k^s – specified activity of y_k. and η_1, η_2 are coefficients, $\lim_{t \to \infty} \eta = 0$, $\lim_{t \to \infty} \sum \eta = \infty$.

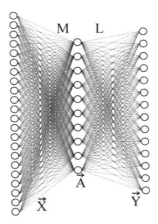

Fig. 5. Neural net structure for $\{X\} \rightarrow \{Y\}$ mapping.

Both learning rules are designed to minimize the distinction in parentheses, but the directions of vectors Mj and Lk changes (with reference to vectors X_i and Ai) are quite different.

Self-organizing maps (SOM) elaborated by Kohonen [3, 4] can solve the mapping task for the input signal space of states. Originally SOM answer is the activity of only one neuron of inner layer, "winner takes all (WTA)". For the Y interpolation purposes better to have several active neurons. SOM with WTA can be easily expanded to SOM with activity center (AC). It is worth mentioning, that SOM learning rule is exactly (1).

Learning rule (2) is used in backpropagation paradigm [5, 6]. Backpropagation technic was successfully implemented for solving different tasks. But the tendency to use purely rule (2) (without rule (1)) resulted in various problems while learning and in some measure useful attempts to solve these problems by means of randomization, normalization, orthogonalization, using multiple layers (deep learning) and so on.

Both rules (1) and (2) are used in counterpropagation network [7]. It was developed in 1986 by Hecht-Nielsen. It is guaranteed to find the correct weights, unlike regular back propagation networks that can become trapped in local minimums during training. Amazingly counterpropagation is less popular, then backpropagation model. Perhaps it's easier to implement pure rule (2), then think about cooperation of rules (1) and (2). Also, original counterpropagation use WTA competitive network (like original SOM), while for purposes of smooth converting $\{X\} \rightarrow \{Y\}$ it's better to apply AC.

Different counterpropagation network extensions were used for manipulation tasks control. The general idea was to convert some data about actual and specified manipulator states X into control signal Y, as shown on Fig. 6.

Two-link manipulator arm is moving in horizontal plane. Two pneumocylinders operate the arm movements. Each pneumocylinder position is controlled by sending pulse-width modulation (PWM) to valves V3 and V4 (Fig. 7). Valve 4 should be opened to move the piston rod to the right. If valves V1 and V2 are switched to the opposite position, short PWM impulses to V3 will cause the slow motion of piston rod leftwards.

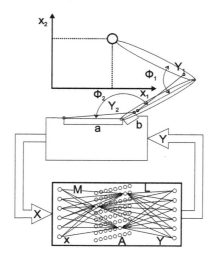

Fig. 6. Neural network manipulator control.

Unpretentiousness of manipulator structure, tasks and control functions was very helpful for detail study and main ideas understanding of neural network manipulator control. It gave possibility to start the work from very simple tasks for neural net and then gradually increase the complexity of the tasks.

First the quasi-static tasks were examined. The simplest task is to control only one manipulator arm link and to be able to stop at specified φ^s starting from arbitrary φ. This task can be easily solved by the instrumentality of standard calculations. The φ_2 angle is controlled by y_2 position. The traditional control formula for Δy_2 to reach specified φ_2^s starting from arbitrary φ_2 is (see Fig. 6 for designations):

$$\Delta y_2 = \sqrt{a^2 + b^2 - 2abcos\varphi_2^s} - \sqrt{a^2 + b^2 - 2abcos\varphi_2} \tag{3}$$

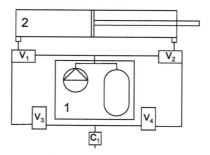

Fig. 7. Pneumocontrol structure. 1 – air pressure source; 2 – pneumocylinder; V1 – V4 – electrically controlled valves; C1 – air flow constrictor.

And Δy_2 is proportional to valves V3 or V4 opening time, T $=k\Delta y_2$, which can be transformed to PWM parameters (frequency, width and number of pulses and valve number (V3 or V4)). In experiments the mean mismatch of single movement was not very high, with minimum about 5 % of $\Delta\varphi_2$ at $\varphi_2^s = \frac{\pi}{2}$. But the displacement could be repeated from closer position and 2–3 movements usually allowed to reach desirable error of link end position less than 2 mm.

Can the results of task solution be improved by means of neural net implementation? Equation (3) gives ideal solution for ideal arm link control model. Real arm link and it's control can slightly differ from ideal model. During traditional control installation and tuning only a, b, φ and multipliers of transformation Δy_2 to PWM parameters could be adjusted, while the function (3) and direct proportionality T = k Δy_2 to PWM parameters are unalterable. Neural nets are representing transformation function in tabular form and are capable for functions form fine tuning during adjustments to real control conditions.

5   Experiments with Neural Net Control

Only one net unit (like shown on Fig. 5) was used in the first experiment. Two angles of the arm link position (actual φ_2 and specified φ_2^s) were forming X for neural net and Y was composed of four PWM parameters. The initial values for vectors Mj and Lk were calculated by (3) and law of transformation Δy_2 to PWM parameters. 1000 input signals X (formed of random φ_2^s and previous φ_2) and learning rules (1) and (2) application resulted in improvement of minimum mean mismatch to 3,6 % for 80 neurons of inner layer. 40 neurons net reached 4,7 % after 350 steps of learning. 20 neurons started from 9 % of minimum mean mismatch and reached reduction to only 7,6 %. So 20 neurons aren't enough for good tabular representation of function (3).

More than 80 neurons of inner layer make the learning time longer (proportionally to N2), but do not lead to sufficient mismatch decrease. 200 and 400 neurons make minimum mean mismatch equal to 3,4 % and 3,3 % respectively. It's meaning, that 3,3 % mismatch take place due to unpredicted random errors.

Second experiment was based on dividing transformation {X} → {Y} into three stages (Fig. 8). On the first stage nonlinear transfers actual φ_2 and specified φ_2^s to y_2 and y_2^s was performed separately on the same net with 15 neurons of inner layer. On the second stage $\Delta y_2 = y_2^s - y_2$ and abs(Δy_2) were calculated. And on the third stage Δy_2 was converted into four PWM parameters (7 neurons for frequency, 10 for width, 10 for number of pulses and 4 for valve number (V3 or V4)). This configuration demonstrated minimum mean mismatch 3,5 %.

The advantage of three stage approach is that all of three stages are simpler, than conversion in one stage. The single transformation was two-dimensional and start working good from 80 inner layers neurons, while three stages transformations are one-dimensional and need less neurons. The second stage (of three stages) was not realized on neural net, but for these linear transformations 10 neurons are more than enough (Fig. 9). The main benefit of neurons number reducing is acceleration of learning process. For three stage approach minimum mean mismatch 3,5 % was reached after processing of 70 input signals.

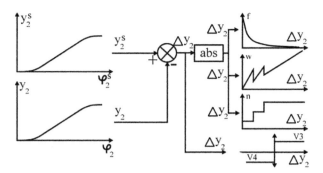

Fig. 8. Tree stages transformation.

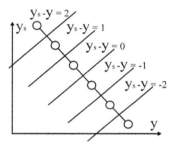

Fig. 9. Isolines $\Delta y\_2 = y\_2{}^\wedge s - y\_2$ and neural net for approximation $f(\Delta y\_2)$.

For small simple tasks the difference between one and several stages transformation isn't sufficient. But for more difficult tasks with dimensionality more than 10 good tabular representation of functions becomes impractical. Too many neurons are needed and the learning time grows unreasonably. The way of solving this problem is to split the complex task into several simpler tasks with dimensionality less than 6–7, 2–3 is desirable and 1 is the best.

After solving the single link quasi-static control task it's time to consider two-link quasi-static control task. The positions of links are defined by angles and, since the dynamic of the process is not analyzed, can be controlled independently. The only difference with training single link quasi-static control task is that the goals aren't angles φ_1^s, φ_2^s, but coordinates x_1^s, x_2^s. The task of conversion coordinates to angles can be easily solved geometrically, but also it can be solved by means of neural net. Anyway specified coordinates x_1^s, x_2^s should be transferred to φ_1^s, φ_2^s. Angles φ_1 and φ_2 are measured by angle sensors. So pairs φ_1^s, φ_1 and φ_2^s, φ_2 can be transformed to operating pneumo cylinders positions PWM parameters. The traditional control formula (3) or one of two neural nets approaches can be used for fulfilling the conversions. Mismatches of all three methods for two-link arm are proportional to results for one link with multiplier 1,41. Learning times for neural nets approaches are approximately four times longer.

6   Conclusion

The best results for two-link arm also showed the three stage transformations method. Minimum mean mismatch 4,8 % of ΔX was reached after processing of 300 input signals. 1–2 movements were necessary to reach desirable less than 2 mm error of two-link arm end position in the center of desk and 2–3 movements at the desk edge.

There are some important questions left to investigate in the quasi-static manipulator control task: balancing of learning rules coefficients, emphasizing of sufficient variables, quick tuning without changing the shape of nonlinear transformations and others. The next step is considering dynamic manipulator control tasks in order to reach the specified points by one movement, without additional correcting movements. The corrections must be made during the movement. These tasks have higher complexity, because except main variables their time derivatives should be taken in account. Also, in contrast to quasi-static control, high transformations performance is needed to have time to make several corrections during quick movement. But any solutions of dynamic manipulator control tasks by neural net means will greatly extend the manipulators application area. The performed experiments have proven the effectiveness of software and hardware tools of intelligent robotics and their correspondence to the tasks. Detailed development of these instruments will provide shorter (faster) worktime of the robots, and improved logical features of the robots. Future plans are developing fully "assembled" intelligent robot-manipulator with planning and playing games abilities based on visual system and neural "nervous" system. The computer technologies that have been developed will be the universal tools for this activity.

References

1. Aliceychik, A., et al.: Intelligent technologies for manipulation tasks. In: Kravets, A., Shcherbakov, M., Kultsova, M., Shabalina, O. (eds.) CIT&DS 2015. CCIS, vol. 535, pp. 23–40. Springer, Heidelberg (2015)
2. Orlov, I.: Synthesis of motions for manipulation systems for spaces with complex relationships and constraints. Ph.D. dissertation, Keldysh Institute of Applied Mathematics, Moscow (2013). (in Russian)
3. Kohonen, T.: Self-organized formation of topologically correct feature maps. Biol. Cybern. **43**, 59–69 (1982)
4. Kohonen, T.: Self-Organizing Maps, 3rd edn. Springer, New York (2001)
5. Rumelhart, D.E., Hinton, G.E., Williams, R.J.: Learning representations by back-propagating errors. Nature **323**, 533–536 (1986)
6. Hinton, G., Deng, L., Yu, D., Dahl, G., Mohamed, A., Jaitly, N., Senior, A., Vanhoucke, V., Nguyen, P., Sainath, T., Kingsbury, B.: Deep neural networks for acoustic modeling in speech recognition - the shared views of four research groups. IEEE Signal Process. Mag. **29**, 82–97 (2012)
7. Hecht-Nielsen, R.: Counterpropagation networks. Appl. Opt. **26**, 4979–4984 (1987)

Dynamic Noise Reduction in the System Measuring Efficiency of Light Emitting Diodes

Galina Malykhina[(✉)] and Yuri Grodetskiy

Peter the Great St. Petersburg Polytechnic University,
Saint Petersburg, Russia
{g_f_malychina, ygrod}@mail.ru

Abstract. Algorithm and method of signal waveform reconstruction in the problem of light emitting diodes efficiency measuring system was developed. Introduced method uses blind sequential signal extraction technique based on generalized skewness and generalized kurtosis allows to solve problem of noise reduction.

Keywords: Light emitting diodes · Neural network · Generalized skewness · Generalized kurtosis

1 Introduction

The problem of a noise suppressing in the channels of information - measuring systems (IMS) cannot be solved always via classical methods of linear low pass filtering or linear adaptive filtering. An example of such system is IMS for measuring parameters of light emitting diodes (LED), which comprises several channels to measure the efficiency of the LEDs and photodiodes. Measurement is performed by comparing two cooling curves, characterizing crystal LED, obtained after exposure of short current pulses of positive and negative polarity I_{forv}, I_{rev} as shown in Fig. 1.

Problem of measurement LED efficiency is especially critical for LEDs used in medicine instruments [1, 2]. The scientific publications [3, 4] consider metrological problems of LEDs. This research is devoted to the problem of measurement of efficiency of LEDs.

Loss associated with heating of crystal dominates among all reasons of the losses of LED capacity. LED is affected by forward pulses of direct current I_+, backward pulses of direct current I_- and small forward bias current to turn-on the LED. The first part $U_+(t)$ of curve characterizes cooling process after effecting current I_+. At this part crystal radiates light and heats simultaneously. The second part $U_-(t)$ characterizes cooling process after effecting current I_-. At this part crystal heats only. Difference between power on the first and the second part of curve allows to calculate power, expended to radiation. The normalized difference of power characterizes LED's efficiency.

Choosing the amplitude of the reverse pulse U_{rev} subject to a constant amplitude of direct pulse U_{forv} we compare two cooling curves $U_+(t)$ and $U_-(t)$. When two LED

© Springer International Publishing Switzerland 2016
L. Cheng et al. (Eds.): ISNN 2016, LNCS 9719, pp. 302–309, 2016.
DOI: 10.1007/978-3-319-40663-3_35

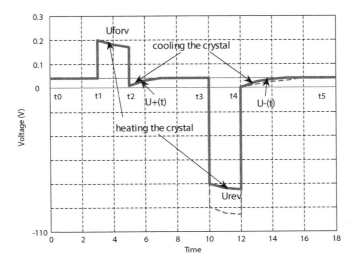

Fig. 1. Curve of heating and cooling the LED crystal, shown without following the scale. Dotted line shows variation of U_{rev} during adjustment of two cooling curves.

cooling curves differ less than a small value ε the compensation is obtained. Condition of compensation is of the form $\|U_+(t) - U_-(t)\| < \varepsilon$. When compensation is reached the value of LED efficiency η may be calculated by the following formula:

$$\eta = \frac{\int\limits_{t1}^{t2} I_{forv} U_{forv} dt - \int\limits_{t3}^{t4} I_{rev} U_{rev} dt}{\int\limits_{t1}^{t2} I_{forv} U_{forv} dt} \tag{1}$$

where I_{forv}, I_{rev} - forward and reverse current of the diode.

Condition of compensation is given by formula:

$$\int\limits_0^T [U_+(t) - U_-(t)]^2 dt \le \varepsilon, \tag{2}$$

where $U_+(t) = U(t_2 + t)$, $U_-(t) = U(t_4 + t)$ - curves of LED crystal cooling as affected by current pulse of positive and negative directivity, ε - acceptable error; $T \le t_3 - t_2, T \le t_5 - t_4$.

Presence of noise in the channels does not allow to determine fact of compensation [5]. Assuming the allowable mismatch error is $\delta = 10^{-2}$ the appropriate SNR can be calculated via equation:

$$SNR = 10 \cdot \lg \frac{u^2}{\sigma_n^2} = 10 - 10\lg\frac{1}{2} - 10\lg\frac{\rho+1}{\rho} \qquad (3)$$

where $u^2 = \sum_{\tau=0}^{N-1} u_+ (t+t_1)^2$, $\rho = r_n/\sigma_n^2$, σ_n is standard deviation of noise.

The cooling part of the signal is the most important for the efficiency evaluation, and therefore must be estimated especially precisely. Also noise should be removed on the flat part of impulses. The technique intends for automatic check of any type of LED.

If the average relative error of the cooling curves mismatch is 1 % and $\rho = 0...0.5$ then value of signal-noise relation (SNR) should be 19.23 dB. If the average relative error of mismatch of 0.1 % is permissible, then SNR = 29.33 dB. Hereby we lay down conditions that measurement of efficiency of LED makes sense if SNR is grater then 30 dB.

In the task of suppressing noise in LEDEMS we suggested to apply method of sequel blind signal extraction [6] which can be considered as nonlinear adaptive filtering. This method is implemented by cascade neural network (NN).

The object of the investigation is to develop combined scenario of signal processing for LEDEMS. The required SNR on the cooling part, as a most important part of signal, must be grater then 30 dB. It is desirable that suggested technique allows boosts SNR grater then 30 dB.

2 Method and Material of Restoring Signal Waveform

Consider simple model of mixing signal components $\mathbf{s}(t) = [s_1(t), s_2(t), \ldots, s_m(t)]^T$ by nonsingular matrix $\mathbf{H}$ [7] $\mathbf{x}(t) = \mathbf{H} \cdot s(t)$. The input of LEDEMS is presented by a mixture of signal and noise. For the correlated signal components advisable to apply decorrelation. The result of the mixing is the observed signal-vector $\mathbf{x}(t) = [x_1(t), x_2(t), \ldots, x_m(t)]^T$. Matrix $\mathbf{W}$ must be such that the estimation $\mathbf{y}(t)$ of the unknown vector-signal $\mathbf{s}(t)$ would be the result of applying the separating matrix to the measured signal $\mathbf{y}(t) = \mathbf{W} \cdot \mathbf{x}(t)$. In other words, the task of restoring signal waveform reduces to estimating vector $\mathbf{y}(t)$ of the original signal by searching the inverse mixing operator $\mathbf{W}$. Since only one component of the mixture - the useful signal is under interest, it was proposed to apply method of the signal extraction. This method may be implemented by cascade neural network with weights $\mathbf{W}$. Matrix $\mathbf{W}$ consists of vectors $\mathbf{W} = [\mathbf{w}_1, \mathbf{w}_2, \ldots, \mathbf{w}_m]$, though every vector is estimated in cascade of NN.

Signal extraction from mixture of signal and noise is possible if signal and noise have distinct statistical characteristics. Approach based height-order statistical moments for signal and noise was used. To justify this approach, we carried out a study of the statistical characteristics of signal and noise. It showed that the probability density function (PDF) of the signal is characterized by high skewness and kurtosis, but the PDF of noise is close to Gaussian, the skewness of the noise is close to zero, the excess of the noise is small. A more detailed study of the PDF of noise was performed using generalized Gaussian PDF, which presents a family of distributions of various shapes, which are characterized by three parameters: the expectation (m_x), standard deviation (σ_x) and a shape parameter (α) [8]. The density of generalized Gaussian PDF is determined according to the expression:

$$f(x) = \frac{\alpha}{2\lambda\sigma_x \cdot \Gamma(1/\alpha)} \cdot \exp\left(-\left|\frac{x - m_x}{\lambda\sigma_x}\right|^{\alpha}\right); \quad \Gamma(a) \equiv \int\limits_0^{\infty} x^{a-1}\exp(-x)dx \quad (4)$$

The estimates of statistical moments of the noise showed that the form of its PDF is close to Gaussian as the parameter a = 2.08 of exponential power distribution close to the parameters of the Gaussian a = 2. Statistical moments of signal and noise were estimated using outputs of electronic circuit of measuring system. The signal has a significant skewness 4.04 and kurtosis 17.3. The essential difference between the statistical properties of the noise and signal in the LEDEMS allows to restore the waveform effectively, using statistical moments of PDF as objective function [9].

The difference between the low orders statistical moments of signal and noise is smaller than the difference between the high order statistical moments, but estimation error of higher order statistical moments is greater. Therefore it is necessary to find a compromise between the difference between the statistical moments and the errors of its assessment.

3 Objective Function Based on Combination of Skewness and Kurtosis

Let the objective function uses skewness - normalized third central moment:

$$J(\mathbf{w}) = -\frac{1}{3}|k_3(y)| = -\frac{\beta}{3}k_3(y) \quad (5)$$

where y is a signal, $k_3(y) = \frac{E\{y^3\}}{E^2\{y^{3/2}\}}$ normalized skewness, β is a parameter that denotes the sign of the skewness. The objective function (8) allowed restore waveform with SNR = 16 dB.

If the objective function is normalized fourth moment− kurtosis then the objective function is calculated in accordance with the expression:

$$J(\mathbf{w}) = -\frac{1}{4}|k_4(y)| = -\frac{\beta}{4}k_4(y), \quad (6)$$

where $k_4(y) = \frac{E\{y^4\}}{E^2\{y^2\}}$ is the normalized kurtosis.

Combining kurtosis and skewness in objective function $J(\mathbf{w}) = -\frac{\beta}{3}k_3(y) - \frac{\beta}{3}k_4(y)$ we obtained the better results. Training of NN was carried out by gradient descent on the following rule:

$$\frac{dw}{dt} = \mu\beta_1 \cdot \left(\frac{m_2^{3/2}}{m_3^3}E\{y^2\mathbf{x}\} - \frac{m_3 \cdot m_2^{1/2}}{m_2^3}E\{y\mathbf{x}\}\right) + \mu\beta_2\left(\frac{1}{m_2^2}E\{y^3\mathbf{x}\} - \frac{m_4}{m_2^3}E\{y\mathbf{x}\}\right)$$

(7)

where m_p, $p = 1,2,3,4$ are statistical moments of order p.

Our objective is to find such vector $\mathbf{w_i}$ that maximizes the non-Gaussianity of output signal. On the other hand the absolute value of normalized kurtosis and normalized skewness may be considered as one of simplest measure of non-Gaussianity of extracted signal. Equation (7) shows the standard gradient descent approach to minimization of normalized kurtosis and skewness.

Activation function of neurons of the network takes the form:

$$\phi(y) = \left[\beta_1 \cdot \left(\frac{m_2^{3/2}}{m_2^3}E\{y^2\} - \frac{m_3 \cdot m_2^{1/2}}{m_2^3}E\{y\}\right) + \beta_2\left(\frac{1}{m_2^2}E\{y^3\} - \frac{m_4}{m_2^3}E\{y\}\right)\right] \quad (8)$$

The activation function in the considering neuron is implicit. It is changing during process of adaptation of neuron weights. Adaptation process of the weights:

$$\mathbf{w}(k+1) = \mathbf{w}(k) + \mu(k)\phi(\mathbf{y}(k))\mathbf{x}(k) \quad (9)$$

where $\mu(k)$- learning rate, providing compromise between accuracy and time of adaptation. In simulation experiment $\mu(k) = 0.01$ Structure of NN may be seen in Fig. 2. The input vector $[x_{11}(t), x_{12}(t)]^T$ represents signals of two channels created via electronic circuit. We implement on-line NN training using gradient descent algorithm (9) and adaptive activation function (8).

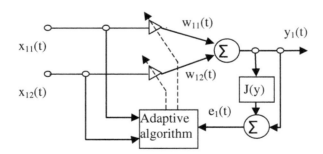

Fig. 2. Block-diagram of one cascade illustrates the sequential signal extraction.

4   Order of Skewness and Kurtosis

Height order statistical moments are more sensitive to the difference between PDF of signal and PDF of noise, but estimation error of height order statistical moments is greater. Therefore it is reasonable to choose generalized skewness A_{pq} and generalized kurtosis k_{pq} of order p and q as objective function, keeping in mind compromise between sensitivity and estimation error. Generalized moments are defined by the following formulas:

$$A_{pq}\{y\} = \frac{E\{\text{sign}(y) \cdot |y|^p\}}{E^q\{y^{p/q}\}}; \quad k_{pq}\{y\} = \frac{E\{|y|^p\}}{E^q\left\{|y|^{p/q}\right\}} \tag{10}$$

The objective function of the combined method based on generalized moments is written as:

$$J_{p1q1p2q2}(\mathbf{w}) = \frac{1}{p_1} A_{p1q1}(\mathbf{w}^T\mathbf{x}) + \frac{1}{p_2} k_{p2q2}(\mathbf{w}^T\mathbf{x}). \tag{11}$$

Where $p1, q1$ - the order of skewness, $p2, q2$ - the order of kurtosis.

The following expression was obtained for the NN training by gradient descent method:

$$\frac{d\mathbf{w}}{dt} = \mu E\left\{\text{sign}(y) \cdot |y|^{p_1-1}\mathbf{x}_1\right\} \frac{1}{m_{p1/q1}^{q1}(y)} + E\left\{|y|^{p2-1}\mathbf{x}\right\} \cdot \frac{1}{m_{p2/q2}^{q2}(y)} - \frac{m_{p1}(y)}{m_{p1/q1}^{q1+1}(y)} \cdot$$
$$\cdot E\left\{\text{sign}(y)|y|^{p1/q1-1}\mathbf{x}\right\} - \frac{m_{p2}(y)}{m_{p2/q2}^{q2+1}\left\{|y|^{p2/q2}\right\}} \cdot E\left\{\text{sign}(y)|y|^{p2/q2-1}\mathbf{x}\right\}. \tag{12}$$

For successful implementation of blind extraction method distance between skewness A_{p1q1}, A_{p2q2} and kurtosis k_{p1q1}, k_{p2q2} of signal and noise should be greater. To select the order of generalized moments we used the Mahalanobis distance between the estimated moments of the signal and noise. Analysis of Mahalanobis distance allowed choose the order of generalized kurtosis and generalized skewness for the implementation in combined method of signal extraction. It has been established that the selection of low order generalized moments is effective since the moments have less estimation error and greater Mahalanobis distance.

Therefore it was suggested to apply the combined method of the second order generalized skewness $(p1 = 2, q1 = 1)$ and first order generalized kurtosis $(p2 = 1, q2 = 1/2)$. Procedure of signal waveform restoration may be formulated by following items:

1. Sampling the sequence of digital impulses under the condition of initial value of LED current.
2. Prewhitening samples by using the standard technique of principal component analyses.
3. Computing the statistical moments by using formulas (5), (6).
4. Computing the activation function by formula (8).
5. Execution the discrete-time adaptation rule (9).
6. Transition to item 4.

7. Checking condition (2).
8. If condition (2) is true, then computing the LED efficiency via formula (1).
9. If condition (2) is false, then alteration the current and transition to item 1.

5   Calculation and Modeling

Simulation model was used for investigation of suggested method. Signal is a deterministic function changing during fitting of cooling curves. Noise is presented as autoregressive process. Using model of signal, noise and the different objective functions we can compare values of SNR.

Sequential blind extraction based on normalized third central moment – skewness (2) allowed to recover signal with SNR = 16 dB.

Normalized forth central moment − kurtosis (6) allowed to obtain SNR = 23 dB.

The combined method, based on skewness and kurtosis, showed that signal may be restored with SNR = 28 dB. Training the NN performed four times faster than training method I (mA) based on a single moment. Waveform restoration by combined method, based on I (mA) skewness of order 3 and kurtosis of order 4, SNR = 28 dB.

Result recovery waveform, SNR = 32 dB

Combined method based on generalized skewness $(p1 = 2, q1 = 1)$ and generalized kurtosis $\left(p2 = 1, q2 = \frac{1}{2}\right)$ was verified. Training a NN is twice as fast as the previous case. Result of signal waveform restoration was SNR = 32 dB. The process of signal extraction is shown in Fig. 3.

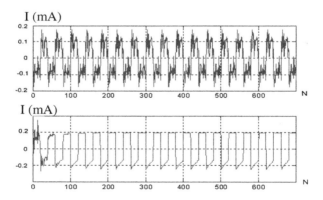

Fig. 3. Waveform restoration by combined method, based on generalized skewness and generalized kurtosis, SNR = 32 dB.

Using a second-order learning algorithm it is possible to achieve SNR = 41 dB at the cooling curves of LED crystal. Investigation of sequential method signal restoration shows that SNR increases with perfecting the method. SNR of LED efficiency measurement system applying gradient descent method is 32 dB, adaptation rate of neural network is two periods approximately.

Occurrence of uncorrelated Gaussian noise in two channels of LEDEMS provides necessity of application the robust signal extraction method. Implementation of robust method requires three channels as minimum. Modeling of robust method shows that signal extraction is possible if ratio of uncorrelated noise and signal is not grater then 7 % on the curves of crystal cooling.

6 Conclusion

Suggested method of signal extraction allows to implement multi-temporal comparison of the crystal cooling curves. It makes available to measure the efficiency of LEDs.

The suggested combined method of signal recovery, taking into account two generalized statistical moments, allows to reach SNR = 32 on critical parts heating and cooling curve.

The future research is supposed to be focused on the second order algorithm of NN adaptation and on suppressing uncorrelated components of noise.

References

1. Belyavskaya, O.A., Vilisov, A.A., Zakharova, G.N., et al.: Type LED device for intracavitary therapy. Electron. Ind. **1**(2), 173–178 (1998). (in Russian)
2. Vilisov, A.A.: LED devices for biology and medicine. Electron. Ind. **1**(2), 19–25 (1998)
3. Gorshkov, T.B., Sapritsky, V.I., Stolyarevskiy, R.I.: Metrological base of light measurements in Russia. Lighting **4**, 48–54 (2011). (in Russian)
4. Lovinsky, L.S.: Spectral photometry uncertainty new methods of measurements and calculations/Lovinsky, L.S. metrology: a monthly. Adjust. Sci. Eng. Zh Measur. Equip. **10**, 34–45 (2001). (in Russian)
5. Lovinsky, L.S., Sorokin, V.I.: The Temperature dependence of the spectral sensitivity of the photodiodes. Measur. Equip. **2**, 30–31 (1992). (in Russian)
6. Liua, W., Mandicb, D.P.: A normalized kurtosis-based algorithm for blind source extraction from noisy measurements. Sig. Process. **86**, 1580–1585 (2006)
7. Cichocki, A., Unbehauen, R.: Robust neural networks with on-line learning for blind indentation and blind separation of sources. IEEE Trans. Circuits Syst.- I: Fundam. Theory Appl. **43**, 894–906 (1996)
8. Kartamyshev, A.V., Malykhina, G.F.: Methods of removing noise in the meter settings reference emitters. Neurocomput. Dev. Appl. **6**, 54–61 (2007). (in Russian)
9. Hyvarinen, J., Karhunen, E.O.: Independent Component Analysis. Wiley, New York (2001)

Neural Network Technique in Some Inverse Problems of Mathematical Physics

Vladimir I. Gorbachenko[1], Tatiana V. Lazovskaya[2], Dmitriy A. Tarkhov[2],
Alexander N. Vasilyev[2(✉)], and Maxim V. Zhukov[1]

[1] Penza State University, 40 Krasnaya Street, 440026 Penza, Russia
`gorvi@mail.ru`, `maxim.zh@gmail.com`
[2] Peter the Great St. Petersburg Polytechnical University,
29 Politechnicheskaya Street, 195251 Saint-Petersburg, Russia
`tatianala@list.ru`, `dtarkhov@gmail.com`, `a.n.vasilyev@gmail.com`

Abstract. The general neural network approach to solving the inverse problems is considered. By applying the developed technique, we solve two different ill-posed problems. The first one is a coefficient inverse problem; the second one is an evolutionary inverse problem with experimental measurements. The neural network approximation parameters are determined by the global minimization of some error functional. Two different minimization algorithms are presented. The regularization of the problem solutions is different for each task: Morozov's condition and the regeneration of the test point ensembles respectively.

Keywords: Boundary value problem · Ill-posed problem · Inverse problem · Artificial neural network · Radial basis function · Learning · Error functional · Global optimization

1 Introduction

The direct problem is to find solutions for a given equation and some initial and boundary conditions. In the inverse problem an equation and/or initial and/or boundary conditions do not succeed completely, but there is some additional information [1,2]. When solving such problems, it is necessary not only to find a solution of the equation but also to restore mathematical model components (for example, sources, boundary or initial conditions, etc.). Inverse problems are as the rule incorrect and require the use of regularization methods for solving [3].

Let us consider the differential problem

$$Lu(\mathbf{x}) = f(\mathbf{x}), \ \mathbf{x} \in \Omega, \tag{1}$$

with boundary conditions

$$Bu(\mathbf{x}) = p(\mathbf{x}), \ \mathbf{x} \in \partial\Omega, \tag{2}$$

and additional conditions

$$Du(\mathbf{z}) = \psi(\mathbf{z}), \ \mathbf{z} \in Z \subset \partial\Omega \cup \Omega, \tag{3}$$

© Springer International Publishing Switzerland 2016
L. Cheng et al. (Eds.): ISNN 2016, LNCS 9719, pp. 310–316, 2016.
DOI: 10.1007/978-3-319-40663-3_36

where u is a required solution, L is a differential operator, B, D are some admissible operators, $f(\mathbf{x})$, $p(\mathbf{x})$ – some functions, Ω is a solution domain, $\partial\Omega$ – a domain boundary, $\psi(\mathbf{z})$ – some function of measurements.

The error of some approximate solution $\hat{u}(\mathbf{x})$ is defined by an error functional. Here, the error functional J is given in the discrete form

$$J = J_1 + \delta_b J_b + \delta_d J_d, \tag{4}$$

where

$$J_1 = \sum_{i=1}^{M_1} \left(L\hat{u}(\mathbf{x}_i) - f(\mathbf{x}_i) \right)^2, \tag{5}$$

$$J_b = \sum_{i=1}^{M_2} \left(B\hat{u}(\mathbf{t}_i) - p(\mathbf{t}_i) \right)^2, \tag{6}$$

$$J_d = \sum_{i=1}^{M_3} \left(D\hat{u}(\mathbf{z}_i) - \psi(\mathbf{z}_i) \right)^2 \tag{7}$$

are the terms corresponding to the differential equation (1), the boundary conditions (2) and the additional data (3) respectively; δ_b, δ_d are positive penalty coefficients; $\hat{u}$ is the approximate solution of the task (1)–(3); $\{\mathbf{x}_i\}_{i=1}^{M_1} \subset \Omega$, $\{\mathbf{t}_i\}_{i=1}^{M_2} \subset \partial\Omega$, $\{\mathbf{z}_i\}_{i=1}^{M_3} \subset Z$ are some test points ensembles.

The solution of the problem (1)–(3) consists in finding the global minimum of the error functional, which is represented here by the discrete form (4). Since exact global extremum seeking is possible only in exceptional cases, we look for the approximate solution $\hat{u}(x)$ of the problem of finding the global minimum of the error functional. Note that the test points can be changed during solving (neural network learning). In this way, we solve the problem of finding an approximation to the global extremum of the functional sequence. This approach allows us to overcome the hazard of getting trapped in local extrema.

Typically, to solve such extremum problems gradient minimization methods are applied. Known methods [1,2] use the functional or the parametric optimizations. In the case of the functional optimization, the gradient of the corresponding integral functional is used, which is found as the solution of a conjugate problem. In the case of parametric optimization, the required parameter of the inverse problem is represented in the form of an expansion in some functional basis, coefficients of this expansion are calculated in the process of solving. Both approaches are complicated and cumbersome and do not allow us to find the global minimum of the functional (4). Moreover, as already mentioned, the inverse problems are incorrect and require a regularization.

Nowadays, meshless methods for solving boundary value problems based on the use of neural networks [4–6,13] are making steady headway.

The authors propose a general neural network approach to solving boundary value problems [5,6], common for direct and inverse problems. The problem

(1)–(3) required components ($u(x)$ and others) are sought as neural network outputs in the form

$$\hat{u} = \sum_{i=1}^{N} c_i v(\mathbf{x}, \mathbf{a}_i),\tag{8}$$

where scalars c_i and vectors $\mathbf{a}_i$ are input network weights; v is selected neural network basis element [5,11].

The network weights are determined by the minimization of the error functional (4). This task is the global nonlinear optimization problem.

We consider two examples of ill-posed problems: the coefficient inverse problem (some equation coefficients are indeterminate) and the evolutionary inverse problem (indeterminate initial conditions). For solving these problems we use the unified neural network approach suggested and different network learning approaches. When the first way, before the solution of the problem, we need to select the type and the number of basis functions. The second approach includes the growing network algorithm with a rejection.

2   Coefficient Inverse Problem for Second Order Elliptic Equation

Let us consider the inverse problem for the Helmholtz equation [7], which describes the steady oscillation processes (mechanical, acoustic, thermal, electromagnetic, quantum, etc.)

$$-\frac{\partial^2 u(x,y)}{\partial x^2} - \frac{\partial^2 u(x,y)}{\partial y^2} + c(y)u(x,y) = 0, \ (x,y) \in \Omega,\tag{9}$$

$$u(x,y) = p(x,y), \ (x,y) \in \partial\Omega,\tag{10}$$

with additional data

$$\frac{\partial u(x,y)}{\partial \mathbf{n}} = \psi(x,y), \ (x,y) \in \partial\Omega,\tag{11}$$

where $\mathbf{n}$ is the external normal to $\partial\Omega$, where $\Omega = \{(x,y) | 0 < x < l_x, 0 < y < l_y\}$, and the coefficient $c(y)$ is required.

The task (9)–(11) is an ill-posed nonlinear inverse problem, and it has the only solution under the restriction (11) [2]. It is typical practice when the additional condition is given with an error. Therefore, we use the approximate equation

$$\frac{\partial u(x,y)}{\partial \mathbf{n}} \approx \psi^\delta(x,y) = \psi(x,y) + \sigma, \ (x,y) \in \partial\Omega,\tag{12}$$

where σ is some random variable on the segment $[-\delta, \delta]$.

We solve the above problem using the neural network approach. The approximate solution $\hat{u}(x)$ and the required coefficient $\hat{c}(y)$ are sought in the form (8). The error functional corresponds in the structure to the general approach form (4), where the expressions (9), (10) and (12) specify operators L, B, D.

To solve the problem, we use the network of radial basis functions (RBFs) [8] with gaussian basis elements. The centers of RBFs initially located randomly in the domain $\bar{\Omega} = \Omega \cup \partial\Omega$. Fixed test points were randomly distributed in the same domain. The number N of basis elements and the number M of test points were chosen experimentally in view of the relation $M \propto N^{1/3}$, where $\propto$ is the proportionality [9].

We modified the trust region method to learn RBF-networks [10,11] and applied it to solving this problem. The principle of trust region method is as follows. At each iteration, the minimized functional is approximated by a polynomial of the second degree in a sufficiently small neighborhood of previous approximation. This neighborhood is called the trust region. Further, the minimum of the polynomial is sought in this region. The corresponding point of minimum is taken for a new approximation to the required global minimum. The iteration process is repeated until the stop condition implementation.

The iterative regularization method (Morozov's condition) was used [12]. Neural network training continues as long as

$$J_d > M_3\delta^2, \tag{13}$$

where δ is the maximum error of the measurement of additional parameters, M_3 is from the expression [7].

We find the coefficient inverse problem (9), (10) and (12) solutions in the case of $c(y) = 10y$, $p(x,y) = 1 + x$ in the unit square Ω ($l_x = l_y = 1$).

The additional data (12) in the discrete form are generated with the use of the direct problem (9), (10) neural network solution if the coefficient $c(y)$ is known.

Next, we solve the inverse problem. The network approach parameters are $M_1 = 100$, $M_2 = 44$, $\delta_b = \delta_d = 1000$, the number of basis RBF for $\hat{u}(x)$ is equal to 14 for both statement, the number of basis RBF for $\hat{c}(y)$ for the inverse statement is equal to 3.

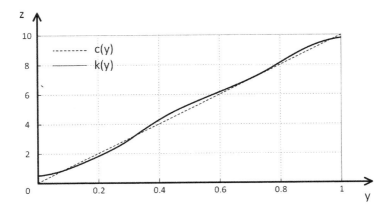

Fig. 1. Neural network approximation for the coefficient $c(y)$ at $\delta = 0$

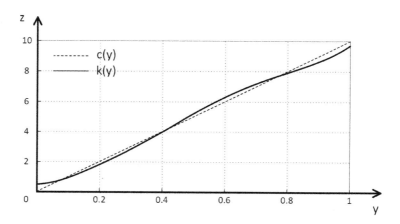

Fig. 2. Neural network approximation for the coefficient $c(y)$ at $\delta = 0.08$

We have solved two inverse problems for data error parameter values $\delta = 0$ and $\delta = 0.08$. The obtained solutions errors are estimated using the ratio mean square error formula. The coefficient $\hat{c}(y)$ errors are equal to 0.031 at $\delta = 0$ and to 0.041 at $\delta = 0.08$ (Fig. 1 and 2 show the results). The solution $\hat{u}(x, y)$ errors are equal to 0.018 and 0.021 correspondingly.

3   Evolutionary Inverse Problem for Heat Equation

Let us consider the problem of defining the temperature field from experimental data with the requirement that the samples have different temperature conductivity coefficients. Here, this initial temperature distribution is equal for all samples.

The problem statement is formulated in the form

$$u_t = ru_{xx}, \ (x, t) \in (0, 1) \times (0, T), \ r \in [r_{min}, r_{max}], \tag{14}$$

with the boundary conditions

$$u(0, t, r) = 0, \ u(1, t, r) = 0, \tag{15}$$

and the additional data

$$u(x_i, t_i, r_i) = f_i, \ i = 1, \ldots, p. \tag{16}$$

The initial condition

$$\phi(x) = u(x, 0, r). \tag{17}$$

is an unknown function.

The problem (14)-(16) is an ill-posed evolutionary inverse problem for the classical heat equation with point experimental data and an interval parameter.

For solving this problem we use the combined algorithm.

At the first stage, the problem solution $u(x, t, r)$ is sought as a neural network approximation in the form (8) with parametric basis functions

$$v(x, t, r, \mathbf{a}) = \exp\left(-a_1(x - a_2)^2 - a_3(x - a_4)(t - a_5) - a_6(t - a_7)^2\right)$$
$$\times \operatorname{th}\left(a_8(r - a_9)\right), \tag{18}$$

where $\mathbf{a} = (a_1, \ldots, a_9)$ is the neural network nonlinear weight vector.

The network weights $(c_i, \mathbf{a}_i)$ are determined by the minimization of the error functional in the general neural network approach form (4), where operators L, B, D are specified by the expressions (14)-(16). We use the algorithm combining RProp and the cloud method [6]. The test point ensembles are assumed as uniformly distributed in the corresponding regions. These test points are randomly regenerated after some number of steps of the optimization algorithm. This regeneration makes it possible to avoid neural network overfitting. In addition, after each test point regeneration we use the following evolutionary algorithm so called the growing neural network with selecting elements. One new neuron (a summand) is added to the sum (8), then new network weights are adjusted and the obtained error functional value is compared with the previous one. If the error increases, a new summand is excluded from the solution.

After terminating the learning process we go over to the next stage and try the weights c_i over again as the solutions of the linear system (in our case, the error functional J is quadratic in c_i)

$$\frac{\partial J}{\partial c_i} = 0, \ i = 1, \ldots, N, \tag{19}$$

where N is the summary number of the neural network basis functions.

We have performed computations on the neural network model with parameters $N = 200$, $N_b = 50$, $N_d = 150$. The experimental data (16) had the random error uniformly distributed on the segment $[-0.01, 0.01]$. The number of the attempts to add a neuron is equal to 200, the summary number of the neural network basis functions is 172. The error of the initial condition recovery does not exceed 10 percent. We have succeeded in capturing the fact that initial temperature distribution is independent of the parameter r.

4   Conclusion

We have developed and applied the iterative neural network technique of media property identification using some additional data. We have solved the coefficient and the evolutionary inverse problems and elaborated two optimization algorithms. The first one is the trust region minimization method modified for RBF-learning. In this case, Morozov's regularization condition makes it possible to avoid neural network overfitting for imperfect data. In the second algorithm, we use the combination of RProp and the cloud methods followed by special parameter matching. The regularization used is the periodical regeneration of the test point ensembles.

The general neural network technique needs no implicit approaches (for instance, construction and solving conjugate problems [2]) that simplifies the solving process of the original problem.

The possibility of the problem extension is consideration of heterogeneous data (other equations and conditions, measurements, etc.), other types of neural network basis elements, piecewise-defined parameters.

Acknowledgements. The work was supported by the Russian Foundation for Basic Research, project numbers 16-08-00906a, 14-01-00660a, and 14-01-00733a.

References

1. Aster, R.C., Borchers, B., Thurber, C.H.: Parameter Estimation and Inverse Problems. Academic Press, New York (2012)
2. Samarskii, A.A., Vabishchevich, P.N.: Numerical Methods for Solving Inverse Problems of Mathematical Physics. Walter de Gruyter, Berlin (2007)
3. Engl, H.W., Hanke, M., Neubauer, A.: Regularization of Inverse Problems. Springer, New York (1996)
4. Yadav, N., Yadav, M., Kumar, M.: An Introduction to Neural Network Methods for Differential Equations. Springer, Netherlands (2015)
5. Vasilyev, A.N., Tarkhov, D.A.: Neural Network Modeling. Principles, Algorithms, Applications. SPbSPU Publishing House, Saint-Petersburg (2009). (in Russian)
6. Tarkhov, D.A.: Neural network models and algorithms. Radiotekhnika, Moscow (2014). (in Russian)
7. Polyanin, A.D., Nazaikinskii, V.E.: Handbook of Linear Partial Differential Equations for Engineers and Scientists. Chapman and Hall/CRC, Boca Raton (2016)
8. Haykin, S.O.: Neural Networks and Learning Machines. Prentice Hall, Upper Saddle River (2008)
9. Niyogi, P., Girosi, F.: On the relationship between generalization error, hypothesis complexity, and sample complexity for radial basis functions. Neural Comput. **8**(4), 819–842 (1996)
10. Conn, A.R., Gould, N.M., Toint, P.L.: Trust regions methods. MPS-SIAM (1987)
11. Gorbachenko, V.I., Zhukov, M.V.: The approaches and methods of radial basis functions networks learning to solve mathematical physic problems. Neurocomputers Dev. Appl. **13**, 12–18 (2013)
12. Morozov, V.A.: Methods for Solving Incorrectly Posed Problems. Springer, New York (1984)
13. Lazovskaya, T.V., Tarkhov, D.A.: Fresh approaches to the construction of parameterized neural network solutions of a stiff differential equation. St. Petersburg Polytechnical Univ. J. Phys. Math. **1**(2), 192–198 (2015). http://dx.doi.org/10.1016/j.spjpm.2015.07.005

The Model of the Robot's Hierarchical Behavioral Control System

A.V. Bakhshiev and F.V. Gundelakh$^{(\boxtimes)}$

Russian State Scientific Center for Robotics and Technical Cybernetics (RTC),
Saint-Petersburg, Russian Federation
{alexab, f.gundelakh}@rtc.ru

Abstract. The paper observes an approach to the implementation of the behavioral control system for robotic systems, based on the modeling both mechanisms of memorization and reproduction of motor acts. The models of neurons and neural networks for motion control are based on some well-known properties of the natural neural networks that controls muscle contraction. A feature of the model is the presence of neural network's structural adaptation and the usage of a dynamic neuron's model, which allows to describe the structure of dendrites and synapses. The hierarchy of motor memory model levels is described. The results of modeling the behavior of the neural motion control network are shown.

Keywords: Neural networks · Behavior control · Bionics · Motor memory · Modeling · Neuron model

1 Introduction

In the development of the robot's behavioral control systems the most commonly used approach is that in which, depending on the chosen final task, the control system is equipped with a set of algorithms that solve individual subtasks (collision avoidance, detour obstacles, recognition of objects from a given set of classes, return to the starting point, and so on), and then a certain decision-making system is used, which integrates solutions of subtasks into a single strategy of robot's behavioral control as part of the problem being solved. However, in dealing with the individual subtasks are often used artificial neural networks (ANN), but their application is generally limited to strictly specified context (a task or list of tasks that should be resolved by the robot). It might be said that in this case, the ANNs are used locally.

In the development of robotic systems with a high degree of autonomy seems promising a "global" approach, in which the neural network receives information from internal and external sensors of the robot and is shorted to the external environment through its effectors. At the same time, the context of a specific problem, which must be solved by the robot, is one of the many possible contexts and not imposed from within the robot's behavioral control system and is transmitted through the mechanism of dialogue with the system by using internal motivation mechanisms.

Among the theories that are used or can be used to build the global behavioral control systems can be distinguished the theory of functional systems (TFS) [1] and the

© Springer International Publishing Switzerland 2016
L. Cheng et al. (Eds.): ISNN 2016, LNCS 9719, pp. 317–327, 2016.
DOI: 10.1007/978-3-319-40663-3_37

theory of neural group selection [2]. Also the large-scale projects of brain modeling should be noted [3].

In the scope of the model which is reviewed in the article, we attempted to focus on solving motion control tasks as the main tasks to be solved by all the animals in the scope of interaction with the environment. Nervous system appeared through the evolution, as a tool for solving problems of increasing complexity, which are directly or indirectly associated with the movement in the environment. The rest of the functionality is an "upgrade" of this basic function. That is, the most important thing is the execution the motor act and the mechanisms of memorization of simple and complex movements as a sets of possible reactions to the environment. The solution to this problem is based on the information about the neural structures with known morphology and function, to create their models, study their functioning, and apply this knowledge to solving problems in the control of the technical systems. However, with the lack of information about the network morphology it seems acceptable to develop own neural network architecture, borrowing the well-known principles of natural neural networks, when it is possible.

2 Description of the Model

Figure 1 shows the functional diagram of neural network control system (CS), receiving information from the sensors and closed on the environment through effectors. Here, the generalization of sensory data refers to the process of memorizing images, the motor memory means the ability of the system to reproduce complex movements, set of which the system was taught earlier. The decision-making block is simplified to the block of reflex activity, meaning that the reflexes can be quite complex. All connections on the scheme are unidirectional as though sensors and effectors do not interact with each other explicitly, but in practice they are fundamentally not always separable. Muscle fiber, which is an effector, is provided with a set of sensors that determine the length of the muscles and produced effort.

Figure 2 shows the levels of hierarchy of the proposed robot control model. Where:

$y^{(s)}$ – data about the current position of the control object;
$u^{(rcn)}$ – the control action on the object;

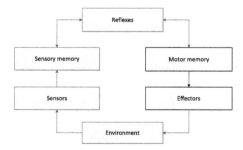

Fig. 1. Detailed circuit CS in terms of memory and reflexes

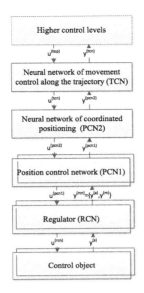

Fig. 2. Structural diagram of motor memory levels of the robot's behavioral control system

$y^{(rcn)}$ – regulator output data;

$y^{(a)}$ – vector of the output values from afferent neurons, corresponding to the information from sensors about the current state of the control object;

$y^{(m)}$ – vector of the motoneuron's activity, providing an information about current target position of the control object;

$u^{(pcn1)}$ – vector of the control actions on the regulator, activating necessary motoneurons for the transition to a new position;

$y^{(pcn1)}$ – vector of the current position of the robot's executive system element from a variety of pre-trained possible positions;

$u^{(pcn2)}$ – vector of the desired position of the robot's executive system element from a variety of pre-trained positions;

$y^{(pcn2)}$ – vector of the current coordinated position of all elements of the robot from a variety of pre-trained coordinated position;

$u^{(tcn)}$ – vector of the desired coordinated position;

$y^{(tcn)}$ – vector describing the current performed by the robot motion path;

$u^{(top)}$ – vector to set the desired trajectory from a variety of pre-trained trajectories.

The neural networks described here are based on the model of a neuron, presented in [5]. A feature of the model is a representation of the neuron membrane, consisting of separate parts, modeling dendrites and soma cells. Each of these sections of the membrane is a dynamic object, and the resulting contribution of synapses formed in different sections, depends both on the parameters of the synapses and on the spatial arrangement of the membrane section relatively to the other sections and the generating area. These features make it easy to implement the interaction between neurons required to implement the interaction, particularly in case of circular structures with positive feedback that are described below.

Level of the regulator provides information on the current state of the control object (for example, the manipulator link) and provides a transition to a predetermined position. Model of the regulator (regulatory control network - RCN) is described below. General view of the regulator structure based on biosimilar pulsed neuron model [4], is shown in Fig. 3. This regulator maintains a predetermined position of the link. The regulator is based on neural network model of spinal level of the muscular contraction control [5]. The entire range of possible positions of the manipulator unit is divided into N segments being in each of which is considered to retention of the target position. In this case the angular size of the segment is actually a positioning error. Increasing the number of segments and thus, the number of controlling neural networks can provide more precise control. Transition of the manipulator to a specified segment is carried out by a simultaneous activation of a corresponding number of control elements. Permanent link location in one of the segments will be called the current spatial configuration of this link.

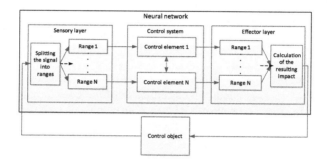

Fig. 3. Functional diagram of the regulator's neural network

Hereinafter, the manipulator, as a control object is chosen for an example. The architecture of the proposed neural networks and methods of its settings do not depend on the model of the control object. For example, instead of the manipulator's drive units we can consider mobile robot's wheel drives without modification of structure of neural network's connections. For this reason, to demonstrate the possibilities of the proposed neural networks, we can abstract from a particular mathematical model of control object.

Control elements, in turn, have the structure shown in Fig. 4.

In this case, we mean that control system has a finite set of possible positions, whose number is determined by the structure of the neural network of the regulator.

The positioning neural network (position control network - PCN), ensures the coordinated positioning of the robot's executive system. This neural network is shown in the scheme twice. PCN1 network allows memorization and reproducing (via controlling the regulator) the specific positions of a single element of the robot control system. PCN2 network unites a set of PCN1 networks for coordinated control of the

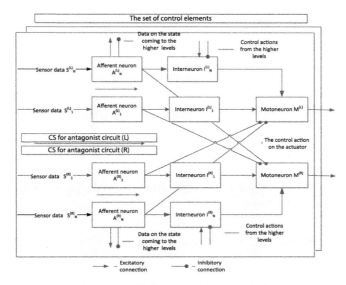

Fig. 4. Structural diagram of the control element

entire executive system. Thus, PCN2, in turn, provides the ability to memorize and reproduce a finite set of coordinated positions of all elements of the robotic system.

The input of the PCN1 level is the data about the activity of afferent neurons of the regulator (RCN), the activity of which describes the parameters of the object in space. Downward output back to the RCN level is an activity supplied to the interneurons, the activity of which determines the change in motoneurons' activity, and thus changes the character of the object's movement.

Such a trainable neural network model for positioning a single actuator is shown in Fig. 5.

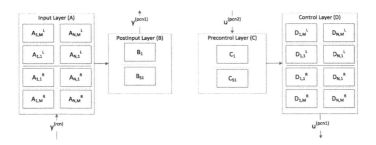

Fig. 5. Scheme of the positioning neural network (PCN1)

Its main function is to fix in the structure of the network connections possible options of the position of the object in space and to be able to reproduce the state by the signal from the higher level.

In such a network, afferent neurons of the control system $R_{i,j}^{A,L}, R_{i,j}^{A,R}$ $(i = \overline{1,N},$ $j = \overline{1,M})$ are connected with the relevant neurons $A_{i,j}^{P(1),L}, A_{i,j}^{P(1),R}$ $(i = \overline{1,N}, j = \overline{1,M})$ from the layer A, which are copies of the afferent neurons, where N - number of afferent channels and M - the number of control elements. Layer A displays the activity of afferent neurons. Each neuron $B_s^{P(1)}$ $(s = \overline{1,S_1})$ of the layer B represents one of the S1 positions of the object and memorizing these positions occurs by establishing excitatory connections between the active neurons of layer A and layer B corresponding neurons, as well as the establishing inhibitory connections between the active neurons in layer A and the previously trained neurons from layer B. Thus, each neuron in the layer B is associated with a single position of an object.

Simultaneously with the creation of connections between the layers A and B, made the creation of appropriate connections between neurons $C_s^{P(1)}$ $(s = \overline{1,S_1})$ of the layer C and neurons $D_{i,j}^{P(1),L}, D_{i,j}^{P(1),R}$ $(i = \overline{1,N}, j = \overline{1,M})$ of the layer D, which provide inverse transformation of the given activity of neurons from layer C, each of which defines the desired position of the control object, similar to B, to the output of the activity of the layer D which comes on the interneurons of the RCN $R_{i,j}^{I,L}, R_{i,j}^{I,R}$ $(i = \overline{1,N}, j = \overline{1,M})$. General view of the structure of the created connections is shown in Fig. 6.

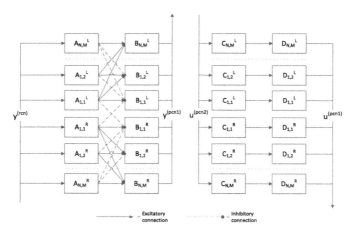

Fig. 6. The structure of the neural connections of the positioning network (PCN1)

Thus, outputs of the B neurons are the outputs of a neural network for the higher control levels, while the input C neurons are its inputs from these levels. The difference between the activity of neurons in B and C defines the difference between the target and current position of the control object.

The maximum number of memorized positions for the PCN1 is:

$$S_{1,\max} = 2 \times M \times N \qquad (1)$$

PCN2 network has a structure similar to PCN1 shown in Fig. 6. Thus, for example, by driving the manipulator multiple units, it can memorize the relative locations of these units. Scheme of for this case is shown in Fig. 7.

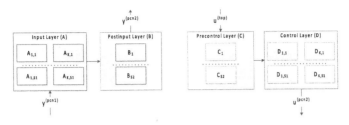

Fig. 7. Scheme of coordinated positioning neural network (PCN2)

In the case of coordinated position neural network for multiple control objects, neural network structure will be similar to PCN1, only the input of layer A will be fed from the outputs of the layers B to lower control levels (PCN1), and the signals from the layer D outputs of the network will be transferred to inputs of the C layers of the lower layers of management. The signals from the output of layer B of the coordinated positioning neural network are the outputs to a higher control level, and the inputs of C layer of the network receive control signals from the outputs of higher control level.

The maximum number of memorized coordinated positions of k actuators for the PCN2 is:

$$S_{2,max} = \prod_{k=1}^{K} S_{1,k} \tag{2}$$

where:

$S_{1,k}$ – number of memorized states in the k-th PCN1 level;
K – number of PCN1 levels (number of actuators).

Depending upon the effector RCN level can be excluded from the model in Fig. 2, for example, in cases where the engine (effector) has its own regulator. In this case, the regulator level degenerates to the module, which converts the data from the sensors to the neural network representation and a module that converts the neural network representation of the desired position in the control action on the regulator. It is a compromise. As part of this approach, it is desirable to use effectors for which the neural network representation of control will be more natural.

Consider the case of a possible realization of a neural network motion control along the trajectory (trajectory control network – TCN) (Fig. 8a). Within the model for the movement along the trajectory we mean consecutive change of positions of the robot's executive system memorized earlier. Thus the movement along the trajectory is reduced to consecutive switching of activity of control neurons from the PCN2 level and downward to the level of RCN level.

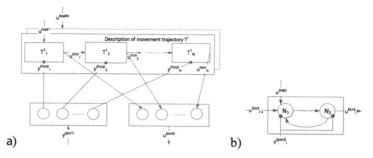

Fig. 8. Description of TCN level. (a) Structural diagram motion control level along the trajectory. (b) Structural diagram implementing trajectory element

Here $T_j^i, i = \overline{1,M}, j = \overline{1,N_i}$ - neural structures for the transition of the system between trajectory sections. Conditionally, we call them trajectory elements. Each trajectory element activates one memorized position on PCN2 level that translates actuation system in this state. On Fig. 8, M – number of memorized trajectories; N_i – number of elements in i-th trajectory. To start the movement along the trajectory the first trajectory element is activated with the higher control level. Further, upon reaching this position, which is determined by the activation of the corresponding neuron on level PCN2, this trajectory element is inhibited and simultaneously activates the next trajectory element and movement process continues until the run for the last trajectory element. Also, the execution of the motion can be interrupted by a signal from a higher level.

Figure 8b shows structural diagram of trajectory element T, satisfying the requirements presented above.

Element represents a ring structure of two neurons with positive feedback. A single excitation of such a structure causes it to a stable state of generation. At a signal from the lower level acting simultaneously on both the neuron in the ring, structure can be inhibited. To the only exciting relationship, which are connected in series such ring structures, does not lead to the fact the simultaneous launch of all elements of the trajectory, the relationship between the elements trajectory is using features of the structural organization of the membrane in the neuron model. At the input of the ring structure $u_{j-1}^{(tcn)}$ several inhibitory and excitatory synapses are formed.

Thus inhibitory synapses on the dendrite are located closer to the soma of the neuron and thus carry more weight, but less time effect on the contribution to the neuron excitation. And excitatory synapses are formed on the dendrites of remote segments and, consequently, have less weight, but a longer time of the influence. Thus at the moment when activity appears $u_{j-1}^{(tcn)}$, neuron N_1 of the next trajectory element becomes inhibited due to a "strong" inhibitory synapses, but with the loss of activity, the inhibitory effect disappears faster than the remaining excitation with dendrites, and launches the next ring structure.

This implementation of the TCN level allows to simulate the following control levels, in turn, more complex trajectories (by introducing, similar to the PCN, additional levels: TCN1, TCN2, etc.), and then, even more abstracting and adding information from sensors (the left side of the diagram in Fig. 3), to simulate the reaction of robotic systems as simple conditioned reflexes, and, subsequently, more complex reflexes, obtained as a combination of simple ones.

3   Experiments

The behavior of RCN network level was researched on the example of the task to control the model of the two-link manipulator. Each regulator must ensure the maintenance of a specified link position in space in its coordinate system. The model of the DC motor was chosen as a model of manipulator unit's drive. In the research, our aim was to demonstrate the possibility of structural adaptation of the neural network and the applicability of the borrowed from biology basic neural structures for solving control tasks in technical systems.

As follows from the above that quality control depends primarily on the number of control elements (CE). Naturally, that the increase in the number of control elements improves handling: reduced amplitude link fluctuations around the equilibrium position. This is confirmed by the dependence shown below (Fig. 9a).

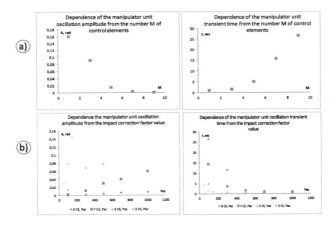

Fig. 9. The results of the regulator settings research. (a) The dependence the manipulator unit oscillation amplitude and transient time from the number M of control elements. (b) The dependence the manipulator unit oscillation amplitude and transient time from the impact correction factor value

With the growing number of control elements considerably reduced the amplitude of link fluctuations in the equilibrium position, but increases during the transition process by changing the link position. However, the growth of transition time occurs much more slowly than the decrease of the oscillation amplitude – thus, the resulting impact correction factor (PAC), can be obtained simultaneously suitable control parameters for oscillation amplitude at the target position and time for transition to a new position (Fig. 9b).

Figure 10 shows the result of the scheme work. It can be seen that a single activation of the first trajectory element leads to the startup sequence of the remaining trajectory elements after receiving confirmation from the lowest level that the shift in position, given the previous trajectory element is made.

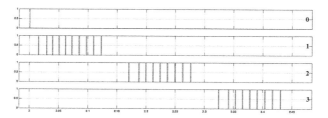

Fig. 10. An example of the functioning of sequential activation of ring structures, responsible for the working off the three trajectory elements. 0 – exciting pulse from the upper level; 1 – activity of the first trajectory element; 2 – activity of the second trajectory element; 3 – activity of the third trajectory element; on the axis of abscissae there is time in seconds

4   Conclusions

It was shown that by changing the structure of the network (structural adaptation) at the level of the regulator, offered by the robot's behavioral control system, the target parameters of the control by accuracy and transient time can be achieved. The increase in the number of the control elements enhances control accuracy, while the increase of the impact correction factor can speed up the system's response. Selection of the optimal values of these two parameters can be carried out at the runtime.

Also the possibility of memorizing states of the target system by reorganizing the structure of the PCN level was shown.

The proposed architecture of TCN network is based on known in neurophysiology ring structures with positive feedback. Such elements are easy to implement on the basis of the dynamic integrate-and-fire neuron model that allows detailed modeling of dendritic structure of the cell.

The proposed architecture of neural network is easily scaled into a large number of the robot's degrees of mobility and does not depend on the type of the robot's executive systems. Also, the network architecture allows further generalizations at high levels, the structure of which is similar to the levels described, and thus complicates the possible behavior of the robot and expands the range of motions available to him.

In the future we plan to integrate the developed system into a single closed structure of neural network that includes sensors, reflex activity and development models.

References

1. Anokhin, P.K.: Biology and Neurophysiology of the Conditioned Reflex and Its Role in Adaptive Behavior. Pergamon, New York (1974)
2. Krichmar, J.L., Edelman, G.M.: Machine psychology: autonomous behavior, perceptual categorization and conditioning in a brain-based device. Cereb. Cortex **12**, 818–830 (2002)
3. McKinstry, J.L., Edelman, G.M., Krichmar, J.L.: A cerebellar model for predictive motor control tested in a brain-based device. PNAS **103**(9), 3387–3392 (2006)
4. de Hugo, G., Chen, S., Ben, G., Lian, R.: A world survey of artificial brain projects, part I: large-scale brain simulations. Neurocomputing **74**, 3–29 (2010)
5. Bakhshiev, A., Gundelakh, F.: Mathematical model of the impulses transformation processes in natural neurons for biologically inspired control systems development. In: 4th International Conference on Analysis of Images, Social Networks and Texts, Yekaterinburg, vol. 1452, pp. 1–12 (2015)

Object Trajectory Association Rules for Tracking Trailer Boat in Low-frame-rate Videos

Jing Zhao[1], Shaoning Pang[1(✉)], Bruce Hartill[2], and Abdolhossein Sarrafzadeh[1]

[1] Department of Computing, Unitec Institute of Technology New Zealand,
Auckland, New Zealand
{jzhao,ppang}@unitec.ac.nz
[2] Fisheries and Marine Ecology, National Institute of Water and Atmospheric
Research New Zealand, Auckland, New Zealand

Abstract. Tracking object accurately in one frame per minute (1-fpm) video is believed to be impossible, because the one-minute discontinuity of object coupled with dynamic background variation implies that the motion and appearance of target is theoretically not predictable. In the context of maritime boat ramps traffic surveillance, we propose in this paper a novel approach to tracking object in the low-frame-rate (LFR) of 1-fpm videos, where the motion discontinuity of object is mitigated by adopting target lifespan path-template and association rules of behavior prediction. The approach has been applied to trailer boat counting at three maritime boat ramps in New Zealand. The obtained accuracy goes above 90 %, with reference to the ground truth manual counting.

1 Introduction

Tracking objects in low-frame-rate (LFR) videos is highly challenging, because the long time interval between frames makes the motion and appearance of target less predictable. Conventional object tracking methods, such as particle filtering [1–4] and kernel tracking [5–8], they are designed for performing object tracking in the rate above 20 frames per second (fps). Also, almost all traditional methods assume that the motion of target is smooth with no abrupt variations [9]. Some existing methods further constrain the target motion to be of constant velocity or constant acceleration based on a priori information. However under LFR conditions, we can not impose such constraints on the motion or appearance of objects, and this makes tracking an object highly challenging in such scenario.

In handling LFR videos, several tracking approaches propose advanced motion model and data association [10–12]. However, these methods are insufficient to cope with the LFR of one frame per minute (1-fpm), because the capturing rate of their images is merely in the range of 1 to 10 fps. At the capturing rate of 1-fpm, targets move abruptly from one frame to another and enter/exit the scene frequently (average 3-5 frames of target lifespan). The huge discontinuity caused by such LFR includes (1) the unpredictable variations of an object on its appearance, scale, and motion. (2) dynamic change of the background. Due to the big time gap, the change on background is likely to be visible.

© Springer International Publishing Switzerland 2016
L. Cheng et al. (Eds.): ISNN 2016, LNCS 9719, pp. 328–337, 2016.
DOI: 10.1007/978-3-319-40663-3_38

For example, a pile of leaves on the road may float in the wind from one to another place within 60 seconds. In the case of maritime boat ramp surveillance, the borderline between the water and land varies over time with the rise and fall of the tide, which causes the background change substantially as levels of illumination, tidal heights, and reflectance vary throughout the day, and throughout the year. This unpredictable variation of targets, coupled with dynamic background makes object tracking in such case extremely difficult.

To increase the motion continuity and make object more traceable under LFR conditions, we investigate the behavior of boats and vehicles passing across boat ramps, and extract lifespan templates to characterize boat launching and retrieval behavior. Based on these templates, a set of association rules for tracking trailer boat are derived to mitigate the motion discontinuity seen in LFR videos. The proposed tracking algorithm is described as follows. Moving object detection is performed and followed by combo object modeling to recognize boat-vehicle combo from tracked objects. With the results of object detection and combo object modeling, single and combo objects are tracked by the multi-feature based matching and data association. The strategies for increasing the motion continuity presented in Sect. 3 are further exploited to improve the tracking results under LFR conditions.

2   Tracking Trailer Boat

2.1   Moving Object Detection

Detecting moving objects from image sequences is a fundamental task for video interpolation. To cope with the dynamic maritime environment, we apply ABM-lw, an adaptive background modeling for land and water composition scenes [13] to perform real-time traffic surveillance at maritime boat ramps. ABM-lw separates land and water scenes by constructing a computing model, in which different learning rates and background updating strategies are exploited for areas of land and water, and the influence of tide, sunrise and sunset is accounted for changes in outdoor luminance. It has been demonstrated that ABM-lw has better object detection and fewer false alarms of surveillance videos with luminance changes, tidal movement and dynamic backgrounds at three different boat ramps.

When a boat is towed, the vehicle is active driving and the trailered boat is passive. The minimum distance between two objects is shown when the front of the trailered boat faces the rear of vehicle. This distance increases when the vehicle turns to either left or right. This distance reaches a maximum when the passive object (i.e., the trailered boat) starts to move in following the active object motion. Let us denote the passive and active object by e^i and e^j respectively. When the front of e^i faces the rear of e^j, the minimum distance is shown between two objects. When the active object e^j turns to the right or left, the two objects distance increases. We simply judge that e^i and e^j form a combo if the following rule is satisfied:

$$\tfrac{1}{2}w_i + \tfrac{1}{2}w_j + \varepsilon \le d_{ij} \tag{1}$$

where d_{ij} is the Euclidean distance between the center of e^i and e^j, w_i and w_j are the width of bounding box of e^i and e^j, respectively, and ε is the gap between two bounding boxes.

2.2   Object Matching

In calculating matching score, we conduct the fusion of multiple features and define the matching score separately for single and combo object. For single object, we simply combine multiple features introduced above. Then, the likelihood for single object matching can be calculated as

$$p(e_{t-1}|e_t) = \prod_{i=1} p_i(e_{t-1}|e_t), \tag{2}$$

where i is the index of features. In our case, three different features are considered, which follows that $p_i(e_{t-1}|e_t)$ can be the likelihood of intensity distribution, average intensity of interest points or texture feature. The score for single object matching can be calculated as

$$\delta_e(e_{t-1}, e_t) = \ln p(e_{t-1}|e_t). \tag{3}$$

We fully consider the characteristics of different types of objects, and choose the most discriminative feature set for each type to measure the similarity between objects. A combination of various useful features, including appearance-based, texture-based and geometry-based features, gives a reliable solution to object tracking in LFR case, since it makes the trackers robust against a wider range of variations, including motion discontinuity, illumination changes, dynamic background, appearance changes, noise etc. To measure the proximity between two objects, three features, namely histogram of intensity, average intensity of interest points: the speeded-up robust features (Surf) [14], and texture feature: the LBP texture operator [15], are investigated in this work.

3   Increasing Motion Continuity in LFR Videos

Consider object tracking in 1 frame per minute videos. During the one-minute interval between two consecutive images, object in observation may change dramatically and jump to next status of interest. For example, for boat lunching at a ramp, the trailer boat may show up at the shoreline in one frame, then disappear afterwards frames; In the case of retrieving a boat, one frame may capture a vehicle in land area, next frame gives a vehicle-boat combo on the way out of the ramp(i.e., the activity of picking up boat at shoreline is not recorded, as it happens within the 1 min interval). Therefore due to LFR of 1 fpm, the status of object is shown very erratic and has discontinuity in the video.

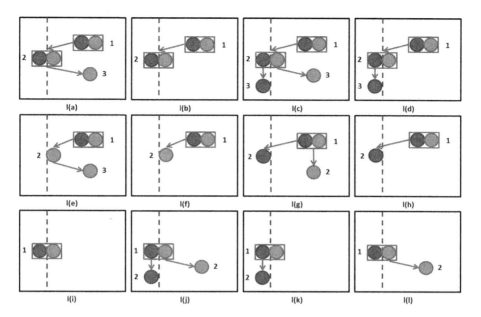

Fig. 1. Illustration of object lifespan templates for boat launching. The dash line represents the boundary between water and land, with its left as water and right as land. The rectangle denotes a combo with its internal component left as boat and right as vehicle.

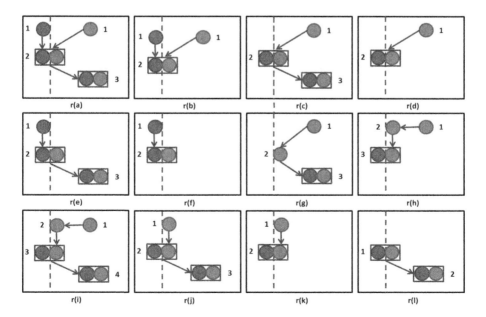

Fig. 2. Illustration of object lifespan templates for boat retrieving.

3.1 Lifespan Template

To increase the motion continuity and make object more traceable under LFR conditions, it's worth noting that the behavior of boats and vehicles passing across boat ramps falls in the list of templates. Figures 1 and 2 present the extracted lifespan templates about the behavior of boat launching and retrieving, respectively. In the figures, the dashed line represents the boundary between water and land, with its left as water and right as land. The rectangle denotes a combo object whose left is boat and right is vehicle. Recall that trailer boat moves always backwards to the water for lunching or retrieving boat, thus the object closer to water is recognized as the boat.

Here, the number denotes the lifespan stage s. The first stage (i.e., $s = 1$) is the beginning of a lifespan, which refers to the first occurrence of an object. The last stage is the lifespan end, after which the object dies out. After the first occurrence of an object, if the observed object changes its status (to be defined below), then stage number increases by one.

The status of a single object is determined by the current location of the object as either @$land$, @$water$, or @$shoreline$. Specifically, if the object bounding box is in water area (see above for the definition of water area), @$water$ is assigned; if the distance from bounding box center to the shoreline is not greater than half width of the bounding box, then @$shoreline$ is the current object status; otherwise the status is assigned as @$land$. For combo object, its status is determined by that of vehicle in the combo. For example, Fig. 1(a) gives one lifespan template of launching boat. A boat-vehicle combo occurs firstly @$land$ in stage 1, then moves to @$shoreline$ in stage 2, vehicle back to @$land$ (after releasing the boat in the water) in stage 3 (final stage). Similarly, Fig. 2(a) demonstrates a lifespan template of retrieving boat. In stage 1, a single object (vehicle) presents @$land$, while another single object (boat) appears @$water$; then these two single objects construct a combo @$shoreline$ in stage 2, and this combo occurs @$land$ in stage 3 (final stage).

3.2 Object Trajectory Association Rule

As the result of above observation on boat launching and retrieving, we derive total eleven rules for trailer boat tracking, mitigating to the maximum motion discontinuity caused by the frame rate of 1 fpm. For simplicity, we explain here the top two most important rules as,

R(a) If a combo object moves from @$land$ to @$shoreline$ in first two stages, then the object is unlikely back to @$land$ for the future. The rule can be summarized as

$$\{(z \in @land) \wedge (s = 1)\} \wedge \{(z \in @shoreline) \wedge (s = 2)\} \\ \rightarrow z \notin @land, \ for \ \forall s >= 3 \tag{4}$$

R(a) is derived from lifespan templates l(a-d) of launching boat. Occasionally, a combo may go from @$land$ to @$shoreline$ for maintenance purpose,

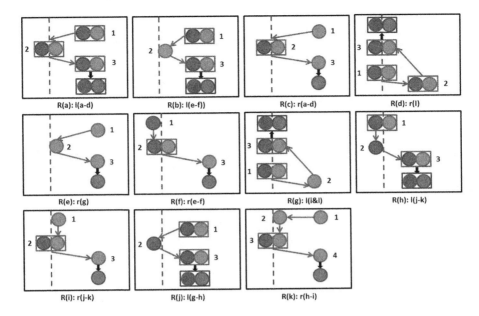

Fig. 3. Illustration of object trajectory association rules for trailer boat tracking.

then back to @*land* and leave, but this probability is very low. As illustrated in Fig. 3(a), one combo goes from @*land* in stage 1 to @*shoreline* in stage 2. If it appears on @*land* in stage 3, apparently, this combo is wrongly recognized and should be identified as a new one.

R(b) If a combo object appears on @*land* in stage 1, and its contained vehicle object presents at @*shoreline* in stage 2, then the combo is unlikely back to @*land* for the future. The rule can be summarized as

$$\{(o \subset z) \wedge (z \in @land) \wedge (s = 1)\} \wedge \{(o \in @shoreline) \wedge (s = 2)\} \tag{5}$$
$$\rightarrow z \notin @land, \, for \, \forall s >= 3$$

R(b) is derived from lifespan templates l(e-f) of launching boat. Occasionally, the driver may change mind and pick up the boat, and they construct a combo again and go back to @*land* and leave, but this probability is very low. As illustrated in Fig. 3(b), one combo goes from @*land* in stage 1 to @*shoreline*, and only the vehicle appears at @*shoreline* in stage 2, since the launching boat already disappears. If the combo appears on @*land* in stage 3, apparently, this combo is wrongly recognized and should be identified as a new one.

The application of rule is described in Algorithm 1, where the set of rules for trailer boat tracking R consists of total eleven rules.

Algorithm 1. Implementation of Tracking Rule

Require: trajectory To_t^j and Tz_t^j, set of rules for trailer boat tracking R.
Ensure: updated trajectory $T'o_t^j$ and $T'z_t^j$
  1: label lifespan stages on trajectory To_t^j as Sect. 3.1;
  2: find current single object trajectory To_t^j aligned rule r^* in R;
  3: $T'o_t^j \leftarrow$ update To_t^j with r^* as in Fig. 3;
  4: label lifespan stages on trajectory Tz_t^j as Sect. 3.1;
  5: find current combo trajectory Tz_t^j aligned rule r^* in R;
  6: $T'z_t^j \leftarrow$ update Tz_t^j with r^* as in Fig. 3;

---

4 Experiments

The image data we used for experiments was collected by New Zealand's National Institute of Water and Atmospheric Research (NIWA). Table 1 describes the experimental data, which includes 2010-2012 image sequences captured at Waitangi, Takapuna and Raglan boat ramp, and which records the number of frames, and the truth number of objects from manual count for each ramp and each year. The frame size of the video is 720×576 pixels, and the frame rate is 1-fpm.

In our experiment, the proposed tracking approach is compared with the state-of-the-art approaches. The same initialization is set to all algorithms for fair comparison, and the parameters of all methods are tuned to achieve the best performance. The proposed system is implemented in Matlab on a PC with a Quad Core 3.4 GHZ CPU and 8 GByte memory. The video interpolation speed of the implemented system is 0.25 seconds per frame.

4.1 Trailer Boat Counting Accuracy

We apply proposed tracking method to conduct computer boat counting at the three boat ramps, and evaluate boat counting accuracy with reference to manual counts. We count on lifespan templates since a trailer boat behavior matches always one of the 12 lifespan templates in either launching or retrieving category. Consider the number of lifespan for launching differs to that of retrieving. We calculate the number boats as the average number of these two categories templates. Boat counting performance of the proposed method was evaluated and

Table 1. Experimental data

Year	No. of Frames (No. of Boats)		
	Waitangi	Raglan	Takapuna
2010	77,760 (1772)	79,200 (783)	86,400 (2844)
2011	83,520 (1422)	82,080 (908)	84,960 (2777)
2012	69,120 (1140)	80,640 (724)	77,760 (2174)
Total	230,400 (4334)	241,920 (2415)	249,120 (7795)

Table 2. Comparison of boat counting performance in terms of NRMSE percentage

Method	PF [4]	EMS [10]	CPF [11]	STEREO [12]	Proposed
Waitangi	87.23	53.61	47.36	45.56	9.71
Raglan	84.49	51.58	45.57	42.31	9.62
Takapuna	89.62	59.32	51.43	48.47	9.83
Average	87.11	54.84	48.12	45.45	**9.72**

compared with one conventional tracking method: particle filter (PF) [16], and three LFR tracking methods: extended mean shift (EMS) [10], cascade particle filter for LFR (CPF) [11] and object tracking in LFR stereo videos (STEREO) [12]. For performance evaluation, we calculate the differences between the ground truth (i.e., daily boat number from manual count), and daily boat number provided by different algorithms by NRMSE, normalized root mean squared error

$$NRMSE = \frac{\sqrt{\frac{\sum_{i=1}^{K}(N_i - C_i)^2}{K}}}{N_{max} - N_{min}}, \tag{6}$$

where N denotes the ground truth of boat number from manual count, and C is the number from one tracking method. K is the total number of days for boat counting, and N_{max} and N_{min} represent the maximum and minimum number of boat from manual count, respectively.

Table 2 presents the NRMSE in percentage for five object tracking methods boat counting at Waitangi, Raglan, and Takapuna, respectively. As seen from the table, PF fails to track over 80 % object of interest, since the conventional

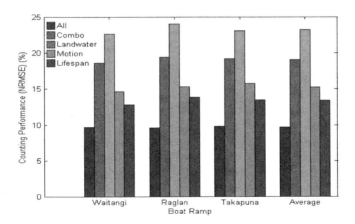

Fig. 4. Boat counting performance in terms of NRMSE versus unselected factors for object tracking. Each color bar denotes one factor that is not considered when tracking objects across frames. (Color figure online)

tracking methods rely on the motion continuity of an object. Some strategies were considered by EMS, CPF and STEREO to deal with abrupt motion in LFR data, they performed relatively better than PF, but their errors still above 40 %, which obviously does not satisfy our application requirement for monitoring recreational fishing efforts over time. In contrast, the proposed method gives average less 10 % counting error for all three maritime boat ramps. This demonstrates that our approach successfully overcomes the extremely dynamic background and severe unpredictability problems compared with the state-of-the-art methods.

For the proposed method, we further investigate its four key factors that impact boat counting performance. These include (1) modeling separately land and water scenes (denoted as Landwater factor), (2) modeling boat-vehicle combo tracking (Combo), (3) increasing motion continuity of objects (Motion), and (4) utilizing lifespan templates of launching and retrieving boat to improve counting accuracy (Lifespan). A sensitivity experiment on these factors is performed using the same data for evaluating boat counting performance, one factor is unselected each time. In this test, we suppose that object counting with all factors obtains the best performance, and removing one of the factors leads to the decrease of performance. The significance of each unselected factor is then indicated by the decrease of counting performance measured in NRMSE.

Figure 4 displays the counting performance without different selected factor as well as that of with all factors. As shown in Fig. 4, the Landwater factor has the greatest influence on the counting performance. If without the consideration of tidal information and modeling land and water scenes separately, the counting accuracy is expected to reduce 13 % in average. This can be explained that if lack of accurate background estimation, the accuracy of object detection may deduct largely. The second important factor is the Combo factor, which causes over 9 % counting accuracy loss as seen in Fig. 4. This indicates that for trailer boat tracking, correlation between vehicle and boat contributes significantly to the success of boat tracking and counting.

5   Conclusions

This paper propose a new algorithm for tracking trailer boat in one frame per minute videos. To handle the information gap between frames, we investigate the behavior of launching and retrieving boats and extract lifespan association rules for target motion prediction. Experimental comparative tests and quantitative performance evaluations on a real-world boat-flow analysis and counting system has demonstrated the benefits of the proposed algorithm.

References

1. Rui, Y., Chen, Y.: Better proposal distributions: Objects tracking using unscented particle filter. In: Proceedings of IEEE Conference on Computer Vision and Pattern Recognition, pp. 786–793 (2001)

2. Zhou, S., Chellappa, R., Moghaddam, B.: Visual tracking and recognition using appearance-adaptive models in particle filters. IEEE Trans. Image Process **13**(11), 1491–1506 (2004)
3. Bruno, M.: Bayesian methods for multiaspect target tracking in image sequences. IEEE Trans. Sig. Process **52**(7), 1848–1861 (2004)
4. Breitenstein, M., Reichlin, F., Leibe, B., Koller-Meier, E., Van Gool, L.: Online multiperson tracking-by-detection from a single, uncalibrated camera. IEEE Trans. Pattern Anal. Mach. Intell. **33**(9), 1820–1833 (2010)
5. Comaniciu, D., Ramesh, V., Meer, P.: Real-time tracking of nonrigid objects using mean shift. In: Proceedings of IEEE Conference on Computer Vision and Pattern Recognition, pp. 142–149 (2000)
6. Comaniciu, D., Ramesh, V., Meer, P.: Kernel-based object tracking. IEEE Trans. Pattern Anal. Mach. Intell. **25**(5), 564–577 (2003)
7. Wang, L., Yan, H., Wu, H., Pan, C.: Forward-backward mean-shift for visual tracking with local-background-weighted histogram. IEEE Trans. Intell. Transp. Syst. **14**(3), 1480–1489 (2013)
8. Shen, C., Kim, J., Wang, H.: Generalized kernel-based visual tracking. IEEE Trans. Circuits Syst. Video Technol. **20**(1), 119–130 (2010)
9. Yilmaz, A., Javed, O., Shah, M.: Object tracking: A survey. ACM Computing Surveys **38**(4), 13–58 (2006). p. article 13
10. Porikli, F., Tuzel, O.: Object tracking in low-frame-rate video. In: Image and Video Communications and Processing 2005, pp. 72–79 (2005)
11. Li, Y., Ai, H., Yamashita, T., Lao, S., Kawade, M.: Tracking in low frame rate video: A cascade particle filter with discriminative observers of different life spans. IEEE Trans. Pattern Anal. Mach. Intell. **30**(10), 1728–1740 (2008)
12. Chuang, M., Hwang, J., Williams, K., Towler, R.: Tracking live fish from low-contrast and low-frame-rate stereo videos. IEEE Trans. Circuits Syst. Video Technol. **25**(1), 167–179 (2015)
13. Zhao, J., Pang, S., Hartill, B., Sarrafzadeh, A.H.: Adaptive background modeling for land and water composition scenes. In: Murino, V., Puppo, E. (eds.) ICIAP 2015. LNCS, vol. 9280, pp. 97–107. Springer, Heidelberg (2015)
14. Bay, H., Ess, A., Tuytelaars, T., Van Gool, L.: Speeded-up robust features (surf). Comput. Vis. Image Underst. **110**, 346–359 (2008)
15. Heikkila, M., Pietikainen, M.: A texture-based method for modeling the background and detecting moving objects. IEEE Trans. Pattern Anal. Mach. Intell. **28**(4), 657–662 (2006)
16. Khan, Z., Balch, T., Dellaert, F.: Mcmc-based particle filtering for tracking a variable number of interacting targets. IEEE Trans. Pattern Anal. Machine Intell. **27**(11), 1805–1918 (2005)

Learning Time-optimal Anti-swing Trajectories for Overhead Crane Systems

Xuebo Zhang[1(✉)], Ruijie Xue[1], Yimin Yang[2], Long Cheng[3],
and Yongchun Fang[1]

[1] Institute of Robotics and Automatic Information System (IRAIS) and Tianjin Key Laboratory of Intelligent Robotics, Nankai University, Tianjin 300071, China
{zhangxuebo,fangyc}@nankai.edu.cn, xueruijie233@163.com
[2] Department of Electrical and Computer Engineering, University of Windsor, Windsor N9B3P4, Canada
eyyang@uwindsor.ca
[3] State Key Laboratory of Management and Control for Complex Systems, Institute of Automation, Chinese Academy of Sciences, Beijing 100190, China
chenglong@compsys.ia.ac.cn

Abstract. Considering both state and control constraints, minimum-time trajectory planning (MTTP) can be implemented in an 'offline' way for overhead crane systems [1]. In this paper, we aim to establish a real-time trajectory planning model by using machine learning approaches to approximate those results obtained by MTTP. The fusion of machine learning regression approaches into the trajectory planning module is new and the application is promising for intelligent mechatronic systems. In particular, we first reformulate the considered trajectory planning problem in a three-segment form, where the acceleration and deceleration segments are symmetric. Then, the offline MTTP is applied to generate a database of minimum-time trajectories for the acceleration stage, based on which several regression approaches including Extreme Learning Machine (ELM) and Backpropagation Neural Network (BP) are adopt to approximate MTTP results with high accuracy. More important, the resulting model only contains a set of parameters, rather than a large volume of offline data, and thus machine learning based approaches could be implemented in low-cost digital signal processing chips required by industrial applications. Comparative evaluation results are provided to show the superior performance of the selected regression approach.

Keywords: Overhead crane systems · Minimum-time trajectory planning · Machine learning · Regression techniques

1 Introduction

Overhead crane systems are one typical kind of engineering machines, which are widely used in construction, manufacturing, harbors, etc., to transport various payloads. With the fast development of factory automation, automatic overhead crane systems are highly demanded in practical applications, and a lot of

© Springer International Publishing Switzerland 2016
L. Cheng et al. (Eds.): ISNN 2016, LNCS 9719, pp. 338–345, 2016.
DOI: 10.1007/978-3-319-40663-3_39

feedback control and motion planning approaches have been proposed in the literature [1–7], wherein two major performance indices, namely the transportation speed and anti-swing performance, are usually the focus of these works. Unfortunately, these two performance indices are somehow contradictive with each other. Therefore, it is important to derive the maximum efficiency of a transportation task given both state and control constraints, including the maximum allowable swing angle, the maximum speed and acceleration, and so on.

In order to address this issue, minimum-time trajectory planning (MTTP) with both state and control constraints, has already been solved in the previous work by applying the quasiconvex optimization theory [1]. However, the optimal solution in [1] is essentially an offline approach which requires several minutes before starting a transportation task. While some other real-time trajectory planning approaches have been reported in the literature [2–5], they cannot guarantee the time optimality under complex state and control constraints. Therefore, how to achieve real-time trajectory planning of overhead crane systems is still an open and important practical problem, and it is necessary to find an efficient approach so that the results of MTTP could be used in a real-time manner in practical applications to improve the transportation efficiency.

By sampling densely all related parameter settings including maximum allowable swing angle, boundary conditions, rope lengths, and so on, the offline MTTP in [1] could be used for each sample to obtain the corresponding results of planned trajectories, based on which a database is established describing the mapping between parameter settings and optimal trajectories. To realize the trajectory planning in a real-time way, the most intuitive method is to use the database as a lookup table or dictionary, from which the time-optimal trajectory could be retrieved fast. However, the memory requirement is high for lookup tables and only sparse discrete samples could be saved. To alleviate these restrictions, we aim to use machine learning regression approaches to establish a model to approximate the offline MTTP results, and in this way, only a regression model with several parameters is involved and it can easily be implemented on a low-cost digital signal processing unit in practical crane control systems.

In this paper, motivated by the practical need of real-time trajectory planning with low-cost and high reliable implementation, machine learning approaches are introduced to learn the results of the offline MTTP in [1] for overhead crane systems, and thus real-time minimum-time anti-swing trajectory planning is achieved. First, we propose a three-segment minimum-time trajectory planning framework which largely decreases the size of the solution space, thus it reduces the difficulty for application of machine learning approaches to achieve high accuracy. Then, machine learning regression algorithms are used to generate two regression models that approximate mapping between parameter settings and optimal trajectories obtained by offline MTTP. Comparative simulation results using Extreme Learning Machine (ELM) [8–10], BP (Back Propagation) [11], PC-ELM (Parallel Chaos Optimization-based ELM) [13], EM-ELM (Expectation Maximization-based ELM) [12] are provided, and the performance evaluation are analyzed. The main contribution of this brief is that: (1) a three-segment planning framework, which is appropriate for machine learning regression, is proposed; (2) machine learning regression approaches are introduced to convert the offline MTTP into an online time-optimal trajectory planning module that is easily implemented on low-cost digital signal processing (DSP) units.

The paper is structured as follows: after the introduction, the problem statement is given in Sect. 2. The overall framework of the proposed approach is presented in Sect. 3. Then, machine learning algorithms are introduced in Sect. 4. Application results of ELM, BP, EM-ELM, PC-ELM are compared and analyzed in Sect. 5. Conclusions are given in the last section.

2   Problem Statement

Consider a two-dimensional underactuated overhead crane systems shown in Fig. 1, the system state $q \in \mathbb{R}^4$ is defined as $q = [x\ v\ \theta\ w]^T$, where $x \in \mathbb{R}$ and $v(t) \in \mathbb{R}$ represents the position and velocity of the cart, respectively; $\theta \in \mathbb{R}$ and $w(t) \in \mathbb{R}$ are the payload swing angle and swing angular velocity, respectively. $l \in \mathbb{R}$ denotes the rope length.

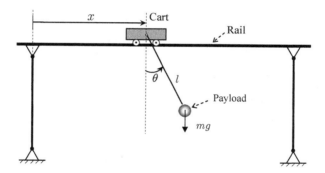

Fig. 1. Overhead crane systems

The objective of Minimum-Time Trajectory Planning (MTTP) in [1] is to find the minimum-time trajectory from the initial configuration $q(0) = [0\ 0\ 0\ 0]^T$ to the terminal configuration $q(t_f) = [x_f\ 0\ 0\ 0]^T$, while guaranteeing the following constraints

$$|v(t)| \leq v_{\max}, \quad |a(t)| \leq a_{\max}, \quad |\theta(t)| \leq \theta_{\max}, \tag{1}$$

where x_f is the terminal position, and t_f denotes the computed terminal time with MTTP. The symbol $a(t)$ denotes the acceleration, and $v_{\max}$, $a_{\max}$ and $\theta_{\max}$ represent the maximum velocity, maximum acceleration and maximum allowable swing angle, respectively.

However, the MTTP in [1] is time-consuming and it can only be conducted offline. In this paper, we will aim to propose a real-time MTTP by learning from the database generated by offline MTTP.

3   The Overall Framework

3.1   MTTP-based Three-segment Trajectory Design

To transport a payload with an overhead crane from the initial configuration $q(0) = [0\ 0\ 0\ 0]^T$ to the terminal configuration $q(t_f) = [x_f\ 0\ 0\ 0]^T$, when the transportation distance x_f is long enough, it is reasonable to split the overall trajectory into three segments:

1. Acceleration stage: MTTP is used so that the cart accelerates to the maximum speed without residual swing in the minimum time, while satisfying the constraints in (1). Accordingly, the initial configuration and terminal configuration are set as $q(0) = [0\ 0\ 0\ 0]^T$ and $q(T_{up}) = [x_{up}\ v_{max}\ 0\ 0]^T$, respectively, where T_{up} and x_{up} are the resultant acceleration time and terminal position of the acceleration stage, respectively, determined by MTTP.
2. Constant velocity stage with $v = v_{max}$.
3. Deceleration stage is symmetric (or inverse) to the acceleration stage.

Based on the uniqueness of solutions, we know that the minimum-time trajectory of the acceleration stage and that of the deceleration stage are symmetrical for each single task from the initial configuration $q(0) = [0\ 0\ 0\ 0]^T$ to the terminal configuration $q(t_f) = [x_f\ 0\ 0\ 0]^T$. Hence, in this paper, we only need to learn the trajectories of the acceleration stage, and then the trajectories of the deceleration stage could be obtained by simple inversion of the corresponding time label of trajectories in the acceleration stage.

3.2   Learning Optimal Trajectories of the Acceleration Stage

In the following, we will give the framework for adoption of machine learning algorithms to learn time-optimal anti-swing trajectories in the acceleration stage in Fig. 2. In this brief, we make an initial trial for learning trajectories obtained with different rope length, supposing that the maximum speed, maximum acceleration, and maximum allowable swing angle are kept constant.

In Fig. 2, by sampling n different values of the rope length $l = l_1, l_2, \cdots, l_n$, we run the offline MTTP in [1] to form a database of the resultant acceleration trajectories. For each rope length l, we use MTTP to obtain the trajectory $x(l, t)$ with $t \in [0, T_{up}(l)]$, where $T_{up}(l)$ is the terminal time of the acceleration stage. Based on such a database of trajectories, machine learning regression approaches are applied in this paper to train two regression models for $T_{up}(l)$ and $x(l, t)$ that could be implemented in low-cost DSP units.

More specifically, the relationship between the rope length l and the acceleration time T_{up}, which is actually a two-dimensional (2-D) curve, is first fitted using machine learning regression approaches. Through the regression model, given any l, we can obtain the predicted $\hat{T}_{up}(l)$. Based on this, we can use another machine learning regression to train the relationship among l, $T_{up}(l)$ and $x(l, t)$, which is a three-dimensional (3-D) surface. After training of regression model $\hat{T}_{up}(l)$ and $\hat{x}(l, t)$, the trajectory can be generated in a real-time manner, as shown in Fig. 3.

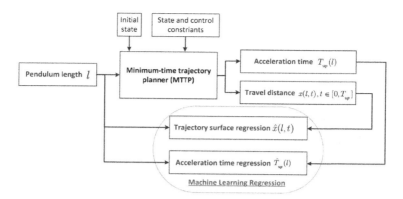

Fig. 2. The framework of machine learning regression with MTTP results

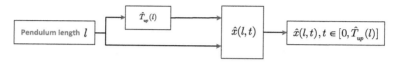

Fig. 3. Real-time prediction for the minimum-time trajectories

Remark 1: In general, the training and testing accuracy for the acceleration time is less than two control periods, which is so small that the trajectory learning is almost not affected by the approximate error of the acceleration time.

4   ELM-based Regression

A lot of regression techniques have been reported in the literature, and in this paper, we choose the recently developed Extreme Learning Machine (ELM), which is a novel single-hidden-layer neural network algorithm proposed by Huang et al. [2]. ELM randomly generates the hidden nodes independent of the training data, and thus the parameters for hidden layers need not to be tuned. Suppose there is one output node, the output function $f_M(\boldsymbol{x})$ of ELM is $f_M(\boldsymbol{x}) = \sum_{i=1}^{M} \beta_i h_i(w_i, b_i, \boldsymbol{x})$, $\boldsymbol{x} \in \mathbb{R}^n$, where $h_i(w_i, b_i, \boldsymbol{x})$ is the output of the ith hidden node, with w_i and b_i being the learning parameters. M denotes the number of hidden nodes, β_i represents the output weight between the hidden layer and the output node. The detailed machine learning algorithm for regression with ELM is ignored in this paper, please refer to [8–10].

5   Performance Evaluation

In this section, we investigate the performance of the ELM-based approach [9] compared with other regression approaches of BP [11], PC-ELM [13], EM-ELM [12], when these approaches are applied into the proposed framework.

Before proceeding to evaluation, we first establish the database for training and testing for the minimum-time trajectory of the acceleration stage. For training data set, the rope length l is sampled from 1 m to 10 m by a step of 0.1 m. For testing data, 91 samples which are different from the training data, are randomly generated between 1 m and 10 m. In the evaluation, to obtain a good performance, the input data were normalized into $[-1, 1]$. These simulations are conducted in Matlab2012a on a desktop of Intel Core I3 and 2GB RAM.

Table 1. Performance evaluation for regression of T_{up}

Methods	Training time (sec)	Training accuracy	Testing accuracy
ELM	0.0000	0.0076	0.0152
PC-ELM	0.5460	0.0076	0.0152
EM-ELM	1.9968	0.0115	0.0168
BP	1.5132	0.0079	0.0151

Table 2. Performance evaluation for regression of $x(l,\ t)$

Methods	Training time (sec)	Training accuracy	Testing accuracy
BP	206.7325	0.0017	0.0029
ELM	3.6660	0.0030	0.0038
PC-ELM	1328.0209	0.0030	0.0038
EM-ELM	68.1412	0.0125	0.0127

It should be note that, though only a number of 91 samples for rope length are used in the cases of training and testing, the number of the resultant trajectory points are more than 15000 in both cases.

Based on the training data and testing data, the comparisons of training and prediction performance by approaches of ELM, BP, PC-ELM, EM-ELM are presented in Tables 1 and 2. To make a fair comparison, parameters of these four approaches are turned artificially to achieve the best performance. From Tables 1 and 2, it is seen that ELM is the most efficient algorithm in terms of the training speed. Regarding the training and testing accuracy (RMSE, root-mean-square error), it is seen that for $T_{up}(t)$, ELM is better than BP, however, ELM is worse than BP for $x(l, t)$. The algorithm of PC-ELM present similar performance with ELM in accuracy but it is slower. To sum up, ELM is the most efficient one and it is a good choice for future learning tasks with multiple parameters and large data. The prediction results using ELM are shown in Figs. 4 and 5, it is seen that the predicted trajectory errors are so small that it could be applied for overhead crane systems.

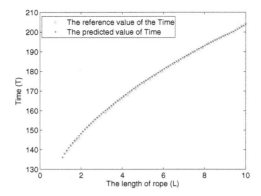

Fig. 4. ELM-based prediction and the testing data for $T_{up}(l)$

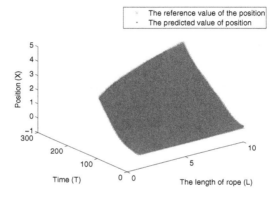

Fig. 5. ELM-based prediction and the testing data for $x(l,\ t)$

6   Conclusion

In this paper, we propose a real-time trajectory planning approach by combing machine learning regression and the previous offline MTTP. Three-segment trajectory planning is introduced so that only the acceleration stage needs to be learned, which largely decreases the size of the solution space for machine learning regression. Several regression approaches are adopted to accomplish the tasks for regression of the acceleration time $T_{up}(l)$ and trajectories $x(l,t)$. Quantitative analysis of four algorithms are given to show that ELM has good performance.

In the future, we will extend the proposed methodology to learn optimal trajectories with multiple inputs of the parameter settings, in addition to the rope length considered in this paper. Afterwards, we will implement the extended algorithm in a DSP unit to accomplish the real-time crane transportation tasks.

Acknowledgments. This work is supported by National Natural Science Foundation of China (NSFC) (61573195, 11372144).

References

1. Zhang, X.B., Fang, Y.C., Sun, N.: Minimum-time trajectory planning for under-actuated overhead crane systems with state and control constraints. IEEE Trans. Ind. Electron. **61**(12), 6915–6925 (2014)

2. Sun, N., Fang, Y.C., Zhang, X.B., Yuan, Y.: Transportation task-oriented trajectory planning for underactuated overhead cranes using geometric analysis. IET Control Theor. Appl. **6**(10), 1410–1423 (2012)

3. Sun, N., Fang, Y.: An efficient online trajectory generating method for underactuated crane systems. Int. J. Robust Nonlinear Control **24**(11), 1653–1663 (2014)

4. Wu, X., He, X., Sun, N.: An analytical trajectory planning method for underactuated overhead cranes with constraints. In: 33rd Chinese Control Conference, pp. 1966–1971. IEEE Press, New York (2014)

5. Sun, N., Fang, Y.C., Zhang, Y., Ma, A.: A novel kinematic coupling-based trajectory planning method for overhead cranes. IEEE/ASME Trans. Mechatron. **17**(1), 166–173 (2012)

6. Cheng, L., Wang, Y., Ren, W., Hou, Z.G., Tan, M.: Containment control of multi-agent systems with dynamic leaders based on a PI^n-type approach. IEEE Trans. Cybern. **45**, 1–14 (2015)

7. Cheng, L., Liu, W., Hou, Z.G., Yu, J.Z., Tan, M.: Neural network based nonlinear model predictive control for piezoelectric actuators. IEEE Trans. Ind. Electron. **62**(12), 7717–7727 (2015)

8. Huang, G.B., Zhu, Q.Y., Siew, C.K.: Extreme learning machine: theory and applications. Neurocomputing **709**(1–3), 489–501 (2006)

9. Huang, G.B., Zhou, H., Ding, X., Zhang, R.: Extreme learning machine for regression and multiclass classification. IEEE Trans. Syst. Man Cybern. Part B: Cybern. **42**(42), 513–529 (2012)

10. Huang, G.B.: An insight into extreme learning machines: random neurons random features kernels. Cogn. Comput. **6**(3), 376–390 (2014)

11. Robert, H.N.: Theory of the backpropagation neural network. Neural Netw. **1**(1), 65–93 (1988)

12. Fang, G., Huang, G.B., Lin, Q., Gay, R.: Error minimized extreme learning machine with growth of hidden nodes and incremental learning. IEEE Trans. Neural Netw. **20**(8), 1352–1357 (2009)

13. Yang, Y., Wang, Y., Yuan, X.: Parallel chaos search based incremental extreme learning machine. Neural Process. Lett. **37**(3), 277–301 (2013)

Attitude Estimation for UAV with Low-Cost IMU/ADS Based on Adaptive-Gain Complementary Filter

Lingling Wang$^{(\boxtimes)}$, Li Fu, Xiaoguang Hu, and Guofeng Zhang

Department of Automatic Science and Electrical Engineering,
Beihang University, Beijing, China
{wangling0908,fuli,xiaoguang,gfzhang}@buaa.edu.cn

Abstract. This paper describes a robust and practical complementary filter (CF) algorithm for unmanned aerial vehicle (UAV) attitude estimation with low-cost inertial measurement unit (IMU) and embedded air data system (ADS). Utilizing a fuzzy logical system, the UAV dynamic modes including different accelerations and turns can be attained. Based on the compensation of acceleration and centrifugal forces in turns using ADS information, the gains of complementary filter adapts to the dynamic modes to yield robust performances. The simulation and experimental results show that the proposed adaptive-gain complementary filter approach can obtain robust and accurate attitude estimation even when the UAV is subject to strong acceleration or in turn mode.

Keywords: Attitude estimation · UAV · Adaptive-gain complementary filter · Fuzzy logic system

1 Introduction

Nowadays, there has been a growing interest in developing UAV for military and civil applications at low altitude [1]. The UAV represents some challenging navigation problems due to its constrains in size, weight and cost, which leads to make smaller and cheaper systems as navigation components equipped on it. The micro electromechanical system (MEMS) with low power, light weight and low cost opens up a wide range of applications for UAV [2]. However, the attitude errors usually accumulate over time due to MEMS inertial raw outputs polluted by different noises under vibration circumstances. So it is an essential task for UAV to develop autonomous attitude estimation algorithms with robustness and simplicity, which can operate in high vibration environment and remain autonomous function with GPS denied.

The filtering techniques make it possible to fuse different kind of information to achieve state estimation. In recent years, a considerable number of filtering techniques for attitude estimations have been studied [3–5], among which the widely used filter technique is Extended Kalman Filter [6–8]. However, there are some several drawbacks for this method. On one hand, its computation due to linearization is very large and it is difficult to choose the filter parameters because of noise affections. On the other hand, the linearization of nonlinear models with less robustness is in general very

© Springer International Publishing Switzerland 2016
L. Cheng et al. (Eds.): ISNN 2016, LNCS 9719, pp. 346–355, 2016.
DOI: 10.1007/978-3-319-40663-3_40

difficult to prove the convergence of the estimation error to zero [9]. In order to remain the nonlinear characteristics of system models, some nonlinear filter techniques, such as Unscented Kalman Filter (UKF) and Particle Filter (PF) are developed [10, 11]. However, these nonlinear filters are computationally demanding and unsuitable for the embedded processors in UAV [12].

An alternative method which has been attracted many researchers is the Complementary Filter (CF) technique to fuse multi-sensor data. In particular, It is widely applied for flying system and attitude observer due to its similarity and effectiveness, which meets the UAV requirements suited for implementation on embedded hardware [13]. The CF is effective when the same physical quantity can be estimated from multiple sensors with different frequency characteristics. So reference [12] combined accelerometer output with integrated gyroscope output to estimate fixed-wing UAV attitude based on CF. However, the result is not optimal due to the CF fixed parameters not adapted to vehicle dynamics. The Full estimation of vehicle attitude as well as gyro biases can be obtained, but fails when the vehicle dynamics are sufficiently large [14].

In this paper, to cope with these situations and obtain accurate attitude for UAV, an adaptive-gain CF, augmented by an acceleration compensation model and motion switching architecture based on fuzzy logical system is proposed to yield robust performance, even when the vehicle is subject to strong dynamic motions. The dynamic modes of UAV are divided into low-acceleration mode and high acceleration mode (including turn) according to the fuzzy logical system based on IMU and ADS. The CF gains can automatically tune and produce robust and accurate attitude estimation.

The body of paper consists of four sections followed by a conclusion and acknowledgement. In Sect. 2, attitude correction equations under different dynamic mode are derived based on the derived UAV kinematics together with the output of IMU and ADS. Section 3 details the design of adaptive-gain CF. The simulations and discusses are provided in Sect. 4. A conclusion and acknowledgement are drawn after all sections.

2 Attitude Measurement with IMUand ADS

2.1 Attitude Estimation Based on IMU

The most commonly attitude description is Euler angles, which is defined as shown in Fig. 1 by yaw (φ), pitch (θ) and roll (γ) in the specific sequence of rotation with respect to Z, Y and X axes of a referred coordinate system. The Euler angles illustrate the relationship of transformation between body frame (b-frame) and navigation frame (n-frame), which is defined as north-east-downward (NED) during the equation derivation. The attitude can be attained by integrating the 3-axis angular rates of gyroscopes respect to its body axis system.

The Euler kinematics describes the relationship between angular rates and time rate of the Euler angles. Based on the direction cosine matrices between b-frame and n-frame, the Euler kinematics about the time derivatives of the attitudes can be derived as follows:

$$\dot{\varphi} = (\omega_y \sin \hat{\gamma} + \omega_z \cos \hat{\gamma})/\cos \hat{\theta}$$
$$\dot{\theta} = \omega_y \cos \hat{\gamma} - \omega_z \sin \hat{\gamma} \tag{1}$$
$$\dot{\gamma} = \omega_x + (\omega_y \sin \hat{\theta} \sin \hat{\gamma} + \omega_z \sin \hat{\theta} \cos \hat{\gamma})/\cos \hat{\theta}$$

where $\omega_x, \omega_y, \omega_z$ are gyroscope measurements from an onboard IMU with respect to b-frame and denote the roll, pitch and yaw rate of the vehicle, respectively. $\dot{\varphi}, \dot{\theta}, \dot{\gamma}$ represent the time derivatives of the Euler angles. $\hat{\theta}, \hat{\gamma}$ are estimated pitch and roll by CF. Eq. (1) will be used in Subsect. 3.1.

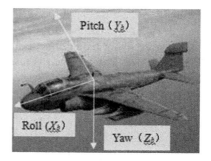

Fig. 1. Attitude definition

The integrated attitude errors will accumulate significantly over time due to sensor noise and biases. To correct the gyroscopic attitude estimations, an attitude observer is required to bound the gyroscopic prediction errors. Besides integrating gyroscope outputs, attitude can also be estimated by accelerometers detecting gravity in non-acceleration motions. Assuming that the UAV is driving in steady-level states or stationary motions, Eq (2) for attitude estimation based on accelerometer can be written as:

$$\begin{bmatrix} f_x \\ f_y \\ f_z \end{bmatrix} = (\boldsymbol{C}_n^b)^{\mathrm{T}} \begin{bmatrix} 0 \\ 0 \\ g \end{bmatrix} \tag{2}$$

where f_x, f_y, f_z denote the specific force measured by strapdown accelerometers in the b-frame. $\boldsymbol{C}_n^b$ is the direction cosine matrix between b-frame and n-frame, whose components are cosine and sine functions of φ, θ and γ. g is the gravitational acceleration expressed in n-frame. Then estimated pitch and roll from accelerometers in steady-level motions can be computed in Eq (3).

$$\theta_{acc} = \arcsin(f_x/g), \ \gamma_{acc} = -\arctan(f_y/f_z) \tag{3}$$

where θ_{acc} and γ_{acc} are estimated attitudes by using the local gravity vector and accelerometer outputs. However, the attitude calculation based on Eq. (3) leads to an incorrect attitude determination under high dynamic conditions due to influences coursed by moving acceleration.

2.2   Acceleration Compensation Based on ADS Data

The accelerometers can not detect gravitational acceleration and centrifugal forces under accelerated motions or in turns. In order to improve the accuracy of accelerometer estimated attitude, the change in transient forward velocity should be known from other aided information. The air data system can provide airspeed, attack angle and slide angle during UAV flight. And the airspeed vector V_{air} can be projected to b-frame from air-frame by the matrix C_a^b defined as

$$C_a^b = \begin{bmatrix} \cos \alpha \cos \beta & -\cos \alpha \sin \beta & -\sin \alpha \\ \sin \beta & \cos \beta & 0 \\ \sin \alpha \cos \beta & -\sin \alpha \sin \beta & \cos \alpha \end{bmatrix}$$

$$V_{airb} = \begin{bmatrix} V_{bx} & V_{by} & V_{bz} \end{bmatrix}' = C_a^b V_{air} \tag{4}$$

where α and β represent attack angle and slide angle, respectively. V_{airb} is the airspeed vector with respect to b-frame. A physical interpretation of pitch in Eq. (3) holds that the change in pitch angle results in different specific forces acting onto the aircraft. However, when the vehicle is moving with apparent acceleration, the pitch estimation extension including the time derivative of the forward velocity component in body frame can be drawn from Eq. (3). The corrected pitch angle is

$$\hat{\theta}_{acc} = \arcsin((f_x - \dot{V}_{bx})/g). \tag{5}$$

When the vehicle turns with a turn radius, another form acceleration experienced by the vehicle is the centripetal acceleration

$$a = \Omega \times V_b \tag{6}$$

where a means the centripetal acceleration vector, Ω is the angular rate vector measured by 3-axis gyroscope on board. Given the actual outputs f_y and f_z from the accelerometers, the corrected roll estimation during turning is given by

$$\hat{\gamma}_{acc} = -\arctan((f_y - a_y)/(f_z - a_z)) \tag{7}$$

3 Complementary Filter Design

3.1 Conventional Complementary Filter Analysis and Design

In general, accelerometer and gyroscope are the basic sensors to measure specific force and angular rate. A fast attitude estimate can be obtained by integrating the gyroscope outputs according to Eq. (1). The gyroscope integrated attitude has a high frequency dynamic characteristics, but is usually disturbed by a growing error caused by gyroscope drifts and bias. On the other hand, the accelerometer attitude estimated using Eqs. (5) and (7) has a low frequency characteristic.

The CF is a frequency-based approach, which consists of a low-pass filter and a high-pass filter and can combine them to give a all-pass filter. Therefore, it allows fusion of independent measurements of the same signal with different spectral characteristics, and can combine slow moving signals from accelerometer and fast moving signals from gyroscope [15]. In order to combine accelerometer output for low frequency attitude estimation with integrated gyroscope output for high frequency estimation, a data fusion algorithm based on CF is adopted in the research. Figure 2 shows a block diagram for the conventional CF attitude filtering estimator.

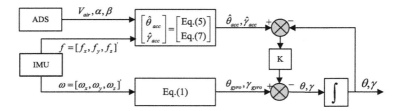

Fig. 2. Conventional CF attitude filtering estimator.

It is obvious that two parallel complementary filters including pitch complementary filter (PCF) and roll complementary filter (RCF) are required in the data fusion. Expression for PCF and RCF are

$$
\begin{bmatrix} \dot{\theta}_{filter}(t) \\ \dot{\gamma}_{filter}(t) \end{bmatrix} = \begin{bmatrix} \dot{\theta}_{gyro}(t) \\ \dot{\gamma}_{gyro}(t) \end{bmatrix} + \begin{bmatrix} K_1[\theta_{acc}(t) - \theta_{filter}(t-1)] \\ K_2[\gamma_{acc}(t) - \gamma_{filter}(t-1)] \end{bmatrix}
$$
$$
\begin{bmatrix} \theta_{filter}(t) \\ \gamma_{filter}(t) \end{bmatrix} = \begin{bmatrix} \theta_{filter}(t-1) \\ \gamma_{filter}(t-1) \end{bmatrix} + \begin{bmatrix} \dot{\theta}_{filter}(t)\Delta t \\ \gamma_{filter}(t)\Delta t \end{bmatrix}
$$

(8)

where K_1 and K_2 are the gain for PCF and RCF, respectively. $\theta_{acc}(t)$ and $\gamma_{acc}(t)$ denote the attitude estimations based on outputs of accelerometer and ADS at current time t. $\dot{\theta}_{gyro}(t)$ and $\dot{\gamma}_{gyro}(t)$ indicate the differential forms of pitch and roll derived from Euler kinematic equation based on angular rate measurement at current time t. $\dot{\theta}_{filter}(t)$ and $\dot{\gamma}_{filter}(t)$ are the rates of estimated attitude changes from previous time $t-1$ to current time t, which will be used for attitude estimation by CF at current time. The attitude estimations of complementary filters at previous time $t-1$ are denoted as $\theta_{filter}(t-1)$ and $\gamma_{filter}(t-1)$. Δt is the sample interval of inertial sensors. Through Eq. (8), the attitudes $\theta_{filter}(t)$ and $\gamma_{filter}(t)$ of complementary filters at current time can be acquired.

3.2 Adaptive-Gain Complementary Filter Design Based on Fuzzy Logical System

The CF designed in Subsect. 3.1 can be regarded as CF with fixed gains to some extent. It is very important to select right gains for both PCF and RCF under any motions. Under steady states, such as station and low dynamic modes, the complementary filter gains can be easily acquired and the estimation angles converge well. However, the gains under steady states are not suitable for high dynamic motions. Under strong forward acceleration or in turn mode, the accumulated errors can not be corrected effectively by the fixed gain filter.

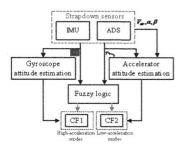

Fig. 3. Adaptive-gain CF structure

To overcome the typical problems and disadvantages of fixed gain complementary filter for attitude estimation, a adaptive-gain complementary filter based on fuzzy logic system is considered in this section. Figure 3 shows the adaptive-gain complementary filter structure with fuzzy logic system. All information for CFs are supplied by gyroscope attitude estimation and accelerator attitude estimation. A block diagram for the attitude filtering estimator augmented by fuzzy logical system for pitch and roll channel is showed in Fig. 4.

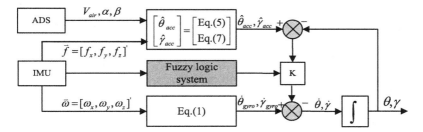

Fig. 4. Realization of adaptive-gain complementary filter

The proposed approach is to exploit fuzzy logic system to generate proper CF gains according to the dynamic conditions. The values and variances of specific forces and angular rates sensed by inertial sensor are inputs to fuzzy logical system, whose results are moving styles including high-acceleration modes and low-acceleration modes (steady mode and stationary mode). The rule storage of Sugeno fuzzy logical system is divided into two levels based on specific force variance and gyroscope outputs of many experiments.

When the square root of x-axis and y-axis specific force is large than 0.1 g, it indicates that the vehicle is in high-acceleration mode. Under this circumstance, if the absolute value of angular rate is not less than 2.0°/s, the vehicle is turning along to corresponding body axis. The filter gains vary from concrete modes with different accelerations in CF1. When the specific force square root is smaller than 0.1 g, the vehicle is in a mode with low-acceleration. In this case, the vehicle is moving without acceleration if the square root value is not less than 0.05 g. Otherwise it is in a stationary mode. The filter gains are also different in steady and stationary modes of CF2.

The threshold values of inertial measurement to identify the moving styles are extracted from amount of experimental data. The attitude estimations based on accelerator and ADS can be computed utilizing Eqs. (5) and (7). The change of attitude over sample interval is acquired based on Eq. (1). In order to reduce the attitude accumulated errors and fuse attitude information effectively adaptive to variant dynamics, the filter gain vector K should be changeable under varying flight dynamics. According to the moving styles judged by fuzzy logical system. K has the following scenarios shown in Table 1:

Table 1. Value of filter gain vector K

Vehicle moving styles	K value
Stationary mode	$K_1 = 1.0$, $K_2 = 1.0$
Steady mode	$K_1 = 0.5$, $K_2 = 0.5$
Turning mode	$K_1 = 0$ or 0.5, $K_2 = 0.5$ or 0
Moving mode	$K_1 = 0$, $K_2 = 0.2$
Others	$K_1 = 0$, $K_2 = 0$

Stationary Mode. The K value is [1.0, 1.0] in this mode. When the vehicle does not move, the attitude estimation based on accelerator is much more accurate than the integrated attitude, whose error accumulate over time due to gyroscope bias and drift. At this time, the filter gain should tend to accelerator attitude estimation.

Steady Mode. The K value is [0.5, 0.5] in this mode. When the vehicle moves with non-acceleration, the parameter weight tending to accelerator attitude estimation should be lowered appropriately due to vibration caused by engine and others.

Turning Mode. The K value varies with the turning mode along body axis. When the vehicle is turning along body x-axis, K is [0.5, 0]. While it is turning along body y-axis, K is [0, 0.5].

Moving Mode. The K value is [0, 0.2] in this mode. When the vehicle moves forward with acceleration, it is not proper to give much trust to accelerator attitude estimation in the fusion process, especially for the pitch estimating channel.

Others. The K value is [0, 0] if the vehicle works in other high dynamic modes not judged by fuzzy logical system.

4   Experimental Results

To verify the proposed complementary filter algorithm implemented with C language in an embedded DSP (digital system processer), a Micro-autonomous AHRS is developed as shown in Fig. 5(a). An ADIS16350 IMU with 3-axis MEMS gyroscopes and accelerators and a small embedded ADS mounted on the AHRS, measuring angular rates, specific force and other parameters. The inertial sensor outputs are collected via SPI serial port and interfaced to the embedded processor at frequency of 100 Hz. Then the proposed attitude estimating algorithm is running and the result of attitude estimation is displayed on PC or other screens. In order to evaluate the performances of proposed algorithm under all motions, a dynamic test with the AHRS fixed on a car is executed along the 5th Ring road in Beijing (Fig. 5b). At the same time, another AHRS with fiber gyroscope is also installed on the car and it can provide high accuracy attitude information as compare reference.

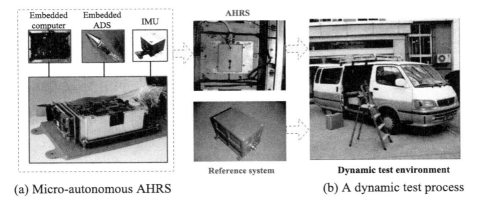

(a) Micro-autonomous AHRS   (b) A dynamic test process

Fig. 5. Experimental process (a) Micro-autonomous AHRS (b) A dynamic test process

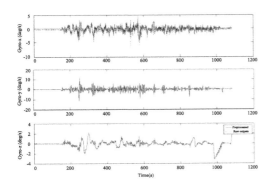

Fig. 6. Gyroscope output during dynamic test

The deterministic errors caused by fixed bias and scale factor can be removed by specific calibration procedures, such as multi-position calibration method [16]. The angular rates measurements during dynamic test are shown in Fig. 6. In stationary motion from 0 s to 150 s, it is very obvious that the angular rates contain bias errors in three directions represented in blue dot curves. In order to decrease the deterministic errors effecting on attitude estimation, a pre-process of error calibration and smoothing is implemented before attitude calculation and complementary filter. The result of pre-process is drawn in red line curves and the processed information is used to obtain attitude estimation.

The dynamic test was carried out by fixing the AHRS on the driving car and then compared the reference AHRS showing in Fig. 7. It is distinct that the accumulated errors of roll and pitch reach about 40° and 6° after 1100 s, even though the deterministic errors of gyroscope have been calibrated by pre-process. The results from reference AHRS is drawn in red. The black curves represent the results based on the proposed adapted-gain CF. Figure 7(b) shows the attitude errors between the AHRS outputs and reference values. From the results of stationary mode in the beginning 150 s, the errors of roll and pitch angles are within 0.1° and 0.4°, respectively. Although the errors in moving mode are larger than that in static mode, the errors of roll and pitch angles in all motions are within the limits of 1°. The results show that the proposed method can improve the attitude accuracy in real-time effectively. It is also very encouraging since the estimation process was performed with low-cost MEMS inertial sensors with a simple and practical complementary filter.

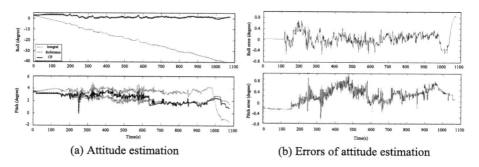

(a) Attitude estimation   (b) Errors of attitude estimation

Fig. 7. Experimental results (a) Attitude estimation (b) Errors of attitude estimation

5   Conclusions

A practical and convenient data fusion method based on adapted-gain CF has been proposed in this paper. The establishment of attitude estimation observer based on accelerator, ADS information is introduced to model estimation observer when the vehicle turns or moves with some acceleration. Based on the measurements from accelerometers and gyroscopes, the fuzzy logic system is used to judge the vehicle's moving styles and adapt the parameters of the filter gain. The proposed method has shown a good performance for depressing the accumulate attitude errors caused by deterministic gyroscope errors as well as the noise corruption from accelerometer measurement in all dynamic circumstances. It is validated that the estimated attitude angles accuracy are within the acceptable region for most of practical implementations. So the proposed adapted-gain complementary filter is much suitable for development of low-cost and autonomous navigation system.

Acknowledgments. This work was supported by a grant from the National Natural Fund of China, Grant No. 61375082 and Buaa blue sky star project.

References

1. Petricca, L., Ohlckers, P., Grinde, C.: Micro-and nano-air vehicles: state of the art. Int. J. Aerosp. Eng. **2011**, 1–17 (2011)
2. Ananda, C.M., Akula, P., Prasad, S.: MEMS sensor suites for Micro Air Vehicle (MAV) autopilot. In: International Conference on Circuits, Communication, Control and Computing, pp. 291–294 (2014)
3. Tian, Y., Wei, H.X., Tan, J.D.: An adaptive-gain complementary filter for real-time human motion tracking with MARG sensors in free-living environments. IEEE Trans. Neural Syst. Rehabil. Eng. **21**(2), 254–265 (2013)
4. Lai, Y.C., Jan, S.S., Hsiao, F.B.: Development of a low-cost attitude and heading reference system using a three-axis rotating platform. Sensors **10**, 2472–2491 (2010)
5. Zhu, R., Sun, D., Zhou, Z.Y., Wang, D.Q.: A linear fusion algorithm for attitude determination using low cost MEMS-based sensors. Measurement **40**, 322–328 (2007)

6. Gebre-Egziabher, D., Hayward, R.C., Powell, J.D.: Design of multi-sensor attitude determination systems. IEEE Trans. Aerosp. Electron. Syst. **40**(2), 627–649 (2004)
7. Bonnabel, S., Rouchon, P.: Symmetry-preserving observers. IEEE Trans. Autom. Control **53** (11), 2514–2526 (2008)
8. Crassidis, J., Markley, F., Cheng, F.: Survey of nonlinear attitude estimation methods. J. Guid. Control Dyn. **30**(1), 12–28 (2007)
9. Bonnabel, S., Deschaud, J.E., Salaun, E.: A simple nonlinear filter for low-cost ground vehicle localization system. In: 50th IEEE Conference on Decision and Control and European Control Conference, pp. 3270–3275 (2011)
10. Harada, T., Uchino, H., Mori, T., Sato, T., Hongo, B.K.: Portable absolute orientation estimation device with wireless network under accelerated situation. In: IEEE International Conference Robot Automation, pp. 21412–21417 (2004)
11. Mahony, R., Hamel, T., Pflimlin, J.M.: Nonlinear complementary filters on the special orthogonal group. IEEE Trans. Autom. Control **53**, 1203–1218 (2008)
12. Euston, M., Coote, P., Mahony, R., Kim, J., Hamel, T.: A complementary filter for attitude estimation of a fixed-wing UAV. In: 2008 IEEE/RSJ International Conference on Intelligent Robots and Systems, pp. 340–345 (2008)
13. Rhoads, G.D., Wagner, N.A., Taylor, B.J., Keen, D.B.: Design and flight rest results for a 24 hour fuel cell unmanned aerial vehicle. In: 8th Annual International Energy Conversion Engineering Conference (2010)
14. Lageman, C., Mahoney, R., Trumpf, J.: State observers for invariant dynamics on a Lie group In: the International Conference on the Mathematical Theory of Networks and Systems (2008)
15. Yoo, T.S., Hong, S.K., Yoon, H.M., Park, S.: Gain-Scheduled complementary filter design for a MEMS based attitude and heading reference system. Sensors **11**, 3816–3830 (2011)
16. Fu, L., Yang, X., Wang, L.L.: A novel calibration procedure for dynamically tuned gyroscope designed by D-optimal approach. Measurement **46**, 3173–3180 (2013)

Hot-Redundancy CPCI Measurement and Control System Based on Probabilistic Neural Networks

Dan Li$^{(\boxtimes)}$, Xiaoguang Hu, Guofeng Zhang, and Haibin Duan

State Key Laboratory of Virtual Reality Technology and Systems,
Beihang University, Beijing 100191, People's Republic of China
lidanbuaa@sina.com

Abstract. Aiming at the requirements of high reliability and high real time for aerospace measurement and control systems, a solution based on Compact PCI (CPCI) bus for hot-redundancy of hardware structure is proposed. And an advanced fault diagnosis method for checking the faults of cards based on the probabilistic neural networks (PNN) is used in this system. A set of hot-redundancy experimental system platforms are developed to perform experimental verification on device-level hot-redundancy technology. Simulation and experiment results show that the system can realize hot-redundancy.

Keywords: Probabilistic neural networks · Fault diagnosis · Hot-redundancy · Hot-swap · Compact PCI bus

1 Introduction

The aerospace measurement and control tasks in the new period have more strict requirements for computer systems. Large Scale Test System (LSTS) applied in aerospace fields has major application requirements for system-level hot-swap and hot-redundancy technology with multiple redundancies and backups. The most representative application of the Beihang University is technology lies on the promotion and reconstruction of hot-swap and hot-redundancy bus systems, and traditional PCI bus products can no longer satisfy the need for high stability, hot-swap and hot-redundancy [1]. As a result, Bus structure calculation control platform with high stability and reliability, which has the feature of hot-swap and hot-redundancy, has become the prerequisite of developing modern measurement and control systems, also the precondition for system-level hot-swap and hot-redundancy technology to be more stable and reliable [2]. In recent years, the measurement and control signal processing systems, based on standard Compact PCI (compact PCI bus, CPCI for short, which can be compatible with PXI) have drawn more attention, especially in the aerospace ground testing fields, the system has occupied an important position.

Fault diagnosis technology is used to detect the running status of the working equipment in the hot-redundancy system. Through the fault diagnosis of the running status in the working equipment, it can decide whether to use the standby device to replace the running equipment [3]. As a typical pattern classification, fault diagnosis can classify the cases that need the diagnosis into the corresponding fault categories; artificial neural network can successfully solve the problems that are difficult for the traditional pattern

© Springer International Publishing Switzerland 2016
L. Cheng et al. (Eds.): ISNN 2016, LNCS 9719, pp. 356–364, 2016.
DOI: 10.1007/978-3-319-40663-3_41

recognition methods. Therefore, fault diagnosis is one of the important application fields in artificial neural network [4, 5]. As a kind of classification neural network with good performance, PNN can directly consider the probability characteristics of sample space. In addition, it has the characteristics of global optimization, so it can be used in the field of fault diagnosis well.

2   The Hardware Technology of Hot-Redundancy

Hot-standby refers that two redundancies work simultaneously under normal circumstances, and when one of them is detected as malfunction, the system will disconnect the fault redundancy and enable single redundant for degradation operation.

Hot-standby technology can be simply divided to single machine hot-standby technology and dual machine hot-redundancy technology. Single machine hot-standby is adopted by this equipment. Both of the main card and standby card of two functional cards can standby each other. Hot-redundancy function was complete by controlling the interrupt response of corresponding board cards.

2.1   Working Principle of Single Machine Hot-Redundancy

Single machine hot-standby technology adopts the way of single access lines hot-redundancy. As shown in the Fig. 1, single machine hot-standby system includes several terminal equipment, authorization host and access device which connects the terminal equipment and authorization host. It is characterized that set main card and standby card to make them as standby mutually.

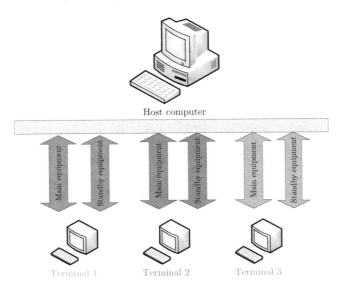

Fig. 1. The mode of single machine hot-redundancy

The operating principle is the inversion for main card and standby card by setting different response time. As shown in the Fig. 2, after reaching the setting response time of the main card, if the main card is normal, the main line card will request signal to connect the line, and complete data transmitting and receiving; in case of the main card failure, it will not response [6]. After reaching the setting response time of the standby card, the standby card will request signal to connect the line, and complete data transmitting and receiving [7].

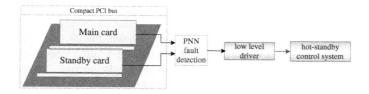

Fig. 2. Work flow of single machine hot-redundancy

2.2 Fault Isolation Technology of Single Machine Hot-Redundancy

Single machine hot-standby adopts main and standby access equipment parallel setting to make them as standby to each other [8].

If the main card is normal operation, the main equipment will respond to the request signal, connect to the data, control the line and complete data transmitting and receiving. In case of the main card failure, the bus of main card board will be in failure and will not respond to the request signal; after reaching the setting response time of the standby card board, it will respond to the request signal after detecting that the main card board is in failure. It will connect to the data, control the line and complete data transmitting and receiving. If the standby card failed to work either, both of them will not respond to the request signal data transmitting and receiving can not be achieved.

Thus, it is possible to detect whether the failure location is on the main card or standby one. In addition, it can cut off the connection between the failure one and the host to realize fault isolation.

3   PNN Fault Detect Technology of Hot-Redundancy

3.1   PNN

The cards are diagnosed by hot-redundancy system with fault diagnosis method based on neural networks. The operation states of cards are related to all failure symptoms. It's hard to express this complicated relation with formulas, while neural networks, due to their specialties, can identify the uncertain mode of operation states. During the test, information from each source wasn't very certain, while neural networks could identify whether the fault should be insulated according to similarities of samples received by the present system.

The faults of cards are checked by probabilistic neural networks mode in this paper. PNN is a parallel computing method, developed from an estimation method of Probability Density Function(PDF), based on Bayes classification rules and Parzen windows [9]. PNN can replace non-linear learning algorithm with linear learning algorithm to finish the work, and keep the high precision of the non-linear learning algorithm. PNN is composed of input layer, mode layer, summation layer and output layer. Its basic structure is as shown in Fig. 3.

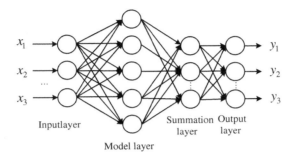

Fig. 3. Topological structure of the PNN

The fault diagnosis method based on PNN is widely accepted in probability statistics. It can be described as follows: supposing there are two known fault modes θA and θB, for the fault feature samples which will be judged $X = (x1, x2, \ldots, xn)$:

If $hAlAfA(X) > hBlBfB(X)$ then $X \in \theta A$;

If $hAlAfA(X) < hBlBfB(X)$ then $X \in \theta B$;

On the formal, hA and hB are the prior probability of θA and θB in failure mode ($hA = NA/N, hB = NB/N$). lA is the cost factor that X (belongs to the fault sample of θA) is mistakenly parted to θB. lB is the cost factor that X (belongs to the fault sample of θB) is mistakenly parted to θA. fA and fB are the Probability Density Function (PDF) of θA and θB. Usually PDF can't be gained with precision but its statistic can be calculated according to fault feature samples.

In 1962, a method of estimating the PDF from some random samples was proposed by Parzen. If there are enough samples, functions gained from this method can smoothly get close to the former PDF. The PDF estimator formula is shown in the following:

$$fA(X) = \frac{1}{(2\pi)^{P/2}\delta^P} \frac{1}{m} \sum \exp\left[-\frac{(X - X_{ai})^T(X - X_{ai})}{2\delta^2}\right] \tag{1}$$

X_{ai} is the i th training vector θA of and m is the training patterns of θA. δ is the smoothing parameter and its value determines the width of the clock-shaped curve centering on the sample point.

3.2   Modeling

In any mode of neural networks, the chosen input value must correctly reflect features of problems. The CPCI has a complicated power environment including power 1.5 V, 3.3 V, 5 V, 12 V and −12 V, in which power 3.3 V, 5 V and 12 V are related to the operation of functional chips. When the board card has faults, current of the stated power will become abnormal. Therefore in the mode of neural networks, the current value of those three electric power sources are regarded as the input feature vector, and the degree of whether the card should be insulated is the output vector. After the following procedures of designing the PNN mode, the flow chart is shown in the Fig. 4.

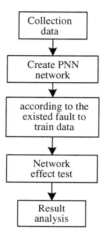

Fig. 4. Design process of PNN

3.3   Implementation by MATLAB and Simulation Analysis

The experimentation used NEWPNN parameters in MATLAB to create PNN, and the selected value of PNN SPREAD was 1.5. In the experimentation, the scale of the matrix was 33 × 4. The first three lines stands for current value and the fourth line stood for the classified output. The number "1" represented the state of acception and the number "2" represented the state of insulation. The first 23 samples were the training samples of PNN and the last 10 samples were the versification samples.

After the training, the training data as input were put into the PNN and judgments of samples were all correct, as shown in Fig. 5. When forecast samples were used to test faults, only one sample got the wrong judgment of faults, as shown in Fig. 6. Therefore it was effective to use the PNN to detect faults of board cards.

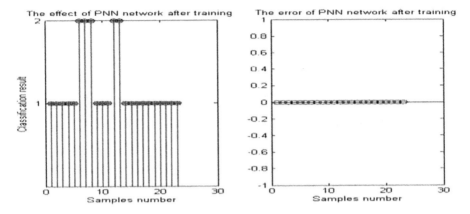

Fig. 5. The effect and error after training

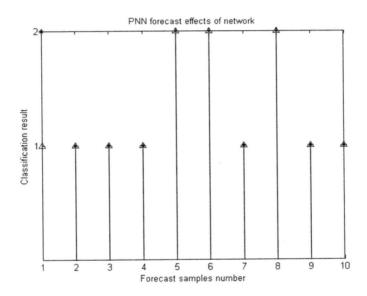

Fig. 6. PNN forecast effects of network

4   Confirmatory Experiment

The purpose of establishing CPCI hot-swap system is to increase the reliability of the system through building multiple standby, namely to build a redundant system. As shown in Fig. 7, at the experiment, insert two serial communication cards and a digital input card. One of the serial communication cards is under operation (main card) and the other one is in hot reservoir system (standby card).

Fig. 7. Insert board of CPCI computer

A group of 32 bit output digital data is controlled by the upper computer to the platform, and the platform to receive high 16 bit data of the 32 bit output digital data output to serial communication card.

Given the main card disable interrupts when the systems is normal working. When the main card completed a full workflow, standby card is automatically reset. If the system detects that the primary card is removed, the standby card enables an interrupt and take over the work. During the system operation, the same kinds of card main card and standby card are in the same working conditions at the same time. Using the serial communication cards transmit to the digital input card these data: 0×10000000, 0×30000000, 0×70000000. In order to verify the reliability of the redundant system, the main card was input much higher voltage from the platform to simulate the failure of the main card during the reading process of the main card. The PNN algorithm which written into the chip, had diagnosed the main card was failure. The main card will be removed safely, and immediately start up the standby card to take over the function of main card. The working condition of standby card is normal.

On the control interface of serial communication card, click on the channel one "read FIFO" button. Read the data, the result is correct. The pulling of card will not cause any impact on reading. Since after the main card read the data, it will upload the data in a timely manner, after starting up the standby card, there is no difference with the main card as shown in Fig. 8.

Insert the main card to the chassis once more. Simulate the status of the main card working continuously after standby card overhaul under ordinary operation. The main card can be identified by the system immediately, replace the standby card, and work continuously.

Through the above experiment, it has been verified that the system is in conformity with the hot-standby system. The reliability of the system can be enhanced by applying the system.

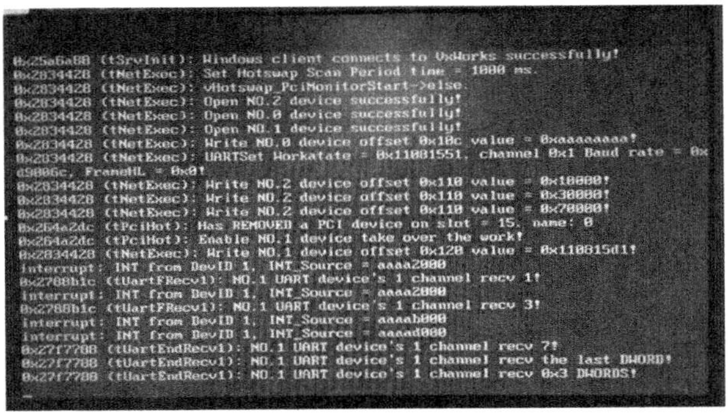

Fig. 8. Standby card taking over main card

5   Conclusion

An in-depth research is carried out on the hot-redundancy technology of widely applicable CPCI bus measurement and control system to satisfy the requirements of high reliability and real-time, and a solution of hot-swap technology based on high available models is proposed. The problem of the card was tested by using the probabilistic neural network. An experimental verification platform system of hot-swap and hot backup technology is constructed. The feasibility of above-mentioned theoretical contents has been verified through simulation and experiment. The results of simulation and experience revealed the feasibility of the technical scheme and proved its stability in performance.

References

1. Xin, L., Jin, L.Z.: Compact PCI bus industrial PC technology present situation and application. Electron. Technol. Appl. **7**(7), 6–8 (2002)
2. He, Z., Men, C., Chen, Y., et al.: Schedulability of fault tolerant real time system based on local optimum checkpoint under priority mixed strategy. Chin. J. Electron. **24**(02), 236–244 (2015)
3. Qian, H., Ma, J., Li, Z.: Navigation sensors fault diagnosis based on wavelet transformation and neural network. In: 2007 IEEE 8th International Conference on Electronic Measurement and Instruments, pp. 592–595. IEEE Press (2007)
4. Chine, W., Mellit, A., Lughi, V., et al.: A novel fault diagnosis technique for photovoltaic systems based on artificial neural networks. Renew. Energy **90**, 501–512 (2016)
5. Jia, F., Lei, Y., Lin, J., et al.: Deep neural networks: a promising tool for fault characteristic mining and intelligent diagnosis of rotating machinery with massive data. Mech. Syst. Sig. Process. **72–73**, 303–315 (2015)
6. Chen, Y., David, K., Yim-Shu, L.: A hot-swap solution for paralleled power modules by using current-sharing interface circuits. IEEE Trans. Power Electron. **21**(6), 1564–1571 (2006)
7. Li, Z., Hu, X.: An extension HA design for the hot-swap application of compact PCI device. In: 8th IEEE Conference on Industrial Electronics and Applications, pp. 646–650. IEEE Press (2013)

8. Hong, G., Bo, Y.: A new hot standby control system for single and communication transmit to protect high speed (350 Km/H) of railway safety. In: IEEE 2010 International Conference on E-Product E-Service and E-Entertainment (ICEEE), pp. 1–3. IEEE Press, Henan (2010)
9. Specht, D.F.: Probabilistic neural networks. Neural Netw. **3**(1), 109–118 (1990)

Individually Adapted Neural Network for Pilot's Final Approach Actions Modeling

Veniamin Evdokimenkov[1], Roman Kim[1], Mikhail Krasilshchikov[1],
and German Sebrjakov[2(✉)]

[1] Moscow Aviation Institute (National Research Centre), Moscow, Russia
evn@netland.ru, romanvkim@yandex.ru, mnkr@mail.ru
[2] State Research Institute of Aviation Systems, Moscow, Russia
sebr@gosniias.ru

Abstract. The individually adapted neural network model for pilot's final approach actions modeling allowing for touchdown accuracy forecast at any time instant on glide path is under consideration. Network model optimization algorithm, based on onboard flight recorder data, is suggested. Final approach test modeling on MIG-AT flight simulator by a group of pilots with different flight experience was executed. Results show pilot's individual flying manner turns out by network architecture, input vector composition, and network parameters.

Keywords: Individually adapted neural network model · MIG-AT flight simulator · Network architecture · Input vector composition · Network parameters

1 Introduction

A pilot professional activity simulation model development is most sophisticated problem in human engineering. Linear model for pilot activity simulation has appeared in the late forties of the 20th century. Further research revealed fundamental patterns of pilot behavior, led to classic control theory models in vicinity of open-loop control system cutoff frequency.

Spectral density noise term for pilot's model was developed to the middle of the 60th. In spite of model demerits, it was very useful for dynamics and applied flight control problems.

Never the less, completely suitable pilot model as a member of a flight control system still absent. Most studied models to date consider a pilot as a member of steering system.

The issue of interest is the so called prognostic-optimization model. The model uses well known human engineering principle: skilled operator doing almost optimally considering all constraints and personal comprehension of the problem. Model formalization is based on assumption the pilot considers gradual control, minimizing expenditure of energy, will be optimal one. Optimal control synthesis is realized with respect to A.A. Krasovsky' generalized work functional (GWF). Let us note the GWF

L. Cheng et al. (Eds.): ISNN 2016, LNCS 9719, pp. 365–372, 2016.
DOI: 10.1007/978-3-319-40663-3_42

model synthesizes an "ideal pilot" optimal control actions. This allows regard it as virtual instructor for real pilot control actions support.

All above mentioned models describe an "averaged" pilot because of its activity fits typical frame. The question at issue is that frame appears wide enough. For example, the widespread pilot model approximation by transfer function has double scattered time constant. Modern military plane dynamics and battle conditions calls for instant response considering pilot' individual character. In other words, a need of individually adapted pilot activity models in consideration of pilot' specificity arises as confirmed by works [1–5].

2 Individually Adapted Neural Network Architecture

Pilot control activity model development is complicated by absence of fundamental theory allowing mathematically describe pilot's individual psycho physiological reaction. The only reliable model development basis is data gathered both from pilot's real final approaches and flight simulator training. Moreover, this information increases from flight to flight in course of pilot professional activity that makes possible continuous model update regarding the change of experience and age-related changes in psycho physiological reactions. Possible approach to experimental pilot's individually adapted model construction follows hereinafter.

Let $Z^T = (z_1,...,z_n)^T$ be a vector of the pilot-aircraft system state. Its components evaluate aircraft state, throttle lever (TL), and control stick (CS) actions. Vector Z can be monitored in final approach trajectory at any distance l from center of the runway. Let a pilot has performed N final approaches to the date. Safe touchdown takes place when vector Z components satisfy the conditions $z_{imin} \leq z_i \leq z_{imax}$, $i = 1,..,n$ at runway end crossing moment L. The given set of conditions outline safety touchdown terminal hyper rectangle in space of pilot-aircraft system state. It stands for reason use of scalar indicator function μ for qualitative assessment of touchdown performance:

$$\mu = \max_i \left| \frac{z_i(L) - 0,5(z_{imax} + z_{imin})}{0,5(z_{imax} + z_{imin})} \right|$$

It is obvious, that $0 \leq \mu \leq 1$. Case $\mu = 0$ stands for aircraft is in the center of safety hyper rectangle, i.e. indicates "ideal" touchdown. Case $\mu > 1$ indicates abnormal touchdown performance. Thereby, after a pilot has performed a certain number of final approaches N for any moment of standard mode performance l we have $Z^k(l)$, $k = 1,...,$ N pilot-aircraft system state samples, "marked" by indicators μ_k, k = 1,...,N (see Fig. 1).

Hereinafter, we understand relation $\mu(Z,l)$ as pilot model which predicts touchdown accuracy μ for any state vector Z at moment l. Sundry pilots with comparable flight experience demonstrate own aircraft control manner, hence the relation $\mu(Z,l)$ has to be adapted to specific pilot control actions. Correctness of this statement has been proved in [1, 2] by statistical treatment of large set of test data gathered from final approaches performed by pilots having comparable qualification. The results statistically confirm specific features of control actions by distinctions in values of pilot-aircraft system state intrinsic to distinct pilots. This agree with aviation medicine and human engineering confirming pilot's personal motor stereotypy formation in course of training [6, 7]. By this, pilot trains apply optimal rate and frequency of control levers movement.

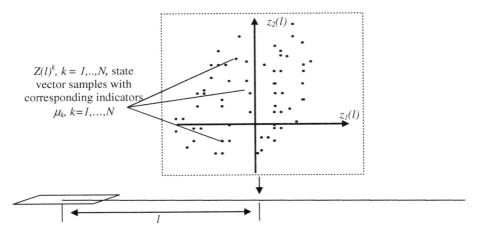

$Z(l)^k$, $k = 1,...,N$, state vector samples with corresponding indicators μ_k, $k=1,...,N$

Fig. 1. Touchdown performance in pilot-airplane state space

Hereby, we need create $\mu(Z,l)$ multidimensional relation using $Z^k(l)$, μ_k, $k = 1,...,N$ samples of pilot's activity. This is a typical regression analysis task. Let us consider possible approaches to creation of such relation. Multidimensional $\mu(Z,l)$ relation forming-up problem based on $Z^k(l)$, μ_k, $k = 1,...,N$ realization is a typical regression analysis task which solution is

$$\mu(Z,l) = a_1f_1(Z,l) + a_2f_2(Z,l) + \ldots + a_kf_k(Z,l) \tag{1}$$

where $f_i(Z,l)$, $i = 1,...,k$ are known scalar functions; a_i, $i = 1,...,k$ are unknown factors. In spite of work [8] specify the problem, basic functions $f_i(Z,l)$, $i = 1,...,k$ in (1) specification selection remains in abeyance. The situation is aggravated with considerable pilot-aircraft system state vector Z length that significantly complicates functions specification on real flight data. Thus, neural network model for $\mu(Z,l)$ is more preferable solution for problem under consideration.

Application of neural network of various types has shown [9, 10] multiple layer perceptron (MLP) sufficiently fits relation $\mu(Z,l)$. This follows from Kolmogorov's theorem proving that MLP fits any continuous function to any desired degree of precision. Individual pilot model $\mu(Z,l)$ adaptation by multilayer perceptron consists in selection of its input parameters, number of layers, number of neurons in each layer and input weights. Let us note that composition of input parameters define a pilot-specific run-time instrumental flight image. Other parameters define pilot's personal run-time image transform into control actions.

The following section describes neural network models of pilots constructed using results of final approach training on MIG-AT aircraft simulator.

3 Pilot Neural Network Model Study

Series of final approaches on MIG-AT aircraft simulator have been used for pilot individually adapted neural network model analysis. Table 1 lists monitored state vector components.

Table 1. The list of recorded state vector components

Component	Parameter
Z_1	V- airspeed
Z_2	P- thrust
Z_3	$\dot{\gamma}$- roll rate
Z_4	$\dot{\psi}$- yaw rate
Z_5	$\dot{\vartheta}$- pitch rate
Z_6	n_X- overload projection onto body axis X
Z_7	n_Y- overload projection onto body axis Y
Z_8	n_Z- overload projection onto body axis Z
Z_9	ϑ- pitch
Z_{10}	γ- roll
Z_{11}	α- attack angle
Z_{12}	β- glide angle
Z_{13}	H- altitude
Z_{14}	ψ- yaw
Z_{15}	V_X- airspeed projection onto X topodetic axis
Z_{16}	V_Y- airspeed projection onto Y topodetic axis
Z_{17}	V_Z- airspeed projection onto Z topodetic axis
Z_{18}	l- distance to runway center
Z_{19}	Z- runway lateral error
Z_{20}	θ- flight path angle
Z_{21}	δZ- lateral glide slope error
Z_{22}	δH - elevator glide slope error

Two operators took part in experiments. The first operator is first-time pilot. The second operator has considerable aircraft control experience. Each operator performed 50 touch-downs. Hereinafter $\mu^1(Z,l)$ and $\mu^2(Z,l)$ denotes neural models of the operators. First of all, a multilayer perceptron optimization calls for minimization fitting error with respect to input vector configuration, the number of neural network layers, and the number of neurons in each layer. Accuracy of $\mu^1(Z,l)$ approximation is shown in Table 2. Measure of accuracy is least mean square with respect to point series $\mu_j(Z^j(l_t) l_t)$, $j = 1,...,50$, $t = 1,...,500$ used to measure the indicator function. The actual indicator function values obtained on aircraft simulator are indicator function "measurements". Table 2 lists perceptron architectures in error descending order. Table 3 shows similar data for relation $\mu^2(Z,l)$.

The results obtained allow to draw the following conclusions:

1. Two-layer perceptron ensures acceptable accuracy of functions $\mu^1(Z,l)$ and $\mu^2(Z,l)$ fitting.
2. Two-layer perceptron reveals operator's individual flying manner with respect to input vector composition, neurons number, and weighing factors.

Table 2. Function $\mu^1(Z,l)$ fitting error with respect to perceptron architecture

MLP architecture	Inputs number	Input vector configuration	Layer/neurons per layer	Fitting error
	15	$Z_1, Z_2, Z_4,$ $Z_7\text{-} Z_9,$ $Z_{11}-Z_{14},$ $Z_{16}-Z_{20},$	1 / 7 2 / 1	0.15
	8	$Z_2, Z_4, Z_{14},$ $Z_{17}, Z_{18},$ Z_{20}, Z_{21}, l	1 / 2 2 / 1	0.13
	22	$Z_1\text{-} Z_{21}, l$	1 / 7 2 / 1	0.08

Let us consider the first operator optimal neural network model structure in details (see Table 2). This two-layer model has 22 input parameters. The first layer has 7 neurons. Single neuron in the second layer output is indicator function value. To provide required fitting accuracy the network was trained on a series of $\mu_j(Z^j(l_t)l_t)$, $j = 1,...,50$, $t = 1,...,500$ using back-propagation algorithm [9]. Obtained optimal synaptic weights $(W_{ij}^1, i = 1,...,22; j = 1,...,7; W_{ij}^2, j = 1,...,7)$ and sigmoid activation functions parameters ensure 0.08 fitting accuracy value. We should note that optimal neural network architecture with two-layer perceptron agree with the neural networks theory stating that for any R - layer network ($R > 2$), a two-layer equivalent exists.

Now consider optimal neural network architecture for the second operator (see Table 3). Two-layer perceptron appears to be optimal in this case as well. Here we have 7 inputs. There are 5 neurons in the first layer. Network output from single neuron of the second layer is indicator function value. At obtained optimal synaptic weights ($W_{ij}^1, i = 1, \ldots, 7; j = 1, \ldots, 5$; $W_{ij}^2, j = 1, \ldots, 5$) and sigmoid activation functions parameters fitting accuracy is 0.09.

It is easy to see that neural network of the second operator much simpler than that for the first operator due to their individual piloting manner. Thus, second operator is mature pilot. The operator holds aircraft on glide path, assiduity doing in horizontal plane. This is proved by input vector is composed of "planar" trajectory parameters. Very parameters affect touchdown accuracy significantly. Moreover, individual manner of the second operator does not vary. This conclusion follows from Fig. 2 which displays his neural networks obtained after $N = 20$, $N = 40$ and $N = 50$ landings.

Table 3. Function $\mu^2(Z,l)$ fitting error with respect to perceptron architecture

MLP architecture	Inputs number	Input vector configuration	Layer/neurons per layer	Fitting error
	4	Z_{14}, Z_{15}, Z_{17}, Z_{20}	1 / 2 2 / 1	0.22
	11	Z_3, Z_4, Z_5, Z_8, Z_9, Z_{13}, Z_{14}, Z_{16}, Z_{17}, Z_{18}, Z_{20}	1 / 9 2 / 1	0.12
	7	Z_1, Z_{11}, Z_{13}, Z_{14}, Z_{17}, Z_{18}, Z_{21}	1 / 5 2 / 1	0.09

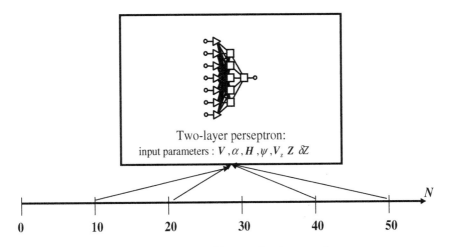

Fig. 2. Neural network architecture for the second operator

Neural network architecture remains constant proving persistent experience of the pilot. In contrast, the first operator is beginner pilot. He has no mature hand to control aircraft neither at final approach, nor at touchdown. The neural network architecture for this operator changes in course of his activity as shown in Fig. 3.

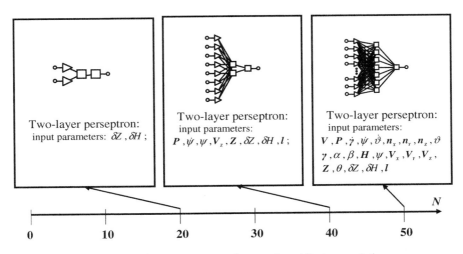

Fig. 3. The first operator neural network architecture variation

Beginner pilot only tries to follow the glide path rely on navigation indicators. He do not try forecast variation of trajectory data. Subsequently the number of parameters analyzed increases, however flying manner remains unchanged due to pilot's low experience.

Thus, on the basis of the received results it is possible to draw the following conclusions:

- expert pilot has own flying manner that develops by the neural network architecture;
- neural network model for an experienced pilot adapts for the expense of its parameters; its architecture remains constant.
- neural network model for the entry-level pilot adapts by both the architecture and parameters adjustment.

4 Conclusion

The individually adapted neural network model for pilot's final approach actions modeling allows for touchdown accuracy forecast at any time instant on glide path. Results of flight modeling on the MIG-AT flight simulator, performed by a group of operators, are given. It was shown that pilot's individual flying manner turns out by network architecture, input vector composition, and network parameters. Composition of input vector, obtained as a result of network optimization, develops pilot's personal instrumental image forming in course of his flight activity.

References

1. Bazlev, D.A., Evdokimenkov, V.N, Krasilshchikov, M.N.: Creation of characteristic sets for the individual adapted support of the pilot when performing the standard flight modes. Russian Academy of Sciences News. Theory and Control Systems 4 (2008)
2. Krasilshchikov, M.N., Evdokimenkov, V.N., Bazlev, D.A.: The Individual Adapted Onboard Monitoring Systems of Technical Plane State and Operating Pilot Actions Support. MAI, Huddersfield (2011)
3. Sebryakov, G.G., Krasilshchikov, M.N., Evdokimenkov, V.N.: Control and intellectual support of the pilot actions when performing the standard flight modes on the basis of the individual adapted approach. Mechatron. Autom. Manag. 8, 16–23 (2014)
4. Sebrjakov, G.G.: Problems of semi-automatic systems development for aircraft pointing. J. Comput. Inf. Technol. 10, 2–8 (2007)
5. Sebrjakov, G.G.: Modelling of a human operator activity in semi-automatic systems for dynamic objects control. Mechatron. Autom. Control 4, 17–29 (2010)
6. Zavalova, N.D., Lomov, B.F., Ponomarenko, V.A.: Image in mental activity regulation system. Science 174, (1986)
7. Zavalova, N.D., Lomov, B.F., Ponomarenko, V.A.: Principle of the active operator and distribution of functions between the person and the machine. Psychological Questions 1971 (3), 3–5 (1971)
8. Draper, N.R., Smith, H.: Appl. Regression Anal. Dialectics, Multiple Regression. M. (2007)
9. Haykin, S.: Neural Networks. 2nd edn. tr from English. Williams publishing house, Chicago (2006)
10. Wasserman, P.: Neural Computing: The Theory and Practice. Mir, Moscow (1992)

Clustering, Classification, Modeling, and Forecasting

On Neurochemical Aspects of Agent-Based Memory Model

Alexandr A. Ezhov[(✉)], Andrei G. Khromov, and Svetlana S. Terentyeva

Troitsk Institute for Innovation and Fusion Research, 142190 Moscow, Troitsk, Russia
ezhov@triniti.ru

Abstract. This study is aimed to discuss neurochemical basis of the earlier proposed agent-based memory model. The model takes into account such sociological factor as inequality and permits to reveal specific memory codes arising in high unequal societies due to phase transition. It also has been offered as a heuristic tool to predict a new competitive pairs in human brain's memory localized in the hippocampus and prefrontal cortex in both hemispheres. Recently new results on dopamine system asymmetry and its influence to the balance of reward-seeking and punishment avoidance behavior together with the role of dopamine level in egalitarian behavior have been obtained. We present arguments in favor of these results can be used as neurochemical basis for this agent-based memory model.

Keywords: Agent-based model · Inequality · Memory codes · Dopamine

1   Introduction

Recently the results of studies of the agent-based model formulated in [1] and describing the emergence of specific binary memory codes [2] in the system characterized by high inequality [3] have been offered as a heuristic tool to predict a new competitive pairs in the human brain's memory localized in the hippocampus and prefrontal cortex in both hemispheres [4]. It has been argued that this model is able to account for the influence of sociological factors e.g. inequality to the brain memory. The original model [1] is based on the hypothesis connecting Lefebvre ethical systems [5] and brain lateral organization. Namely, the first and the second ethical systems introduced by Lefebvre on the base of opposite interpretations of two basic algebra operations (summation and multiplication) correspond to dominance of left and right brain hemispheres correspondingly. This assumption permitted to derive the rules of agent interaction. Using ultrametric data analysis [6] of the set of agent histories the specific memory codes for agents with right and left brain dominance had been derived [3]. These results have been used to propose the hypothesis about possible organization of some structures and functions of human memory [4]. In particular a new competitive pairs in human brain's memory localized in the hippocampus and prefrontal cortex in both hemispheres have been hypothesised. However both the agent-based and also memory model have no neural level basis. In this presentation some recent results relating to brain dopamine system asymmetry and its influence to the balance of reward-seeking and punishment avoidance behavior together with the role of dopamine level in egalitarian behavior have been used

L. Cheng et al. (Eds.): ISNN 2016, LNCS 9719, pp. 375–384, 2016.
DOI: 10.1007/978-3-319-40663-3_43

for formulating desired neurochemical basis [1]. In the first section we describe the problem connected with the study of the influence of economic and social inequality to human brain and memory. In the second section we review the agent-based model [1] which has been developed to take into account an inequality of resource presentation in different social (identity) niches. In the third section the agent-based model has been used as heuristic tool to propose hypothesis on possible memory structures [4].

The main result is presented in the fourth section. It is demonstrated that the general model basis described in [1–4] can be justified by consideration of neural processes, especially by recent findings in dopamine system asymmetry and its influence to human behavior. In conclusion we outline and compare briefly different levels of the agent-based memory model.

2   Brain in Unequal Society

One of the important problems of cognitive science is the influence of economic and social inequality to human brain and memory. It is well-known that social and economic inequality describes the difference both between people and countries. In recent years many papers devoted to the question of people reaction to inequality have been published.

It is remarkable that the economic mainstream – neoclassical economics – has failed to explain the form of the wealth distribution curve. This has been successfully described by physicists only in 21st century [7]. Besides, the human behavior in situations of an unjust distribution of income is contrary to the main feature of a rational agent of neoclassical economics – *homo economicus* – the utility maximization. This was confirmed by the results of studying the classical *Ultimatum Game* [8]. When playing in reality, the players reject unjust splitting of a potential prize by giving preference to the case of receiving nothing but not an unjust part. It is also notable that this rejection of any unequal sharing can be suppressed by applying a magnetic field to the right prefrontal cortex of player brain [9]. In 2010 the study of frontal parts of striatum revealed a strong aversion to a highly unequal payment for equivalent job: the aversion was expressed by human who received both excessive and insufficient payments [10]. By studying the key factors of the economic growth, it was found that any improvement of law systems, property rights, or democracy is of no use if a sharp inequality in society exists [11]. Therefore, when building realistic economic and social models, it is extremely important that the inequality level should be taken into account. Such models are being developed in the growing field of *econophysics* [12].

3   Agent-Based Model of Unequal Society

In the model proposed in [1] the artificial world consists of cells populated by an arbitrary number of agents. The crucial feature of an artificial agent is it has two kinds of resources: physical and mental ones. The physical resource is necessary for the agent to be alive physically, so the resource can be treated as food, money, etc. The second one corresponds to the agent's identity – the complex of its self-identification as belonging to a given nation, religion, profession, country, etc. The agent dies if any of its resources expires. Its death can be either physical, or mental. The latter corresponds to full loss of the agent's identity. It is supposed that the initial cell where the agent starts its life corresponds to its given identity.

It is also assumed that if an agent changes the cell it occupies its mental resource gradually degrades. On the other hand, the agent is provoked to change its cell because the food needed to survive physically can be presented in another cell (it is also supposed that physical resource degrades in time). So, to survive mentally the agent should try to hold its cell, if possible, but to survive physically it should change its cell, because the necessary food can appear in other cells [1]. The point is that the probability to receive a proposal for food depends on the cell and the distribution of this probability across the cell is highly non-uniform, in general. Preventing both resources from vanishing (mental and physical surviving) turns out to be a contradictory problem. The crucial model feature which connects this with the field of statistical physics is that inequality in the food proposal in a different cell can be described by an effective temperature of environment in such a manner that its high value corresponds to a low level in inequality of food presentation in different cells, while a low value – to a highly unequal presentation of it across these cells [1, 3].

We interpret the appearance of food in the cell being free of a specific agent as the environment proposal either to change the identity or to pay a unit of mental resource for the food. For non-interacting agents there are only *two reasonable strategies* to survive. Following the first one, the agent keeps its mental resource regardless of the cell in which the food is proposed. Following the second one, the agent always accepts environmental proposals for increasing its physical resource.

These two strategies were called in [1] the *right* brain and *left* brain strategies, correspondingly. There is a real basis to relate such kinds of behavior with the dominance of corresponding brain hemisphere. It follows from the exponential form of logical functions representing these strategies which were used by Lefebvre in his algebra of consciousness [5], actually from the absence and presence of logical negation in them. Indeed, there exists some experimental evidence that in general all logical operations are functions of the left hemisphere only. Other arguments in favor of this interpretation are available, e.g. [13]. The most interesting behavior of agents can be in the situation where they can interact mentally when making decision. Concrete forms of this interaction were obtained by making a proposal that the left brain dominance corresponds to the first (Western) ethical system, while the right one – to the second (Soviet, Eastern) ethical system. Lefebvre has demonstrated that it is possible to construct two algebraic models of ethic which differ in interpretation of two basic operations – *adding* and *multiplication*. They can be interpreted as *confrontation* and *cooperation*, correspondingly, or *vice versa* [5]. These two variants of ethic based on these interpretations represent the results of the experiments with emigrants from Europe (1st system) and the Soviet Union (2nd system).

By relating two strategies with two ethical systems it was possible to show that the cooperative right brain dominant agents effectively *attract* each other, while the competitive left brain dominant ones effectively *repulse* each other (in the case of self-interaction an agent changes its dominant hemisphere). As follows from [1] the equilibrium state of the collective of cooperative right brain agents is described by statistics of Bose-Einstein. The equilibrium state of the left brain dominant agents is described by a variant of the Fermi-Dirac statistics.

More information about the model properties can be revealed not from equilibrium distributions, but from the study of *fluctuations*, i.e. temporal deviations of agent-in-cell numbers from their equilibrium mean values. This demands to look at the dynamics of agent movement through world cells and to record the *histories* of agents in the systems under condition of thermodynamic equilibrium. Let us call the choice of the destination cell where the material resource (food) is offered to the agent as "the proposal" to go to this cell. The agent can either *accept* or *reject* this proposal. According to its decision the agent can either stay in the initial cell or go to the destination cell. Now we can define three different situations which can occur:

- The destination cell differs from the current cell of the agent and the agent accepts the proposal to go to a new cell. We can denote this situation using a pair of characters AG (*accept* and *go*).
- The destination cell differs from the current cell of the agent, but the agent rejects this proposal and stays in the previous cell. We can denote this situation using a pair of characters RS (*reject* and *stay*).
- The destination cell coincides with the current agent cell. In this case the agent unconditionally accepts the proposal and stays in the previous cell. We can denote this situation using a pair of characters AS (*accept* and *stay*).

By collecting the events happened to the given agent we can form its history which can be represented by a three-symbol ternary code (h1, h2, h3) assuming that h1 = AG, h2 = RS, h3 = AS. The next step is connected with the derivation of two memories from a single history. The memories are obtained by partial forgetting this history. Indeed, we can easily derive two special binary memory codes by omitting the second or the first character in the ternary history code. The first choice gives characters A and R – we call this code {A, R}. The second choice leaves characters S and G and we call this code {S, G} [2]. If we describe the memory of the given agent using the {A, R} or {S, G} codes, we may evidently lose some information about the true history. In particular, if any agent accepts the proposal (A), it can either stay in the initial cell or go to another one. On the other hand, if the agent stays in the initial cell, this implies that it either accepts the proposal to stay there or rejects the proposal to go to another cell. Therefore, it is evident that such two binary coding schemes for memory are not equivalent: it is impossible to reconstruct the {A, R} agent memory given the {S, G} memory and vice versa.

These two binary codes can define *memory codes* because there is some evidence that the brain can really work *with the binary codes only* [14]. As it was shown in [2], the possibility of coding the agent memories by use of two different schemes opens the way to relate memories of the right and left brain dominant agents by calculating the degree of ultrametricity [6] for sets of their incomplete binary coded memories. Briefly speaking to calculate the degree of ultrametricity of a set of agent memories it is necessary to form *triples* of these memories and to study triangles with their sides equal to distances between these memories. It turns out that some of these triangles will be proper, or isosceles with the small base. The portion of such triangles is just the ultrametricity degree. If this degree is equal to one, the set has a simple tree-like hierarchical structure. So the degree of ultrametricity is characterizing the degree of hierarchical organization of the memory set. By comparing ultrametrical characteristics of the memory sets [3], it was found, that for two kinds of agents, the both coding schemes were *complementary* if the inequality in presenting food to different cells is not too high. This reflects a high symmetry between memory codes. But if the inequality exceeds some threshold value, phase transition occurs and two coding schemes preserve their symmetry *only* if the {A, R} scheme is attributed to the right brain agents and the {S,G} scheme to the left brain agents.

One can give an interesting interpretation of this phenomenon and explain why the {A,R} coding becomes specific for the right brain agents which attract each other, and the {S,G} coding − for the left brain agents which repulse each other. Recall that the agents solve a contradictory problem: they try to hold the both resources. By definition, the left brain agents are those which try to accept the environment proposal i.e. to enhance the first (physical) resource. On the other hand, the right brain agents try to hold their current identity cell in order to save their second (mental) resource.

In this sense they possess built-in automatic strategies which permit them to hold the first and the second resources, correspondingly. By doing so, they, however, risk to lose their *uncontrolled* resource (the second for the left brain agents and the first for the right brain ones). So it will be extremely useful for the agents to *memorize* the cases of loosing or receiving the second (for the left brain agents) and the first (for the right brain ones) resources or *count* them in order to control their quantity.

It is of interest enough, that for the right brain agents this corresponds to the act of *memorizing the cases* where they accepted or rejected the environment proposal (to consume or not to consume food). On the contrary, for the left brain agents it corresponds to the act of *memorizing the cases* where they stayed in the current cell or went to another cell. It just corresponds to the use of the {A,R} code in the memory of the right brain agents and the {S,G} code in the memory of the left brain agents.

4   Memory Structures

The agent-based model [1] has been used as heuristic tool to propose hypothesis on possible memory structures [4]. Specific codes can correspond to the codes of two hemispheres of the single model brain. It is widely recognized that human memory is not unitary but composed of distinct systems localized in different brain structures and, at least, it can have dual character. These dual-memory theories introduced different dichotomies such as a declarative/procedure knowledge, explicit/implicit memories, etc. Now it is clear that the picture is more complicated and memory has a multiple character [15]. The dual theory with a declarative/procedure dichotomy attributed them to the hippocampus of the medial temporal lobe and to the striatum of the basal ganglia, correspondingly [16]. These memories are functionally independent but can interact with each other [17]. Different theories presume that this interaction have rather competitive than cooperative character. In fact, memory mechanisms in the hippocampus and striatum competitively interact with each other at the response level after the process of learning has already occurred [17].

Also different memory systems differ not only in the locus of the competitive inter-action, but in their ability to be modulated by external factors such as distraction. This suggests that the locus of competition, be it at the level of acquisition or response, could influenced by external factors as well [17] Moreover, "...current research is limited in that it does not explain how various social, environmental, and physiological parameters influence the interaction between the two competing memory systems [18] ". The model [1] is able, in principle, to offer a tool that allows this influence to be taken into account. In fact it employs two memory systems – the first uses the code {S,G} and can be treated as the novelty detective ("Go" corresponds to transition to a new niche) and the second one registers accepting and rejection ("Reject") of the proposal. We can accept the hippocampus as the candidate brain structure for implementing the first function. On the other hand, the prefrontal cortex that is involved in the decision-making [19] can be treated as a structure for the recording acception or rejection. Indeed the experimental studies carried by D. Knoch et al. [9] showed that just the magnetic field stimulation of the right dorsolateral prefrontal cortex (DLPFC) suppresses the rejection of an unfair proposal in the ultimatum game. Since the {A,R} code becomes specific for the right brain dominant agent, we can speculate that in certain conditions of high inequality the right DLPFC can be the site for the {A,R} coding [4]. The hypothesis about presentation of memory in the prefrontal cortex (PFC) can be supported by considerations also presented in [19]. Wood and Grafman [19] argued that apart from these other processes performed by PFC it also plays important representational role in brain. These repre-sentations are "memories that are localized in neural networks that encode information and, when activated, enable access to this stored information".

This indicates a possible relation between PFC and memory representations consid-ered in the agent-base model. In [4] it has been also suggested that hippocampus can be considered as a location of memory coded by using the {S,G} code. Though DLPFC

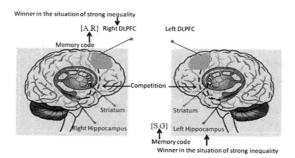

Fig. 1. Two competitive systems in two brain hemispheres are shown. The known hippocampus-striatum (short black interrupted arc) and the hypothetical hippocampus-PFC (longer black interrupted arc). Under high inequality in this hypothetical competitive pair PFC dominates over the hippocampus in the right hemisphere and uses the specific {A,R} code to represent memory records. On the other hand, the hippocampus dominates over PFC in the left hemisphere and uses the {S,G} code to represent memory records.

has no direct connection to the hippocampus it can interact with the last via ventromedial PFC. The model [1] suggests that in the situation of high inequality (unfairness) the {A,R} code be specific for the right brain agents while the {S,G} code – for the left brain ones. Then one can guess that in this situation the right DLPFC becomes a winner in the competition with the right hippocampus, while the left hippocampus dominates over the left DLPFC (Fig. 1). Some analogy of this situation is observed, e.g., in a major depressive disorder characterized by a hypoactivity of the left DLPFC and also by a hyperactivity of the right DLPFC [20].

5   Neural Basis: Role of Dopamine

Is it possible to present some neural basis to the model [1]? Really, we can note that the existence of just two types of agents can be inferred from two cases of asymmetry between dopamine receptors in brain hemispheres. Recently it has been shown by Tomer et al. [21] that relative sensitivity to *reward* and *punishment* reflects lateralization of dopamine signaling. It was found that individual differences in baseline asymmetric binding to D2-like receptors in the putamen and frontal cortex predict greater sensitivity to positive, approach-related versus aversive experiences. Specifically, relatively higher D2 receptor binding in the left hemisphere is associated with preference for rewarding events, whereas stronger tendency to avoid aversive outcomes is predicted by relatively higher binding in the right hemisphere.

It is quite naturally to connect *approach tendency* with the tendency *to change cell* and to consume food and *avoidance tendency* with the decision *to hold previous cell*. Then higher D2 receptor binding in the left hemisphere directly corresponds to the concept of left hemisphere dominance (left brain agent strategy) in the model [1], while higher D2 receptor binding in the right hemisphere corresponds to the concept of right

Table 1.   Different levels of the agent-based model basic features.

Model	Social level	Brain laterality	Neuron level
2 types of agents 2 types of resources 2 rational strategies	Western and Eastern type of behavior	Hemispheric dominance interpretation of Boolean form of rational strategies	Two cases of dominances of D2 dopamine receptors Reward approach – right brain excess of D2 receptors Risk aversion – left brain excess of D2 receptors
Agents interaction: attraction and repulsion Bose and Fermi-like statistics	Egalitarian behavior Competitive behavior Lefebvre ethical systems	Hemispheric dominance	Dopamine level increase leads to egalitarian behavior (from competition to cooperation)
Irrationality description Quantum implication Fractional statistics	Connection of religiosity and irrationality	Modification of right brain dominant strategy	Dopamine level decrease in right hemisphere leads to weakening religiosity

hemisphere dominance (right brain agent strategy). The authors of [21] stress that absolute binding values in each hemisphere were not associated with either measure of motivational bias, and asymmetry in D2 receptor binding was not related to absolute levels of approach or avoidance tendencies but predicted the direction and relative strength of the motivational bias. However dopamine value can be connected with *the type of interaction* between agents – *cooperative or competitive*. In the model [1] an interaction between agents has been considered as collaborative or competitive and by using the correspondence between left brain dominance and 1st ethical system and right brain dominance with 2nd ethical system [5] the character of interaction (repulsion and attraction) has been derived. Neural basis for the choice of the type of interaction can be found in Saez et al. [22]. It was found that pharmacologically manipulated dopamine levels provides causal evidence for the role of dopamine in *egalitarian behavior*. Increased dopamine tone led to decisions that prioritized egalitarian motives. At the neural level, there is substantial evidence suggesting a link between dopamine and social behavior. Also dopaminergic augmentation via catechol-O-methyl transferase (COMT) inhibition increased egalitarian tendencies in participants who played an extended version of the dictator game. Computational modeling of choice behavior shows that tolcapone exerted selective effects on inequity aversion. In particular, it has been found that enhancing dopaminergic tone via COMT inhibition is sufficient to increase inequity-averse behavior. This role of dopamine can be connected with the transition of agent interaction from competition to cooperation. It should been also mentioned that the model [1] has been generalized to quantum domain to simulate irrational (e.g. religious) behavior [23]. This generalization is based on the usage of some quantum analog of implication function and leads to fractional statistics of agent in-cell distributions [23]. Again, it can be related to dopamine level in brain. Really in [24] it is noted that hallucinogenic drugs such as Psilocybin and LSD, which indirectly stimulate dopamine activity in the brain's frontal lobes, can produce religious experience even in the avowedly non-religious. These hallucinogens produce vivid imagery, sometimes along with near psychotic breaks or intense spiritual experience, all tied to stimulation of dopamine receptors on neurons in the limbic system, and in the prefrontal cortex. It has been also hypothesized that when dopamine in the limbic and prefrontal regions of the brain was high, but not too high, it would produce the ability to entertain unusual ideas and associations, leading to heightened creativity, inspired leadership and profound religious experience. When dopamine was too high it would produce mental illness in genetically vulnerable individuals. In those who had been religious before, fanaticism could be the result [24]. What is interesting, laterality plays important role in these phenomena. It has been discovered that for patients with religious leanings prior to getting sick, only a subgroup lost religious fervour after illness set in. These were patients with *'left-onset disease'* – meaning that their muscle problems had begun on the *left side* of the body, correlating with dysfunction in the *right prefrontal regions* of the brain. Those with left-onset disease reported significantly lower scores on all dimensions of religiosity compared to those with right-onset disease. McNamara surmised it was loss of dopamine in the right half of the brain [24]. It is interesting that in quantum generalization of original model [1] aiming to take into account irrational (including religious) behavior [23] the strategy of right brain agent changes.

6 Conclusion

Earlier, the agent-based model proposed in [1] was based on the binary nature of agent resources (model level) leading to two rational agent strategies and two types of agents, correspondingly. These two types of agents have been related to different hemisphere dominance (brain laterality level). The agent interaction rules have been derived using the hypothesis about the relation of two hemisphere dominance with two Lefebvre ethic systems (Western and Eastern) – social level. So, the model has no neural basis. In this presentation we add some neurochemical basis to the model connected with recent finding in asymmetry of dopamine receptors (defines agent type) and also dopamine level (defines character of interactions). We summarize these and other model basic features in the table below. New neurochemical features are presented in the last column. We hope, that agent-based memory model presented in the Sect. 3 and able to take into account the influence of such social parameter as inequality is also receives some neural basis. It can open the way to develop in future neural networks models of social brain using agent-based heuristics (Table 1).

References

1. Ezhov, A.A., Khrennikov, A.Y.: Agents with left and right dominant hemispheres and quantum statistics. Phys. Rev. E **71**(1), 016138, 1–8 (2005)
2. Ezhov, A.A., Khrennikov, A.Y.: On ultrametricity and a symmetry between Bose-Einstein and Fermi-Dirac systems. In: p-Adic Mathematical Physics: Proceedings of AIP Conference 826, pp. 55–64 (2006)
3. Ezhov, A.A., Khrennikov, A.Y., Terentyeva S.S.: Indication of a possible symmetry and its breaking in a many-agent model obeying quantum statistics. Phys. Rev. E **77**(3), 031126, 1–12 (2008)
4. Ezhov, A.A., Terentyeva, S.S.: Agent-based model heuristics in studying memory mechanisms. Psychology **5**, 369–379 (2014)
5. Lefebvre, V.A.: Algebra of Conscience. Kluwer Academic Publisher, Dordrecht (2001)
6. Murtagh, F.: From data to the p-Adic or ultrametric model. p-Adic numbers. Ultrametric Anal. Appl. **1**, 58–68 (2009)
7. Scafetta, N., Picozzi, S., West, B.J.: An out-of-equilibrium model of the distribution of wealth. Quant. Financ. **4**, 353–364 (2004)
8. Thaler, R.H.: Anomalies: the ultimatum game. J. Econ. Perspect. **2**(4), 195–206 (1988)
9. Knoch, D., Gianotti, L.R., Pasquale-Leone, A., Treyer, V., Regard, M., Hohmann, M., Brugger, P.: Disruption of right prefrontal cortex by low-frequency repetitive transcranial magnetic simulation induces risk-taking behavior. J. Neurosci. **26**, 6469–6472 (2006)
10. Tricomi, E., Rangel, A., Camerer, C.F., O'Doherty, J.P.: Neural evidence for inequality-averse social preferences. Nature **463**, 1089–1091 (2010)
11. Easaw, J., McKay, A., Savoia, A.: Inequality, democracy and institutions. World Dev. **38**, 142–154 (2010)
12. Mantegna, R.N., Stanley, H.E.: An Introduction to Econophysics. Cambridge University Press, Cambridge (2000)
13. Rotenberg, V.S., Arshavsky, V.V.: Right and Left Brain Hemispheres Activation in the Representatives of Two Different Cultures. Homeostasis **38**, 49–57 (1997)

14. Lin, L., Osan, R., Tsien, J.: Organizing principles of real-time memory encoding: neural clique assemblies and universal neural codes. Trends Neurosci. **29**(1), 48–57 (2006)
15. Squire, L.R.: Memory systems of the brain: a brief history and current perspective. Neurobiol. Learn. Mem. **82**, 171–177 (2004)
16. Squire, L.R.: Memory and the hippocampus: a synthesis from findings with rats, monkeys, and humans. Psychol. Rev. **99**, 195–231 (1992)
17. Krupa, A.K.: The competitive nature of declarative and nondeclarative memory systems: converging evidence from animal and human brain studies. UCLA Undergraduate Sci. J. **22**, 39–46 (2009)
18. Poldrack, R.A., Packard, M.J.: Competition among multiple memory systems: converging evidence from animal and human brain studies. Neuropsychologia **41**, 245–251 (2003)
19. Wood, J.N., Grafman, J.: Human prefrontal cortex: processing and representational perspectives. Nat. Rev. Neurosci. **4**, 139–147 (2003)
20. Grimm, S., Beck, J., Schuepbach, D., Hell, D., Boesiger, P., Bermpohl, F., Niehaus, L., Boeker, H., Northoff, G.: Imbalance between left and right dorsolateral prefrontal cortex in major depression is linked to negative emotional judgment: an fMRI Study in severe major depressive disorder. Biol. Psychiatry **63**, 369–376 (2008)
21. Tomer, R., Stagter, H.A., Christian, B.T., Fox, A.S., King, C.R., Murali, D., Gluck, M.A., Davidson, R.J.: Love to win or hate to lose? Asymmetry of dopamine D2 receptor binding predicts sensitivity to reward versus punishment. J. Cogn. Neurosci. **26**, 1039–1048 (2014)
22. Saez, I., Zhu, L., Set, E., Kayzer, A., Hsu, M.: Dopamine modulates egalitarian behavior in humans. Curr. Biol. **25**, 912–919 (2015)
23. Ezhov, A.A., Khromov, A.G., Terentyeva, S.S.: On the quantum implication function and strategies for multi-agent models. In: PRIMA2014 International Conference on Principles and Practice of Multi-Agent Systems. Queensland, Australia (2014)
24. McNamara, P.: The God Effect. Religion Spawns Both Benevolent Saints and Murderous Fanatics. Could Dopamine Levels in the Brain Drive That Switch? https://aeon.co/essays/the-dopamine-switch-between-atheist-believer-and-fanatic (2014)

Intelligent Route Choice Model for Passengers' Movement in Subway Stations

Eric Wai Ming Lee[(✉)] and Michelle Ching Wa Li

Department of Architecture and Civil Engineering, City University of Hong Kong,
Kowloon Tong, Hong Kong S.A.R., China
`ericlee@cityu.edu.hk`, `chingwalimic@gmail.com`

Abstract. Current practice of designing subway stations usually based on relevant design guidebooks and experiences of the designers. Improper station design may lead to bottleneck areas which may reduce the efficiency of the passenger flow. In Hong Kong, microscopic pedestrian movement models have been adopted to predict the pedestrian flow patterns inside subway stations. However, the route choice decisions are required to be pre-defined by the designers. In reality, a passenger should make the decision based on the visual information he/ she received. This study collected the actual pedestrian behaviors from subway stations and adopted support vector machine to simulate the decision making on route choice. The results showed that, with 95 % confidence level, the percentage of correct prediction achieved almost 80 %.

Keywords: Route choice · Support vector machine · Subway station

1 Introduction

Hong Kong is a high population city. The major transportation is the mass transport railway (MTR) which is a underground railway system with total 87 numbers subway stations covering most of the major districts in Hong Kong. During rush hours (i.e. 7:00–9:00 and 17:50–19:30), over 90 % capacity of the MTR system is utilized. In subway stations of the system, any bottleneck area may lead to congestion and hence reduces the efficiency of the subway station. In order to investigate the pedestrian flow pattern of a subway station and identify the location of the bottleneck areas, microscopic pedestrian movement models are adopted to simulate the dynamicpedestrian flow pattern. The microscopic models are pedestrian movement model to simulate the movements of every individual pedestrians and the interaction between them and also the obstacles. These computer simulation models usually adopt agent-based social force model [1] simulating pedestrians' movement by Newton's Second Law or cellular automata model [2] to simulate the interaction between pedestrians-pedestrians and pedestrians-obstacles by predefining different rules for different scenarios to guide the pedestrians moving between grids. Currently, the route choice decision of the pedestrian is pre-defined by the user before the simulation. Usually, minimum travelling-time or minimum travelling distance is adopted as the criterion of the decision. These approaches may oversimplify the route choice decision making process.

© Springer International Publishing Switzerland 2016
L. Cheng et al. (Eds.): ISNN 2016, LNCS 9719, pp. 385–392, 2016.
DOI: 10.1007/978-3-319-40663-3_44

Route choice in pedestrian behaviors is a complicated decision process. A pedestrian makes a route choice decision based on the information he/she received. The information includes, for example, the paths of the routes in front of a passenger, the pedestrian traffic condition, time constraint of the passenger, shop attraction, etc.

2 Modeling of Passengers' Decision Making in Subway Station

Making the route choice decision based on the visual information received by a passenger is a human involved behavior which could be highly nonlinear. We investigated the feasibility to adopt intelligent approach to simulate the decision making on the choice of automatic fair collection gate group of a passenger. Each gate group may consist of different number of gates as shown in Fig. 1. The gate groups are situated at different locations of the stations. When a passenger reaches an entrance of a subway station from street, he/she will choose one of the gate groups to enter into the paid area. Figure 2 illustrates the layout plans of the gate groups and the possible routes to approach.

Fig. 1. Example illustrates the two gate groups that passengers should choose for entering into the paid area.

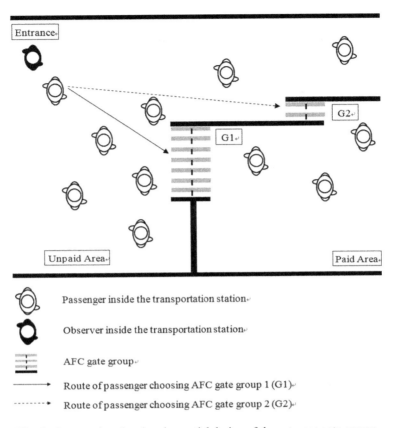

Fig. 2. Layout plan showing the spatial design of the entrance gate groups.

2.1 Problem Identification

The pedestrian movement pattern in a subway station is governed by the route choice decision of the passengers. The choice of gate groups in subway station is one of the decisions in their movement. In the practice of subway station design, the gate groups should be separated apart from each other for serving passengers coming from different entrances.

In this study, we are going to simulate the passenger's decision of choosing either one of the nearest two gate groups in front of the passenger. The nearest two gate groups are denoted here below as G1 and G2. This binary decision is the output of the decision model. The input parameters are the considerations of the passengers when he/she making the decision. With reference to our previous study [3], the following factors were identified in choosing gate groups inside a subway station. They are (i) numbers of gates of the two nearest gate groups; (ii) travelling distances from the entrance of the passenger entering into the station to the two nearest gate groups and (iii) numbers of direction changes along the paths to the two nearest gate groups. In this study, we introduce a new input parameter which is the numbers of passengers waiting in front of the two nearest gate groups. The input and output parameters of the classification model are summarized in Table 1.

Table 1. Input and output parameters of the proposed SVM.

	Parameter
Input	1. Number of gates of G1
	2. Number of gates of G2
	3. Travelling distance from entrance of the passenger to G1
	4. Travelling distance from entrance of the passenger to G2
	5. Number of direction change along the route to G1
	6. Number of direction change along the route to G2
	7. Number of passengers waiting in front of G1
	8. Number of passengers waiting in front of G2
Output	The gate group that the passenger chosen

2.2   Data Collection and Modeling

The methodology of data collection is described as follows. When a passenger enters into a subway station through an entrance, an observer (shown in black color in Fig. 2) will follow the path of the passenger until the passenger passing through the gate group. The eight input parameters stated in were collected by the observer as the sample input. The sample output is the gate group (i.e. either G1 or G2 as shown in Fig. 2) that the passenger chosen to enter into the paid area. In this study, only two gate groups are considered since, in the practice of subway station design, it is seldom to have three of more gate groups located adjacent to each other. Therefore, the route choice decision task in this study is a binary classification.

We adopted the above methodology and collected total 4420 samples from 6 subway stations during rush hours. Support vector machine (SVM) [4] was adopted as the decision model in this study. SVM is a robust classifier. It determines a decision hyperplane in high-dimensional space demarcating different classes of samples such that the spaces between the decision hyperplane and different classes of samples (i.e. margin) are maximized as shown in Eq. (1) where are the training samples, is the radial basis function in form of taken to be the kernel function. The term is the margin to be maximized, is the slack variable and is the regularization parameter. The literature [4] describes the detail development and mathematical formulation of the SVM model.

$$\min_{w,b,\xi} \frac{1}{2} w^T w + C \sum_{i=1}^{l} \xi_i \tag{1}$$

Subject to $y_i(w^T \phi(x_i) + b) \geq 1 - \xi_i, \xi_i > 0$

We implemented the SVM toolbox LIBSVM [5] in the MATLAB platform for this study. The value of the kernel function was taken to be 1/(number of input parameters) (i.e. $1/8 = 0.125$). The regularization parameter was taken to be 1. The model training and performance evaluation process are detailed as follows. Total 1000 numbers of trials with different random data extractions were carried out in this study. In each trial, 10 % of the total samples (i.e. 442 samples) were randomly extracted from the data pool as test samples for evaluating the performance of the trained SVM model which was trained by the other 90 % of the samples (i.e. 3978 samples). The trained model was applied to the test samples to evaluate the percentage of correct classification. Upon the completion of all trials, 1000 numbers of percentages of correct classification were collected.

In order to benchmark the performance of the SVM in this application, multi-layered perceptron (MLP) [6] model is also adopted. Single hidden layer was selected. Sigmoid function (i.e.) was taken to be the activation function of the hidden neurons. The number of hidden neurons was determined by the rule of thumb [7] as shown in Eq. (2) of which is the number of hidden neuron, is the number of training samples, and are, respectively, the numbers of input and output parameters of the model.

$$N_h = \sqrt{N_t} + \frac{N_i + N_0}{2} \tag{2}$$

Similar to the SVM training, 1000 trials were performed. In each trial, the available 4420 samples were randomly divided into training samples (70 %), validation samples (20 %) and testing samples (10 %). The number of hidden neurons was estimated to be 60 (i.e. $(4420 \times 0.7)^{0.5} + (8 + 1)/2 \approx 60$). Early-stop validation approach was adopted to stop the model training if no further reduction in the validation error over 200 epochs. The trained MLP model was applied to the test samples to evaluate the percentage of correct classification which would be compared to that of the SVM model to benchmark their performances. Upon the completion of all trials, 1000 numbers of percentages of correct classification were obtained.

3   Results and Discussion

After the completion of the 1000 trials for each SVM and MLP as described in Sect. 2.2, the results are plotted in the histogram as shown in Fig. 3. The percentage of correct classification of SVM is ranged from 76.1 % to 87.4 % while that of MLP is ranged from 76.1 % to 87.4 %. From the distribution profiles, it shows that the SVM performs better than the MLP. We further analyzed quantitatively the performances of them by statistical approach. Since the valid range of the percentage of correct classification is bound between 0 % and 100 %, we adopted beta distribution to approximate the distribution of the results. By considering the 95 % confidence level as shown in Fig. 4, it shows that the percentage of correct classification of SVM and MLP are 79 % and 77 % respectively. We may statistically conclude that the performance of the SVM model is better than the MLP model in this application.

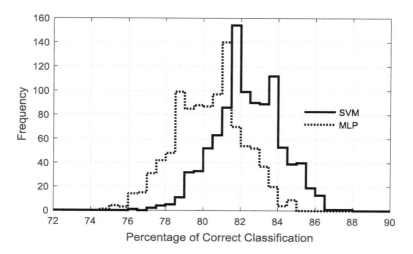

Fig. 3. Histogram summaries the results of the trials from SVM and MLP.

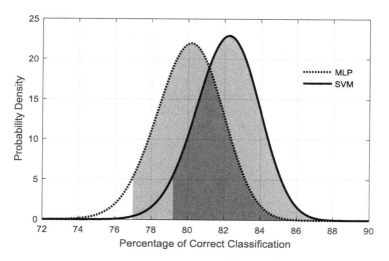

Fig. 4. Considering the one-tailed 95 % confidence level (shaded areas) of the beta distributions, the % of correct classification of SVM and MLP are respectively 79 % and 77 %.

We also compare the result of this study with the similar study [8] carried out by our research team in 2013. It is a pilot study on using intelligent approach to simulate the decision making of passenger in choosing a gate group to approach. That study adopted MLP as the decision was found to be 60 % with 95 % confidence level. In this study, we found that the performance of the MLP model can be dramatically improved from 60 % to 77 % by introduction of the new parameter as one of the model inputs. Since the problem of this study is human-behavior related, we consider the SVM model performed reasonably well and it should be applicable for practical engineering design.

We further analyze the performance of classification of the SVM and MLP by comparing their confusion matrices as shown in Fig. 5. The matrices show that the correct prediction of GG2 is the higher than the correct prediction of GG1. It can be explained by the sample distribution that, among the total 4430 samples, the numbers of samples for GG1 and GG2 are respectively 154 and 289 respectively. The ratio of GG1 samples to GG2 samples is 0.53 (i.e. 154/289 = 0.53). Since the samples for GG2 is more than GG1, the percentage of correct classification for GG2 is reasonably higher than that of GG1 in both MLP and SVM models. The GG1/GG2 ratio of the correct prediction for MLP model is 0.37 (i.e. 95/257 = 0.37) while that of the SVM model is 0.47 (i.e. 119/252 = 0.47). It shows that the ratio of SVM is comparatively more close to the ratio of the GG1/GG2 samples. Also, the percentages of misclassification by SVM model (i.e. 8.3 % and 7.9 %) are more close to each other comparing to the MLP model (i.e. 8.4 % and 12.2 %). We may conclude that the SVM model is able to provide less prejudice result comparing to the MLP model.

MLP Model		Actual gate group		MLP Model		Actual gate group	
		GG1	GG2			GG1	GG2
Predicted gate group	GG1	21.4% (95)	8.4% (37)	Predicted gate group	GG1	21.4% (95)	8.4% (37)
	GG2	12.2% (54)	58.0% (257)		GG2	12.2% (54)	58.0% (257)

Fig. 5. Confusion matrices of the results predicted by the SVM and MLP models.

Currently, different microscopic pedestrian movement models (e.g. EXODUS [9], SIMULEX [10], AnyLogic®, etc.) are available in the market for simulating the pedestrian movement. The route choice decision of some of these pedestrian movement models require user to predefine the probabilities of different routes by which the model distributes the pedestrians to the routes accordingly. Some models simply use shortest distance as the criterion to determine the route choice which may oversimply the decision making process. This study collected actual real data of the route choice behavior from the site surveys inside subway stations. The SVM model successfully built a nonlinear mapping between the visual information and the route choice decision. This approach is considered more reasonable than the traditional approach since it did not make any assumption on the human behavior but captured the route choice behavior directly from the collected data. Therefore, replacing the traditional route choice engine of the pedestrian movement model by this SVM model would provide a more realistic pedestrian movement patterns and also the bottleneck areas.

4   Conclusions

This study adopted SVM to simulate the decision making of a passenger in subway station on choosing a gate group to approach. The percentage of correct classification of the SVM is 79 % which is higher than that of the MLP with 77 %. It was also found that the introduction of the parameter "number of passengers waiting in front of gate group" can dramatically improve the model performance. This study demonstrated the feasibility of adopting the intelligent approach in route choice decision making. We expect that the same technique can also be adopted in different passenger activities in subway stations (e.g. choices of escalator or staircase, choice of gates of a gate group, etc.). More on-site surveys will be carried out to collect data for model training. These are the future works in front of our research team.

Acknowledgments. The work described in this paper was fully supported by a grant from the Research Grant Council of the Hong Kong Special Administrative Region [Project No. City U 11206714].

References

1. Helbing, D., Farkas, I., Vicsek, T.: Simulating dynamical features of escape panic. Nature **407**, 487–490 (2000)
2. Burstedde, C., Klauck, K., Schadschneider, A., Zittartz, J.: Simulation of pedestrian dynamics using a two-dimensional cellular automaton. Physics A **297**, 507–525 (2001)
3. Yuen, J.K.K., Lee, E.W.M., Lo, S.M., Yuen, R.K.K.: An intelligence-based optimization model of passenger flow in a transportation station. IEEE Trans. Intell. Transp. Syst. **14**, 1290–1300 (2013)
4. Cortes, C., Vapnik, V.: Support-vector network. Mach. Learn. **20**, 1–25 (1995)
5. Chang, C.C., Lin, C.J.: LIBSVM: a library for support vector machines. ACM Trans. Intell. Syst. Technol. **2**, 1–27 (2011)
6. Rosenblatt, F.: Principles of Neurodynamics. Spartan Books, New York (1962)
7. Ward: Neuroshell 2 Manual, Ward Systems Group Inc., Frederick, MA (2008)
8. Yuen, J.K.K., Lee, E.W.M., Lo, S.M., Yuen, R.K.K.: An intelligence-based optimization model of passenger flow in a transportation station. IEEE Trans. Intell. Transp. Syst. **14**, 1290–1300 (2013)
9. Galea, E.R., Galparsoro, J.M.P.: A computer-based simulation-model for the prediction of evacuation from mass-transport vehicles. Fire Saf. J. **22**, 341–366 (1994)
10. Thompson, P.A., Marchant, E.W.: A computer-model for the evacuation of large building populations. Fire Saf. J. **24**, 131–148 (1995)

A Gaussian Kernel-based Clustering Algorithm with Automatic Hyper-parameters Computation

Francisco de A.T. de Carvalho[1]($\boxtimes$), Marcelo R.P. Ferreira[2],
and Eduardo C. Simões[1]

[1] Centro de Informatica, Universidade Federal de Pernambuco,
Avenue Jornalista Anibal Fernandes s/n - Cidade Universitaria,
Recife, PE 50740-560, Brazil
{fatc,ecs4}@cin.ufpe.br
[2] Departamento de Estatistica, Centro de Ciências Exatas e da Natureza,
Universidade Federal da Paraiba, Joo Pessoa, PB 58051-900, Brazil
marcelo@de.ufpb.br

Abstract. The clustering performance of the conventional gaussian kernel based clustering algorithms are very dependent on the estimation of the width hyper-parameter of the gaussian kernel function. Usually this parameter is estimated once and for all. This paper presents a gaussian c-Means with kernelization of the metric which depends on a vector of width hyper-parameters, one for each variable, that are computed automatically. Experiments with data sets of the UCI machine learning repository corroborate the usefulness of the proposed algorithm.

Keywords: Kernel clustering · Kernelization of the metric · Width hyper-parameter

1 Introduction

Clustering is a useful tools to explore data structures and has been widely applied in various areas, including taxonomy, image processing, data mining, and information retrieval. Clustering means the task of organizing a set of patterns into clusters such that patterns within a given cluster have a high degree of similarity, whereas patterns belonging to different clusters have a high degree of dissimilarity [1].

Distance measures are an important component of a clustering algorithm and the Euclidean distance is most commonly used in partitioning clustering algorithms. However, when the data structure is complex (i.e., clusters with non-hyper-spherical shapes and/or linearly non-separable patterns), these algorithms may have poor performance. Kernel-based clustering algorithms have been proposed to tackle these limitations.

Two main approaches have guided the development of kernel-based clustering algorithms: kernelization of the metric, where the cluster centroids are obtained in the original space and the distances between patterns and cluster centroids are

© Springer International Publishing Switzerland 2016
L. Cheng et al. (Eds.): ISNN 2016, LNCS 9719, pp. 393–400, 2016.
DOI: 10.1007/978-3-319-40663-3_45

computed by means of kernels; and clustering in feature space, in which cluster centroids are obtained in the feature space [2,3].

In kernel-based clustering algorithms it is possible to compute Euclidean distances by using kernel functions and the so-called distance kernel trick [3]. The most popular kernel function in applications is the Gaussian kernel. However, the clustering performance of the conventional Gaussian kernel-based clustering algorithms are very dependent on the estimation of the width hyper-parameter of the Gaussian kernel function. This parameter is estimated once and for all.

This paper aims to present a Gaussian c-Means with kernelization of the metric which depends on a vector of width hyper-parameters, one for each variable, that are computed automatically.

The paper is organized as follows. Section 2 recalls the conventional kernel c-means algorithm with kernelization of the metric. Section 3 presents the Gaussian c-Means with kernelization of the metric with automatic computation of width hyper-parameters. In Sect. 4, experiments with data sets of the UCI machine learning repository corroborate the usefulness of the proposed algorithm. Section 5 gives the final remarks of the paper.

2   Kernel C-means with Kernelization of the Metric

This section briefly recalls the basic concepts about kernel functions and the conventional kernel c-means algorithm with kernelization of the metric. Let $E = \{e_1, \ldots, e_n\}$ be a set of n objects described by p real-valued variables. Let $\mathcal{D} = \{\mathbf{x}_1, \ldots, \mathbf{x}_n\}$ be a non-empty set where for $k = 1, \ldots, n$, the kth object e_k is represented by a vector $\mathbf{x}_k = (x_{k1}, \ldots, x_{kp}) \in \mathbb{R}^p$. A function $\mathcal{K} : \mathcal{D} \times \mathcal{D} \to \mathbb{R}$ is called a positive definite Kernel (or Mercer kernel) if, and only if $\mathcal{K}$ is symmetric (i.e., $\mathcal{K}(\mathbf{x}_k, \mathbf{x}_l) = \mathcal{K}(\mathbf{x}_l, \mathbf{x}_k)$) and if the following inequality holds [4]:

$$\sum_{l=1}^{n} \sum_{k=1}^{n} c_l c_k \mathcal{K}(\mathbf{x}_l, \mathbf{x}_k) \geq 0, \forall n \geq 2 \tag{1}$$

where $c_r \in \mathbb{R}, r = 1, \ldots, n$.

Let $\Phi : \mathcal{D} \to \mathcal{F}$ be a nonlinear mapping from the input space $\mathcal{D}$ to a high dimensional feature space $\mathcal{F}$. By applying the mapping Φ, the inner product $\mathbf{x}_l^T \mathbf{x}_k$ in the input space is mapped to $\Phi(\mathbf{x}_l)^T \Phi(\mathbf{x}_k)$ in the feature space. The basic notion in the kernel approaches is that the non-linear mapping Φ does not need to be explicitly specified because each Mercer kernel can be expressed as $\mathcal{K}(\mathbf{x}_l, \mathbf{x}_k) = \Phi(\mathbf{x}_l)^T \Phi(\mathbf{x}_k)$ [5].

One the most relevant implications is that it is possible to compute Euclidean distances in $\mathcal{F}$ without knowing explicitly Φ, by using the so-called distance kernel trick [3]:

$$\begin{aligned}
||\Phi(\mathbf{x}_l) - \Phi(\mathbf{x}_k)|| &= (\Phi(\mathbf{x}_l) - \Phi(\mathbf{x}_k))^T (\Phi(\mathbf{x}_l) - \Phi(\mathbf{x}_k)) \\
&= \Phi(\mathbf{x}_l)^T \Phi(\mathbf{x}_l) - 2\Phi(\mathbf{x}_l)^T \Phi(\mathbf{x}_k) + \Phi(\mathbf{x}_k)^T \Phi(\mathbf{x}_k) \\
&= \mathcal{K}(\mathbf{x}_l, \mathbf{x}_l) - 2\mathcal{K}(\mathbf{x}_l, \mathbf{x}_k) + \mathcal{K}(\mathbf{x}_k, \mathbf{x}_k)
\end{aligned}$$

2.1   Kernel C-means with Kernelization of the Metric

The kernel c-means with kernelization of the metric (hereafter named KCM-K) provides a partition $\mathcal{P} = \{P_1, \ldots, P_c\}$ of E into c clusters, and a matrix of representatives (called hereafter a prototypes) $\mathbf{G} = (\mathbf{g}_1, \ldots, \mathbf{g}_c)$ of the clusters in the partition $\mathcal{P}$. The prototype of cluster P_i $(i = 1, \ldots, c)$ is represented by the vector $\mathbf{g}_i = (g_{i1}, \ldots, g_{ip}) \in \mathbb{R}^p$.

From an initial solution, the matrix of prototypes $\mathbf{G}$ and the partition $\mathcal{P}$ are obtained iteratively in two steps (representation and allocation) by the minimization of a suitable objective function, here-below denoted as J_{KCM-K}, that gives the total homogeneity of the partition computed as the sum of the homogeneity in each cluster:

$$J_{KCM-K}(\mathbf{G}, \mathbf{U}) = \sum_{i=1}^{c} \sum_{e_k \in P_i} ||\Phi(\mathbf{x}_k) - \Phi(\mathbf{g}_i)||^2 \tag{2}$$

where, by using the so-called distance kernel trick [3], $||\Phi(\mathbf{x}_k) - \Phi(\mathbf{g}_i)||^2 = \mathcal{K}(\mathbf{x}_k, \mathbf{x}_k) - 2\mathcal{K}(\mathbf{x}_k, \mathbf{g}_i) + \mathcal{K}(\mathbf{g}_i, \mathbf{g}_i)$.

Hereafter we consider the Gaussian kernel, the most commonly used in the literature: $\mathcal{K}(\mathbf{x}_l, \mathbf{x}_k) = \exp\left\{-\frac{||\mathbf{x}_l - \mathbf{x}_k||^2}{2\sigma^2}\right\} = \exp\left\{-\frac{1}{2}\sum_{j=1}^{p}\frac{1}{\sigma^2}(x_{lj} - x_{kj})^2\right\}.$, where σ^2 is the width hyper-parameter of the Gaussian.

Then, $\mathcal{K}(\mathbf{x}_k, \mathbf{x}_k) = 1, \forall k$, $\mathcal{K}(\mathbf{g}_i, \mathbf{g}_i) = 1, \forall i$, and $||\Phi(\mathbf{x}_k) - \Phi(\mathbf{g}_i)||^2 = 2 - \mathcal{K}(\mathbf{x}_k, \mathbf{g}_i)$ and thus, the objective function J_{KMC-K} becomes:

$$J_{KCM-K}(\mathbf{G}, \mathbf{U}) = 2\sum_{i=1}^{c} \sum_{e_k \in P_i} (1 - \mathcal{K}(\mathbf{x}_k, \mathbf{g}_i)) \tag{3}$$

During the representation step, the partition $\mathcal{P}$ is kept fixed. The objective function J_{KMC-K} is optimized with respect to the prototypes. Thus, from $\frac{\partial J_{KMC-K}}{\partial \mathbf{g}_i} = 0$ and after some algebra, the cluster prototypes are obtained as follows:

$$\mathbf{g}_i = \frac{\sum_{e_k \in P_i} \mathcal{K}(\mathbf{x}_k, \mathbf{g}_i)\mathbf{x}_k}{\sum_{e_k \in P_i} \mathcal{K}(\mathbf{x}_k, \mathbf{g}_i)}, \ i = 1, \ldots, c \tag{4}$$

In the allocation step, the cluster prototypes are kept fixed. The clusters P_i $(i = 1, \ldots, c)$, that minimizes the clustering criterion given in Eq. 2, are updated according the following allocation rule:

$$P_i = \{e_k \in E : 1 - \mathcal{K}(\mathbf{x}_k, \mathbf{g}_i) \leq 1 - \mathcal{K}(\mathbf{x}_k, \mathbf{g}_h), \forall h \neq i, h = 1, \ldots, c\} \tag{5}$$

These two steps are repeated until the convergence of KCM-K.

3   KCM-K with Automatic Computation of Width Hyper-Parameters

The kernel c-means with kernelization of the metric and automatic computation of width hyper-parameters (hereafter named KCM-K-H) provides a partition

$\mathcal{P} = \{P_1, \ldots, P_c\}$ of E into c clusters, a vector of width hyper-parameters (one for each variable) $\mathbf{s} = (s_1^2, \ldots, s_p^2)$ and a matrix of prototypes $\mathbf{G} = (\mathbf{g}_1, \ldots, \mathbf{g}_c)$ of the clusters in the partition $\mathcal{P}$.

From an initial solution, the matrix of prototypes $\mathbf{G}$, the vector of hyper-parameters $\mathbf{s}$ and the partition $\mathcal{P}$ are obtained iteratively in three steps (representation, computation of the width hyper-parameters and allocation) by the minimization of a suitable objective function, here-below denoted as $J_{KCM-K-H}$, that gives the total homogeneity of the partition computed as the sum of the homogeneity in each cluster:

$$J_{KCM-K}(\mathbf{G}, \mathbf{s}, \mathbf{U}) = \sum_{i=1}^{c} \sum_{e_k \in P_i} ||\Phi(\mathbf{x}_k) - \Phi(\mathbf{g}_i)||^2 \tag{6}$$

where

$$||\Phi(\mathbf{x}_k) - \Phi(\mathbf{g}_i)||^2 = \mathcal{K}^{(\mathbf{s})}(\mathbf{x}_k, \mathbf{x}_k) - 2\mathcal{K}^{(\mathbf{s})}(\mathbf{x}_k, \mathbf{g}_i) + \mathcal{K}^{(\mathbf{s})}(\mathbf{g}_i, \mathbf{g}_i) \tag{7}$$

with

$$\mathcal{K}^{(\mathbf{s})}(\mathbf{x}_l, \mathbf{x}_k) = \exp\left\{ -\frac{1}{2} \sum_{j=1}^{p} \frac{1}{s_j^2}(x_{lj} - x_{kj})2 \right\}$$

Because $\mathcal{K}^{(\mathbf{s})}(\mathbf{x}_k, \mathbf{x}_k) = 1, \forall k$, $\mathcal{K}^{(\mathbf{s})}(\mathbf{g}_i, \mathbf{g}_i) = 1, \forall i$, and $||\Phi(\mathbf{x}_k) - \Phi(\mathbf{g}_i)||^2 = 2 - \mathcal{K}^{(\mathbf{s})}(\mathbf{x}_k, \mathbf{g}_i)$, the objective function $J_{KMC-K-H}$ becomes:

$$J_{KCM-K-H}(\mathbf{G}, \mathbf{s}, \mathbf{U}) = 2\sum_{i=1}^{c} \sum_{e_k \in P_i} (1 - \mathcal{K}^{(\mathbf{s})}(\mathbf{x}_k, \mathbf{g}_i)) \tag{8}$$

During the representation step, the vector of width hyper-parameters $\mathbf{s}$ and the partition $\mathcal{P}$ are kept fixed. The objective function $J_{KMC-K-H}$ is optimized with respect to the feature weights. Thus, from $\frac{\partial J_{KMC-K-H}}{\partial \mathbf{g}_i} = 0$ and after some algebra, the cluster prototypes are obtained as follows:

$$\mathbf{g}_i = \frac{\sum_{e_k \in P_i} \mathcal{K}^{(\mathbf{s})}(\mathbf{x}_k, \mathbf{g}_i)\mathbf{x}_k}{\sum_{e_k \in P_i} \mathcal{K}^{(\mathbf{s})}(\mathbf{x}_k, \mathbf{g}_i)}, \; i = 1, \ldots, c \tag{9}$$

In the computation of the width hyper-parameters step, the matrix of prototypes $\mathbf{G}$ and the partition $\mathcal{P}$ are kept fixed. First, we use the method of Lagrange multipliers with the restriction that $\prod_{j=1}^{p} \left(\frac{1}{s_j^2} \right) = \gamma$, where γ is a suitable parameter, and obtain

$$L_{KCM-K-H}(\mathbf{G}, \mathbf{s}, \mathbf{U}) = 2\sum_{i=1}^{c} \sum_{e_k \in P_i} (1 - \mathcal{K}^{(\mathbf{s})}(\mathbf{x}_k, \mathbf{g}_i)) - \omega \left(\prod_{j=1}^{p} \frac{1}{s_j^2} - \gamma \right) \tag{10}$$

Then, we compute the partial derivatives of $L_{KCM-K-H}$ w.r.t $\frac{1}{s_j^2}$ and ω, and by setting the partial derivatives to zero, and after some algebra we obtain

$$\frac{1}{s_j^2} = \frac{\gamma^{\frac{1}{p}} \left\{ \prod_{h=1}^{p} \sum_{i=1}^{c} \sum_{k=1}^{n} \mathcal{K}^{(\mathbf{s})}(\mathbf{x}_k, \mathbf{g}_i)(x_{kh} - g_{ih})^2 \right\}}{\sum_{i=1}^{c} \sum_{k=1}^{n} \mathcal{K}^{(\mathbf{s})}(\mathbf{x}_k, \mathbf{g}_i)(x_{kj} - g_{ij})^2}, \; j = 1, \ldots, p \tag{11}$$

In the allocation step, the cluster prototypes and the width hyper-parameters are kept fixed. The clusters P_i $(i = 1, \ldots, c)$, that minimizes the clustering criterion given in Eq. 2, are updated according the following allocation rule:

$$P_i = \left\{ e_k \in E : 1 - \mathcal{K}^{(\mathbf{s})}(\mathbf{x}_k, \mathbf{g}_i) \leq 1 - \mathcal{K}^{(\mathbf{s})}(\mathbf{x}_k, \mathbf{g}_h), \forall h \neq i, h = 1, \ldots, c \right\} \quad (12)$$

These three steps are repeated until the convergence of KCM-K-H. The K-means algorithm can be viewed as an Expectation-Maximization (EM) algorithm and it is convergent because each EM algorithm is convergent [2]. As the KCM-K-H algorithm is a modified version of the classical K-means algorithm, its convergence can be proved.

4 Empirical Results

This section discusses the performance and the usefulness of the proposed algorithm in comparison with the KCM-K algorithm. All experiments ware conducted on the same machine (OS: Windows 7 Professional 64-bits, Memory: 16 GiB, Processor: Intel Core i7-X990 CPU @ 3.47 GHz).

Thirteen datasets from the UCI Machine learning Repository [6], namely, Breast tissue, Ecoli, Image segmentation, Iris plants, Isolet, Leaf, Libras Movement, Multiple features, Seeds, Thyroid gland, Urban land cover, Breast cancer wisconsin (diagnostic), and Wine, with different number of objects, variables and a priori classes, were considered in this study. Table 1 (in which n is the number of objects, p is the number of real-valued variables and K is the number of a priori classes) summarizes these data sets.

Table 1. Summary of the data sets.

Data sets	n	p	K	Data sets	n	p	K
Breast tissue	106	9	6	Multiple features	2000	649	10
Ecoli	336	7	8	Seeds	210	7	3
Image segmentation	2100	19	7	Thyroid gland	215	5	3
Iris	150	4	3	Urban land cover	675	148	9
Isolet	6238	617	26	Brest cancer winsconsin	569	30	2
Leaf	310	14	36	Wine	178	13	3
Libras Movement	360	90	15				

KCM-K and KCM-K-H were executed on these data sets 100 times. In each execution, they were run also 100 times and the best results, according to the respective objective functions, were selected.

The parameter *gamma* of the KCM-K-H algorithm was set as $\gamma = (\sigma^2)^p$, where σ is the optimal width hyper-parameter used in the conventional KCM-K algorithm that is estimated as the average of the 0.1 and 0.9 quantiles of $||\mathbf{x}_l - \mathbf{x}_k||^2$, $l \neq k$ [7].

To compare the quality of the partitions provided by these algorithms, the adjusted Rand index (ARI) [8], and the mutual normalized information (MNI) [9] were considered.

Table 2 shows the average and standard deviation (in parenthesis) of the ARI and NMI indices for the 100 executions of the KCM-K and KCM-K-H algorithms. They were computed to the partitions provided by these algorithms on the data sets of Table 1, in comparison with the a priori partitions. It can be observed that whatever the considered index, the KCM-K-H algorithm outperforms the KCM-K algorithm on the majority of the data sets. Moreover, the application of the Welch's t-test shows that, for all the indices considered, the KCM-K-H algorithm is significantly better than the KCM-K algorithm (with $\alpha = 0.01$) in eight out of the thirteen data sets considered int this study.

Table 2. Performance of the algorithms.

Data sets	ARI		NMI	
	KCM-K	KCM-K-H	KCM-K	KCM-K-H
Breast tissue	0.160 (0.037)	0.338 (0.030)	0.314 (0.033)	0.545 (0.026)
Ecoli	0.419 (0.011)	0.386 (0.066)	0.611 (0.009)	0.562 (0.031)
Image segmentation	0.374 (0.007)	0.475 (0.049)	0.501 (0.006)	0.629 (0.026)
Iris	0.742 (0.004)	0.734 (0.172)	0.765 (0.002)	0.785 (0.096)
Isolet	0.496 (0.027)	0.475 (0.020)	0.725 (0.009)	0.715 (0.010)
Leaf	0.303 (0.018)	0.380 (0.022)	0.667 (0.009)	0.713 (0.009)
Libras Movement	0.305 (0.013)	0.271 (0.027)	0.588 (0.012)	0.555 (0.025)
Multiple features	0.482 (0.015)	0.528 (0.078)	0.615 (0.009)	0.657 (0.056)
Seeds	0.705 (0.000)	0.679 (0.002)	0.678 (0.001)	0.681 (0.002)
Thyroid gland	0.164 (0.003)	0.328 (0.239)	0.219 (0.001)	0.377 (0.187)
Urban land cover	0.088 (0.010)	0.213 (0.162)	0.169 (0.011)	0.314 (0.236)
Brest cancer winsconsin	0.069 (0.067)	0.733 (0.003)	0.128 (0.105)	0.645 (0.002)
Wine	0.371 (0.000)	0.833 (0.046)	0.428 (0.000)	0.813 (0.035)

Table 3 shows the average and standard-deviation (in parenthesis) of the execution time of the KCM-K and KCM-K-H algorithms on the data sets of Table 1, as well as the ration between the average execution time of KCM-K-H algorithm and KCM-K algorithms. It can be observed that this ration was at most two on the majority of the data sets considered in this study.

Table 4 shows the best results (according to the respective objective functions) of the KCM-K and KCM-K-H algorithms on the data sets of Table 1, according to the ARI and NMI indices. It can be observed that whatever the considered index, the KCM-K-H algorithm outperforms the KCM-K algorithm in ten out of the thirteen the data sets of this this study. Moreover, the application of the Wilkoxon signed-ranks test shows that, for all the indices considered, the KCM-K-H algorithm is significantly better than the KCM-K algorithm (with $\alpha = 0.01$) on these data sets.

Table 3. Average and standard-deviation of the execution time of the algorithms.

Data sets	Time (min)		Average ratio
	KCM-K	KCM-K-H	
Breast tissue	0.54 (0.29)	0.98 (0.54)	1.81
Ecoli	3.50 (2.04)	4.00 (2.27)	1.14
Image segmentation	173.74 (102.06)	735.49 (418.03)	4.23
Iris	0.33 (0.22)	0.35 (0.19)	1.06
Isolet	160089.0 (90386.6)	289138.8 (161862.5)	1.80
Leaf	10.31 (6.03)	18.06 (10.57)	1.75
Libras Movement	182.67 (108.64)	276.21 (158.23)	1.51
Multiple features	68662.0 (38010.3)	107343.6 (60659.6)	1.56
Seeds	1.33 (0.73)	2.49 (1.22)	1.87
Thyroid gland	0.55 (0.32)	1.54 (0.94)	2.8
Urban land cover	880.84 (500.30)	1246.39 (703.75)	1.41
Brest cancer winsconsin	15.40 (8.93)	28.86 (17.03)	1.87
Wine	1.62 (0.90)	2.18 (1.20)	1.34

Table 4. Performance of the algorithms.

Data sets	ARI		NMI	
	KCM-K	KCM-K-H	KCM-K	KCM-K-H
Breast tissue	0.1501	0.2944	0.3066	0.5515
Ecoli	0.4215	0.4105	0.6171	0.6049
Image segmentation	0.3774	0.4221	0.5070	0.6083
Iris	0.7436	0.8856	0.7660	0.8641
Isolet	0.5221	0.4756	0.7302	0.7234
Leaf	0.2934	0.3532	0.6668	0.7120
Libras Movement	0.3097	0.3236	0.5959	0.6101
Multiple features	0.4900	0.6759	0.6221	0.7512
Seeds	0.7056	0.6789	0.6792	0.6804
Thyroid gland	0.1652	0.6050	0.2194	0.5374
Urban land cover	0.0958	0.3560	0.1815	0.5106
Brest cancer winsconsin	0.0066	0.7297	0.0348	0.6416
Wine	0.3711	0.8348	0.4287	0.8215

5 Final Remarks and Conclusions

The clustering performance of the conventional KCM-K, the gaussian kernel-based clustering algorithm is very dependent on the estimation of the width hyper-parameter of the Gaussian kernel function, that is estimated once and for all. In this paper we proposed KCM-K-H, a Gaussian c-Means with kernelization of the metric and automatic computation of a vector of width hype-parameters, one for each variable.

Experiments with thirteen data sets from UCI machine learning repository, with different number of objects, variables and a priori classes, showed the performance of the proposed algorithm. Based on the computation of the adjusted Rand index and normalized mutual information index, we can conclude that in the majority of the data sets considered in this study, the proposed KCM-K-H algorithm is significantly better then the conventional KCM-K algorithm.

Acknowledgments. The authors are grateful to the anonymous referees for their careful revision, and CNPq and FACEPE (Brazilian agencies) for their financial support.

References

1. Jain, A.K.: Data clustering: 50 years beyond k-means. Pattern Recogn. Lett. **31**, 651–666 (2010)
2. Camastra, F., Verri, A.: A novel Kernel method for clustering. IEEE Trans. Neural Netw. **27**, 801–804 (2005)
3. Filippone, M., Camastra, F., Masulli, F., Rovetta, S.: A survey of Kernel and spectral methods for clustering. Pattern Recogn. **41**, 176–190 (2008)
4. Mercer, J.: Functions of positive and negative type, and their connection with the theory of integral equations. Philos. Trans. R. Soc. London Series A Containing Papers Math. or Phys. Charact. **209**, 415–446 (1909)
5. Mueller, K.R., Mika, S., Raetsch, G., Tsuda, K., Schoelkopf, B.: An Introduction to Kernel-based learning algorithms. IEEE Trans. Neural Netw. **12**, 181–202 (2001)
6. Blake, C.L., Merz, C.J.: UCI Repository of Machine Learning databases, Irvine, CA, University of California, Department of Information and Computer Science (1998). http://www.ics.uci.edu/mlearn/MLRepository.html
7. Caputo, B., Sim, K., Furesjo, F., Smola, A.: Appearence-based object recognition using SVMS: which Kernel should i use?. In: Proceedings of NIPS Workshop on Statistical Methods for Computational Experiments in Visual Processing and Computer Vision (2002)
8. Hubert, L., Arabie, P.: Comparing partitions. J. Classif. **2**, 193–218 (1985)
9. Manning, C.D., Raghavan, P., Schuetze, H.: Introduction to Information Retrieval. Cambridge University Press, Cambridge (2008)

Network Intrusion Detection with Bat Algorithm for Synchronization of Feature Selection and Support Vector Machines

Chunying Cheng[1(✉)], Lanying Bao[2], and Chunhua Bao[1]

[1] College of Computer Science and Technology,
Inner Mongolia University for Nationalities, Tongliao 028043, China
chengchunying_80@163.com
[2] College of Mathematics, Inner Mongolia University for Nationalities,
Tongliao 028043, China

Abstract. In order to improve the detection rate of network intrusion, this paper proposes a kind of bat algorithm (BA), which can optimize the intrusion detection model of support vector machine (BA-SVM). In this algorithm, parameters of the SVM support vector machine are coded as individual bats first, and the detection rate of network intrusion is put as the parameter objective function. Then, the optimum parameter of support vector machine is found by simulating the bat flight. Finally, a network intrusion detection model is established based on optimal parameters, and simulation experiments are performed with KDD CUP99 dataset. The results show that this model could not only improve the detection rate of network intrusion, but also reduce the training time, and therefore improve the effect of network intrusion detection.

Keywords: Bat algorithm · Network intrusion · Feature selection · Support vector machine

1 Introduction

With the growing size and openness of the network, the number of cyber attacks and the degree of damage showed an upward trend. The traditional security measures can not satisfy the requirements of modern network security. As an active network security measure, network intrusion detection system has become a research focus in the field of network security [1].

Essentially, network intrusion detection is a multi-classification problem and mainly includes key procedures such as feature selection, classification parameters optimization etc. In practical applications, in order to solve SVM parameter optimization problems better, researchers have proposed genetic algorithms [2], PSO [3], AFSA [4] and other parameters to perform optimization of SVM parameters. In order to solve the conundrum of classification parameters and features set selection in network intrusion, this paper proposes a network intrusion detection method of synchronization of feature selection and SVM parameters based on bat algorithm (BA-SVM) and verify the effectiveness of BA-SVM with KDD CUP 99 dataset.

© Springer International Publishing Switzerland 2016
L. Cheng et al. (Eds.): ISNN 2016, LNCS 9719, pp. 401–408, 2016.
DOI: 10.1007/978-3-319-40663-3_46

2 Bat Algorithm

Bat algorithm (BA) is a new type of stochastic optimization algorithm for the optimal solution of swarm intelligence search proposed by Yang from University of Cambridge in 2012 [5]. It is simple and easy to implement this algorithm, and it has become a hotspot of heuristics in recent years. Bat algorithm is a kind of optimization technique based on iteration. In the algorithm, all solutions are initialized to a group of random ones, and then the optimal solution is found through iterative search. And part of the new solution is generated around the optimal solution with randomly flight, which can enhance the local search. Compared with other algorithms, bat algorithm is superior to them in terms of accuracy and validity.

Bats can explore for preys and avoid obstacles with its echolocation function, and they can find their habitats in the darkness. These bats emit loud sound pulses, and then listen to echoes bounced back from objects around. In bat algorithm, let's assume x_i is the position of a bat, and it flies arbitrarily at a speed of u_i, searching targets with a frequency of f_i, a variable wavelength of λ, and a responsiveness of A_o.

When the bat is searching for preys, it is using a pulse frequency of:

$$f_i = f_{\min} + (f_{\max} - f_{\min}) \times rand \tag{1}$$

The update formula of flying velocity of bats in search of preys is:

$$u_{id}^t = u_{id}^{t-1} + (x_{id}^t - x_d^*) \times f_i \tag{2}$$

The update formula of the location of bats in search of preys is:

$$x_i^t = x_i^{t-1} + u_i^t \tag{3}$$

In the formula, f_i is the pulse frequency used by the i th bat in search of preys, $[f_{\min}, f_{\max}]$ is the searching range of pulse frequency, and $rand \in [0, 1]$ is a random factor uniformly distributed. x^* is the position of current global optimal solution, it is the best location after searching and comparing positions of all the bats. For local search, once a solution is selected among current optimal solutions, new local solutions generated by random walk of each bat are shown as formula (4):

$$x_{new} = x_{old} + \varepsilon A^t \tag{4}$$

In the formula, $\varepsilon \in [-1, 1]$ is a random number, and $A^t = A_i^t$ is the average response of all bats in the same time period. When a bat finds its prey, the volume will be reduced, while the pulse rate will increase. The volume will change at any convenient value, and the responsiveness A_i and pulse rate r_i will be updated according to formula (5).

$$A_i^{t+1} = \alpha A_i^t, \quad r_i^{t+1} = r_i^0[1 - \exp(-\gamma t)] \tag{5}$$

In the formula, α and γ are constants, r_i^0 indicates the maximum pulse frequency of the i th bat, r_i^{t+1} indicates pulse frequency of the i th bat at $t+1$ time, γ is the increase factor of pulse frequency, $\gamma > 0$, A_i^t indicates the sound intensity when the i th bat generate pulses at t time, $0 \le \alpha \le 1$ is the attenuation coefficients of pulse sound intensity.

3   SVM Algorithm

Vapnik et al. proposes SVM algorithm. According to the VC dimension theory and structural risk minimization principle, this algorithm studies the limited sample information, finds the best compromise point in complexity and ability of learn model, and get the best generalization. Let's assume the training set is $\{x_i, y_i\}$, and the hyperplane equation for the SVM is:

$$wX + b = 0 \tag{6}$$

The corresponding SVM classification decision function is:

$$f(x) = \text{sgn}(wX + b) \tag{7}$$

For linear classification problems, the risk of experience can be zero. Based on the principle of structural risk minimization, the hyperplane problem for maximum classification interval can be transformed into:

$$\min \frac{1}{2} w \cdot w + C \sum_{i=1}^{n} \xi_i \tag{8}$$
$$s.t. \quad y_i (w \cdot x_i + b) \ge 1 - \xi_i \quad \xi_i \ge 0 \quad i = 1, 2, \cdots, n$$

By introducing Lagrange multiplier, Eq. (8) is transformed into the dual form:

$$\min \frac{1}{2} \sum_{i,j=1}^{n} a_i a_j y_i y_j (\varphi(x_i) \cdot \varphi(x_j)) + \sum_{i=1}^{n} a_i \tag{9}$$

In the formula, the point corresponding to $a_i > 0$ is called support vector. Thus, the optimal hyperplane decision function of support vector machine can be:

$$f(x) = \text{sgn}[\sum_{i=1}^{n} a_i y_i (X_i \cdot X) + b] \tag{10}$$

For nonlinear problems, by introducing the kernel function φ, the input vector is transformed into a high dimensional feature space and summarized for linear mapping, namely:

$$\phi(x) = [\varphi_1(X), \varphi_2(X), \cdots, \varphi_n(X)]^T \tag{11}$$

According to Hilbert-Schmidt theory, as long as it satisfies Mercer conditions, then it may correspond to the inner product of a transformation space, namely:

$$K(X_i, X_j) = \phi^T(X_i)\phi(X_j) \tag{12}$$

If the appropriate kernel function is selected, the decision function of SVM for nonlinear classification problems can be expressed as:

$$y = \text{sgn}[\sum_{i=1}^{n} a_i y_i k(X_i \cdot X) + b] \tag{13}$$

In SVM, kernel functions have an important influence on the algorithm. Among many kernel functions, RBF kernel functions have fewer parameters to be optimized, and have good analysis ability for high-dimensional data. Therefore RBF kernel function is selected as the kernel function for support vector machine.

4 Network Intrusion Feature Selection and SVM Parameters for BA

Feature selection will aim at a set of feature data, according to a certain selection rules to remove redundant or irrelevant features, and leave only some of the most effective feature subset. In network intrusion detection, under the premise of not reducing the detection accuracy, the goal of feature selection is to effectively eliminate redundant or irrelevant features, to decrease dimensions of features, to reduce the computing time, and to improve the efficiency of the detection algorithm. Bat algorithm is an efficient random search and optimization method, so it can be used to synchronize selected subset of intrusion features and support vector machine parameters.

(1) BA-SVM Design

The goal of feature selection and SVM parameter optimization is to select as few features and optimal parameters as possible, and obtain higher accuracy network intrusion detection. Therefore let's put intrusion detection rate as the target of SVM parameter optimization, which can be expressed as:

$$f = w_a \times Acc + w_f(\sum_{i=1}^{N_f} f_i)^{-1} \tag{14}$$

In the formula, w_a represents the weight of classification correct rate, Acc is used to verify the accuracy of the set in network intrusion detection, w_f indicates the weight of the number of the selected feature, the i th feature is selected, N_f represents the total number of features, f_i represents the i th full-feature selection status, that is:

$$f_i = \begin{cases} 1, \text{The ith feature is not selected;} \\ 0, \text{The ith feature is selected} \end{cases} \tag{15}$$

Encoding of bat position is composed of three segment groups, nuclear parameter C, nuclear parameter γ and the mask of feature subset f, and all of them are coded in binary, which will get the encoding structure of bat position as shown in Fig. 1.

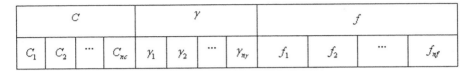

Fig. 1. Encoding method of bat position.

The 1st and 2nd segments, $C_1 \sim C_{nc}$ and $\gamma_1 \sim \gamma_{n\gamma}$ represent support vector machine parameters C and γ respectively, and need to be decoded when returning the optimal solution, which will decode the binary to the actual value. C And the decoding function of γ is:

$$\Gamma(R, L) = \min_{R} + \frac{\max_R - \min_R}{2^l - 1} \times d \tag{16}$$

In the formula, d is a decimal value, $\min_R$ and $\max_R$ are the minimum and maximum values of R respectively, Γ is the actual value of the parameter R.

The 3rd segment of bat position, $f_1 \sim f_{nf}$ represent feature selection status in the nf th network intrusion detection respectively, "1" indicates that the corresponding feature is selected, "0" indicates that nothing is selected.

2. Network intrusion detection process based on BA-SVM:

Step 1. Collects historic data of network status, and they will be pre-treated correspondingly.

Step 2. Initializes related parameter values of bat algorithm, including the number of individual bats N, the volume A and the pulse frequency r of each bat, and its position vector x and velocity vector u, frequency range f and the position x_i of each bat.

Step 3. Initializes the position of the bat population, and the position of each bat of solution is composed of C, γ, f.

Step 4. Decodes the position of bats, gets support vector machine parameters and feature subset, and then establishes a detection model based on support vector machine parameter values and the selected feature subset, and calculates the correct rate of classification of test samples, and finally gets the fitness value of individual bats by calculating according to formula (14).

Step 5. Finds the current optimal solution based on the merits of fitness values, and updates the search pulse rate, speed and position of bats according to the formula (1), (2) and (3).

Step 6. (1) Generates a uniformly distributed random number *rand*, and if *rand* $> r_i$, puts random perturbations on the current optimal solution, produces a new solution, and carries out cross-border processing on the new solution.

(2) Generates a uniformly distributed random number $rand$, and if $rand < A_i$ and $f(x_i) < f(x^*)$, and the new solution generated in step (6) is accepted, then updates A_i and r_i according to formula (5).

Step 7. Sorts the fitness values of all bats, and finds out the current optimal solution.

Step 8. Repeats steps (4) to (8), until the maximum number of iterations is found.

Step 9. Finds out the global optimal solution, and obtains SVM parameter (C, γ) values by decoding optimal solutions, processes the training samples according to the optimal feature subset, enters the optimum parameters in SVM and builds models, and inspects the test samples, analyses optimal performance based on test results.

5 Simulation Result

To test the performance of BA-SVM, let's perform simulation experiments using data in standard test set of network intrusion KDD CUP 99 database, and process classification fields into four categories of intrusion detection: DOS, U2R, U2L and Probe. Let's select 10,000 records randomly from the file KDD CUP99. Data.gz to train data, select 2000 records randomly from the file corrected. gz to test data, and the first data record contains 41 feature attributes. In order to make BA-SVM test results comparable, compare them with the two models: one is SVM-1 model, which selects network intrusion detection feature with BA alone, SVM parameters is determined randomly; the other is SVM-2 model, which restricts network intrusion detection feature with BA first, and then optimize SVM parameter with BA. Each algorithm uses correct detection rate (%) and average detection time (s) as evaluation indices. The experiment was repeated 5 times, the same number of training set and test set are randomly selected to test each time, and then stat test results of the five feature selections.

In order to determine the differences generated by different factors of characteristic properties upon the classification results and operational efficiency, select three groups of characteristic properties for test. Through several experiments, it shows that the first set of features in the classification accuracy and the computational time is the best on overall performance, and therefore select it as the optimal feature subset in network intrusion. For the corresponding optimal SVM parameters $C = 100$, $\gamma = 0.191$, we build network detection feature models with optimal feature subset and optimal parameters of SVM, then compare network intrusion detection model of BA-SVM and those of SVM-1, SVM-2 models, and verify the validation set, test results are shown in Table 1, the average detection time of different test samples for each model is shown in Fig. 2.

It could be learned from Table 1, when we use BA-SVM in the intrusion detection, the detection rate was significantly higher than the comparison models of SVM-1, SVM-2, and the detection efficiency is higher, and the training time is significantly reduced. It is shown in simulation results that there are a natural link between network intrusion detection feature selection and SVM parameters. They split this link, while BA-SVM can enable the optimization of network feature selection and SVM parameters and they can reach optimization at the same time, which could improve network intrusion detection correct rate and it is also a good way to ensure network security.

Table 1. Comparison of detection performance of each model.

Intrusion Type	Detect correct rate (%)			Average detection time (s)		
	SVM-1	SVM-2	BA-SVM	SVM-1	SVM-2	BA-SVM
DOS	87.01	92.15	94.47	1.12	0.81	0.48
U2R	86.94	90.56	93.91	0.92	0.76	0.45
R2L	85.92	86.11	91.36	0.84	0.29	0.21
Probe	83.15	90.13	92.04	1.79	1.31	0.59

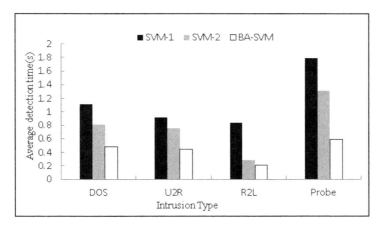

Fig. 2. Comparison of the average detection time between different models.

6   Conclusion

During the process of Network intrusion detection, the main problems we faced are the selection of features subset and the optimization of classifier parameters. With the linkages between them, this paper proposed a network intrusion detection method based on bat algorithm synchronization selection of features subset and support vector machine parameters. It is shown in the result of the simulation test that BA-SVM can get optimal feature subset selection, effectively streamline network data features, and improve the classification algorithm network intrusion detection speed and detection rate, which has broad application prospects in the field of network security.

Acknowledgments. This work was supported by the Inner Mongolia University for Nationalities Funds of China under Grant No. NMDYB15079.

References

1. Mao, X.G.: Computer and information network security problems and strategies. Sci. Technol. Inf. **3**, 65–66 (2010)

2. Chen, G., Wang, H.Q., Sun, X.: Model selection for SVM classification based on kernel prototype and adaptive genetic algorithm. J. Graduate Univ. Chin. Acad. Sci. **29**, 62–69 (2012)
3. Shan, L.L., Zhang, H.J., et al.: Parameters optimization and implementation of mixed kernels ε-SVM based on improved PSO algorithm. Appl. Res. Comput. **30**, 1636–1639 (2013)
4. Gao, L.F., Zhao, S.J., Gao, J.: Application of artificial fish-swarm algorithm in SVM parameter optimization selection. Comput. Eng. Appl. **49**, 86–90 (2013)
5. Yang, X.S.: A new metaheuristic bat-inspired algorithm. Nat. Inspired Coop. Strat. Optim. Sci. **284**, 65–74 (2012)

Motion Detection in Asymmetric Neural Networks

Naohiro Ishii[1]($\boxtimes$), Toshinori Deguchi[2], Masashi Kawaguchi[3],
and Hiroshi Sasaki[4]

[1] Aichi Institute of Technology, Toyota, Japan
ishii@aitech.ac.jp
[2] Gifu National College of Technology, Gifu, Japan
deguchi@gifu-nct.ac.jp
[3] Suzuka College of Technology, Mie, Japan
masashi@elec.suzukact.ac.jp
[4] Fukui University of Technology, Fukui, Japan
hsasaki@ccmails.fukui-ut.ac.jp

Abstract. To make clear the mechanism of the visual movement is important in the visual system. The prominent feature is the nonlinear characteristics as the squaring and rectification functions, which are observed in the retinal and visual cortex networks. Conventional model for motion processing in cortex, is the use of symmetric quadrature functions with Gabor filters. This paper proposes a new motion sensing processing model in the asymmetric networks. To make clear the behavior of the asymmetric nonlinear network, white noise analysis and Wiener kernels are applied. It is shown that the biological asymmetric network with nonlinearities is effective and general for generating the directional movement from the network computations. The qualitative analysis is performed between the asymmetrical network and the conventional quadrature model. The results are applicable to the V1 and MT model of the neural networks in the cortex.

Keywords: Asymmetrical neural networks · Directional movement · Nonlinear visual pathway · Wiener kernels · Motion detection

1 Introduction

In the biological neural networks, the sensory information is processed effectively and speedily. Reichard [1] evaluated the sensory information by the auto-correlations in the neural networks. The nonlinear characteristics as the squaring function and rectification function, which are observed in the retina [12, 13] and visual cortex networks [4–7], respectively. Conventional model for cortical motion sensors is the use of symmetric quadrature functions with Gabor filters, which is called energy model [2, 3]. Recent study by Hess and Bair [3] discusses quadrature is not necessary or sufficient under certain stimulus condition. Then, minimal models for sensory processing are expected. This paper proposes a new motion sensing processing model in the biological asymmetric networks. The nonlinear function exists in the asymmetrical neural networks.

L. Cheng et al. (Eds.): ISNN 2016, LNCS 9719, pp. 409–417, 2016.
DOI: 10.1007/978-3-319-40663-3_47

To find out cells function in the biological networks, white noise stimulus [8–13] are often used in physiological experiments. In this paper, to make clear the behavior of the asymmetric network with nonlinearity, white noise analysis and Wiener kernels are applied. It is shown that the asymmetric network with nonlinearities is effective and superior for generating the directional movement detection from the network computations. We analyze the asymmetric network based on the retinal circuit of the catfish [8, 9, 15, 16], from the point of the optimization of the network model. It is shown that the asymmetric network with nonlinearity has the ability for the directional movement of the stimulus, which can be written in directional equations by Wiener kernels. Then, it is shown that the directional equations obtained are selective for the preferred and null direction stimulus in the asymmetric network. It is shown that the quadrature model works with Gabor functions [2, 3], while the asymmetric network does not need their conditions.

2   Biological Neural Networks

In the biological neural networks, the structure of the network, is closely related to the functions of the network. The network suggests the biological function of the organism. Naka et al. [13] presented a simplified, but essential networks of catfish inner retina as shown in Fig. 1. Visual perception is carried out firstly in the retinal neural network as the special processing between neurons. The following asymmetric neural network is extracted from the catfish retinal network [13]. The asymmetric structure network with a quadratic nonlinearity is shown in Fig. 1, which composes of the pathway from the bipolar cell B to the amacrine cell N and that from the bipolar cell B, via the amacrine cell N with squaring function to the N cell.

Figure 1 shows a network which plays an important role in the movement perception as the fundamental network. It is shown that N cell response is realized by a linear filter, which is composed of a differentiation filter followed by a low-pass filter. Thus, the asymmetric network in Fig. 1 is composed of a linear pathway and a nonlinear pathway. Here, the stimulus with Gaussian distribution is assumed to move from the left side to the right side in front of the network in Fig. 1, as shown in Fig. 2. $x''(t)$ is mixed with $x(t)$. Then, we indicate the right stimulus by $x'(t)$. By introducing a

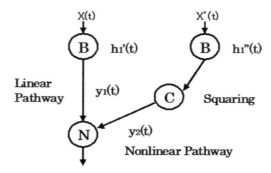

Fig. 1. Asymmetric network with linear and squaring nonlinear pathways.

B1 cell   **B2 cell**

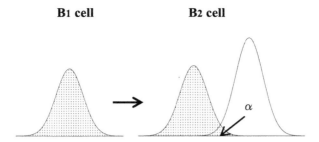

Fig. 2. Stimulus movement from the left to the right side.

mixed ratio, α, the input function of the right stimulus, is described in the following equation, where $0 \leq \alpha \leq 1$ and $\beta = 1 - \alpha$ hold. Then, Fig. 2 shows that the moving stimulus is described in the following equation

$$x'(t) = \alpha x(t) + \beta x''(t) \tag{1}$$

Let the power spectrums of $x(t)$ and $x''(t)$, be p and p'', respectively an equation $p'' = kp$ holds for the coefficient k, because we assumed here that the deviations of the input functions are different in their values. Figure 2 shows that the slashed light is moving from the receptive field of B_1 cell to the field of the B_2 cell. The mixed ratio of the input $x(t)$, α is shown in the receptive field of B_2 cell. The stimulus on both cells in Fig. 2 is shown in the schematic diagram as shown in Fig. 3.

B1cellB2cellα

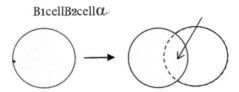

Fig. 3. Schematic diagram of the preferred stimulus direction.

First, on the linear pathway of the asymmetrical network in Fig. 1, the input function is $x(t)$ and the output function is $y(t)$, which is an output after the linear filter of the bell N.

$$y(t) = \int h_1'''(\tau)(y_1(t - \tau) + y_2(t - \tau))d\tau + \varepsilon \tag{2}$$

where $y_1(t)$ shows the linear information on the linear pathway $y_2(t)$ shows the non-linear information on the nonlinear pathway and ε shows error value. The $y_1(t)$ and $y_2(t)$ are given, respectively as follows,

$$y_1(t) = \int\limits_0^\infty h_1'(\tau)x(t-\tau)d\tau \tag{3}$$

$$y_2(t) = \int\limits_0^\infty \int\limits_0^\infty h_1''(\tau_1)h_1''(\tau_2)x'(t-\tau_1)x'(t-\tau_2)d\tau_1 d\tau_2 \tag{4}$$

We assume here the linear filter N to have only summation operation without in the analysis. Thus the impulse response function $h_1'''(t)$ is assumed to be value 1 without loss of generality.

2.1 Directional Equations from Optimized Conditions in the Asymmetric Networks

Under the assumption that the impulse response functions, $h1'(t)$ of the cell B_1, $h1''(t)$ of the cell B_2 and moving stimulus ratio α in the right to be unknown, the optimization of the network is carried out. By the minimization of the mean squared value ξ of ε in Eq. (2), the following necessary equations for the optimization of equations are derived,

$$\frac{\partial \xi}{\partial h_1'(t)} = 0, \frac{\partial \xi}{\partial h_2''(t)} = 0, \frac{\partial \xi}{\partial \alpha} = 0 \tag{5}$$

Then, the following three equations are derived for the optimization satisfying the Eq. (5).

$$
\begin{aligned}
&E[y(t)x'(t-\lambda)] = \alpha p h_1'(\lambda) \\
&E[(y(t)-C_0)x'(t-\lambda_1)x'(t-\lambda_2)] = 2\{(\alpha^2 + k\beta^2)p^2 h_1''(\lambda_1)h_1''(\lambda_2)\} \\
&E[(y(t)-C_0)x(t-\lambda_1)x(t-\lambda_2)] = 2\alpha^2 p^2 h_1''(\lambda_1)h_1''(\lambda_2) \\
&E[(y(t)-C_0)x''(t-\lambda_1)x''(t-\lambda_2)] = 2\beta^2 (kp)^2 h_1''(\lambda_1)h_1''(\lambda_2)
\end{aligned} \tag{6}
$$

where C_0 is the mean value of, $y(t)$ which is shown in the following. Here, the Eq. (6) can be rewritten by applying Wiener kernels, which are related with input and output correlations method developed by Lee and Schetzen [14]. From the necessary optimization equations in (5), the following Wiener kernel equations are derived as shown in the following [8–10]. First, we can compute the 0-th order Wiener kernel $C0$, the 1-st order one and $C11(\lambda)$ the 2-nd order one $C21(\lambda1, \lambda2)$ on the linear pathway by the cross-correlations between $x(t)$ and $y(t)$. The suffix i, j of the kernel, $Cij(\bullet)$ shows that i is the order of the kernel and $j = 1$ means the linear pathway, while $j = 2$ means the nonlinear pathway. Then, the 0-th order kernel under the condition of the spatial interaction of cell's impulse response functions $h1'(t)$ and $h1''(t)$ becomes,

$$C_{11}(\lambda) = \frac{1}{p}E[y(t)x(t - \lambda)] = h_1'(\lambda) \tag{7}$$

since the last term of the second equation becomes zero. The 2-nd order kernel is also derived from the optimization Eq. (8) as follows,

$$C_{21}(\lambda_1, \lambda_2) = \frac{1}{2p^2}E[(y(t) - C_0)x(t - \lambda_1)x(t - \lambda_2)]$$
$$= \alpha^2 h_1''(\lambda_1)h_1''(\lambda_2) \tag{8}$$

From Eqs. (1), (7) and (8), the ratio, α which is a mixed coefficient of $x(t)$ to, is $x'(t)$ shown by α^2 as the amplitude of the second order Wiener kernel. Second, on the nonlinear pathway, we can compute the 0-th order kernel, $C0$ the 1-st order kernel $C12(\lambda)$ and the 2-nd order kernel by the $C22(\lambda1, \lambda2)$ cross-correlations between $x(t)$ and $y(t)$ as shown in the following, which are also derived from the optimization Eq. (6).

$$C_{12}(\lambda_1, \lambda_2) = \frac{1}{p(\alpha^2 + k\beta^2)}E[y(t)x'(t - \lambda)]$$
$$= \frac{\alpha}{\alpha^2 + k(1 - \alpha)^2}h_1'(\lambda) \tag{9}$$

and

$$C_{22}(\lambda_1, \lambda_2) = h_1''(\lambda_1)h_1''(\lambda_2) \tag{10}$$

The motion problem is how to detect the direction of the stimulus in the increase of the ratio α in Fig. 3. This implies that for the motion of the light from the left side circle to the right one, the ratio α can be derived from the kernels described in the above, in which the second order kernels C_{21} and C_{22} are abbreviated in the representation of Eqs. (8) and (10).

$$(C_{21}/C_{22}) = \alpha^2 \tag{11}$$

holds. Then, from the Eq. (11) the ratio α is shown as follows

$$\alpha = \sqrt{\frac{C21}{C22}} \tag{12}$$

The Eq. (12) is called here α-equation, which implies the directional stimulus on the network and shows the detection of the movement by the α without it's direction. This shows that the α-equation is determined by the second order kernels on the linear pathway and the nonlinear one in the network. From the first order kernels $C11$ and $C12$, and the second order kernels in the above derivations, the directional equation from the left to the right, holds as shown in the following,

$$\frac{C12}{C11} = \frac{\sqrt{\frac{C21}{C22}}}{\frac{C21}{C22} + k(1 - \sqrt{\frac{C21}{C22}})^2} \tag{13}$$

The Eq. (13) shows the direction of the moving stimulus from the left to the right.

2.2 Algorithm for Movement Detection

When α increases to α' in the preferred direction, the movement will take place. Then, necessary conditions of the movement is to satisfy the directional equation (13) at α and that at α', respectively. The detection of the directional movement is carried out in the following steps.

① In the case of the directional movement from the left to the right, the Eq. (12) for α at time t, $\alpha(t)$, i.e., the root of kernel correlations ratio in the left side of the Eq. (13) is computed. Then, the Eq. (13) at time t is checked whether the equation holds. $(t + \Delta t)$

② α' at time $(t + \Delta t)$, $\alpha'(t + \Delta t)$ is computed similarly. The Eq. (13) is checked whether the equation holds.

③ Assume here that the following holds,

$$\alpha(t) < \alpha'(t + \Delta t) \tag{14}$$

④ When ① and ② are satisfied, the directional movement is written as

$$\alpha \rightarrow \alpha' \tag{15}$$

In the null direction from the right to left side stimulus, often called the null direction, the schematic diagram of the directional stimulus, is shown in Fig. 4.

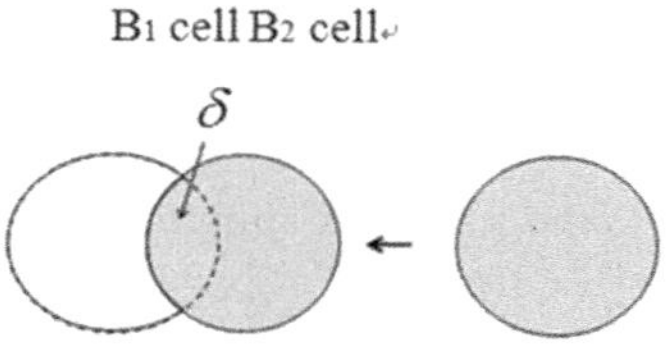

B1 cell B2 cell

Fig. 4. Schematic diagram of stimulus from right to left (null direction).

Figure 4 are derived similarly as the Sect. 2.1.

$$C_{11}(\lambda) = h_1'(\lambda)$$

$$C_{21}(\lambda_1, \lambda_2) = \frac{k^2 \delta^2}{(\alpha^2 + k\delta^2)^2} h_1''(\lambda_1) h_1''(\lambda_2) \tag{16}$$

Similarly, the following equations are derived on the nonlinear pathway,

$$C_{12}(\lambda) = \delta h_1'(\lambda)$$
$$C_{22}(\lambda_1, \lambda_2) = h_1''(\lambda_1)h_1''(\lambda_2) \tag{17}$$

From Eqs. (16) and (17), the ratio δ is derived, which is abbreviated in the notation.

$$\delta = \frac{C_{12}}{C_{11}} \tag{18}$$

and the following directional equation is derived.

$$\frac{C_{11}}{C_{12}} = \frac{k\sqrt{\frac{C_{22}}{C_{21}}}}{(1 - \frac{C_{12}}{C_{11}})^2 + k(\frac{C_{12}}{C_{11}})^2} \tag{19}$$

It is proved that any set of Wiener kernels values satisfying the both Eqs. (13) and (19) does not exist. Thus, Eqs. (13) and (19) are different.

3 Comparison with Conventional Quadrature Models

Motion detection of the conventional quadrature models under the same conditions in this paper is analyzed. The quadrature model in Fig. 5 is well known as the energy model for motion detection [2, 3], which is a symmetric network model. The model with Gabor filters are used as the functions $h_1(t)$ and $h_1'(t)$ in the models [2, 3]. Under the same stimulus conditions in the asymmetric network in Figs. 1 and 3, the Wiener kernels are computed in the symmetric quadrature model in Fig. 5. In Fig. 5, the first order kernels disappear on the left and the right pathways. Only the second order kernels are computed in Fig. 5. On the left pathway in Fig. 5, the second order kernel $C_{21}(\lambda_1, \lambda_2)$ is computed as follows,

$$C_{21}(\lambda_1, \lambda_2) = \frac{1}{2p^2} \iint h_1(\tau)h_1(\tau')E[x(t-\lambda)x(t-\lambda')x(t-\lambda_1)x(t-\lambda_2)]d\tau d\tau'$$
$$+ \frac{1}{2p^2} \iint h_1'(\tau)h_1'(\tau')E[x'(t-\lambda)x'(t-\lambda')x(t-\lambda_1)x(t-\lambda_2)]d\tau d\tau' \tag{20}$$
$$= h_1(\lambda_1)h_1(\lambda_2) + \alpha^2 h_1'(\lambda_1)h_1'(\lambda_2)$$

On the right pathway in Fig. 5, the 2nd order kernel $C_{22}(\lambda_1, \lambda_2)$ is computed similarly,

$$C_{22}(\lambda_1, \lambda_2) = \frac{\alpha^2}{(\alpha^2 + k\beta^2)^2} h_1(\lambda_1)h_1(\lambda_2) + h_1'(\lambda_1)h_1'(\lambda_2) \tag{21}$$

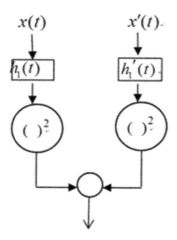

Fig. 5. Quadrature energy model with Gabor filters.

In the conventional energy model of motion [2, 3], the Gabor functions are given as

$$h_1(t) = \frac{1}{\sqrt{2\pi\sigma}} \exp(-\frac{t^2}{2\sigma^2}) \sin(2\pi\omega t) \quad h_1'(t) = \frac{1}{\sqrt{2\pi\sigma}} \exp(-\frac{t^2}{2\sigma^2}) cos(2\pi\omega t) \quad (22)$$

When Gabor functions in the quadrature model are given as the Eq. (22), the motion parameter α and the motion equation are computed as follows,

$$\alpha = \sqrt{\frac{C_{21}(\lambda_1, \lambda_2) - h_1(\lambda_1)h_1(\lambda_2)}{h_1'(\lambda_1)h_1'(\lambda_2)}} \quad (23)$$

$$C_{22}(\lambda_1, \lambda_2) = \frac{\alpha^2}{(\alpha^2 + k\beta^2)^2} h_1(\lambda_1)h_1(\lambda_2) + h_1'(\lambda_1)h_1'(\lambda_2) \quad (24)$$

Note that the conventional quadrature model generates the motion Eq. (24) under the condition of the given Gabor functions (22), while the asymmetric network in Fig. 1 generate the motion Eq. (13) without the condition of the Gabor functions. Thus, the asymmetric network has also general ability compared to the conventional quadrature model.

4   Conclusion

The neural networks are analyzed to make clear functions of the biological asymmetric neural networks with nonlinearity. This kind of networks exits in the biological network as retina and brain cortex of V1 and MT areas. In this paper, the behavior of the asymmetrical network with nonlinearity, is analyzed to detect the directional stimulus from the point of the neural computation. The conditions for the directive movement of the stimulus, are derived based on the optimization of the network. The qualitative motion ability is compared between the asymmetric network and the conventional quadrature model. It was shown that the quadrature model works with Gabor filters, while the asymmetrical network does not need their conditions. These results will suggest functions of the detection behavior of the movement in the cortex, V1 and MT areas.

References

1. Reichard, W.: Autocorrelation, a principle for the evaluation of sensory information by the central nervous system. Rosenblith edn. Wiley, New York (1961)
2. Adelson, E.H., Bergen, J.R.: Spatiotemporal energy models for the perception of motion. J. Opt. Soc. Am. A 2(2), 284–298 (1985)
3. Heess, N., Bai, R.W.: Direction opponency, not quadrature, is key to the 1/4 cycle preference for apparent motion in the motion energy model. J. Neurosci. 30(34), 11300–11304 (2010)
4. Chubb, C., Sperling, G.: Drift-Balanced Random Stimuli, A General Basis for Studying Non-Fourier Motion. J. Optical Soc. of America A, 1986–2006 (1988)
5. Taub, E., Victor, J.D., Conte, M.: Nonlinear preprocessing in short-range motion. Vis. Res. 37, 1459–1477 (1997)
6. Simonceli, E.P., Heeger, D.J.: A model of neuronal responses in visual area MT. Vis. Res. 38, 743–761 (1996)
7. Heeger, D.J.: Normalization of cell responses in cat striate cortex. Vis. Neurosci. 9, 181–197 (1992)
8. Marmarelis, P.Z., Marmarelis, V.Z.: Analysis of Physiological Systems – The White Noise Approach. Plenum Press, New York (1978)
9. Marmarelis, V.Z.: Nonlinear Dynamic Modeling of Physiological Systems. Wiley-IEEE Press, New Jersey (2004)
10. Marmarelis, V.Z.: Modeling methodology for nonlinear physiological systems. Ann. Biomed. Eng. 25, 239–251 (1997)
11. Wiener, N.: Nonlinear Problems in Random Theory. The MIT Press, Cambridge (1966)
12. Sakuranaga, M., Naka, K.I.: Signal transmission in the catfish retina. III. Transmissioto type-C cell. J. Neurophysiol. 53(2), 411–428 (1985)
13. Naka, K.I., Sakai, H.M., Ishii, N.: Generation of transformation of second order nonlinearity in catfish retina. Ann. Biomed. Eng. 16, 53–64 (1988)
14. Lee, Y.W., Schetzen, M.: Measurements of the Wiener kernels of a nonlinear by cross-correlation. Int. J. of Control 2, 237–254 (1965)
15. Ishii, N., Ozaki, M., Sasaki, H.: Correlation computations for movement detection in neural networks. In: Negoita, M.G., Howlett, R.J., Jain, L.C. (eds.) KES 2004. LNCS (LNAI), vol. 3214, pp. 124–130. Springer, Heidelberg (2004)
16. Ishii, N., Deguchi, T., Kawaguchi, M.: Neural computations by asymmetric networks with nonlinearities. In: Beliczynski, B., Dzielinski, A., Iwanowski, M., Ribeiro, B. (eds.) ICANNGA 2007. LNCS, vol. 4432, pp. 37–45. Springer, Heidelberg (2007)

Language Models with RNNs for Rescoring Hypotheses of Russian ASR

Irina Kipyatkova[1,2(✉)] and Alexey Karpov[1,3]

[1] St. Petersburg Institute for Informatics and Automation
of the Russian Academy of Sciences (SPIIRAS), St. Petersburg, Russia
{kipyatkova, karpov}@iias.spb.su
[2] St. Petersburg State University of Aerospace Instrumentation (SUAI),
St. Petersburg, Russia
[3] ITMO University, St. Petersburg, Russia

Abstract. In this paper, we describe a research of recurrent neural networks (RNNs) for language modeling in large vocabulary continuous speech recognition for Russian. We experimented with recurrent neural networks with different number of units in the hidden layer. RNN-based and 3-gram language models (LMs) were trained using the text corpus of 350M words. Obtained RNN-based language models were used for N-best list rescoring for automatic continuous Russian speech recognition. We tested also a linear interpolation of RNN LMs with the baseline 3-gram LM and achieved 14 % relative reduction of the word error rate (WER) with respect to the baseline 3-gram model.

Keywords: Recurrent neural networks · Language model · Automatic speech recognition · Russian speech

1 Introduction

For automatic speech recognition (ASR) a language model (LM) is needed. The most widely used model is n-gram model which estimates posterior probability of the word consequence in a text. Commonly 3-gram model is employed. The usage of n-gram LMs with longer context can lead to the data sparseness problem. LMs based on recurrent neural networks (RNN) estimate probabilities based on all previous history that is their advantage over n-gram models.

In our research we used RNN LM for N-best list rescoring of automatic speech recognition (ASR) system. In Sect. 2 we give a survey of using NNs for LM creation, in Sect. 3 we describe RNN LM, in Sect. 4 we present our baseline LM, Sect. 5 gives a description of our RNN LMs, experiments on using RNN LM for N-best list rescoring for Russian speech recognition are presented in Sect. 6.

2 Related Work

The use of NN for LM training was firstly presented in [1]. RNN for language modeling was firstly used in [2]. In [3], a comparison of LMs based on feed-forward and recurrent NN was made. On the test set RNN LM showed 0.4 % absolute word error rate (WER) reduction comparing to feed-forward NN.

© Springer International Publishing Switzerland 2016
L. Cheng et al. (Eds.): ISNN 2016, LNCS 9719, pp. 418–425, 2016.
DOI: 10.1007/978-3-319-40663-3_48

In [4], the strategies for NN LM training on large data sets are presented: (1) reduction of training epochs; (2) reduction of number of training tokens; (3) reduction of vocabulary size; (4) reduction of size of the hidden layer; (5) parallelization. It was shown that when data are sorted by their relevance the fast convergence during training and the better overall performance are observed. A maximum entropy model trained as a part of NN LM that leads to significant reduction of computational complexity was proposed. 10 % relative reduction was obtained comparing to the baseline 4-gram model.

In [5] it was proposed to call RNN LM to compute LM score only if newly hypothesized word has a reasonable score. Also cache based RNN inference was proposed in order to reduce runtime. Three approaches for exploiting succeeding word information in RNN LMs were proposed in [6]. In order to speed up training noise contrastive estimation training was investigated in [7] for RNNLMs. Noise contrastive estimation does not require normalization at the output layer and thereby allows speeding up training. A novel RNN LM dealing with multiple time-scale contexts was presented in [8]. Several lengths of contexts were considered in one LM. In [9], paraphrastic RNN LMs, which use multiple automatically generated paraphrase variants, were investigated. In [10] Long Short-Term Memory (LSTM) NN architecture was explored for modeling English and French languages. Investigation of the jointly trained maximum entropy and RNN LMs for Code-Switching speech is presented in [11]. It was proposed to integrate part-of-speech and language identifier information in RNN LM. In [12] the discriminative method for RNN LM was proposed. As a discriminative criterion the log-likelihood ratio of the ASR hypotheses and references was used.

RNN LM for Russian was firstly used in [13]. RNN LM was trained on the text corpus containing 40M words with vocabulary size of about 100K words. An interpolation of the obtained model with the baseline 3-gram and factored LMs was carried out. The resulted LM was used for rescoring 500-best list that demonstrated 7.4 % relative improvement of WER.

Despite of the increasing popularity of usage NNs for language modeling there are only a few studies on NN-based LMs for Russian. We made a research of implementation RNNs for Russian LM creation.

3 Artificial Neural Networks for Language Modeling

We used the same structure of RNN LM as in [2]; it is presented in Fig. 1. RNN consists of an input layer x, a hidden (or context) layer s, and an output layer y. The input to the network in time t is vector $x(t)$. The vector $x(t)$ is a concatenation of vector $w(t)$, which is a current word in time t, and vector $s(t-1)$, which is output of the hidden layer obtained on the previous step. Size of $w(t)$ is equal to vocabulary size. The output layer $y(t)$ has the same size as $w(t)$ and it represents probability distribution of the next word given the previous word $w(t)$ and the context vector $s(t-1)$. The size of the hidden layer is chosen empirically and usually it consists of 30–500 units [2].

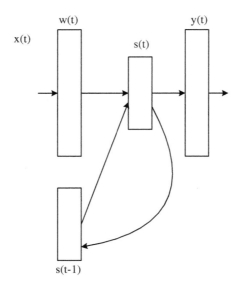

Fig. 1. General structure of the recurrent neural network.

Input, hidden, and output layers are as follows [2]:

$$x(t) = w(t) + s(t-1)$$

$$s_j(t) = f\left(\sum_i x_i(t)u_{ji}\right)$$

$$y_k(t) = g\left(\sum_j s_j(t)u_{kj}\right),$$

where $f(z)$ is sigmoid activation function:

$$f(z) = \frac{1}{1+e^{-z}}$$

$g(z)$ is softmax function:

$$g(z_m) = \frac{e^{z_m}}{\sum_k e^{z_k}}$$

NN training is carried out in several epochs. Usually, for training the back propagation algorithm with the stochastic gradient descent is used.

In order to speed up training in [14] it was suggested to perform factorization of the output layer. Words were mapped to classes according to their frequencies. At first, probability distribution over classes was computed. Then, probability distribution for the words that belong to a specific class was computed. In this case, word probability is computed as follows:

$$P(w_i|h_i) = (P(c_i)|s(t))P(w_i|c_i, s(t)),$$

where c_i is a class of the given word, h_i is a history of the previous word.

4 Training Textual Corpus and Baseline Language Model

For the language model creation, we collected and automatically processed a Russian text corpus of a number of on-line newspapers. The procedure of preliminary text processing and normalization is described in [15]. At first, texts were divided into sentences. Then, a text written in any brackets was deleted, and sentences consisting of less than six words were also deleted. Uppercase letters were replaced by lowercase letters, if a word began from an uppercase letter. If a whole word was written by the uppercase letters, then such change was made, when the word existed in a vocabulary only. The size of the corpus after text normalization is over 350M words, and it has above 1M unique word-forms.

For the statistical text analysis, we used the SRI Language Modeling Toolkit (SRILM) [16]. During LMs creation we used the Kneser-Ney discounting method, and did not apply any n-gram cutoff. We created various 3-gram LMs with different vocabulary sizes, and the best speech recognition results were obtained with 150K vocabulary [17]. The perplexity measure of the baseline model was 553. So this vocabulary was chosen for further experiments with N-best list rescoring.

5 Creation of Language Models Based on Recurrent Neural Networks

For creation of RNN LM we used Recurrent Neural Network Language Modeling Toolkit (RNNLM toolkit) [18]. We made factorization of the output layer of RNN and created LMs with the number of classes equal to 100 and 500. We created models with different number of units in the hidden layer: 100, 300, and 500 [19, 20].

Then we have made a linear interpolation of the RNN LMs with the baseline 3-gram model. In this case, the probability score was computed as follows:

$$P_{IRNN}(w_i|h_i) = \lambda P_{RNN}(w_i|h_i) + (1 - \lambda)P_{BL}(w_i|h_i)$$

where $P_{RNN}(w_i|h_i)$ is a probability computed by the RNN LM; $P_{BL}(w_i|h_i)$ is a probability computed by the baseline 3-gram model; λ is an interpolation coefficient.

LMs are evaluated by perplexity which is computed on held-out text date. Perplexity can be considered to be a measure of on average how many different equally most probable words can follow any given word. Lower perplexities represent better LMs [21]. Perplexities of the obtained models computed on the text corpus of 33M words are presented in Table 1. The interpolation coefficient of 1.0 means only

Table 1. Perplexities of RNN LMs interpolated with 3-gram LM.

Language model	Number of classes	Interpolation coefficients			
		0.4	0.5	0.6	1.0
RNN with 100 hidden units + 3-gram LM	100	457	465	482	981
	500	471	482	500	1074
RNN with 300 hidden units + 3-gram LM	100	457	467	484	997
	500	432	436	446	843
RNN with 500 hidden units + 3-gram LM	100	394	392	396	766
	500	417	419	428	870

RNN LM was used. In the table, we can see RNN LMs have smaller perplexities than the 3-gram LM.

6 Experiments

Architecture of the Russian ASR system with developed RNN LMs is presented on Fig. 2. The system works in 2 modes [15]: training and recognition. In the training mode, acoustic models of speech units, LMs, and phonemic vocabulary of word-forms that will be used by recognizer are created.

For training the speech recognition system we used our own corpus of spoken Russian speech Euronounce-SPIIRAS [22]. The database consists of 16,350 utterances pronounced by 50 native Russian speakers (25 male and 25 female). Each speaker pronounced more than 300 phonetically-balanced and meaningful phrases. Total duration of speech data is about 21 h. For acoustic modeling, we applied continuous density Hidden Markov Models (HMMs).

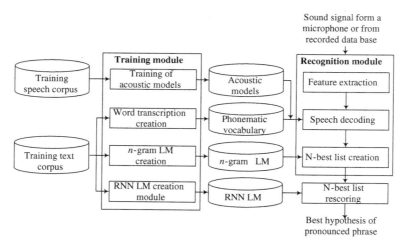

Fig. 2. Architecture of Russian ASR system with RNN LMs.

To test the ASR system we used a speech corpus that contains 500 phrases pronounced by 5 different speakers (each speaker said the same 100 phrases). The phrases were taken from the materials of an on-line newspaper that were not used in the training data.

For automatic speech recognition, we applied the open-source Julius engine ver. 4.2 [23]. At speech decoding stage, the baseline 3-gram language models were used, and N-best list of hypotheses was created. Then RNN LM was applied for rescoring obtained N-best list of hypotheses and for selection of the best recognition hypothesis for pronounced phrase.

The WER obtained with the baseline 3-gram LM was 26.54 %. We produced a 50-best list and made its rescoring using RNN LMs as well as RNN LMs interpolated (+) with the baseline model using various interpolation coefficients. Obtained results are summarized in Table 2.

Table 2. WER obtained after rescoring N-best lists with RNN LMs (%).

Language model	Number of classes	Interpolation coefficients			
		0.4	0.5	0.6	1.0
RNN with 100 hidden units + 3-gram LM	100	24.72	24.91	24.98	26.72
	500	24.78	24.83	24.83	27.45
RNN with 300 hidden units + 3-gram LM	100	24.10	24.18	24.51	25.49
	500	23.88	23.84	24.25	25.24
RNN with 500 hidden units + 3-gram LM	100	23.24	22.87	22.96	23.97
	500	23.91	23.60	23.73	24.12

In the table we can see that in the most cases the rescoring decreased the WER in comparison with the baseline model excepting the case of using RNN LMs with 100 hidden units without interpolation with the baseline model. Application of RNNs with 100 classes gave better results than RNNs with 500 classes. The lowest WER = 22.87 % was achieved using RNN LM with 500 hidden units and 100 classes interpolated with 3-gram model using the interpolation coefficient of 0.5.

Our results are consistent with those obtained in [13]. But we used training set of 350 million words that is 10 times larger set than in [13]. WER obtained in [13] with help of RNN was equal to 32.9 %. Our results are better and support the hypothesis that RNN-based LMs improve speech recognition accuracy.

7 Conclusion

In the paper, we have described the implementation of RNN LMs for rescoring N-best hypotheses lists of the ASR system. The advantage of RNN LMs over n-gram LMs is that they are able to store arbitrary long history of a given word. We have tried RNNs with various number of units in the hidden layer, also we tested the linear interpolation of the RNN LM with the baseline 3-gram LM. And we achieved 14 % relative reduction of WER using RNN LM with respect to the baseline model.

Acknowledgments. This research is partially supported by the Council for Grants of the President of Russia (projects No. MK-5209.2015.8 and MD-3035.2015.8), by the Russian Foundation for Basic Research (projects No. 15-07-04415 and 15-07-04322), and by the Government of the Russian Federation (grant No. 074-U01).

References

1. Schwenk, H., Gauvain, J.-L.: Training neural network language models on very large corpora. In: Proceedings of the Conference on Empirical Methods on Natural Language Processing. Association for Computational Linguistics, Vancouver, B.C., Canada, pp. 201–208 (2005)
2. Mikolov, T., Karafiat, M., Burget, L., Cernocky, J., Khudanpur, S.: Recurrent neural network based language model. In: Proceedings of INTERSPEECH 2010, vol. 2, pp. 1045–1048. Makuhari, Chiba, Japan (2010)
3. Sundermeyer, M., Oparin, I., Gauvain, J.-L., Freiberg, B., Schluter, R., Ney, H.: Comparison of feedforward and recurrent neural network language models. In: Proceedings of International Conference on Acoustics, Speech and Signal Processing (ICASSP), Vancouver, B.C., Canada, pp. 8430–8434 (2013)
4. Mikolov, T., Deoras, A., Povey, D., Burget L., Černocký, J.: Strategies for training large scale neural network language models. In: Proceedings of ASRU 2011, Hawaii, pp. 196–201 (2011)
5. Huang, Z., Zweig, G., Dumoulin, B.: Cache based recurrent neural network language model inference for first pass speech recognition. In: Proceedings of International Conference on Acoustics, Speech and Signal Processing (ICASSP) 2014, Florence, Italy, pp. 6404–6408 (2014)
6. Shi, Y., Larson, M., Wiggers, P., Jonker, C.M.: Exploiting the succeeding words in recurrent neural network. In: Proceedings of INTERSPEECH 2013, Lyon, France, pp. 632–636 (2013)
7. Chen, X., Liu, X., Gales, M.J.F., Woodland, P.C.: Recurrent neural network language model training with noise contrastive estimation for speech recognition. In: Proceedings of IEEE International Conference on Acoustics, Speech and Signal Processing (ICASSP), Brisbane, Australia, pp. 5411–5415 (2015)
8. Morioka, T., Iwata, T., Hori, T., Kobayashi, T.: Multiscale recurrent neural network based language model. In: Proceedings of INTERSPEECH 2015, Dresden, Germany, pp. 2366–2370 (2015)
9. Liu, X., Chen, X., Gales, M.J.F., Woodland, P.C.: Paraphrastic recurrent neural network language models. In: Proceedings IEEE International Conference on Acoustics, Speech and Signal Processing (ICASSP), Brisbane, Australia, pp. 5406–5410 (2015)
10. Sundermeyer, M., Schlüter, R., Ney, H.: LSTM neural networks for language modeling. In: Proceedings of INTERSPEECH 2012, pp. 194–197 (2012)
11. Vu, N.T., Schultz, T.: Exploration of the impact of maximum entropy in recurrent neural network language models for code-switching speech. In: Proceedings of 1st Workshop on Computational Approaches to Code Switching, Doha, Qatar, pp. 34–41 (2014)
12. Tachioka, Y., Watanabe, S.: Discriminative method for recurrent neural network language models. In: Proceedings of IEEE International Conference on Acoustics, Speech and Signal Processing (ICASSP), Brisbane, Australia, pp. 5386–5390 (2015)

13. Vazhenina, D., Markov, K.: Evaluation of advanced language modeling techniques for Russian LVCSR. In: Železný, M., Habernal, I., Ronzhin, A. (eds.) SPECOM 2013. LNCS, vol. 8113, pp. 124–131. Springer, Heidelberg (2013)

14. Mikolov, T., Kombrink, S., Burget, L., Černocký, J.H., Khudanpur, S.: Extensions of recurrent neural network language model. In: Proceedings of International Conference on Acoustics, Speech and Signal Processing (ICASSP), pp. 5528–5531 (2011)

15. Karpov, A., Markov, K., Kipyatkova, I., Vazhenina, D., Ronzhin, A.: Large vocabulary Russian speech recognition using syntactico-statistical language modeling. Speech Commun. **56**, 213–228 (2014)

16. Stolcke, A., Zheng, J., Wang, W., Abrash, V.: SRILM at sixteen: update and outlook. In: Proceedings of IEEE Automatic Speech Recognition and Understanding Workshop ASRU 2011, Waikoloa, Hawaii, USA (2011)

17. Kipyatkova, I., Karpov, A.: Lexicon size and language model order optimization for Russian LVCSR. In: Železný, M., Habernal, I., Ronzhin, A. (eds.) SPECOM 2013. LNCS, vol. 8113, pp. 219–226. Springer, Heidelberg (2013)

18. Mikolov, T., Kombrink, S., Deoras, A., Burget, L., Černocký, J.: RNNLM-recurrent neural network language modeling toolkit. In: ASRU-2011, Demo Session (2011)

19. Kipyatkova, I., Karpov, A.: A comparison of RNN LM and FLM for Russian speech recognition. In: Ronzhin, A., Potapova, R., Fakotakis, N. (eds.) SPECOM 2015. LNCS, vol. 9319, pp. 42–50. Springer, Heidelberg (2015)

20. Kipyatkova, I., Karpov, A.: Recurrent neural network-based language modeling for an automatic Russian speech recognition system. In: Proceedings of International Conference AINL-ISMW FRUCT, St. Petersburg, Russia, pp. 33–38 (2015)

21. Moore, G.L.: Adaptive Statistical Class-Based Language Modelling. Ph.D. thesis, Cambridge University (2001)

22. Jokisch, O., Wagner, A., Sabo, R., Jaeckel, R., Cylwik, N., Rusko, M., Ronzhin, A., Hoffmann, R.: Multilingual speech data collection for the assessment of pronunciation and prosody in a language learning system. In: Proceedings of SPECOM 2009, St. Petersburg, Russia, pp. 515–520 (2009)

23. Lee, A., Kawahara, T.: Recent development of open-source speech recognition engine julius. In: Proceedings of Asia-Pacific Signal and Information Processing Association Annual Summit and Conference (APSIPA ASC) 2009, Sapporo, Japan, pp. 131–137 (2009)

User-Level Twitter Sentiment Analysis
with a Hybrid Approach

Meng Joo Er[1(✉)], Fan Liu[2], Ning Wang[1], Yong Zhang[2],
and Mahardhika Pratama[3]

[1] Marine Engineering College, Dalian Maritime University, Dalian, China
emjer@ntu.edu.sg, n.wang.dmu.cn@gmail.com
[2] School of Electrical and Electronic Engineering,
Nanyang Technological University, Singapore, Singapore
{liuf0024,yzhang067}@e.ntu.edu.sg
[3] Department of Computer Science and IT,
La Trobe University, Melbourne, Australia
m.pratama@latrobe.edu.au

Abstract. With the objective of extracting useful information from the opinion-rich data on Twitter, both supervised learning-based and unsupervised lexicon-based methods for sentiment analysis on Twitter corpus have been studied in recent years. However, the unique characteristics of tweets such as the lack of labels and frequent usage of emoticons poses challenges to most of the existing learning-based and lexicon-based methods. In addition, studies on Twitter sentiment analysis nowadays mainly focus on domain specific tweets while a larger amount of tweets are about personal feelings and comments on daily life events. In this paper, a hybrid approach of augmented lexicon-based and learning-based method is designed to handle the distinctive characteristics of tweets and perform sentiment analysis on a user level, providing us information of specific Twitter users' typing habits and their online sentiment fluctuations. Our model is capable of achieving an overall accuracy of 81.9 %, largely outperforming current baseline models on tweet sentiment analysis.

Keywords: Twitter · Social media · Date mining · Sentiment analysis

1 Introduction

Social media services such as Twitter and Face book have become increasingly popular platforms for people to share personal experience and opinions towards events, celebrities, politics and products. Extensive studies have been focusing on discovering useful information from domain specific tweets with the help of sentiment analysis. However, instead of studying tweets on a specific topic, in this paper, we exploit methods for sentiment analysis and study tweets on a user level.

In the recent years, much research on computation intelligence has been reported [1, 2]. User-level Twitter sentiment analysis has seldom been studied despite the fact that Twitter is frequently used as a channel for its users to record and express their thoughts towards daily events. As demonstrated in [3], tweets can be classified into five

© Springer International Publishing Switzerland 2016
L. Cheng et al. (Eds.): ISNN 2016, LNCS 9719, pp. 426–433, 2016.
DOI: 10.1007/978-3-319-40663-3_49

categories according to the content, namely pointless babble, conversation, self-promotion and news and spam, while pointless babble and conversation take up 78.1 % of all tweets. Investigating Twitter corpus on a user level allows us to understand the sentiment fluctuation and online behavior of Twitter users. These insights are particularly helpful for social and psychological studies [4]. In this context, we propose a hybrid approach that combines lexicon-based and learning-based methods to perform unsupervised Twitter sentiment analysis on a user level. Our approach can be generalized into five steps, namely data cleansing, lexicon augmentation, generation of training data, subjectivity classification and polarity classification. During lexicon augmentation, polarity words in the internet language that are often used by an individual Twitter user, such as "lol" and "wtf", can be identified and added into the original lexicon list. The enriched lexicon list is then used to extract training data for subsequent classifiers.

Compared to traditional supervised methods of sentiment analysis, this unsupervised model is more efficient and less time consuming. The fact that text data crawled from Twitter are unlabeled makes supervised learning a difficult and challenging task. To avoid the trouble of manual labeling, lexicon-based methods are often used. However, lexicon-based methods suffer from low recall. Using a combination of lexicon-based and learning-based methods not only eliminates the manual labeling process necessary for learning-based method, but also improves the low recall of lexicon-based method. In addition, the augmented lexicon method of [5] used in this approach is particularly suitable for user-level sentiment analysis. The augmented lexicon method was used to identify words in internet language that are frequently used to describe a specific entity. However, even though people may express the same sentiment towards an entity, it is common that people use different words and spell them differently. The consistency in word choice and spelling is stronger in tweets written by the same user than those written by different users.

The remainder of this paper is organized as follows. In Sect. 2, we review existing literature related to our work. In Sect. 3, we present our proposed method in details. In Sect. 4, we apply our model on empirical data and evaluate its performance. Conclusions and future works are given in the last section.

2   Related Works

Generally speaking, classification models for sentiment analysis fall into two categories, namely supervised and unsupervised learning-based model. In early research of sentiment analysis using supervised learning method, the performance of several machine learning models including Naïve Bayes and SVMs is evaluated [6]. Since then, researchers have been developing more sophisticated learning-based models to perform subjectivity and polarity classification on short sentences. However, due to the lack of labeled data, unsupervised learning has become more and more important in real-world social media applications. One of the most popular ways to perform unsupervised sentiment analysis is the lexicon-based method [7] which employs a lexicon list to determine the overall polarity of a sentence.

Combining lexicon-based method and learning-based method to perform sentiment analysis is not a new idea. In [8], a subjectivity lexicon is used to auto-generate training data for subjectivity classification on text corpus. A similar idea was also applied to sentiment classification of reviews [9]. However, the approach of [9] does not consider the characteristics of tweets and only classifies reviews into positive and negative ones while our approach is able to classify tweets into three classes (positive, negative and neutral). Subsequently, an unsupervised approach with augmented lexicon method was developed to perform subjectivity classification on tweets [5]. The augmented lexicon method enriches the original lexicon list by finding potential polarity indicators. The proposed approach of this paper is based on augmented lexicon method with modifications and improvements to cater for the characteristics of tweets such as the extensive usage of emojis. In addition, unlike most previous studies on Twitter sentiment analysis which are mainly based on domain specific data, we focus on the application of sentiment analysis on a user level.

3   Proposed Model

In this section, we present the proposed hybrid approach for Twitter sentiment analysis.

3.1   Data Cleansing

Unlike traditional text corpus such as movie reviews, tweets contain large amount of emoticons, emojis and internet slangs. Noisy and unstructured Twitter data significantly affect the performance of essential NLP tools such as sentence tokenizer and POS tagger. Therefore, data cleansing is necessary before we perform further analysis. In the cleansing process, URLs and "@username" are replaced by "httpaddr" and "Username". "#" is removed but the following letters are kept, for example, "#tgif" becomes "tgif". All capital letters are converted to lowercase except for "I" and "U" in "Username". We also restitute some abbreviated words commonly used in internet language, for example, we change "u" to "you". Another important function of the cleansing process is the extraction emoji/emoticon from raw tweets. Emoticons are combinations of ASCII characters that resemble facial expressions and emojis are ideograms, each with a unique corresponding Unicode. A raw tweet string will be transformed into a 2-tuple; the first element stores the text part of the tweet and the second element stores the emojis and emoticons in the tweet.

Table 1. Examples of emoji/emoticon lexicon list.

	Positive	Negative
Strong Subjective:	=) =D <3 😊 😊 😊 💖	:(:< =/ 😒 😩 😶 😣
Weak Subjective:	📷 🖐 💅 🎨	🤕 💊 😿 💩
Objective:	👀 👤 ⚙ 1	

3.2   Lexicon Augmentation and Generation of Training Data

In this step, we identify potential subjectivity indicators that are not in our original lexicon list and enrich our list so that when extracting tweets using a lexicon-based method in the following steps, the problem of low recall can be alleviated to some extent. The idea of lexicon augmentation is that if the occurrence of a word appears more frequently in subjective sentences, the word is more likely to be a subjectivity indicator. In the context of tweets, these potential indicators can be words in internet language such as "lol" or words spelled in abbreviated ways such as "luv" for "love".

A widely used lexicon list for sentiment analysis, MPQA subjectivity lexicon [8], is used here as our original lexicon list. In this manually labeled list of words in Standard English, there are 5,569 strong subjective words and 2,653 weak subjective words with their respective polarity. Besides MPQA lexicon, we create a lexicon for emojis and emoticons as well. There are a total of 751 emoji characters commonly supported across most platforms [10]. We have manually selected 78 strong subjective emojis and 34 weak subjective ones to form the emoji/emoticon lexicon. We then assign a score of +2 and −2 to strong positive and strong negative words/emojis respectively and +1 and −1 to weak positive and weak negative words/emojis respectively. Table 1 shows some examples in the manually labeled emoji/emoticon lexicon list.

To perform lexicon augmentation, we first apply baseline lexicon-based method to identify subjective tweets with our original word lexicon and emoji lexicon. To do this, we adopt the approach of [11] with slight modifications. In this approach, tweets are split into several micro-phrases $m_1 \cdots m_n$, according to splitting cues which consist of punctuations and conjunction words. The subjectivity and polarity of tweets $S(T)$ and $P(T)$ are then calculated by summing the polarity score of each micro-phrase. If a negation word appears in the micro-phrase, the polarity of the phrase is reversed. To take into consideration of emojis and emoticons, we treat the emojis and emoticons in a tweet as one micro-phrase. The subjectivity and polarity of the tweets can be calculated as follows:

$$P(T) = \sum_{i=1\ldots n} \sum_{j=1\ldots k} \mathrm{score}\,(w_j) \tag{1}$$

$$S(T) = \sum_{i=1\ldots n} \left| \sum_{j=1\ldots k} score(w_j) \right| \tag{2}$$

where n is the number of micro-phrases, k is the number of words in i^{th} micro-phrase and $\mathrm{score}\,(w_j)$ is the polarity score of word w_j.

After extracting subjective tweets with lexicon-based method, we use the Chi-square test to identify potential subjective words among these subjective tweets. Since the expected count of a word is often less than five in small datasets, the Yates corrected Chi-square method is used. The first step of the Yates' Chi-square test is to

Table 2. Contingency table.

	With w	Without w	Row Total
Subjective tweets	f_{11}	f_{12}	$f_{11}+f_{12}$
Objective tweets	f_{21}	f_{22}	$f_{21}+f_{22}$
Column total	$f_{11}+f_{21}$	$f_{12}+f_{22}$	

compute the contingency table (Table 2) for each unique unigram w in the corpus where f_{ij} is the frequency of the corresponding cell as follows:

$$\chi^2(w) = \sum_{i=1,2} \sum_{j=1,2} \frac{\left(\left|f_{i,j} - E_{i,j}\right| - 0.5\right)^2}{E_{i,j}} \tag{3}$$

where $E_{i,j}$ is the expected frequency given by

$$E_{i,j} = \frac{(f_{i,1} + f_{i,2}) \times (f_{1,j} + f_{2,j})}{f_{11} + f_{12} + f_{21} + f_{22}} \tag{4}$$

However, words that have high chi-square score are not necessarily polarity indicators; they can be entities that the author often comments on positively or negatively. Therefore, manual inspection is needed when deciding which word to add into the lexicon list as a subjectivity indicator. After identifying potential subjective words, we perform a similar test using subjective tweets to determine the polarity of these words.

We iterate this process for a few rounds until no more polarity indicators can be found to add to the lexicon list. To generate training data for subsequent classification, we apply the aforementioned lexicon-based method one last time with the most updated lexicon list.

3.3 Subjectivity and Polarity Classification

In this step, we perform polarity classification based on a two-step approach [12]. In this approach, polarity classification is done in two steps: first, perform subjectivity classification and then further classify subjective data into positive and negative data. To ensure high precision, tweets with $|P(T)| \geq 2$ are selected as subjective training data and tweets with $S(T) = 0$ are selected as objective training data. Our basic features are unigrams, bigrams and emojis/emoticons. In addition, we also include two more handcrafted features: (1) Whether the tweet (before cleansing) contains words in all capital letters; (2) Whether the tweet contains lengthening words such as "cooool".

We build our subjectivity and polarity classifiers using neural networks with two hidden layers. The input dimension of our classifiers depends on the number of unigram and bigram features from the user's tweets, but the number of neurons at each hidden layer and at the output layer is fixed. We use *Adam* as our objective function optimizer of [13]. However, parameters such as the number of neurons at each level and training batch size have not fine tuned.

Table 3. List of words with a high chi-square score.

Confidence interval	Potential subjective indicators
99.5 % (>7.879)	ass, thanks
99 % (>6.635)	tf, lol, luv
97.5 % (>5.024)	netflix, food, craving, god, lafayette

4   Performance Evaluation

To evaluate the performance of our proposed method, we collected a random list of Twitter user IDs and selected three users whose account satisfies the following two conditions: (1) The account is maintained by a real person for personal use; (2) This user has more than 2,000 tweets. We crawled 2,000 tweets from each of the selected user (user 227386****, user 256999**** and user 141277****) using Twitter API. A total of 1,785 tweets from each user are used as training data and the remaining 215 are used for testing. Testing data are manually labelled by three different persons to avoid bias.

After data cleansing, we apply augmented lexicon method to identify potential subjectivity indicators for each user and create their user-specific augmented lexicon list. Table 3 shows the list of words with high chi-square scores (excluding words already in the lexicon list and stop words) for user 227386****. However, when deciding which word should be added into the lexicon list, manual inspection is needed. In this case, for example, words such as "netflix", "food" and "lafayette" represent entities that the author often comments positively and negatively on, and these word by themselves, unlike "ass", "thanks", "tf" and "luv", do not emit any positive and negative sentiment. Thus, we should not include "netflix", "food" and "lafayette" into lexicon list.

After we have determined the polarity of these indicators, we add them to the original lexicon list and use this user-specific lexicon list to extract subjective and objective tweets for the subsequent classification processes. In the case of user 227386****, we have identified 910 subjective tweets and 784 objective tweets among training data. However, these labelled tweets are not sufficient for training. To collect more data for training, we use the same user-specific lexicon list to label 5,000 tweets randomly selected from Stanford Twitter Sentiment (STS) dataset. The STS dataset contains 1,600,000 labelled tweets with emojis and emoticons removed. As a result, we have identified 1,512 subjective tweets and 1,623 objective tweets using user 227386****'s augmented lexicon. We use these 3,135 tweets with the 1,694 from user 227386**** for training and apply our trained model to classified testing data for this user. We repeat this process for user 256999**** and user 141277**** and perform polarity classification for these two users as well.

The average accuracy rate of the classification model for these three different users is 81.9 %. To demonstrate the effectiveness of our proposed model, we compare the performance of the following four models on the three sets of user data above. Table 4 shows the average precision, recall and F-score for positive, negative and objective tweets.

- Learning-based method using neural network trained by the STS dataset;
- Lexicon-based method proposed by Musto et al. (2014) with MPQA lexicon list and Emoji/Emoticon lexicon list;
- A hybrid of lexicon-based and learning-based method without augmented lexicon;
- A hybrid of lexicon-based and learning-based method with augmented lexicon.

From Table 4, we can see that the learning-based model has the worst performance with an overall accuracy of 57.4 %. One of the main reasons is that emojis and emoticons are removed from the tweets in the STS dataset. Emojis and emoticons are often strong indicators of the sentiment of tweets. In many cases, tweets appear objective if emojis are removed, for instance, "Ready for Friday ☺". After we take into consideration of the effect of emoji/emoticon, we are able to achieve 75.5 % accuracy using the lexicon-based method. Applying the learning-based method with augmented lexicon method can further improve the performance since the learning model is able to pick up other sentence features and user-specific texting habits can be learned.

Table 4. Precision, recall and F-score results

	Learning-based method			Lexicon-based method		
	Precision	Recall	F-score	Precision	Recall	F-score
Positive	0.584	0.623	0.603	0.799	0.751	0.774
Negative	0.562	0.420	0.480	0.540	0.891	0.673
Objective	0.657	0.764	0.706	0.853	0.771	0.810
Accuracy	57.4 %			75.5 %		
	Hybrid method without augmented lexicon			Hybrid method with augmented lexicon		
	Precision	Recall	F-score	Precision	Recall	F-score
Positive	0.808	0.790	0.799	0.851	0.838	0.844
Negative	0.734	0.747	0.740	0.737	0.817	0.775
Objective	0.766	0.829	0.796	0.814	0.855	0.834
Accuracy	78.6 %			81.9 %		

5   Conclusions

Text data on Twitter are short and fast involving, which poses challenges to Twitter sentiment analysis. In this paper, we proposed an unsupervised method, which is able to deal with the frequent usage of emojis/emotions and identify polarity words frequently used in informal English and internet language to perform sentiment analysis on a user level. Through evaluation using empirical data, we show that the proposed method indeed can successfully identify potential polarity indicators according to user-specific texting habit. More importantly, it outperforms traditional learning-based method and lexicon-based method. For the next step, we are interested in coupling tweet polarity with other user information such as date, time and location when the tweet was posted so that we can further investigate the user's sentimental fluctuation and its possible cause.

Acknowledgement. The authors would like to acknowledge the funding support from the National Natural Science Foundation of P. R. China (under Grants 51009017 and 51379002), Applied Basic Research Funds from Ministry of Transport of P.R. China (under Grant 2012-329-225-060), and Program for Liaoning Excellent Talents in University (under Grant LJQ2013055).

References

1. Wang, N., Er, M.J., Han, M.: Generalized Single-hidden layer feed forward networks for regression problems. IEEE Trans. Neural Netw. Learn. Syst. **26**(6), 1161–1176 (2015)
2. Wang, N., Er, M.J., Han, M.: Parsimonious extreme learning machine using recursive orthogonal least squares. IEEE Trans. Neural Netw. Learn. Syst. **25**(10), 1828–1841 (2014)
3. Pears Analytics: Twitter Study August – 2009. http://www.pearanalytics.com/wpcontent/uploads/2009/08/Twitter-Study-August-2009.Pdf
4. De Choudhury, M., Counts, S., Horvitz, E.: Social media as a measurement tool of depression in populations. In: 7th International Conference on Weblogs and Social Media, pp. 128–137. ACM, New York (2013)
5. Zhang, L., Ghosh, R., Dekhil, M., Hsu, M., Liu, B.: Combining lexicon-based and learning-based methods for Twitter sentiment analysis. HP, Technical report HPL-2011-89 (2011)
6. Pang, B., Lee, L., Vaithyanathan, S.: Thumbs up? Sentiment classification using machine learning techniques. In: ACL-02 Conference on Empirical Methods in Natural Language Processing, vol. 10, pp. 79–86. Association for Computational Linguistics, Stroudsburg (2002)
7. Ding, X., Liu, B., Yu, P.S.: A holistic lexicon-based approach to opinion mining. In: 2008 International Conference on Web Search and Data Mining, pp. 231–239. ACM, New York (2008)
8. Wiebe, J., Riloff, E.: Creating subjective and objective sentence classifiers from unannotated texts. In: Gelbukh, A. (ed.) CICLing 2005. LNCS, vol. 3406, pp. 486–497. Springer, Heidelberg (2005)
9. Tan, S., Wang, Y., Cheng, X.: Combing learn-based and lexicon-based techniques for sentiment detection without using labeled examples. In: 31st Annual International ACM SIGIR Conference on Research and Development in Information Retrieval, pp. 743–744. ACM, New York (2008)
10. Novak, P.K., Smailović, J., Sluban, B., Mozetič, I.: Sentiment of Emojis. PLOS ONE **10** (12), e0144296 (2015)
11. Musto, C., Semeraro, G., Polignano, M.: A comparison of lexicon-based approaches for sentiment analysis of microblog posts. In: 8th International Workshop on Information Filtering and Retrieval, DART, vol. 1314, pp. 59–68, Pisa, Italy (2014)
12. Barbosa, L., Feng, J.: Robust sentiment detection on Twitter from biased and noisy data. In: 23rd International Conference on Computational Linguistics: Posters, pp. 36–44. Association for Computational Linguistics, Stroudsburg (2010)
13. Kingma, D., Ba, J.: Adam: a method for stochastic optimization. In: The International Conference on Learning. Representations (ICLR), San Diego (2015)

The Development of a Nonlinear Curve Fitter Using RBF Neural Networks with Hybrid Neurons

Michael M. Li$^{(\boxtimes)}$

Centre for Intelligent Systems and School of Engineering and Technology,
Central Queensland University, Rockhampton, QLD 4701, Australia
m.li@cqu.edu.au

Abstract. This paper investigates a new method using radial basis function (RBF) neural networks with an additional linear neuron for solving nonlinear curve fitting problem. The complicated unknown function to be fitted is approximated by a set of Gaussian basis function with a linear term correction. The proposed new technique is first used to evaluate two benchmark examples and subsequently applied to fit several heavy ion stopping power datasets (MeV energetic projectiles in aluminium). Due to the linear correction effect, the proposed approach significantly improves accuracy of fitting without adding much computational complexity. The developed method can be served as a standalone curve fitter or implemented as a proprietary software module to be embedded in an intelligent data analysis package for applications in regression analysis.

Keywords: Curve fitting · RBF · Neural networks · Stopping power

1 Introduction

Curve fitting is not a new concept but it takes on a new life when it is actually used in many modern scientific domains. By fitting a set of measured data points to a smooth curve, relevant analytical properties of the curve such as the rate of change, the local minimum and maximum points, and the asymptotic behavior can be conveniently calculated by using calculus. Thus underlying characteristics produced by the dataset can be determined accurately.

The process of curve fitting is to construct a curve with a mathematical function or an equation. The constructed curve should match the shape of data point distribution through a certain measure of the closeness between the data and the model function. To perform a curve fitting job, it usually involves two correlative tasks: (i) To find an appropriate regression function with a few parameters so that the function can reliably describe the phenomenon or the system represented by the data points; (ii) To resolve the values of parameters by a numerical technique. For the first task, it relies on a good understanding of the underlying science or phenomenon, the observation of data point distributions, and the researcher's experience in data analysis. The candidate's regression functions are often available from some practical handbooks and scientific computing software packages. For example, Arlinghaus' Practical Handbook of Curve

© Springer International Publishing Switzerland 2016
L. Cheng et al. (Eds.): ISNN 2016, LNCS 9719, pp. 434–443, 2016.
DOI: 10.1007/978-3-319-40663-3_50

Fitting [1] is a comprehensive manual containing numerous parametric functions with illustrations of application examples ranging from atmospheric science to epidemiology. Some popular statistical and computing software packages such as MATLAB, R, and GraphPad etc. also provide a built-in function list that allows users to configure a fitting with a set of given input data. However, these functions or their combinations may not always best describe a complicated process or a system in real world application due to the complexity of the problem. As for the second task, the common practice is to use the least squares approximation to determine the fitting parameters, where the sum of squares between the raw data points and those modeled by the fitting function is min-imized through a numerical analysis algorithm, commonly Levenberg-Marquartd algorithm. Since the used parametric function often has a nonlinear relation with the unknown parameters (for example, a three-parameter Gompertz function), determina-tion of the optimal parameters could be difficult and actually requires a lengthy iterative process starting from an initial guess. To guarantee its convergence, the selection of a good initial value may be critical and it also may need a few times of human intervention if the iteration process does not work towards a correct direction.

To address the above challenges, we consider using neural networks to provide a novel technique for finding the fitting function and determining the corresponding parameter values. In terms of the neural network method, the raw data are fitted by a set of basis function denoted by neurons. Each neuron has a definite pre-set function. This makes the proposed method generic without concerning an exhaustive searching for a specific regression function in a particular application, while the parameters are determined by a training algorithm based on the statistical learning theory. Mathe-matically, curve fitting is one type of classical problems in function approximation theory, where the function with the unknown analytical form is approximated by a group of simple function like polynomial or exponential, or by a set of basis function such as sigmoid function or radial basis function (RBF). Since RBF neural networks have been proved to possess universal approximation capacity [2, 3], it is considered as a precedence option. However, in the practical applications, one technique issue becomes obvious. To ensure its best approximation property, the number of basis function should be chosen to be sufficiently large in a network configuration. This situation is not favorable in the curve fitting problems since the primary goal of curve fitting often is to find a relatively simple empirical formula with as few as possible number of coefficients. If only a few of nodes in a network is dealt with, the boundary behavior of the basis function may deteriorate due to the localized effect from the superposition of limited number of Gaussian function. To overcome this difficulty, we introduce an additional linear neuron in the RBF network architecture. The extra linear term plays a correction role through an adaptive training process. This effectively balances the requirement of number of basis function and the fitting accuracy.

In this article, the proposed RBF neural network with an additional linear neuron is used to study for a few benchmark examples and subsequently it is applied to fit stopping power curves versus the energy of projectile (a group of MeV energetic projectiles Li, B, C, N, Al, Si, Ni, and Cu in the elemental target Al). Stopping power data have extreme important applications in two fields - ion beam analysis technique and radiation therapy [4]. The proposed approach is a global method, since it contains the weighted linear superposition from a series of Gaussian basis function at different centers spreading the data spectrum and each center constructs an approximation through a Gaussian with a linear correction based on samples in their neighborhood.

The organization of this paper is as follows. In Sect. 2, the proposed method is described; and the benchmark numerical examples are tested. Next, computer simulations and results for stopping power data are discussed in Sect. 3. Finally, Sect. 4 concludes the paper.

2   The Proposed Method

2.1   RBF Neural Networks with Additional Linear Neuron

Mathematically, an RBF network is a linear combination of a set of weighted radial basis functions. It typically consists of three layers – an input layer, a hidden layer with activation functions and an output layer. The network represents a nonlinear mapping that transforms the input vector into the output scalar through a set of basis functions. When it is applied to the problem of curve fitting, the unknown function $f(\mathbf{x})$ to be fitted can be directly approximated by an RBF as below,

$$f(\mathbf{x}) \approx \sum_{i=1}^{N} w_i \varphi(||\mathbf{x} - \mathbf{c}_i||) \tag{1}$$

where $\mathbf{x}$ is the input vector, w_i are weights, φ is a set of N basis functions with centers $\mathbf{c}_i$ $(i = 1...N)$ and $||.||$ denotes the Euclidian distance.

A number of functions have been tested as basis functions. The Gaussian function is typically selected as the basis function in many applications [5, 6], due to its outstanding analytical properties such as smoothness and infinite differentiability. In the case of Gaussian as the basis function, we have the following form RBF network,

$$y(\mathbf{x}) = \sum_{i=1}^{N} w_i \exp(-\frac{(||\mathbf{x} - \mathbf{c}_i||)^2}{2\sigma^2}) \tag{2}$$

where σ is the width, and y is the output of the network.

The Gaussian function is a type of localized function. It decays quickly from the locations nearby the centre. More generally, its influence decreases according to the Mahalanobis distance from the centre. This suggests that data points far from centers with a large Mahalanobis distance will fail to activate that basis function. Particularly, provided the centers sorted in ascending, basis functions at the margins of lower boundary of first center and upper boundary of last center will not be able to efficiently to deal with the behavior of the function to be approximated since other Gaussians influence little in the margin regions. In addition, in the vicinities near centers, the Gaussian might overestimate a bit for the function. To overcome these problems, we proposed to add an extra linear term to the usual RBF network so that it not only can help efficiently to reproduce the global behavior of function but also plays a role of corrections.

With adding the extra linear term, the RBF network has the following form,

$$y(\mathbf{x}) = \sum_{i=1}^{N} w_i \varphi(||\mathbf{x} - \mathbf{c}_i||) + k\mathbf{x} + h \tag{3}$$

where k is the linear coefficient, and h is the constant term.

The training of RBF network requires determinations of parameters for centers and weights. It usually is a two-stage process. In first the stage, the number of centers and their locations are determined and then follow a process to evaluate the weights. There are a few ways to determine the centers which include the random selection from subset of training data, clustering techniques and a supervised learning process. We only consider the relative simple one - the random selection from the training set for our applications. Following the determination of centers, a simple and straightforward equation for evaluating weights can be derived as below. By re-writing the Eq. (3) in the form of matrix,

$$\begin{aligned} y(\mathbf{x}) &= \sum_{i=1}^{N} w_i \varphi(||\mathbf{x} - \mathbf{c}_i||) + k\mathbf{x} + h \\ &= \mathbf{\Phi}^T(r)\mathbf{W} \end{aligned} \tag{4}$$

where

$$r = ||\mathbf{x} - \mathbf{c}_i||$$
$$\varphi_i(r) = \varphi(||\mathbf{x} - \mathbf{c}_i||)$$

$$\mathbf{\Phi}^T(r) = [\varphi_1(r), \varphi_2(r), \ldots \ldots \varphi_N(r), \mathbf{x}, 1]$$

$$\mathbf{W} = [w_1, w_2, \ldots \ldots w_N, k, h]^T$$

Considering each point in the training set $\{\mathbf{x}_j, y_j\}_{j=1}^{M}$ for Eq. (4), we obtain

$$\sum_{i=1}^{N} w_i \varphi_i(r_j) + k\mathbf{x}_j + h = y_j \quad j = 1, 2, \ldots M \tag{5}$$

The set of linear equations of (5) can be re-written in terms of matrix form,

$$\mathbf{\Phi}^T \mathbf{W} = \mathbf{Y} \tag{6}$$

$$\text{where } \mathbf{\Phi} = \begin{bmatrix} \varphi_1(r_1) & \varphi_2(r_1) & \cdots & \varphi_N(r_1) & x_1 & 1 \\ \varphi_1(r_2) & \varphi_2(r_2) & \cdots & \varphi_N(r_2) & x_2 & 1 \\ \cdot \\ \cdot \\ \cdot \\ \varphi_1(r_M) & \varphi_2(r_M) & \cdots & \varphi_N(r_M) & x_M & 1 \end{bmatrix}$$

$$\varphi_i(r_j) = \exp(-\frac{\|\mathbf{x}_j - \mathbf{c}_i\|^2}{2\sigma^2})$$

$$Y = [y_1, y_2, \ldots \ldots y_M]^T$$

The solution of the matrix Eq. (6) is

$$W = (\mathbf{\Phi}^T \mathbf{\Phi})^{-1} \mathbf{\Phi}^T Y$$
$$= \mathbf{\Phi}^+ Y \tag{7}$$

where $\mathbf{\Phi}^+$ is the pseudoinverse of matrix $\mathbf{\Phi}$; it is defined as

$$\mathbf{\Phi}^+ = (\mathbf{\Phi}^T \mathbf{\Phi})^{-1} \mathbf{\Phi}^T \tag{8}$$

The computation of the pseudoinverse $\mathbf{\Phi}^+$ can be conveniently performed by using the algorithm known as singular-value decomposition (SVD).

2.2 Numerical Examples

Before the proposed method is applied to actual stopping power datasets, two numerical examples are used to demonstrate the linear term effects in the developed approach. As the first example, the zero-order spherical Bessel function is examined,

$$J_0(t) = \frac{\sin(t)}{t}$$

Due to its smoothness, this is a relatively simple case where the smooth nonlinear function $J_0(t)$ is sampled in the interval $[-10,10]$ with 50 data points and subsequently are fitted using the proposed RBF neural network technique. Figure 1 compares the fitting result with the linear term and without the linear term. It can be seen that the linear term indeed improves the fitting at both end positions where it offsets the gradually weakening impact of Gaussian terms.

The second test example is a typical nonlinear regression problem where the MacKay's Hermit polynomial [7] is added 10 % random noise and sampled in the interval $[-4, 4]$. The data might be difficult to fit using a conventional numerical method due to its noise. Considering a 3-centered RBF neural network with a linear term for this application, the dataset is fitted well as shown in Fig. 2. However, if the linear term is removed, surprisingly the RBF neural network gives an unacceptable fitting result. It completely fails the fitting as clearly seen in Fig. 2. The results show

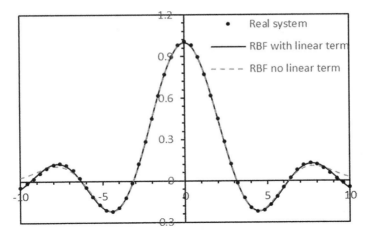

Fig. 1. Testing $J_0(t)$ dataset using the developed RBF fitting (Color figure online)

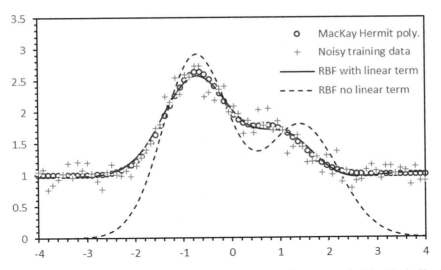

Fig. 2. Comparison of RBF fitting with linear and without linear term for MacKay's Hermit polynomial (Color figure online)

that the proposed method is considerably robust in the case of difficult data when the moderate level noises exist.

3 Experimental Results and Discussion

In addition to numerical examples in last section, our method has been used to perform the empirical fitting of stopping power curves for heavy ions Li, B, C, N, Al, Si, Ni, and Cu in the aluminium foils with incident energies from 0.01 meV/per nucleon up to

4.5 meV/per nucleon. Aluminium is chosen as the target material because it is a typical metal substance with wide use in industry. Its stopping power knowledge is representative for understanding the interaction mechanism of electrons with substance during the collision process. The data considered are mainly from the data collections published by Nuclear Data Services, International Atomic Energy Agency (IAEA) [8], with significant personal contribution from Paul and Schinner [9]. Since the measurements of these data have spanned over half a century at different laboratories, errors or inconsistencies may be unavoidable. A simple filtering principle has been applied to exclude individual data points with greater uncertainties or outliers.

We adopted the procedure of random subset selection to determine the center locations of Gaussians. In the problems of curve fitting or regression analysis, it should be desirable if there is a minimal amount of number centers while not deteriorating the fitting accuracy. Our empirical tests have found that a set of 4 centers is minimal for our problem and a different set of centers selected randomly has little impact to the fitting outcomes but gives the different weight values. The center positions along the energy axis for the following fittings are set to {0.01572, 0.06107, 0.1944, 0.6103}, and parameter σ is set to 1. After this, it is straightforward to achieve weights from a computation according to Eqs. (7) and (8).

As the result of fitting, a group of values of parameters w_1, w_2, w_3, w_4, k, and h along with the normalized root mean squared error $\sqrt{\varepsilon^2}$ are obtained and listed in Table 1 for each projectile. In all cases, the coefficient of determination of R^2 are greater than 0.99, which indicates the data points close to the fitted curve. The resultant parameters then can be conveniently used as a set of coefficients thus a simple empirical equation describing the relationship of stopping power (S) upon the projectile energy (E) can be expressed as

$$S(E) = \sum_{i=1}^{4} w_i e^{-(E-E_i)^2/2} + kE + h \tag{9}$$

where E_i ($i = 1,2,3,4$) are constants, and values of coefficients w_i, k and h can be found from the tabulation for various projectiles.

Constrained to the pages of this paper, only partial figures of fitting curves are presented. Here two representative figures are illustrated. Figures 3(a, b) are plots of

Table 1. Fitting parameters for projectiles Li, B, C, N, Al, Si, Ni and Cu in target Al.

Projectile	w_1	w_2	w_3	w_4	k	h	$\sqrt{\varepsilon^2}$
Li	29.80	−56.44	71.84	40.60	−5.04	3.31	0.0551
B	−12.59	1.85	12.62	−16.21	0.016	10.65	0.0751
C	47.65	−131.58	164.55	−99.01	12.95	26.64	0.0376
N	−124.60	206.26	−239.52	134.53	−20.48	−2.69	0.1672
Al	−77.62	167.79	−228.42	156.72	−14.25	−17.72	0.1734
Si	261.37	−470.09	564.62	−317.37	49.25	33.10	0.2344
Ni	−1014.8	984.03	−945.05	279.25	−198.24	262.14	0.4500
Cu	198.99	−13.70	−21.56	136.69	53.40	−119.29	0.2566

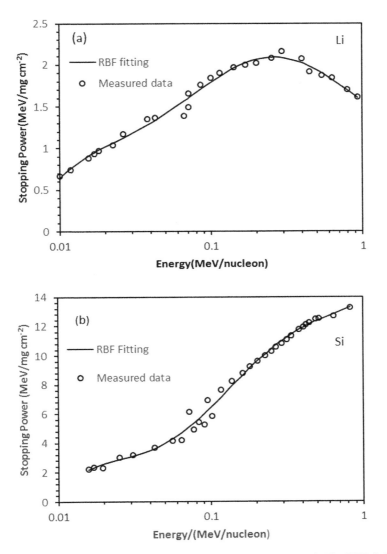

Fig. 3. (a)–(b). Stopping powers for Li and Si projectiles in the target Al. The RBF fitting with linear term are compared with the measured data.

fitted stopping power curves along with the original measured data for Li and Si in Al. From these figures, it can be seen that the agreement is excellent; the fitted curves exhibit the general shape of data points and typical features including peaks. The peak of curves reflects the underlying interaction of projectile particles against the atoms of target matter where the shift of the peak position depends on the atomic number of the projectile. This is consistent with the theoretical modelling from the sophisticated quantum physics. On average, the deviations of fitted curves from the corresponding

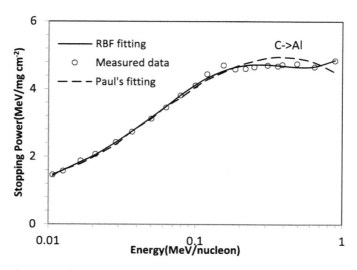

Fig. 4. Stopping power for C ions in aluminium versus incident energy. The measured data are compared to RBF fitting and Paul's empirical formula prediction.

experimental data are within 3.8 %. However, we noticed that systematic error of measured data ranges between 2.5 % and 10 %. Only a few isolated points have a relative large discrepancy.

Finally, to compare our fitting with those from other methods, we consider a representative numerical-based method - Paul's empirical fitting [9]. Figure 4 compares our result with Paul's fitting for the carbon projectiles in the aluminium target. It can be clearly seen that at the low energy end our result and Paul's empirical fitting are consistently close to the measured points while at the high energy region, our fitting matches the experiment points well but the Paul's result deviates them in a moderate degree.

4   Conclusions

A novel method using RBF neural networks has been proposed for nonlinear curve fitting. By introducing an additional linear neuron, the constructed neural network enriches the function behavior, where a global approximation is constructed by a series of superposition of Gaussian with a linear correction. The fitting parameters are determined by the learning algorithm. Benchmark numerical examples and several heavy ions stopping power data fitting have been demonstrated with satisfactory results. Since many physical systems exhibit the underlying behaviors to be close to but not exactly Gaussian, particularly stopping phenomenon of energetic particles in materials which typically reflects this characteristic, the proposed approach is well suited for fitting stopping power curves. The proposed method has two outstanding advantages over the conventional numerical based technique. First, it is a generic method with a relative few number of parameters. This allows a simple and fast computation for an automated data fitting task where high volume data are required to deal with. Second, this adaptive method performs well with the noisy data since the noise in data is efficiently averaged during the training. A further investigation and more tests from diverse datasets should allow the implementation of this method as a proprietary software module to be embedded into a practical intelligent data analysis package for various applications in curve fitting.

References

1. Arlinghaus, S.L.: Practical Handbook of Curve Fitting. CRC Press, Boca Raton (1994)
2. Park, J., Sandberg, I.W.: Universal approximation using radial basis function. Neural Comput. **3**, 246–257 (1991)
3. Poggio, T., Girosi, F.: Networks for approximation and learning. In: Proceeding of IEEE 78, pp. 1481–1497 (1990)
4. Konac, G., Klatt, Ch., Kalbitzer, S.: Universal fit formula for electronic stopping power of all ions in Carbon and Silicon. Nucl. Instrum. Meth. Phys. Res. B **146**, 106–113 (1998)
5. Haykin, S.: Neural Networks: A Comprehensive Foundation, 2nd edn. Prentice Hall (1998)
6. Li, M.M., Verma, B., Tickle, K.: RBF neural networks for solving the inverse problem of backscattering spectra. Neural Comput. Appl. **17**, 391–399 (2008)
7. MacKay, D.J.C.: Bayesian interpolation. Neural Comput. **4**, 415–447 (1992)
8. Stopping Power of Matter for Ions. https://www-nds.iaea.org/stopping/index.html
9. Paul, H., Schinner, A.: Empirical stopping power tables for ions from $_3$Li to $_{18}$Ar and from 0.001 to 1000 MeV/nucleon in solids and gases. At. Data Nucl. Data Tables **85**, 377–452 (2003)

Networks of Coupled Oscillators for Cluster Analysis: Overview and Application Prospects

Andrei Novikov$^{(\boxtimes)}$ and Elena Benderskaya

Peter the Great St. Petersburg Polytechnic University,
St.-Petersburg, Russia195251
spb.andr@yandex.ru, helen.bend@gmail.com

Abstract. Rapid growth of data is caused by evolution of information technologies in various areas and as a result there is a need to have approaches and
methods that are able to ensure faster and efficient solution of data mining
problems. Oscillatory networks can be considered as a way that is proposed by
nature to solve problems of data mining and especially for cluster analysis. The
article presents analysis of oscillatory networks based on Kuramoto model and
chaotic dynamic. Those models are representatives of two classes of oscillators:
harmonic and chaotic oscillators respectively. Additionally considers how to
apply the oscillatory networks for general problems of cluster analysis are
examined.

Keywords: Oscillatory network · Chaotic network · Phase oscillator ·
Synchronization · Kuramoto model · Cluster analysis · Parallel algorithm

1 Introduction

Cluster analysis is one of the most relevant and demanded scientific area of knowledge
that is rapidly evolving. Its formalization leads to a mathematical formulation of the
problem of clustering in almost all subject areas of science [10, 16]. Researches from
various disciplines contribute to development of cluster analysis methods when try to
take into account specifics of problem solution for data from its own domain of
knowledge and also when use features of data from various sources [6, 10, 14, 21]. In
this paper we do not present full review of the new clustering methods, but identify
some trends in the development of cluster analysis and special attention is given to area
of bio-inspired cluster analysis methods.

2 Methods Based on Oscillatory Networks

Basis of oscillatory methods constitute oscillatory elements that are coupled with each
other in a system. The oscillatory element can generate regular dynamic (periodic
oscillations) and chaotic dynamic (chaotic oscillations). The key property in oscillatory
systems is self-synchronization that leads to one of the system states: global, partial or
local and desynchronization. Synchronization also can be divided into three types:
absolute, antiphase and phase synchronization [12]. Usually the phase synchronization

© Springer International Publishing Switzerland 2016
L. Cheng et al. (Eds.): ISNN 2016, LNCS 9719, pp. 444–453, 2016.
DOI: 10.1007/978-3-319-40663-3_51

is used for solving clustering problem where each synchronous ensemble of oscillators encodes only one cluster.

Kuramoto model is one of the successful model that ensures various states of synchronization between oscillators. The model is flexible and can be adapted to solve various practical problems because of simplicity of the differential equation. It allows to study synchronization processes that depend on structure and parameters of oscillatory network [1, 4]. The Kuramoto model consists of a population of N coupled phase oscillators that is described by the following differential equation [18]:

$$\dot{\theta}_i = \omega_i + \frac{K}{N} \sum_{j=1}^{N} \sin\left(\theta_j - \theta_i\right) \tag{1}$$

Phase coordinate θ_i is a basic state variable of oscillator i distributed in the interval from 0 to 2π. Intrinsic frequency of i oscillator ω_i can be considered as a bias that is randomly initialized in line with some probability distribution. Couplings between oscillators are defined by coupling strength K that is important parameter that affects processes of synchronization. The oscillators are said to synchronize if [8]:

$$\lim_{t \to \infty} \left|\dot{\theta}_i(t) - \dot{\theta}_j(t)\right| = 0 \tag{2}$$

Synchronization processes play important role in oscillatory networks where each ensemble of synchronized oscillators corresponds to a single encoded feature from input data space. For example, ensemble of synchronized oscillators can be considered as a single cluster in case of cluster analysis. Networks that are based on Kuramoto model ensure general states of synchronization: global, local or partial synchronization and desynchronization. These states can be set by tuning of coupling strength K between oscillators. When oscillatory network contains inhomogeneous coupling strength that is less or greater than critical coupling strength then local synchronization can be observed in the network. The state of local synchronization implies existence of more than one ensemble of synchronized oscillators.

State of synchronization can be defined by estimation of degree of global synchronization r in network [18]:

$$r = \left|\frac{1}{N \exp(i\varphi)} \sum_{j=1}^{N} \exp\left(i\theta_j\right)\right|, \quad \varphi = \frac{1}{N} \sum_{j=1}^{N} \theta_j \tag{3}$$

Here φ denotes average phase of oscillators. Global synchronization is observed when degree of synchronization $r \to 1$ and desynchronization is observed when synchronization degree $r \to 0$.

Nevertheless, order r does not coverages to zero if coupling strength between oscillators is less than critical value K_c when system is not able to reach global or partial synchronization state, this issue has been deeply investigated and discussed in paper [1]. For example, rough estimate indicates that order r is bounded between $(16 - \pi^2)^{0.5}/4$ and $3^{0.5}/2$ for critical value of coupling strength when the fixed-point is in the $(-\pi/2, \pi/2)$ region [17]. $K > K_c$ is necessary condition of synchronization and sufficient condition is existence $T \geq 0$ for each $t \geq T$, $|\theta_i(t) - \theta_j(t)| < \pi/2 - \varepsilon$, where $0 < \varepsilon < \pi/2$.

Phase distribution from 0 to 2π is undesirable because of unstable behavior that follows from the equation of the Kuramoto model in case of odd number of oscillators equidistant from each other in this range because phase locking cannot be reached [23]. Detailed study of stability can be found in paper [25].

An important advantage of oscillatory networks based on the Kuramoto model is ability to ensure described states of synchronization in various network structures. We have performed experimental study and can confirm that global and local synchronization can be successfully ensured in networks with structures such as grid, star, unidirectional list and bidirectional list [21]. Our experimental study has shown that convergence time of synchronization depends on type of network structure and also on number of oscillators and coupling strength between them. Oscillatory network with a grid structure where each oscillator has connections with four neighbors (right, left, top and bottom) has quadratic dependence of convergence rate $O(n^2)$. For comparison, network with a list structure, where each oscillator has connection with two neighbors (right and left), has cubic dependence $O(n^3)$ [22].

One of the ways to increase convergence rate of synchronization is an adaptive coupling strength between oscillators that can be used instead of static K [27]. We have found out in our experiments that usage of adaptive coupling strength significantly reduces time of synchronization between oscillators, for example, convergence rate of the network with grid structure has liner dependence $O(n)$ instead of $O(n^2)$ and the network with bidirectional list structure has quadratic dependence instead of cubic $O(n^3)$ [21]. Another approach to improve convergence rate is to reduce input feature space by encoding using self-organized feature map (Sync-SOM algorithm) [22].

Informally, the clustering problem is considered as a decomposition problem of a given sample of objects into disjoint subsets that are called clusters. Each cluster should consist of the most similar objects. Clustering allows determining significant attributes, identifying structures or some distribution in feature space. In the oscillatory networks each ensemble of synchronous oscillators corresponds to one cluster of data.

Consider the problem of clustering N-dimensional data that uses oscillatory network based on modified Kuramoto model. The dynamic of each oscillator in the network is described by the modified Kuramoto model of the phase oscillator:

$$\dot{\theta}_i = \frac{K}{N_i} \sum_{j \in N_i} \sin(\theta_j - \theta_i) \tag{4}$$

Phase of oscillator is denoted by θ_i and it is a key state variable that is distributed from 0 to π. All oscillators have the same value of frequency ω therefore it is omitted in this equation due to permanent offset that reduces time of synchronization. Coupling strength K between oscillators is also the same. N_i denotes set of oscillators that have connections with oscillator i.

The degree of local synchronization can be used as an estimation of the end of clustering process that can be considered as stop condition:

$$r_c = \left| \sum_{i=1}^{N} \frac{1}{N_i} \sum_{j \in N_i} e^{\theta_j - \theta_i} \right| \tag{5}$$

Clustering process can be considered complete when the degree of local synchronization reaches $r_c \rightarrow 1$. Oscillators whose phase coordinate converge to a single point encode the same cluster and since each oscillator corresponds to only one object then allocated clusters and objects can be easily extracted from the output dynamic of the network.

The Fundamental Clustering Problems Suite has been used for experimental study of the oscillatory network. Experiments have shown that clusters successfully allocated for almost all samples, except sample EngyTime where two uncertain clusters with equal density intersect in two-dimensional space. The deviation is explained by the fact that clusters are formed by local oscillatory networks whose structures are defined by connectivity radius that should be chosen in accordance with input data like in density based clustering algorithms. Thus EngyTime is considered as a one cluster.

There is another one clustering algorithm based on the Kuramoto model that is known as the hierarchical Sync algorithm [26] that uses the same oscillator dynamic (4). General idea of the algorithm is to substitute radius-connectivity parameter by amount of clusters that should be allocated. On the first step, the smallest connectivity-radius is set and simulation is performed until local synchronization is reached. On the second step, if amount of allocated clusters is less than specified than the connectivity-radius is increased and simulation is repeated. This process is repeated until required amount of clusters is allocated. The advantage of this algorithm is lack of connectivity-radius, but the disadvantage is performance degradation (execution time becomes significantly bigger).

We have compared clustering results and execution times of bio-inspired algorithms Sync (order parameter $r_c = 0.99$) [20, 21] and Sync-SOM [22] (results of hSync [26] are not presented because it requires a lot computational resources, but at the same time clustering results are similar to Sync) with well-known algorithms: agglomerative (with single link) [3], BIRCH [28], CURE [15], DBSCAN [11], K-Means [19], ROCK [14], X-Means (with Bayesian information criterion) [24]. A Python, C++ data mining library PyClustering has been used for that purpose (only python implementation of mentioned algorithms) on PC Intel® Core™ i5-3320M CPU @ 2.60 GHz, 2.60 GHz, 4.00 GB RAM. Results of the comparison are presented in Table 1. The oscillatory based algorithms provide the same features for cluster analysis like DBSCAN and ROCK. Sync algorithm requires much more time than others in case of elongated clusters that are close to each other like WingNut, TwoDiamonds, Chainlink. Sync-SOM algorithm significantly reduces execution time by encoding input feature space, but still has the same problem with elongated clusters like WingNut – it takes ~ 106 s to allocate clusters. Nevertheless, oscillatory based algorithms can be executed in parallel and parallel implementations of these algorithms increases their performance.

To be executed in parallel is one of the most important advantages of neural network and the oscillatory networks based on Kuramoto model are not exception. We have developed multi-core implementation of the network for studying of the mentioned feature using multi-core station HP ProLiant BL460c Generation 7 with four

Table 1. Comparison of clustering results and execution times of the algorithms. Each cell consists of execution time and clustering correctness: (+) means correct result and (−) means incorrect cluster allocation.

Sample (amount of objects)	Clustering algorithms								
	Sync (r = 0.99)	Sync-SOM	Agglom. (single-link)	BIRCH	CURE	DB-SCAN	K-Means	ROCK	X-Means (BIC)
Atom (800 objs)	2.757 (+)	11.41 (+)	631.9 (+)	0.943 (−)	4.815 (−)	2.525 (+)	0.0106 (−)	218.5 (+)	0.7211 (−)
Chainlink (1000 objs)	315.4 (+)	26.87 (+)	1492 (+)	0.305 (−)	5.477 (−)	2.998 (+)	0.0167 (−)	419.8 (+)	0.0935 (−)
Hepta (212 objs)	6.189 (+)	2.867 (+)	13.33 (+)	0.173 (+)	0.303 (+)	0.1472 (+)	0.00817 (+)	3.905 (+)	0.0527 (+)
Lsun (403 objs)	223.4 (+)	10.46 (+)	82.14 (+)	0.783 (−)	0.878 (+)	0.3793 (+)	0.0593 (−)	25.48 (+)	0.0424 (−)
Target (770 objs)	282.5 (+)	15.94 (+)	553.5 (+)	0.849 (−)	3.034 (+)	1.789 (+)	0.0201 (−)	196.1 (+)	0.0965 (−)
Tetra (400 objs)	50.68 (+)	7.493 (+)	89.93 (−)	0.206 (−)	0.972 (+)	0.4188 (+)	0.00355 (+)	26.27 (+)	0.0648 (+)
TwoDiamonds (800 objs)	847.7 (+)	66.87 (+)	698.6 (−)	0.133 (+)	3.282 (+)	1.463 (+)	0.00641 (+)	220.3 (+)	0.0454 (+)
WingNut (1016 objs)	1410 (+)	106.5 (+)	1292 (+)	0.350 (−)	2.763 (+)	2.407 (+)	0.00778 (−)	448.2 (+)	0.0861 (−)

processors Intel® Xeon™ X5660 2.80 GHz/6-core/12 MB/95 W DDR3-1333, HT, Turbo 2/2/2/2/3/3, 8 MB shared L3 cache. Each oscillator of the network is considered as a single unit that is represented by separate thread that has own context where current state of corresponding oscillator is stored such as phase coordinate, the context of oscillator i is read-only for neighbors and available to write by owner – Fig. 1.

In our multi-core implementation each thread can be executed by one of the 24 core. Simulation of the network has been performed with fixed number of iteration using 100000 steps. Average execution time of the multi-core implementation with

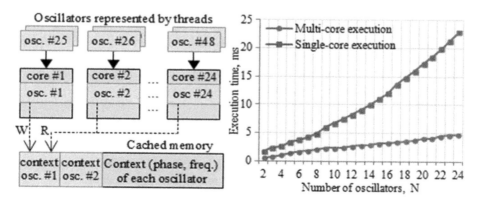

Fig. 1. (Left) The scheme of the multi-core implementation of the oscillatory network. (Right) The comparison of execution time of multi-core implementation and single-core implementation with sizes from 2 to 24 oscillators during 100000 iterations.

sizes from 2 to 24 oscillators (in order to obtain results of completely parallel execution using workstation with 24 cores) has linear character in comparison with the single-core implementation as it is shown on Fig. 1 (right). Results of simulation of the network with bigger sizes are presented in Table 2.

Table 2. Comparison of execution times with various number of oscillators.

Oscillators, N	200	300	400	500	600	700	800	900	1000
Single-core., ms	915	2233	3618	5166	7110	10931	14279	16873	20101
Multi-core., ms	123	295	458	628	810	1211	1561	1775	2111
Difference, times	7.43	7.56	7.9	8.22	8.77	9.02	9.14	9.50	9.52

The proposed multi-core implementation of the network ensures general hallmark of neural networks to be executed in parallel and provides opportunity to use oscillatory networks based on Kuramoto model for real practical problems such as cluster analysis or image segmentation [23].

3   Methods Based on Chaotic Neural Networks

The chaotic neural network based on the logistic transformation deserves special attention among models of chaotic oscillator systems. Dynamic states of the chaotic neural network allow to solve clustering problems when there is a minimum priori information. Amount of clusters is determined during simulation of the network [2]. The best result of clustering is determined by results of post-processing of output dynamic of the network.

The structure of the chaotic oscillatory network is similar to the recurrent network with a single layer where all elements are connected with each other (all-to-all type structure) [2, 6]. Learning algorithm of the chaotic neural network is based on weight formation. The weight defines strength of interaction between neurons. Each neuron corresponds to only one object from input data set.

State of each neuron is initialized by a random value $y_i \in [-1, 1]$. State of neuron is defined by following equation:

$$y_i(t+1) = \frac{1}{\sum\limits_{i \neq j} w_{ij}} \sum\limits_{i \neq j} w_{ij} f(y_i(t)) \tag{6}$$

where $f(y_i)$ is defined by the logistic function that is used as a transfer function of neuron $f(y) = 1-2y^2$. The chaotic neural network is initialized by input pattern and as a result matrix of weights between neurons is computed. Weights are adjusted only once in line with following rule:

$$w_{ij} = \exp\left(-\frac{\|\mathbf{x}_i - \mathbf{x}_j\|}{2a^2}\right) \qquad (7)$$

where $\|\mathbf{x}_i - \mathbf{x}_j\|$ is Euclidian distance between two objects from input data and a is scale constant that is defined by algorithm [5] and described as follows. The Delaunay triangulation is formed in line with input data set, after that vector B_i is formed for each point that contains amount of nearest neighboring points in accordance with formed triangulation, $B_i \subset L$, where L – set of numbers of all N objects in input data set. List of distances V_i between point i and its neighbors B_i is formed for each point and using this list average distance is calculated for each V_i. Finally the scale constant a is computed as arithmetic average of all average values V_i. Detection methods of phase synchronization are deeply examined in paper [7].

Ability of the chaotic neural network to perform clustering is determined by the network dynamic that is based on the nonlinear logistic map. The nonlinear logistic map is defined by network states and synchronization processes. Various states of synchronization can occur in the chaotic neural network as it has been investigated in paper [6, 7]. The fragmentary chaotic synchronization occurs when complex input patterns are processed by the network when each cluster characterized by a unique "melody", for example, four groups of objects can be allocated from the network dynamic as it is presented in Fig. 2.

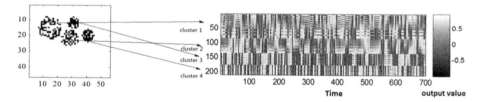

Fig. 2. The fragmentary chaotic synchronization of the chaotic neural network where four groups of objects have the same "melody" and as a result can be allocated into four clusters.

Experimental study has shown that chaotic neural networks are able to successfully preform clustering without any information about input data set using various combinations of parameters and using method for detection of fragmentary chaotic synchronization. The chaotic neural network is not able to perform clustering for data sets without ambiguous allocated clusters with sparseness of feature like traditional algorithms, for example, several intersecting sets that have the same density [13]. The clustering method that is based on the chaotic neural network has high computational complexity. Example of cluster allocations with different values of phase difference for fragmentary chaotic synchronization is presented in Fig. 3.

The estimation of resource intensity of the chaotic neural network requires to take into account following components: the first one is to calculate amount of nearest neighbors for each neuron using Delaunay triangulation – complexity is $O(N \log(N))$. The second one is to calculate weights between neurons – complexity is $O(N^2)$. The

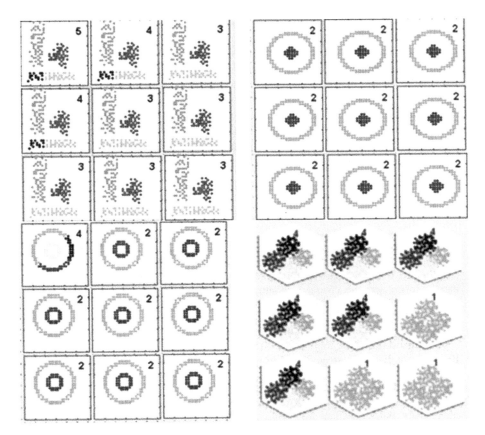

Fig. 3. Results of clustering using combination of the parameters for detection of fragmentary chaotic synchronization.

third one is to calculate dynamic of the chaotic network – complexity is $O(N^2)$. The forth one is post-processing of the whole dynamic of the network, and the fifth one is comparison of each neuron output and other outputs of neurons – complexity is $O(N^2)$.

The limit ε and the percent of phase's differences are key parameters for clustering because these parameters are used for post-processing of dynamic of the network. The chaotic network does not require any parameters due to self-organization processes that are inherent by many nonlinear dynamic systems and as a result this property allows to achieve qualitative results.

4   Conclusions

The oscillatory networks based on Kuramoto model take an intermediate position between traditional methods and chaotic methods. Comparison of the oscillatory-based methods and traditional algorithms is studied. The oscillatory networks require choice of parameter of similarity or amount of cluster like traditional methods. Advantage of the oscillatory networks is opportunity to parallelize process of simulation and the multi-core implementation is discussed. The chaotic neural network does not require any parameter of similarity due to automatic analysis of input data set.

Advantage of oscillatory and chaotic networks is versatility: stability of the methods to any peculiarities in topology of input patterns. Analysis of dynamic of the oscillatory networks can be used to resolve ambiguities in input patterns and as consequence achieve qualitative result of clustering unlike traditional methods. Also analysis of dynamic of network helps reduce a priori information about input data set and as result it helps automate process of clustering.

References

1. Acebron, A.J., Bonilla, L.L., Vincente, C.J.P., Rotort, F., Spigler, R.: The Kuramoto model: a simple paradigm for synchronization phenomena. Rev. Mod. Phys. **77**, 137–185 (2005)
2. Angelini, L., Carlo, F., Marangi, C., Pellicoro, M., Nardullia, M., Stramaglia, S.: Clustering data by inhomogeneous chaotic map lattices. Phys. Rev. Lett. **85**, 78–102 (2000)
3. Anil, K., Dubes, J.C., Dubes, R.C.: Algorithms for Clustering Data. Prentice Hall, Englewood Cliffs (1998)
4. Arenas, A., Diaz-Guilera, A., Kurths, Y., Moreno, Y., Changsong, Z.: Synchronization in complex networks. Phys. Rep. **469**, 93–153 (2008)
5. Benderskaya, E.N., Zhukova, S.V.: Clustering by chaotic neural networks with mean field calculated via Delaunay triangulation. In: Corchado, E., Abraham, A., Pedrycz, W. (eds.) HAIS 2008. LNCS (LNAI), vol. 5271, pp. 408–416. Springer, Heidelberg (2008)
6. Benderskaya, E.N., Zhukova, S.V.: Large-dimension image clustering by means of fragmentary synchronization in chaotic systems. Pattern Recogn. Image Anal. **19**, 306–316 (2009)
7. Benderskaya, E.N., Zhukova, S.V.: Chaotic Clustering: Fragmentary Synchronization of Fractal Waves. In: Tlelo-Cuautle, E. (ed.) Chaotic Systems, pp. 187–202. InTech (2011)
8. Chopra, N., Spong, M.W.: On exponential synchronization of Kuramoto oscillators. IEEE Trans. Autom. Control **54**, 353–357 (2009)
9. Dimitriadou, E., Weingessel, A., Hornik, K.: Voting-merging: an ensemble method for clustering. In: Dorffner, G., Bischof, H., Hornik, K. (eds.) ICANN 2001. LNCS, vol. 2130, pp. 217–224. Springer, Heidelberg (2001)
10. Eidswick, J.A.: On some fundamental problems in cluster set theory. Proc. Am. Math. Soc. **39**, 163–168 (1973)
11. Ester, M., Kriegel, H.P., Sander, J., Xu, X.: A density-based algorithm for discovering clusters in large spatial databases with noise. In: KDD-96 Proceedings, pp. 226–231 (1996)
12. Fujisaka, H., Shimada, T.: Phase synchronization and nonlinearity decision in the network of chaotic flows. Phys. Rev. E **55**, 2426–2433 (1997)
13. Gan, G., Ma, C., Wu, J.: Data Clustering Theory, Algorithms, and Applications. ASA-SIAM Series on Statistics and Applied Probability (2007)
14. Guha, S., Rastogi, R., Shim, K.: ROCK: a robust clustering algorithm for categorical attributes. In: Proceedings of the 15th International Conference on Data Engineering, pp. 512–521 (1999)
15. Guha, S., Rastogi, R., Shim, K.: CURE: an efficient clustering algorithm for large databases. Inf. Syst. **26**, 35–58 (2001)

16. Han, J., Kamber, M.: Data Mining: Concepts and Techniques (The Morgan Kaufmann Series in Data Management Systems). Morgan Kaufmann, Burlington (2005)
17. Jadbabaie, A., Motee, N., Barahona, M.: On the stability of the Kuramoto model of coupled nonlinear oscillators. In: Proceedings of the 2004 American Control Conference Boston, Massachusetts, 30 June–2 July 2004
18. Kuramoto, Y.: Chemical Oscillations Waves, and Turbulence. Springer, Berlin (1984)
19. MacQueen, J.B.: Some methods for classification and analysis of multivariate observations. In: Proceedings of Fifth Berkley Symposium on Mathematical Statistics and Probability, pp. 281–297 (1967)
20. Miyano, T., Tsutsui, T.: Data synchronization as a method of data mining. In: Proceedings of the 2007 International Symposium on Nonlinear Theory and its Applications, NOLTA 2007, pp. 224–227 (2007)
21. Novikov, A.V., Benderskaya, E.N.: Oscillatory neural networks based on the Kuramoto model. Pattern Recogn. Image Anal. **27**, 365–371 (2014)
22. Novikov, A.V., Benderskaya, E.N.: SYNC-SOM: double-layer oscillatory network for cluster analysis. In: Proceedings of 3rd International Conference on Pattern Recognition Applications and Methods, pp. 305–309 (2014)
23. Novikov, A., Benderskaya, E.: Oscillatory network based on Kuramoto model for image segmentation. In: Malyshkin, V. (ed.) PaCT 2015. LNCS, vol. 9251, pp. 210–221. Springer, Heidelberg (2015)
24. Pelleg, D., Moore, A.: X-means: extending K-means with efficient estimation of the number of clusters. In: Proceedings of the Seventeenth International Conference on Machine Learning, pp. 727–734 (2000)
25. Rogge, J.A., Aeyels, D.: Existence of partial entrainment and stability of phase locking behavior of coupled oscillators. Prog. Theor. Phys. **112**, 921–942 (2004)
26. Shao, J., He, X., Bohm, C., Yang, Q., Plant, C.: Synchronization-inspired partitioning and hierarchical clustering. IEEE Trans. Knowl. Data Eng. **25**, 893–905 (2013)
27. Xin, B.L., Qin, B.Z.: Adaptive cluster synchronization in coupled phase oscillators. In: International Conference on Information Engineering and Computer Science, ICIECS 2009 (2009)
28. Zhang, T., Ramakrishnan, R., Livny, M.: BIRCH: an efficient data clustering method for very large databases. In: Proceedings of the SIGMOD 1996, pp. 103–114, June 1996

Day-Ahead Electricity Price Forecasting Using WT, MI and LSSVM Optimized by Modified ABC Algorithm

H. Shayeghi[1(✉)], A. Ghasemi[1], and M. Moradzadeh[2]

[1] Technical Engineering Department,
University of Mohaghegh Ardabili, Ardabil, Iran
hshayeghi@gmail.com, ghasemi.agm@gmail.com
[2] Systems Engineering Research Group, Department of Engineering
and Technology, University of Huddersfield, Huddersfield, UK
m.moradzadeh@hud.ac.uk

Abstract. This paper presents a novel hybrid algorithm to forecast day-ahead prices in the electricity market. Seeking for more accurate price forecasting techniques, this hybrid price-forecasting algorithm works based on Mutual Information (MI), Discrete Wavelet Transform (DWT), Least Squares Support Vector Machine (LSSVM) optimized by a Interactive Artificial Bee Colony (IABC) technique. The numerical simulation results show that the proposed hybrid algorithm improves the accuracy of electricity price forecasting in Spanish electricity market in comparison to previously-known classical and intelligent methods.

Keywords: Price forecasting · Modified ABC algorithm · Wavelet packet transform · SVM · Mutual information

1 Introduction

Since the antecedent of the electricity generation in 1882 at New York Pearl Street (NYPS) power station, the electric industries followed the non-competitive price, traditional or regulated industrial frameworks for control, power generation and marketing. At this time, many industrialized countries restructured their electric power industries to be competitive for increase survivability, efficiency, security affordability and sustainability in a competitive environment [1]. In the recent years, the electric power in all countries obvious deregulation processes [2]. Forecast processes play a major role in both operation and planning of today power systems. The main goal for electricity-market participants is to have a clear and cost-effective market without any market's forces. Hereby, to maximize their profits, they need to exact and robust estimate of future electricity price to make their bidding policies for the real market. As results, consumers can derive a plan to maximize their purchased electricity from the pool, or use self production capability to protect themselves against high prices.

The importance of electricity price forecasting for optimal operation of the power systems and its complexity on the other hand, motivated many research works in the

© Springer International Publishing Switzerland 2016
L. Cheng et al. (Eds.): ISNN 2016, LNCS 9719, pp. 454–464, 2016.
DOI: 10.1007/978-3-319-40663-3_52

recent years. These techniques are Dynamic Regression Models (DRM) [3], Auto-Regressive Integrated Moving Average (ARIMA) [4], Neural Network (NN) [5, 6], Generalized Auto-Regressive Conditional Heteroskedastic (GARCH) [7], Support Vector Machines (SVMs) [8], neural network + cuckoo search algorithm [9], Fuzzy Inference System and Least Squares Estimation (FIS-LSE) [10], Wavelet Neural Network (WNN) [11], Least Squares Support Vector Machine (LSSVM) [12], Panel Cointegration and Particle Filter (PCPF) [13], Fuzzy NNs (FNN) [14], and Sensitivity based Dynamic Model (SDM) [15] and etc. Notwithstanding of all mentioned researches in the area of electricity price forecasting, there is still a gap for more accurate, less prediction errors and robust price forecast method which make maximum profits for market players. Specially, there is an essential need of robust feature selection algorithm for designing the input vector data of electricity price forecast, which can consider the nonlinear modeling of the price signal. The important difficulty of ANN can be found in its model that cannot overcome the characteristics of high volatility in electricity price series. In the other words, this problem is the broad lack of understanding of how they actually solve an electricity price forecasting problem. Therefore, to conquest this shortcoming, Least Squares Support Vector Machine (LSSVM) was applied for forecasting [8]. The important reason behind the choice of LSSVM is its high accuracy and global solution. When the control parameters of LSSVM aren't suitably adjusted then it cannot produce accurate forecast result. The main reason is that the selection of the parameters in LSSVM structure has a main effect on the forecasting accuracy. Thus, a novel modified Artificial Bee Colony (ABC) algorithm is proposed to optimize the LSSVM parameters. The main disadvantage of the standard ABC technique is the fact that it is not powerful enough to maximize the exploitation capacity [16]. To improve the overall performance and effectiveness of the modified ABC algorithm, the theory of universal gravitation is introduced into the consideration of the affection between worker and scout bees on searching process to improve the global and local search ability. Moreover, to decrease price signal noise, DWT has been used. In the other hand, with increasing of input data dimension into learning machine, LSSVM become accordingly harder to utilize. Also, high-dimensional data is a serious difficulty for many classification models due to its high computational cost and memory usage. Hence, Mutual Information (MI) is used as a special tool (to minimize relative entropy) in evaluating the approximation of the joint probability with the product of marginal probabilities.

2 Proposed Forecast Method Tools

Before explains how the proposed hybrid algorithm will work in price forecast, to the readers convenience, their details and mathematical formulations have expressed. These applied tools for price forecasting are DWT, MI, LSSVM and modified ABC algorithm.

2.1   Wavelet Transform (WT)

In this section, the WT is briefly reviewed (for more details see Ref. [8]). The main goal of the WT is its ability to represent any function as a superposition of a set of wavelets or basis functions. These functions are small waves located in different times; the wavelet transform can provide information about both the time and frequency domains as follows [8]:

$$\psi_{j,k}(t) = \frac{1}{\sqrt{s_0^j}} \psi(\frac{t - k\tau_0 s_0^j}{s_0^j}) \tag{1}$$

where, j and k are integers, $s_0 > 1$ and τ_0 are fixed dilation step and dilation step, respectively. The φ and ψ of DWT can be calculated as:

$$\varphi(2^j t) = \sum_{i=1}^{k} h_{j+1}(k)\varphi(2^{j+1} t - k), \psi(2^j t) = \sum_{i=1}^{k} g_{j+1}(k)\varphi(2^{j+1} t - k) \tag{2}$$

Then, a signal $f(t)$ can be written as:

$$f(t) = \sum_{i=1}^{k} \lambda_{j-1}(k)\varphi(2^{j-1} t - k) + \sum_{i=1}^{k} \gamma_{j-1}(k)\varphi(2^{j-1} t - k) \tag{3}$$

2.2   Mutual Information (MI)

MI can be seen as a simplistic optimizing voting algorithm, which includes and excludes attributes in order to maximize value of the evaluation function. There are some profits for these methods before applying forecasting methods, such as decrease the pattern complexity, improve the forecast accuracy, avoiding over-fitting and lessening the computational time. Let $X = X_1, X_2, ..., X_n$ and $P(X) = P(X_1), P(X_2), ..., P(X_n)$ be vector of variable and probability distribution then the entropy theory $H(X)$ can be approximated by [8]:

$$H(X) = -\sum_{i=1}^{n} P(X_i) \log_2(P(X_i)) \tag{4}$$

For universalization, X and Y denote discrete random variables with a joint probability distribution $P(X, Y)$, then the joint entropy function $H(X, Y)$ is given by:

$$H(X, Y) = -\sum_{i=1}^{n} \sum_{j=1}^{m} P(X_i, Y_j) \log_2\left(P(X_i, Y_j)\right) \tag{5}$$

Also, when certain variables are known and others are not, the remaining uncertainty is calculated using the conditional entropy [15] as follows:

$$H(Y/X) = \sum_{i=1}^{n} P(X_i)H(Y/X = X_i) = -\sum_{i=1}^{n} P(X_i) \sum_{j=1}^{m} P(Y_j/X_i) \log_2\left(P(Y_j/X_i)\right)$$

$$= -\sum_{i=1}^{n} \sum_{j=1}^{m} P(X_i, Y_j) \log_2\left(P(Y_j/X_i)\right) \tag{6}$$

Then $H(X, Y)$ and the $H(X/Y)$ or $H(Y/X)$ is defined by Eq. (7) known as chain-rule:

$$H(X, Y) = H(X) + H(Y/X) = H(Y) + H(X/Y) \tag{7}$$

Thus, the mathematical formulation of mutual information $MI(X, Y)$ between two variables X and Y can be expressed as:

$$MI(X, Y) = \sum_{i=1}^{n} \sum_{j=1}^{m} P(X_i, Y_j) \log_2\left(P(X_i, Y_j)/P(X_i)P(Y_j)\right) \tag{8}$$

Mutual information is a measure of the amount of information that one random variable contains about another random variable. It is the reduction in the uncertainty of one random variable due to the knowledge of the other. When the MI is large value then X and Y are closely related and vice versa. Consequently, the two variables aren't closely related and dependent when MI becomes zero. Let Y_1, Y_2, ..., Y_N denote input features and Y_m (candidate input) have high relate with the target variable X based MI (X, Y_m) then it is a good candidate as input feature, since by employing Y_m the uncertainty of X reduces more than using the other candidate inputs.

2.3   Least Squares Support Vector Machine (LSSVM)

The main difficulty of SVM can be found in the burden time calculation for training when input data has high dimension [17]. Therefore, an improved structure of original SVM, named LSSVM is employed. Assume that the input and output training data with x_i, y_i ($i = 1,2, ..., l$), respectively. Thus, LSSVM defines the regression function as:

$$\min J(w, e) = \frac{1}{2}w^T w + \frac{1}{2}\gamma \sum_{i=1}^{l} e_i^2, \quad y_i = w^T \phi(x_i) + b + e_i, i = 1, 2, ..., l \tag{9}$$

where, w, γ, b, e_i and $\phi(0)$ are the weight vector, the penalty parameter, bias term, approximation error and the nonlinear mapping function, respectively. Matching lagrange function can be calculated by:

$$L(w, e, \alpha, b) = J(w, e) - \sum_{i=1}^{l} \alpha_i w^T \phi(x_i) + b + e_i - y_i \tag{10}$$

where, α_i denote the Lagrange multiplier. By the Karush-Kuhn-Tucker (KKT) conditions can be obtianed [20]:

$$\frac{\partial L}{\partial w} = 0 \Rightarrow w = \sum_{i=1}^{l} \alpha_i \phi(x_i) \quad \frac{\partial L}{\partial b} = 0 \Rightarrow \sum_{i=1}^{l} \alpha_i = 0$$

$$\frac{\partial L}{\partial e_i} = 0 \Rightarrow \alpha_i = \gamma e_i \quad \frac{\partial L}{\partial \alpha_i} = 0 \Rightarrow \alpha_i w^T \phi(x_i) + b + e_i - y_i = 0 \tag{11}$$

Now, reduces the w and e_i, one gets:

$$\begin{bmatrix} b \\ \alpha \end{bmatrix} = \begin{bmatrix} 0 & I_v^T \\ I_v & \vartheta + \gamma^{-1} I \end{bmatrix}^{-1} \begin{bmatrix} 0 \\ y \end{bmatrix}, \; that \begin{cases} y = [y_1, y_2, \ldots, y_l]^T \\ I_v = [1, 1, \ldots, 1]^T \\ \alpha = [\alpha_1, \alpha_2, \ldots, \alpha_l]^T \end{cases} \tag{12}$$

Based on mercer condition and kernel function $(K(x, x_i))$ [20] on matrix ϑ with $\vartheta_{km} = \phi(x_k)^T \phi(x_m), m, k = 1, 2, \ldots, l$, the LSSVM regression is defined by:

$$y(x) = \sum_{i=1}^{l} K(x, x_i) \times \alpha_i + b \tag{13}$$

2.4   Interactive ABC Algorithm

Although, the original ABC algorithm (See Ref. [16] for more details) is effective in finding of the best solution along search space in the optimization problems, but its mechanism is based on the relation between employed bee and selected bee by the roulette wheel technique. The process effects with a random factor which leads to it can not use full exploitation capacity. To cope with this shortage, the Newtonian law of universal gravitation is employed. In the IABC algorithm the Eq. (14) is considered which shows the universal gravitations between the selected employed bees are exploited by the roulette wheel selection and the onlooker bee.

$$F_{12} = G \frac{mass_1 mass_2}{d_{21}^2} \hat{d}_{21} \tag{14}$$

$$\hat{d}_{21} = \frac{d_2 - d_1}{|d_2 - d_1|} \tag{15}$$

where, $mass_1$, $mass_2$, d_{21}, $\hat{d}_1$, F_{12} and G are mass of the object 1 and object 2, the separation between the objects, the unit vector defined with Eq. (15), the gravitational force heads from the object 1 to the object 2 and the universal gravitational constant, respectively. In the proposed method, the mass $mass_1$ and $mass_2$ replaced by parameters: $F(\delta_i)$ and $F(\delta_k)$, they are fitness value of the employed bee that picked by applying the roulette wheel selection and the randomly selected employed bee, respectively. We can drive similar formulation for universal gravitation:

$$F_{ik_j} = G \frac{F(\delta_i) \times F(\delta_k)}{(\delta_{kj} - \delta_{ij})^2} \cdot \frac{\delta_{kj} - \delta_{ij}}{|\delta_{kj} - \delta_{ij}|} \qquad (16)$$

$$x_{ij}(t+1) = \delta_{ij}(t) + F_{ik_j} \cdot [\delta_{ij}(t) - \delta_{kj}(t)] \qquad (17)$$

where, $F_{ik} \cdot [\delta_i - \delta_k]$ considered universal gravitation between the employed bee, which is hand-picked by the onlooker bee, and more than one employed bees. F_{ik} plays factor controlling in the roulette wheel selection. By developing and considering the gravitation between the picked employed bee and n selected employed bees, the results can be expressed as:

$$x_{ij}(t+1) = \delta_{ij}(t) + \sum_{k=1}^{n} F_{ik_j} \cdot [\delta_{ij}(\tilde{t}) - \delta_{kj}(t)] \qquad (18)$$

where, $\tilde{F}_{ik_j}$ is the normalized gravitation force.

3   Applying the IABC for Price Forecasting Problem

In general, process of the day-ahead price forecasting by IABC algorithm can be summarized as follows:

Step 1. Sort the historical price data and divide them into training and testing sets, namely T_r and I_n, respectively.

Step 2. Set up the optimal IABC control parameters to arrive all searching ability of it, hereby, the maximum cycle number, colony size and *limit value* are tuned by 50, 40 and 10 values, respectively.

Step 3. Decompose price signal (historical data up to 24 h of day d-1) via DWT in a set of four constitutive subset which is known as a_h, b_h, c_h and d_h, $h = 1,2, \ldots, T$, that the range of T is between (1, 8) weeks. a_h, b_h, c_h are detail with small adjustments and d_h is approximation subset which is the key subset of the DWT output. Thus, employing the DWT to the original prices signal can be calculated as follows:

$$W(p_h; h = 1, \ldots, T) = \{a_h, b_h, c_h, d_h; h = 1, \ldots, T\} \qquad (19)$$

To perform forecasting, the target and candidate inputs variable (T_r and I_n) are normalized between 0 and 1. Then, the proposed MI technique based on the feature selection is used to pick the best input variables for artificial machine learning.

Step 4. Sort the initial population and all their related data in the descending order of fitness.

$$Fitness = \frac{1}{N} \sum_{i=1}^{N} \frac{act_i - forc_i}{y_i} \qquad (20)$$

where, act_i and $forc_i$ represent the actual and forecast values, respectively.

Step 5. Use the IABC algorithm to find the best agent. If the best solution is better than the previous food source then replaced.

Step 6. Employ the inverse DWT form to guess the hourly prices for d^{th} day by means of the estimates d^{th} day of the constitutive subset. In other words, this form is used in turn in order to reconstruct the estimate signal for prices, i.e.,

$$W^{-1}(\{a_h^{est}, b_h^{est}, c_h^{est}, d_h^{est}; h = T+1, \ldots, T+24\}) = P_h^{W,est} \qquad (21)$$

Step 7. Update velocity and position based IABC scheme. In other words, move the agents to the search area for discovering new solution.

Step 8. If Algorithm receives the maximum number of iterations then finished. Otherwise, go to step 2.

The graphical process of the proposed scheme is depicted in Fig. 1.

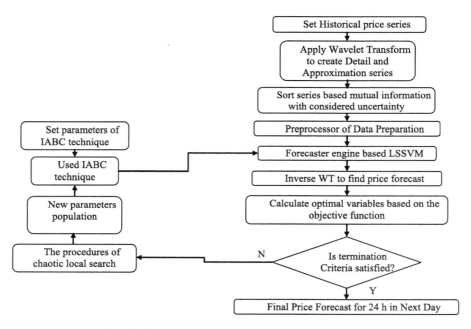

Fig. 1. Flowchart of the proposed forecasting scheme

4   Experimental and Discussion

The usefulness of the proposed method for the solution of the day-ahead price forecasting problem is investigated on Spain test system [18]. To evaluate and compare the performance of the proposed technique than other methods, Mean Absolute Percentage Error (MAPE) index is adopted in this study as:

$$MAPE = \frac{1}{N}\sum_{i=1}^{N}\frac{|P_{iACT} - P_{iFOR}|}{P_{AVE-ACT}}, P_{AVE-ACT} = \frac{1}{N}\sum_{i=1}^{N}P_{iACT} \qquad (22)$$

where, P_{iACT} and P_{iFOR} are actual and predict value and $P_{AVE-ACT}$ denotes the average of actual prices. The suggested forecast algorithm is evaluated for the Spain electricity market compare to available forecast method in the literature. In this way, the target and candidate inputs variable are linearly normalized in the range (0,1) for using by MI. After that the selected training inputs and validation samples are constructed based on the data of 50 days ago. Each of 24 forecasters output is trained through its respective 49 training trials and validated via one validation trial. The proposed price forecast approach has price series data up to the midnight of last night. Commonly, if the kth forecaster engine acquires the hourly prices of the forecast day, the predicted price values for these hours by its earlier forecasters are used. For example, P_{h-1} for the second forecaster is the predicted price by the first forecaster. The obtained numerical results for some test days and weeks are tabulated in Tables 1 (MAPE index) & 2 (variance of the prediction errors, as a measure of uncertainty) and the fourth week of February, May, August, and November are chosen for all seasons.

To reader convenience, the graphical view of price data for the Spanish electricity market is shown in Fig. 2. Moreover, this graphical outlook about the forecast accuracy obtained from the proposed algorithm, its results for the Spanish electricity market and the fall's test week are depicted in Figs. 3 and 4, respectively. The Total computational set-up time of the MI + LSSVM including the execution of data preparation part, training of LSSVM with consideration of the cross-validation (fine-tuning of parameters), and training of IABC is about 20 min, respectively on a Intel Core 2 Duo Processor P8700 2.53 GHz computer with 4 GB RAM memory.

According to Table 1, the proposed hybrid forecast algorithm has partly better weekly MAPE than other available forecast methods in all seasons. Moreover, the average value of the MAPE index of the proposed hybrid forecast algorithm is considerably less than all other forecast methods (last row of Table 1). Table 2 shows the variance of the forecasting errors, as a measure of uncertainty. It can be seen that this index for the proposed hybrid algorithm is moderately less than the other forecasted algorithms for all weeks.

Table 1. Weekly MAPE values in terms of percentage (%) for 4 weeks of the Spanish electricity market

Test week	ARIMA [3]	Wavelet-ARIMA [8]	FNN [14]	NN [8]	Mixed model [19]	MI + CNN [20]	RBFN [21]	Proposed method
Winter	6.32	4.78	4.62	5.23	6.15	4.51	4.27	4.450
Spring	6.36	5.69	5.30	5.36	4.46	4.28	4.58	4.526
Summer	13.39	10.70	9.84	11.40	14.90	6.47	6.76	5.478
Fall	13.78	11.27	10.32	13.65	11.68	5.27	7.35	5.522
Average	9.96	8.11	7.52	8.91	9.30	5.13	5.74	4.993

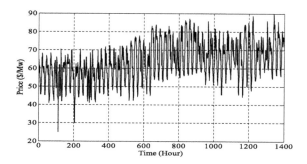

Fig. 2. Price data of Spanish electricity market

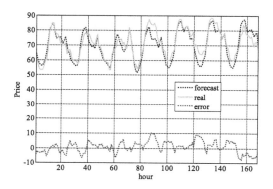

Fig. 3. Hourly prices (green), price forecasts (blue), and absolute value of forecast errors (red) for the fall's test week of the Spanish electricity market

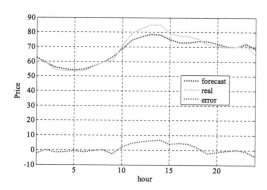

Fig. 4. Hourly prices (green), price forecasts (blue), and absolute value of forecast errors (red) the Spanish electricity market

Table 2. Error variance for 4 weeks of the Spanish electricity market

Test week	ARIMA [3]	Wavelet-ARIMA [8]	FNN [14]	MI + CNN [20]	RBFN [21]	Proposed method
Winter	0.0034	0.0019	0.0018	0.0014	0.0015	0.0013
Spring	0.0020	0.0025	0.0019	0.0014	0.0019	0.0014
Summer	0.0158	0.0108	0.0092	0.0033	0.0047	0.0032
Fall	0.0157	0.0103	0.0088	0.0022	0.0049	0.0021
Average	0.0092	0.0064	0.0054	0.0021	0.0033	0.0020

5 Conclusion

Day-ahead price forecasting of electricity market is volatile based on the uncertainties from the power market. In this paper, a new prediction strategy integrated of IABC algorithm based mutual information and LSSVM as forecast engine is proposed for day-ahead price forecast of electricity markets. This proposed hybrid forecast algorithm is applied on well-known Spanish market and compared with most recently published forecasting algorithms. It can be seen from figures and Tables; the comparing index in all four test weeks is lower than the results in all other methods. The main feature of the proposed hybrid forecast method is that it can take the complex dynamic pattern of electricity price series with simplicity in implementation.

References

1. Xiao, L., Wang, J., Hou, R., Wu, J.: A combined model based on data preanalysis and weight coefficients optimization for electrical load forecasting. Energy **82**, 524–549 (2015)
2. Ghofrani, M., Ghayekhloo, M., Arabali, A., Ghayekhloo, A.: A hybrid short-term load forecasting with a new input selection framework. Energy **81**, 777–786 (2015)
3. Conejo, A.J., Plazas, M.A., Espinola, R., Molina, A.B.: Day-ahead electricity price forecasting using the wavelet transform and ARIMA models. IEEE Trans. Power Syst. **20**, 1035–1042 (2005)
4. Zhou, M., Yan, Z., Ni, Y.X., Li, G., Nie, Y.: Electricity price forecasting with confidence-interval estimation through an extended ARIMA approach. IEE Proc. Gener. Transm. Distrib. **153**, 187–195 (2006)
5. Zhang, L., Luh, P.B.: Power market clearing price prediction and confidence interval estimation with fast neural network learning. In: IEEE Power Engineering Society Winter Meeting, vol. 1, pp. 268–273 (2002)
6. Hong, Y.-Y., Hsiao, C.-Y.: Locational marginal price forecasting in deregulated electricity markets using artificial intelligence. Inst. Elect. Eng. Gen. Transm. Distrib. **149**, 621–626 (2002)
7. Garcia, R.C., Contreras, J., Akkeren, M.V., Garcia, J.B.C.: A GARCH forecasting model to predict day-ahead electricity prices. IEEE Trans. Power Syst. **20**, 867–874 (2005)
8. Shayeghi, H., Ghasemi, A.: Day-ahead electricity prices forecasting by a modified CGSA technique and hybrid WT in LSSVM based scheme. Energy Convers. Manag. **74**, 482–491 (2013)

9. Taherian, H., Nazer, I., Razavi, E., Goldani, S.R., Farshad, M., Aghaebrahimi, M.R.: Application of an improved neural network using cuckoo search algorithm in short-term electricity price forecasting under competitive power markets. J. Oper. Autom. Power Eng. **1**, 136–146 (2013)
10. Li, G., Liu, C.C., Mattson, C., Lawarree, J.: Day-ahead electricity price forecasting in a Grid environment. IEEE Trans. Power Syst. **22**, 266–274 (2007)
11. Aggarwal, S.K., Saini, L.M., Kumar, A.: Electricity price forecasting in Ontario electricity market using wavelet transform in artificial neural network based model. Int. J. Control Autom. Syst. **6**, 639–650 (2008)
12. Suykenns, J.A.K., Vandewalle, J.: Least squares support vector machine. Neural Process. Lett. **9**, 293–300 (1999)
13. Li, X.R., Yu, C.W., Ren, S.Y., Chiu, C.H., Meng, K.: Day-ahead electricity price forecasting based on panel cointegration and particle filter. Electr. Power Syst. Res. **95**, 66–76 (2013)
14. Amjady, N.: Day-ahead price forecasting of electricity markets by a new fuzzy neural network. IEEE Trans. Power Syst. **21**, 887–896 (2006)
15. Bompard, E., Ciwei, G., Napoli, R., Torelli, F.: Dynamic price forecast in a competitive electricity market. IET Gener. Transm. Distrib. **1**, 776–783 (2007)
16. Shayeghi, H., Ghasemi, A.: A modified artificial bee colony based on chaos theory for solving non-convex emission/economic dispatch. Energy Convers. Manag. **79**, 344–354 (2014)
17. Vapnik, V.N.: The Nature of Statistical Learning Theory. Springer, New York (1995)
18. http://www.ree.es/cap03/pdf/Inf_Oper_REE_99b.pdf
19. Garcia-Martos, C., Rodriguez, J., Sanchez, M.J.: Mixed models for short-run forecasting of electricity prices: application for the Spanish market. IEEE Trans. Power Syst. **22**, 544–551 (2007)
20. Keynia, F.: A new feature selection algorithm and composite neural network for electricity price forecasting. Eng. Appl. Artif. Intell. **25**, 1687–1697 (2012)
21. Shafie-Khah, M., Moghaddam, M.P., Sheikh-El-Eslami, M.K.: Price forecasting of day-ahead electricity markets using a hybrid forecast method. Energy Convers. Manag. **52**, 2165–2169 (2011)

Categorization in Intentional Theory of Concepts

Dmitry Zaitsev[1(✉)] and Natalia Zaitseva[2]

[1] Department of Philosophy, Lomonosov Moscow State University,
Moscow, Russia
zaitsev@philos.msu.ru
[2] Russian Foreign Trade Academy, Moscow, Russia
natvalen@list.ru

Abstract. The recent rapid growth of empirical results in neuroscience has widened a proverbial explanatory gap between a first-person introspective experience and objective third-person data. Given different competing approaches, phenomenological framework proved an adequate methodological tool to bridge this gap. The present paper aims at a phenomenological representation of categorization by means of a modified functional theory of concept, termed Intentional Theory of Concept. The 1st section hereof serves as an introduction into the subject matter. In the 2nd section, we will focus on key phenomenological ideas of intentionality and analogous apperception (appresentation) and briefly describe our approach. The 3rd section contains a short summary of the modern version of the classical theory of concepts and its functional variant. Integrating the substance of the previous sections, the 4th one will offer an interpretation of categorization by means of a functional theory of concepts formally presented by typed lambda calculus. The final section provides a summary of findings and prospects for a follow-up study.

Keywords: Phenomenology · Analogous apperception · Categorization · Theory of concepts · Lambda calculus

1 Introduction

This paper contributes to a new approach to an examination of cognitive activity from a phenomenological perspective. More specifically, we will zero on the operation of categorization and introduce its novel interpretation in terms of the modernized theory of concepts.

One of the most important challenges is the venerable mind-body problem and in particular its modern reincarnation the 'hard problem of consciousness' coined by David Chalmers. Very briefly, it can be explicated as the problem of explaining how individual experiences are possible, why and how we experience anything. These phenomenal aspects of our mental life, which can be accepted only introspectively, are instantiated by intrinsic qualities known as qualia. Contrary to popular belief, the hard problem of consciousness appeared to be not a whim of a beautiful scholastic mind but a very specific task, which can be

© Springer International Publishing Switzerland 2016
L. Cheng et al. (Eds.): ISNN 2016, LNCS 9719, pp. 465–473, 2016.
DOI: 10.1007/978-3-319-40663-3_53

articulated as finding how physical processes in the brain give rise to subjective experience?

However, conrming the status of a true perpetual problem, many attempts to solve it met serious embarrassment of the so-called 'explanatory gap'.

"For no matter how deeply we probe into the physical structure of neurons and the chemical transactions which occur when they fire, no matter how much objective information we come to acquire, we still seem to be left with something that we cannot explain, namely, why and how such-and-such objective, physical changes, whatever they might be, generate so-and-so subjective feeling, or any subjective feeling at all. This is the famous explanatory gap for qualia" [15].

Recent findings in neuroscience have only exacerbated these difficulties. According to R.Yuste, it was an impressive advance in multineuronal recording techniques together with progress in optical probe design and synthesis accompanied by a revolution in optical hardware, that caused 'not just a quantitative change in the amount of data acquired but a qualitative modification in the mindset with which neuroscientists approach neural circuits' [19]. All that essentially deepens the gap between objective data of natural sciences and subjective experience both related to the same phenomena. In other words, this explanatory gap may be characterized in terms of coordination of first-person data gained by introspection with third-person empirical data gathered by external observer.

To bridge the explanatory gap, the pains of neuroscientists alone are evidently insufficient, it is exactly the interdisciplinary field that strongly needs collaborative efforts of specialists (experts) in different areas, including philosophy as a fundamental science of reality and mind. The revived interest in phenomenal consciousness and advent of embodied approaches to cognition, accompanied by salient progress in neuroscience, resulted in growing criticism of computational theory of mind and search for competitive alternatives. It was phenomenology that recently manifested itself as a prominent framework of an adequate methodology for philosophy of mind and study of consciousness (for more detail consult [6, 14]).

This paper is essentially interdisciplinary, that is why it contains two preliminary parts designed to introduce some relevant ideas and findings of phenomenology and modern theory of concept. In the 2nd section, we will briefly describe our approach and emphasize key phenomenological ideas of intentionality and analogous apperception (appresentation). The 3d section contains a short summary of modern version of classical theory of concepts and its functional variant. Putting it all together, in the 4th section we will offer an interpretation of categorization by means of functional theory of concepts formally presented by typed lambda calculus. The final section of the paper summarizes the results and delineates the horizonts for follow-up study.

2   Intentionalaty and Analogous Appresentation

Phenomenology is typically considered as the study of human conscious experience from the first-person point of view, and is definitely one of the key disciplines in philosophy among ontology, epistemology, logic etc. At the same time, its scope is not limited to classical philosophical problems, and ranges from the social interactions and behavior of economic agents to embodied subjectivity, neural correlates of consciousness (NCC) and other intriguing fundamental problems par excellence related not to a scientific progress in a certain field as such, but rather to perpetual mysterium cosmographicum.

There are at least three possible strategies to close proverbial explanatory gaps in phenomenological fashion [5]. One way is to reduce first-person data to third-person empirical records. This strategy is labeled as 'heterophenomenology', and the main problem with it is such a reduction emasculates the phenomenological content by replacement the first-person datum of personal experience by indirect contexts of the form "The subject X reports that he/she experiences Y". The second alternative is known as 'naturalized phenomenology', with "neurophenomenology" of F.Valera as the most telling example. He introduced this term as the name 'to designate a quest to marry modern cognitive science and a disciplined approach to human experience', the latter being Husserlian phenomenology. He claimed the goal of this quest to respond to the issues raised by D. Chalmers and address directly the hard problem of consciousness 'by gathering a research community armed with new pragmatic tools enabling them to develop a science of consciousness' [16].

Our vision of the field is apparently different and can be qualified under the third phenomenological strategy of "front-loaded phenomenology". In the words of Gallese, "we should phenomenologize Cognitive Neuroscience rather than naturalizing phenomenology" [7]. The underlying idea is to use methodological and worldview insights that have already been worked out in phenomenology as the basis for experimental design and keynote idea specifying the direction for subsequent interpretation of experimental data in neuroscience. Put more transparently, we see the advantage of phenomenology in cognitive (neuro-) science in providing a well-defined and deep-laid pattern for interpretation of empirical findings, that, as we hope, may in after years spring to a new theoretical framework for a science of consciousness.

In what follows we will restrict our consideration of phenomenology to two concepts playing key roles in the current research, namely intentionality and analogous appresentation. Intentionality is in truth a central phenomenological concept, which corresponds, according to E. Husserl, to a fundamental property of human consciousness. We proceed further with a standard philosophical definition of intentionality as "a peculiarity of human consciousness which consists in its directedness to an object, in 'ideal' positing of an object in thought, thereby becoming an 'ideal' object" [12]. So defined, intentionality is a characteristic of human consciousness, which presupposes, as stated by [11], its (1) "existence-independence" and (2) "conception-dependence". The former is derived from intentionality as posing an ideal object of directedness. It evidently makes intentional cognitive act to be independent of the existence of its object and, hence, puts forward one more important phenomenological concept of intentional object. Intentional objects do not exist in an ordinary sense, as, say chairs, tables or animals; rather they exist as an integral part of phenomenological content of the acts that represent them. In a natural way these considerations leads to a conclusion that "the intentionality of an act depends not just on which object the act represents but on a certain conception of the object represented" [11], and in this way intentionality turns to be closely related with conceptualization and categorization as fundamental operation in any theory of concepts.

The idea of apperceptive transfer, or analogizing apprehension, appears, when Husserl runs into a problem of inresubjectivity in the Fifth Cartesian Meditation [8]. Phenomenological method presupposes that any object is to be considered only with respect to a subject, as meaningful for a certain subject. At this rate, in order to avoid solipsism, we must justify the existence of 'intesubjective' surrounding world (shared by different subjects). In so doing, a problem of the Other arises. We do not have a direct access to the other consciousness, and to assure oneself of existence of the other mind Husserl appeals to universal cognitive device – analogous appresentation, or appresentation.

Analogous appresentation (co-presentation) presupposes the recognition of an object on the basis of already experienced 'similar' object. This procedure is in turn based on the principle of Pairing, that is a kind of passive synthesis in which a pair is constructed. It makes possible a so called apperceptive transfer of meaning from the object we already apprehended and categorized to a new case. Applied to the perception of the Other, it results in my body and the other body making up a pair. In this ground the other body is treated as bearer of mental life. According to Husserl, this procedure is universal in the sense that it is applied by an agent everyday and every moment perceiving our surroundings. Here goes his famous example of external experience, when the directly visible front side of any physical object always appresents a rear side of this object. In his own words: "Even the physical things of this world that are unknown to us are, to speak generally, known in respect of their type. We have already seen like things before, though not precisely this thing here. Thus each everyday experience involves an analogizing transfer of an originally instituted objective sense to a new case, with its anticipative apprehension of the object as having a similar sense" [8].

There are several important comments to be made. First of all, pairing and corresponding appresentaion are often interpreted (maybe due to a tradition of English translation of Husserls writings) in terms of 'similarity' or 'likeness'. However such an interpretation may turn out to be misleading. Apperceptive transfer is not a relation but rather a process of *likening to* (more close to homoosis than to similarity of two things 'equal in right'). Secondly, as noted by Husserl himself, it is not analogical inference, not a thinking act. Appresentation does not presuppose premises and conclusion, there is no act of inference involved, it is a pure (embedded) cognitive mechanism. Thirdly, it is connected with the procedure of typification. Apperceptive transfer enables to typify new object, to attribute it certain type. Thus we find ourselves in the space of typified, meaningful objects, and the process of conceptualization, recognition of an object appears as an embedding into a meaning structure.

3   Theory of Concepts (Modern Version)

Categorization is one of the most basic mental operations which provides classification of objects and events and forms a ground for recognition and understanding the surrounding world and its inhabitants. Quite predictably the problem of adequate interpretation of this phenomenon has been for decades a point at issue in cognitive science. Many successful attempts to meet the challenge were made within the context of differently constructed theories of concept. This paper is no exception.

Below we will consider in more detail the so called modern version of the classical theory of concepts developed in Russia, mostly presented in Russian, and for this reason remains unknown to academic community. It originates in seminal monograph of E.K. Voishvillo published in 1967 [17], and was further developed by him in its extended 1989th edition [18]. More up-to-date version of this theory may be found in [2], and the paper [10] is specifically devoted to importance and prospects of 'Voishvillian doctrine of concept'. The main idea underlying this theory is that of formal representation of concept (and its logical form) as $\langle \alpha_1, ... \alpha_n \rangle A^n(\alpha_1, ... \alpha_n)$, where $\alpha_1, ... \alpha_n$ are variables and $A^n(\alpha_1, ... \alpha_n)$ is n-ary predicate with $\alpha_1, ... \alpha_n$ free. Hereinafter, to simplify the consideration, we will deal with only concept-constructions of the type $\alpha A(\alpha)$. This construction added as a well-defined expression to second-order language is not a term for definite description because it does not contain any operator of description, rather it may be interpreted as a specified variable which narrows the range on unspecified variable $\alpha - \{\alpha : A(\alpha)\}$. Such formal presentation allows defining strictly such 'classical' characteristics of a concept as genus, extension and intension. A genus for a concept $\alpha A(\alpha)$ is a set corresponding to a range of a variable α. Above introduced set $\{\alpha : A(\alpha)\}$ is exactly an extension of a corresponding concept $\alpha A(\alpha)$ and is labeled as $W\alpha A(\alpha)$. An intension of a concept $\alpha A(\alpha)$ is not a collection of properties but a complex property $A(\alpha)$ represented by a predicate formula.

The next step in a feasible direction was made in [20, 21], where a concept is presented in Fregian manner as a certain function and formalized by means of lambda-calculus. The straightforward idea is to consider relevant propositional functions, where 'relevant' applied to function means 'really depend on its argument', and propositional functions correspond to predicates taking individuals as their arguments and returning truth-values. Thereafter, a concept as a function when applied to individuals from its extension takes the True as its value.

4   Categorization in More Detail

In full generality, our approach to cognitive activity rests upon the following guiding principles.

   1. *Intentionality is considered as a universal fundamental characteristics of cognition shared by animated bodies of various kinds.*
   We do not associate intentionality necessarily with human consciousness and see it as an essential property of embedded (and embodied) cognitive faculty, to be interpreted broadly in terms of interaction with surrounding world, aimed at adaptation.
   2. *Intentionality may be presented as functional relations from stimuli (taken as intended objects) to recognized (meaningful) individuals, relativized to a particular subject.*
   Typically intentionality is described in terms of subject-object interaction. However, if one occupies a first-person position towards the surrounding world and phenomenologically interprets it as experienced by individual consciousness, then all cognitive acts of an agent can be identified as sharing the same source. It makes implicit the meaning-bestowal function of intentionality: as we noticed before, it transforms a stimulus into an ideal meaningful (intentional) object.
   3. *Thus understood, intentionality may be considered as a concept function from stimuli into intentional objects.*
   The latter claim opens a possibility to develop a conceptual framework for analyzing cognitive activity, and makes the corresponding theory of concept intentional. In the context of current research we will focus on the conceptual presentation of categorization as providing a basic level of cognitive activity.
   In so doing, a categorization will be modeled via analogous apperception-like function. The core feature of this function is that it represents the above mentioned process of likening stimulus to a sample (reference standard). Such a likening is based on an identification of certain moments, sides or parts of perceived object and sample object. The process of likening is not homogeneous and can be very roughly divided into two phases, in accordance with Edelmans *reentry* [4] and Ivanitskys *informational synthesis* [9]: primary sensations recognition stage and subsequent synthesis of a perceived object as a whole.

   "Objects and their properties are perceived to be unitary, despite the fact that a given perception results from parallel activity in the brain of many different maps, each with different degrees of functional specialization. A striking case is the extrastriate visual cortex, in which different areas specialized for color, motion, and form act together to yield a coherent response to an object" [3].

   We propose a formal presentation of thus interpreted categorization by means of simply typed lambda calculus. More precisely categorization corresponds to beta-conversion.
   Typed lambda calculus is formal system (calculus) designed to implement a process of computation via application of functional abstractions. Its typed version is sensitive to a type of computed data. Formally, it is based on two operations: lambda-abstraction λ and application $\bullet$. If $M(x)$ means that Mx depends on x, then $\lambda x.M(x)$ is a functional abstraction expression denoting a map $x \longmapsto M(x)$. Together application (of a function to an argument) and

abstraction are enough to present a simple calculation, as illustrated by the following example: $\lambda x.(x + 1) \bullet 2 = (2 + 1) = 3$. In general the form of such calculation can be expressed as $(\lambda x.M(x)) \bullet N = M(N)$, which is known as β-conversion. For more detail consult [1].

There is a different approach to formal explication of calculation known as combinatory logic. The underling idea is to present the same applicative system not by taking β-conversion as a deductive postulate but rather by 'simulation' of its working via special terms. Keeping in mind our idea of relevant functions together with two possible types of application connected with individual categorization and property categorization we arrive at the BCI combinatory logic as a logic of concepts.

We proceed further with a formal language containing sets of variables $\{x\}$, constants $\{a\}$, functional variables $\{f\}$ and constants f, type-variables $\{\tau\}$ and three special symbols of application ($\bullet$), lambda-abstraction λ and functional mapping ($\rightarrow$). In a natural inductive manner we define notions of term and type, and add a specific definition of a concept-term:

(1) if F is a functional constant and v – variable, then $F\bullet v$ is a concept-term;
(2) if F is a functional variable and k – constant, then $F \bullet k$ is a concept-term.

If t is a term and A – type, $t : A$ is formula.

Now we are in a position to formulate the rules to simulate categorization. For that purpose we need:

one basic 'typed' rule of application: $AR \; \frac{t_1 : A \rightarrow B[\Gamma] \; t_2 : A[\Delta]}{t_1 \bullet t_2 : B[\Gamma, \Delta]}$,

two rules of permutation: $PerB \; \frac{\lambda v. F \bullet v \bullet t : B[\Gamma]}{F \bullet (v \bullet t) : B[\Gamma]}$

$PerC \; \frac{\lambda V. V \bullet k \bullet t : B[\Gamma]}{(V \bullet t) \bullet k) : B[\Gamma]}$

and Identity, or apperception rule: $IR \; \frac{\theta \bullet (\alpha \bullet t) \bullet \delta : B[\Gamma]}{\theta \bullet t \bullet \delta : B[\Gamma]}$

Two following examples illustrate two clauses for categorization. First consider 'property typing schema'.

1. $\lambda V.V \bullet a : A \rightarrow B[\Gamma]$
2. $F: A[\Delta]$
3. $\lambda V.V \bullet a \bullet F: B[\Gamma, \Delta]$ AR, 1,2
4. $(V \bullet F) \bullet a : B[\Gamma, \Delta]$ $PerC$, 3
5. $F \bullet a : B[\Gamma, \Delta]$ IR, 4

Repeated for every 'part and moment' of perceived stimulus, this procedure results in a complex whole.

At the second stage, this aggregation of typed components is in turn recognized as an intended object.

1. $\lambda x.F \bullet x : A \rightarrow B[\Gamma]$
2. $a : A[\Delta]$
3. $\lambda x.F \bullet x \bullet a : B[\Gamma, \Delta]$ AR, 1,2
4. $F \bullet (x \bullet a) : B[\Gamma, \Delta]$ $PerB$, 3
5. $F \bullet a : B[\Gamma, \Delta]$ IR, 4

5   Conclusion

This paper provides only the first step on the road of development a new Intentional Theory of Concepts which in turn appears to be only a part, albeit very important, of a complex phenomenological conception of cognitive activity. The main intermediate result we obtained here, from our point of view, consists in explication of categorization as a special case of universal cognitive intentional mechanism. Thereby we extend this mechanism to similar but non-conscious processes. To that end, 'theory of concepts' is just a customary convenient term to denote the formal presentation of this mechanism. Reference to conceptual activity does not necessarily mean reflection as its feasibility condition, it may carry an embedded (built-in) unconscious process.

Possible directions of future work concern both the further development of Intentional Theory of Concepts and the elaboration of intentional (phenomenological) theory of cognitive activity. One interesting perspective of research is to apply concept representation to higher-order cognitive functions. In order to grasp the concept-generation procedure, one can add to our formalism a rather straightforward rule of concept introduction:

$$IntroC \quad \frac{t:B[\Gamma,\alpha:A]}{\lambda\alpha.t:(A\rightarrow B)[\Gamma]}, \text{ provided } \alpha \text{ is free in } t.$$

Another promising line of research concerns embedded reasoning. If we look in more detail at the Rule of Application AR, it is easy to note that typed suffixes of premises and conclusion are directly reminiscent of the Modus Ponens schema of inference. We do not think that it is a matter of random coincidence. This argument form unlike some of the other inference rules or deductive postulates can be treated as a kind of embedded reasoning due to underling operation of categorization.

Surprisingly, categorization appears to have much in common with spatial navigation as presented by recent findings in neuroscience (discovery of so called 'place cells' and 'grid cells'). Developing a Quiroga's idea of "striking similarities between concept cells in the human Medial Temporal Lobe and place cells in the rodent hippocampus' [13], we see hopeful prospects in treating categorization as a 'navigation' in a meaning space, but it is a very different story.

References

1. Barendregt, H.: Lambda calculi with types. In: Abramsky, S., Gabbay, D.M., Maibaum, T.S.E. (eds.) Handbook of Logic in Computer Science, vol. II, pp. 118–309. Oxford University Press (1992)
2. Bocharov, V., Markin, V.: Introduction to Logic. Forum, Moscow (2008). In Russian
3. Edelman, G.M.: Neural darwinism: selection and reentrant signalling in higher brain function. Neuron. **10**, 115–125 (1993)
4. Edelman, G.M., Gally, J.A.: Reentry: a key mechanism for integration of brain function. Front. Integr. Neurosci. **7**, 63 (2013)

5. Gallagher, S.: Fantasies and facts: epistemological and methodological perspectives on first and third-person perspectives. Phenomenol. Mind. **1**, 40–46 (2011)
6. Gallagher, S., Zahavi, D.: The Phenomenological Mind: An Introduction to Philosophy of Mind. Routledge, London (2008)
7. Gallese, V.: Corpo vivo, simulazione incarnata e intersoggettivit: una prospettiva neurofenomenologica. In: Cappuccio, M., (ed.) Neurofenomenologia, Bruno Mondadori Milano, pp. 293–326 (2006)
8. Husserl, E.: Cartesian Meditations: An Introduction to Phenomenology. Martinus Nijhoff, The Hague (1960). Cairns, D. (trans.)
9. Ivanitsky, A.M., Ivanitsky, G.A., Sysoeva, O.V.: Brain science: on the way to solving the problem of consciousness. Int. J. Psychophysiol. **73**, 101–108 (2009)
10. Markin, V.: Voishvillo's Doctrine of concept: importance and prospects. Log. Inv. **20**, 60–77 (2014). In Russian
11. McIntyre, R., Smith, D.W.: Theory of intentionality. In: Mohanty, J.N., McKenna, W.R. (eds.) Husserl's Phenomenology: A Textbook, pp. 147–179. Center for Advanced Research in Phenomenology and University Press of America, Washington, D. C. (1989)
12. Motroshilova, N.: New Encyclopedia of Philosophy, 2nd edn. Intentionality, Moscow (2000)
13. Quiroga, R.Q.: Concept cells: the building blocks of declarative memory functions. Nat. Rev. Neurosci. **13**, 587–597 (2012)
14. Schmicking, D., Gallagher, S. (eds.): Handbook of Phenomenology and Cognitive Science. Springer, Cambridge (2010)
15. Tye, M.: "Qualia", The Stanford Encyclopedia of Philosophy (Fall 2015 Edition). In: Zalta, E.N., (ed.). http://plato.stanford.edu/archives/fall2015/entries/qualia/
16. Varela, F.: Neurophenomenology: a methodological remedy for the hard problem. J. Conscious. Stud. **3**, 330–349 (1996)
17. Voishvillo, E.: Concept. Isd. Moskovskogo Universiteta, Moscow (1967). In Russian
18. Voishvillo, E.: Concept as a Form of Thought. Isd. Moskovskogo Universiteta, Moscow (1989). In Russian
19. Yuste, R.: From the neuron doctrine to neural networks. Nat. Rev. Neurosci. **16**, 487–497 (2015)
20. Zaitsev, D.: Concept as a relevant function. In: Materials of the Workshop the Logic Center of the Institute of Philosophy, Russian Academy of Science, vol. 16, pp. 46–53 (2002). In Russian
21. Zaitsev, D.: Relevant logic of concepts. In: Zaitsev, D., Il'in, A., Markin, V. (eds.) Logic and Century, Sovremennii Tetradi, Moscow, pp. 130–138 (2003). In Russian

A Novel Incremental Class Learning Technique for Multi-class Classification

Meng Joo Er[1($\boxtimes$)], Vijaya Krishna Yalavarthi[2], Ning Wang[1], and Rajasekar Venkatesan[2]

[1] Marine Engineering College, Dalian Maritime University, Dalian, China
emjer@ntu.edu.sg, n.wang.dmu.cn@gmail.com
[2] School of Electrical and Electronic Engineering,
Nanyang Technological University, Singapore, Singapore
{vijayakr001,raja0046}@e.ntu.edu.sg

Abstract. In this paper, a novel technique for multi-class classification, which is independent of the number of class constraints and can learn the new classes it encounters, is developed. The developed technique enables remodelling of the network to adapt to the dynamic nature of non-stationary input samples. It not only can learn the new classes, but also the new patterns created in the input. The proposed algorithm is evaluated using various benchmark datasets and a comparative study of classification performance shows that the proposed algorithm is superior.

Keywords: ELM · OS-ELM · Incremental class learning · Sequential learning · Multi-class classification

1 Introduction

Classification problems are one of the age-old problems in the computational intelligence community. In classification, the problem of grouping input sequences into one of more than two disjoint classes is called multi-class classification. Based on learning paradigms, multi-class classification problems are divided into two categories. The first one is batch learning where the entire input data are available for training which takes place throughout the availability of data once. Another one is sequential learning where the initial training takes place with available data and later, training happens continuously whenever new datum arrives.

In the recent years, much research on computation intelligence pertaining to sequential learning has been reported [1–3]. In multi-class classification, sequential learning algorithms can learn the new data of a fixed number of classes which it has been trained initially. However, whenever a new class is encountered, the network is to be retrained with the entire set of data, which makes it very time consuming. In applications like cognitive robotics, the number of classes varies and the trained data may not be available to retrain the network with the new class encountered. This makes the existing sequential learning algorithms unsuitable for these applications.

In this paper, a novel algorithm, which can adapt to the new classes it encounters and changes to the input pattern by automatically changing the architecture of the

© Springer International Publishing Switzerland 2016
L. Cheng et al. (Eds.): ISNN 2016, LNCS 9719, pp. 474–481, 2016.
DOI: 10.1007/978-3-319-40663-3_54

network, is proposed. The proposed algorithm updates the network without losing the knowledge of the previous training. The proposed algorithm is different from the CIELM of [4] where it does not consider changes in the input patterns created during the addition of a new class. The Class-Incremental Learning was proposed using the support vector machines [5] where the problem is considered as binary classification. A new class is considered as one class and all the existing classes as another. The major disadvantage of this method is that the data will be unbalanced and accuracy is reduced.

The proposed algorithm is developed from the Online Sequential Extreme Learning Machine (OS-ELM) of [1]. The OS-ELM originates from the batch learning Extreme Learning Machine (ELM) of [6]. Several versions of OS-ELM and ELM are proposed in the literature [7–13]. The performance of the OS-ELM [1] is evaluated on several benchmark datasets and the results are compared with various sequential learning techniques. Simulation results show that the OS-ELM gives better generalization performance at very high learning speed. The performance of the proposed algorithm is evaluated on several standard datasets and it is compared with the CIELM and OS-ELM. In the proposed methodology, we systematically increase the number of output neurons to facilitate the learning of new classes and the hidden layer neurons to adapt to changes in the input patterns.

The paper is organised as follows: In Sect. 2, a brief review on the ELM and OS-ELM is discussed. This is followed by the proposed algorithm in Sect. 3. Simulation results are discussed in Sect. 4 and conclusions are drawn in the last section.

2   Background

In this section, a brief review of the ELM and OS-ELM is presented.

2.1   Extreme Learning Machine

Let us assume that the data have N samples represented by $\{(x_i, y_i)\}_{i=1}^{N}$ where x_i is the input vector and y_i is the output vector. Consider the hidden layer with P neurons. Then, the output can be given as follows:

$$f_P(x) = \sum_{j=1,}^{P} \beta_j G(w_j, b_j, x) = y \qquad (1)$$

As per the theory of ELM [6], input weights w and hidden neurons bias b are randomly determined and the output weights β are obtained by $\beta = H^+ Y$ where H^+ is the Moore-Penrose generalized inverse of the hidden layer output matrix and $Y = [y_1, .., y_N]^T$.

2.2   Online Sequential Extreme Learning Machine

In the OS-ELM, it is assumed that the data arrive sequentially. The OS-ELM has two phases of training [1].

Initialisation Phase: In this phase, the OS-ELM works like the batch learning ELM with an initial data N_0, represented by $\{(x_{0_i}, y_{0_i})\}_{i=1}^{N_0}$ and H_0 is the hidden layer output matrix for the initial training section N_0. The output weight vector β^0 is calculated using

$$\beta^0 = M_0^{-1} H_0^T Y_0 \qquad (2)$$

where $M_0 = H_0^T H_0$ is stored for the sequential learning phase.

Sequential Learning Phase: After the initialisation phase, data arriving sequentially are used to train the network. In this phase, the output weight vector β^{k+1} is calculated as the function of the previous output weight vector β^k, partial hidden layer output matrix H_{k+1} and the output vector Y_{k+1}. Let k be 0 for the first block. The output weight vector β^{k+1} is calculated as follows:

$$\beta^{k+1} = \beta^k + M_{k+1}^{-1} H_{k+1}^T \left(Y_{k+1} - H_{k+1}\beta^k \right) \qquad (3)$$

where $M_{k+1} = M_k + H_k^T H_k$.

3   Proposed Algorithm

The proposed algorithm works like OS-ELM until a new output class arrives. Whenever a new output class or set of classes arrive, the output layer of the network will be updated with the addition of nodes based on the number of new classes. Along with the output layer, the hidden layer is also updated with the addition of new nodes to adapt to the pattern changes in the input data.

Consider an initial dataset containing N_0 samples represented by $Q = \left\{ \left(x_Q^{(i)}, y_Q^{(i)} \right) \right\}_{i=1}^{N_0}$ where $x_Q^{(i)} \epsilon X$ is the input vector and $y_Q^{(i)}$ is the output vector with m classes. The proposed algorithm works like the OS-ELM until new classes arrive. Hence, β^0 is determined using (2). Later, the data arrive block by block and a block may have one or more samples. Let us assume that there is no new class until k^{th} block. The output weight matrix β^k is determined using (3).

Later, $(k+1)^{th}$ block with a new data of N_1 samples and Δm new classes represented by $K = \left\{ \left(x_K^{(i)}, y_K^{(i)} \right) \right\}_{i=1}^{N_1}$ arrive. To facilitate the new classes in the output, we add Δm new nodes in the output layer. This makes a total of m^* nodes in the output layer. There will be a change to the input pattern whenever a new class arrives. To accommodate this change, we increase the number of neurons in the hidden layer. Let ΔP nodes be added to the hidden layer and the input weights and bias are generated randomly as per the theory of OS-ELM [1]. The total number of nodes in the hidden layer be P^*. Due to the addition of new hidden nodes, M_k will be modified to M_k^* as follows:

$$M_k^* = I_{P^* \text{x} P} * M_k * I_{P^* \text{x} P}^T \tag{4}$$

where $I_{P^* \text{x} P} = \begin{bmatrix} I_{P \text{x} P} \\ 0_{\Delta P \text{x} P} \end{bmatrix}$ is a rectangular identity matrix and $0_{\Delta P \text{x} P}$ is a zero matrix of order $\Delta P \text{x} P$.

All the newly added hidden layer neurons are connected to all the m^* output nodes with zero weights. Due to the introduction of new hidden layer neurons and output neurons, the output weight vector is modified to β^{k^*} which is given by

$$\beta^{k^*} = I_{P^* \text{x} P} * \beta^k * I_{m \text{x} m^*} \tag{5}$$

where $I_{m \text{x} m^*} = [I_{m \text{x} m} \ 0_{m \text{x} \Delta m}]$ is a rectangular identity matrix and $0_{m \text{x} \Delta m}$ is a zero matrix of order $m \text{x} \Delta m$.

Then β^{k+1} is calculated using

$$\beta^{k+1} = \beta^{k^*} + M_{k+1}^{-1} H_{k+1}^T (Y_{k+1} - H_{k+1} \beta^{k^*}) \tag{6}$$

where $M_{k+1} = M_k^* + H_{k+1}^T H_{k+1}$.

Subsequently it will continue as the OS-ELM until new classes arrive.

In the proposed algorithm, the network restructure itself allowing the changes in the data without losing the previous information.

4   Simulation Results and Discussions

The proposed algorithm works in the dynamic environment where the number of classes is often unknown. The algorithm automatically redesigns the network for adapting new classes and the variations to the input patterns without losing the previous knowledge. The proposed algorithm is tested on five benchmark datasets which are Waveform, Image Segmentation, Satellite Image, Letter Recognition and Pen-Based Recognition of Handwritten Digits given in Table 1. All the datasets are taken from the UCI machine learning repository.

We assume that the data are balanced and uniformly distributed. Each dataset is divided into 70 % for training and 30 % for testing. Without loss of generality and for the sake of simplicity, we demonstrate the addition of one new class. The training dataset is split into two subsets where the latter has one more class of the data that were not present in the former. The first subset is used for the initial training. The number of hidden neurons is found using the training and validation method where 80 % of the data are used for training and 20 % for validation. Experiments are repeated over 50 trials and the optimal number of hidden neurons which gives best validation accuracy is used for training. We consider the ratio of hidden to output neurons as constant and whenever a new class is presented, hidden neurons are added along with the output neurons to keep the ratio same. After initial training, the number of hidden layer neurons and output neurons are increased and trained for the new class. The distribution of datasets is given in Table 1.

Table 1. Specifications of datasets used

Dataset	Attributes	Train data	Test data	Range	No. of classes	Hidden nodes
Waveform	21	3500	1500	1–705	2	100
				706–3500	3	150
Image segmentation	19	1617	963	1–416	6	150
				417–1617	7	175
Satellite image	36	4504	1931	1–1033	5	150
				1034–4504	6	180
Letter recognition	16	3247	1392	1–811	5	350
				812–3247	6	420
Pen-based recognition of handwritten digits	17	3659	1583	1–953	6	240
				953–3659	7	280

The testing accuracy curve of the Waveform data is shown in Fig. 1. Initially, the network is trained with two classes. When the third class is introduced at the 706th sample, the network updated itself by increasing the number of neurons in the output and hidden layers to adapt to the new class and input variations. Testing accuracy suddenly rises when a new class has arrived and settled down after a few iterations.

Figure 2 shows the testing accuracy for the letter recognition data in which A, B, C, D and E are the base classes. Initially, the network is trained with these five base classes. It can be seen that for in 811th sample F class is introduced and learning the new class did not affect the performance of the existing classes significantly.

Consistency is an important factor every algorithm should have. The proposed algorithm is experimented several times on each of the aforementioned datasets and the testing accuracy of all the trails is shown in Fig. 3.

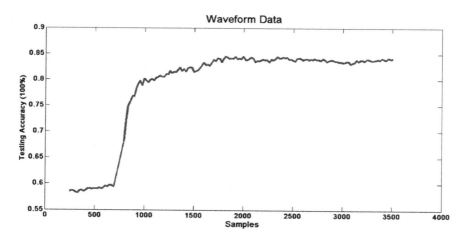

Fig. 1. Testing accuracy of waveform data

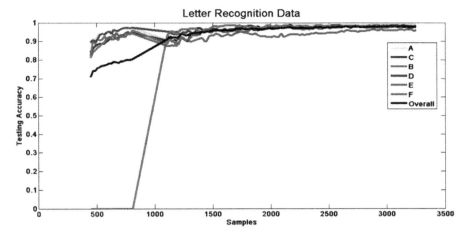

Fig. 2. Overall and individual testing accuracy for letter recognition data

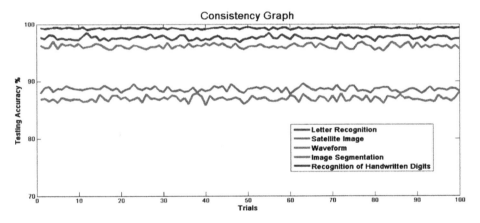

Fig. 3. Consistency of the proposed algorithm in testing accuracy

It can be observed from the graph that the proposed algorithm is consistently accurate in each of the attempts. The deviation of the testing accuracy in each case is in the order of 0–2 % of the mean value which is nominal.

The performance of the proposed algorithm is compared with the OS-ELM and CIELM algorithms and simulation results are given in Table 2. Testing accuracy along with the standard deviations for all the data sets mentioned are compared. A computer with Intel core i7 processor and 8 GB RAM was used for all the simulation works.

It can be observed from the simulation results that the testing accuracy of the proposed algorithm is better than the CIELM and at par with the OS-ELM. In summary, the proposed algorithm can learn the new class it encounters without losing the previous knowledge and it can adapt to pattern variations at the input.

Table 2. Performance comparison

Data type	Proposed algorithm testing accuracy	OS-ELM testing accuracy	CIELM testing accuracy
Waveform	85.40 ± 0.67	85.30 ± 0.68	83.90 ± 0.71
Image segmentation	95.27 ± 0.34	95.07 ± 0.73	94.94 ± 0.37
Satellite image	88.60 ± 0.37	88.93 ± 0.57	88.17 ± 0.42
Letter recognition	98.03 ± 0.31	98.50 ± 0.34	97.62 ± 0.34
Pen-based recognition of handwritten digits	99.28 ± 1.20	99.39 ± 1.10	99.18 ± 1.30

5 Conclusions

In this paper, a novel algorithm for incremental class learning is proposed. This algorithm adapts to the new output classes it encounters and variations to the input patterns by modifying the network without losing previous knowledge. The proposed algorithm is well suited to real-time applications like cognitive robotics where the number of output classes is often unknown. The performance of the proposed algorithm is compared with the OS-ELM and CIELM and the proposed algorithm shows promising results.

Acknowledgement. The authors would like to acknowledge the funding support from the National Natural Science Foundation of P. R. China (under Grants 51009017 and 51379002), Applied Basic Research Funds from Ministry of Transport of P. R. China (under Grant 2012-329-225-060), and Pro-gram for Liaoning Excellent Talents in University (under Grant LJQ2013055).

References

1. Liang, N.-Y., Huang, G.-B., Saratchandran, P., Sundararajan, N.: A fast and accurate online sequential learning algorithm for feedforward networks. IEEE Trans. Neural Netw. **17**, 1411–1423 (2006)
2. Wang, N., Er, M.J., Han, M.: Parsimonious extreme learning machine using recursive orthogonal least squares. IEEE Trans. Neural Netw. Learn. Syst. **25**, 1828–1841 (2014)
3. Wang, N., Er, M.J., Han, M.: Generalized single-hidden layer feedforward networks for regression problems. IEEE Trans. Neural Netw. Learn. Syst. **26**, 1161–1176 (2015)
4. Zhao, Z., Chen, Z., Chen, Y., Wang, S., Wang, H.: A class incremental extreme learning machine for activity recognition. Cogn. Comput. **6**, 423–431 (2014)
5. Zhang, B., Su, J., Xu, X.: A class-incremental learning method for multi-class support vector machines in text classification. In: 2006 International Conference on Machine Learning and Cybernetics, pp. 2581–2585. IEEE (2006)
6. Huang, G.-B., Zhu, Q.-Y., Siew, C.-K.: Extreme learning machine: theory and applications. Neurocomputing **70**, 489–501 (2006)
7. Li, M.-B., Huang, G.-B., Saratchandran, P., Sundararajan, N.: Fully complex extreme learning machine. Neurocomputing **68**, 306–314 (2005)

8. Zhu, Q.-Y., Qin, A.K., Suganthan, P.N., Huang, G.-B.: Evolutionary extreme learning machine. Pattern Recogn. **38**, 1759–1763 (2005)
9. Huang, G.-B., Chen, L., Siew, C.-K.: Universal approximation using incremental constructive feedforward networks with random hidden nodes. IEEE Trans. Neural Netw. **17**, 879–892 (2006)
10. LI, B., Wang, J., Li, Y., Song, Y.: An improved on-line sequential learning algorithm for extreme learning machine. In: Liu, D., Fei, S., Hou, Z.-G., Zhang, H., Sun, C. (eds.) ISNN 2007, Part I. LNCS, vol. 4491, pp. 1087–1093. Springer, Heidelberg (2007)
11. Rong, H.-J., Huang, G.-B., Sundararajan, N., Saratchandran, P.: Online sequential fuzzy extreme learning machine for function approximation and classification problems. IEEE Trans. Syst. Man Cybern. B Cybern. **39**, 1067–1072 (2009)
12. Huang, G.-B., Wang, D.H., Lan, Y.: Extreme learning machines: a survey. Int. J. Mach. Learn. Cybern. **2**, 107–122 (2011)
13. Wang, X., Han, M.: Improved extreme learning machine for multivariate time series online sequential prediction. Eng. Appl. Artif. Intell. **40**, 28–36 (2015)

Basis Functions Comparative Analysis in Consecutive Data Smoothing Algorithms

F.D. Tarasenko and D.A. Tarkhov$^{(\boxtimes)}$

Saint-Petersburg Polytechnic University of Peter the Great,
Saint-Petersburg, Russia
oudi@mail.ru, dtarkhov@gmail.com

Abstract. In this paper, we investigate algorithms for constructing experimental data dependence based on sequential processing of the points one by one. Four algorithms are reviewed, comparative analysis for different basis functions, a level of noise and other options is made. In addition to static data, there was an investigation of dynamic data case. The sine with variable frequency is used as an approximative function. Numerical experiments led to the conclusions about the comparative efficiency of algorithms and basis functions. The recommendations for the use of the algorithms are given.

Keywords: Consecutive algorithms · Data processing · RBF network · Spline

1 Introduction

Problems of high neural network training costs stimulate the search for ways to accelerate this process. The main directions of the acceleration are

(1) Finding a good initial approximation to the required weights of the neural network, which allows to significantly accelerate global optimization algorithms for functional errors.
(2) Parallelization of the learning process that allows you to accelerate it, using multiprocessor computer systems and graphics cards.
(3) The use of specialized processors.

The work deals with algorithms to realize the first two approaches. If we solve the problem of constructing RBF-network on data set, you can use the algorithms discussed below to quickly build a good approximation, using equidistant from the fixed centers and the wide range of Gaussian basis functions. Further, it is possible to clarify the position of the centers, and the width of the basis functions are weighting coefficients using any global optimization algorithm. Less smooth functions can be used in the first step. In the next step, using a pre-built approximation of basis functions and scaling operations can be obtained neural network approximation with the required activation functions - Gaussians, sigmoid, etc.

Algorithms considered in the work allow for effective parallelization. Thus, procedures for distribution of the points for intervals and updates the weights of the basis functions can be separated. Wherein, the interval, into which the new point may be

© Springer International Publishing Switzerland 2016
L. Cheng et al. (Eds.): ISNN 2016, LNCS 9719, pp. 482–489, 2016.
DOI: 10.1007/978-3-319-40663-3_55

determined simultaneously with the specification of the weights caused by the previous point. Since the basis functions are non-zero only in a small neighborhood center, it is possible to update the weight of several functions simultaneously and in parallel. Another option is to parallelize the simultaneous use of the considered algorithms and evolutionary algorithm of neural network training. This updated with the help of another portion of the data processing network includes an evolving population.

Separately mention studied the possibility of applying the algorithms we have considered building a model for dynamic data. The data obtained from the function, which varies with time. This situation is common to observe the object whose behavior changes in the process of building the model. The possibility of restructuring the model in accordance with these changes.

These algorithms can be used in hybrid neural network procedures for constructing approximate solutions of differential equations. In the first phase of operation of such procedures by the classical method of finite difference constructed approximation on points, the second - with the help of the algorithms considered in this paper the neural network approximation is built, on the third - it is specified by the methods discussed in [8–10].

Algorithms of consecutive data smoothing were studied in several works [11, 12].

In this paper, we review the methods [1–3] of finding dependence $y = f(x)$ on experimental data $(x_1, y_1), (x_2, y_2), \ldots, (x_N, y_N)$ in the situation when the points (x_i, y_i) are received and processed one by one, which may be associated with the need to process data in real-time. Let the required dependence provide in view if $y = \sum_{j=1}^{n} c_j \varphi_j(x)$, where $\varphi_j = \varphi(\alpha_j |x - z_j|)$ or $\varphi_j = \varphi(\alpha_j (x - z_j)^2)$.

Such types of functions are called RBF-nets [3–5]. Finding parameters c_j, α_j and z_j is called network learning. In this paper considers the case, where only the parameter c_j is finding. We set other parameters ourselves.

As it known, every piecewise linear function can be represented in the sum form $\sum_{j=1}^{n} c_j \varphi (\alpha_j (x - z_j))$, if we chose triangular cap as basis function (Fig. 1).

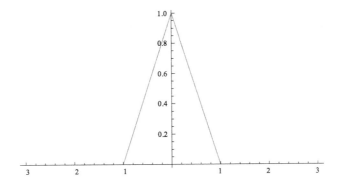

Fig. 1. The plot of the triangular cap.

$$y = \varphi(x) = (1 - |x|)_+ = \begin{cases} 1 - |x|, & |x| \leq 1 \\ 0, & |x| > 1 \end{cases}$$

In a such way any spline can be decomposed in the sum of basis functions, which match to its degree of smoothness [6].

In the numerical experiments we used following basis splines [6]: parabola, cubic parabola, gaussian.

We studied several algorithms for data smoothing:

(1) Processing points one by one with the adaptation of the weights of the basis functions with the nearest center.
(2) same, but with the adaptation of the weights of the two nearest functions.
(3) same as (2) but assumes a linear dependence of the speed of adaptation of the weights from the distance of the abscissa of a point to the center of the basis function [1].
(4) Finding the optimal coefficients of decomposition in basis functions the solution of the linear system.

For approaches 1 and 3 was also studied the variation of the algorithm with a previously determined law of variation of step.

In approaches 1–3 as the basis for adaptation of the coefficients is the minimization of a quadratic functional of the error.

2 Approaches

2.1 Approach 1

The initial values of coefficients $c_j = 0$. In the step from N-1 point to N the changing of weight coefficient of the basis function with the nearest center, written as $\Delta_k(N)$, is equal

$$\Delta_k(N) = \frac{Q_k}{S_k}, \tag{1}$$

where

$$S_k = \sum_{i=1}^{N} \varphi_k^2(x_i), \quad Q_k = \sum_{i=1}^{N} \varphi_k(x_i)\delta_i(N) \tag{2}$$

Moving to the next step we need to recalculate the error according to the formula:

$$\delta_i(N+1) = \delta_i(N) - \Delta_k \varphi_k(x_i). \tag{3}$$

2.2 Approach 2

Unlike the first approach, we select two basis functions between which centers is the received point.

The changing of weights is made according to the formulas

$$\Delta_k = \frac{Q_k S_{k+1} - Q_{k+1} P_k}{S_k S_{k+1} - P_k^2}, \Delta_{k+1} = \frac{Q_{k+1} S_k - Q_k P_k}{S_k S_{k+1} - P_k^2} \tag{4}$$

where

$$P_k = \sum_{i=1}^{N} \varphi_k(x_i)\varphi_{k+1}(x_i). \tag{5}$$

When the denominator of (2) turns to the zero we use formulas (7) and (8). Moving to the next step the terms in the sum (2) and (5) are added for all basis functions, for which $\varphi_k(x_i) \neq 0$.

Error δ_i when moving to the next step you need to recalculate according to the formula:

$$\delta_i(N+1) = \delta_i(N) - \Delta_k \varphi_k(x_i) - \Delta_{k+1} \varphi_{k+1}(x_i) \tag{6}$$

2.3   Approach 3

Let x_N is between the centers of the functions with numbers k and k + 1. In previous notation we use the formula [1]:

$$\Delta_k(N) = \Delta(N)\lambda, \Delta_{k+1}(N) = \Delta(N)(1 - \lambda), \tag{7}$$

where $\lambda = \frac{z_{k+1} - x_N}{z_{k+1} - z_k}$ [1].

We find the number $\Delta_k(N)$ by the minimization of the error function and so it is equal

$$\Delta(N) = \frac{\lambda Q_k + (1 - \lambda)Q_{k+1}}{\lambda^2 S_k + 2\lambda(1 - \lambda)P_k + (1 - \lambda)^2 S_{k+1}}. \tag{8}$$

Moving to the next step we need to recalculate an error according to the formula Error when moving to the next step you need to recalculate according to the formula

$$\delta_i(N+1) = \delta_i(N) - \Delta(N)\left(\lambda\varphi_k(x_i) + (1 - \lambda)\varphi_{k+1}(x_i)\right). \tag{9}$$

For approaches 1 and 3 decreasing by a given law $\Delta_k(N)$ can be used instead of formulas (1), (7) and (8), for example, $\Delta(N) = \left(1 - \frac{2}{T}\right)^N \delta_k(N)$, where $T = 5n + N_{max}$ [1]. This significantly reduces the amount of computation.

2.4   Approach 4

This approach involves the construction of linear regression and involves solving systems of linear equations, which can be obtained by a recurrent way when we get a new point [3, 4, 7].

3 Results of Numerical Experiments

Comparative testing of algorithms of the approaches 1–4 for the above basis functions for different values of the error dimensions, the number of points and number of basis functions was done. Also, the cases of dynamic (variable in time) data were analyzed at the different sets of parameters and the choices of basis functions. The interval of the argument [0;1]. Below are illustrations of the algorithms results, some of the most characteristic data of numerical experiments and conclusions obtained in the analysis of the results of these experiments. The first picture shows the result of processing experimental dependence, obtained by generate points around the function $\sin(\pi x) + 0.1 \sin(10\pi x)$, using the first algorithm and the formulas from [1] to 20 basis functions, a 400-hundred experimental points, the error of "measurement" – 0.1 and 10 runs of the algorithm.

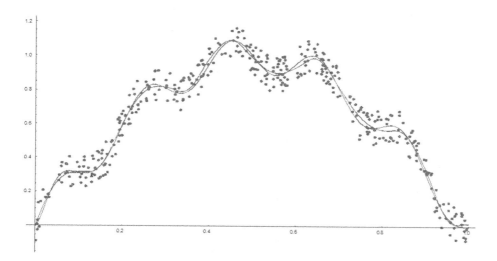

Fig. 2. The result of the first algorithm for static data.

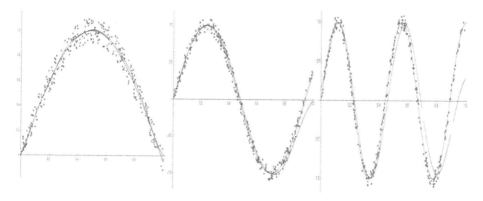

Fig. 3. Results of the second approach for dynamic data.

Figure 3 shows the results of applying this second approach to dynamic data. We approximate the function $\sin(w\pi x)$, where w varies over the computational time from 1 to 4.5. Basis function - cubic parabola. 300 experimental points. 10 basis functions. Measurement error - 0,1. There are three moments in which the original function has the forms, respectively: $\sin(\pi x)$, $\sin(2,25\ \pi x)$, $\sin(\pi x\ 4,5)$ (Fig. 3, Table 2).

Table 1. Mathematical expectation of error for the eight tests of the algorithm with 50 points. Static data. Approximative function - $\sin(\pi x)$. The number of basis functions - 10. The number of experimental points - 50.

ε	m	Approach	Triangular cap	Parabola	Cubic parabola	Gaussian
0.01	10	1 E.	0.0045	0.0359	0,0251	0,0337
1	10		0.2880	0.3115	0,0404	0,0308
0.01	1		0.0247	0.0363	0,0536	0,0209
0.01	10	1 A.	0,0047	0,0050	0,0056	0,3342
1	10		0,0320	0,0406	0,0361	0,0355
0.01	1		0,1888	0,1126	0,1130	0,1313
0.01	10	2	0,0039	0,0039	0,0054	0,0038
1	10		0,0331	0,0330	0,0281	0,0343
0.01	1		0,0057	0,0150	0,0195	0,0146
0.01	10	3 E.	0,0132	0,0125	0,0238	0,0217
1	10		0,0292	0,0378	0,0387	0,0398
0.01	1		0,0464	0,0403	0,0365	0,0487
0.01	10	3 A.	0,0044	0,0044	0,0053	0,0422
1	10		0,0278	0,0342	0,0358	0,0579
0.01	1		0,1016	0,0975	0,0945	0,1041
0.01		4	0,0031	0,0035	0,0040	0,0080
1			0,2838	0,2542	0,2633	0,2901

Here and next, ε - error of observations, m - the number of iterations of an algorithm, n - the number of points, A. is for adjustable step in accordance with the following formula from [1], E. is for exact step in accordance with the formulas given above for each algorithm.

Table 2. Mathematical expectation of error for the four tests of the algorithm. Dynamic data. Approximative function - $\sin(w\pi x)$, where w varies over the computational time from 1 to 2 and up to 4.5. The number of basis functions is 10. Measurement error is 0.1.

m	n	w	Approach	Triangular cap	Parabola	Cubic parabola	Gaussian
100	300	4,5	1	0,4252	0,4084	0,4060	0,4360
50	100	2		0,1908	0,1910	0,1755	0,1811
100	300	4,5	2	0,2880	0,2624	0,2481	0,2348
50	100	2		0,1454	0,1361	0,1321	0,1233
100	300	4,5	3	0,4176	0,4296	0,4422	0,4767
50	100	2		0,1854	0,1815	0,1779	0,1817

4 Conclusions

(1) Reviewed algorithms and basis functions showed good performance in the considered task.
(2) None of the algorithms does not have a decisive advantage over others. This allows us to recommend the simplest of them – the first.
(3) The advantages of exact computation of step over the formula from [1] could not compensate for a substantial increase in computational complexity, if the number of points is large enough, or duplicated as necessary.
(4) For large errors and a small number of experimental points, all the methods are unsatisfactory. Some better than others, the results of smoothing using cubic basis function for a large number of points and the gaussian for small.
(5) The results we got using different basis functions approximately the same. The choice of specific function is dictated by conditions on the smoothness. If such conditions exist, then the preferred is a triangular function because of the minimality of computational complexity. If these conditions are not known in advance or sufficiently rigid, it is preferable gaussian, as having infinite smoothness.
(6) The best of the considered algorithms for dynamic data smoothing is approach 2, as it handles much better the end of the interval.

The paper is based on research carried out with the financial support of the grant of the Russian Science Foundation (Project No. 14-38-00009, the program-targeted management of the Russian Arctic zone development). Peter the Great St. Petersburg Polytechnic University.

References

1. Hakimov B.V.: Modeling Correlation Dependences with Splines on the Examples in Geology and Ecology, p. 144. Moscow State University, Neva, St. Petersburg (2003). (in Russian)
2. Hakimov, B.V., Mikheev, M.I.: Nonlinear model of a neuron - multi-dimensional spline. Neurocomputers Dev. Appl. **7**, 36–40 (2012). (in Russian)
3. Tarkhov, D.A.: Consecutive algorithms for data smoothing. Neurocomputers Dev. Appl. **3**, 11–18 (2015). (in Russian)
4. Tarkhov, D.A.: Neural network models and algorithms. In: Radio Engineering, Moscow, p. 352 (2014). (in Russian)
5. Haykin, S.: Neural Networks: A Comprehensive Foundation, p. 823. Prentice Hall, Upper Saddle River (1999)
6. Svinyin, S.F.: Basis splines in the theory of samples of signals. Science 118 (2003). St. Petersburg (in Russian)
7. Albert, A.: Regression and the Moor-Penrose Pseudoinverse, p. 179. Academic Press, New York (1972)
8. Vasilyev, A.N., Tarkhov, D.A.: Mathematical models of complex systems on the basis of artificial neural networks. Nonlinear Phenom. Complex Syst. **17**, 327–335 (2014)

9. Kainov, N.U., Tarkhov, D.A., Shemyakina, T.A.: Application of neural network modeling to identification and prediction in ecology data analysis for metallurgy and welding industry. Nonlinear Phenom. Complex Syst. **17**, 57–63 (2014)

10. Lazovskaya, T.V., Tarkhov, D.A.: Fresh approaches to the construction of parameterized neural network solutions of a stiff differential equation. St. Petersburg Polytechnical Univ. J. Phys. Math. **1**, 192–198 (2015)

11. Ueno, T., Kawanabe, M., Mori, T., Maeda, S., Ishii, S.: A semiparametric statistical approach to model-free policy evaluation. In: Proceedings of the 25th International Conference on Machine Learning, pp. 1072–1079 (2008)

12. Bottou, L., Lecun, Y.: On-line learning for very large datasets. Appl. Stoch. Models Bus. Ind. **21**, 137–151 (2005)

FIR as Classifier in the Presence
of Imbalanced Data

Solmaz Bagherpour, Àngela Nebot$^{(\boxtimes)}$, and Francisco Mugica

SOCO Research Group, Universitat Politècnica de Catalunya,
UPC Barcelona-Tech, Barcelona, Spain
{sbagherpour,angela,fmugica}@cs.upc.edu

Abstract. In this paper we are investigating the potentiality of the Fuzzy
Inductive Reasoning (FIR) methodology as classifier applied to real world
dataset. FIR is a modeling and simulation methodology that is best suited for
dealing with regression and time series prediction. It has been shown in previous
works that FIR methodology is a powerful tool for the identification and prediction of real systems, especially when poor or non-structural knowledge is
available. FIR methodology falls under rule base supervised learning techniques. In this study we are studying the performance of the basic FIR classifier
applied to imbalanced classification problems and comparing its results with
well-known instance based and rule based approaches.

Keywords: Fuzzy Inductive Reasoning · FIR classifier · Imbalance datasets ·
Supervised learning

1 Introduction

There are several applications for Machine Learning, the most significant of which is
predictive data mining. If every instance in any dataset is given with a known label then
the learning is called supervised. Supervised methods attempt to discover the relationship between input attributes and a target attribute. The relationship discovered is
represented in a structure referred to as a model. Usually models describe and explain
phenomena, which are hidden in the dataset and can be used for predicting the value of
the target attribute knowing the values of the input attributes. Several different well
known approaches exist for classification tasks. In this study we want to compare the
basic Fuzzy Inductive Reasoning (FIR) classifier with some of the well known
rule-based and instance-based approaches as C4.5 [1], CN2 [2], PART [3], RISE [4],
kNN [5], RIPPER [6] and Modlem [7]. We chose them mainly due to their similarity to
FIR regarding to their comprehensibility and human-readability. High imbalance due to
rare cases or rare classes is an inseparable part of nature and real world problem
domains. Decision systems that aim to detect these rare but important cases have
always faced difficulty detecting them because of different reasons. The most obvious
problem is the associated lack of data for rare cases among training data. Even if they
are detected they still can be problematic because it is still hard to make any generalization out of few data points. Rare cases might be obscured by common cases; this is
especially more problematic when the algorithms rely on greedy search that examine

© Springer International Publishing Switzerland 2016
L. Cheng et al. (Eds.): ISNN 2016, LNCS 9719, pp. 490–496, 2016.
DOI: 10.1007/978-3-319-40663-3_56

one variable at a time. Another problem arises from the metrics used for evaluation of data mining algorithms. For example, the use of classification accuracy will cause classifier induction algorithms focus more on common cases rather than on rare cases. Another problem can arise from the bias of those data mining systems that employ a maximum-generality bias, since it can impact their ability to mine rare cases. In this work we are interested in study the performance of the basic FIR classifier, i.e. without including specific techniques to deal with imbalanced data sets such are algorithm or data level methods, when applied over imbalanced real world problems and compare its performance with common rule based and instance based approaches [8].

2 Basic FIR Classifier

The FIR methodology has been improved considerably in the past years, and it has proven to be useful for modelling and predicting soft sciences systems, such as biomedical and ecological applications [9]. FIR is a data-driven methodology that is based on systems behavior rather than on structural knowledge. Therefore it is well suited for systems for which no a priori structural knowledge is available, such as biomedical, biological and ecological systems [9]. FIR has been mainly applied to regression and time series prediction problems, and it is usually viewed as a modelling and simulation methodology. FIR structure is described in Fig. 1. The raw training data is converted (through a discretization pre-process) to qualitative fuzzy information by a fuzzification function. This function converts each quantitative (real-valued) data point into one qualitative triple. The first element of the triple is the class value, the second element is the fuzzy membership value, and the third element is the side value. The side value indicates whether the qualitative value is to the left or to the right of the peak value of the associated membership function. In order to convert quantitative values into qualitative triples, it is necessary to provide to the function the number of classes into which the space is going to be discretized, as well as the discretization algorithm. The modelling function (Learning Optimal Mask in Fig. 1) is responsible for finding causal relations between variables, i.e. a mask in the FIR terminology. A mask is a matrix representation of the spatial and temporal relations between the selected variables, which from machine learning terminology it can be viewed as a feature selection strategy. The optimal mask is the one that maximizes the forecasting power of the qualitative model. Each mask is evaluated by mean of a quality measure, mainly based on the Shannon entropy. The obtained mask is used to form the behavior matrix of the system, which can be interpreted as a special kind of fuzzy finite state machine relating the mask inputs to the mask output (fuzzy rules). Once the mask and the behavior matrix are available, predictions can take place using the FIR classification algorithm, employing a specialization of the 'k-nearest neighbors rule' commonly used in pattern recognition. For a deeper insight into the FIR methodology, the reader is kindly referred to [9].

Algorithm 1 shows the main steps of the KNN-based classifier of FIR methodology.

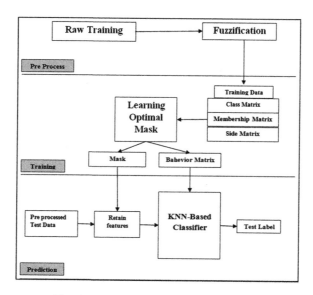

Fig. 1. Fuzzy inductive reasoning scheme

Algorithm 1. *KNN-Based Classifier of FIR*

Apply Selected Mask to the instance to be predicted
Same_Patterns = Find_Patterns(P,BM)
  # finds all the instances of Pattern P in the Behavior Matrix (BM) of the selected Mask
if Same_Patterns = NULL
    Apply lower complexity MASK with its related BM
    # finds a new sub-optimal MASK
else
    Same_Pattern_Distances = Find_Same_Patterns_Distances(P, BM)
    # finds distances between P and all the same Pattern instances found in the BM
if All Same_Pattern_Distances = 0
    Predicted_Output_Class = Majority_Output(Same_pattern_Distances)
    # predicts as output the class that is more represented in the patterns with distance 0
else
    Neighbours = FindNeighbours(K,Same_Pattern_Distances)
    # finds K nearest Neighbours among Same Pattern Instances
    Predicted_Output_Class = Majority_Output(Neighbours)
    # predicts as output the class that is more represented in the K nearest patterns

3   Experimental Results on Real World Datasets

The main goal of the experiment is to evaluate the classification ability of basic FIR classifier in presence of class imbalance. We compared FIR with other standard rule induction classifiers presented in [8], where the authors proposed binary classification problems. Following their approach, from datasets with multiclass domain the smallest class is selected as minority class and aggregated the remaining classes into one majority class. When dealing with imbalanced data sets, it is essential to use different performance measures in order to reflect different aspects of the learning algorithm. It is important to take into account that if the choice of metrics doesn't value the minority class then the imbalance problem is not issued well. The most commonly used metric which is the accuracy is not a suitable metric for imbalanced datasets since it is biased towards the majority class. There are other metrics that can be used and are less affected by imbalance as precision, recall, F-measure, Sensitivity, Specificity, geometric mean, ROC curve, AUC and precision-recall curve. In this research, in order to compare the performance of all classifiers three measures are used: Sensitivity of the minority class, G-mean and F-measure.

From the confusion matrix it is possible to obtain:

(1) $TruePositiveRate = TP/(TP + FN)$
(2) $TrueNegativeRate=TP/(TP + FN)$
(3) $FalsePositiveRate=FP/(TN + FP)$
(4) $Precision=TP/(TP + FP)$

(5) $F\text{-measure} = 2 \cdot \dfrac{Precision.Recall}{Precision + Recall}$

(6) $G\text{-mean} = \sqrt{Sensitity.Specificity}$

True Positive Rate is called *Sensitivity* (also known as *Recall*), while True Negative Rate is called *Specificity*. The F-measure or balanced F-score (F_1 score) is the harmonic mean of precision and recall. An important, useful property of G-mean is that it is independent of the distribution of examples between classes. The selected datasets are described in Table 1. They have different imbalance ratios, sizes and number of attributes.

The Experiments were run with a stratified 10 fold cross validation. The results of the 8 algorithms are shown in Table 2. For each data set and algorithm, the Sensitivity, G-Mean and F-Measure of the results obtained are presented. From Table 2 it can be seen that FIR obtains the best sensitivity result for one data set, the best G-Mean for

494   S. Bagherpour et al.

Table 1. Description of the datasets. Obtained from UCI machine learning repository

Data set	Num. instances total	Num. inst. minority class	% Imbalance	Num. attributes
Abalone	4,177	335	8.02	8
Balance	625	49	7.84	4
Breast cancer	699	241	34.47	9
CMC	1,473	333	22.61	9
Ecoli	336	35	10.42	7
Haberman	306	81	26.47	3
Ionosphere	351	126	35.89	34
New-Thyroid	215	35	16.28	5
Pima	768	268	34.89	8
Transfusion	748	178	23.80	4
Vehicle	846	199	23.52	18

three data sets and the best F-Measure for one data set. It has, also, the second sensitivity score for Ecoli, Haberman and Vehicle. Moreover, FIR performance measures are in general close to the best ones, except for Breast cancer, Ionosphere and CMC data sets. The Balance data set is quite special and FIR gets very bad results as almost the rest of the algorithms do.

We use a statistical approach to compare the differences in performance between all classifiers. We apply non-parametric Friedman test to the three measures to globally compare the performance of the 8 classifiers on the 11 selected datasets. The null-hypothesis in these tests is that all the classifiers perform equally well. We find that the null hypothesis can be rejected for Sensitivity and G-Mean results, with p values of 0.0009 and 0.019, respectively. However, in the case of F-Measure the null hypothesis cannot be rejected because the p value is greater than $\alpha = 0.05$, i.e. 0.1102. Table 3 presents the mean average of the Sensitivity, G-Mean and F-Measure of all the datasets for each algorithm. Inside parenthesis is the ranking of each algorithm. As it can be seen from Table 3, C4.5 is the algorithm that performs better on average taking into account all the measures. Then, PART and FIR algorithms have almost the same performance, i.e. PART has slightly better sensitivity and F-Measure whereas FIR has better G-Mean. Both algorithms have performances very close to the ones obtained by C4.5.

Table 2. Sensitivity, G-mean and F-measure obtained for each algorithm and data set. The best results for each measure and data set is shaded.

Data sets	Measure	Algorithms							
		RISE	kNN	C4.5	CN2	PART	RIPPER	Modlem	FIR
Abalone	Sensitivity	0.128	0.137	0.339	0.160	0.188	0.184	0.245	0.218
	G-Mean	0.345	0.358	0.568	0.396	0.419	0.421	0.484	0.457
	F-Measure	0.192	0.208	0.393	0.253	0.269	0.282	0.326	0.293
Balance	Sensitivity	0.000	0.004	0.018	0.018	0.000	0.000	0.000	0.000
	G-Mean	0.000	0.009	0.019	0.019	0.000	0.000	0.000	0.000
	F-Measure	0.000	0.007	0.019	0.019	0.000	0.000	0.000	0.000
Breast	Sensitivity	0.959	0.968	0.917	0.886	0.947	0.896	0.887	0.892
	G-Mean	0.963	0.969	0.929	0.929	0.950	0.928	0.926	0.929
	F-Measure	0.949	0.957	0.912	0.915	0.932	0.910	0.910	0.914
CMC	Sensitivity	0.293	0.308	0.404	0.096	0.377	0.071	0.256	0.283
	G-Mean	0.507	0.517	0.586	0.258	0.543	0.255	0.472	0.490
	F-Measure	0.351	0.358	0.434	0.140	0.361	0.124	0.311	0.320
Ecoli	Sensitivity	0.505	0.578	0.597	0.185	0.420	0.445	0.400	0.583
	G-Mean	0.638	0.701	0.717	0.284	0.554	0.587	0.568	0.727
	F-Measure	0.517	0.592	0.593	0.244	0.450	0.473	0.465	0.570
Haberman	Sensitivity	0.224	0.181	0.244	0.184	0.334	0.180	0.240	0.272
	G-Mean	0.375	0.334	0.426	0.345	0.468	0.355	0.401	0.464
	F-Measure	0.240	0.214	0.300	0.235	0.349	0.233	0.262	0.315
Ionosphere	Sensitivity	0.902	0.629	0.837	0.779	0.840	0.818	0.824	0.754
	G-Mean	0.928	0.780	0.878	0.870	0.888	0.874	0.890	0.827
	F-Measure	0.913	0.747	0.847	0.850	0.864	0.848	0.872	0.785
New-Thyroid	Sensitivity	0.928	0.867	0.850	0.866	0.933	0.855	0.812	1.000
	G-Mean	0.951	0.921	0.901	0.915	0.953	0.911	0.878	0.986
	F-Measure	0.947	0.895	0.843	0.906	0.918	0.879	0.848	0.935
Pima	Sensitivity	0.551	0.558	0.507	0.408	0.591	0.377	0.485	0.556
	G-Mean	0.666	0.681	0.649	0.600	0.679	0.581	0.641	0.687
	F-Measure	0.577	0.599	0.567	0.512	0.596	0.484	0.550	0.607
Transfusion	Sensitivity	0.297	0.319	0.386	0.150	0.429	0.088	0.371	0.355
	G-Mean	0.507	0.529	0.579	0.342	0.602	0.266	0.529	0.555
	F-Measure	0.354	0.385	0.443	0.214	0.462	0.149	0.354	0.407
Vehicle	Sensitivity	0.831	0.865	0.867	0.329	0.883	0.874	0.859	0.880
	G-Mean	0.895	0.914	0.911	0.513	0.919	0.919	0.916	0.910
	F-Measure	0.855	0.877	0.867	0.433	0.875	0.885	0.892	0.853

Table 3. Mean average of sensitivity, G-mean and F-measure of all the datasets for each algorithm. Inside parenthesis is the ranking of each algorithm.

	RISE	kNN	C4.5	CN2	PART	RIPPER	Modlem	FIR
Sensitivity	0.51 (4)	0.49 (5)	0.54 (1)	0.37 (8)	0.54 (2)	0.43 (7)	0.49 (6)	0.53 (3)
G-Mean	0.615 (4)	0.610 (5)	0.65 (1)	0.50 (8)	0.63 (3)	0.55 (7)	0.609 (6)	0.64 (2)
F-Measure	0.53 (4)	0.53 (5)	0.56 (1)	0.43 (8)	0.55 (2)	0.48 (7)	0.53 (6)	0.54 (3)

4 Conclusions

In this paper we studied the performance of the basic FIR classifier when applied to binary imbalanced classification problems and compared its results with well-known instance based and rule based approaches. The main question motivating the study was if the performance of FIR applied as classifier is highly affected by imbalance or not. The results obtained are very encouraging since the basic FIR classifier is able to perform as well as PART algorithm and only slightly worse than C4.5. FIR clearly outperforms RISE, kNN, CN2, RIPPER and Modlem classifiers. Our conclusion based on the carried experiments is that the performance of FIR on these datasets is not necessarily affected by degree of imbalance present in the datasets. The next step is the enhancement of the basic FIR classifier focusing on data level methods (as for example a re-sampling method) or algorithm level methods.

References

1. Quinlan, J.R.: C4.5: Programs for Machine Learning. Morgan Kaufmann Publishers Inc., San Francisco, USA (1993)
2. Clark, P., Niblett, T.: The CN2 induction algorithm. Mach. Learn. **3**, 261–283 (1989)
3. Frank, E., Witten, I.: Generating accurate rule sets without global optimization. In: Proceedings of the Fifteenth International Conference on Machine Learning, pp. 144–151 (1998)
4. Domingos, P.: Unifying instance-based and rule-based induction. Mach. Learn. **24**, 141–168 (1996)
5. Weinbergerl, K.Q., Saul, L.K.: Distance metric learning for large margin nearest neighbor classification. J. Mach. Learn. Res. **10**, 207–244 (2009)
6. Cohen, W.: Learning trees and rules with set-valued features. In: AAAI 1996 Proceedings of the Thirteenth National Conference on Artificial Intelligence, pp. 709–716 (1996)
7. Stefanowski, J.: Rough set based rule induction techniques for classification problems. In: Proceedings of 6th European Congress on Intelligent Techniques and Soft Computing, pp. 109–113 (1998)
8. Napierala, K., Stefanowski, J.: BRACID: a comprehensive approach to learning rules from imbalanced data. J. Intell. Inf. Syst. **39**, 335–373 (2012)
9. Nebot, A., Mugica, F.: Fuzzy inductive reasoning: a consolidated approach to data-driven construction of complex dynamical systems. Int. J. Gen. Syst. **41**, 645–665 (2012)

Neural Network System for Monitoring State of a High-Speed Fiber-Optical Linear Path

I.A. Saitov, O.O. Basov$^{(\boxtimes)}$, A.I. Motienko, S.I. Saitov, M.M. Bizin, and V.Yu. Budkov

SPIIRAS, 39, 14th Line, St. Petersburg 199178, Russia
oobasov@mail.ru

Abstract. The paper presents a methodology for the synthesis of systems for monitoring the state of a high-speed fiber-optical linear path, based on the information and measuring control system that implements neural network recognition algorithms with synthesis by dominance. The proposed information-measuring system processes levels of the average intensity of the optical signal received at various carriers in a certain retrospective over a defined period of time. Neural network algorithms are synthesized by optimizing the parameters of a neural network that allow to make a conclusion about the state of a high-speed fiber-optical linear path.

Keywords: Optical transport networks · Main optical path · Neurocomputers · Neural network · Information and measuring control system

1 Introduction

In the last decades, the optical transport networks (OTN), serving an average of 90 % of volume of long-distance and international traffic, have become a basis of national OTS. It is possible to mark out the following features of the current state of the art in a subject domain [1]: (1) productivity of the optical fiber baseband transmission paths (OFBTP) with wavelength-division multiplexing (WDM) has increased many times and reached tens of terabits per second; (2) transmission distance of nonregenerative signaling exceeded 1000 km; (3) synchronous transmission technologies are succeeded by the whole group of asynchronous technologies: from the known operator options *Ethernet* to perspective *OTN (Optical Transport Network)* and *GMPLS* [2]. It has become possible to transmit signals of various formats in spectral channels of one OFBTP with WDM (OFBTP have become heterogeneous).

These factors significantly complicate monitoring process of the baseband transmission path (BTP) state of the real time. Standard control facilities of OFBTP of von Neumann type [3] have no sufficient productivity for processing the incoming massifs of optical signal intensity measurements. Meanwhile there are results of researches that show a possibility for application of neural network approaches to formation of a system for monitoring the state of heterogeneous OFBTP with WDM [4–6].

© Springer International Publishing Switzerland 2016
L. Cheng et al. (Eds.): ISNN 2016, LNCS 9719, pp. 497–504, 2016.
DOI: 10.1007/978-3-319-40663-3_57

2   The Essence of the Approach

Let heterogeneous OFBTP with WDM be an object of control, in which there are R carriers with wavelengths $\lambda_1, \ldots, \lambda_R$. For realization of continuous control of the state of OFBTP, in subject domain it is often proposed to organize a measurement channel (Fig. 1) via which the part of group signal energy will be fed through a coupler to the Information and Measuring Control System (IMCS).

Research has shown that it is expedient to use a specialized optical neural network as a compute kernel of IMCS [3].

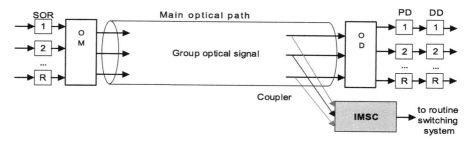

Fig. 1. The organization of a measurement channel in OFBTP

3   Estimability Analysis of OFBTP Parameters

Let τ_{MP} be a maintenance period (MP) of OFBTP during which all meaning characteristics of all components of the main optical path (MOP) can be measured, and τ_{split} be some time period, during which there is a transition of a certain spectral channel (SC) from operating state Ω_1 to a state of parametric failure Ω_2. Then it is possible to allocate the following two groups of factors of operating conditions of MOP [3, 4]: (1) irreversible factors, for which $\tau_{split} > \tau_{MP}$ is true, connected with phenomena of wear of the OFBTP elements; (2) factors (reversible and irreversible), causing changes of intensity parameters of an optical signal at the input of a photodetector with $\tau_{split} < \tau_{MP}$.

Therefore, it is necessary to choose such intensity parameters of an optical signal, which, on the one hand, are capable to reflect the dynamics of change of OFBTP state with the required accuracy, and on the other hand, allow using neurooptical information processing systems.

A study [7] has showed that the characteristics, defining a state of the r-th spectral channel $\left(r = \overline{1, R}\right)$ of incoherent OFBTP with a passive pause regarding reliability of information transfer, are the average number of photons of a signal n_{avgr} and background noise n_{nr} observed during measurement time at the input of a photodetector. Numerical and full-scale experiments showed that for the existing OFBTP, a measurement of sample average number of photons with the given accuracy and reliable estimation $\psi = 0,99$ will require observation of several thousand impulses, and observation time τ_{obs} will be units of milliseconds.

4 Preparation Data Array for Processing in Neural Network Information and Measuring System

Further, it is proposed to identify the following sets of states of OFBTP that reflect the suitability of their R spectral channels [3].

Fault-free (FF) set, when all spectral channels provide transfer of a flow with the required speed B_r^{RQRD} and necessary reliability $p_{err\,r}$ of transmission

$$Q_{FF}^{LP} = \{R, B_r^{RQRD} | p_{errr} \leq p_{errr}^{perm}\}; \tag{1}$$

where $p_{err\,r}^{perm}$ is a permissible value of a bit mistake in the channel.

Operational (Op) (intermediate), when most spectral channels R_{FF} are fault-free, and in other R_{Op} spectral channels there can be a certain (limited) decrease in reliability of transfer (not worse than $p_{err\,r}^{al}$) regarding the maximum transmission speed:

$$Q_{Op}^{LP} = \{\exists r \in R_{Op}, B_r^{RQRD} | p_{err\,r}^{al} \geq p_{err\,r} > p_{err\,r}^{perm}\}, \tag{2}$$

where $R_{Op} \cup R_{FF} = R$, $\forall r \in R_{FF}(B_r^{RQRD} | p_{errr} \leq p_{err\,r}^{perm})$; $R_{FF} \geq R_{thld}, R_{FF} + R_{Op} \geq R_{crit}$.

Alarmed (Al) (inoperable), when a number of fault-free spectral channels appears to be less than a threshold value R_{thld}, or the sum $R_{FF} + R_{Op}$ of fault-free and operational spectral channels becomes less than a value R_{crit}:

$$Q_{Al}^{LP} = \{\exists r \in R_{Op}, B_r^{RQRD} | R_{FF} < R_{thld} \text{ or } R_{FF} + R_{Op} < R_{crit}\}. \tag{3}$$

Requirements for values $B_r^{RQRD}, p_{err\,r}^{perm}, p_{err\,r}^{al}, R_{FF}, R_{Op}, R_{thld}, R_{crit}$ are considered to be set by a metasystem.

For unambiguous reference of a state of OFBTP to the sets Q_{FF}^{LP}, Q_{Op}^{LP} or Q_{Al}^{LP}, during functioning of spectral channels it is necessary to have some combinations of impulses with the determined characteristics in a group signal at all bearing wavelengths [8]. Such impulses combinations (IC), having length N_{IC} of symbols can periodically be entered into TOM by methods of temporary division of channels (TDM) and branch off in IMCS (Fig. 2) together with an information optical signal.

In the developed system, by processing N_{IC} of symbols at each of the carriers a sample average value of a number of photons per bit μ_{rk} in the r-*th* spectral channel in the k-*th* moment of time will be received. It follows that IMCS for a state of OFBTP has to process the values of average intensity of an optical signal received at various carriers in a certain retrospective in some period (Fig. 2).

Suppose that in the presence of several classes of states of spectral channels of OFBTP $Y = (\Omega_z), w = \overline{1, W}$ K measurements are carried out on N_{IC} impulses, and estimates $\{\mu_{rk}, r = \overline{1, R}; k = \overline{1, K}\}$ are obtained for each spectral channel. It is required to determine an estimate $y_{rK+1} = f(\mu_{rK+1}), y_r = 1, \ldots, W$ of processes of parameters alteration $\{\mu_{rk}\}$ according to the criterion $c_r(y_r, \mathring{y}_r)$, set by a penalty function $c_r(y_r, \vartheta) = 1 - \delta(y_r; \mathring{y}_r), \mathring{y}_r = 1, \ldots, W$, where δ is the Kronecker symbol; $\mathring{y}_r$ is a true value of a sought estimate.

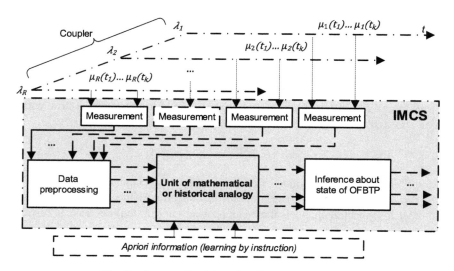

Fig. 2. A generalized functional structure of IMCS

Each image $y_{rk}(k = 1, \ldots, K, K + 1)$ can be put in correspondence with its distribution $\varpi(\mu_r | \dot{y}_r)$ in space of the observed R-dimensional random process $(\mu_{r1}, \ldots, \mu_{rK+1})$. Knowing some a priori distribution $p(\dot{y}_r)$, it is possible to solve the problem by the usual Bayesian methods. However, during analysis of MOP the situation is complicated by the fact that the variables μ_{rk} are not independent, and the distributions $\varpi(\mu_r | 1), \ldots \varpi(\mu_r | W)$ and $p(\dot{y}_r)$ are a priori unknown. Their a priori knowledge can be replaced by the process of learning by instruction, i.e. the process of supervisory communication of additional information to IMCS. For this purpose, during measurements and tests of MOP a set of training examples for each IC is formed.

Everything mentioned above can be ensured with the help of means of neurooptic that allow for realization of parallelized analysis of parameters of the whole MOP on the time interval $\tau_P = [t_1, t_k]$, i.e., when an extrapolation argument is presented by a volume optical image – a parameter matrix of the form:

$$X'_{FF} = \begin{pmatrix} \mu_{11} & \mu_{12} & \cdots & \mu_{1k} \\ \mu_{21} & \mu_{22} & \cdots & \mu_{2k} \\ \cdots & \cdots & \cdots & \cdots \\ \mu_{R1} & \mu_{R2} & \cdots & \mu_{Rk} \end{pmatrix}, \tag{5}$$

where μ_{rk} is the mean number of photons in the r-th spectral channel by N_{IC} in the k-th moment of time. It is necessary to create a type of a matrix-prediction according to the given argument by methods of mathematical and/or historical analogy X''_{Pred}. Analysis of this matrix allows us to make a conclusion about the projected state of OFBTP for taking control actions.

5   The Technique for Synthesis of Neural Network Information-Measuring Control System for OFBTP State

Let the vector X_e in the selected feature space be represented by the magnitudes and directions of changes of the matrix elements X'_p. A vector of the governing parameters $X_{\text{gov}\,e}$ of MOP combines characteristics of SOR and OFA in the e-th situation, $e = \overline{1, E}$, known to the control system by values of the corresponding pumping currents of SOR and OFA: $I_{Se}(t_k)$ and $I_{\text{Amp}\,e}(t_k)$ respectively.

Thus a case of the functioning of MOP can be written in the form:

$$C_e = (X_e(t_k); Y_e(t_k + \tau_{\text{Cs}})) = (X_{\text{Op}\,e}(t_k),\ I_{\text{FF}\,e}(t_k),\ I_{\text{Amp}\,e}(t_k); Y_e(t_k + \tau_{\text{Cs}})),$$

where $X_e = (X_{\text{Pe}}(t_k),\ I_{Se}(t_k),\ I_{\text{Amp}e}(t_k))$ is the vector of causes (Cs) of the e-th situation; $Y_e(t_k + \tau_{\text{Cs}})$ is the vector taken as a consequence of the e-th situation; τ_{Cs} is a time interval between the cause and the consequence.

Obtained in the course of traffic control of MOP, a training set cannot contain the full volume of information characterizing all the possible consequences $Y(t_k + \tau_{\text{Cs}})$ of all the possible set of causes $X(t_k)$. So extrapolating functional, modeling on its basis an operator F of cause-effect relation $Y(t_k + \tau_{\text{Cs}}) = F(X(t_k))$, will not be predetermined, i.e., $Y(t_k + \tau_{\text{Cs}}) = F_M^{(1)}(X(t_k))$, where $F_M^{(1)}$ is a model of operator F of the first approximation.

Using the assumption that all tests are conducted at one point in time t_0, from the study of governing parameters in the time space we can pass to their analysis in the feature space. For this purpose, the test report can be written in the form $R_{test} = \{\,S_{ij}\,\}$, where each situation S_{ij} corresponds to the expression:

$$S_{ij} = (X_{\text{test}\,ij}; Y_{ij}) = (X_{\text{Op}ij},\ I_{\text{FF}i},\ I_{\text{Amp}i}; Y_{ij}), \tag{6}$$

where Y_{ij} is the reaction of the state of the spectral channel of a regeneration section in the i-th situation to the j-th effect of an algorithm for MOP testing; $X_{\text{test}\,ij} = (X_{\text{Op}\,ij},\ I_{\text{FF}\,i},\ I_{\text{Amp}\,i})$ is a set of parameters describing a situation in which this reaction to the j-th impact is carried out; $j = 1,\ldots, G$. Therefore, in interests of control system training, the information array can be obtained, where each i-th case of exploitation contains G_i observation situations by which the functionality $F_M^{(2)}$ of the operator $Y(t_i + \tau_{\text{Cs}}) = F_M^{(2)}(X'(t_k))$ of the second approximation is synthesized.

Then the task of developing the control system is reduced to a task of synthesis of a set of neural network algorithms for formation F_M by optimizing the neural network parameters allowing one to draw an inference:

$$Y_M : Y = F(X) \rightarrow Y_M = F_M(X), \tag{7}$$

$$|Y - Y_M| \leq \varepsilon^*, \tag{8}$$

where ε^* is the specified value, reflecting the requirements to the adequacy of a generated model.

The training set for formation of model structure must be presented by a set of cases of the form $R_a = \{C_a\} = \{(X_{OPa}, I_{FFa}, I_{Ampa}; Y_a)\}$, where $(X_{Opa}, I_{FFa}, I_{Ampa}) = X_a$ is the cause of the consequence Y_a.

The process of obtaining F_M with the known neural network structure G by available experimental data means the adjustment of transmission coefficients of interneuronal communication with the aim of minimizing the model error. However, direct measurement of this error in practice is not achievable, so we use the estimate

$$\xi_1 = \sum_{X \in X_a} |F_M(X) - Y_a|.$$

The unknown error $\xi_2 = \sum_{X \in X_b} |F_M(X) - Y_b|$ made by the model F_M on data not previously used in training is called an error of model integration. Since the true value of the integration error ξ_2 is inaccessible, in practice its estimation is used, which is obtained from the analysis of a part of the examples X_b for which system responses Y_b are known, but were not used in the training $X_b \in C_b = \{X_b, Y_b\}, X_b \cap X_a = \emptyset$.

A sample $C_b = \{X_b, Y_b\}$ is hereinafter referred to as a test sample (verification).

The actual *method of the synthesis of control system* for MOP state is considered as a sequence of the following stages.

1. The synthesis of an artificial neural network that reproduces the logic of MOP functioning, i.e., the structure definition of G_l and parameters $\Phi = \{\phi_{lh}\}$ of a neural network, modeling an operator $F : Y_a = F(X_a), F \to F_M$.
2. The choice of a learning algorithm for neural networks allowing realization of information-measuring elements transformation of a set $\{X_a\}$ to the elements of a set $\{Y_a\}$ in accordance with the criterion of suitability:

$$\Delta = \left|Y_a - F_M(X_{Pai}, X_{Sai}, X_{Ampai})\right| \leq \Delta^*, \tag{9}$$

where Δ^* is a permissible value of error Δ of representation of a precedent C_a, i.e., recovery of the consequence Y_{Ma} of the cause X_a, where $Y_{Ma} = F_M(X_a)$.
3. Based on processing of vectors $\{X_b\}$, the search for such a set of vectors $\{Y_{Mb}\}$, which would reflect the predicted states of $\{Y_b\}$ spectral channels. The reliability criterion of prediction is the expression

$$D_{Pred} = \left|Y_b - F_M(X_{Pbi}, X_{Sbi}, X_{Ampbi})\right| \geq D^*_{Pred}. \tag{10}$$

A neural network, synthesized in such a way after training, will ensure with the required accuracy displaying of the observed vector X_c in the inference about the state of MOP Y_{Mc}. If in the report on the current observations R_c there is a vector X_c similar to the well-known one of the neural network, $X_c = X_a = (X_{Opai}, X_{FFai}, X_{Ampai})$, then the functions of IMCS are limited to associative search (by historical analogy) of a given precedent and the restoration of its consequence as a sought-for prediction, i.e., $Y_c = Y_{Ma}$.

During the research, we have developed the algorithm [3] to identify the state of the MOP on the long-living intervals with the *synthesis by dominance*. We have also received a patent [9] for invention of neurooptical controller that implements the above-mentioned approach on the components of integrated and nonlinear optics.

6 Conclusion

The presented approach to creation of the control system for a state of OFBTP based on application of a control combination entered into a group optical signal, as well as the information and measuring system realizing neural network algorithms of recognition with synthesis by dominance, allows one to monitor a state of OFBTP in real time. At present, based on computer simulation, the following is being carried out: (1) testing of neural networks of different structures to ensure maximum reliability of OFBTP state identification and minimize training time; (2) search for ways to optimize the characteristics of the control system such as the depth of retrospection (τ_r), frequency rate of current measurements (τ_m) and a range (τ_l) of lead, which depend on the characteristics (a number of spectral channels and speed of information transmission in them) of the specific OFBTP. The developed approach for optical path monitoring is oriented to implementation in smart environments and cyberphysical systems to support safe and trust connection between distributed embedded modules, robots, cloud services, user devices and users [10–15].

Acknowledgment. This work is partially supported by the Russian Foundation for Basic Research (grant No. 16-08-00696-a) and the Council for Grants of the President of Russia (Projects No. MK-7925.2016.9).

References

1. Hudgings, J., Nee, J.: WDM all-optical networks. EE228 Project Report, Oslo, 29 (1996)
2. Rosen, E., Viswanathan, A., Callon, R.: Multiprotocol label switching architecture. RFC 3031 (2001)
3. Saitov, I.A., Muzalevskii, D.Y.: Continuous monitoring of a fiber-optical baseband transmission path based on intellectual optical-signal processing facilities. Telecommun. Radio Eng. **70**, 1501–1508 (2011). (in Russian)
4. Toge, K., Ito, F.: Recent research and development of optical fiber monitoring in communication systems. Photonic Sens. **3**, 304–313 (2013)
5. Saitov, I.A., Myasin, N.I.: A model of a fiber-optical baseband transmission path with wavelength-division multiplexing and fiber-optical amplifiers. Telecommun. Radio Eng. **70**, 1729–1738 (2011). (in Russian)
6. Delavaux, J.-M.P., Nagel, J.A.: Multi-stage erbium-doped fiber amplifier design. Lightwave Technol. **135**, 703–720 (1995)
7. Robert, M., Gagliardi, S.K.: Optical communication: transl. from Eng. Ed. A.G. Sheremetyev. M.: Svyaz' (1978) (in Russian)
8. Maamoun, K., Mouftah, H.: Survivability Issues in Optical and Optical Wireless Access Networks: Monitoring Trail Deployment for Fault Localization in All-Optical Networks and Radio-over-Fiber Passive Optical Networks (2012)

9. Saitov, I.A., Myasin, N.I., Muzalevskii, D.Y.: Device for continuous monitoring operating capacity of fibre-optic linear channel. Patent of RF for an invention No. 2400015 from 20.09.2010. Application No. 2009102711/28(003451) from 27.01.2009
10. Yusupov, R.M., Ronzhin, A.L.: From smart devices to smart space. Herald Russ. Acad. Sci. **80**, 45–51 (2010). MAIK Nauka
11. Budkov, V., Prischepa, M., Ronzhin, A.: Dialog model development of a mobile information and reference robot. Pattern Recognit. Image Anal. **21**, 458–461 (2011). Pleiades Publishing
12. Saveliev, A., Basov, O., Ronzhin, A., Ronzhin, A.: Algorithms for low bit-rate coding with adaptation to statistical characteristics of speech signal. In: Ronzhin, A., Potapova, R., Fakotakis, N. (eds.) SPECOM 2015. LNCS, vol. 9319, pp. 65–72. Springer, Heidelberg (2015)
13. Karpov, A.A., Ronzhin, A.L.: Information enquiry kiosk with multimodal user interface. Pattern Recognit. Image Anal. **19**, 546–558 (2009). MAIK Nauka/Interperiodica, Moscow
14. Basov, O., Ronzhin, A., Budkov, V., Saitov, I.: Method of defining multimodal information falsity for smart telecommunication systems. In: Balandin, S., Andreev, S., Koucheryavy, Y. (eds.) NEW2AN/ruSMART 2015. LNCS, vol. 9247, pp. 163–173. Springer, Heidelberg (2015)
15. Ronzhin, A., Prischepa, M., Budkov, V.: Development of means for support of comfortable conditions for human-robot interaction in domestic environments. In: Botía, J.A. et al. (Eds.) Workshop Proceedings of the 8th International Conference on Intelligent Environments, pp. 221–230. IOS Press (2012)

Pattern Classification with the Probabilistic Neural Networks Based on Orthogonal Series Kernel

Andrey V. Savchenko[(✉)]

Laboratory of Algorithms and Technologies for Network Analysis,
National Research University Higher School of Economics,
Nizhny Novgorod, Russia
avsavchenko@hse.ru

Abstract. Probabilistic neural network (PNN) is the well-known instance-based learning algorithm, which is widely used in various pattern classification and regression tasks, if rather small number of instances for each class is available. The known disadvantage of this network is its insufficient classification computational complexity. The common way to overcome this drawback is the reduction techniques with selection of the most typical instances. Such approach causes the shifting of the estimates of the class probability distribution, and, in turn, the decrease of the classification accuracy. In this paper we examine another possible solution by replacing the Gaussian window and the Parzen kernel to the orthogonal series Fejér kernel and using the naïve assumption about independence of features. It is shown, that our approach makes it possible to achieve much better runtime complexity in comparison with either original PNN or its modification with the preliminary clustering of the training set.

Keywords: Pattern classification · Small sample size problem · Probabilistic neural network (PNN) · Orthogonal series kernel · Nonparametric density estimates

1 Introduction

Pattern classification task is one of the most important application of artificial neural networks [1]. Unfortunately, such classifiers as multilayer perceptron or support vector machines (SVM) do not show the highest accuracy, if the number of training samples is not sufficient to learn complex neural network structures [2, 3]. In such a case it is possible to apply the instance-based learning methods [4], e.g. the k-nearest neighbor (k-NN) or the Bayesian approach [5]. In the latter case the nonparametric techniques, e.g., the Parzen kernel, can be used to estimate the unknown probabilistic densities of each class. The widely known parallel implementation of this nonparametric approach is the probabilistic neural network (PNN) [6, 7]. This feedforward network with two hidden layers is characterized by extremely fast training procedure. The PNN is commonly used in many pattern classification tasks, such as optical character recognition [8], face recognition [9], speech recognition [10], medical diagnostics [11], text authorship attribution [12], variance change point estimation in the control charts [13],

© Springer International Publishing Switzerland 2016
L. Cheng et al. (Eds.): ISNN 2016, LNCS 9719, pp. 505–512, 2016.
DOI: 10.1007/978-3-319-40663-3_58

etc. However, the computational complexity of the online classification with the PNN is rather low due to the requirement of having one neuron for each training sample in the pattern layer [6, 9]. The known way to overcome this drawback is the reduction techniques with selection of the most typical instances based on the clustering of the training set [4, 14]. However, such an approach causes the loss of the PNN advantages. In fact, the training procedure includes the clustering algorithm and becomes rather difficult [5]. Moreover, the resulted decision is not optimal in the Bayesian sense, especially, when the sizes of the clusters are quite different.

Thus, the purpose of our research is to develop the modification of the PNN, which possesses all its advantages, but is much faster. The rest of the paper is organized as follows. In Sect. 2 we introduce our modification of the PNN, in which the Parzen window in the pattern layer is replaced to the orthogonal series kernel [7, 15, 16]. In Sect. 3 experimental study results are presented for rather simple wine dataset from the UCI repository. Finally, concluding comments are given in Sect. 4.

2   Materials and Methods

Classification is the task of assignment a new observation $\mathbf{x} = [x_1, \ldots, x_M]$ to one of $C > 1$ classes. Here M is the size of the feature vector $\mathbf{x}$. Each class is specified by a training set $\{\mathbf{x}_r(c)\}, r \in \{1, \ldots, R_c\}$ of $R_c \geq 1$ instances $\mathbf{x}_r(c) = [x_{r;1}(c), \ldots, x_{r;M}(c)]$. The total size of the training set is equal to $R = \sum_{c=1}^{C} R_c$. According to statistical approach, it is assumed that the feature vectors from one class have the same probability distribution, and the task is reduced to the testing of C hypothesis $W_c, c = \overline{1, C}$ about distribution of the observation $\mathbf{x}$ [5]. The Bayesian criterion is written as follows

$$v = \underset{c \in \{1, \ldots, C\}}{argmax} \frac{R_c}{R} \cdot f(\mathbf{x}|W_c), \qquad (1)$$

where the c-th class prior probability is estimated as R_c/R [6], and $f(\mathbf{x}|W_c)$ is the conditional probability density function (likelihood) of the c-th class. This likelihood is estimated with the Parzen kernel $K(\mathbf{x}, \mathbf{x}_r(c))$ [6]. In this case, the criterion (1) is transformed to the PNN-based rule:

$$v = \underset{c \in \{1, \ldots, C\}}{argmax} \frac{1}{R} \sum_{r=1}^{R_c} K(\mathbf{x}, \mathbf{x}_r(c)). \qquad (2)$$

The runtime complexity of such instance-based classifier is equal to $O(MR)$. Moreover, it is necessary to store all instances; hence, the memory requirements are also rather high. It is known, that the unknown densities in (2) can be estimated with the orthogonal series kernel [15] instead of the Parzen kernel. According to this approach, an appropriate orthogonal basis is chosen (e.g., the Fourier, Hermite or

Legendre series [7]), and each feature is normalized. However, the computational complexity remains identical. To overcome this drawback, the naïve Bayesian rule with an assumption about conditional independence of features can be used [5]:

$$v = \underset{c \in \{1,...,C\}}{argmax} \frac{R_c}{R} \prod_{i=1}^{M} \hat{f}_c(x_i), \tag{3}$$

It is known that the canonical form of the density estimates (2) with the orthogonal series can be replaced to the simplified form. For example, the Fejér kernel can be used [15]:

$$\hat{f}_c^{(OS)}(x_i) = 0.5 + \sum_{k=1}^{m} \left(a_{i;k}^{(1)}(c) \cdot \psi_k^{(1)}(x_i) + a_{i;k}^{(2)}(c) \cdot \psi_k^{(2)}(x_i) \right), \tag{4}$$

where

$$\psi_k^{(1)}(x_i) = cos(k \cdot \pi \cdot x_i), \ \psi_k^{(2)}(x_i) = sin(k \cdot \pi \cdot x_i). \tag{5}$$

The coefficients are estimated as follows

$$a_{i;k}^{(l)}(c) = \frac{m-k+1}{R_c \cdot (m+1)} \sum_{r=1}^{R_c} \psi_k^{(l)}(x_{r;i}(c)), l = \overline{1,2}, \tag{6}$$

Here m is the number of terms in the series (4). It is known [15], that the estimation (4) converges to the real probability distribution with the convergence rate $O(R^{-1/3})$, if the number of terms m is chosen as follows: $O(R^{1/3})$[17]. Hence, in the proposed method we use the following value of this parameter: $m = R^{1/3}$. Thus, if the original PNN (2) is modified in a described way (3), (4), (5) and (6), its runtime complexity will be defined as $O(MR^{1/3})$. Moreover, the memory requirements also decrease to $O(MR^{1/3})$, because only coefficients (6) need to be stored. Thus, such a modification does not implement the memory-based classifier anymore [4].

Though the computation of the basis functions (5) of the new observation can be quite complex, it can be simplified by using the known trigonometric rules:

$$\psi_k^{(1)}(x_i) = \psi_{k-1}^{(1)}(x_i) \cdot \psi_1^{(1)}(x_i) - \psi_{k-1}^{(2)}(x_i) \cdot \psi_1^{(2)}(x_i)$$
$$\psi_k^{(2)}(x_i) = \psi_{k-1}^{(1)}(x_i) \cdot \psi_1^{(2)}(x_i) + \psi_{k-1}^{(2)}(x_i) \cdot \psi_1^{(1)}(x_i), \tag{7}$$

with the initialization $\psi_1^{(1)}(x_i) = cos(\pi \cdot x_i), \ \psi_1^{(2)}(x_i) = sin(\pi \cdot x_i)$.

The proposed orthogonal series-based PNN (OS-PNN) is presented in Fig. 1. To simplify this figure we do not show the upper indices 1 and 2 of the basis functions (5) and coefficients (6). In this modification the new kernel layer is added to calculate the

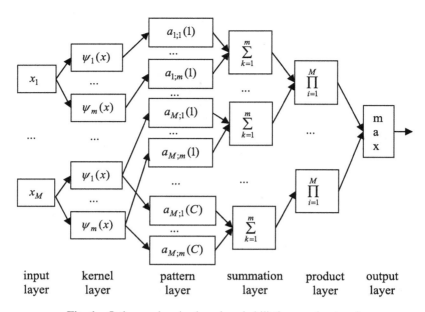

input   kernel   pattern   summation   product   output
layer   layer   layer   layer   layer   layer

Fig. 1. Orthogonal series-based probabilistic neural network

basis functions (5) and (7). The pattern layer does not include anymore the comparison of new observation and all instances with the Parzen kernel (2). On the contrary, only $2mMC$ products with the coefficients (6) are computed in this layer in Fig. 1. The output (or decision) layer produces the class label with the maximal posterior probability (3).

The proposed OS-PNN saves all advantages of the classical PNN [6] over other neural network-based classifiers [1]. First, it converges to the optimal Bayesian decision. Second, unlike many networks, the PNN (Fig. 1) does not contain recursive connections from the neurons back to the inputs. Hence, it can be implemented completely in parallel. But the most important advantage is an excellent training speed in comparison with back propagation [5]. The new sample can be added even in real time applications. The network begins to generalize each new observed set of patterns causing the decision boundary to become closer to the optimal one. In fact, the following recurrent equation can be applied to refine the model, if the new instance $\mathbf{x}_{R_c+1}(c)$ of the c-th class becomes available:

$$\tilde{a}_{i;k}^{(l)}(c) = \left(R_c \cdot a_{i;k}^{(l)}(c) + \psi_k^{(l)}\left(k \cdot \pi \cdot x_{R_c+1;i}(c)\right) \right) / (R_c + 1), l = \overline{1,2}, \qquad (8)$$

The most important property of our modification is the improved performance: the online classification (3), (4), (7) is approximately $R^{2/3}$ –times faster, than the original PNN (2). The next section experimentally supports this claim.

3   Experimental Study

In this section we use the popular Wine dataset from UCI repository ($T = 178$ total instances, $M = 12$ features, $C = 3$ classes). The proposed modification (3), (4), (5), (6), (7) and (8) of the PNN (Fig. 1) is experimentally compared with: (1) the original PNN (2); (2) the traditional reduction of the PNN [14] with the k-median clustering (no more, than 10 clusters); (3) the k-NN with $k = 3$ neighbors [5]; and (4) multi-class one-versus-all linear SVM from libSVM library. In the OS-PNN (Fig. 1) we implemented either most widely-used Hermite kernel [18] or the Fejér kernel [15] discussed in the previous section. The mean and standard deviation of each feature are evaluated on the basis of the available training sample, and the features are normalized.

We compute the error rate (in %) and the average time (in ms) to recognize an observation with a laptop (4 core i7, 6 Gb RAM) and Visual C ++ 2013 Express compiler (x64 environment) and optimization by speed. As it was stated in introduction, we concentrate on the small sample size problem. Hence, the modified version of the q-fold cross-validation is applied. Namely, the original dataset is randomly partitioned into $q \in \{2, \ldots, 10\}$ equal sized subsamples. Of the q subsamples, a single subsample is retained as the *training* data of the size $R = T/q$. The number of terms m is chosen to be equal to $max\left((T/q)^{1/3}, 5 \right)$ in order to obtain an accurate estimation of the probability density for a very small training sample. The remaining $(q - 1)$ subsamples are used as *testing* data. This process is then repeated q times. The average error rates and classification times are presented in Figs. 2 and 3, respectively.

Based on these results, we can draw the following conclusions. First, the instance-based classifiers (PNN, k-NN) are more accurate than SVM for very small training samples (Fig. 2). Nevertheless, SVM is 15–45 times faster (Fig. 3). The performance of the instance-based classifiers dramatically decreases, when the size of the training set becomes larger.

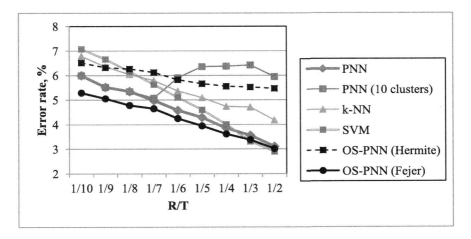

Fig. 2. Classification error rate, %

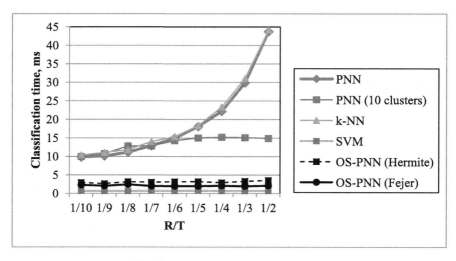

Fig. 3. Average classification time, ms

Second, the reduction of the PNN with the clustering obviously improves its classification speed, when the number of clusters is much less than the size R of the complete training set. However, such clustering significantly increases the error rate, because the resulted decision looses the property of the Bayesian optimality (see Fig. 2, $R/T \geq 1/6$).

Thirdly, the application of the Fejér kernel (4), (5) and (6) in the proposed OS-PNN (Fig. 1) leads to the 0.5–1.3 % more accurate decisions, than the traditional Hermite kernel. Moreover, our implementation is 30–70 % faster (Fig. 3). Finally, the proposed OS-PNN is only 3-times slower, than the very fast linear SVM, though the accuracy of our modification is 0.5–1.8 % higher, when the training sample size is rather small.

In the next experiment we tested our network (Fig. 1) with the large Skin Segmentation dataset [19] from UCI repository. This dataset contains $C = 2$ classes (skin/non-skin pixel) and 245057 instances (RGB pixels, i.e., $M = 3$ features); out of which 50859 are the skin samples and 194198 are non-skin samples. To make the assumption about feature independence realistic, the principal component analysis (PCA) was applied to the training set, and all feature vectors were converted to their M principal components. We used the cross-validation procedure described earlier with $q = 10$, i.e., the training set contains $R = 24506$ instances. The number of terms in the proposed OS-PNN is chosen as follows: $m = R^{1/3} \approx 30$. The number of features here is very low. Hence, in addition to the linear SVM we tested non-linear SVM with RBF kernel. The average error rate and classification time are presented in Table 1.

Table 1. Experimental results, skin segmentation dataset

	k-NN	Linear SVM	Non-linear SVM	PNN	PNN (10 clusters)	OS-PNN
Error rate, %	3.22 ± 2.59	34.61 ± 18.5	3.29 ± 2.66	3.26 ± 1.17	6.19 ± 7.33	3.57 ± 1.2
Classification time, ms	4838.6 ± 105.2	0.6 ± 0.1	10.9 ± 0.8	4491.5 ± 95.8	3.1 ± 0.5	1.1 ± 0.1

Here, first, the instance-based learning methods (k-NN, PNN) are characterized by the lowest error rates, but their performance is inappropriate. Second, linear SVM cannot distinct the classes. However, non-linear kernel made it possible to obtain rather accurate decision, though it is 18-times slower, than linear SVM. Third, though the clustering in the PNN is rather fast, it is 3 % less accurate than the PNN. Finally, the OS-PNN with Fejér kernel (3), (4), (5), (6), (7) and (8) is again one of the most promising techniques. The proposed classifier is 4000-times faster, then the original PNN, but the difference in their error rates is not meaningful.

4   Conclusion and Future Work

In this paper we proposed a modification of the PNN by replacing the Parzen estimate of the class probability density (2) to the orthogonal expansions with the Fourier series and the Fejér kernel (4), (5) and (6). We demonstrated, that such approach saves all advantages of the classical PNN including the convergence to the optimal Bayesian decision. Our preliminary experimental studies showed that the proposed modification significantly improves the classification performance of the PNN (Fig. 3). Moreover, the density estimate in our network is parameters-free. Hence, there is no need to optimize the smoothing parameter of the Gaussian kernel in the PNN. Unfortunately, our OS-PNN is based on the naïve assumption about independence of features (3), because, the estimation of the multivariate probability distribution with the orthogonal series kernel can be rather complex [7]. Thus, the main direction for further research is the implementation of the Bayesian networks [5] in our classifier in order to evaluate the joint probability density functions only for several (2–3) dependent features. Another important research direction is the thorough experimental study of our approach in real applications, such as image recognition, authorship attribution, etc. [9].

Acknowledgements. The work is partially supported by Laboratory of Algorithms and Technologies for Network Analysis, National Research University Higher School of Economics.

References

1. Haykin, S.O.: Neural Networks and Learning Machines. Prentice Hall, Harlow (2008)
2. Deng, W., Liu, Y., Hu, J., Guo, J.: The small sample size problem of ICA: a comparative study and analysis. Pattern Recogn. **45**, 4438–4450 (2012)
3. Savchenko, A.V., Belova, N.S.: Statistical testing of segment homogeneity in classification of piecewise–regular objects. Int. J. Appl. Math. Comput. Sci. **25**, 915–925 (2015)
4. Wilson, D.R., Martinez, T.R.: Reduction techniques for instance-based learning algorithms. Mach. Learn. **38**, 257–286 (2000)
5. Theodoridis, S., Koutroumbas, K.: Pattern Recognition. Academic Press, Burlington (2008)
6. Specht, D.F.: Probabilistic neural networks. Neural Netw. **3**, 109–118 (1990)
7. Rutkowski, L.: Adaptive probabilistic neural networks for pattern classification in time-varying environment. IEEE Trans. Neural Netw. **15**, 811–827 (2004)

8. Romero, R.D., Touretzky, D.S., Thibadeau, R.H.: Optical Chinese character recognition using probabilistic neural networks. Pattern Recogn. **30**, 1279–1292 (1997)
9. Savchenko, A.V.: Probabilistic neural network with homogeneity testing in recognition of discrete patterns set. Neural Netw. **46**, 227–241 (2013)
10. Li, X., Zhang, S., Li, S., Chen, J.: An improved method of speech recognition based on probabilistic neural network ensembles. In: 11th International Conference on Natural Computation (ICNC), pp. 650–654. IEEE (2015)
11. Hirschauer, T.J., Adeli, H., Buford, J.A.: Computer-aided diagnosis of Parkinson's disease using enhanced probabilistic neural network. J. Med. Syst. **39**, 1–12 (2015)
12. Savchenko, A.V.: Statistical recognition of a set of patterns using novel probability neural network. In: Mana, N., Schwenker, F., Trentin, E. (eds.) Artificial Neural Networks in Pattern Recognition (ANNPR 2012), LNCS, vol. 7477, pp. 93–103. Springer, Heidelberg (2012)
13. Amiri, A., Niaki, S.T.A., Moghadam, A.T.: A probabilistic artificial neural network-based procedure for variance change point estimation. Soft Comput. **19**, 691–700 (2015)
14. Kusy, M., Kluska, J.: Probabilistic neural network structure reduction for medical data classification. In: Rutkowski, L., Korytkowski, M., Scherer, R., Tadeusiewicz, R., Zadeh, L. A., Zurada, J.M. (eds.) ICAISC 2013, Part I. LNCS, vol. 7894, pp. 118–129. Springer, Heidelberg (2013)
15. Devroye, L., Gyorfi, L.: Nonparametric Density Estimation: The L1 View. Wiley, Hoboken (1985)
16. Greblicki, W., Pawlak, M.: Classification using the Fourier series estimate of multivariate density functions. IEEE Trans. Syst. Man Cybern. **11**, 726–730 (1981)
17. Schwartz, S.C.: Estimation of probability density by an orthogonal series. Ann. Math. Stat. **38**, 1261–1265 (1967)
18. Greblicki, W.: Asymptotic efficiency of classifying procedures using the Hermite series estimate of multivariate probability densities. IEEE Tran. Inf. Theory **27**, 364–366 (1981)
19. Bhatt, R.B., Sharma, G., Dhall, A., Chaudhury, S.: Efficient skin region segmentation using low complexity fuzzy decision tree model. In: India Conference (INDICON), pp. 1–4. IEEE (2009)

Neural Network Methods for Construction of Sociodynamic Models Hierarchy

Ekaterina A. Blagoveshchenskaya[1,2], Aleksandra I. Dashkina[2],
Tatiana V. Lazovskaya[2(✉)], Viktoria V. Ryabukhina[1],
and Dmitriy A. Tarkhov[2]

[1] Petersburg State Transport University, 9 Moskovsky pr.,
190031 Saint Petersburg, Russia
kblag2002@yahoo.com, vicennea@bk.ru
[2] Peter the Great St. Petersburg Polytechnical University,
29 Politechnicheskaya Street, 195251 Saint Petersburg, Russia
wildroverprodigy@yandex.ru, tatianala@list.ru, dtarkhov@gmail.com

Abstract. The article includes the following modern approaches to modelling sociodynamic processes: Kolmogorov equation system for Markov process with a discrete set of states, Fokker-Planck equation, multiagent systems, etc. As an example, one demographic task of predicting is solved. We compare the simplest neural network approach with an approach based on a special evolutionary model. The article also justifies applying the neural network modelling for producing solutions to the above mentioned equations, determination of their coefficients on the basis of observations and making more precise models, including the dependence of human behaviour on a psychological type. A possibility of making models more precise as new data come in has been discussed.

Keywords: Sociodynamics · Artificial neural networks · Markov processes · Kolmogorov equation · Fokker-Planck equation · Hierarchical systems · Multiagent systems

1 Introduction

Over the past years the problem of predictive models for sociodynamic processes has been gaining growing importance. Thus, a decision was adopted in the European Union to build a global system in 2013–2020 for prediction of social, economic, and environmental occurrences in their interrelation [1]. A similar task was stated at the Plenary Session of 16th St.-Petersburg International Economic Forum [2].

In order to solve this problem it is necessary to construct a hierarchy of interrelated mathematical models in economics, sociology, politology, etc. that can approach in their accuracy and predictive power to physical models. The statistical physics performs transition from molecular dynamics to equations for continuous medium. When developing sociodynamic models it is expedient

L. Cheng et al. (Eds.): ISNN 2016, LNCS 9719, pp. 513–520, 2016.
DOI: 10.1007/978-3-319-40663-3_59

to take initial data, as individual behaviour, including psychological and other similar factors, and then proceed to economical, social, mental (mems), etc. interrelating fields.

The main scientific problem, which must be dealt with when creating this kind of mathematical and information models, consists in indispensable transition from opinion-based and semi-empirical approach to application of modern mathematic simulation methods, justified planning, and optimal management typical of science and engineering.

It is essential to select classes of mathematic models and a method for their application. The requirements for these models can be stated as follows:

- Possibility to construct a model with any degree of complexity model using standard rules and standard elements
- Robustness of model against errors in initial data and a priori assumptions
- Possibility to deal with contradictory assumptions
- Possibility to adjust the model after receiving new data

All these requirements are met in neural network models [3,4].

2   Modern Approaches to Sociodynamic system modelling

Designing a comprehensive programme of developing a particular region, for example, the Arctic zone of the Russian Federation, involves tackling a great number of economic, environmental and social problems. Major mineral extraction projects are not only supposed to be cost-effective, but also take into account their influence on the environment and the population as well as the consequences for the Russian economy. As a result, interaction between experts in different fields is necessary. These experts should formalize their knowledge into uniform mathematical models in order to attain mutual understanding.

A system of Kolmogorov differential equations for Markov process with a discrete set of states was proposed by [5] in the capacity of such models. Let us describe this model in simplest case, when system state is assigned by vector $\mathbf{n}$ with integer coordinates $\{n_i\}$. The examples of such vectors are the sizes of particular national, social, etc. groups of population, the numbers of companies involved in different types of business and varying in size. Let $P(\mathbf{n}, t)$ be a probability of being in state $\mathbf{n}$ at time point t. We denominate as $\mathbf{n}_{ji}$ a vector, which differs from $\mathbf{n}$ by increase of coordinate j and decrease of coordinate i by 1. Then $\mathbf{n}_{i+}$ is a vector with coordinate i increased by 1, and $\mathbf{n}_i$ is a vector with coordinate i decreased by 1. Then "Main Equation" will look as follows:

$$
\frac{dP(\mathbf{n}, t)}{dt} = \sum_{i \neq j} w_{ij}(\mathbf{n}_{ji}, t) P(\mathbf{n}_{ji}, t) - \sum_{i \neq j} w_{ji}(\mathbf{n}, t) P(\mathbf{n}, t) +
$$

$$
\sum_i w_{i+}(\mathbf{n}_{i-}, t) P(\mathbf{n}_{i-}, t) - \sum_i w_{i+}(\mathbf{n}, t) P(\mathbf{n}, t) +
$$

$$
\sum_i w_{i-}(\mathbf{n}_{i+}, t) P(\mathbf{n}_{i+}, t) - \sum_i w_{i-}(\mathbf{n}, t) P(\mathbf{n}, t), \tag{1}
$$

where

$$w_{ij}(\mathbf{n}, t) = \left. \frac{\partial P(\mathbf{n}_{ij}, \tau | \mathbf{n}, t)}{\partial \tau} \right|_{\tau=t}, \quad w_{i\pm}(\mathbf{n}, t) = \left. \frac{\partial P(\mathbf{n}_{i\pm}, \tau | \mathbf{n}, t)}{\partial \tau} \right|_{\tau=t}$$

are intensities of transitions from the current state to the state with an increased or decreased coordinate i.

If the process under consideration does not involve transitions $\mathbf{n}_{i+}$ and $\mathbf{n}_{i-}$ (if the demographic dynamics is considered, it means that there is neither natality, nor mortality, or, to be precise, they are disregarded in comparison with the migratory processes) the Eq. (1) looks simpler:

$$\frac{dP(\mathbf{n}, t)}{dt} = \sum_{i \neq j} w_{ij}(\mathbf{n}_{ji}, t) P(\mathbf{n}_{ji}, t) - \sum_{i \neq j} w_{ji}(\mathbf{n}, t) P(\mathbf{n}, t). \tag{2}$$

Further analysis requires assumptions with regard to intensities of transitions $w_{ij}(\mathbf{n}, t)$. Different variants are considered by [5], e.g.

$$w_{ij}(\mathbf{n}, t) = \mu_{ij} n_j \exp[u_i(\mathbf{n}_{ij}, t) - u_j(\mathbf{n}, t)], \tag{3}$$

where u_i is a utility function of state i. This function is significant because its increase raises the probability of transition into state i and decreases the probability of reverse transition. The utility function is assumed to be linear relative to $\mathbf{n}$. Different variants of dynamics in the system under examination are considered further (in [5]), according to Eq. (1) or (2), and depending on ratio of coefficients in this function. The case of quadratic dependence of utility function on $\mathbf{n}$ is examined by [6]. On the basis of dynamics observed in real systems, coefficients of these dependencies are estimated that may allow generating prediction for the future. In addition, dependence of these coefficients on other factors is examined (e.g. dependence of migration properties on social and economic factors, which is also supposed to be linear). Evidently, real dependencies are usually non-linear; so, if linear approximation has poor accuracy, search of non-linear dependencies should be performed. For the above reasons, neural network functions are suggested as desired non-linear ones.

In [5], for the sake of simplifying the analysis, examination of stationary case $w_{ij}(\mathbf{n}, t) = w_{ij}(\mathbf{n})$ was recommended, and proceeding to quasi-averages, transition rates being averaged according to probabilities. Then (2) is changed by equation system as:

$$\frac{d\hat{n}_i}{dt} = \sum_{i \neq j} w_{ij}(\hat{\mathbf{n}}) - \sum_{i \neq j} w_{ji}(\hat{\mathbf{n}}). \tag{4}$$

Besides Markov processes with a discrete set of states, processes with a continuous set of states are examined by [6]. Conditional distribution density $f(\mathbf{y}, \tau | \mathbf{x}, t)$ in such processes complies to the equation system:

$$\begin{cases} \dfrac{\partial f}{\partial t} + \sum_{i=1}^{n} a_i \dfrac{\partial f}{\partial x_i} + \dfrac{1}{2} \sum_{i,j=1}^{n} b_{ij} \dfrac{\partial^2 f}{\partial x_i \partial x_j} = 0; \\[2ex] \dfrac{\partial f}{\partial \tau} + \sum_{i=1}^{n} \dfrac{\partial (a_i f)}{\partial y_i} - \dfrac{1}{2} \sum_{i,j=1}^{n} \dfrac{\partial^2 (b_{ij} f)}{\partial y_i \partial y_j} = 0, \end{cases} \qquad (5)$$

where

$$a_i = \lim_{\tau \to t} \frac{1}{\tau - t} \mathrm{M}[Y_i - X_i | \mathbf{X} = \mathbf{x}], \ b_{ij} = \lim_{\tau \to t} \frac{1}{\tau - t} \mathrm{M}[(Y_i - X_i)(Y_j - X_j) | \mathbf{X} = \mathbf{x}].$$

In particular, these equations are satisfied with solutions of differential equations system

$$\frac{dU_i}{dt} = a_i(U_1, \ldots, U_n, t) + \sum_{m=1}^{n} g_{im}(U_1, \ldots, U_n, t) \xi_m(t), \ b_{ij} = n g_i g_j,$$

where $\xi_m(t)$ are mutually independent, and represent a white noise.

In one-dimensional case these equations are reduced to

$$\frac{\partial f}{\partial t} + a \frac{\partial f}{\partial x} + \frac{1}{2} b \frac{\partial^2 f}{\partial x^2} = 0; \ \frac{\partial f}{\partial \tau} + \frac{\partial (af)}{\partial y} - \frac{1}{2} \frac{\partial^2 (bf)}{\partial y^2} = 0. \qquad (6)$$

The second equation is commonly called a Fokker-Planck equation. The corresponding differential equation is the following

$$\frac{\partial U}{\partial t} = a(U, t) + g(U, t) \xi(t), \ b = g^2. \qquad (7)$$

The problem of identifying a and b functions by real data arises again in these models. Besides, in order to solve these particular equations it is necessary to know the initial and boundary conditions. Tendency to zero on the infinity seems to be appropriate as boundary conditions, but it does not correspond to the practical problems, in which it is known that all the components x or some of them are not negative. Initial conditions are even more difficult. Commonly the results of monitoring the process are known instead of them, which means that the classical methods of solving mathematical physics problems, such as finite differences method, finite element method, etc., cannot be applied. The methods set forth in the scholarly works [3,4] allow formulating approximate values (5), (6) taking into consideration all the information available.

Using data on the distribution of the population of Russia at the $m = 14$ age groups for 2010–2014, decide to test the prediction task. Data for the first three years are used to construct an evolutionary neural network model, in the last two – to check the results of the projection of this model.

The first approach does not use Fokker-Planck equation. The population distribution by ages is looking as an radial basis function (RBF) neural network approximation in the form

$$u_1(x, t, \mathbf{a}) = \sum_{j=1}^{n} a_{j1} \exp \left(- a_{j2}(x - a_{j3})^2 \right) \exp \left(- a_{j4}(t - a_{j5})^2 \right), \qquad (8)$$

where $n = 10$ is the number of neurons. The variable x is the age interval, the variable t is time, in our case, a year; $\mathbf{a} = \{a_{ji}\}$ is the matrix of neural network weights. The network weights are determined by the minimization of the error functional, built on the basis of available data. Here, it has the form

$$J = \sum_{k=1}^{3} \sum_{i=1}^{m} \left(u_1(x_i, t_k, \mathbf{a}) - n_{ki} \right)^2, \tag{9}$$

where n_{ki} is the normalized ($\sum_{i=1}^{m} n_{ki} = 1$) quantity index of the age group i in year t_k.

In the second approach we take into account the evolving nature of the required distribution function. As the basis functions neural network uses the fundamental solution of the Fokker-Planck equation [7]. The distribution function is sought as neural network outputs in the form

$$u_2(x, t, \mathbf{b}) = \sum_{j=1}^{n} \frac{b_{j1} b_{j3}}{\sqrt{t + b_{j2}^2}} \exp \frac{-b_{j3}^2 (x - b_{j4} + b_{j5}(t + b_{j2}^2))^2}{t + b_{j2}^2}, \tag{10}$$

where $n = 10$ is the number of neurons, weight matrix $\mathbf{b} = \{b_{ji}\}$ components are determined by the minimization of the error functional similar to the functional (9).

The graphs of neural network approximate models of population distributions in 2010 year are shown at Fig. 1.

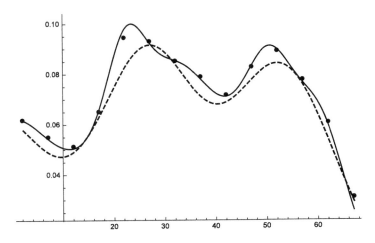

Fig. 1. Neural network approximate models for population distribution, 2010. Dots are real data, dashed line is the RBF-model, solid line is the model based on the Fokker-Planck equation

As we can see, neural network approximation description of the data in both models is good enough. These models predictions for 2014 are shown at the Fig. 2.

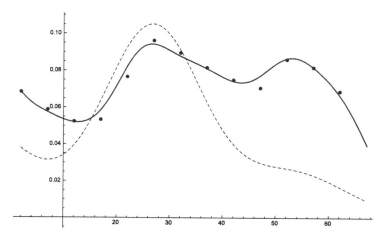

Fig. 2. Neural network models predictions for population distribution, 2014. Dots are real data, dashed line is the RBF-model prediction, solid line is the prediction of the model based on the Fokker-Planck equation

Obviously, the prediction of the model based on the Fokker-Planck equation is precise enough, while the RBF-model does not catch even the character the predicted distribution.

It is necessary to point out that the approach [3] to making stable neural network models on the basis of dissimilar data including differential data implies selecting not only free parameters, but also the structure of the model. It is particularly vital for sociodynamic models given the complexity of the object which is modelled. We can account additional conditions in the neural network models by adding to the error functional the corresponding terms.

For specific algorithms of this kind refer to [3] and the source literature quoted there.

3   Activity and Memory as Human Element

Besides their hierarchical structure, real engineering and socio-economic systems are characterised by activity and memory due to the human element. In order to optimise their operation, mathematical models of these phenomena are necessary to develop, as well as value scales, collective behaviour, etc. connected with them. Neuronal nets were already applied [8] for more simple models of such occurrences. It is advisable to combine this approach with methods for structural and parametrical identification of such systems on the basis of observed parameters [3,4], and management of such systems, including value scale management. For modeling we apply multiagent systems.

Particularly, the dynamics of the agent is assumed to follow minimisation task for functionals hierarchy, which may be interpreted as a value scale. At that, dynamics of the agent can be managed through changing internal parameters, adjusting external conditions (optimised by functionals) and rebuilding the value system (the position of the functional on a corresponding scale), eliminating some functionals and inserting other ones (that can optimise functionals of ambient system or managing the subject of the respective level).

If the agent simulates a human, particular dependencies are constructed based upon psychical features and specific interaction with environment.

Personality type 1: concentrated on model of reality as a whole, and on long-term prediction of its dynamics. Accuracy of model and its predictions makes a great contribution to quality functional. Activity is mainly directed at improvement of the model and realisation of desired development scenario.

Personality type 2: differs from Personality type 1 by substitution of reality as a whole for the nearest social environment.

Personality type 3: differs from Personality type 2 by obtaining quality functional evaluation from the environment. Need in adjusting activity according to variable functionals requires proceeding from long-time prediction to short-time ones.

Personality type 4: differs from Personality type 3 by obtaining not only quality evaluation, but also reality models from environment. High intensity of information exchange leads to reducing quantity of social contacts. Differences in models being obtained require significant efforts for their concordance and building a consistent picture. Main part in quality functional belongs to evaluation of one's own place in foreign reality models.

Personality type 5: differs by obtaining reality models from society as a whole that leads to reducing information exchange with the nearest environment, and long process of comparison between reality models and reality as such. Decrease of this discrepancy contributes the greatest part into desired change of quality functional.

Personality type 6: differs from Personality type 5 by striving for influence on reality as a whole, in order to bring it to concordance with one's internal model. Awareness of difficulty to make such influence alone by oneself leads to directing either at local possible changes or participation in great communities able to making global changes.

Personality type 7: differs from Personality type 6 by substitution of influence on reality for searching of own optimal place in reality, including optimal environment. Rapid changes in ambient reality lead to employing short-time predictions.

Personality type 8: differs from Personality type 7 by substitution of movement through reality for organisation of reality according to one's internal model and short-time predictions. For such organisation, social opportunities are actively employed, first of all, the nearest environment and accessible levels of social hierarchy.

Personality type 9: differs from Personality type 8 by substitution of changing ambient reality for changing one's own internal model according to ambient reality and opinions in the nearest environment. Such adjustment requires long concordance of different sub-models that leads to reduction of activity due to duration of such process.

4 Conclusion

Creating sociodynamics intellectual management systems is a problem the immediate future. We have presented an overview of the models of such systems. The analysis presented demonstrates the desirability of using the neural network technologies for modeling sociodynamics. We are confident that in the coming years will be to create a real system of prediction and control using the above approach.

Acknowledgments. The paper is based on research carried out with the financial support of the grant of the Russian Science Foundation (Project No. 14-38-00009, The program-targeted management of the Russian Arctic zone development).

References

1. FuturICT. http://www.futurict.eu/
2. Putin's speech at the plenary session of the 16th St. Petersburg international economic forum. http://kremlin.ru/transcripts/15709
3. Tarkhov, D.A., Vasilyev, A.N.: Mathematical models of complex systems on the basis of artificial neural networks. Nonlin. Phenom. Complex Syst. **17**(3), 327–335 (2014)
4. Kainov, N.U., Tarkhov, D.A., Shemyakina, T.A.: Application of neural network modeling to identification and prediction in ecology data analysis for metallurgy and welding industry. Nonlin. Phenom. Complex Syst. **17**(1), 57–63 (2014)
5. Weidlich, W.: Sociodynamics: A Systematic Approach to Mathematical Modelling in the Social Sciences. Harwood Academic, Amsterdam (2000)
6. Helbling, D.: Quantative Sociodynamics. Springer, Heidelberg, Dordrecht, London, New York (2010)
7. Polyanin, A.D.: Spravochnik po linejnym uravneniyam matematicheskoj fiziki. FizMatLit, Moscow (2001)
8. Mosalov, O.P., Red'ko, V.G., Prokhorov, D.V.: Simulation of the evolution of autonomous adaptive agents. Math. Models Comput. Simul. **1**(1), 156–164 (2009)

Application of Hybrid Neural Networks for Monitoring and Forecasting Computer Networks States

Igor Saenko[1(⊠)], Fadey Skorik[2], and Igor Kotenko[1]

[1] Saint-Petersburg Institute for Informatics and Automation of the
Russian Academy of Sciences, 14-th Liniya, 39, Saint-Petersburg 199178, Russia
{ibsaen,ivkote}@comsec.spb.ru
[2] Military Telecommunication Academy, Tikhoretsky 3,
Saint-Petersburg 194064, Russia
f.skorik@bk.ru

Abstract. Nowadays, monitoring and forecasting computer networks states are the most important components of network administration processes. For their realization, it is necessary to use the means having high adaptability and resistance to external noise. Hybrid neural networks possess such properties. The paper considers possibilities of application of hybrid neural networks as a basis for models of monitoring and forecasting of computer networks states. Hybrid neural networks have high adaptability and resistance to external noise. Results of the experiments showed that the offered models possess rather high precision of classification of the current and predicted states of computer networks.

Keywords: Hybrid neural networks · Kohonen map · Computer networks · Forecasting · Monitoring · State indicator

1 Introduction

Nowadays, monitoring and forecasting computer networks states are the most important components of network administration processes. They allow reaching high stability and safety of computer networks. Improvement and expansion of software and hardware tools for modern computer networks cause continuous complication of their administration and management processes. At the same time, essential restrictions of intellectual, technical and economic type are imposed on these processes. Because of high dynamics and nonlinearity of these processes in modern computer networks, and complexity of structural communications between nodes, it is necessary to use for monitoring and forecasting computer network states the means having high adaptability and resistance to external noise. Such means are fuzzy logical inference algorithms and artificial neural networks. However, application of fuzzy logic methods for these purposes is limited because of impossibility of independent acquisition of new knowledge, which then could be used in the inference mechanisms. Therefore, from this point of view, artificial neural networks are more preferable. However, the analysis of the results received with their assistance can be significantly complicated.

© Springer International Publishing Switzerland 2016
L. Cheng et al. (Eds.): ISNN 2016, LNCS 9719, pp. 521–530, 2016.
DOI: 10.1007/978-3-319-40663-3_60

It is possible to avoid the problems mentioned above if to use hybrid neural networks as a basis to create state monitoring and forecasting systems for computer networks. The hybrid neural network is understood as an artificial neural network in which symbolical calculations are used [1, 2]. The hybrid neural network has the following advantages: (a) it carries out a logical inference based on fuzzy logic, transparent for the end users; (b) it keeps opportunity to get new knowledge by means of artificial neural networks and to use them in further work.

Application of hybrid neural networks in state monitoring and forecasting systems for computer networks allows: (a) to simplify significantly information processing and analysis; (b) to automate processes of detecting the arising emergency situations; (c) to reveal prerequisites to their emergence in the future. Thereby it is possible to create the adaptive system that will be capable to make fuzzy logic references and to acquire new knowledge by training. Similar systems are successfully used in many areas (machine learning, face recognition, dynamic behaviors, medicine, signal processing etc.) [3–6]. They allow receiving the necessary results without participation of the human and with small computing expenses.

The paper suggests using hybrid neural networks to solve a problem of state monitoring and forecasting for computer networks with any topology. The initial data of the problem is a set of indicators, which characterize states of controlled network nodes and is collected in real time. It is required to draw in the shortest time a conclusion about the current and predicted states of the network in general by processing of values of these indicators.

The main theoretical contribution of the paper consists in the development of original models for state monitoring and forecasting for computer networks. The models are based on a combination of self-organizing and hybrid neural networks and do not require an essential time and computing expenses for their functioning.

The paper has the following structure. In Sect. 2 the review of relevant works is presented. Section 3 defines a set of indicators necessary for state assessment of computer networks. Sections 4 and 5 consider the models of hybrid neural networks for state monitoring and forecasting of computer networks. In Sect. 6, the experimental results are discussed. Conclusions about the received results and the directions of further researches are presented in Sect. 7.

2   Related Work

The analysis of the works devoted to using hybrid neural networks as the elements of monitoring, remote control and administration systems showed that in most cases the areas of their application are strongly limited.

[7] suggests to realize the expert system for network monitoring based on fuzzy logic and artificial neural networks. The core of this system is the knowledge base with a set of rules. The rules define network states. As the recognition mechanism, a combination of a neural network and a fuzzy inference system is offered. However, an essential lack of this system is the impossibility of its self-learning.

[8] considers features of using hybrid neural networks to realize self-learning system containing fuzzy logic rules. However, correct learning of these networks

demands essential time, intellectual and computing expenses. So, it is not seemed as possible and expedient in our cases.

[9–11] demonstrate systems intended to detect intrusions and abnormal user behavior. The systems use models of hybrid neural networks. They are rather simple in realization and do not demand considerable computing resources for their functioning. However, the general shortcoming of these systems is an impossibility to predict the consequences arising from influence of the revealed anomalies.

[12] suggests an approach to forecast the states of the network elements based on a combination of probabilistic and usual multilayered neural networks. The combination of neural networks of various types allows reducing significantly time for training of neural network systems. However, the main shortcoming of this approach is low stability in the conditions of noise impact on initial data. This impact is important characteristic of real computer networks.

In [13], a nonlinear identification system based on the self-organizing Kohonen map is described. Because of realization specifics, it is convenient for application in state monitoring systems for computer networks. It allows data processing with high noise level and with low value of the error coefficient. However, a lack of this system is insufficient opportunities for interpretation of received results.

[14] suggests to use particle swarm algorithms for training artificial neural networks. However, on our opinion, it is very difficult to introduce these results in our work due to the complexity of forming the goal functions.

[15] suggests a dynamic evolutionary system with fuzzy logic that implements adaptive learning in near-real time. The system is able to predict the trend of the input parameters. However, in our opinion, this approach is rather complicated and not always comprehensible.

Thus, the analysis of related work shows that hybrid neural networks possess great opportunities to solve the problem defined in the paper. At the same time, known models of hybrid neural networks cannot be directly applied for this purpose. Besides, it is advisable to use Kohonen maps [16] for solving the problem.

3 Computer Network State Indicators

To receive a complex assessment of computer network states, it is expedient to consider a set of indicators, which characterize accuracy, quality, reliability and productivity of computer networks. Thus, the formed system of indicators has to meet the following requirements: (1) provide simplicity of control, (2) supply the total and not excess assessment of computer networks, (3) allow carrying out complex research of the network environment, (4) make it possible carrying out an assessment and forecasting of computer network states. The offered system of indicators, answering to the requirements mentioned above and sufficient, in our opinion, for state assessment of computer networks, is presented in Table 1. These indicators, as a rule, are dimensionless and demand the minimum volume of data for their calculation. Thus, the coverage of the functional properties for computer networks remains complete.

Table 1. Network state indicators system

Indicator	Formula	Comments
Accuracy	$D = 1 - P_{er}$	P_{er} – probability of a mistake
Data operating time on a mistake	$Q = \frac{W}{\mu}$	W – the volume of the processed data, μ – average of mistakes at data processing
Average time of delivery of the message	$\bar{T}(k, r)$	k – message priority, r – delivery mode
Average delay of packets	$M = \sum\limits_{i=1}^{N} T_i$	N – number of the transferred packets, T_i – transferring time for i-th packet
Reliability	$\bar{T} = \frac{1}{\lambda}$	λ – frequency of mistakes
Coefficient of operational readiness	$K = k_r P(t)$	k_r – coefficient of readiness, $P(t)$ – probability of no-failure operation in time t
Time of reaction to user request	$T_a = T_{at} +$ $+ T_{re} + T_{an} +$ $+ T_p + T_{pa}$	T_{at} – preparation of request, T_{re} – transfer of the request to the server, T_{an} –preparation of the answer, T_p – transfer of the answer, T_{pa} – answer processing
Effective speed of data transfer	$V = \frac{N_b}{T_s}$	N_b – amount of useful information, T_s – communication session time

Besides, the offered system of indicators promotes essential simplification of preparation of the initial data used further in the models offered for state monitoring and forecasting of a computer network.

4 A Hybrid Neural Network Model for Computer Network State Monitoring

The assessment of computer network state is a difficult and multistage process. The values of functioning indicators of both separate nodes and subsystems, and all network in general are estimated in this process. In addition, interrelations and dependences between network elements are defined. Besides, comparison of the obtained data with standard values and expert opinions should be done.

Definition of interrelations and data formalization of a real computer network represents the problem that is almost insolvable by usual mathematical methods. The reason is that the structure of information exchange in network is formed dynamically under the influence of diverse, often external and internal factors uncoordinated among themselves. Thus, it is necessary to consider that during operating of fuzzy expert and control systems the set of fuzzy rules formed by the expert can have contradictions. It can be incomplete and not fully display the real state. However, the combination of algorithms of fuzzy logical inference with opportunities of an artificial neural network allows constructing the model, which is overcoming these difficulties and adequately estimating current state of computer networks.

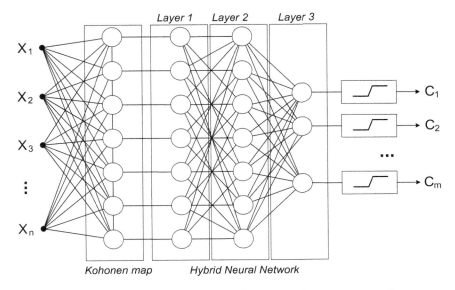

Fig. 1. Hybrid neural network model for monitoring of computer network states

The model of the hybrid neural network suggested for monitoring of computer network states is represented in Fig. 1.

This model represents combining two artificial neural networks which are a self-organizing Kohonen map and a three-layer hybrid neural network. For a filtering the values received at the outputs of the hybrid neural network and definition of an output class (C_1, C_2, ..., C_m), corresponding to the current state of the computer network, the blocks realizing step function with the set activation threshold are used.

The values of indicators received from client applications, the results of the network traffic analysis and the saved-up statistical information are considered as initial data (X_1, X_2, X_3, ..., X_n).

Kohonen map is used for initial sorting and clustering of the arriving values of indicators. Its mission is to structure the initial data for the hybrid neural network [16].

The hybrid neural network incorporates three layers. The first layer is intended for definition of accessory degree of input values to the corresponding fuzzy sets given earlier. The second layer consists of neurons, which signals x_i and weights w_i are united by means of triangular co-norms as follows:

$$p_i = S(w_i, x_i), \quad i = 1, 2 \ldots n,$$

where n – number of neuron inputs.

The outputs of each of neurons of the second layer are calculated with application of the triangular norm and are defined as

$$y = T(p_1, \ldots, p_n) = T(S(w_1, x_1), \ldots, S(w_n, x_n)).$$

The third layer includes the standard neurons with sigmoid-based activation function, which inputs are formed as the linear combinations of the second layer neuron outputs. Outputs of the third layer neurons are defined as degree of accessory of the considered object to a certain class given earlier.

Process of model functioning assumes the following sequence of actions: (1) clustering of the indicators values, (2) processing of the received values by means of the hybrid neural network, (3) filtering of the received values and allocation of the target class defining current state of the computer network.

Therefore, having a certain set of indicator values on the input of the considered model, it is possible to interpret unambiguously its output as an assessment of the current state of the computer network.

5 A Hybrid Neural Network Model for Computer Network State Forecasting

The offered hybrid neural network model for computer network state forecasting is a logical continuation of the model considered earlier for the computer network state monitoring. In the offered model, the multilayered perceptron is connected to outputs of the hybrid neural network. The use of this neural network is needed because it has simple structure and is easily trained. In addition, it has rather high accuracy of the output data. The structure of the model offered for the computer network state forecasting is presented in Fig. 2. The perceptron plays a role of a forecasting module in the model. It receives the outputs of the hybrid neural network determining the current state of the computer network. Then it forms at the outputs the value defining degree of belonging of the network state to the predetermined class of states through the given time interval. The values of indicators used as initial data $(X_1, X_2, X_3, \ldots, X_n)$ are similar to the values considered in the model for the computer network state

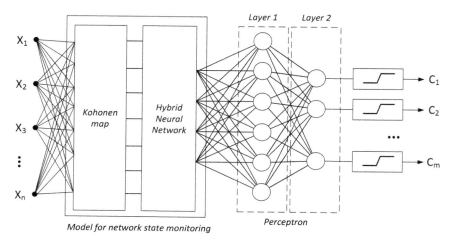

Fig. 2. Hybrid neural networks model for forecasting of computer network states

monitoring. The results of forecasting are filtered by the blocks realizing a step function with the set activation threshold. Thereby the definition of one of the resultant classes $(C_1, C_2, \ldots, C_m)$ is provided defining the predicted state of the computer network. Functioning of the hybrid neural network model for the computer network state forecasting assumes the following sequence of actions: (1) clustering of the indicators values, (2) processing of the received values by means of the hybrid neural network, (3) creation of the forecast based on the output values of the hybrid neural network, (4) filtering of the received values and allocation of the target class defining predicted state of the computer network.

As a result, it is possible to interpret unambiguously the predicted state of the computer network through the given time interval processing a set of indicators values on the model input.

6   Experimental Results

The experimental assessment of the offered models for monitoring and forecasting of computer network states was carried out for the computer network including the database server, the file server and 10 workstations. Models were realized on C# and included in the administrator software. Experiments were made according to the following scheme. Accumulation of statistical information about values of the chosen indicators was originally carried out. The indicators remained in the form of temporary series. The volume of the series was sufficient for correct training of the artificial neural networks, which were a part of the models. Then the structure of the models was formed in compliance with a configuration of the network and with the number of controlled indicators. The artificial neural networks, which are the parts of the models, were subsequently trained. The outputs of each of the models were associated with a certain number of classes. Each of the classes was corresponded to a certain, strictly given state of a computer network.

The self-organizing Kohonen map was a part of both models and allowed to reduce dimensions of vectors of the arriving values of indicators. It transformed multidimensional vectors to two-dimensional ones that allowed carrying out visualization of all available states of the computer network on the two-dimensional plane. The example of such visualization of the results of 320 measurements by a neural network of dimension 100 * 100 neurons is represented in Fig. 3. It is possible to allocate conditionally three main areas on the received chart: (1) all nodes operate, and exchange of information on the network is minimal or is absent; (2) the intensive network exchange of local nodes with the database server takes place; (3) the hard work of users with the file server is observed.

Nevertheless, there are a number of states on the plane, which cannot be unambiguously defined and are not related to one of the considered areas.

The hybrid neural networks are served to the purposes of their identification. They also are the parts of both models. The results of the operation of the Kohonen maps are used as the training selections. It significantly simplifies the process of training and the structure of the hybrid neural network. Thereby, it reduces the needs for time and computing resources. Training of the hybrid neural network and multilayered

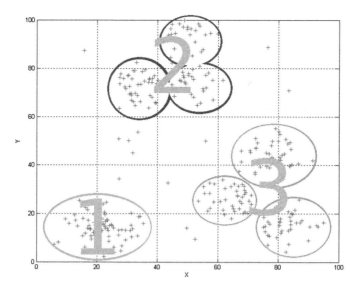

Fig. 3. Projection of the set of available computer network states to the two-dimensional plane

Table 2. Mean square errors for the hybrid neural network and the multilayered perceptron

Name	Hybrid neural network		Perceptron	
	Value	Epoch number	Value	Epoch number
Minimal	0.0025	800	0.041	6500
Middle	0.0032	9600	0.067	22000
Maximal	0.0044	4200	0.084	800

perceptron was carried out using the return distribution algorithm. The test results are represented in Table 2, where the values of the mean square errors for the hybrid neural network and the multilayered perceptron are presented.

As we can see from Table 2, the hybrid neural network has higher precision and requires carrying out smaller amount of epochs in comparison with a multilayered perceptron on the training stage. After finishing the training stage for artificial neural networks, the models of monitoring and forecasting of computer network states are ready to the decision making. Table 3 represents the results of state monitoring and forecasting for the computer network after carrying out 320 tests.

As we can see from Table 3, at the decision-making stage the offered hybrid neural networks for monitoring and forecasting of computer network states provide sufficiently high accuracy. The maximum classification error for the computer network state monitoring does not exceed 4.8 %, and for state forecasting – no more than 7.2 %. At the same time, we can observe the following regularity. The state class including bigger number of tests corresponds to the bigger classification error. All these results testify to high efficiency of solving the problem of monitoring and forecasting of computer network states by the application of hybrid neural networks.

Table 3. Experimental results of monitoring and forecasting of computer network states

State classes	State monitoring		State forecasting	
	Tests number	Error, %	Tests number	Error, %
Class 1	87	1.7	91	2.5
Class 2	122	4.8	128	7.2
Class 3	111	2.2	101	3.4

7 Conclusion

The paper suggests the new approach to the solution of the problem of monitoring and forecasting of computer network states based on the application of the models containing hybrid neural networks. Kohonen maps are the parts of the offered models and used as a source of initial data for hybrid neural networks. It allows reducing significantly costs of training these hybrid neural networks. At the same time, possibility to use the fuzzy logic inference mechanisms completely remains. Application of a multilayered perceptron as means of forecasting allows defining possible changes of the state of the controlled computer network that will happen eventually.

The experimental assessment of the offered models on the computer network of small dimension showed their rather high efficiency. Considering simplicity of structures of the offered models and the minimum requirements to the computing resources used by them, it is possible to draw a conclusion that these models can receive successful application in systems of remote control and monitoring for computer networks of any configuration.

The directions of further researches relates with features of the application of hybrid neural networks in monitoring systems for computer networks having multilevel structure and complex hybrid topology.

Acknowledgment. This research is being supported by The Ministry of Education and Science of The Russian Federation (contract # 14.604.21.0147, unique identifier RFMEFI60414X0147).

References

1. Awan, Z.K., Khan, A., Iftikhar, A.: Hybrid Neural Networks: From Application Point of View. LAP Lambert Academic Publishing, Saarbrücken (2012)
2. Wermter, S., Sun, R.: An overview of hybrid neural systems. In: Wermter, S., Sun, R. (eds.) Hybrid Neural Systems. LNCS, vol. 1778, pp. 1–13. Springer, Heidelberg (2000)
3. Chen, Y., Kak, S., Wang, L.: Hybrid neural network architecture for on-line learning. Intell. Inf. Manag. 2(4), 253–261 (2008)
4. Lawrence, S., Giles, C.L., Tsoi, Ah.Ch., Back, A.D.: Face recognition: a hybrid neural network approach. Technical report (1996)
5. Wan, L., Zhu, L., Fergus, R.: A hybrid neural network-latent topic model. In: Proceedings of 15th International Conference on Artificial Intelligence and Statistics (AISTATS), La Palma, Canary Islands, pp. 1287–1294 (2012)

6. Psichogios, D.C., Ungar, L.H.: A hybrid neural network-first principles approach to process modeling. AIChE J. **38**(10), 1499–1511 (1992)
7. Azruddin, A., Gobithasan, R., Rahmat, B., Azman, S., Sureswaran, R.: A hybrid rule based fuzzy-neural expert system for passive network monitoring. In: Proceedings of Arab Conference on Information Technology ACIT, Dhaka, pp. 746–752 (2002)
8. Mishra, A., Zaheeruddin, Z.: Design of hybrid fuzzy neural network for function approximation. J. Intell. Learn. Syst. Appl. **2**(2), 97–109 (2010)
9. Bahrololum, M., Salahi, E., Khaleghi, M.: Anomaly intrusion detection design using hybrid of unsupervised and supervised neural network. Int. J. Comput. Netw. Commun. (IJCNC) **1** (2), 26–33 (2009)
10. Garcia-Teodoro, P., Diaz-Verdejo, J., Macia-Fernandez, G., Vazquez, E.: Anomaly-based network intrusion detection: techniques, systems and challenges. Comput. Secur. **28**(1–2), 18–28 (2009)
11. Zhang, Z., Manikopoulos, C.: Neural networks in statistical anomaly intrusion detection. J. Neural Netw. World. **3**, 305–316 (2001)
12. Kotenko, I., Saenko, I., Skorik, F., Bushuev, S.: Neural network approach to forecast the state of the internet of things elements. In: Proceedings of XVIII International Conference on Soft Computing and Measurements (SCM 2015), pp. 133–135. IEEE Xplore, St. Petersburg (2015)
13. Souza, L.G.M., Barreto, G.A.: Nonlinear system identification using local ARX models based on the self-organizing map. Learning and Nonlinear Models - Revista da Sociedade Brasileira de Redes Neurais (SBRN) **4**(2), 112–123 (2006)
14. Kasabov, N., Hamed, H.N.A.: Quantum-inspired particle swarm optimization for integrated feature and parameter optimization of evolving spiking neural networks. Int. J. Artif. Intell. **7** (11), 114–124 (2011)
15. Kasabov, N.K., Song, Q.: Dynamic evolving neural-fuzzy inference system and its application for time-series prediction. IEEE Trans. Fuzzy Syst. **10**(2), 144–154 (2002)
16. Kohonen, T.: The self-organizing map. Proc. IEEE **78**(9), 1464–1480 (1990)

Multiclass Ensemble of One-against-all SVM Classifiers

Catarina Silva[1,2]($\boxtimes$) and Bernardete Ribeiro[1]

[1] Center for Informatics and Systems, University of Coimbra, Coimbra, Portugal
{catarina,bribeiro}@dei.uc.pt
[2] School of Technology and Management, Polytechnic Institute of Leiria, Leiria, Portugal

Abstract. Ensembles of classifiers have recently received a resounding interest due to their successful application in different scenarios. In this paper, our main focus is on using ensembles of one-against-all classifiers in multiclass problems. Current approaches in multiclass problems are often focused in dividing the problem but seldom focus on cooperating strategies between classifiers. We propose a framework based on Support Vector Machines (SVM) one-against-all baseline ensemble classifiers that includes a Multiclass Ensemble Function (MEF) to heuristically incorporate both the predictions of individual classifiers as well as the confidence margin associated with those predictions to determine the final ensemble output. The results achieved with the renown Iris and Wine datasets show the performance improvement achieved by the proposed multiclass ensemble of one-against-all SVM classifiers.

Keywords: Ensembles · Multiclass classification · SVM

1 Introduction

The ubiquitous data that is generated everyday by the huge amount and variety of devices and sensors spread all over the internet is overwhelming. From this data we need to extract information. Machine learning and pattern recognition are fields whose techniques are well-established for knowledge mining or scientific discovery. However, in many real applications patterns that can represent an object, a process or an event are rarely binary. For a given problem, classification consists of determining to which region a pattern consisting of a feature vector **x** belongs to. While for binary classification many approaches have been tackled with great success, the multiclass approach that occurs often in the real world is a less-studied problem. One straightforward goal is to obtain the sought decision boundaries that separate decision regions among all possible for the problem at hand. In the case of one-against-all approach we achieve an absolute separation which corresponds to a hierarchical classification. With the proviso that by obtaining the correct hyperplane separability one can solve the multiple class problem, a good tradeoff of the bias-variance dilemma should be obtained.

L. Cheng et al. (Eds.): ISNN 2016, LNCS 9719, pp. 531–539, 2016.
DOI: 10.1007/978-3-319-40663-3_61

Having a single model that covers all the input space might not be the best solution. In particular, for complex problems a divide and conquer strategy may be more appropriate. Thus, it is worthy to consider creating several localized models that can take advantage of the specific characteristics of their operating domain and by combining them together they can better mimic the model governing the true data distribution. The rationale is to obtain a model that resembles as close as possible to the subjacent model that governs the real data distribution. Therefore, if we take one step further and use instead of a classifier, an ensemble of classifiers which combine several sub-models and often present better generalization performance than their constituent models, still the above tradeoff can be attained provided that the integrated classifiers are diverse and accurate [1]. In addition if we resort our classifiers in the ensemble to be Support Vector Machines (SVMs) one can expect that by properly choosing their parameters the machine capacity is low although the combination might still yield a large bias and low variance. Classifiers such as decision trees, k nearest neigbhors, neural networks are also possible. However, they are high capacity classifiers with large variance and low bias.

SVMs are large margin classifiers that can provide good generalization independent of the training set's distribution by using the principle of structural risk minimization [2]. Support Vector Machines transform the input vectors nonlinearly into a high-dimensional feature space through a kernel function so that the data can be separated by linear models. The principle of SVM is to search an optimal hyperplane able to classify the two classes and maximize the margin of the separation. In the last decade SVMs have gained wide popularity due to the good generalization performance on high-dimensional and a relatively small amount of data.

In this contribution we propose a framework based on Support Vector Machines (SVM) one-against-all baseline ensemble classifiers that includes a Multiclass Ensemble Function (MEF) to incorporate both the predictions of individual classifiers as well as the confidence margin associated with those predictions to determine the final ensemble output. We further show that the framework presents performance improvements in a benchmark dataset.

The rest of the paper is organized as follows. In the next section, we set the foundations and background for multiclass approaches and ensemble strategies. In Sect. 3, we introduce our approach of Multiclass Ensemble of One-Against-All SVM Classifiers. Experiments and results are described and analyzed in Sect. 4. Finally, Sect. 5 addresses conclusions and future work.

2   Background and Related Work

2.1   Multiclass Classification

Multiclass learning is the problem of assigning labels to instances where the labels are drawn from a finite set of elements and is being increasingly required by modern applications, such as text classification, protein function classification, speech recognition, music categorization and semantic scene classification.

In multiclass classification the challenges are numerous. Feature selection and dimensionality reduction methods must take into account the relevance of features not only to a particular class, as in the binary setting, but to their impact on all classes.

In [3] a multiclass framework is introduced using boosting by constructing symmetric functions, which contrasts with the usual AdaBoost-type boosting algorithms using linear separators. In [4] a robust classification framework is described for tongue-movement ear pressure signals using an ensemble voting methodology. The ensemble members are comprised of different combinations of sensor inputs and a relevant error reduction is achieved using majority voting. This is achieved through a combination of rejection based on ambiguity in the ensemble and diversity in the misclassified instances across the ensemble members.

In [5] the problem in focus is multiclass classification problems with imbalanced classes, and ensembles are shown to be an effective technique that has increasingly been adopted to combine multiple learning algorithms to improve overall prediction accuracy and may outperform any single sophisticated classifiers. In this case the proposed framework combines simple nearest neighbour and Naive Bayes algorithms and uses a tree-based ensemble combination function.

In [6] a hierarchical ensemble method of neural networks is proposed to tackle multiclass problems. Experimental testing shows that the proposed method can reduce the uncertainty of the decision.

2.2   Ensembles

Ensemble based systems (also known under various other names, such as multiple classifier systems, committee of classifiers, or mixture of experts) have shown favorable results compared to those of single-expert systems for a broad range of applications requiring automated decision making under a variety of scenarios.

In matters of great importance that have financial, medical, or other implications, we often seek a second opinion before making a decision, sometimes more. In doing so, we analyze each one, and combine them using some implicit process to reach a final decision that is apparently the best informed one. This process of consulting several experts before making a final decision is perhaps second nature to us; yet, the extensive benefits of such a process in classification systems is still being discovered by the computational intelligence community [7]. In [8], a sample of the vast literature on classifier combination can be found, on both the theory and implementation of ensemble based classifiers. In [9] a thorough survey can be found on ensemble classification and regression.

According to [10] there are three main reasons to use ensembles: statistical, representational and computational. Regarding statistical ones, either there is not sufficient data to find the optimal hypothesis or there are many different hypothesis with limited data. In the representational set of reasons, unknown functions may not be present in the hypotheses space or even a combination of present hypotheses may expand it. For the computational issue, the learning algorithms may stuck in local minima, therefore using an ensemble might be useful. Hence, in generating the multiclass ensembles, diversity and accuracy are both important for each member classifier [11]. In summary, an ensemble rationale is that, given a task that requires specific knowledge, k experts may perform better than one, given that their individual responses are duly combined.

A classifier committee is then characterized by (i) a choice of k classifiers, and (ii) a choice of a combination function [12], usually denominated voting algorithm. The classifiers should be as independent as possible to guarantee a large number of inductions on the data [13]. A common voting algorithm is majority voting, where each base classifier (expert) votes on the class the example should belong to and the majority wins (in two-class problems an odd number of classifiers should be used). In [14], a comparison of methods for multiclass Support Vector Machines can be found, emphasizing the extension of SVM to multiclass classification, namely by comparing different approaches: one-against-all, one-against-one, and directed acyclic graph SVM (DAGSVM).

3   Multiclass Ensemble of One-against-all SVM Classifiers

Figure 1 generically depicts the proposed ensemble approach that constitutes the base of the framework for multiclass ensemble of one-against-all SVM classifiers. The training set is used to construct T one-against-all baseline SVM classifiers that will later be conjugated using a MultiClass Ensemble Function (MEF), f.

Considering a $\mathcal{D}$ training dataset, $\mathcal{D} = \{\mathbf{x}_n, \mathbf{y}_n\}_{n=1}^N$ with N examples, where $\mathbf{x}_n$ are the input feature vectors and $\mathbf{y}_n$ are the label vectors, a set of one-against-all T SVM classifiers $\mathcal{C} = \{\mathbf{C}_i(.)\}_{i=1}^T$ can be constructed to serve as baseline classifiers.

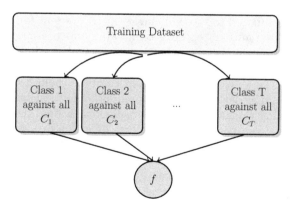

Fig. 1. Ensemble approach.

Having the baseline SVM classifiers and a testing set $\mathcal{S} = \{\mathbf{x}_n^t, \mathbf{y}_n^t\}_{n=1}^M$ with M examples, one can define the Multiclass Ensemble Function (MEF) as shown in Algorithm 1.

This combination function acts as a cooperation function that uses the SVM margins with which each testing example is classified to untie the controversial classifications. When a testing example is classified as positive by only one of the baseline one-against-all classifiers MEF assumes that as the correct output, but that is a trivial scenario. Otherwise, a different approach is followed depending on one of two hypothesis: (i) either the testing example is deemed negative by all baseline classifiers or (ii) it is considered positive by more than one.

In (i) the margins are compared and the example is sought as belonging to the class whose classifier considered it negative by a lesser difference to the average positive examples. In this scenario the average is evidently taken from the training set.

Whereas in (ii), to untie by deciding which of the competing base classifiers gets the testing example, it is sought as of the class whose classifier considered it positive by a margin with larger distance to the average negative examples. Again, this margin is taken from the training set.

The rationale for this procedure of defining the MEF is that the average margins with which positive and negative training examples were classified are representative of respectively the positive and negative testing examples of the same class and used accordingly.

In the next section we are going to present the deployment of the proposed approach to a specific scenario, showing its effectiveness in enhancing classification performance in multiclass settings.

Algorithm 1. MEF - Multiclass Ensemble Function.

Consider:

the trainset $\mathcal{D} = \{\mathbf{x}_n, \mathbf{y}_n\}_{n=1}^{N}$ with N examples,

the testset $\mathcal{S} = \{\mathbf{x}_n^t, \mathbf{y}_n^t\}_{n=1}^{M}$ with M examples,

$\mathbf{x}_n$ and $\mathbf{x}_n^t$ the input feature vectors,

$\mathbf{y}_n$ and $\mathbf{y}_n^t$ the label vectors.

Ensemble: set of one-against-all T SVM classifiers $\mathcal{C} = \{\mathbf{C}_i(.)\}_{i=1}^{T}$.

for each classifier $\mathbf{C}_i$ **do**

    calculate the average margin of the predicted positive examples: PM_i

    calculate the average margin of the predicted negative examples: NM_i

end for

for each testing example $\mathbf{x}_n^t$ **do**

    let each classifier C_i define the predicted class p_n^i of the

    let each classifier C_i define the SVM margin in that prediction v_n^i

    if only one classifier C_i predicts the example to be in the given class **then**

        assign class i to $\mathbf{x}_n^t$

    else if no classifier predicts the example to be in the given class **then**

        assign the class of the classifier that minimizes $|v_n^i - PM_i|$

    else if more that one classifiers predict the example to be in the given class **then**

        assign the class of the classifier that maximizes $|v_n^i - NM_i|$

    end if

end for

---

4   Experimental Setup

To test the proposed multiclass ensemble of one-against-all SVM classifiers app-roach we first define the performance metrics and introduce two multiclass case studies: the Iris and the Wine datasets, available for instance at https://ics.uci.edu/ml/datasets/Iris and https://archive.ics.uci.edu/ml/datasets/Wine respectively and described in following. The experimental setup also includes the construction ans evaluation of the baseline SVM ensemble classifiers.

4.1   Dataset and Performance Metric

The *Iris* dataset [15] is one of the best known datasets, extensively used in machine learning research. The data set contains 3 classes of 50 instances each, where each class refers to a type of the iris plant. The numeric attributes, including the class, are:

1. sepal length in cm;
2. sepal width in cm;
3. petal length in cm;
4. petal width in cm;
5. Class variable: Iris Setosa; Iris Versicolour; Iris Virginica.

The division in train and test sets was carried using a random 60/40 segmented split, i.e. maintaining proportion among the 3 classes.

The Wine dataset includes the results of a chemical analysis of wines grown in the same region in Italy but derived from three different cultivars. The analysis determined the quantities of 13 constituents found in each of the three types of wines.

The numeric attributes, including the class, are:

1. Class variable: Alcohol (1,2,3)
2. Malic acid
3. Ash
4. Alcalinity of ash
5. Magnesium
6. Total phenols
7. Flavanoids
8. Nonflavanoid phenols
9. Proanthocyanins
10. Color intensity
11. Hue
12. OD280/OD315 of diluted wines
13. Proline

There are several metrics that can be used in multiclass scenarios [16]. In this approach we will be using a simple accuracy measure as defined in:

$$Accuracy = \frac{\# \text{ of testing examples well classified}}{\# \text{ of testing examples}}, \tag{1}$$

that permits the direct comparison between different approaches and also between baseline and ensemble classifiers.

4.2   Ensemble Classifiers and Performance Results

In Table 1 baseline classification results using SVM for both datasets are presented. These are typical results for both datasets. To apply the MEF combining ensemble function, we start by calculating the average margins for positive and negative examples in training, resulting in the values presented in Table 2.

Table 1. Baseline results on Iris and Wine datasets.

Iris	Accuracy (# of errors)	Wine	Accuracy (# of errors)
Setosa	100.0% (0)	Alcohol 1	94.37% (4)
Versicolour	98.33% (1)	Alcohol 2	95.77% (3)
Virginica	83.33% (10)	Alcohol 3	98.59% (1)

Table 2. Average positive and negative margins for each classifier.

	C_1^{iris}	C_2^{iris}	C_3^{iris}	C_1^{wine}	C_2^{wine}	C_3^{wine}
Avg. Positive	1.02	1.10	0.55	1.17	1.34	1.62
Avg. Negative	-1.29	-0.95	-1.11	-1.29	-1.67	-1.45

At this point it is possible to apply the MEF to the non-trivial examples in both datasets. A trivial example is a testing example that is classified as positive by only one of the baseline one-against-all classifiers.

The data for the non-trivial examples is presented in Tables 3 and 4 and, as can be easily verified, only one remains wrongly classified in each dataset dataset. This latter error generally happens because it was wrongly considered as trivial.

Table 3. Classification margins for each classifier in the non-trivial examples - Iris.

Example	Correct class	C_1	C_2	C_3
1	Versicolour	-1.43	-0.43	-0.07
2	Virginica	-1.43	-0.02	-0.05
3	Virginica	-1.68	-0.81	0.12
4	Virginica	-1.67	-0.99	0.24
5	Virginica	-1.27	-0.04	-0.14
6	Virginica	-1.63	-0.65	0.25
7	Virginica	-1.47	-1.00	-0.01
8	Virginica	-1.67	-0.94	0.26
9	Virginica	-1.52	-0.64	0.00
10	Virginica	-1.58	-0.81	0.16
11	Virginica	-1.41	-0.65	-0.02

Table 4. Classification margins for each classifier in the non-trivial examples - Wine.

Example	Correct class	C_1	C_2	C_3
1	Alcohol 1	0.29	−0.72	−1.26
2	Alcohol 1	0.06	−0.07	−1.83
3	Alcohol 1	0.07	−1.16	−0.62
4	Alcohol 1	0.18	−0.49	−1.77
5	Alcohol 2	−1.86	−0.18	0.32
6	Alcohol 2	−0.77	−0.40	−0.37
7	Alcohol 2	−1.35	−0.15	−0.19

5   Conclusions and Future Work

In this paper we presented a framework for multiclass classification that is based on one-against-all SVM that form the baseline classifiers of an ensemble and are combined using a Multiclass Ensemble Function (MEF) to incorporate both the predictions of individual classifiers as well as the confidence margin associated with those predictions to determine the final ensemble output. This combination supports the evidence that a good trade-off of bias-variance dilemma was attained by reducing the variance term without affecting the bias term.

The results achieved with the experiments have shown that potential errors from simple majority voting combination functions can be avoided by the use of the proposed MEF with streamlined easiness.

Future research is expected to deal with different datasets that can present different challenges, namely with different scenarios in confidence margins.

References

1. Džeroski, S., Panov, P., Ženko, B.: Ensemble methods in machine learning. In: Encyclopedia of Complexity and Systems Science, pp. 5317–5325 (2009)
2. Vapnik, V., Cortes, C.: Support-vector networks. Mach. Learn. **20**(3), 273–297 (1995)
3. Lefaucheur, P., Nock, R.: Robust multiclass ensemble classifiers via symmetric functions. In: Proceeding of the 18th International Conference on Pattern Recognition, pp. 136–139 (2006)
4. Mace, M., Abdullah-Al-Mamun, K., Wang, S., Gupta, L., Vaidyanathan, R.: Ensemble classification for robust discrimination of multi-channel, multi-class tongue-movement ear pressure signals. In: International Conference of the IEEE Engineering in Medicine and Biology Society, pp. 1733–1736 (2011)
5. Sainin, M.: A direct ensemble classifier for imbalanced multiclass learning. In: IEEE Conference on Data Mining and Optimization (DMO), pp. 59–66 (2012)
6. Oong, T., Isa, N.: One-against-all ensemble for multiclass pattern classification. Appl. Soft Comput. **12**(4), 1303–1308 (2012)
7. Polikar, R.: Ensemble based systems in decision making. IEEE Circ. Syst. Mag. **6**(3), 21–45 (2006)

8. Kuncheva, L.: Combining Pattern Classifiers - Methods and Algorithms. Wiley, New York (2004)
9. Ren, Y., Zhang, L., Suganthan, P.: Ensemble classification and regression - recent developments, applications and future directions. IEEE Comput. Intell. Mag. **11**(1), 41–53 (2016)
10. Dietterich, T.G.: Ensemble methods in machine learning. In: Kittler, J., Roli, F. (eds.) MCS 2000. LNCS, vol. 1857, pp. 1–15. Springer, Heidelberg (2000)
11. Gu, S., Chen, R., Jin, Y.: Multi-objective ensemble generation. WIREs Data Min. Knowl. Discov. **5**(5), 234–245 (2015)
12. Silva, C., Ribeiro, B.: Inductive Inference for Large Scale Text Classification. Kernel Machines for Text Classification. SCI, vol. 255, pp. 31–48. Springer, Heidelberg (2010)
13. Gacquer, D., Piechowiak, S., Delmotte, F., Delcroix, V.: A genetic approach for training diverse classifier ensembles. In: International Conference on Information Processing and Management of Uncertainty in Knowledge-Based Systems, pp. 798–805 (2008)
14. Hsu, C.W., Lin, C.J.: A comparison of methods for multi-class support vector machines. IEEE Trans. Neural Netw. **13**(2), 415–425 (2002)
15. Fisher, R.A.: The use of multiple measurements in taxonomic problems. Ann. Eugenics **7**(2), 179–188 (1936)
16. Sokolova, M., Lapalme, G.: A systematic analysis of performance measures for classification tasks. Inf. Process. Manage. **45**, 427–437 (2009)

Long Exposure Point Spread Function Modeling with Gaussian Processes

Ping Guo$^{(\boxtimes)}$, Jian Yu, and Qian Yin

Image Processing and Pattern Recognition Laboratory,
Beijing Normal University, Beijing 100875, China
{pguo,yinqian}@bnu.edu.cn

Abstract. The long exposure point spread function (PSF) model is commonly used to improve signal-noise-ratio of astronomical object imaging and reduce the effect of atmospheric turbulence. In this paper, a move-and-superposition method of modeling the long exposure PSF based on Gaussian process is proposed. Experimental results show that the proposed modeling method can obtain more accurate estimation of final PSF for astronomical object imaging process than that of the simple shift-and-add PSF model.

Keywords: Long exposure · Modeling PSF · Gaussian process

1 Introduction

In ground-based observations, there are obviously distinct assumptions in modeling of the astronomical object imaging degradation. Which can be categorized roughly two cases: one is the short exposure point spread function (PSF) (exposure time about milliseconds, usually less than 15 ms), the other is the long exposure PSF (exposure time about minutes even to hours). After the atmospheric turbulent phase screen is effectively simulated, the short exposure PSF can be easily described by the pupil function of a telescope system.

The research work on the modeling the long exposure PSF is comparatively rare in the literature. Some researchers [1–4] just simply add all the short exposure PSF together to generate the long exposure PSF, which is called simple superposition model in this paper. These simple shift-and-add models are not accurate enough to represent the long exposure PSF, because it is an extended speckle rather than the same size of short exposure PSF in fact. Consequently, the simple shift-and-add model of long exposure PSF can not describe the real astronomical imaging processing and will cause imperfect astronomical signal processing results.

As we known, a Gaussian process is a statistical distribution where observations occur in a continuous domain. In a Gaussian process, every point in some continuous input space is associated with a normally distributed random variable. Gaussian processes are important in statistical modeling because of properties inherited from the normal. Inspired by this statistical modeling technique,

© Springer International Publishing Switzerland 2016
L. Cheng et al. (Eds.): ISNN 2016, LNCS 9719, pp. 540–546, 2016.
DOI: 10.1007/978-3-319-40663-3_62

in this paper, we propose a method that can obtain a more accurate model of long exposure PSF, which is that a set of simulated short exposure PSFs are considered as the normal distribution of random variable in the Gaussian processes.

2   Short Exposure PSF

The short exposure images can be assumed to be affected by a "frozen pattern" introduced by the Taylor hypothesis of frozen turbulence [5,6] that is moved across the aperture of the telescope. The exposure time of the short exposure images is usually less than 15 ms. The short exposure PSF function can be expressed in the following Eqs. (1) and (2):

$$S(x, y) = \left| F\left\{ P(x, y) e^{iO(x, y)} \right\} \right|^2 . \tag{1}$$

where $S(x, y)$ expresses the short exposure PSF function and $F\{\cdot\}$ represents the Fourier transform, $O(x, y)$ is the simulated atmospheric turbulent phase screen computed with Fast Fourier Transformation (FFT) based method [7–9]. $P(x, y)$ is the ideal pupil function of value 1 inside and 0 outside the telescope aperture, which is defined as follows.

$$P(x, y) = \begin{cases} 1 & for \quad \sqrt{x^2 + y^2} < \frac{D}{2} \\ 0 & otherwise \end{cases} \tag{2}$$

where D is used to simulate the diameter of the telescope.

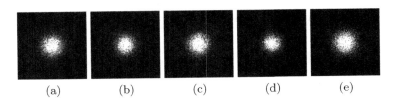

(a)   (b)   (c)   (d)   (e)

Fig. 1. Some samples of the short exposure PSFs

Some samples of the short exposure PSFs are shown in Fig. 1, which illustrate the different sample component of the short exposure PSF. From Fig. 1, we can see that among these short exposure PSFs, the size is not the same and shift a little from the center of the image also.

3   Long Exposure PSF

Long exposure PSF can be considered to be formed as the result of integrating several short exposure PSFs, each of which contains information in different spatial frequency ranges.

3.1   Simple Shift-and-Add Model

The atmospheric turbulence not only distorts the distribution of the short exposure PSFs but also shifts the centroid of them randomly. In the literature, the
long exposure PSF is estimated by the simple shift-and-add model [1–4], which
is by means of the average of a time series of simulated short exposure PSFs
under the assumption that the centroid of the short exposure PSFs are shifted
to the same point and added together. It can be described as follows:

$$L(x,y) = \frac{1}{T} \sum_{t=1}^{T} S(x(t) - x_{centroid}(t), y(t) - y_{centroid}(t)). \tag{3}$$

where $L(x,y)$ is the long exposure PSF, T is the number of the total sample
intervals during the exposure time, $(x_{centroid}(t), y_{centroid}(t))$ is the centroid of
the short exposure PSFs.

The simple shift-and-add long exposure model can improve the signal-noise-
ratio of astronomical imaging process with some limitation, and usually be
utilized to astronomical image reconstruction problem. However, it does not
describe the actual long exposure process properly and cannot be applied to
estimate the PSF to get better model, therefore it is not suitable for astronomical spectrum extraction also [10].

3.2   Gaussian Process Move-and-Superposition Model

It is important for the accuracy estimation of the long exposure PSF in order to
reduce the spectrum extraction errors, we need to denote not only the movement
of the short exposure PSFs' centroid, but also the distortion of the short exposure
PSF if we take the short exposure PSF as a component in long exposure PSF
estimation.

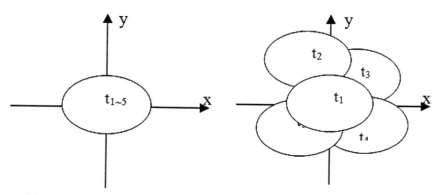

(a) Simple Shift-and-Add Model   (b) GP Move-and-Superposition Model

Fig. 2. The long exposure PSF models

The short exposure PSFs move and superimpose randomly over time. The centroid of the short exposure PSFs are unpredictable random vectors, therefore we can simply regarded them as Gaussian distribution random series. In other words, the process of the short exposure PSFs movement and superposition over time is assumed as a Gaussian Process (GP), this is our main idea.

The simple superposition model is shown in Fig. 2(a) and the proposed Gaussian process move-and-superposition model is shown in Fig. 2(b) while t_1 to t_5 represent the different short exposure PSFs for five sample component.

3.3 Squared Exponential GP

GP is defined by its mean function and covariance function. In this paper, squared exponential GP is used for its merit of smooth curvature. The mean function μ and covariance function k of squared exponential GP are defined as Eq. (4).

$$\mu(x) = 0, \quad k(x, y) = exp(-a\,|x - y|^2), a > 0. \tag{4}$$

Figure 3 shows a sample of squared exponential GP.

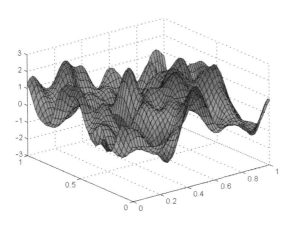

Fig. 3. A sample of Squared Exponential GP.

3.4 Long Exposure PSF Based on GP Move-and-Superposition Model

The proposed GP move-and-superposition model of long exposure PSF formation can be expressed in the following Eq. (5):

$$\begin{cases} L(x, y) & = \quad \frac{1}{T}\sum_{t=1}^{T} S(x(t), y(t)), \\ f(x_{centroid}) = GP(\mu(x_{centroid}), k(x_{centroid}, x'_{centroid})), \\ f(y_{centroid}) = GP(\mu(y_{centroid}), k(y_{centroid}, y'_{centroid})), \\ r/2 & \geq \quad \sqrt{(x - x_{centroid})^2 + (y - y_{centroid})^2}. \end{cases} \tag{5}$$

In the equation above, T is the number of the total sample intervals during the exposure time, GP represents the Gaussian process with mean variable μ and covariance function k, r is the radius of dithering. In this paper, mean variable μ is set to 0 and covariance function k uses the squared exponential function which has the merit of smooth curvature.

4   Experimental Results and Analysis

In this section, we experimental evaluate the proposed method with following experiment: simulation of the speckle formed on CCD image when LAMOST acquiring observation data [11] using two long exposure PSF models described on above. The oval-shaped spot of the arc lamp image in LAMOST can be considered to reflect the shape of the actual long exposure PSF. The length of the semi-major axis of the oval-shaped spot ranges from 4 to 5 pixels, while the length of the semi-minor axis ranges from 3 to 4 pixels. Therefore, r, the radius of dithering, is used to bind the long exposure PSF. Using the above models, i.e. Eqs. (3)–(5), the simulation results are shown in Fig. 4.

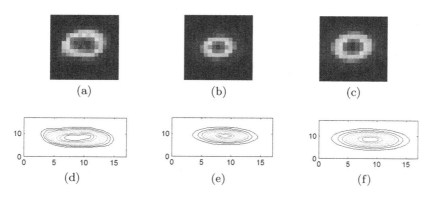

Fig. 4. The long exposure PSF models. (a) actual long exposure PSF in LAMOST. (b) simulation of simple shift-and-add model. (c) simulation of GP move-and-superposition model. (d) contour of "a". (e) contour of "b". (f) contour of "c".

A sample of the CCD spot with long exposure acquisition time which can be assumed as actual PSF of LAMOST imaging is illustrated in Fig. 4(a), the simple shift-and-add model formation result of the long exposure PSF model is shown Fig. 4(b) and the GP move-and-superposition model of the long exposure PSF is shown in Fig. 4(c). The contour illustration of Fig. 4(a)–(c) are shown in Fig. 4(d)–(f), respectively.

From Fig. 4, we can see that long exposure PSF modeled by the proposed method is an extended speckle and it is more similar than that of the simple shift-and-add model.

To objectively assess the accuracy and effectiveness of the proposed method, we adopt the root mean square error (RMSE) criterium to evaluate the error introduced by simulation model.

The RMSE of the simulation models is defined as follows:

$$RMSE = \sqrt{\frac{\|psf^{simulation} - psf^{actual}\|}{n}}. \tag{6}$$

In the Eq. (6), $psf^{simulation}$ is the simulation model and psf^{actual} is the actual long exposure PSF in LAMOST, n is the amount of the pixels in the model. Let T be 10, 50 and 100 min, using the above models, we can get the results as shown in Table 1.

Table 1. RMSE of the simulation models

T	Simple shift-and-add model	GP move-and-superposition model
10	1.4756	0.9882
50	1.3756	0.9569
100	1.2641	0.9268

From Table 1, we can see that the RMSE value of the GP based move-and-superposition model is lower than that of the simple shift-and-add model. In addition to other experimental results, we can conclude that it is apparent that the proposed model has the capacity to obtain better result than that of the simple shift-and-add model. The more accurate astronomical object PSF model will generate more accurate astronomical signal processing results consequently [10].

5   Summary

In this paper, modeling long exposure PSF during the astronomical object imaging is studied. And the GP based move-and-superposition model is proposed to represent the actual long exposure PSF formation. Experimental results with real world LAMOST imaging data show that the proposed model is more accurate than that of the simple shift-and-add model. Consequently, with proposed model, we can obtain more accurate astronomical signal processing results, especial for spectrum extraction task.

Acknowledgments. The authors would like to thank Ms. Min Yan for her help in typesetting work of this manuscript. The research work described in this paper was fully supported by the grants from the National Natural Science Foundation of China (Project No. 61375045 and 61472043), the joint astronomic fund of the national natural science foundation of China and Chinese Academic Sinica (Project No. U1531242), and Beijing Natural Science Foundation (Project No. 4142030 and Project No. 4162027). Prof. Qian Yin and Ping Guo are the authors to whom all correspondence should be addressed.

References

1. Fried, D.L.: Optical resolution through a randomly inhomogeneous medium for very long and very short exposures. J. Orient. Soc. Australia **56**, 1372–1379 (1966)
2. Knox, K.T., Thompson, B.J.: Recovery of images from atmospherically degraded short-exposure photographs. Astrophys. J. **193**, L45–L48 (1974)
3. Véran, J.P., Rigaut, F., Maître, H., Rouan, D.: Estimation of the adaptive optics long-exposure point-spread function using control loop data. J. Opt. Soc. Am. A **14**, 3057–3069 (1997)
4. Marino, J.: Long exposure point spread function estimation from solar adaptive optics loop data. Ph.D. Dissertation, New Jersey Institute of Technology (2007)
5. Taylor, G.I.: The spectrum of turbulence. Proc. R. Soc. London A Math. Phys. Eng. Sci. **164**, 476–490 (1938). The Royal Society Press
6. Zaman, K.B.M.Q., Hussain, A.K.M.F.: Taylor hypothesis and large-scale coherent structures. J. Fluid Mech. **112**, 379–396 (1981)
7. Welsh, B.M.: Fourier-series-based atmospheric phase screen generator for simulating anisoplanatic geometries and temporal evolution. In: Optical Science, Engineering and Instrumentation 1997, pp. 327–338. International Society for Optics and Photonics Press (1997)
8. Sedmak, G.: Implementation of fast-Fourier-transform-based simulations of extra-large atmospheric phase and scintillation screens. Appl. Opt. **43**, 4527–4538 (2004)
9. Von Karman, T.: Progress in the statistical theory of turbulence. Proc. Nat. Acad. Sci. U.S.A. **34**, 530–539 (1948)
10. Yu, J., Yin, Q., Guo, P., Luo, A.: A deconvolution extraction method for 2D multi-object fibre spectroscopy based on the regularized least-squares QR-factorization algorithm. Mon. Not. R. Astron. Soc. **443**, 1381–1389 (2014)
11. Luo, A., Zhang, H.T., Zhao, Y.H.: Data release of the LAMOST pilot survey. Res. Astron. Astrophys. **12**, 1243 (2012)

Neural Network Technique for Processes Modeling in Porous Catalyst and Chemical Reactor

Tatiana A. Shemyakina, Dmitriy A. Tarkhov, and Alexander N. Vasilyev[✉]

Peter the Great St. Petersburg Polytechnical University,
29 Politechnicheskaya Street, 195251 Saint-Petersburg, Russia
sh_tat@mail.ru, dtarkhov@gmail.com, a.n.vasilyev@gmail.com

Abstract. We study neural network methods for building parametric neural network models for heterogeneous data. As the data are used differential equations, boundary conditions, and numerical approximations. We will carry out inspection of methods on the example of two practically interesting problems. We consider the neural network models for heat and mass transfer in porous catalyst granule and in non-isothermal chemical reactor. In the second example, the range of the parameter is wider than the existence area of the differential equation. Some specified hybrid method for the neural network construction is in use; it allows achieving higher precision.

Keywords: Catalyst grain · Non-isothermal chemical reactor · Initial boundary value problem · Diffusion equation · Neural network modeling · Artificial neural network · Training · Error functional · Global optimization · Hybrid method

1 Introduction

Modeling chemical and physical reactions, environmental and biological processes, etc. [1–7,9–14] leads to the solution of boundary value problems for nonlinear ordinary and partial differential equations. In practice, for parameters included in the formulation of the problem, we do no know exact values. We only know the intervals of their change.

Often there is a need to study the dependence of the solution found concerning these parameters, choose the optimum settings based on some criteria arising from the application tasks, etc. This requires a functional dependence of the solution not only on the variables of the problem but also on the parameters mentioned above.

With rare exceptions, it is impossible to find the exact solution of such problem. Classical approximate methods (net-point method, finite element method, Galerkin's method, etc.) require solving a series of tasks on a sufficiently representative set of parameter values. We have developed [5–7,10–14] the unified

© Springer International Publishing Switzerland 2016
L. Cheng et al. (Eds.): ISNN 2016, LNCS 9719, pp. 547–554, 2016.
DOI: 10.1007/978-3-319-40663-3_63

neural network approach to the construction of approximate models for complex systems; this approach allows us to simplify the solution to the problem. We were able to construct a neural network for which the parameters are the input variables. This circumstance allows using one neural model for all sets of parameter values.

Such an approach could lead to a paradigm change in supercomputing. Now a user sends on a supercomputer some problem which is solved to a certain set of parameters. As a result of calculations, a supercomputer generates and gives a great set of numbers. This set of numbers requires a significant effort to analyze, interpret and use. When implementing our approach supercomputing will issue a parametric neural network model. The end user will be convenient to use this model for further research.

In this paper, we test these methods on two tasks. In the first of them we construct a parametric neural network model of processes of heat and mass transfer in a porous catalyst granule. The second one is related to a macrokinetic model of the non-isothermal chemical reactor. This model is constructed in a wider range of the parameter than the region of the existence of the exact solution of the equation considered. Such a situation can occur in applications where the range of the parameters for which the boundary value problem is adequate to the object being modeled and has an exact solution is not known in advance.

2 Neural Network Model of a Catalyst

The analysis of the balance of heat and mass in a flat pellet of the porous catalyst at a catalytic reaction leads to the following boundary value problem for an ordinary differential equation of second order [8]. We want to find $y(x)$ a solution of the ordinary differential equation

$$\frac{d^2y}{dx^2} = \alpha(1+y)\exp\left[-\frac{\gamma\beta y}{1-\beta y}\right],\tag{1}$$

which satisfies boundary conditions

$$\frac{dy}{dx}(0) = 0,\ y(1) = 0.\tag{2}$$

We considered the procedure of constructing neural network that gives the solution of the problem at values from certain intervals. We carried out computations for the following intervals of parameters change: $\alpha \in (0.05; 0.15)$, $\beta \in (0.4; 0.6)$, and $\gamma \in (0.8; 1.2)$. The approximate solution of this problem we build in the form of an artificial neural network of specified architecture [15]

$$y(x,\alpha,\beta,\gamma) = \sum_{i=1}^{N} c_i v(x,\alpha,\beta,\gamma,\mathbf{a}_i),\ \mathbf{a}_i = (a_{1i}, a_{2i}, \ldots, a_{8i}).\tag{3}$$

As basis neuro elements we choose the Gaussian distributions:

$$v(x,\alpha,\beta,\gamma,\mathbf{a}_i) = \exp\{-a_{1i}(x-a_{2i})^2 - a_{3i}(\alpha-a_{4i})^2 - a_{5i}(\beta-a_{6i})^2 - a_{7i}(\gamma-a_{8i})^2\}.$$

The weights of the neural network are parameters c_i and $\mathbf{a}_i$. They are defined in the process of training the network by minimizing the error functional that includes the residuals in the satisfaction equation and boundary conditions:

$$J(y) = \sum_{j=1}^{M} \left(\frac{d^2 y}{dx^2} - \alpha(1+y) \exp\left[-\frac{\gamma \beta y}{1 - \beta y} \right] \right)^2 (x_j, \alpha_j, \beta_j, \gamma_j) + \tag{4}$$

$$\delta \sum_{j=1}^{M} \left(\left(\frac{dy}{dx} \right)^2 (0, \alpha_j, \beta_j, \gamma_j) + y^2(1, \alpha_j, \beta_j, \gamma_j) \right).$$

The error functional is calculated in the set of sample points $(x_j, \alpha_j, \beta_j, \gamma_j)$, $(0, \alpha_j, \beta_j, \gamma_j)$, and $(1, \alpha_j, \beta_j, \gamma_j)$, $j = 1, \ldots, M$; $\{x_j\}_{j=1}^{M}$ is a set of periodically regenerated test points on the segment $[0, 1]$; δ is a positive penalty coefficient. We used a modified polyhedron method and a dense cloud method [5] for the selection of optimal neural network weight values for the approximate solutions $y(x, \alpha, \beta, \gamma)$ based on the minimization of the functional $J(y)$. The latter method proved to be more effective. The values at the control points for a network of $N = 30$ neuro elements differ from those data specified in the monograph [8] in less than 2 percent.

Next, we significantly expanded the range of variation of the parameter $\alpha \in (0; 0.25)$ (dependence on this one is most interesting from the point of view of applications). In this case, we have options $\beta = 0.5$ and $\gamma = 1$. The underlying neuro elements use functions of the form:

$$v(x, \alpha, a_{1i}, a_{2i}, a_{3i}, a_{4i}) = \exp\{-a_{1i}(x - a_{2i})^2\} \text{th}\{-a_{3i}(\alpha - a_{4i})\}, \ i = 1, \ldots, N.$$

We trained neural network on the basis of minimizing the error functional $J(y)$ by a method RProp with the regeneration of sample points (x_j, α_j), $j = 1, \ldots, M$, every 5 steps of the algorithm.

There have been three series of computational experiments for networks of 10, 30 and 100 neurons. In the first series, we applied the above approach with no changes. In the second and third series, we used a hybrid method. The hybrid method is to add in the functional some elements of the type $\delta \sum_{i=1}^{m} \left(y(x_i', \alpha_i) \right)^2$ where x_i' are fixed points, in them the values of the sought function $y(x_i', \alpha_i) = y_i$ are known. For instance, those values are found by some approximate numerical method. We compared the difference of neural network solutions and the solutions found in the package *Mathematica*.

First series.
We consider an acceptable error of 0.005 for a network of 10 neurons with $\alpha = 0.1$, what can not be said about the error at $\alpha = 0.01$ and $\alpha = 0.25$. Although the absolute error is small enough, the nature of the solution for small values of the parameter α is reflected by neural network model not accurately. Assume that the problem is in an insufficient number of neurons.

The calculation results show that for a network of 30 neurons with $\alpha = 0.25$ and $\alpha = 0.1$ we really managed to achieve a significant reduction of the error. In $\alpha = 0.01$ the problem of approximation remains.

We have not received a substantial reduction of the error for a network of 100 neurons.

For a better neural network approximation we use a hybrid method. For this, we construct approximate pointwise solution on a uniform one-dimensional grid with $\alpha = 0.01$. The discrepancy of the sought-for solution with the pointwise solution is included in the error functional in the form of a term with a positive penalty (weighting) factor. We treat the pointwise values of the solution as additional data for a neural network solution.

In **the second series** of numerical experiments the weighting multiplier of the summand corresponding to the difference between the network output and the pointwise solution has not changed in the learning process.

The calculation results show that 10 neurons are not enough to assimilate all the available information – the equation, boundary conditions, and pointwise approximation to the solution. The results obtained when training a network of 30 neurons significantly better results for a network of 10 neurons. The approximate error of neural network solution at $\alpha = 0.01$ is roughly equivalent to the error of the pointwise approximation used. The use of additional information at $\alpha = 0.01$ can significantly increase the accuracy and at $\alpha = 0.1$ as well. Using the approximate solution for a network of 100 neurons has allowed us much better to teach a neural network for small values of the parameter, increasing the accuracy of approximation and at $\alpha = 0.1$. When another $\alpha \in (0; 0.25)$, a neural network provides an approximation of similar quality. Of particular note is that at $\alpha = 0.01$ the error, which gives the neural network, is fewer errors of additional data used.

To further improve the solution in **the third series** of computational experiments, we reduce the weight of the term in the error functional, responsible for the misalignment of the neural network model and the discrete approximation, multiplying this weight by 0.95 each time we regenerate trial points.

For a network of 10 neurons, the considered approach does not increase the accuracy of the obtained neural network approximation.

For a network of 30 neurons, the result is noticeably better than under the previous approach (with a constant weight factor). This circumstance is explained by the fact that at the initial stage of training the neural network, the error is significantly higher than that used a pointwise approximation, and it allows one to speed up neural network training. At the stage when the neural network gives an error comparable to the error of the pointwise approximation, the use of the latter becomes impractical and its impact on the learning process decreases by the above adjustment weight for the corresponding term in the error functional.

The increase in the number of neurons to 100 improved the accuracy of neural network model.

The specified hybrid algorithm of the neural network training allows getting the approximation that is significantly better not only the neural network approximation constructed without additional information, but pointwise approximation used. The results obtained show that 100 neurons is sufficient to digest all pieces of information used when learning – equations, boundary conditions, and pointwise approximation.

Built in an explicit form the approximate neural network solution models very accurately the joint processes of heat and mass transfer in the particle of the porous catalyst not only for specific values of parameters but in intervals of change of these parameters; the model is given by a single neural network.

3   Neural Network Model of a Chemical Reactor

A special case of the catalyst model is a model of chemical reactor. With the help of neural network modeling, one of the variants of the model of chemical reactors was investigated in the works [3,4]. The steady-state problem of a thermal explosion in a plane-parallel case was considered. This problem is interesting because the exact solution is known, the region of solution existence and the parameter values under which the solution of the problem does not exist are known as well ($\alpha > \alpha_{cr} \approx 0.89$). We consider this one as a model for learning algorithms of neural networks. The absence of a stationary solution indicates a thermal explosion. Let us search for the approximate solution of the boundary value problem:

$$\frac{d^2y}{dx^2} + \alpha \exp(y) = 0, \ \frac{dy}{dx}(0) = 0, \ y(1) = 0. \tag{5}$$

in the form of neural network output

$$v(x,\alpha) = \sum_{k=1}^{N} c_k \exp\{-a_k(x - \tilde{x}_k)^2\}\text{th}\{b_k(\alpha - \tilde{\alpha}_k)^2\}, \tag{6}$$

the parameters of which are found in the phase of network training by minimizing the error functional that includes the residuals in the satisfaction differential equation, boundary conditions, and additional conditions:

$$J(y) = \sum_{j=1}^{M} \left(\frac{d^2y}{dx^2} + \alpha \exp(y)\right)^2 (x_j, \alpha_j) + \tag{7}$$

$$\delta_1 \sum_{j=1}^{M} \left(\left(\frac{dy}{dx}(0, \alpha_j)\right)^2 + \left(y(1, \alpha_j)\right)^2\right) + \delta_2 \sum_{i=1}^{m} \left(y(x_i', \alpha') - y_i\right)^2,$$

where α_j are test points from the segment $[\alpha_{min}; \alpha_{max}]$; x_i' are fixed points, such that $y(x_i', \alpha') = y_i$; $\delta_1, \delta_2 > 0$ are penalty parameters.

We have identified the $y(x, \alpha)$ were neural network solution to the network elements of N, $\hat{y}(x, \alpha)$ analytical solution, and the function $\tilde{y}(x, \alpha)$ found in the package *Mathematica*. The parameter α can accept the values at which the exact solution is missing.

Neural network approach was used with (hybrid method) and without considering additional information derived from measurement or rough numerical solution at a fixed set of points considered in the last term of the error functional. To apply the hybrid method, we constructed an approximate pointwise solution at values of: $\alpha' = 0.5$, $\alpha' = 0.8$, $\alpha' = 0.87$, which was seen as additional data to solve the problem. It differs significantly from the analytical solution for all variants of computational experiments.

We used a network of N neuro elements: $N = 30$, $N = 50$, $N = 100$, and chose the following intervals of the parameter change: $\alpha \in [0.4; 1]$, $\alpha \in [0.85; 0.9]$. We conducted two numerical experiments.

In **the first experiment**, we apply the neural network approach when the values of the parameter $\alpha \in [0.4; 1]$. Thus, consider the following cases:

(a) without additional conditions;
(b) with additional condition at the value of the parameter $\alpha' = 0.5$;
(c) with additional condition at the value of the parameter $\alpha' = 0.8$.

In the case (a) the accuracy of the neural network solution is not sufficient - the relative error is about 10 percent. In the case of (b) and (c) the output of the neural network $y(x, \alpha)$ coincides with the analytical solution.

For a small value of the parameter $\alpha = \alpha_{min} = 0.4$ ($\alpha = \alpha_{min} = 0.85$) the result of the calculation turns out better (the approximation error is less) for the case of (b), than options (a) and (c) of the parameter $\alpha' = 0.8$ ($\alpha' = 0.87$).

When a sufficiently large parameter $\alpha = \alpha_{max} = 1$ ($\alpha = \alpha_{max} = 0.9$) the exact solution of the problem does not exist, and we get the approximate function of the $\tilde{y}(x, \alpha)$ were. The resulting approximation turns out bad for all cases, much worse than with the parameter $\alpha = \alpha_{min} = 0.4$ ($\alpha = \alpha_{min} = 0.85$). In this case, the *Mathematica* function gives the $\tilde{y}(x, \alpha)$, which does not satisfy the boundary condition on the right end of the segment. At the same time, the neural network $y(x, \alpha)$ gives a more balanced function. We note that it is possible to control the accuracy of satisfying the boundary conditions of the problem by using penalty multipliers.

In **the second experiment**, we are narrowing the interval of the parameter $\alpha \in [0.85; 0.9]$. The value α of the parameter closes to the critical value $\alpha \approx \alpha_{cr}$. As before, consider three cases:

(a) without additional conditions;
(b) with additional condition at the value of the parameter $\alpha' = 0.5$;
(c) with additional condition at the value of the parameter $\alpha' = 0.87$.

For the case of (c) the output of the neural network $y(x, \alpha)$ practically coincided with the analytical solution $\hat{y}(x, \alpha)$.

For the case of (b) the output of the neural network $y(x, \alpha)$ differs significantly from the analytical solution $\hat{y}(x, \alpha)$. It is noted a sharp increase in error after a critical value of the parameter that may be indicative of the absence of the exact solution.

The results of two experiments showed that in the case (c) a relative error of calculations is quite small: within 10^{-3} percent.

4   Conclusion

Numerical experiments showed:

Methods for constructing neural network models of complex systems significantly improve neural network solution if you are using additional information. It can be obtained, for example, using the approximations based on traditional numerical methods (even not very accurate).

The neural network allows constructing an approximate solution of the parametric problem on the interval for which the exact solution does not exist. The effect of the lack of solutions manifested by a sharp increase in the errors in satisfying the equation.

In applications where an unknown interval of the existence of a solution, neural networks allow specifying the approximate solution. For this purpose, additional data – measurements or numerical solution for one value of the parameter – can be used. The effect of refinement is lost when approaching the critical value of the parameter.

The trained neural network can be used to determine the parameters based on the measurements by these parameters minimizing the discrepancy between the measurement data and the output of the neural network.

The results obtained allow to suggest that such hybrid algorithms can be effective for a sufficiently representative class of problems on the construction of approximate solutions for ordinary and partial differential equations.

Acknowledgements. This work was supported by the Russian Foundation for Basic Research, project numbers 14-01-00660A, 14-01-00733A.

References

1. Vasilyev, A.N., Tarkhov, D.A., Shemyakina, T.A.: Neural network model for the solution of the catalyst problem. Hybrid Method. Problems of Informatics in Education, Management, Economics, and Technics, pp. 58–62 (2014). (in Russian)
2. Vasilyev, A.N., Tarkhov, D.A., Shemyakina, T.A.: Hybrid method in construction of a neural network model of the catalyst. Mod. Inf. Technol. IT-Educ. **1**(10), 476–484 (2014). (in Russian)
3. Vasilyev, A.N., Tarkhov, D.A., Shemyakina, T.A.: Model of non-isothermal chemical reactor based on parametric neural networks. Problems of Informatics in Education, Management, Economics and Technics, pp. 96–99 (2015). (in Russian)
4. Vasilyev, A.N., Tarkhov, D.A., Shemyakina, T.A.: Model of non-isothermal chemical reactor based on parametric neural networks. Hybrid Method Mod. Inf. Technol. IT-Educ. **2**(11), 271–278 (2015). (in Russian)
5. Vasilyev, A.N., Tarkhov, D.A.: Neural Network Methods and Algorithms of Mathematical Modeling. SPb Publishing House of the Polytechnical University, St. Petersburg (2014). (in Russian)
6. Vasilyev, A.N., Tarkhov, D.A.: Principles and Techniques of Neural Network Modeling. SPb Publishin House "Nestor-History"(2014). (in Russian)

7. Vasilyev, A.N., Tarkhov, D.A., Shemyakina, T.A.: A Neural Network Approach to Problems of Mathematical Physics. SPb Publishing House "Nestor-History" (2015). (in Russian)

8. Na, C.: Computational Methods for Solving Applied Boundary Value Problems. The Publishing World, Singapore (1982)

9. Vasilyev, A., Tarkhov, D., Guschin, G.: Neural Networks Method in Pressure Gauge Modeling. In: 10th IMEKO TC7 International Symposium on Advances in Measurement Science vol. 2, pp. 275–279 (2004)

10. Tarkhov, D.A., Vasilyev, A.N.: New neural network technique to the numerical solution of mathematical physics problems. I Simple problems, Optical Memory and Neural Networks (Information Optics) **14**(1), 59–72 (2005)

11. Tarkhov, D.A., Vasilyev, A.N.: New neural network technique to the numerical solution of mathematical physics problems. II Complicated and nonstandard problems Optical Memory and Neural Networks (Information Optics) **14**(2), 97–122 (2005)

12. Vasilyev, A.N., Tarkhov, D.A.: Mathematical models of complex systems on the basis of artificial neural networks. Nonlinear Phenomena in Complex Systems **17**(3), 327–335 (2014)

13. Lazovskaya, T.V., Tarkhov, D.A.: Fresh approaches to the construction of parameterized neural network solutions of a stiff differential equation. St. Petersburg Polytechnical Univ. J. Phys. Math. **1**, 192–198 (2015). http://dx.doi.org/10.1016/j.spjpm.2015.07.005

14. Kainov, N.U., Tarkhov, D.A., Shemyakina, T.A.: Application of neural network modeling to identification and prediction problems in ecology data analysis for metallurgy and welding industry. Nonlin. Phenom. Complex Syst. **17**(1), 57–63 (2014)

15. Haykin, S.O.: Neural Networks and Learning Machines. Prentice Hall, Upper Saddle River (2008)

Fine-Grained Real Estate Estimation
Based on Mixture Models

Peng Ji[1], Xin Xin[1], and Ping Guo[1,2(✉)]

[1] School of Computer Science, Beijing Institute of Technology,
Beijing, 100083, China
{paulj,xxin}@bit.edu.cn
[2] Image Processing and Pattern Recognition Laboratory,
Beijing Normal University, Beijing, 100875, China
pguo@ieee.org

Abstract. People always want to know how much does the apartment they select really value, which is more useful than the whole trend of the housing price in one area for buyers. However, the transaction records about one certain kind of apartments are few or often lacked. In this paper, a novel estimation concept, fine-grained real estate estimation, is proposed and aims at evaluating the price of an apartment in one garden. One problem of the fine-grained estimation is that data is sparse. To deal with this sparsity problem, the strategy of applying the mixture model is put forward to alleviating the data sparsity. The experiments are all conducted in the real data from six districts second-hand housing transaction records (Between January 1st and July 15th, 2015) in Beijing, China. We find that proposed mixture model method has double better accuracy compared with other five traditional approaches in MAE and RMSE metrics.

Keywords: Regression · Mixture models · Real estate estimation

1 Introduction

Along with the accelerating pace of urbanization and the continuous improvement of the urban infrastructure, the population is gradually gravitating towards the cities in China, which makes the housing problem become the public focus. To people who want to settle in a city, purchasing a house is a rigid demand. And many people need professional guidance and reasonable reference to know how much is the selected house worth before paying for it. However, only a few people can get useful reference information. Most of others select only through constant comparison between houses and houses due to lack of the transaction information of the same apartments. What's more, the vast majority of researches about the real estate are based on a large-scale region, such as the house price action of a district or the average price of a sub-district. We call it coarse-grained. The finest information customers can get are the average price of a certain garden and the marked price of an apartment provided by real estate agents. But they may be still unaware of the cost of a specific apartment layout per square meter. Our work is aimed at estimating the unit price of a specific apartment in gardens.

© Springer International Publishing Switzerland 2016
L. Cheng et al. (Eds.): ISNN 2016, LNCS 9719, pp. 555–564, 2016.
DOI: 10.1007/978-3-319-40663-3_64

It is well known that house price is influenced by many factors. To coarse-grained analysis of house price tendency, the garden location, transportation, surrounding public facilities and service etc. will exert an influence. All of these factors' impact on the garden will reflect on the unit price level. For fine-grained apartment estimation, based on the average price level of a garden, apartments' own properties (such as house type, apartment layout and orientation, etc.) can be utilized. In addition, the average unit price present continuously rise in a fluctuant way. Generally, the average price of each month is 2 %–4 % higher than the last month. Hence, time dimension should be considered when estimating.

Beijing is the target city of our research. 45903 second-hand housing transaction records of 4760 gardens in 13 districts (ChangPing, ChaoYang, DaXing, DongCheng, FangShan, FengTai, HaiDian, MenTouGou, ShiJingShan, ShunYi, TongZhou, XiCheng, YiZhuang) from January 1st to July 15th are achieved, accounting for 40 % transaction records of Beijing, 2015. In experiment, six districts are selected as experiment data.

In this task, the noisy data is difficult to filter out. Besides, data sparsity is another challenge to alleviate, which is also a common problem in general urban computing tasks [1, 9]. Unfortunately, this task has not been thoroughly explored yet. For solving this problem, a mixture model is proposed and our work has three primary contributions:

- Fine-grained real estate estimation. This task aimed at estimating the price of one apartment, which has never been studied. Compared with previous works about the coarse-grained real estate forecasting, the fine-grained provided more valuable reference information.
- Dealing with data sparsity. The method proposed in this paper cluster the data into a few classes, and each class of data has the similar properties. Data density will increase by this way and the data sparsity problem will alleviate.
- Real data estimation. The experiments conducted in real transaction data of Beijing, and the estimation values are compared with true values. We verified that the proposed method outperforms traditional approaches double in MAE and RMSE.

2   Related Works

2.1   Coarse-Grained Real Estate Evaluation

In paper [1], this article first introduces the concept of urban computing, discussing its general framework and key challenges from the perspective of computer sciences. Real estate evaluation is widely studied and commonly considered as a branch of urban computing. Fu et al. [2, 7] proposed geographic dependencies and mixed land-use latent models for real estate appraisal. In their papers, potential resale and investment value (in a long term thing) of communities was predicted. Park et al. 2015 [3] proposed an improved housing price prediction model by using machine learning algorithms such as C4.5, RIPPER, Naive Bayesian and AdaBoost to assist a house seller or a real estate agent make better informed decisions based on house price valuation.

Differences. Most previous coarse-grained real estate evaluation was concentrated on the overall trend of a community or a large area. However our paper is focused on the unit price of an apartment in one garden. Unlike the coarse-grained evaluation focusing on communities' surrounding environment, the fine-grained pays more attention on the properties of the apartment itself. It is a brand-new application, which has not been thoroughly studied before.

2.2 Existing Methods of Real Estate Estimation

The typical methods, such as linear regression model [8] and artificial neural network [10] have performed well in traditional estimations. Approaches based on geographic information account for the largest. For different applications, multiple methods have been developed, such as the spatial statistical model [4], coupled with GIS as well as GWR model. A hybrid genetic algorithm (HGA) approach to instance selection in artificial neural networks (ANNs) [5] is another method to predict house price. In paper [6], the result of M5 model trees was demonstrated to outperform commonplace multiple linear regression models and artificial neural networks.

Differences. Existing methods have been demonstrated work-well in the corresponding applications. However, few of them can be directly employed in our task. In order to predict unit price of each type of apartments, data sparsity is the first problem to solve. Therefore, we design a specific mixture model for this unique task. Then, compare our method with the traditional approaches in the same data.

3 Estimation Framework

3.1 Problem Statement

We consider real estate estimation as a regression problem. Given a set of apartments $E = \{e_u\}(u = 1, \ldots, M)$ for sale, our objective is to estimate the price of a place of real estate e. Each real estate e can be represented as a feature vector $\mathbf{f}(e) = (f_1(e), \ldots, f_K(e))$, where K indicates the number of features, such as layout, orientation and the apartment's belonging garden.

$$p = \omega \cdot \mathbf{f}(e) \tag{1}$$

where ω is the corresponding parameters.

Suppose real estates for sale in the training set are represented as $E = \{e_u\}$, and $Y = \{y_u\}$ denotes actual sale prices of e_u, where u = 1,..., M, and M is the quantity of training instances. Therefore, our objective is to minimize the prediction error in Eq. (2).

$$\sum_{i=1}^{M} \left\| \omega \cdot \mathbf{f}(e_i) - y_i \right\|^2 \tag{2}$$

3.2 Regularization

By minimizing the quadratic sum in Eq. (2), a set of ω can be learnt from the training data. After we have ω, we can predict the prices of given apartments by calculating Eq. (1). Nevertheless, this regression model is too ideal to achieve satisfactory performance since noisy data in the training set is unavoidable. In addition, the model is possible to overfit the training data.

Regularization is a common strategy to solve an ill-posed problem or to prevent overfitting, which refers to a process of introducing additional information in the objective function. We can employ regularization techniques to improve the proposed regression model.

In real estate estimation, a lot of fine-grained prior knowledge is available. We summarize them into three categories of information about the target apartment: layout, orientation and floor. For example, with respect to the layout, the apartments with multiple bedrooms are more expensive than those with single bedroom. Besides, high-rise apartments usually possess high sale prices than low-rise apartments. These knowledge can be utilized as regularization terms in our final objective function.

According to practical requirements, we can represent the prior knowledge as several constraints. Obviously, these constraints are functions of parameter ω. Suppose we summarize N constraints, and each constraints can be denoted as a regularization term $c_j(\omega)$. Then the objective function becomes

$$\sum_{i=1}^{M} \left\| \omega \cdot \mathbf{f}(e_i) - y_i \right\|^2 + \sum_{j=1}^{N} \lambda_j \cdot c_j(\omega) \qquad (3)$$

where λ_j is to adjust the weight of the jth regularization term in the objective function. As we mentioned before, each apartment can be represented as a vector of features $\mathbf{f}(e) = (f_1(e), \dots, f_K(e))$. Each dimension of $\mathbf{f}(e)$ denotes a property of the target apartment. Let us suppose $f_a(e)$ represents that the target apartment has multiple bedrooms, and $f_b(e)$ has single bedroom. Then we can conclude that $f_a(e)$ should have larger weight than $f_b(e)$ in final objective function, i.e. $\omega_a > \omega_b$.

We can employ logistic function to represent this constraints, thus the objective function becomes:

$$\sum_{i=1}^{M} \left\| \omega \cdot \mathbf{f}(e_i) - y_i \right\|^2 + \lambda_1 \log \frac{1}{1 + exp^{(\omega)}} \qquad (4)$$

3.3 Mixture Regression Model

An intuitive strategy to learn the prediction function is to minimize the objective function using existing training data, which is named as global model (GM) in this paper. GM can achieve an overall optimal performance for all the real estates in the data set.

In GM, a fixed set of combination weights (i.e., ω) are learned to optimize the overall performance for all apartments. However, the best combination parameter for a given apartment is not always the best for others. The apartments are extremely diverse in several properties, including layout, orientations, locations, etc. For instance, a south-oriented apartment usually has a higher price for sale than a north-oriented one if they are competitive on other properties in China real estate market. Besides, apartments located in some high-quality gardens are desirable than those apartments of low-quality gardens. It is unreasonable to apply identical regression parameter for apartments belonging to different gardens/courts.

We know that some high-price apartments share some common properties, correspondingly, low-price apartment share some common properties too. We introduce a latent class to capture the apartment information in the learning framework. The apartments in one class will obtain identical combination weights in the prediction function.

A latent variable z is utilized to indicate which class the combination weights $\omega_z = (\omega_{z1}, \ldots, \omega_{zK})$ are drawn from. The choice of z depends on the target apartment e. The probability of choosing hidden class z given apartment can be calculated with softmax function in Eq. (5),

$$P(z|e) = \frac{1}{Z_d}\exp\left(\sum_{j=1}^{L} \alpha_j^z \mathbf{f}(e)\right) \qquad (5)$$

where α is the corresponding parameter, and Z_d is the normalization factor that scaled the exponential function to be a probability distribution. L is the number of latent classes.

The prediction function can be re-written as

$$p = \frac{1}{Z_d}\exp\left(\sum_{j=1}^{L} \alpha_j^z \mathbf{f}(e)\right) \cdot \omega^z \mathbf{f}(e) \qquad (6)$$

In comparison to GM, the mixture model can exploit the following advantages: (a) the combination weights are able to change across apartments and hence lead to a gain of flexibility; (b) it offers probabilistic semantics for the latent apartment classes and thus each apartment may be associated with multiple classes.

4   Experiments

4.1   Data Description and Processing

Experimental verifications are conducted in the second-hand housing transaction data of Beijing, China. Six major urban districts (ChaoYang, DongCheng, XiCheng, HaiDian, FengTai, ShiJingShan) are selected as evaluated area, due to other districts only account for a small percentage of transaction data and their average price per square meter are much lower than the average price of Beijing (Fig. 2).

As shown in Fig. 1, the price rose with fluctuations in the last 18 months. Firstly, in order to estimate with greater accuracy, we select six major districts transaction data between January 1st and July 15th, 2015 and divide them into 12 periods (Each two weeks are regarded as a period and February is regarded as a single period due to that month is Chinese Spring Festival). Secondly, delete abnormal trading records. After filtering, our experiment data contains 32968 records of 3680 gardens in six districts which account about for over 70 % in all transaction data. Finally, in each district, 20 % of the data is divided into the test set and 80 % of the data is divided into the training set.

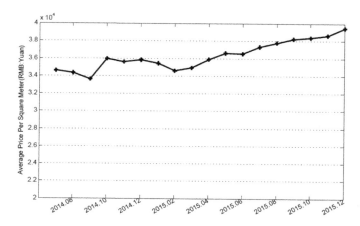

Fig. 1. House price trend of Beijing.

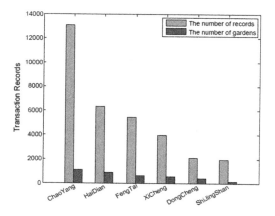

Fig. 2. Transactions of 6 districts in Beijing.

4.2 Feature Selection

In this step, some significant features are selected for distinguishing between different apartments in the same garden. In our work, three features are extracted to estimate, including (a) apartment floor, (b) apartment layout, (c) apartment orientation. In most gardens, the buildings are generally divided into two kinds: slab-type buildings (less than 7 floors, without elevator) and tower buildings (more than 7 floors, with elevator) and each room in the building with three types of floors: low, middle and high.

Apartment layout often represents the number of bedrooms and living-rooms. However, it is not precise to decide which layout an apartment belongs to solely based on the room number of it. The apartment square is also a factor which influences the unit price. Hence, room number and square should be considered together as the feature of apartment layout. Based on the analysis of statistical results, the category is showed in Table 1:

Table 1. Apartment layout classification

Category	Description	Average price per m^2 in HaiDian (yuan/m^2)
Compact apartment	One bedroom and square =<40 m^2; More than one bedroom and square =<50 m^2	52896
One bedroom with one dining-room	One bedroom and 40 <square <=50 m^2	50687
One bedroom with one living-room	One bedroom and 50 <square <=70 m^2	45226
Two bedrooms with one dining-room	Two bedrooms and 50 <square <=65 m^2	49824
Two bedrooms with one living-room	Two bedrooms and 65 <square <=90 m^2	45172
Three bedrooms with one dining-room	Three bedrooms and 50 <square <=80 m^2	52713
Three bedrooms with one living-room	Three bedrooms and 80 <square <=110 m^2	46394
Capacious apartment	One bedroom and square >70 m^2; Two bedrooms and square >70 m^2; Three bedrooms and square >110 m^2; Others	46035

As can be seen in this table, the greater the area is, the lower the unit price when the apartments have the same bedroom.

Apartment orientations are divided into four kinds: (a) orientations include south and north; (b) north, northeast and northwest; (c) the rest orientations which include south; (d) the rest orientations which do not include south. So far, each apartment can be categorized into one of 192 types according to three features.

4.3   Evaluation Metrics and Baselines

In the experiments, set L = 5 in Eq (6). It is illustrated in Fig. 3 that MAE and RMSE are less when L is equal to 5.

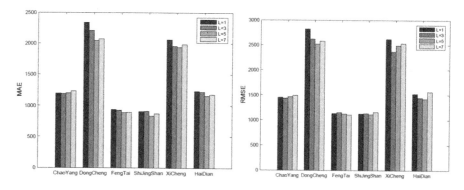

Fig. 3. MAE and RMSE in different L.

We utilize two metrics, the Mean Absolute Error (MAE), and the Root Mean Square Error (RMSE) for evaluations. A smaller MAE or RMSE value indicate a better performance. They are defined as

$$MAE = \frac{\sum_{i=1}^{n} |y_i - \hat{y}_i|}{n}, RMSE = \sqrt{\frac{\sum_{i=1}^{n} (y_i - \hat{y}_i)^2}{n}},$$

where $\hat{y}_i$ is the prediction, and y_i is the ground truth.

We compare our framework with the following baselines.

1. Average: Regard the average unit price of each garden as the estimation value.
2. KNN-3: Regard the average unit price of each garden and its three neighbors as the estimation value by computing distance through latitude and longitude.
3. KNN-5: Regard the average unit price of each garden and its five neighbors as the estimation value by computing distance through latitude and longitude.
4. Tensor Factorization: Regard the gardens, features and time periods as three dimensionalities of a tensor and the unit prices as values in the tensor. Use tensor factorization to estimate the missing data.
5. Neural Network: Encode three features as inputs of the neural network, and the unit price as the output and train a 3-layer neural network for prediction.

Based on the same training set and test set, the above 5 methods and our method are verified.

4.4 Overall Performance

Table 2 shows the performance of different methods in the whole data and the content in the parentheses is the percentage of MAE and RMSE in overall average price. The bottom bolded method is the one proposed in this paper. It is observed that the estimation error (MAE and RMSE) is reduced to less than half of the Neural Network method, which has the best results of the five baselines. The KNN-5 method performs the worst and the KNN-3 underlies it. This result shows that the average unit price of one garden has few similarities with its neighbors. The Tensor Factorization and Neural Network almost achieve the same result, but their accuracy improves a little compared with the Average method which is the most simple and intuitive approach. For Tensor Factorization and Neural Network, the data is too sparse due to their feature numbers are large. The Mixture Model alleviates the sparsity problem by clustering the apartments with similar properties, and considers them as a class. Due to the number of class is far less the number of apartments, the data looks not sparse relatively.

Table 2. Overall performance of different methods

Methods	MAE	RMSE
KNN-5	5043.427(11.0 %)	6688.717(14.5.%)
KNN-3	4186.667(9.1 %)	5661.704(12.3 %)
Average	3044.474(6.6 %)	4095.168(8.9 %)
Tensor factorization	2778.913(6.0 %)	3971.91(8.6 %)
Neural network	2659.791(5.8 %)	3906.203(8.5 %)
Mixture model	**1283.216(2.8 %)**	**1585.176(3.4 %)**

The two sub-figures in Fig. 4 show the six methods' MAE and RMSE at six districts data. The Mixture Model can always achieve the best result. For each district, higher its average unit price is, worse the estimation result get. The average prices (unit: yuan/m2) of these districts respectively are: ChaoYang (37088), DongCheng (51310), FengTai (32341), ShiJingShan (30498), XiCheng (58607), HaiDian (48060). This satisfies the common sense that the unit price difference is usually large when purchasing high-price apartments, thus the variance become large. This conclusion can be demonstrated that FengTai and ShiJingShan both achieve the best estimation result in two evaluation metrics of all methods.

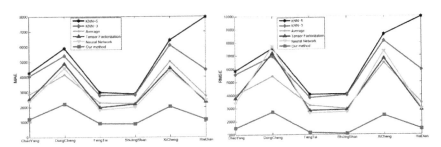

Fig. 4. MAE and RMSE of different methods.

5 Conclusions

In this paper, we investigate a novel task, fine-grained real estate estimation, which has never been studied before due to previous works often consider the gardens as the finest granularity. By utilizing the proposed mixture model, the data sparsity challenge has been alleviated. The results of experiments conducted in real data demonstrate that our model works. MAE and RMSE respectively are 5.8 % and 8.5 % in neural network which gets the best result of the five baselines. Nevertheless, the mixture model obtain over 100 % improvements in that two evaluation metrics when compared with the neural network.

Acknowledgements. The work described in this paper was mainly supported by the National Natural Science Foundation of China (No. 61300076, No. 61375045), the Ph.D. Programs Foundation of Ministry of Education of China (No. 2013110112-0035), the joint astronomic fund of the national natural science foundation of China and Chinese Academic Sinica (Project No. U1531242), Beijing Natural Science Foundation (Project No. 4162054, 4142030, and 4162027), and the Excellent young scholars research fund of Beijing Institute of Technology.

References

1. Zheng, Y., Capra, L., Wolfson, O., Yang, H.: Urban computing concepts, methodologies, and applications. ACM Trans. Intell. Syst. Technol. **5**(3) (2014)
2. Fu, Y., Xiong, H., Ge, Y., Yao, Z., Zheng, Y., Zhou, Z.-H.: Exploiting geographic dependencies for real estate appraisal: a mutual perspective of ranking and clustering. In: 20th ACM SIGKDD International Conference, pp. 1047–1056. ACM Press, New York (2014)
3. Park, B., Bae, J.-K.: Using machine learning algorithms for housing price prediction: the case of Fairfax county, Virginia housing data. Expert Syst. Appl. **42**(6), 2928–2934 (2014)
4. Geng, J., Cao, K., Yu, L., Tang, Y.: Geographically weighted regression model (GWR) based spatial analysis of house price in Shenzhen. In: 19th International Conference on Geoinformatics, pp. 1–5. IEEE Press, New York (2011)
5. Shi, H., Li, W.: Fusing neural networks, genetic algorithms and fuzzy logic for analysis of real estate price. In: 2009 Information Engineering and Computer Science, pp. 1–4. IEEE Press, New York (2009)
6. Del, C.C.: A comparison of data mining methods for mass real estate appraisal. University Library of Munich, Munich (2010)
7. Fu, Y., Liu, G., Papadimitriou, S., Xiong, H., Ge, Y., Zhu, H., Zhu, C.: Real estate ranking via mixed land-use latent models. In: 21th ACM SIGKDD International Conference on Knowledge Discovery and Data Mining, pp. 299–308. ACM Press, New York (2015)
8. Bailey, M., Muth, R., Nourse, H.: A regression method for real estate price index construction. J. Am. Stat. Assoc. **58**(304), 933–942 (1963)
9. Fu, Y., Ge, Y., Zheng, Y., Yao, Z., Liu, Y., Xiong, H., Yuan, N.J.: Sparse real estate ranking with online user reviews and offline moving behaviors. In: 14th IEEE International Conference on Data Mining (ICDM 2014), pp. 120–129. IEEE Press, New York (2014)
10. Peterson, S.: Neural network hedonic pricing models in mass real estate appraisal. J. Real Estate Res. **31**(2), 147–164 (2009)

Intellectual Analysis System
of Big Industrial Data for Quality Management
# of Polymer Films

Tamara Chistyakova[1(✉)], Mikhail Teterin[1], Alexander Razygraev[1],
and Christian Kohlert[2]

[1] State Institute of Technology Saint Petersburg Russia, Saint-Petersburg, Russia
`nov@technolog.edu.ru`
[2] Klöckner Pentaplast Europ GmbH & Co. KG, Montabaur, Germany
`c.kohlert@kpfilms.com`

Abstract. The article is devoted to application of methods and technologies of the intellectual analysis of the large production data for the large-tonnage production of a polymeric film being characterized complex structure, existence of systemic linkages, semi-structured information, power consumption and a large number of the controlled and accumulated industrial information. The novelty consists in application of methods and data mining technologies in production of polymeric films. The article describes the application of neural networks for big data analysis. Application of neural networks allowed to carry out the forecast of quality of production that in turn allowed to reduce flaws. Also, this article describes the application of other data mining algorithms, which enabled to identify the causes of the decreasing quality of products and to find the values of controls in the system, which provide the specified quality and output as recommendations to the operator. Thus, the application of the developed intellectual system for analysis of large industrial data provided to reduce defects and increase productivity. Approbation of the developed system took place at plants of Russia and Germany and confirmed its working efficiency for management and change-over of productions of a polymeric film on various types of product.

Keywords: Classification · Neural network · Forecasting · Optimization · Data mining · Polymer films

1 Introduction

Modern production of polymer films is characterized by high cost of raw materials, high complexity of the technological process caused by the use of a large number of production stages, the use of additional devices, the monthly change in the product range and a variety of equipment types used [1, 2]. Customers' high demands for production quality cause annual losses of companies in the production of polymeric materials in the order of several billion euros due to the lack of time for analysis of the data and perceptions obtained in real time and the absence of a monitoring system satisfying the needs of the production line operators and of the quality engineers [3]. The most important indicators of the polymer film quality are: the absence of defects on

© Springer International Publishing Switzerland 2016
L. Cheng et al. (Eds.): ISNN 2016, LNCS 9719, pp. 565–572, 2016.
DOI: 10.1007/978-3-319-40663-3_65

the surface – black dots, destructive (brown) streaks, inclusions of unmelted polymer and modifier particles, cracks (the bursting of air bubbles) – the color values, the film width and film thickness. The polymer film quality depends on quality of raw material and technological parameters of production [4].

The source information is the data from the extrusion-calender lines (Fig. 1) for the production of polymer films, characterized by: being multi-stage (production line includes the following stages of production: mixing; powder (extruder); moulding (calender); cooling; coiling); being multiproduct; large-tonnage (the equipment recycles 1000 kg per hour); continuity; huge volume of accumulated expertise in the conditions of their production (billion records); a large scope of controlled information (250 sensors); high power consumption [5]; complex structure and the presence of system linkages that describe weakly formalized information systems and complex application objects of study that leads to considerable complication of the rules for the construction of the formalized information-analytical models describing the potential relationships in the data.

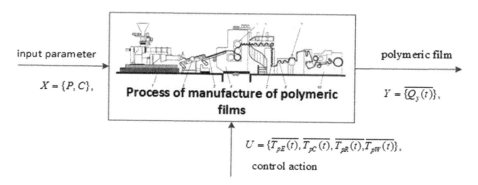

Fig. 1. Description of the production process for polymer films

Under these conditions, traditional approaches to information processing become ineffective. As a result there is an actual problem to develop and apply new multi-method approaches to data analysis. An actual direction of improving the efficiency of analysis of large volumes of semi-structured information, and building of analytic and information models describing the behavior of complex systems is the use of a systematic multi-method technology of intellectual analyze which is called "Data Mining".

The aim of this project is to create software, which will allow the analysis of production data received from the monitoring system and identify the process parameters that caused the deviations of the actual values of quality thus providing a chance to improve the efficiency of production.

The use of this complex allowed to identify the complex structural relations between technological parameters, as well as to increase the quality of the products.

Technical implementation is to create a visualization system that allows us to build trends of technological parameters and quality indicators for the management production staff. This system is related to existing industrial SCADA systems. Information from the SCADA-system is downloaded to a staging database, where it interacts with the developed intelligent system.

2   The Formalized Description of the Process

The formalized description of the polymer film materials production process can be represented in the form (Fig. 1): where $Y(t)$ – is the vector of output variables, representing quality indicators of the polymer films; $X(t) = \{P(t)$ – type of polymer film, $C(t)$ – component composition of the raw material$\}$; $U(t)$ – vector of control actions, where $T_{pE}(t)$ – vector of technological parameters of the extruder; $T_{pC}(t)$ – vector of technological parameters of the calender operation; $T_{pR}(t)$ – vector of technological parameters of the pulling device; $T_{pW}(t)$ – vector of technological parameters of the winding machines, where $t = t_1 \div t_{set}$.

The technological parameters of the extruder are $Pr^e(t)$ – the output of the extruder, m/min; S – extruder screw rotation speed, rev/min; $T^s(t)$ – the temperature of the screw, $°C$; $T^h(t)$ – temperature of the heaters, $°C$.

The process parameters of four-roll calender include: $T^W(t)$– the temperature of rolls, $°C$; $Pr^W(t)$ – the velocity of the rolls, m/min; $To^W(t)$– torque rolls, N·m; $P^{concave}(t)$ – pressure concave bending, Pa; $P^{convex}(t)$ – pressure convex curve, Pa; $P^{bend}(t)$ – pressure blend curve, Pa; $P^{offset}(t)$ – pressure offset curve, Pa; $FL(t)$ - the level of filling of gaps, mm.

Technological parameters of the pulling device: $T^{tor}(t)$– the temperature of pulling devices, $°C$; $Pr^{tor}(t)$ – the performance of pulling devices, m/min; $To^{tor}(t)$ – the torque pulling devices, N · m; $T^{pr}(t)$ – temperature pressure rollers, $°C$; $To^{pr}(t)$ – the torque pressure rollers, N·m; $T^{cr}(t)$ – temperature of the cooling rollers, $°C$; $To^{cr}(t)$ – the torque cooling rollers, N·m; $T^{sr}(t)$ – temperature hardening of rolls, $°C$; $To^{sr}(t)$ – the torque hardening of rolls, N·m; $T^t(t)$ – the temperature of the stretch rollers, $°C$;

Technological parameters of the pulling device: $T^{wi}(t)$ – the temperature of the winding machines, $°C$; $S^{wi}(t)$ – the speed of winding, m/min; $S^{wis}(t)$ – winding speed drawing, m/min; $L(t)$ – the length of the wound roll, m; $FT(t)$ – the tensioning force on a winding machine, N.

3   Problem Definition and Solving

The computer support system allows solving the following tasks:

- the definition of input parameters $X(t)$ and control actions $U(t)$, which led to abnormal situation $Y_{set1} < Y(t) < Y_{set2}$, occurred in the period of time from t_1 to t_2 [5];
- the prediction of output parameters $Y(t)$ for given input parameters $X(t)$ and control actions $U(t)$ during a period time from t_2 to t_3;
- the definition of control actions $U(t)$ and input parameters $X(t)$, which allow to obtain the best film quality Y_{opt} [5].

The algorithm of the intellectual analysis for big data (up to 200 million records) allowing to issue recommendations to production staff on prediction of behavior of an object, on clarification of the true reasons, on determination of the best values of the control values to obtain the given quality of production (Fig. 2) is developed for the solution of objectives. At the first stage of managing staff export data from database of industrial parameters and characteristics to support computer system and input the threshold restriction on the film quality parameters [6]. The second stage is the design of a database and export production data into a data mining system database. The experts select the most significant parameters (Fig. 2) and make a request to the data mining system database to form an array of significant information for experts. The array is split into 3 data sets: training data set (70 %), evaluation data set (15 %) and test data set (15 %). Training data set is analyzed using one of the proposed data mining methods, and evaluates the adequacy of the model generated using the evaluation data set. Production quality is predicted using the test data set.

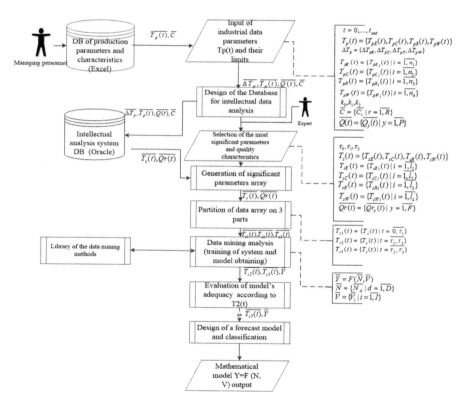

Fig. 2. The data mining algorithm for industrial production of the polymer film

The software solution (Fig. 3) is developed for realization of this algorithm and includes the following components: infoware (database of equipment parameters, production data, and knowledge base of emergency situations); mathware (mathematical model based on neural network, decision trees, support vector machines, and naive Bayesian classifier) and optimization algorithms [6].

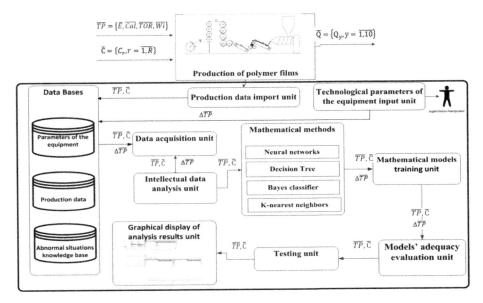

Fig. 3. The functional structure of a software solution

Using the wide range of algorithms is caused by the fact that each method solves the problem with the given accuracy and it has disadvantages. The decision trees allow to identify cause of the defect on the film surface, have a high degree of interpretability and high speed training.

The neural network was used for solving the prediction task. Rumelhart multilayer perceptron was used ideally for predicting film quality in the production. The input of neural network receives input parameters $X(t)$ and control actions $U(t)$ in normalized form, and the at output we get the predicted output parameter values $Y_{pred}(t)$. The neural network was trained by backpropagation. This iterative gradient method minimizes the error of the neural network and obtains the goal output value [7–9] (Fig. 4).

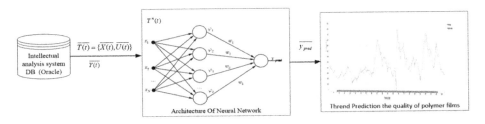

Fig. 4. The scheme of the forecasting unit

The linear regression model provides the search for the relationship between the input and the output variable based on the regression equation. The regression model learns quickly and can be easily implemented, but in case of lack of data, model would be inadequate, and in case of the large amount of data, the model can be overfitting [10]. Regression model significance is evaluated using Fischer's test.

There also was implemented a classification algorithm based on Random Forest search technology. The algorithm implements subsequent procedure for building machine learning algorithm composition and generates the ensemble of classifiers using random sampling with replacement, that means selection of several random subsets while sampling. This algorithm is used when high accuracy of classification is required. The algorithm tends to overfitting, especially on noisy data and requires large amounts of memory for storage of mathematical models [11–13].

Testing the software solution performance was conducted on data obtained from industrial plants in Russia and Germany Corporation "Klöckner Pentaplast". The data contain 200 million rows for 250 different data processing parameters. Testing considered one of the major defects – black dots. Regression model, algorithm of Random Forests, a neural network (Rumelhart multilayer perceptron), a decision tree and knowledge bases about abnormal situations were used.

As a test example the case in which the neural network predicted abnormal situation is demonstrated. Neural network training occurred the extrusion-calender data for line № 14 of the corporation "Klöckner Pentaplast". Data is read from the database in a sequential chronological order. The input layer of the neural network receives 7 parameters, the number of neurons in the hidden layer is 15. The forecast trend is shown on Fig. 5. The trend for the black dots number in case of abnormal situation exceeding the permissible value of black dots quantity is shown [14].

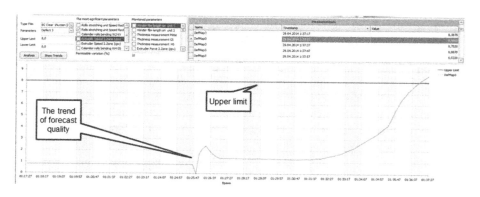

Fig. 5. Prediction trend for the number of black dots

To find the causes of the predicted abnormal situation the regression model and Random Forests algorithm were used. An array of the most important data such as the speed of the calender rolls, the speed of the tempering rolls, the speed of removable rolls, the roll bending of calender rolls, the screw speed in the extruder, the temperature of the screw in the extruder, the temperature of the mixture in a hot mixer including

1000 measurements (2 h) were formed. Testing was conducted for the film brand "BC Clear 1Nutzen" with thickness 100 μm.

Based on the obtained statistical quality and process parameters data simple regression models were built. Models' significance is estimated using Fischer's test dispersion method. Estimation showed that the highest correlation is observed between the number of black dots and the following technological parameters: velocity of the 3rd calender roll, velocity of the removable roll, velocity of the tempering rolls.

On the next step the array, obtained by experts, is divided into 3 parts. Training, evaluation and testing used data mining method – Random Forest. Random Forest showed that the most important parameters are almost all the speeds of the production line rolls. Thus, the results of Random Forests analysis and Fisher's test match (Table 1).

Table 1. The cause of the defect extracted from the knowledge base

Cause	Action	Probability
Insufficient amount of stabilizer	To increase the stabilizer not more than valid 1.1 %	55 %
The irregular shape of the melt	The melt should leak out. To adjust roll bending. To follow the profile	30 %
Mixture temperature is too high	The temperature of the mixture in a hot mixer or the residence time in the hot mixer should be reduced	6 %
Configuration of OCS is wrong	OCS must be configured	6 %
Dots look matte black	To clean the knife head and the body of kneader	2 %
The black dots on the edges	Clean cheeks and a locking ring	1 %

Testing proved the efficiency of developed algorithms for intellectual analysis of industrial data in the production of polymer films, the algorithm of data reduction, the algorithm for finding the reasons of black dots occurrence and the whole software solution.

4   Conclusion

Thus, the intelligent system for quality control of polymer films, including a database of technological parameters, a library of mathematical methods (neural network, decision trees, Random Forests, k-nearest neighbors), the interfaces of visualization of technological and production quality parameters, allows to perform neural network forecast quality of polymer films, to identify the true causes of flaws, to predict, to find combinations of variables, which is basis for resource- and energy-saving management, as it leads to a rational use of energy and reduce of flaws.

The basis of intelligent system is a neural network that adapts to the type of the polymer film using the input parameters to be analyzed, control actions and the values of quality indicators. Neural network with 7 neurons in the input layer, 15 neurons in the hidden layer and 1 neuron on the output layer is trained by back propagation method. Neural network has proven its predictive capability and the accuracy of classification reached 93 %, which meets the requirements of the company where the testing a program complex was done.

Acknowledgments. This research is partially supported by the Russian Foundation for assistance to small innovative enterprises in science and technology (Projects No. 5263GU1/2014) and by the German Academic Exchange Service (DAAD) (Grant of Leonard Euler).

References

1. Kohlert, M., König, A.: Advanced polymeric film production data analysis and process optimization by clustering and classification methods. Front. Artif. Intell. Appl. **243**, 1953–1961 (2012)
2. Kohlert, M., Chistyakova, T.: Advanced process data analysis and on-line evaluation for computer-aided monitoring in polymer film industry. J. Saint Petersburg State Inst. Technol. (Tech. Univ.) **29**, 83–88 (2015)
3. Optical Control Systems GmbH. http://www.ocsgmbh.com
4. Kohlert, M., König, A.: High dimensional, heterogeneous multi-sensor data analysis approach for process yield optimization in polymer film industry. Neural Comput. Appl. Mater. **26**(3), 581–588 (2015)
5. Chistyakova, T., Polosin, A., Razygraev, A., Kohlert, K.: Software system to control the color of thin rigid polymeric materials. J. Autom. Ind. **7**, 12–18 (2012)
6. Kohlert, M., Teterin, M., Konig, K., Chistyakova, T.: Computer support in production polymeric films on data mining. Math. Methods Technics Technol. **3**, 20–25 (2012)
7. Verikas, A., Bacauskiene, M.: Feature selection with neural networks. Pattern Recogn. Lett. **23**(11), 1323–1335 (2002)
8. Haykin, S.S.: Neural Networks and Learning Machines. Pearson Education, Upper Saddle River (2009)
9. Segal, E., Pe'er, D., Regev, A., Koller, D., Friedman, N.: Learning module networks. J. Mach. Learn. Res. **6**, 557–588 (2005)
10. Seber, G., Lee, A.: Linear Regression Analysis. Wiley, New York (2003)
11. Biau, G., Devroye, L., Lugosi, G.: Consistency of random forests and other averaging classifiers. J. Mach. Learn. Res. **9**, 2015–2033 (2008)
12. Breiman, L.: Bagging predictors. Mach. Learn. **24**, 123–140 (1996)
13. Breiman, L.: Random forests. Mach. Learn. **45**, 5–32 (2001)
14. König, A., Gratz, A.: Advanced methods for the analysis of semiconductor manufacturing process data. In: Pal, N.R., Jain, L. (eds.) Advanced Techniques in Knowledge Discovery and Data Mining, pp. 27–74. Springer, London (2005)

Some Ideas of Informational Deep Neural Networks Structural Organization

Vladimir Smolin[(⊠)]

Keldysh Institute of Applied Mathematics, RAS, Moscow, Russia
smolin@keldysh.ru

Abstract. Deep Learning (DL) is a branch of machine learning based on multiple processing layers with complex structure, or otherwise composed of multiple non-linear transformations. Diverse DL models are used for solving different tasks, but have some common features and identic problems. Eight ideas for DL features organization and problems solving are outlined, the DL main goal is specified. The elements and structure of deep learning neuro informational model are discussed.

Keywords: Deep learning · Multiple processing layers · Curse of dimensionality

1 Introduction

Multilayer Deep Neural Networks (DNN) and DL have achieved outstanding performance in comparison to 2–3 layer Neural Networks (NN). It can be referred to such important problems as computer vision, speech recognition, and so on [1]. The BackPropagation (BP) algorithm [2] is thought to be a "basic swing", the basis for learning in most articles on neural networks. But the idea of gradient descent is rather obvious ("shallow") for optimization; some "more deep" ideas are usually used to improve BP.

The analysis of optimization process shows, that we shouldn't only improve BP, several different ideas are necessary to use for complex tasks solving. Ideas under consideration are well known, but they aren't usually applied for NN DL. It seems these ideas that can work better than only improved BP.

2 Reflex and Conscious Activity

Human and higher animals Nervous System (NS) have reflex and conscious forms of activity. Reflexes give fast reaction to any external signal, while conscious activity assumes some delay for modelling and estimation of different variants of behavior before realizing the best one.

Present-day DNNs are aimed to simulate NS reflex activity. In the sake of simplicity it is really better to start from reflexes. But the best is to estimate the applicability of DNNs under investigation for modelling conscious forms of NS activity.

© Springer International Publishing Switzerland 2016
L. Cheng et al. (Eds.): ISNN 2016, LNCS 9719, pp. 573–582, 2016.
DOI: 10.1007/978-3-319-40663-3_66

Idea #1: <The development of DNNs for reflex activity will be much more promising, if to consider DNNs capable of building percepted objects models. These models could be used for the estimation of different variants of behavior before their realization>.

It's better to discover universal DNNs, suitable for objects modelling of various classes. One variant of such NN (and DNN) will be sketched out below.

3 Nonlinear Table Functions

Multiple non-linear transformations for DL can be realized on any NN, converting input signal $\vec{X}$ into output $\vec{Y}$. NN parameters, mainly weights of connections matrix, define function $\vec{Y} = F(\vec{X}, \vec{A})$, where $\vec{A} = \{a_i\}$ is NN elements activity vector. Connection weights and other coefficients matrix is a table of parameters, so $F(\vec{X}, \vec{A})$ is a table function.

Idea #2: <The point of view, that all NN and DNN implement transformations $\vec{Y} = F(\vec{X}, \vec{A})$ by basic vectors $\{\vec{X}_i, \vec{Y}_i\}$ interpolation, makes it possible to compare different neural models>.

What's the use of idea #2? Comparison of nonlinear function interpolation capabilities can be done theoretically, without expensive programming and supercomputers. While "the best results of tests" race looks like a "justice for rich".

4 Neuron as a Memory Cell

Formal Neuron (FN) structure is shown on Fig. 1a. The evident way of FN utilization is to use it as a memory cell for basic vectors $\{\vec{X}_i, \vec{Y}_i\}$ interpolation, making $\vec{M}_i = \vec{X}_i$ and $\vec{L}_i = \vec{Y}_i$, where $\vec{M}_i$ and $\vec{L}_i$ are input and output connection weight vectors. FN activity a_i defines vector $\vec{Y}_i$ contribution to interpolation result. And a_i is proportional to proximity of $\vec{M}_i$ to $\vec{X}$,

$$a_i = max(0, (\vec{X} * \vec{M}_i - b_i - k_i \sum_j a_j)) \tag{1}$$

$$\vec{Y} = f(a_i) * \vec{L}_i \tag{2}$$

where b_i and k_i are bias and back inhibition parameters of i-th FN. $\sum_j$ here and below means summing by all j. All values in (1) and (2) are only non-negative. Formula (1) matches neural layer with back inhibition. For lateral inhibition

$$a_i = max(0, (\vec{X} * \vec{M}_i - b_i - \vec{A} * \vec{K}_i)) \tag{3}$$

where $\vec{A} = \{a_i\}$ is activity vector and $\vec{K}_i$ is i-th FN vector of lateral connection weights.

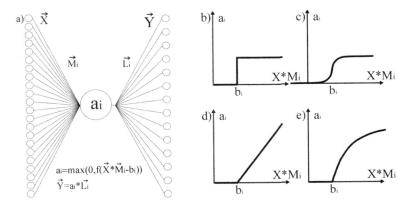

Fig. 1. FN structure (a) and activation functions (b–e).

The $f(a_i)$ selection (see Fig. 1b–e) effects interpolation results. For the sake of simplicity the linear threshold function 1d is considered below, so

$$\vec{Y} = a_i * \vec{L}_i \qquad (4)$$

There is no problem to set input $\vec{M}_i = \vec{X}_i$ and output $\vec{L}_i = \vec{Y}_i$ if only one FN is under consideration. But if NN consists of two or more layers the same connections emanating for the first are incoming for the second layer and so on. There is an alternative: to adjust connection weights according to activity of the first or the second layer. The results will be different, excluding some special cases. This problem makes nontrivial the next idea:

Idea #3: <The structure and learning rules for NNs and DNNs should allow to utilize FNs like a "memory cell" for basic vectors $\{\vec{X}_i, \vec{Y}_i\}$ interpolation>.

For example, CounterPropagation (CP) networks [3] comply the idea #3 and BP networks [2] conflict with it.

5   2 and 3 Layers NN Models

We are going to start from two layers to study connection weights adaptation. Then we will find out what advantages we can get from applying the third layer. Signals always propagate from the left ($\vec{X}$) to the right ($\vec{Y}$), but we can choose, will we adjust the input connection weights ω_{ij} to remember previous layer activity or output connections weights to form required next layer activity. The results of learning will be different.

Designations. Signal transfer direction in interlayer connections matrix is shown by wide arrow. Thin arrow inside wide arrow designate the direction of NN layers mapping. If activity of previous layer is mapped on the next layer input connection weights, then thin arrow has the same direction with wide arrow. If output connection weights are mapping next layer activity, thin arrow direction is opposite to wide arrow.

Letter near the arrows should help to understand, what connections are better to consider. Figure 2 includes 4 symbolic designations of NNs. For 3 layers NN 2 more symbolic designations are possible, but only 4 NNs, shown on Fig. 2 will be considered below.

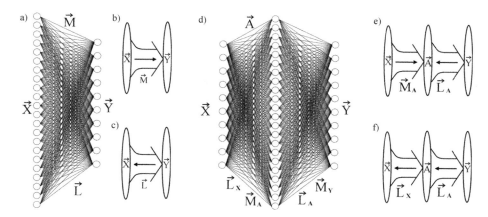

Fig. 2. NN models: structures are (a), (d), designations are (b) SOM, (c) BP, (e) CP, (f) BP.

First let's consider NN with two layers X and Y (Fig. 2a) with activity $\vec{X}$ and $\vec{Y}$. To remember previous layer activity we should use mapping learning algorithm [4]:

$$\Delta \omega_{ij} = \eta_1 (x_i - \omega_{ij}) y_j \Delta t \tag{5}$$

Where η_1 is coefficient and Δ - sign of increment. For FN j from layer Y ω_{ij} is input connection weight, so $\omega_{ij} = m_{ij}$ and for $\vec{M}_j = \{m_{ij}\}$ we can write:

$$\Delta \vec{M}_j = \eta_1 (\vec{X} - \vec{M}_j) y_j \Delta t \tag{6}$$

Formula (6) is quite understandable: vector $\vec{M}_j$ is always changing in the direction to vector $\vec{X}$ and tends to the mean $\vec{X}$ while $y_j > 0$. So (6) is preferable way to describe and explore in vector form connection weights changes during the process of previous layer activity remembering. That is why (6) is widely used in investigations on Self Organizing Mapping (SOM) [4].

If we want to form required next layer activity, BP algorithm [2] should be employed:

$$\Delta \omega_{ij} = \eta_2 (y_j^d - y_j) x_i \Delta t \tag{7}$$

Where y_j^d is desirable value of y_j. For FN i from layer X ω_{ij} is output connection weight, so $\omega_{ij} = l_{ij}$ and for $\vec{L}_i = \{l_{ij}\}$ we can write:

$$\Delta \vec{L}_i = \eta_2 (\vec{Y}^d - \vec{Y}) x_i \Delta t \tag{8}$$

Idea #4: <For adaptational processes understanding it is better to use vector form of adjusting functions like (6) and (8), because the remembering layer activity vector dimensionality matches to connection weights vector dimensionality>.

Nonlinear Table Functions. 2 layers NN can be used for nonlinear table functions approximation. But nonlinearity of 2 layers NN is weak, because the main nonlinear NN properties originate from switching FN on and off, and in 2 layers NN there are no layers for arbitrary switching.

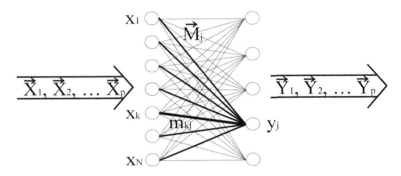

Fig. 3. Reproduction of signals $\vec{Y}_i = \{y_j^i\}$ on FNs of output layer.

Let's use BP (8) and try to remember desirable $\vec{Y}_i^d$ (Fig. 2c). For 2 layers NN the number p of pairs $\{\vec{X}_i, \vec{Y}_i\}$ is restricted by vector $\vec{X}$ dimensionality. To form output signals $\vec{Y}_i = \{y_j^i\}$ we should adjust connection weights $m_{kj} \in \vec{M}_j$ for each y_j^i to satisfy system of p equations:

$$\begin{cases} \sum_{k=1}^{N} m_{kj} x_k^1 - b_j = y_j^1 \\ \sum_{k=1}^{N} m_{kj} x_k^2 - b_j = y_j^2 \\ \quad \dots\dots\dots \\ \sum_{k=1}^{N} m_{kj} x_k^p - b_j = y_j^p \end{cases} \tag{9}$$

Where b_j is a bias of FN j in layer Y, for other designations see Fig. 3. System of p linear Eq. (9) has solutions for the most feck of arbitrary x_k^i and y_j^i if $p \le (N+1)$, where N - number of vector $\vec{X}$ (and $\vec{M}_j$) components. Then interlayer connections matrix can convert $\vec{X}_i$ to $\vec{Y}_i$ for all p pairs $\{\vec{X}_i, \vec{Y}_i\}$. By changing connection weights only linear dependence between layer X activity and layer Y excitation may be adjusted. Using any standard method of system (9) solution the required weights of connections can be easily found.

But ω_{ij} adjustments by adaptive laws like (5) or (7) don't come to linear systems solutions. As a result much less pairs $\{\vec{X}_i, \vec{Y}_i\}$ can be remembered, 5–20 times less than $N + 1$.

Not all arbitrary pairs can be remembered. If system (9) determinant is equal or close to zero, then there is no solution for ω_{ij} or values of ω_{ij} should be unacceptably large. The problem is caused by the attempt to build one hyperplane for representation of all pairs $\{\vec{X}_i, \vec{Y}_i\}$. The problem can be easily solved by piecewise linear approximation of $\vec{Y} = F(\vec{X})$. But for piecewise linear approximation we need some Hidden Layers (HL).

Idea #5: <The way to the wide classes of nonlinear functions $\vec{Y} = F(\vec{X}, \vec{A})$ is directed away from single hyperplane to piecewise linear approximation>.

3 layers NN is shown on Fig. 2d. It seems, that 2 linear conversions by connection matrixes will also result in linear transformation. But if changes of input signal cause switching HL FNs on and off, conversion $\vec{Y} = F(\vec{X})$ becomes nonlinear. FNs nonlinear threshold function (Fig. 1c–e) will cause nonlinearity of $\vec{Y} = F(\vec{X})$ without switching. The example is shown on Fig. 4. Very simple NN structure will help to understand its functioning. Only one input and one output allow to draw graphs, and five FNs of HL are more than enough to show nonlinearity (Fig. 4a). Just setting biases b_i proportional to input connection weights values m_i we can cause FNs with larger m_i to switch on later. Activity a_i of FN is proportional to $\vec{X} * \vec{M}_i$ (see (1)), so each of HL FNs will have some X axis segment, where its activity is larger, than other HL FNs activity (Fig. 4b). It's already enough to get nonlinear properties.

Idea #6: <For getting nonlinear properties it's advisable to set biases of HLs FNs proportional to $|\vec{M}_i|$>.

To make learning process quicker, we should use back inhibition $(-k_i \sum_j a_j$ in (1)). FNs picture of activity with back inhibition depends on $-k_i$ values, see on Fig. 4d. HL with properly adjusted biases and back inhibition works like radial basis function networks [6].

The last we need for piecewise linear function approximation (Fig. 4c) is to adjust output connection weights l_i to suitable values. There is a wide choice of adjustment methods, including BP. Back inhibition allows to make adjustment of output connection weights independently, without spoiling results of previous pairs $\{\vec{X}_i, \vec{Y}_i\}$ learning.

Idea #7: <Back inhibition in HLs will help to speed up learning process>.

Input connections $\vec{M}_i$ also should be adjusted. We should spread HL input connection vectors $\vec{M}_i$ in the input signal states space $X = \{\vec{X}\}$. BP doesn't solve this task, while SOM learning algorithm (6) will be quite relevant [4]. Utilization SOM algorithm (6) for input and BP (8) for output HL connections (Fig. 2e) results in HL FMs using as a "memory cell" and allows to organize piecewise linear function approximation (Fig. 4d).

For more precise approximation the number of HL FNs should be made lager. Applying BP algorithm to all 3 layer NN connections (Fig. 2f) doesn't satisfy

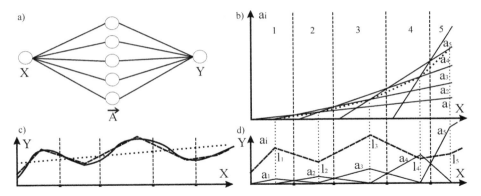

Fig. 4. (a) NN structure; (b) HL A FNs activity a_i without inhibition; (c) nonlinear function and its linear and piecewise linear approximations; (d) HL A FNs activity a_i with inhibition and piecewise linear approximation.

"memory cell" requirements. Adding of HL FNs can help BP to remember more pairs $\{\vec{X}_i, \vec{Y}_i\}$, but only in the case, if systems (9) determinants are fitting for their solving.

SOM algorithm is defined not only by (6). Neighbors by lateral connections of active (with $a_i > 0$) FNs also adjust their input connections according to (6), but with smaller increments. This rule results in vectors $\vec{M}_i$ equable allocation in input signal states space X, but requires initial structure of lateral connections. Equable allocation can be adjusted by means of biases b_i, (6) for this purpose should be added by:

$$\Delta b_i = \eta_3 a_i \Delta t, \quad i : a_i > 0; \quad \Delta b_j = (-\sum_i \Delta b_i)/N, \forall j \tag{10}$$

where N is number of FNs in the layer.

Rule (10) can be applied in addition to lateral connections algorithm or alone, but it allows to execute mapping of the space with a priory unknown dimensionality. For functions approximation we don't need uniform spreading of vectors $\vec{M}_i$ equable allocation in input signal states space X. Allocation should depend on accuracy of approximation. Areas with bad accuracy need more FNs for function approximation. This goal can be easily achieved by making coefficients η_1 and η_3 (in (6) and (10)) proportional to mismatch between function and its current approximation.

Idea #8: <Equable allocation of vectors $\vec{M}_i$ in input signal states space X should provide equal function approximation accuracy, not just uniform spreading>.

6 Functions Approximation on SOM as Universal Object Model

Input signal states space X of any observable object can be mapped by SOM algorithm. 3 layers NN (Fig. 2e) can approximate functions of very broad classes (in restricted X range and with restricted Y derivatives). The map of input signal states space $X = \{\vec{X}\}$ and output function $\vec{Y} = F(\vec{X}, \vec{A})$ can be considered as an information about different object states and ways to react on these states. To complete the model information about object dynamics is needed. Lateral excitatory connections may be used to store information about object dynamics. Functions approximation assumes interpolation on the base of more than one pair $\{\vec{X}_i, \vec{Y}_i\}$. So more than one FN is active $(a_i > 0)$ in a HL. Let's sign FN j vector of incoming lateral excitatory connections as $\vec{H}_j = \{h_{ij}\}$. Adaptation law

$$\Delta\vec{H}_j = \eta_4(\vec{A}(t - \tau) - \vec{H}_j)a_j(t)\Delta t \tag{11}$$

where $\vec{A} = \{a_i\}$ is activity vector of HL FNs, τ and η_4 are parameters, will form connections weights, reflecting the directions of activity spot translocation.

So we can make a conclusion, that 3 layer NN (Fig. 2e) with back inhibition and lateral excitatory connections of HL and adjustment laws (6), (8) and (11) can form models of a wide class of objects. Due to lateral excitatory connections adjustment these models can exhibit inner activity without input signal $\vec{X}$, simulating activity spot translocation during reception of the objects. In the higher levels of DNNs hierarchical structure such models could be used for the estimation of behavior variants before the realization.

7 Multilayered DNNs Structure Elements

Function $\vec{Y} = F(\vec{X}, \vec{A})$ approximation can be implemented by BP, CP or some other NN. The comparison of BP and CP NN structures is shown on Fig. 5 (a) and (b). Why BP needs several (6–10 are recommended) layers for function $\vec{Y} = F(\vec{X}, \vec{A})$ approximation while CP NN can restrain 3 levels? Nonlinear properties are located in HLs FNs, for both, BP and CP NNs. But CP nonlinear properties characteristics are controlled by (10) and can be properly adjusted. BP nonlinear characteristics are randomly allocated. More BP layers are needed to get higher probability to achieve desirable nonlinear properties.

In previous sections an attempt was made to demonstrate, that that 3 layer CP NN (Fig. 2e) with back inhibition and lateral excitatory connections of HL and adjustment laws (6), (8), (10) and (11) can form dynamic models of objects. Can we solve all tasks with the use of such development of CP [3] NN? Of course, no. The main problem is curse of dimensionality, impracticability to describe high-dimensional functions (and objects) by means of table functions. Only low-dimensional (less, than 15–20 degrees of freedom) objects can be described by table. And for adjustment of such table at reasonable time, dimensionality should be even less (6–12). But this development of CP NN can form the main "bricks" for multilayered DNNs hierarchical structure. There are two main methods to build multilayered hierarchical structure of such "bricks".

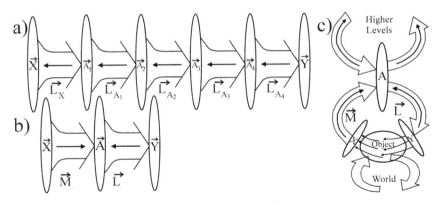

Fig. 5. (a) BP NN structure; (b) CP NN structure; (c) CP NN element for multilayered DNN.

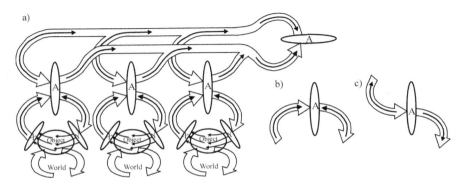

Fig. 6. (**a**) Hierarchical CP NN structure; (b) Mapping layer; (c) Screening layer.

First (Fig. 5c and 6), the activity of HL can be used as an input for the next hierarchical level of multilayered CP NNs structure. This higher level can receive input from several lower "bricks". Similarly output of higher level can be projected on several lower "bricks". The lowest level "bricks" should percept input from object (throw receptors) and project output on affectors. This method follows established in [7] idea, that higher levels are controlling lower layers in the same way, like the lowest layers control muscles. The number of simple hierarchy levels is also limited with curse of dimensionality. 3–4 steps of low-dimensional models uniting will result in high-dimensional model forming. For complex tasks scenes should be decomposed to a set of low-dimensional objects.

Second way to connect levels is to make interface between levels via special "screens", NN layers collecting projections from several higher and lower layers, for decomposition, filtering and prediction of situation development. These "screens" could create the base for conscious forms of activity analogue.

These two main methods to build DNNs are much more complex in comparison to BP algorithm. But BP does nothing to circumvent the curse of dimensionality, multiple levels are used to accumulate enough nonlinearity to be able to approximate nonlinear functions. On the contrary, approximating nonlinear function on the base of topologically correct mapping seems to be the way to deal with the curse of dimensionality.

8   Conclusions

The consideration of multiple non-linear transformations on NNs, assumed to be "bricks" of DNNs, gives some ideas of features organization and problems solving for multilayered NN. Few simple ideas, like #6 "to set biases of HLs FNs proportional to $|\vec{M}_i|$" can be recommended for DNNs. Some ideas can be used in more complex NNs. And other may help to understand some characteristics of information processing in NNs.

Many questions of DNN structural organization, some of paramount importance, like NN dimensionality forming, goal functions for learning, spatial-temporal mapping, were not even mentioned in consideration above. Solutions of these questions seem to be compatible with outlined ideas, but each question is too complex to be added to a short article.

The principal conclusion is: the main problem of nonlinear functions table approximation is curse of dimensionality, and all ideas should be directed to deal with it, diminishing dimensionality of mapping layer structure by all obtainable means.

References

1. Nielsen, M.: Neural Networks and Deep Learning, draft book in preparation (2016). http://neuralnetworksanddeeplearning.com
2. Rumelhart, D., Hinton, G., Williams, R.: Learning representations by back-propagating errors. Nature **323**, 533–536 (1986)
3. Hecht-Nielsen, R.: Counterpropagation networks. Appl. Opt. **26**(23), 4979–4984 (1987)
4. Kohonen, T.: Self-organized formation of topologically correct feature maps. Biol. Cybern. **43**(1), 59–69 (1982)
5. Karp, G.: Cell and Molecular Biology: Concepts and Experiments, 4th edn. Wiley, Hoboken (2005)
6. Broomhead, D., Lowe, D.: Multivariable functional interpolation and adaptive networks. Complex Syst. **2**, 321–355 (1988)
7. Hecht-Nielsen, R.: Confabulation Theory. Springer, Heidelberg (2007)

Meshfree Computational Algorithms
Based on Normalized Radial Basis Functions

Alexander N. Vasilyev[1]([✉]), Ilya S. Kolbin[2], and Dmitry L. Reviznikov[3]

[1] Peter the Great St. Petersburg Polytechnical University,
29 Politechnicheskaya Street, 195251 Saint-petersburg, Russia
a.n.vasilyev@gmail.com
[2] Federal Research Center "Computer Science and Control",
Russian Academy of Sciences, 44/2 Vavilov Street, 119333 Moscow, Russia
iskolbin@gmail.com
[3] Moscow Aviation Institute, 4 Volokolamskoye Shosse, 125993 Moscow, Russia
reviznikov@inbox.ru

Abstract. In this paper, new methods for solving mathematical modelling problems, based on the usage of normalized radial basis functions, are introduced. Meshfree computational algorithms for solving classical and inverse problems of mathematical physics are developed. The distinctive feature of these algorithms is the usage of moving functional basis, which allows us to adapt to solution particularities and to maintain high accuracy at relatively low computational cost. Specifics of neural network algorithms application to non-stationary problems of mathematical physics were indicated. The paper studies the matters of application of developed algorithms to identification problems. Analysis of solution results for representative problems of source components (and boundary conditions) identification in heat transfer equations illustrates that the elaborated algorithms obtain regularization qualities and allow us to maintain high accuracy in problems with considerable measurement errors.

Keywords: Modelling · Boundary value problem (BVP) · Identification problem · Meshfree method · Artificial neural network (ANN) · Normalized radial basis function (NRBF) · Training · Error functional · Optimization

1   Introduction

Neural network technology is one of the most actively developing branches of applied scientific research. Various neural-network-based methods are successfully used in different areas, such as: control problems, forecasting, pattern recognition, multidimensional data fitting, data compression, etc.

Ultimately, a considerable interest has aroused in application of meshfree methods to mathematical modelling problems. This is mainly due to the typical difficulties that rise when grid methods are applied to the solution of multidimensional problems on areas with complex geometry, problems with inexact

© Springer International Publishing Switzerland 2016
L. Cheng et al. (Eds.): ISNN 2016, LNCS 9719, pp. 583–591, 2016.
DOI: 10.1007/978-3-319-40663-3_67

coefficients, inverse problems when there is measurement error, etc. As a rule, in these cases it is necessary to specifically adapt computational algorithms to the considered problem. On the other hand, application of neural network methods allows us to overcome a significant part of these difficulties and to use uniform approaches for solving different types of problems. An important advantage of the considered algorithms from this class is its regularization properties that allow us to apply the elaborated methods to identification problems.

In this paper we develop an approach suggested by A.N. Vasilyev and D.A. Tarkhov [1,2], which is based on using a neural functional basis. The peculiar feature of the method proposed in this paper is that the solution is sought in the form of normalized radial basis function networks [3–6].

2   Neural Network Modelling Algorithms for Stationary Transfer Processes

The starting point for describing the object model is a boundary value problem stated in [1]: over a domain Ω, find a solution $u = u(x)$ to the equation:

$$A(u) = f(x), \quad x \in \Omega \subset R^d,$$

which satisfies the following condition on its boundary Γ:

$$B(u) = g(x), \quad x \in \Gamma,$$

where x is an input vector, A, B are integro-differential operators, f, g are some given functions. The operators can be nonlinear or discontinuous, change type on subdomains, etc. There are no specific restrictions for the boundary.

The solution is sought in the form of a normalized radial basis function (NRBF) network. An output of a network with n neurons can be given by

$$\tilde{u}(x, \psi) = \frac{\sum_{i=1}^{n} \omega_i \phi(||x - x_i^c||, \sigma_i)}{\sum_{i=1}^{n} \phi(||x - x_i^c||, \sigma_i)},$$

where ψ is a parameter vector, ω, σ, $x^c \in \psi$ are weights, "breadths" and coordinates of the neuroelements' centers, ϕ is the radial basis function. A possible approach to the parameter adjustment of a neural network model is to minimize the quadratic error functional, which is formed by substituting the neural network representation into the initial equation and the boundary condition. The error functional is given by a discrete form:

$$J(\tilde{u}) = J(\psi, \Xi) = \sum_{j=1}^{m_\Omega} w_\Omega \left[A(\tilde{u}(x_j^\Omega, \psi)) - f(x_j^\Omega) \right]^2 +$$
$$\sum_{j=1}^{m_\Gamma} w_\Gamma \left[B(\tilde{u}(x_j^\Gamma, \psi)) - g(x_j^\Gamma) \right]^2,$$

where m_Ω, m_Γ is a number of reference points on the domain and on the boundary, respectively, $\{x_j^\Omega\} \subset \Omega$ is a set of reference points on the domain, $\{x_j^\Gamma\} \subset \Gamma$ on the boundary, and $\varXi = \{x_j^\Omega\} \cup \{x_j^\Gamma\}$. Here w_Ω, w_Γ are some coefficients that "level" the inputs of inner and boundary components. The adjustment of w_Ω and w_Γ largely depends on problem specification. Minimization is performed using multidimensional optimization procedures [7]. A considerable advantage of this approach is its generality, since it does not place any restrictions on domain configuration, smoothness of coefficients, etc. The adjustment of parameters ψ of the approximating model $\tilde{u}$ is performed by minimizing the build up functional:

$$J(\psi, \varXi) \xrightarrow{\psi} \min.$$

Applications of the developed method to one-, two- and three-dimensional problems are considered, problems with nonlinear source, with boundary-layer nature of the solution were evaluated, and neural network approximations for problems with curvilinear boundary of the domain were obtained.

To demonstrate the work of the suggested method we consider a typical boundary value test problem as a benchmark [5]. We are looking for a solution to Poisson's equation over a rectangular domain: $\Delta u(x, y) = f(x, y), \Delta = \frac{\partial^2}{\partial x^2} + \frac{\partial^2}{\partial y^2}$ is the Laplace operator, $x \in [0, \pi]$, $y \in [0, \pi]$; satisfying homogeneous boundary conditions: $u(0, y) = u(\pi, y) = u(x, 0) = u(x, \pi) = 0$. The source component is given by $f(x, y) = \sin x \sin y$. The solution is sought in the form of NRBF network $\tilde{u}((x, y), \psi)$, to do so we compose the error functional of the form:

$$J(\tilde{u}) = \sum_{j=1}^{m_\Omega} w_\Omega \left[\Delta\tilde{u}((x, y)_j^\Omega, \psi) - f((x, y)_j^\Omega) \right]^2 +$$

$$\sum_{j=1}^{m_\Gamma} w_\Gamma \left[\tilde{u}((x, y)_j^\Gamma, \psi) \right]^2.$$

The adjustment of parameters ψ is performed using the conjugate gradient method CG_DESCENT [8].

We solved this problem for networks of different sizes. Figure 1 shows NRBF network solutions at different sections $y = $ const, and analytical solution (denoted by dots). Dashed line shows the output of the NRBF network containing 4-neuron, solid line – one of the 8-neuron NRBF network.

To evaluate the accuracy of the obtained approximations we calculated the mean squared error ε on a grid, the results are given in Table 1. We compared the results with the known analytical solutions.

From Table 1 we can see that increasing the dimensionality of approximating network results in higher accuracy of resulting approximation. It is important to note that the error functional and approximation error decrease simultaneously.

In Fig. 1 we can see that the output of the 8-neuron NRBF network almost coincides with the analytical solution.

Table 1. Calculation results

Number of neuroelements, n	Resulting error functional, J	Mean squared error, ε
2	0.0027	0.02513
4	0.0014	0.02212
6	$5.61 \cdot 10^{-5}$	0.00119
8	$1.17 \cdot 10^{-5}$	0.00067

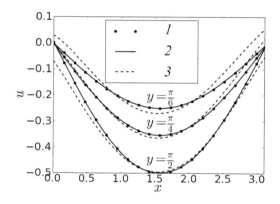

Fig. 1. Analytical and neural network solutions to Poisson's equation. 1 – exact solution, 2 – NRBF-8, 3 – NRBF-4

In what follows we perform a comparative study of neural network methods, based on classical (RBF) and normalized radial basis function networks, for a set of typical test problems [9]. We consider a one-dimensional equation of the form: $u'' = k^2 u$, to which we need to obtain a solution that satisfies boundary conditions $u(0) = u(1) = 1$. The particularity of this problem is the boundary-layer nature of its solution, what becomes more prominent as the coefficient k increases. In this paper we take $k = 27.79$, and the effect is clearly pronounced. Figure 2 illustrates outputs of the obtained two-element neural networks. It can be seen that NRBF-approximation more adequately depicts the exact solution better than the non-normalized network.

We considered a three-dimensional boundary value problem for a stationary heat equation with nonlinear source: $\Delta u = \sin(u^2) + f(x, y, z)$, computational domain Ω is a unit cube, boundary conditions: $u|_\Gamma = \sin(x + y + z)$. Figure 3 depicts problem approximations with 4-neuroelement networks at different sections. It can be seen that the solution obtained using NRBF-approximation almost coincides with the analytical one.

Conducted numerical experiments show that on the average, NRBF-based method approximates with higher accuracy, but parameter adjustment time is longer. At that, for some tasks NRBF-based method converges faster.

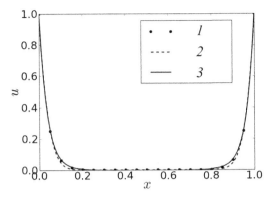

Fig. 2. Solution to one-dimensional problem with boundary-layer nature of the solution. 1 – exact solution, 2 – RBF-2, 3 – NRBF-2

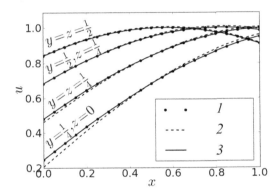

Fig. 3. Solution to three-dimensional problem with nonlinear source. 1 – exact solution, 2 – RBF-4, 3 – NRBF-4

For example, the three-dimensional problem with nonlinear source for 4-neuron networks is solved using NRBF-based method about twice as fast as using RBF networks, and the approximation accuracy is almost a hundred times better.

We compared the neural network method with the finite difference method. For the finite difference implementation, we used the immersed boundary method with fictitious cells, iterative solution to systems of linear algebraic equations was obtained using preconditioned biconjugate gradient stabilized method. We considered a number of model problems with curvilinear domain boundary. A comparative study showed that the methods have similar accuracy and convergence rate. It is important to note that the transition from problems with rectangular boundaries to problems with curvilinear boundaries hardly changed the neural network solution algorithm. This fact shows a significant advantage of the neural network method as compared with the finite difference method, which requires considerable modification in order to become applicable to solving problems with curvilinear boundaries.

Neural network modelling algorithms for non-stationary transfer processes could be realized in two ways: *direct* method, when the time component is placed into the input vector, and *hybrid* method, with finite-difference time partition. Using the hybrid method, at the output we obtain a set of neural networks that carry out spatial approximation at the respective layers of time. The adjustment of network parameters at each time step is performed by minimization of the error functional.

3 Neural Network Algorithms for Solving Inverse Problems (Identification Problems)

We consider an inverse problem stated in this way: we need to find a solution $u = u(x)$ to the equation:

$$A(u) = f(x), \ x \in \Omega,$$

which satisfies the following condition on its boundary :

$$B(u) = g(x), \ x \in \Gamma,$$

where A, B known integro-differential operators, f, g are some functions. The operators can be nonlinear or discontinuous, change type on subdomains, etc. There are no specific restrictions for the boundary. We consider two types of inverse problems: identification of the right-hand side *f(x)* or identification of the boundary condition *g(x)*, functions are not specified (or incompletely specified).

A distinguishing characteristic of considered problems is that instead of the missing data we know an additional set of m_s pointwise values of the function in question (measurement data), and measured values contain a certain error ξ (statistical properties of noises are known):

$$\mathrm{v}_j = u(x_j^\nu) + \xi, \quad j = 1, m_s.$$

We want to obtain a solution in the neural network representation. To do so, we use a pair of neural networks. The first network approximates the function in question, and the second one, depending on type of the problem, approximates the right-hand side or the boundary condition. The adjustment of network parameters is performed by optimization of a discrete quadratic error functional:

$$J(\tilde{u}) = \sum_{j=1}^{m_\Omega} w_\Omega \left[A(\tilde{u}(x_j^\Omega, \psi_u)) - f(x_j^\Omega) \right]^2 +$$

$$\sum_{j=1}^{m_\Gamma} w_\Gamma \left[B(\tilde{u}(x_j^\Gamma, \psi_u)) - g(x_j^\Gamma) \right]^2 +$$

$$\sum_{j=1}^{m_s} w_s \left[\tilde{u}(x_j^\nu, \psi_u) - \nu_j \right]^2 \xrightarrow{\psi} \min,$$

where ψ is the parameter vector of neural network model; $\tilde{u}$ is the approximation of the exact solution to the boundary value problem u; $\psi_u \subset \psi$ are parameters of the neural network $\tilde{u}$; m_Ω, m_Γ is a given number of reference points; w_Ω, w_Γ, w_s are weight coefficients, that level out the input of different components of the functional. If we consider a *problem of the right-hand side identification*, the source component of the first summand is substituted by a neural network approximation $\tilde{f}(x_j^\Omega, \psi_f)$, where $\psi_f \subset \psi$ is the parameter vector, specified as a result of learning. Similarly, if we want to *retrieve the boundary condition*, the unknown summand $g(x_j^\Gamma)$ is substituted by $\tilde{g}(x_j^\Gamma, \psi_g)$, $\psi_g \subset \psi$. Optimizing the functional, as the result we obtain approximations of solution to the boundary value problem $\tilde{u}$ itself and identified functions $\tilde{f}$ or $\tilde{g}$ (depending on the problem). To demonstrate the efficiency of suggested algorithms we explored application of the computational technology to a problem of heat source intensity identification in a stationary heat conduction equation [6, 11]:

$$\begin{cases} -u_{xx} - u_{yy} = f(x), \ 0 < x < \pi, \ 0 < y < +\infty, \\ u(0, y) = u(\pi, y) = 0, \ 0 \le y < +\infty, \\ u(x, 0) = 0, \ 0 \le x \le \pi. \end{cases}$$

The function $f(x)$ was unknown at this problem statement, and we needed to retrieve it from additional "measurements". As results of "measurements" we took noisy exact values of the function in points $u(x, 1) = g(x)$ (quasi-real experiment): $g(x) = \sum_{l=1}^{\infty}(1 - e^{-l})e^{-l}l^{-2} \sin lx$, $g_i^\delta = g(hi) + \delta\theta$, $h = \pi/m_s$, $i = 0, m_s$, where θ is a standard normal random variable. Analytical solution to this problem is known: $f(x) = \sum_{l=1}^{\infty} e^{-l} \sin lx$. To evaluate how measurement errors influence the accuracy of resulting approximations we added to measurements noises δ with amplitudes: 10^{-4}, 10^{-3}, 10^{-2}. Figure 4 illustrates the result of the source component retrieval using a 32-neuroelement NRBF network for $\delta = 10^{-2}$. The exact solution is denoted by dots, dashed line shows the solution

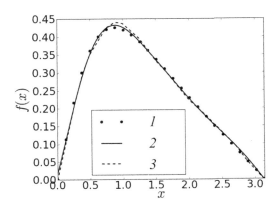

Fig. 4. Result of the right-hand part retrieval. 1 – exact solution, 2 – NRBF-32, 3 – Xie-Zhao

obtained using a modification of Tikhonov's regularization method. It can seen that the neural network method gave us the result with similar accuracy.

4   Conclusion

In this paper the following results are presented:

1. Methods of building neural network models for stationary transfer processes in physical systems containing mixed data were developed. The reasonable ANN-models, based on the usage of normalized radial basis functions, were given.
2. Meshfree computational algorithms for solving elliptic problems of mathematical physics with adaptation of normalized functional basis to solution particularities were elaborated. It was shown that the developed algorithms allow us to efficiently solve problems with boundary-layer nature of the solution, problems on areas with complex geometry, multidimensional problems.
3. A modification of the suggested neural network algorithms that can be applied to inverse problems of mathematical physics was developed. Problems of source component and boundary condition identification were considered. It was shown that regularization properties of neural network algorithms allow us to efficiently solve identification problems with considerable measurement errors.
4. A software package for mathematical modelling of transfer processes in physical systems using normalized radial basis function networks was created [10].

Acknowledgements. The work was supported by the Russian Foundation for Basic Research, project numbers 14-01-00660 and 14-01-00733.

References

1. Vasilyev, A.N., Tarkhov, D.A.: Neural Network Modelling Principles, Algorithms, Applications. SPbSPU Publishing House, Saint-Petersburg (2009). 528 (in Russian)
2. Vasilyev, A.N., Osipov, V.P., Tarkhov, D.A.: Unified modelling process for physicotechnical objects with distributed constants. St. Petersburg State Polytechnical Univ. J. **3**, 39–52 (2010). (in Russian)
3. Bugmann, G.: Normalized radial basis function networks. Neurocomputing (Special Issue on Radial Basis Function Networks) **20**, 97–110 (1998)
4. Haykin, S.: Neural Networks: A Comprehensive Foundation, 2nd edn. Williams Publishing House, Jerusalem (2006). 1104 (in Russian)
5. Kolbin, I.S., Reviznikov, D.L.: Solving mathematical physics problems using normalized radial basis functions neural-like networks. Neurocomputers: Developing, Applications. Radiotekhnika vol. 2, pp. 12–19 (2012). (in Russian)
6. Kolbin, I.S., Reviznikov, D.L.: Solving inverse mathematical physics problems using normalized radial basis functions neural-like networks. Neurocomputers: developing, applications. Radiotekhnika **9**, 3–11 (2013). (in Russian)

7. Panteleyev, A.V., Letova, T.A.: Optimization Techniques Through Examples and Tasks. Vysshaya shkola, Moscow (2005). 544 (in Russian)

8. Hager, W., Zhang, H.: A new conjugate gradient method with guaranteed descent and an efficient line search. SIAM J. Optim. **16**, 170–192 (2005)

9. Chai, Z., Shi, B.: A novel lattice boltzmann model for the poisson equation. Appl. Math. Mech. **32**, 2050–2058 (2008)

10. Kolbin, I.S.: Software package for the solution of mathematical modelling problems using neural networks techniques. Program Eng. **2**, 25–30 (2013). (in Russian)

11. Xie, O., Zhao, Z.: Identifying an unknown source in the Poisson equation by a modified Tikhonov regularization method. Int. J. Math. Comput. Sci. **6**, 86–90 (2012)

Evolutionary Computation

An Experimental Assessment of Hybrid Genetic-Simulated Annealing Algorithm

Cong Jin$^{(\boxtimes)}$ and Jinan Liu

School of Computer, Central China Normal University,
Wuhan 430079, People's Republic of China
jincong@mail.ccnu.edu.cn

Abstract. Genetic algorithm (GA) is a way of solving problems by mimicking the same processes mother nature uses, and has been widely used in many fields. However, it also has some limitations. In this paper, an improved GA is proposed for overcoming these limitations, which is based on the simulated annealing (SA) technology. In binary code, the disadvantageous of selecting crossover gene bit with equal probability is analyzed in depth. Based on these analysis, a crossover operator is proposed, whose crossover probability being adaptive changed with gene bits. The experimental results show that the proposed improved GA algorithm has greater convergence performance than classical GA.

Keywords: Genetic algorithm · Crossover probability · Convergence performance · Simulated annealing · Equal probability

1 Introduction

In the field of artificial intelligence, a genetic algorithm (GA) is a search heuristic that mimics the process of natural selection. This heuristic (also sometimes called a metaheuristic) is routinely used to generate useful solutions to optimization and search problems [1–5]. GA generates solutions using techniques inspired by natural evolution, such as inheritance, mutation, selection, and crossover.

In these basic operations, the crossover operation plays an important role in ensuring the GA optimization process to converge to the global optimum and improve the convergence speed [1]. Common crossover operations include single point crossover [1], multiple point crossovers [4], uniform crossover [3], and so on. These crossover operations can achieve very good results for some certain specific issues, but in general, it is possible that there is the premature convergence and fall local minimum when fitness value very low. In this paper, a new improved GA is proposed based on the most common single point crossover operation and simulated annealing (SA) algorithm [6], called GASA. The proposed improving algorithm can solve the contradiction between local search and global samples.

© Springer International Publishing Switzerland 2016
L. Cheng et al. (Eds.): ISNN 2016, LNCS 9719, pp. 595–602, 2016.
DOI: 10.1007/978-3-319-40663-3_68

2 Improved GA

2.1 The Classical GA

GA [1, 2] represents every possible solution of the problem into an individual for obtaining a group. At the beginning, GA always randomly generates some individuals according to the principle of the fittest individual has more survival chance, GA crosses and combinations these individuals to produce offspring using genetic operations. Because inherited some good traits of the parents, the offspring was significantly better than the previous generation. Thus, the individual groups will gradually toward the evolution of a more optimal solution.

Classical GA is briefly described as follows.

Step 1 Randomly establish an initial population of individuals.
Step 2 Evaluate the fitness F of each individual.
Step 3 Select a mating pair, #1 parent and #2 parent, for reproduction.
Step 4 Create new offspring by performing crossover and mutation operations.
Step 5 Generate a population for the next generation.
Step 6 If the number of generations equals a threshold or termination condition is met, then the current best individuals are presented as a solution; otherwise go back to Step 2.

Although GA has successfully applied in many optimization problems [2–5], there are still some shortcomings. For example, local search ability is poor, and there is a premature convergence and so on, which results in the poor convergence performance, takes a long time to find the optimal solution. These insignificancies hinder the application of GA. How to improve the search capability and convergence performance of GA is a very important, because which is a prerequisite for GA applied to practical solving problem.

To overcome these shortcomings, SA technique is introduced into the GA.

2.2 Improving Method of GA

In this paper, SA will be introduced into the improvement of GA, and the details of SA please reference [7].

In the classical GA, the probabilities of selecting any one gene bit as crossover positions are equal. However, to change these gene bit values with equal probability, will not effectively search the optimal solution in the optimization space. This is because that the change degrees of individual fitness values caused by different gene bit change are different. In other words, the importance of each gene bits is not the same. Therefore, to select crossover position with equal probability will influence the optimization ability of GA. We use an example to illustrate it.

In binary encoding, suppose #1 parent and #2 parent are $X = x_1x_2...x_q$, $Y = y_1y_2...y_q$, where x_i and y_i are i-th bit value of X and Y respectively. If using the single point crossover operation in the i-th bit, we may obtain two offspring, namely $X' = x_1x_2...x_{i-1}y_ix_{i+1}...x_q$, $Y' = y_1y_2...y_{i-1}x_iy_{i+1}...y_q$. For instance, suppose $X = (1001)_2$, $Y = (0010)_2$, when crossover position is in the 3-th bit, then $X' = (1011)_2 = 11$, $Y' = (0000)_2 = 0$. The Hamming distance between parent and offspring is 2 or 9. When crossover position is in the 4-th bit, then $X' = (1000)_2 = 8$, $Y' = (0011)_2 = 3$. The distance between parent and offspring is 1 or 6. So, as fitness function F is continuous, offspring obtained by different crossover position has different influence for the fitness function. Therefore, to select bit with equal probability as crossover position will influence the searching capability of GA. Thus, for the t-th generation population, suppose the crossover probability of the i-th gene bit of each individual is $p_c^{(i)}(t)$, its initial value $= p_c^{(i)}(0)$, the current optimal solution X^* is retained in the algorithm running process. In each generation, for selected parent X_1 and Y_1, two offspring X_2 and Y_2 can be obtained after single-point crossover in the i-th gene bit. Let

$$\bar{F}_i = \frac{F(X_2) + F(Y_2)}{2} \tag{1}$$

where F is fitness function. So, at each generation t, the crossover probability of i-th gene bit may be modified as:

$$p_c^{(i)}(t+1) = \exp\left(\frac{F(X^*(t)) - \bar{F}_i}{F(X^*(t))}\right) \cdot p_c^{(i)}(t), \quad i = 1, 2, \ldots, q \tag{2}$$

where $\bar{F}_i$ is the average fitness of two offspring by performing crossover operation, and q is the bit number of the individual. We notice that, in Eq. (2), the crossover probability $p_c^{(i)}(t)$ is adaptive.

Up now, the improved GA can be described as follows:

Step 1 Randomly establish an initial population of individuals $P(0) = \{X_1(0), X_2(0), \ldots, X_N(0)\}$, where N is the size of solution population. The initial crossover probability of each gene bit is given by the users.

Step 2 According to the definition of the fitness function, calculate the fitness function value $F(X_i(t))$, $1 \leq i \leq N$ for each individual. If the fitness value of the individual is $F_{\max}$, then this individual is unconditionally copied into next generation $P(t+1)$.

Step 3 Select the individuals $X_i(t)$ and $X_j(t)$ for generating next generation population according to selection probability p_s from $P(t)$.

Step 4 According to crossover probability $p_c^{(i)}(t)$, $1 \leq i \leq q$, $X_i(t)$ and $X_j(t)$ are reconstructed. So, the individuals $X_i'(t)$ and $X_j'(t)$ may be obtained. After $X_i'(t)$ and $X_j'(t)$ are reconstructed continually according to mutation probability p_m, the individuals $X_i''(t)$ and $X_j''(t)$ may also be obtained. Their fitness function values $F(X_i''(t))$ and $F(X_j''(t))$ are calculated respectively. If $F(X_s''(t)) \geq \min(X_i(t), X_j(t))$, $s = i, j$, then insert $X_s''(t)$ into $P(t+1)$; otherwise, $X_s''(t)$ is discarded. Modify crossover probability of the i-th gene bit according to Eq. (2).

Step 5 If $P(t+1)$ does not end, then go to *Step* 3; otherwise, go to *Step* 6.

Step 6 Let two average fitness values of the individuals in solutions groups of t-th and $(t+1)$-th generations be $F_{avg}(t)$ and $F_{avg}(t+1)$ respectively. If

$$d = \left\| F_{avg}(t) - F_{avg}(t+1) \right\| < \varepsilon \tag{3}$$

then $X^* = X_{max}(t+1)$, the global optimal solution is obtained, GA end; otherwise, $t = t+1$, go to *Step* 2. Where, ε is a predefined threshold.

3   Experiment Analysis

In this section, through the examples of function optimization, we test the convergence performance of proposed improved GA, and the experimental results are also compared with the results of classical GA and SA. In these experiments, seven typical high dimensional complex functions [8] are selected as follows:

(1) Rosenbrock function

$$f_1(X) = \sum_{i=1}^{n-1}[100(x_i^2 - x_{i+1})^2 + (1-x_i)^2], \quad |x_i| \le 30$$

(2) Ackley function

$$f_2(X) = -20\exp(-0.2\sqrt{\frac{1}{n}\sum_{i=1}^{n}x_i^2}) - \exp(\frac{1}{n}\sum_{i=1}^{n}\cos 2\pi x_i) + 20 + e, \quad |x_i| \le 32$$

(3) Schwefel function

$$f_3(X) = \sum_{i=1}^{n}|x_i| + \prod_{i=1}^{n}|x_i|, \quad |x_i| \le 10$$

(4) Griewank function

$$f_4(X) = \frac{1}{4000}\sum_{i=1}^{n}x_i^2 - \prod_{i=1}^{n}\cos(\frac{x_i}{\sqrt{i}}) + 1, \quad |x_i| \le 600$$

(5) Generalized Penalized function

$$f_5(X) = \frac{\pi}{n}\{10\sin^2(\pi y_i) + \sum_{i=1}^{n-1}(y_i - 1)^2[1 + 10\sin^2(3\pi y_{i+1})] + (y_n - 1)^2\}$$

$$+ \sum_{i=1}^{n}u(x_i, 10, 100, 4),$$

where,

$$y_i = 1 + (x_i + 1)/4, \quad |x_i| \le 600,$$

$$u(x_i, a, k, m) = \begin{cases} k(x_i - a)^m, & x_i > a \\ 0, & -a \le x_i \le a \\ k(-x_i - a)^m, & x_i < -a \end{cases}$$

(6) Generalized Schwefel function

$$f_6(X) = \sum_{i=1}^{n} -x_i \sin(\sqrt{|x_i|}), \quad |x_i| \leq 500$$

(7) Generalized Rastrgin function

$$f_7(X) = \sum_{i=1}^{n} (x_i^2 - 10\cos(2\pi x_i) + 10), \quad |x_i| \leq 5.12$$

In these seven functions, the optimal solution of functions $f_1 \sim f_5$ and f_7 is 0, and the optimal solution of function f_6 is -12569.5. The fitness function of GASA is these functions itself. The parameters of the GASA used for these experiments are shown in Table 1.

For achieving optimal solutions of the seven functions, GA, SA and GASA algorithms are implemented respectively. The average results of 50 independent runs are summarized in Table 2.

Table 1. GASA parameters.

Population size:	50
Maximum generation:	500
Bit number q:	48
$p_c^{(i)}(0)$:	0.04
Selection operation:	Roulette rule
p_m	0.001
ε	0.001

In these experiments, if the absolute error between the final solution and actual optimal solution is greater than 0.5, then we consider algorithm to fall into local optimization. From these experimental results, we can get the following conclusions:

(1) For high dimensional complex functions, the performance of SA and GA is unsatisfactory, and it is easy to fall into local optimization. The number of SA and GA falling into local optimization is much higher than the GASA's.

(2) GASA combines the advantages of SA and GA, which shows desirable performance for high dimensional complex functions. The variances of SA and GA are far greater than GASAs, which shows that GASA is a reliable function optimization algorithm.

In addition to these experiments, we also compare our algorithm with other similar algorithms. The average results of 50 independent runs of GASA are summarized in Table 3.

Table 2. All results have been averaged over 50 runs.

Function	Algorithm	Average optimal solution	Best optimization solution	Worst optimization solution	Variance of optimization solutions	# of falling into local optimization
$f_1(X)$	SA	1.0150721	0.0591347	2.0119472	1.151359	9
	GA	0.8933014	0.1074014	1.7204820	0.818837	10
	GASA	0.0040617	0.0031462	0.0061473	0.000202	0
$f_2(X)$	SA	2.0862463	0.0003195	6.1533578	3.165015	15
	GA	2.1495302	0.1021689	4.0895854	1.358674	12
	GASA	0.0000160	0.0000103	0.0000201	0.000003	0
$f_3(X)$	SA	1.5662748	1.1014635	2.0488136	0.196267	10
	GA	1.6063272	0.9957341	2.1408240	0.926216	18
	GASA	0.0000156	0.0000120	0.0000211	0.000015	0
$f_4(X)$	SA	0.0009235	0.0000011	0.0040413	0.001210	25
	GA	0.0413124	0.0000213	0.0705225	0.020725	21
	GASA	0.0000000	0.0000000	0.0000000	0.000000	0
$f_5(X)$	SA	6.5185036	0.0000002	11.049263	6.613022	31
	GA	4.7830314	0.0000041	8.6458351	6.159527	22
	GASA	0.0000000	0.0000000	0.0000000	0.000000	0
$f_6(X)$	SA	−12075.31	−12495.60	−11472.76	215.71	42
	GA	−12328.28	−12506.87	−12046.97	113.65	46
	GASA	−12569.50	−12569.50	−12569.50	0.000000	0
$f_7(X)$	SA	3.5156967	0.0529483	7.9484165	4.795031	31
	GA	2.6140528	0.0648135	5.1058144	1.739538	35
	GASA	0.0000000	0.0000000	0.0000000	0.000000	0

Table 3. Comparison between other algorithms.

Function	Algorithm	Mean best	Standard deviation
$f_1(X)$	FEP [8]	5.06	5.87
	ABC-PS [9]	8.7094	5.3352
	ARSAGA [10]	/	/
	GASA	0.0031462	0.014213
$f_2(X)$	FEP [8]	0.018	0.012
$f_2(X)$	ABC-PS [9]	2.5021e-12	1.25782-12
	ARSAGA [10]	0.961244	4.18928
	GASA	0.0000103	0.001732
$f_3(X)$	FEP [8]	0.0081	0.00077
	ABC-PS [9]	/	/
	ARSAGA [10]	0.0	0.0
	GASA	0.0000120	0.000015
$f_4(X)$	FEP [8]	0.016	0.022
	ABC-PS [9]	1.0229e-08	3.6532e-06
	ARSAGA [10]	0.0	0.0
	GASA	0.0	0.0
$f_5(X)$	FEP [8]	0.0000096	0.0000036
	ABC-PS [9]	/	/
	ARSAGA [10]	/	/
	GASA	0.0	0.0
$f_6(X)$	FEP [8]	−12554.5	52.6
	ABC-PS [9]	/	/
	ARSAGA [10]	−12569.47	0.0
	GASA	−12569.50	0.0
$f_7(X)$	FEP [8]	0.046	0.012
	ABC-PS [9]	7.7497e-05	5.5547e-04
	ARSAGA [10]	0.0	0.0
	GASA	0.0	0.0

From these experimental results, we may find that the proposed GASA can achieve better performance than the other compared optimization algorithms about mean best and standard deviation two aspects, which shows that the proposed improving crossover operation is effective to improve the performance of GA.

4   Conclusions

In this paper, we investigated the improvement of GA. The proposed GA algorithm can improve convergence performances to avoid falling into local optimization. The main advantages of proposed GASA are as follows: (1) To improve the performance of GA, SA is introduced into GA, which can combine the advantages of SA and GA. Experiment results confirm that the GASA is very effective for complex function optimization. (2) In the process of complex function optimization, the proposed GASA is almost no falling into local optimization. In other words, it can obtain ideal solution, which shows that the proposed GASA has good convergence performance.

The direction of future research is to solve multi-objective optimization problem.

Acknowledgments. This work was supported by Natural Social Science Foundation of China (Grant No.13BTQ050).

References

1. Melanie, M.: An Introduction to Genetic Algorithms. MIT Press, Cambridge (1996)
2. Jin, C., Wang, S.H.: Robust watermark algorithm using genetic algorithm. J. Inf. Sci. Eng. **23**(2), 661–670 (2007)
3. Šetinc, M., Gradišar, M., Tomat, L.: Optimization of a highway project planning using a modified genetic algorithm. Optimization **64**(3), 687–707 (2015)
4. Moradi, M.H., Abedini, M.: A combination of genetic algorithm and particle swarm optimization for optimal DG location and sizing in distribution systems. Int. J. Electr. Power Energy Syst. **34**(1), 66–74 (2012)
5. Elhaddad, Y., Sallabi, O.: A new hybrid genetic and simulated annealing algorithm to solve the traveling salesman problem. World Congr. Eng. **1**, 11–14 (2010)
6. Lombardi, A.M.: Estimation of the parameters of ETAS models by simulated annealing. Scientific reports, 5, Article number: 8417 (2015). doi:10.1038/srep08417
7. e Oliveira Jr., H.A.: Global optimization and its applications. In: Aguiar e Oliveira Junior, H., Ingber, L., Petraglia, A., Rembold Petraglia, M., Augusta Soares Machado, M. (eds.) Stochastic Global Optimization and Its Applications with Fuzzy Adaptive Simulated Annealing. ISRL, vol. 35, pp. 11–20. Springer, Heidelberg (2012)
8. Yao, X., Liu, Y., Lin, G.: Evolutionary programming made faster. IEEE Trans. Evol. Comput. **3**(2), 82–102 (1999)
9. Wang, C.F., Liu, K., Shen, P.P.: Hybrid artificial bee colony algorithm and particle swarm search for global optimization. Mathematical Problems in Engineering, Article ID 832949, 8 p. (2014)
10. Hwang, S.F., He, R.S.: A hybrid real-parameter genetic algorithm for function optimization. Adv. Eng. Inform. **20**(1), 7–21 (2006)

When Neural Network Computation Meets Evolutionary Computation: A Survey

Zonggan Chen, Zhihui Zhan[✉], Wen Shi, Weineng Chen, and Jun Zhang

School of Computer Science and Engineering, South China University of Technology,
Guangzhou, 510006, China
zhanapollo@163.com, junzhanghk@qq.com

Abstract. Neural network (NN) and evolutionary computation (EC) are two of the most popular and important techniques in computational intelligence, which can be combined together to solve the complex real world problems. This paper represents a review of the researches that combined NN and EC. There are 3 main research focuses as follows. In the first research focus, EC algorithms have been widely used to optimize the structure and parameter of the NN, including weight, structure, learning rates, and others. In the second research focus, lots literatures have witnessed that EC based NNs are widespread in the applications such as classification, automatic control, prediction, and many other fields. These two kinds of researches into combining NN and EC are mainly focuses on using EC algorithms to optimize NN, to enhance the NN performance and the NN application ability. Our survey finds that particle swarm optimization is the most popular EC algorithm that the researchers choice to optimize NN in recent year, while genetic algorithm and differential evolution are also widely used. In the third research focus, there are also researches adopted NN as a tool to enhance the performance of EC algorithms. Although the literatures in this focus are not as many as the above two focuses, the existing results show that NN has great potential in enhancing EC algorithms. The survey shows that when NN and EC meet, combining them would result in an effective way to deal with the real world application. This interesting research topic has become more and more significant in the field of computer science and still has much room for development.

Keywords: Neural network · Evolutionary computation · Network optimization

1 Introduction

Neural network (NN), also known as artificial neural network (ANN), is inspired by biological neurons and nervous system [1]. From the perspective of information processing, NN is constructed based on the human's nervous system. As an important branch of computational intelligence, NN has become a very popular cross-disciplinary, which is widely used in biology, electronics, computers, mathematics and physics and other disciplines.

With the rapid development of NN, more and more scientists participated in the research of NN. Since 21st century, NN was widely used in classification, automatic control, artificial intelligence, and many other fields with its great robustness and highly

© Springer International Publishing Switzerland 2016
L. Cheng et al. (Eds.): ISNN 2016, LNCS 9719, pp. 603–612, 2016.
DOI: 10.1007/978-3-319-40663-3_69

parallel distribution. But the optimization of NN is a hard problem and there is no explicit formulation so that it cannot be solved by some traditional method such as dynamic programing. Moreover, in the dawn of "era of big data", the problem that NN met became more and more complex and the scale of NN is more and more complex so that using enumeration algorithm to solve this problem is unacceptable due to the poor efficiency. Therefore, the training of the NN, structure optimization, and the parameter optimization are all great challenges.

With the development of evolutionary computation (EC) [2], using EC algorithms to optimize NN has been popular. EC algorithms aim to solve the optimization problems by means of human intelligence or the art of nature. EC algorithms show great ability in optimizing the NP-hard problems and high time complexity problems. Particle swarm optimization (PSO) [3], genetic algorithm (GA) [4, 5], ant colony optimization (ACO) [6, 7], and differential evolution (DE) [8] are the typical and widely-used EC algorithms.

In this paper, we give a categorization and summarization of the recent research that combine NN and EC. The detailed categorization is shown in Fig. 1. Most of the researches focus on using EC to optimize NN. In the figure, we can see that PSO, GA, and DE are the most widely used algorithms while these algorithms always used to optimize the structure and parameter of NN. Moreover, the applications of these EA-NNs include automatic control, classification, prediction, etc. There are also researches applying NN to enhance the fitness evaluation and search capacity of EC, but not as popular as using EC to optimize NN.

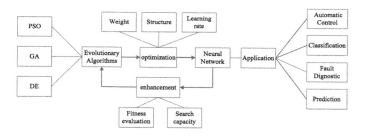

Fig. 1. The categorization of the recent researches.

The rest of this paper is organized as follows. Section 2 presents the optimization of the parameter and structure of NN. Section 3 presents the recent researches about the application of NN. Section 4 presents the EC based NN while Sect. 5 presents some methods that adopted NN to enhance the EC.

2 Optimization of NN

An objective function describing the NN training problem is going to be multimodal. Therefore, algorithms based on gradient methods can easily get stuck in local minima. To avoid this problem, EC algorithms are extensively employed as a global optimization technique. Among all of the parameters of the NN, weight [9–20], structure [11, 14, 19, 21–26, 29, 30], and learning rates [27, 28] are most frequently optimized by EC algorithms.

2.1 Weight

Weight is one of the most vital parameters of NN which has a huge influence of the performance. To use EC algorithms to optimize the weight of NN, each individual x_i consists of k genes (where k represents the number of weights in the trained NN). In Fig. 2(a), a part of NN with neurons n and $n + 1$ is shown. Additionally, in Fig. 2(b), the coding scheme for weights in an individual x_i connected to neurons from Fig. 2(a) is shown.

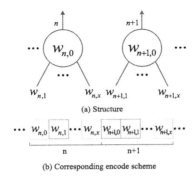

(a) Structure

(b) Corresponding encode scheme

Fig. 2. Part of (a) ANN, corresponding to its (b) chromosome containing the weight values, weights $w_{i,0}$ represent bias weights [9].

2.2 Structure

Structure (or architecture) is one of the core parameters of the NN, which determines the main characteristic of a NN. To improve the performance of the NN by optimizing its structure, some of the papers encode all of the information about NN topology in each individual in EC algorithms [11, 19, 21–26], as the way we do in Sect. 2.1, while others just encode some of the parameters which determines the structure of the neural network such as number of hidden nodes [29], corresponding centers [30], radii [14, 30].

2.3 Learning Rates

In most cases, the learning rates of the online-training weights are usually selected by trial-and-error method, which is time-consuming. Therefore, adopt EC algorithms to adjust the learning rates to further improve the online learning capability of the NNs is a distinguished attempt.

2.4 Others

In addition to the parameters mentioned above, EC algorithms are available to adapt other parameters of NNs such as, timescale [16, 18], parameter of transfer function [19], smoothing factor [31]. Moreover, EC algorithms can also optimize a method that is applied to improve an NN. For example, the paper [32] proposes an orthogonal projection pursuit method to enhance the performance of NN and PSO is used to optimal this method.

However, most papers use EC to optimize more than two parameters of NNs, commonly optimizing the structure and weights simultaneously [11, 19].

3   Applications of NN

This section gives a detailed discussion of the application fields in which EC-NNs have been used and puts a special emphasis on EC-NNs-FUZZ systems, that is, the combination of EC, NN, and fuzzy (FUZZ) approaches.

3.1 Prediction Problems

Prediction techniques, including qualitative forecasts such as Delphi method and quantitative forecasting which includes time series forecasting methods and cause and effect (fishbone diagram), have been a hot spot in the recent years. Publish works have introduced NNs methodologies as efficient cause and effect techniques for prediction problems. Therefore, NN sare extensively applied in a variety of prediction problems such as Micro grids load prediction [29], optimal thermal performance prediction [24], time series prediction [13], and wind power prediction [26].

3.2 Control System Designing

Many techniques that have been used to design control systems, the combination of EC-NNs-FUZZs, which bring together the low-level learning and computational power of neural networks with the high-level human-like thinking and reasoning of fuzzy systems, and apply powerful EC optimization and adaptation, have proven to be one of the most effective approaches. Thus they are applied in the nonlinear system control [21], electric power steering system control [27], predictive control [12], linear synchronous motor drive [28], and evolving gaits of a hexapod robot [18].

3.3 Classification Problem

The classification problem, trained by a set of data containing observations whose categories are already known, determines which set of categories a new observation belongs. It is one of the core and challenging tasks in machine learning and statistics, EC- NNs and EC-NNs-FUZZ systems are diffusely employed to pattern classification [30], classification of semiconductor defects [10], brain-computer interface classification [17], hypoglycemia detection system [19] and power transformer differential protection [31]. Moreover, some NNs are used to optimized the classification problem, for instance, [11] proposes a new multi classification algorithm using multilayer perceptron neural network models, and [23] proposes a novel approach to improve the classification performance.

3.4 Fault Diagnosis Problem

Fault diagnosis, which is used to determine the causes of symptoms, mitigations and solutions in systems engineering and computer science, is another field where EC-NNs are widely used not only in theory improvement [22] but also in real-world problem, such as power transformers fault diagnostic [14].

3.5 Others

Apart from the fields mentioned above, EC-NNs are used in other fields as well, where they can be applied to forecast energy consumption of cloud computing [15], reconstruct the topology of gene regulatory network [16], design custom-made fractal antennas [33], optimize large scale problem [20] as well as employed in Spatio–Temporal system identification [32], and an online modeling algorithm [25].

Note that, used by itself separately from NNs, EC algorithms is famous for its great ability in a variety of optimization problems such as resource scheduling [35, 36], power system [37], and so on.

Anyway, there are also some papers which just use some kinds of EC algorithms to improve the performance of NNs or propose a new NN. Recent EC-NN algorithms worth mentioning include [9], which uses differential evolution algorithm with multiple trial vectors to NN training and [34], which introduces a new granular NN.

4 Popular EC Algorithms for NN Optimization

Based on our survey, we find that PSO, GA, and DE are three most widely used EC algorithms for optimizing NNs. Among them, PSO is the most popular algorithm due to its simple programming implementation and great search capability.

4.1 Optimize NN by PSO

PSO is proposed by Eberhart and Kennedy in 1995, and has fast developed in recent years [38–40]. PSO is inspired by the foraging of birds. Particles in PSO fly in the search space based on the personal experience and learning from other particles in the population.

Many researches use the original PSO to optimize the structure and parameters. The paper [34] proposed a novel granular neural network and using PSO to adjust the allocation of granularity. Moreover, PSO is applied to optimize some novel structures and method in [31, 32].

608   Z. Chen et al.

However, when using original PSO to optimize NN, it may easily converge to local optima due to lack of diversity. Therefore, some improved PSO algorithms is applied in NNs. Dynamic weight is the popular strategy to balance between the local and global search ability. Therefore, the papersb [28] all use this strategy to optimize the NN. For the sake of avoiding local optima, PSO with wavelet mutation (HPSOWM) is used in [19], while the paper [20, 26] proposed enhanced PSO(EPSO) that combines a novel strategy for the selection of *pbest* and *gbest*, which can improve the search capacity.

With the increase of the complexity of the problem, some more improved strategy is proposed. The paper [12] employed chaotic map and Gaussian function to balance the exploration and exploitation. The papers [14, 25] applied QPSO (quantum PSO) to optimize parameters and structure respectively. In order to deal with those complex problems such as brain–computer interface classifier for wheelchair commands, the paper [17] proposed PSO with fuzzy inertia weight ω to enhance searching quality. Moreover, the paper [18] proposed species-based PSO (SSPSO), which uses multi swarm method and a swarm optimize a single neural node.

Hybrid algorithm is another way to enhance the performance of PSO. The paper [16] proposed a novel hybrid ant colony optimization (ACO) and PSO method to reconstruct the topology of gene regulatory networks. The paper [13] proposed CCPSO which combine culture algorithm for accelerating the search and enhancing global search capacity.

4.2 Optimize NN by GA

GA, proposed by Holland, is a search heuristic that mimics the process of natural selection. It can be applied to optimization problems according to selection, crossover, and mutation.

The crossover and mutation of GA can preserve the good combination of different index values. Therefore, GA has a great global search capacity. Many researches use GA to optimize NNs, [22, 23] are some of the typical applications. There are some researchers used real-coded GA (RCGA), such as [23].

4.3 Optimize NN by dE

DE is first proposed by Storn and Price in 1997. Like PSO and GA, DE is inspired in Nature, focusing on a specific, simple algorithm for assigning traits for the next generation based on differences in performance and genes among the population. Moreover, DE is much more simple and straightforward to implement and Main body of the algorithm takes four to five lines to code in any programming language [41], which make it convenient for researchers to use it.

DE is also widespread in optimizing NN. Trial vector is a popular optimization object. In [9], a novel DE with multiple trail vectors is proposed for the training of NNs. To adjust the parameter of DE is another important issue. The paper [21] proposed modified DE which not only adjust scalar number but also provides a cluster-based mutation scheme. The paper [29] used an enhanced version of DE that modified the recombination operator according to the optimization problems. The paper [27] employed improved DE with dynamic scalar number and crossover rate.

5   EC Aided by NN

Although most EC-NN research focuses on using EC to training NNs and to optimize their structure and parameters, a few research projects have addressed the reverse situation whereby neural networks are used to optimize ECs.

A novel multi frequency vibrational PSO (MVPSO) is proposed in [42], which adopted radial basis function (RBF) NNs to enhance controlled diversity. The paper [33] applied NNs as a part of fitness evaluation to enhance the performance of PSO in the field of custom-made fractal multi-band antennas. In order to optimize passive components in high frequency RF ICs, [43] proposed the machine learning-based DE (MMLDE) while NN was used to construct surrogate models online to evaluate the performance of the algorithm.

6   Conclusion and Future Research

In this paper, we have looked over the recent research that combines EC algorithms and NNs. With the growing complexity of real world problems, ECs show great ability in optimizing the structure and parameter such as weight, learning rate and so on. On the other hand, PSO, GA, and DE are 3 popular EC algorithms that extensively used in optimizing NN structure and parameters, such as weight, learning rate and so on. PSO, GA, and DE are the 3 most popular EC algorithms used for EC-NNs. Among them, PSO is the hottest one.

Many researches have used the original version of EC algorithms, or they have adopted some simple enhancements of the standard versions. In the future, researchers should pay more attention to develop specific strategy or hybrid algorithm targeted for different NN or real world problems. Meanwhile, although very few researches employed NN to optimize the EC algorithms but the existing researches obtain great performance, which showed that NNs have great potential in the enhancement of EC algorithms.

Deep learning neural network is a popular research field in recent years, hybrid deep learning neural network and evolutionary computation should raise more attention. The paper [44] applied genetic algorithm to optimize the structure of deep learning neural network. The paper [45] proposed a multi objective sparse feature learning model for deep neural networks and used a self-adaptive multi objective differential evolution (DE) based on decomposition (Sa-MODE/D) to solve the model. With the development of deep learning neural network, hybrid deep learning neural network and evolutionary computation still has much room for development.

References

1. McCulloch, W.S., Pitts, W.: A logical calculus of the ideas immanent in nervous activity. Bull. Math. Biophys. **5**(4), 115–133 (1943)
2. Fogel, D.B.: Evolutionary Computation: The Fossil Record. Wiley-IEEE Press, Hoboken (1998)
3. Kennedy, J., Eberhart, R.: Particle swarm optimization. In: Proceedings of the 6th IEEE International Conference on Neural Networks, pp. 1942–1948 (1995)
4. Holland, J.H.: Genetic algorithms. Sci. Am. **267**, 44–50 (1992)
5. Holland, J.H.: Adaptation in Natural and Artificial Systems: An Introductory Analysis with Applications to Biology, Control, and Artificial Intelligence. University of Michigan Press, Ann Arbor (1975)
6. Dorigo, M., Mauro, B., Thomas, S.: Ant colony optimization. IEEE Comput. Intell. Mag. **1**(4), 28–39 (2006)
7. Dorigo, M., Vittorio, M., Alberto, C.: Ant system: optimization by a colony of cooperating agents. IEEE Trans. Syst. Man Cybern. B Cybern. **26**(1), 29–41 (1996)
8. Storn, R., Price, K.: Differential evolution-a simple and efficient heuristic for global optimization over continuous spaces. J. Global Optim. **11**(4), 341–359 (1997)
9. Slowik, A.: Application of an adaptive differential evolution algorithm with multiple trial vectors to artificial neural network training. IEEE Trans. Industr. Electron. **58**(8), 3160–3167 (2011)
10. Tan, S.H., Watada, J., Ibrahim, Z., Khalid, M.: Evolutionary fuzzy ARTMAP neural networks for classification of semiconductor defects. IEEE Trans. Neural Netw. Learn. Syst. **26**(5), 933–950 (2015)
11. Caballero, J.C.F., Martínez, F.J., Hervás, C., Gutiérrez, A.: Sensitivity versus accuracy in multiclass problems using memetic pareto evolutionary neural networks. IEEE Trans. Neural Netw. Learn. Syst. **21**(5), 750–770 (2010)
12. Han, M., Fan, J., Wang, J.: A dynamic feedforward neural network based on gaussian particle swarm optimization and its application for predictive control. IEEE Trans. Neural Netw. Learn. Syst. **22**(9), 1457–1468 (2011)
13. Lin, C.J., Chen, C.H., Lin, C.T.: A hybrid of cooperative particle swarm optimization and cultural algorithm for neural fuzzy networks and its prediction applications. IEEE Trans. Syst. Man Cybern. Part C Appl. Rev. **39**(1), 55–68 (2009)
14. Meng, K., Dong, Z.Y., Wang, D.H., Wong, K.P.: A self-adaptive rbf neural network classifier for transformer fault analysis. IEEE Trans. Power Syst. **25**(3), 1350–1360 (2010)
15. Foo, Y.W., Goh, C., Lim, H.C., Zhan, Z.H., Li, Y.: Evolutionary neural network based energy consumption forecast for cloud computing. In: Proceedings of International Conference on Cloud Computing Research and Innovation, pp. 53–64 (2015)
16. Kentzoglanakis, K., Poole, M.: A swarm intelligence framework for reconstructing gene networks: searching for biologically plausible architectures. IEEE/ACM Trans. Comput. Biol. Bioinf. **9**(2), 358–371 (2012)
17. Chai, R., Ling, S.H., Hunter, G.P., Tran, Y., Nguyen, H.T.: Brain-computer interface classifier for wheelchair commands using neural network with fuzzy particle swarm optimization. IEEE J. Biomed. Health Inf. **18**(5), 1614–1624 (2014)
18. Juang, C.F., Chang, Y.C., Hsiao, C.M.: Evolving gaits of a hexapod robot by recurrent neural networks with symbiotic species-based particle swarm optimization. IEEE Trans. Industr. Electron. **58**(7), 3110–3119 (2011)

19. San, P.P., Ling, S.H., Nguyen, H.T.: Industrial application of evolvable block-based neural network to hypoglycemia monitoring system. IEEE Trans. Industr. Electron. **60**(12), 5892–5901 (2013)
20. Miranda, V., da Hora Martins, J., Palma, V.: Optimizing large scale problems with metaheuristics in a reduced space mapped by autoencoders—application to the wind-hydro coordination. IEEE Trans. Power Syst. **29**(6), 3078–3085 (2014)
21. Chen, C.H., Lin, C.J., Lin, C.T.: Nonlinear system control using adaptive neural fuzzy networks based on a modified differential evolution. IEEE Trans. Syst. Man Cybern. Part C Appl. Rev. **39**(4), 459–473 (2009)
22. Tan, Y., He, Y., Cui, C., Qiu, G.: A novel method for analog fault diagnosis based on neural networks and genetic algorithms. IEEE Trans. Instrum. Meas. **57**(11), 2631–2639 (2008)
23. Lin, C.T., Prasad, M., Saxena, A.: An improved polynomial neural network classifier using real-coded genetic algorithm. IEEE Trans. Syst. Man Cybern. Syst. **45**(11), 1389–1401 (2015)
24. Horng, J.T., Chang, S.F., Wu, T.Y., Chen, P.L., Hung, Y.H.: Thermal optimal design for plain plate-fin heat sinks by using neuro-genetic method. IEEE Trans. Compon. Packag. Technol. **31**(2), 449–460 (2008)
25. Chen, H., Gong, Y., Hong, X.: Online modeling with tunable RBF network. IEEE Trans. Cybern. **43**(3), 935–947 (2013)
26. Amjady, N., Keynia, F., Zareipour, H.: Wind power prediction by a new forecast engine composed of modified hybrid neural network and enhanced particle swarm optimization. IEEE Trans. Sustain. Energ. **2**(3), 265–276 (2011)
27. Hung, Y.C., Lin, F.J., Hwang, J.C., Chang, J.K., Ruan, K.C.: Wavelet fuzzy neural network with asymmetric membership function controller for electric power steering system via improved differential evolution. IEEE Trans. Power Electron. **30**(4), 2350–2362 (2015)
28. Lin, F.J., Chen, S.Y., Teng, L.T., Chu, H.: Recurrent functional-link-based fuzzy neural network controller with improved particle swarm optimization for a linear synchronous motor drive. IEEE Trans. Magn. **45**(8), 3151–3165 (2009)
29. Amjady, N., Keynia, F., Zareipour, H.: Short-term load forecast of mic rogrids by a new bilevel prediction strategy. IEEE Trans. Smart Grid **1**(3), 286–294 (2010)
30. Gutiérrez, P.A., Hervás-Martínez, C., Martínez-Estudillo, F.J.: Logistic regression by means of evolutionary radial basis function neural networks. IEEE Trans. Neural Netw. **22**(2), 246–263 (2011)
31. Tripathy, M., Maheshwari, R.P., Verma, H.K.: Power transformer differential protection based on optimal probabilistic neural network. IEEE Trans. Power Delivery **25**(1), 102–112 (2010)
32. Wei, H.L., Billings, S., Zhao, Y., Guo, L.: Lattice dynamical wavelet neural networks implemented using particle swarm optimization for spatio-temporal system identification. IEEE Trans. Neural Netw. **20**(1), 181–185 (2009)
33. Patnaik, A., Sinha, S.N.: Design of custom-made fractal multi-band antennas using ANN-PSO. IEEE Antennas Propag. Mag. **53**(4), 94–101 (2011)
34. Song, M., Pedrycz, W.: Granular neural networks: concepts and development schemes. IEEE Trans. Neural Netw. Learn. Syst. **24**(4), 542–553 (2013)
35. Chen, Z.G., Du, K.J., Zhan, Z.H, Zhang, J.: Deadline constrained cloud computing resources scheduling for cost optimization based on dynamic objective genetic algorithm. In: 2015 IEEE Congress on Evolutionary Computation, pp. 708–714 (2015)
36. Chen, Z.G., Zhan, Z.H., Li, H.H., Du, K.J., Zhong, J.H., Foo, W.Y., Li, Y., Zhang, J.: Deadline constrained cloud computing resources scheduling through an ant colony system approach. In: 2015 International Conference on Cloud Computing Research and Innovation, pp. 112–119 (2015)

37. AlRashidi, M.R., El-Hawary, M.E.: A survey of particle swarm optimization applications in electric power systems. IEEE Trans. Evol. Comput. **13**(4), 913–918 (2009)
38. Shen, M., Zhan, Z.H., Chen, W.N., Gong, Y.J., Zhang, J., Li, Y.: Bi-velocity discrete particle swarm optimization and its application to multicast routing problem in communication networks. IEEE Trans. Industr. Electron. **61**(12), 7141–7151 (2014)
39. Zhan, Z.H., Li, J., Cao, J., Zhang, J., Chung, H., Shi, Y.H.: Multiple populations for multiple objectives: a coevolutionary technique for solving multiobjective optimization problems. IEEE Trans. Cybern. **43**(2), 445–463 (2013)
40. Li, Y.H., Zhan, Z.H., Lin, S., Zhang, J., Luo, X.N.: Competitive and cooperative particle swarm optimization with information sharing mechanism for global optimization problems. Inf. Sci. **293**(1), 370–382 (2015)
41. Das, S., Suganthan, P.N.: Differential evolution: a survey of the state-of-the-art. IEEE Trans. Evol. Comput. **15**(1), 4–31 (2011)
42. Pehlivanoglu, Y.V.: A new particle swarm optimization method enhanced with a periodic mutation strategy and neural networks. IEEE Trans. Evol. Comput. **17**(3), 436–452 (2013)
43. Liu, B., Zhao, D., Reynaert, P., Gielen, G.G.: Synthesis of integrated passive components for high-frequency RF ICs based on evolutionary computation and machine learning techniques. IEEE Trans. Comput. Aided Des. Integr. Circuits Syst. **30**(10), 1458–1468 (2011)
44. Shinozaki, T., Watanabe, S.: Structure discovery of deep neural network based on evolutionary algorithms. In: 2015 IEEE International Conference on Acoustics, Speech and Signal Processing, pp. 4979–4983 (2015)
45. Gong, M., Liu, J., Li, H., Cai, Q., Su, L.: A multiobjective sparse feature learning model for deep neural networks. IEEE Trans. Neural Netw. Learn. Syst. **26**(12), 3263–3277 (2015)

New Adaptive Feature Vector Construction Procedure for Speaker Emotion Recognition Based on Wavelet Transform and Genetic Algorithm

Alexander M. Soroka[✉], Pavel E. Kovalets, and Igor E. Kheidorov

Department of Radiophysics and Digital Media Technologies,
Belarusian State University, Minsk, Belarus
soroka.a.m@gmail.com

Abstract. In this paper new original method of adaptive feature vector construction based on wavelet transform and genetic algorithm is proposed. Wavelet-based original feature vector is designed using genetic algorithm and support vector machine as classifier in order to provide better speaker emotion recognition. It was shown that the usage of the proposed adaptive feature vector lets to improve emotion recognition accuracy.

Keywords: Emotion recognition · Feature extraction · Wavelet transformation · Support vector machine · Genetic algorithm

1 Introduction

Feature vector choice is one of the most important stages for audio signal processing, classification and indexing. There are different feature estimation algorithms used for emotion recognition based on mel-frequence coefficients (MFCC), linear prediction coefficients(LPC), energy-based, wavelet-transforms, etc. [7,11] which are widely used for audio signal description and provide good accuracy. Further recognition performance improvement is possible by adaptation of signal description parameters for the task to be solved [12,17]. There is a series of methods for feature parameters optimization but one of the best way to optimize audio feature description is the usage of genetic algorithm [5,10,17] Most of approaches for audio description improvement are to use a combination of different audio characters that leads to significant growth of feature vector dimension. It becomes too sparse as the result and that impairs the accuracy of the ensuring statistical analysis and recognition [3]. To solve the problem mentioned above it is a good idea to decrease the dimension of feature space based on the assumption that a number of parameters have low information significance [3,6,8,13,15]

Such approach leads to the following possible issues:

1. feature vector reduction procedure possibly causes lost of some significant information for following analysis;

© Springer International Publishing Switzerland 2016
L. Cheng et al. (Eds.): ISNN 2016, LNCS 9719, pp. 613–619, 2016.
DOI: 10.1007/978-3-319-40663-3_70

2. considerable cluster overlapping of feature vectors for the same audio signals requires usage of complex nonlinear classifiers.

In order to overcome these drawbacks for speaker emotion recognition in the paper it was proposed to apply new approach to feature vector construction. The developed algorithm assumes usage of wavelet-like transform with adapted basic function designed using genetic algorithm. Within the framework of this approach basic wavelet function form is optimized directly based on training speech data unlike the optimization of separate parameters. The proposed method let to achieve the feature vector which is optimal from the speaker emotion classification viewpoint, in such case it is possible to improve the recognition accuracy and simplify the classifier.

2   Adaptive Wavelet Function Construction Method Based on Genetic Algorithm and Support Vector Machines

Wavelet transform is an integral transform [11], defined by the function of two variables:

$$W\left(\alpha, \tau\right) = \int_{-\infty}^{+\infty} \frac{s\left(t\right)}{\sqrt{\alpha}} w_{p}\left(\frac{t - \tau}{\alpha}\right) dt \qquad (1)$$

where $s\left(t\right)$ - signal to be analyzed, $w_{p}\left(t\right)$ - base wavelet function specified by the set of parameters p, α - scale, τ - time shift.

This function has to satisfy the following condition [11]:

$$\int_{-\infty}^{+\infty} w_{p}\left(t\right) dt = 0 \qquad (2)$$

Let assume that it is possible to find the set of parameters p which allows to localize significant coefficients for the best separation of the given classes in the created feature space. The best optimization criteria for base wavelet function design is to use classifier in order to achieve parameters provided best emotion recognition accuracy. Linear support vector machine is a good choice to estimate vector distribution in the feature space [16]. This method allows to find optimal separating hyperplane $\langle b, x \rangle - b_0 = 0$ in case of binary classification by maximizing the distance between the separating hyperplane and boundaries of the classes, where x - point on the hyperplane, b - the normal vector to the hyperplane, b_0 - the offset of the hyperplane from the origin along the normal vector b. Decision rule for such classifier is described by the following expression:

$$a\left(x\right) = sign\left(\langle b, x \rangle - b_0\right) \qquad (3)$$

Let $X = \{x_i | i = 1...N\}$ - speech feature vectors built using wavelet transform 1 and belong to two different classes. The exact classification of these vectors $c\left(x\right) = \{-1, 1\}$ is known. Decision rule (3) is built using k-fold cross-validation.

The X can be divided into K disjoint subsets: $X_i | X_i \bigcup X_i ... \bigcup X_k; X_i \bigcap X_j = \emptyset; i = 1...K; j = 1...K; i \neq$ ɪ and each subset X_i contains n_k feature vectors. The set of classifiers $\{a_k(\boldsymbol{x}) | k = 1...K\}$ is then trained using the set of feature vectors $X^l = X \backslash X_k$. Let use average classification accuracy for classifiers set as an objective function for feature space optimization:

$$f = \frac{1}{K} \sum_{k=1}^{K} \frac{1}{n_k} \sum_{x \in X_k} \frac{|a_k((x)) + c((x))|}{2} \tag{4}$$

Main task is to find vector $\boldsymbol{p} = (p_1, p_2 ... p_n), p_i \in R$ which determines base wavelet function $w_{\boldsymbol{p}}(t)$ providing $f \rightarrow max$ for all X_i. Time representation of wavelet function $w_{\boldsymbol{p}}(t)$ is defined by Akima spline [1] with $\boldsymbol{p}$ as control points. This choice of a parametric curve is caused firstly by the fact that the resultant curve passes through all sample points. Secondly by resistance of the Akima spline to local overshoots this spline practically has no curve oscillations near overshoot points, as distinct from cubic splines [1]. This property of the Akima spline is quite significant as any additional oscillations make worse the localization of the wavelet function in the frequency domain.

In order to find $\boldsymbol{p} = (p_1, p_2 ... p_n), p_i \in R$ numerical optimization methods are to be used. Genetic algorithm (GA) [2] is the best choice of optimization method due to a number of significant advantages:

- possibility of finding a global extreme of the objective function;
- conditional optimization;
- convergence speed;
- algorithm parallelism that allows to significantly reduce calculation time with the help of modern parallel data processing systems.

The genetic algorithm operates with vector $\boldsymbol{p}$ indirectly through sequence of code symbols $\boldsymbol{q} = (q_1, q_2 ... q_n)$, which is accepted to be called a chromosome in the theory of evolutionary optimization methods. Vector of parameters $\boldsymbol{p}$ is unambiguously defined by the chromosome $\boldsymbol{q}$, while:

1. Each parameter $p_i, i \in 1...N$ is described by the corresponding gene $q_i, i \in 1...N$;
2. Each gene $q_i, i \in 1...N$ consists of M alleles, selected from a finite set. The finite set $\{0, 1\}$ is used for the convenience of the genetic algorithm implementation.

The Akima spline is defined by a set of control points $p(n) = \{p_i, i = 1...N\}$ where p_i value is coded by q_i gene, and $p_i \in [-1, 1]$.

To comply with the condition (5) let us build the Akima spline $s_{\boldsymbol{p}}(x)$ and normalize the acquired set of parameters:

$$\widetilde{p}_i = p_i - \frac{1}{N} \int_0^{+N} s_{\boldsymbol{p}}(x) \, dx, \tag{5}$$

where N - the size of vector $\boldsymbol{p}$. The required wavelet function is built as Akima spline [1] using the set of parameters $\widetilde{p}_i$.

616   A.M. Soroka et al.

As the genome consists of the finite number of alleles, the values are coded with some accuracy ϵ accordingly. In this case the maximum number of values which can be coded by gene, is described by following expression:

$$K_{max} = \frac{max\,(p_i) - min\,(p_i)}{\epsilon} \qquad (6)$$

the minimum number of alleles in a gene, necessary for coding all values, makes:
$L\,(q_i) = \lceil log_2\,(K_{max}) \rceil$

Operations of mutation and crossover in this case are trivial because any combination of alleles is valid. The elitism strategy is used as an individuals selecting algorithm to generate next population. It provides higher convergence speed when solving the problem of base wavelet function optimization in comparison with the pure methods of roulette wheel or tournament selections [2].

On each GA iteration the value of the objective function (4) (classification accuracy) is calculated for each individual. Let use distance measure between best objective function value and average for population as estimation of GA convergence. The GA iteration statistics for training data is shown on Fig. 1 as well as classification accuracy for target testing data using proposed algorithm. As it can be noticed from Fig. 1 the classification accuracy for target testing data increases during GA optimization.

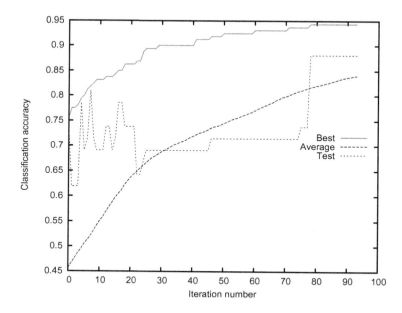

Fig. 1. GA iteration statistics for the emotion "fear" (Color figure online)

One of the advantages of the constructed adaptive wavelet function is the fact that it considers dynamics of a signal throughout all its length [14]. At each scale the wavelet function depending on all signal is constructed and it respectively comprise information about all changes in a studied signal. Delta coefficients are traditionally used for the accounting of acoustic signal's dynamic characteristics at a stage of features description creation in classic methods of the primary analysis. They allow to consider only local changes of a signal, in difference from proposed adaptive wavelet function.

3 Experimentation

3.1 Dataset Description

All speaker emotions recognition experiments were performed on the EMO-DB [4]. This database contains speech recordings made by ten actors (5 female and 5 male) simulated seven emotions: anger, boredom, fear, disgust, happiness, neutral, sadness. The speech material comprises about 800 sentences.

3.2 Experimental Results

The proposed algorithm based on GA and SVM was used to create adaptive wavelet function for all emotions from EMO-DB. In order to do this all records were divided to train and test data subsets: train data which were used to perform wavelet function search and test data which were not evolved into learning process and were used for evaluation only. Database was segmented to isolated words which were used as minimum speech unit for emotion recognition. Training and testing subsets were chosen in proportion approximately one to five. The number of optimized parameters p_i was $N = 128$, the population size 5000. All calculations were done on an MPI cluster consisted of 50 nodes, on each node four Dual Core AMD Opteron 875 HE processors were used, 2.2 GHz and 1024 kB cache using k52 libraries [9].

During experiments the linear support vector machine classifier was used to avoid the influence of non-linear classification effects on the classification accuracy. The SVM penalty parameter C for each classification task was found with the help of the grid search. As the result of training an adaptive base wavelet function was designed for each of seven emotions from data base. Emotions recognition experiment was done using linear SVM classifier and "one-against-all" scheme. In order to estimate the effectiveness of the developed adaptive feature vector (AFV) construction algorithm for emotion recognition it was done the comparison with MFCC-based feature vectors. Experimental results can be found in the Table 1.

As it can be noticed from the table the proposed approach to adaptive speech parameters description provides better emotion recognition performance for all emotions in comparison with widely used MFCC-based vector.

Table 1. Various emotions classification accuracy, %

Emotion	AFV	MFCC
anger	81,1	74,2
boredom	88,7	85,6
disgust	91,5	91,4
fear	91,7	91,7
happiness	90,5	89,7
neutral	90,7	89,2
sadness	89,1	85,1

4   Conclusion

In this paper the algorithm for adaptive wavelet function construction for speaker emotions recognition was proposed based on the genetic algorithm and support vector machines as classifier. It was shown that usage of base wavelet functions generated by the proposed algorithm to construct feature description allows to achieve the steady increase of the classification accuracy. The improvement of emotion recognition accuracy vs mel-frequency cepstral coefficients varies up to 7 %. The proposed method for feature construction can be generalized for other audio processing tasks like speech and music recognition, acoustic events detection, speaker identification etc.

Acknowledgement. Authors acknowledge financial support from Belarusian Republican Foundation for Fundamental Research, project F14-052.

References

1. Akima, H.: A new method of interpolation and smooth curve fitting based on local procedures. J. ACM **17**, 589–602 (1970)
2. Back, T.: Evolution Algorithms in Theory and Practice. Oxford University Press, Oxford (1996)
3. Bellman, R.: Adaptive Control Processes Guided Tour. Princeton University Press, Princeton (1961)
4. Burkhardt, F., Paeschke, A., Rolfes, M., Sendlmeier, W., Weiss, B.: A database of german emotional speech. In: Proceedings of Interspeech, Lissabon, pp. 1517–1520 (2005)
5. Goldberg, D.: Genetic Algorithms in Search, Optimization and Machine Learning, 1st edn. Addison-Wesley Longman Publishing Co., Boston (1989)
6. Huang, C.L., Wang, C.J.: A GA-based feature selection and parameters optimization for support vector machines. Expert Syst. Appl. **31**(2), 231–240 (2006)
7. Huang, X., Acero, A., Hon, H.W.: Spoken Language Processing: A Guide to Theory, Algorithm, and System Development. Prentice Hall, Upper Saddle River (2001)

8. Kiranyaz, S., Raitoharju, J., Gabbouj, M.: Evolutionary feature synthesis forcontent-based audio retrieval. In: 2013 1st International Conference on Communications, Signal Processing, andtheir Applications (ICCSPA), pp. 1–6, February 2013
9. Kovalets, P., AlexanderSoroka: k52: v0.1 (2016). http://dx.doi.org/10.5281/zenodo.49181
10. Lixia, H., Evangelista, G., Zhang, X.: Adaptive bands filter bank optimized by genetic algorithm for robust speech recognition system. J. Cent. South Univ. Technol. **18**, 1595–1601 (2011)
11. Mallat, S.: A Wavelet Tour of Signal Processing: The Sparse Way. Academic Press, San Diego (2008)
12. Murthy, D.S., Holla, N.: Robust speech recognition system designed by combining empirical mode decomposition and a genetic algorithm. Int. J. Eng. Res. Technol. (IJERT) **2**, 2056–2068 (2013)
13. Schuller, B., Reiter, S., Rigoll, G.: Evolutionary feature generation in speech emotion recognition. In: 2006 IEEE International Conference on Multimedia and Expo, pp. 5–8 (2006)
14. Soroka, A., Kovalets, P., Kheidorov, I.: New method of speech signals adaptive features construction based on the wavelet-like transform and support vector machines. In: Ronzhin, A., Potapova, R., Delic, V. (eds.) SPECOM 2014. LNCS, vol. 8773, pp. 308–314. Springer, Heidelberg (2014)
15. Srinivasan, V., Ramalingam, V., Sellam, V.: Classification of normal and pathological voice using GA and SVM. Int. J. Comput. Appl. **60**, 34–39 (2012)
16. Vapnik, V.: The Nature of Statistical Learning Theory. Springer, New York (2000)
17. Vignolo, L.D., Rufiner, H.L., Milone, D.H., Goddard, J.C.: Evolutionary splines for cepstral filterbank optimization in phoneme classification. EURASIP J. Adv. Signal Process. **2011**, 1–14 (2011)

Integration of Bayesian Classifier and Perceptron for Problem Identification on Dynamics Signature Using a Genetic Algorithm for the Identification Threshold Selection

Evgeny Kostyuchenko$^{(\boxtimes)}$, Mihail Gurakov, Egor Krivonosov,
Maxim Tomyshev, Roman Mescheryakov, and Ilya Hodashinskiy

Tomsk State University of Control Systems and Radioelectronics,
Lenina Street 40, 634050 Tomsk, Russia
{key,mrv,hia}@keva.tusur.ru, g.mishell@gmail.com,
{egor-yrga,unreal-max}@mail.ru
http://www.tusur.ru

Abstract. An approach to the integration of multiple methods of user authentication and example of multi-classifier Bayesian and neural network is presented. The approach offers to find the convolution of outputs from multiple classifiers based on the complementary functions and to carry out the selection of the identification thresholds for each of the users. A number of complementary functions that use fundamentally different mathematical functions is analyzed. It is shown the practical need in metaheuristic algorithms for selecting the identification thresholds by comparison with the classic gradient method. The effectiveness some of the proposed series of multi-function, compared with the single use Bayes classifier and neural network is showed.

Keywords: Signature identification · Bayesian · Neural network · Complex method · Identification threshold

1 Introduction

The problem of identification of biometric characteristics of the user as a reinforcing factor is actual. One approach to achieve this is using of the dynamics signature identification [1]. Signature features are selected by using of sensors on the pen graphic tablet. In previous research authors have conducted a study that confirmed the possibility of using neural networks and Bayesian classifier for user authentication based on signature dynamics [2,3].

In this paper we consider the possibility of combining these approaches to improve accuracy. Preliminary works in this direction have already been carried out [4], this paper presents a study of applicability of many functions for complexation. In addition, a comparative study for the selection of authentication threshold algorithms is presented. Consider these stages in more detail.

© Springer International Publishing Switzerland 2016
L. Cheng et al. (Eds.): ISNN 2016, LNCS 9719, pp. 620–627, 2016.
DOI: 10.1007/978-3-319-40663-3_71

2   Parameters Preparation

For preparing of parameters involved in carrying out the authentication procedure, the following actions was performed:

1. measuring dependencies position of the pen on the x tablet (t) and y (t), the height z (t), pressure on the plate p (t), the inclination angle of the pen to the tablet ϕ (t) and angle between the stylus and the plane formed by the axes and y and z pen θ (t), from the time t (total six characteristics);
2. signature was normalized to a fixed size using by a linear transformation, and the dependencies from step one was recalculated, taking into account the normalization;
3. the calculation of the velocity and acceleration of characteristics in time (along with the originals - total we have 18 characteristics);
4. the use of the Fourier transform and the measuring of the amplitudes of the constant component and the first seven harmonics of the time dependence from step three - result is eight amplitudes, 18*8=144 values. Output parameters are recorded in the database and is used by classifiers in the analysis.

Previous studies carried out a separate processing of the parameters using the neural network (perceptron was used) [2] and Bayes classifier [3]. The use of these approaches is not unique, they have been used previously to solve the identification problem by other researchers [5–8].

Current database contains more than 1200 signatures from 8 users.

3   Model of Combining Classifiers Outputs

The following steps were used to combine the outputs of classifiers:

1. carried out the convolution of the output values of the neural networks and Bayes classifier using a monotonic function. This feature further comprises a plurality of coefficients - the parameters of the convolution;
2. optimization of the resulting convolution in terms of selection of the optimum parameters of the convolution and optimization of the decision threshold for the classification. Classification thresholds are selected independently for each user, and may vary among themselves.

To implement this approach, the whole sample divided into three parts - classifiers training (within 60 % of the experiment), optimization training (20 %) and testing for evaluating the quality of the combined classifier (20 %). Convolution function used are shown in Table 1. The first two functions are in fact neural network and Bayesian classifier working separately and used for comparison.

4   Comparison of Combined Classifiers that Use Gradient and Genetic Optimization

As indicators of performance recognition systems type I errors (failures of authorized users) and type II errors (tolerance in unauthorized user) were adopted. Three criteria have been constructed on the basis of indicators:

Table 1. List of functions for combinations of outputs recognition systems. N - outputs of the neural network, B - outputs naive Bayes classifier, α, β - coefficients

#	Function	#	Function
1.	$f(\alpha, \beta, B, N) = B$	18.	$f(\alpha, \beta, B, N) = exp(B^{\alpha}) + exp(N^{\beta})$
2.	$f(\alpha, \beta, B, N) = N$	19.	$f(\alpha, \beta, B, N) = \sqrt{\alpha \cdot B} + \sqrt{\beta \cdot N}$
3.	$f(\alpha, \beta, B, N) = \alpha \cdot B + \beta \cdot N$	20.	$f(\alpha, \beta, B, N) = \sqrt{B^{\alpha}} + \sqrt{N^{\beta}}$
4.	$f(\alpha, \beta, B, N) = lg(\frac{\alpha \cdot N}{\beta \cdot B})$	21.	$f(\alpha, \beta, B, N) = exp(sinh(B + \alpha) + sinh(N + \beta))$
5.	$f(\alpha, \beta, B, N) = lg(B^{\alpha} \cdot N^{\beta})$	22.	$f(\alpha, \beta, B, N) = exp(sinh(B + \alpha)) + exp(sinh(N + \beta))$
6.	$f(\alpha, \beta, B, N) = lg(B^{1/\alpha} \cdot N^{\beta})$	23.	$f(\alpha, \beta, B, N) = sinh(\alpha \cdot B + \beta \cdot N)$
7.	$f(\alpha, \beta, B, N) = B \cdot \alpha \cdot N \cdot \beta$	24.	$f(\alpha, \beta, B, N) = B^{\alpha} + \beta \cdot N$
8.	$f(\alpha, \beta, B, N) = B^{\alpha} + N^{\beta}$	25.	$f(\alpha, \beta, B, N) = lg(B^{\alpha} + \beta \cdot N)$
9.	$f(\alpha, \beta, B, N) = lg(B \cdot \alpha \cdot N \cdot \beta)$	26.	$f(\alpha, \beta, B, N) = sinh(B + \alpha) \cdot sinh(N \cdot \beta)$
10.	$f(\alpha, \beta, B, N) = lg(B^{\alpha} + N^{\beta})$	27.	$f(\alpha, \beta, B, N) = sinh(B^{\alpha}) + sinh(N \cdot \beta)$
11.	$f(\alpha, \beta, B, N) = sinh(B + \alpha) + sinh(N + \beta)$	28.	$f(\alpha, \beta, B, N) = tanh(B + \alpha) \cdot tanh(N \cdot \beta)$
12.	$f(\alpha, \beta, B, N) = sinh(B \cdot \alpha) + sinh(N \cdot \beta)$	29.	$f(\alpha, \beta, B, N) = tanh(B^{\alpha}) + tanh(N \cdot \beta)$
13.	$f(\alpha, \beta, B, N) = sinh(B^{\alpha}) + sinh(N^{\beta})$	30.	$f(\alpha, \beta, B, N) = \sqrt{B^{\alpha}} + sinh(N \cdot \beta)$
14.	$f(\alpha, \beta, B, N) = tanh(B + \alpha) + tanh(N + \beta)$	31.	$f(\alpha, \beta, B, N) = tanh(B \cdot \alpha) + sinh(N \cdot \beta)$
15.	$f(\alpha, \beta, B, N) = tanh(B \cdot \alpha) + tanh(N \cdot \beta)$	32.	$f(\alpha, \beta, B, N) = tanh(B \cdot \alpha) \cdot sinh(N \cdot \beta)$
16.	$f(\alpha, \beta, B, N) = tanh(B^{\alpha}) + tanh(N^{\beta})$	33.	$f(\alpha, \beta, B, N) = tanh(B + \alpha) + sinh(N \cdot \beta)$
17.	$f(\alpha, \beta, B, N) = exp(B \cdot \alpha) + exp(N \cdot \beta)$	34.	$f(\alpha, \beta, B, N) = tanh(B + \alpha) \cdot sinh(N \cdot \beta)$

1. the sum of the probabilities of errors of type I and type II $K = m_1 + m_2$ Allowed that errors are equally important as indicators of the recognition system working. Analog is accuracy - $K = 1 - (m_1 + m_2)/2$;
2. the probability of error without division into Type I and type II errors $K = m$;
3. the sum of the probabilities of errors of the first and second kind $K = m_1 + 10m_2$. Allowed that type II error is ten times more important type I errors.

On current stage optimization was carried out on the first criterion. As a method of optimizing gradient optimization has been selected. This is done on the assumption that the surface of the objective function has a simple shape with the most likely single extremum in each dimension.

4.1 The Gradient Search Algorithm Threshold Values

As a result of testing of the gradient optimization method the unsuitable for practical application of the classifier value was regularly obtained (in particular, the error probabilities greater than 0.5), and the resulting values are very different from each other. This suggests that the assumption of a simple form of the objective function surface was wrong - there is a significant number of local extrema. Based on it two reports made:

1. for best results, the gradient algorithm must be replaced by metaheuristic optimization algorithm which less prone to a halt in the local extremes. At this stage the genetic optimization algorithm was selected;
2. to evaluate the obtained results the selection of the best 50 % of them to exclude obvious errors was held. The obtained gradient optimization test results are presented in Table 2. Count of tests was 100.

   The result of optimization was recognized as unsatisfactory. Given that in a particular case, a search for the global minimum of a complex surface is carried out, the genetic optimization was selected as an alternative optimization method.

Table 2. The results of the gradient optimization applying

#	$m_1 + m_2$		#	$m_1 + m_2$	
	mean	min		mean	min
1.	0,0550	0,0360	18.	0,0798	0,0314
2.	0,0998	0,0314	19.	0,0692	0,0274
3.	0,0972	0,0360	20.	0,0870	0,0292
4.	0,0473	0,0326	21.	0,0608	0,0314
5.	0,0867	0,0274	22.	0,0868	0,0514
6.	0,0660	0,0274	23.	0,1163	0,0581
7.	0,0460	0,0258	24.	0,0602	0,0406
8.	0,0550	0,0431	25.	0,1368	0,0366
9.	0,0901	0,0443	26.	0,0948	0,0492
10.	0,1225	0,0523	27.	0,1097	0,0517
11.	0,0565	0,0403	28.	0,0483	0,0274
12.	0,0964	0,0338	29.	0,1598	0,0280
13.	0,0611	0,0338	30.	0,0347	0,0261
14.	0,0513	0,0354	31.	0,0378	0,0255
15.	0,1456	0,0369	32.	0,1493	0,0274
16.	0,0750	0,0427	33.	0,0827	0,0544
17.	0,1020	0,0360	34.	0,0645	0,0335

4.2   Genetic Algorithm for Finding the Threshold Values

The result of testing the genetic optimization (50 % of the best results) is shown in Table 3. It can be seen that the results are significantly higher than the results of the gradient method. In addition, on the test results can be concluded that the best effective functions are 17, 30 and 11 $(exp(B{\cdot}\alpha)+exp(N{\cdot}\beta), \sqrt{B^\alpha}+sinh(N{\cdot}\beta)$

and $sinh(B + \alpha) + sinh(N + \beta)$). Furthermore, it is seen that the result of the application of these functions for convolution neural network and Bayesian classifier achieves substantial gain for criterion of identification accuracy in comparison with separated using of neural network and Bayesian classifier. For visibility, the test results are shown in Fig. 1. Count of tests was 100.

Table 3. The results of the genetic optimization applying

#	$m_1 + m_2$		#	$m_1 + m_2$	
	mean	min		mean	min
1.	0,0395	0,0120	18.	0,0358	0,0108
2.	0,0344	0,0197	19.	0,0431	0,0268
3.	0,0432	0,0215	20.	0,0430	0,0209
5.	0,0405	0,0181	22.	0,0430	0,0175
6.	0,0326	0,0098	23.	0,0378	0,0185
7.	0,0409	0,0175	24.	0,0396	0,0246
8.	0,0395	0,0194	25.	0,0452	0,0228
9.	0,0431	0,0246	26.	0,0376	0,0252
10.	0,0382	0,0194	27.	0,0363	0,0169
11.	0,0324	0,0083	28.	0,0338	0,0151
12.	0,0434	0,0166	29.	0,0489	0,0221
13.	0,0337	0,0126	30.	0,0307	0,0172
14.	0,0433	0,0117	31.	0,0381	0,0163
15.	0,0468	0,0252	32.	0,0407	0,0138
16.	0,0409	0,0234	33.	0,0408	0,0194
17.	0,0301	0,0185	34.	0,0357	0,0120

5   Comparison with Similar Systems Work

There are other methods of solving on-line identification by signature task. This task can be solved with help of graph theory method [9]. This method is based on weighted signature graph creating. Each signature graph has fearues such as an overall graph weight, graph norm. Graph norm is a single real number that represents the whole graph. In the identification process the given signature graph is created, the graph norm is calculated and compared with the base. If graph norm is equal with graph norm of just one user, it suggests that user as a potential signatory and the identification process is over. In another case user graph is compared with all the graphs of potential users.

Another signature recognition method is based on multi-section vector quantization [10] - an improved version of the classical vector quantization approach. The results of work were compared with the results obtained with other works,

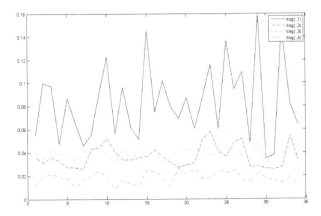

Fig. 1. The values of the three identification quality criteria for the gradient and genetic optimization methods for every function of convolution. itog1 - mean of gradient method, itog2 - minimum of gradient method, itog3 - mean of genetic method, itog4 - minimum of genetic method.

explores identification on user's signature and the dynamics of its affixing, by the accuracy of classification value. Works [9–14] were chosen for this operations. The comparison results are presented in Table 4.

It can be seen that the results obtained are comparable with the results of similar recognition systems, and in some cases even superior to some of them. Thus the proposed approach combining classifiers based on neural networks with the classical naive Bayes classifiers by the convolution results of their work can be considered as applicable. There is a principle problem with testing - very small size of database compare with analogs testing. On next stage this problem will be solved using the same database.

Table 4. Comparison of the results with the results of similar systems of identification by signature

Article	Method	Accuracy	User count
Article [9]	Basic concepts of graph theory	94,25 %	27
Article [10]	Multi-section vector quantization	98 %	330
Article [11]	Statistical analysis	97,75 %	200
Article [12]	Principal Component Analysis +Neural Network	89,475 %	200
Article [13]	Principal Component Analysis +Neural Network	93,1 %	200
Article [14]	Neural Network	95 %	10
Current	Neural Network+Bayes	98,5 %	8

6 Conclusion

In this work, the check of the applicability of a combined approach to the problem of identifying the user on the dynamics of the signature on the basis of joint application of neural networks and Bayesian classifier was performed. It was shown that this approach produces results that are superior to the results of the application of neural networks and Bayesian classifier working separately. Selected classifier outputs convolution function, allowing to obtain the best results. It is functions $exp(B\cdot\alpha)+exp(N\cdot\beta), \sqrt{B^\alpha}+sinh(N\cdot\beta)$ and $sinh(B+\alpha)+sinh(N+\beta)$. The inapplicability of classical gradient method for the selection of the parameters of the convolution function and thresholds to identify each of the users was shown. Testing of the applicability of one of metaheuristic algorithms - genetic algorithm - to solve this problem was held. The results are compared with similar results on the criterion of the identification accuracy. On the basis of comparison, the conclusion the applicability of the proposed approach was done. The total classification accuracy was on average 98,5 % with a minimum at 99,5 %. The next phase of the study is to compare the various quality evaluation classifiers criteria for the proposed approach, as well as obtaining more information about the functions that allow to obtain the best results when the convolution and attempting to summarize this information to select classes of such functions.

Acknowledgments. This work was supported by the Ministry of Education and Science of the Russian Federation within 1.3 federal program research and development in priority areas of scientific-technological complex of Russia for 2014-2020 (grant agreement 14.577.21.0172 on October 27, 2015; identifier RFMEFI57715X0172).

References

1. Tolosana, R., Vera-Rodriguez, R., Ortega-Garcia, J., Fierrez, J.: Pre-processing and feature selection for improved sensor interoperabilityin online biometric signature verification. IEEE Access **3**, 478–489 (2015)
2. Doroshenko, T.Y., Kostyuchenko, E.Y.: The authentication system based on dynamic handwritten signature. In: Proceedings of Tomsk State University of Control Systems and Radioelectronics, vol. 3, pp. 219–223 (2014)
3. Gurakov, M.A., Krivonosov, E.O., Kostyuchenko, E.Y.: User authentication on the signature dynamics based on naive Bayes classifier. In: 11th International Scientific Conference Electronic Instrumentation and Control Systems, pp. 155–158. V-Spectr, Tomsk (2015)
4. Gurakov, M.A., Krivonosov, E.O., Kostyuchenko, E.Y.: Quality parameters of dynamic signature recognition systems based on naive Bayes classifier and neural network. Trudi MAI (2016). accepted for publication
5. Iranmanesh, V., Ahmad, S.M.S., Adnan, W.A.W., Malallah, F.L., Yussof, S.: Online signature verification using neural network and pearson correlation features. In: 2013 IEEE Conference on Open Systems (ICOS), pp. 18–21. Sarawak, Malaysia (2013)
6. Meshoul, S., Batouche, M.: A novel approach for online signature verification using fisher based probabilistic neural network. In: Proceedings - International Symposium on Computers and Communications, pp. 314–319. Riccione, Italy (2010)

7. Kachaykin, E., Ivanov, A.: Identification of authorship of handwritten images using neural network emulator of quadratic forms high dimension. Cybersecurity **12**, 42–47 (2015)
8. Lozhnikov, P.S.: Human identification of the dynamics of writing words in computer systems. Success Mod. Sci. **4**, 129–130 (2004)
9. Fotak, T., Bača, M., Koruga, P.: Handwritten signature identification using basic concepts of graph theory. WSEAS Trans. Sig. Process. **7**, 117–129 (2011)
10. Faundez-Zanuy, M., Gaspar, J.M.P.: Efficient on-line signature recognition based on multi-section vector quantization. Formal Pattern Anal. Appl. **14**, 37–45 (2011)
11. Nilchiyan, M.R., Yusof, R.B., Alavi, S.E.: Statistical on-line signature verification using rotation-invariant dynamic descriptors. In: The 10th Asian Control Conference, ASCC 2015, Kota kinabalu, Malaysia (2015)
12. Adnan, W.A.W., Malallah, F.L., Mumtazah, S., Yussof, S.: Online handwritten signature recognition by length normalization using up- sampling and down-sampling. Int. J. Cyber-Secur. Digital Forensics (IJCSDF) **4**, 302–313 (2015)
13. Iranmanesh, V., Ahmad, S.M.S., Adnan, W.A.W., Yussof, S., Arigbabu, O.A., Malallah, F.L.: Research article online handwritten signature verification using neural network classifier based on principal component analysis. Sci. World J. **2014**, 1–9 (2014). doi:10.1155/2014/381469
14. Babita, P.: Online signature recognition using neural network. Electr. Electron. Syst. **4**, 155 (2015). doi:10.4172/2332-0796.1000155

Cognition Computation and Spiking Neural Networks

Tracking Based on Unit-Linking Pulse Coupled Neural Network Image Icon and Particle Filter

Hang Liu[1,2,3] and Xiaodong Gu[1,2,3(✉)]

[1] Department of Electronic Engineering,
Fudan University, Shanghai 200433, China
xdgu@fudan.edu.cn
[2] Key Laboratory for Information Science of Electromagnetic
Waves (Ministry of Education), Fudan University, Shanghai 200433, China
[3] Research Center of Smart Networks and Systems,
Fudan University, Shanghai 200433, China

Abstract. Visual tracking is a challenging problem in computer vision. Many visual trackers either rely on luminance information or other simple color representations for image description. This paper introduces a tracking algorithm using unit-linking PCNN (Pulse Coupled Neural Network) image icon and particle filter. This approach has the translation, rotation, and scale invariance for using unit-linking PCNN image icon as the features. The experimental results show the proposed approach is with 16.43 % higher median distance precision than the color gradient-based tracker. This unit-linking PCNN image icon-based particle filter tracker can better solve the problems caused by partial occlusions, or out-of-plane rotation, or scale variation, or non-rigid object deformation, or fast motion.

Keywords: Object tracking · Unit-linking PCNN image icon · Particle filter

1 Introduction

Object tracking is one of the most important problems in the field of computer vision with applications ranging from surveillance and human-computer interactions to medical imaging. Given the initial state (e.g., position and extent) of a target object in the first image, the goal of tracking is to estimate the states of the target in the subsequent frames [1]. Although object tracking has been studied for a long time, there are still many problems remain unsolved. Numerous factors, such as illumination variation, occlusion, and background clutters, affect the performance of a tracking algorithm, and if the tracker wants to have a better performance, it is very important to find a stable and typical feature.

So far, most of the tracking algorithms select luminance information or simple color as their features [2–6]. These features do not restrict the operation speed of tracking algorithms, however, they are too simple to give a precise description of the tracked object and lack some in variance. In this paper, we use a feature called unit-linking PCNN (Pulse Coupled Neural Network) image icon. This feature can make up for the

© Springer International Publishing Switzerland 2016
L. Cheng et al. (Eds.): ISNN 2016, LNCS 9719, pp. 631–639, 2016.
DOI: 10.1007/978-3-319-40663-3_72

weak points of the features we describe above and has translation, rotation, and scale invariance. Particle filter algorithm is a very classic target tracking algorithm, so we present a particle filter tracking algorithm which use unit-linking PCNN (Pulse Coupled Neural Network) image icon.

2 The Basic Particle Filter

Particle filtering was developed to deal with the situation that the posterior density $p(X_t|Y_t)$ and the observation density $p(Y_t|X_t)$ are non-Gaussian. The state vector X_t means the quantities of a tracked object while the vector Y_t denotes all the observations $\{Y_1, \ldots \ldots . Y_t\}$ up to time t. Calculating the particles' weight is the most critical step of the particle filter. Using a known probability density function $q(X_t|Y_t)$ generates the sample particles. And then a weighted sample can approximate the probability distribution of the object state.

$$W_i(x_i) = \frac{p(y_{1:i}|x_i)p(x_i)}{q(x_i|y_{1:i})} \propto \frac{p(x_i|y_{1:i})}{q(x_i|y_{1:i})} \qquad (1)$$

Each sample includes an element and it determines the probability that the target appears. The evolution of the sample set is calculated by propagating each sample according to a system model. Finally, to reduce the influence of a smaller weighted particle and to prevent particle degradation, we need a resampling procedure [7]. In this paper, we use adaptive resampling (Fig. 1).

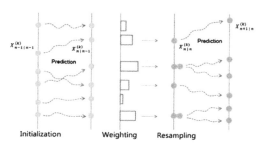

Fig. 1. Basic particle filter with adaptive resampling.

Particle filtering provides a robust tracking framework, as it models uncertainty. It can keep its options open and consider multiple-state hypotheses simultaneously. Since less likely object states have a chance to temporarily remain in the tracking process, particle filters can deal well with short-lived occlusions [8].

3 Unit-Linking PCNN Image Icon

In 1990, Eckhorn introduced the linking field model [9] exhibiting synchronous pulse bursts in the cat visual cortex [10]. In 1993, Johnson called the linking model Pulse Coupled Neural Network (PCNN) by introducing the linking strength to the linking model [11]. PCNN can be applied in many fields, such as image processing, optimization [12–15]. Unit-linking PCNN [12], easy to implemented by hardware, is a simplified version of PCNN retaining PCNN main characteristics. Literature [16] uses unit-linking PCNN to produce image icons as image features.

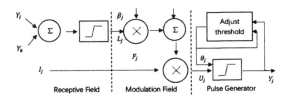

Fig. 2. A unit-linking PCN j [16].

A unit-linking PCN (Pulse Coupled Neuron) consists of three parts: the receptive field, the modulation field, and the pulse generator. Figure 2 illustrates a unit-linking PCN j.

Unit-linking PCNN is a single layer two-dimensional array of laterally linked neurons. The number of neurons in the network is equal to the number of pixels in the input image. Unit-linking PCNN image icon is a 1-dimensional time series, each element of which is equal to the sum of neurons that fire at each iteration. In this paper, the numbers of the iteration are 12, then the unit-linking PCNN image icon includes 12 elements, and if 100 neurons fire at 1st iteration in the whole network, then the 1st element of image icon equals to 100. Obviously the unit-linking PCNN image icon has the translation and rotation invariance because all neurons are identical and have the same connection mode. Unit-linking PCNN image icon also has the scale invariance by using vector-angular similarity or normalized Euclidean similarity because for the identical-content images with different sizes, their unit-linking PCNN image icons are similar [16]. Specific algorithm process will be given in later chapters.

4 Tracking Algorithm Based on Unit-Linking PCNN Image Icon

Our algorithm combines the particle filter and the unit-linking PCNN image icon. The calculation of particles' weight is based on the unit-linking PCNN image icon. So, first of all, we need to know how to calculate the feature of every particle in this paper.

At first, we should turn each frame into Y_i ($i = 1, 2, \ldots, N$), where Y_i is a binary neuron output matrix in each iteration and N is the number of iteration. And then, all we need is only a part of the Y_i. If the position of a particle X_i is (a, b) and the size of target box is (h, w), we can find a corresponding area Y'_i in Y_i based on (a, b) and (h, w). The algorithm is as follows [16].

```
Input: L = 0, U = 0, θ = 1, Y = 0, β = 0.2 , δ
Output: Yᵢ
for i=1 to N do
    L = Step(Y⊗K)
    while(1)
        Inter = Y, U = F.*(1 +βL), Y = Step(U - θ )
        If Y = Inter then
            break
        else
            L = Step(Y⊗K)
        end
end
    Y = Y.
    Θ = θ - δ + 10000*Y.
    for j=(a-w/2) to (a+w/2) do
        for k=(b-h/2) to (b+h/2) do
Yᵢ = Y (j,k)
end
    end

    end
```

In this algorithm, K is a 3*3 kernel, $\begin{bmatrix} 1 & 1 & 1 \\ 1 & 1 & 1 \\ 1 & 1 & 1 \end{bmatrix}$, and it means each neuron is connected in its 3*3 neighboring field and the neighboring field includes the neuron itself. F is the intensity of the pixel of the image and it is normalized between 0 and 1. Besides, the times of iteration are 12 and red dots denote particles (Fig. 3).

Fig. 3. $Y_1 - Y_{12}$ can be calculated after 12 iterations. In a neighborhood of X_i (the yellow dot), $Y'_1 - Y'_{12}$ can be taken from $Y_1 - Y_{12}$ in the green box. (Color figure online)

Then we can calculate the sum of neurons that fire based on each Y'_i and get the unit-linking PCNN image icon of this particle X_i. The size of X_i's neighborhood is the same as the target area.

In the tracking process, we can get the position of target in the first frame. Using the algorithm mentioned above gets the unit-linking PCNN image icon in target area. This is our target template (Icon target) (Fig. 4).

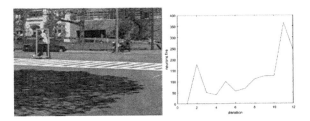

Fig. 4. The PCNN image icon of target template.

The unit-linking PCNN image icon of every particle in each frame will be compared with the target template. If they are closer, the probability that the target appears is higher. Thus, this particle will get a larger weight.

In this paper, we can calculate the particles weight by Eqs. (2) and (3).

$$w_i = \log\left(\sqrt{2\pi} * R\right) + \frac{D^2}{2 * R^2} \tag{2}$$

$$D = E(\text{Icon\_X}_i, \text{Icon\_target})_{diff} \tag{3}$$

In Eq. (2), R is the measurement noise covariance, and D means the difference between the unit-linking PCNN image icon of X_i and the target template $(E(a,b)_{diff}$ means the Euclidean distance of a and b).

Then each particle has different weight, and this process is called importance sampling in the particle filter. In order to solve degradation issues, the particles with high weights may derive more particles to replace the particles with low weights, and this process is called resampling in the particle filter (Fig. 5).

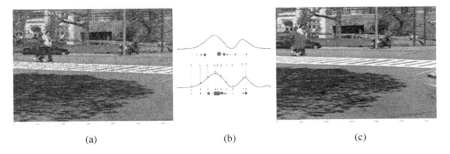

(a)   (b)   (c)

Fig. 5. The particles will be generated randomly in the first frame and they will be distributed randomly in the picture, just like (a). (b) is the process of importance sampling and resampling. After resampling, the particles distribute like (c) and most of them are concentrated in the target area.

After resampling, the central region of the particle is the position of target. So in the framework of particle filtering, the whole process of this tracking algorithm based on unit-linking PCNN image icon is as follows (Fig. 6).

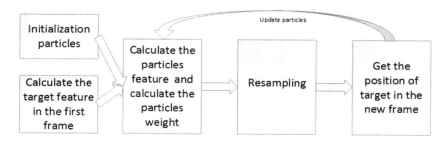

Fig. 6. The process of particle filter algorithm based on unit-linking PCNN image icon.

5   Experiments

We evaluate the proposed particle filter tracking algorithm based on unit-linking PCNN image icon using a Couple and a Girl sequence with challenging factors including heavy occlusion, drastic illumination changes, pose and scale variation, and motion blur. We compare this proposed tracker with a particle filter tracking algorithm based on RGB color gradient, and an algorithm based on KCF [17].

The Couple sequence has 140 frames, and it has out-of-plane rotation, scale variation, non-rigid object deformation, fast motion, background Clutters problems. We can see the performance of three tracking algorithms in Fig. 7.

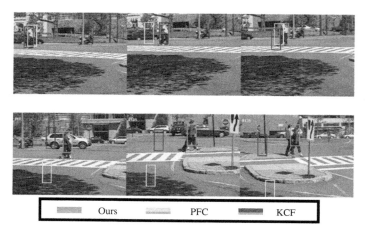

Fig. 7. The results of three tracking algorithms (Couple). (Color figure online)

In Fig. 7, our results are marked with green bounding boxes, the results of particle filter based color gradient are marked with yellow bounding boxes and the results of KCF are marked with blue bounding boxes. Through these six frames at different times, we can see that the results of our algorithm are better than the other two algorithms. Figure 8 is Precision plot of the Couple sequence, we can get the same conclusion through this evaluation criterion.

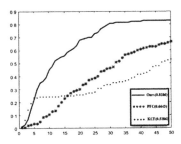

Fig. 8. Precision plot.

In Fig. 8, our tracker improves the PF based on RGB color gradient tracker by 16.43 % in mean distance precision, and our tracker improves the KCF tracker by 30 % in mean distance precision. In this case our approach performs favorably to state-of-the-art tracking methods.

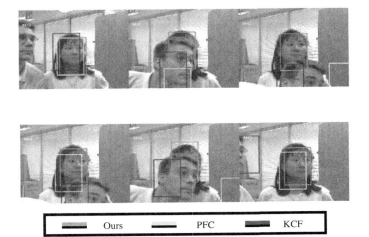

Fig. 9. The results of three tracking algorithms (Girl). (Color figure online)

In Fig. 9, our results are marked with green bounding boxes, the results of particle filter based color gradient are marked with yellow bounding boxes and the results of KCF are marked with blue bounding boxes. The target objects in the Girl sequence are partially occluded at times (See 435, 458 of the girl sequence in Fig. 9) which makes the tracking tasks more difficult. We can see from the Fig. 9 that only our algorithm is able to track the objects successfully in most frames of this sequence. It can be concluded that our algorithm can successfully solve the occlusion well.

6   Conclusion

The particle filter tracking based on unit-linking PCNN image icon, can handle non-rigid and fast moving objects with changes well. Moreover, using this tracker can well track targets with occlusion or clutter. In particle filter trackers, the unit-linking PCNN image icon performs better than color gradient.

Acknowledgments. This work was supported by National Natural Science Foundation of China under grant 61371148.

References

1. Smith, T.F., Waterman, M.S.: Identification of common molecular subsequences. J. Mol. Biol. **147**, 195–197 (1981)
2. Dinh, T.B., Vo, N., Medioni, G.: Context tracker: exploring supporters and distracters in unconstrained environments. In: CVPR (2011)
3. Hare, S., Saffari, A., Torr, P.: Struck: structured output tracking with kernels. In: ICCV (2011)
4. Kalal, Z., Matas, J., Mikolajczyk, K.: P-n learning: bootstrapping binary classifiers by structural constraints. In: CVPR (2010)
5. Sevilla-Lara, L., Learned-Miller, E.G.: Distribution fields for tracking. In: CVPR (2012)
6. Zhang, K., Zhang, L., Yang, M.-H.: Real-time compressive tracking. In: Fitzgibbon, A., Lazebnik, S., Perona, P., Sato, Y., Schmid, C. (eds.) ECCV 2012, Part III. LNCS, vol. 7574, pp. 864–877. Springer, Heidelberg (2012)
7. Hue, C., Cadre, J.L., Pérez, P.: Tracking multiple objects with particle filtering. IEEE Trans. Aerosp. Electron. Syst. **38**(3), 791–812 (2002)
8. Nummiaro, K., Koller-Meier, E., Van Gool, L.: Object tracking with an adaptive color-based particle filter. In: Van Gool, L. (ed.) DAGM 2002. LNCS, vol. 2449, pp. 353–360. Springer, Heidelberg (2002)
9. Eckhorn, R., Reitboeck, H.J., Arndt, M., Dicke, P.W.: Feature linking via synchronization among distributed assemblies: simulation of results from cat cortex. Neural Comput. **2**, 293–307 (1990)
10. Eckhorn, R., Bauer, R., Jordan, W., Brosch, M., Kruse, W., Munk, M., Reitboeck, H.J.: Coherent oscillations: a mechanism of feature linking in the visual cortex? Multiple electrode and correlation analyses in the cat. Biol. Cybern. **60**(2), 121–130 (1988)
11. Johnson, J.L., Ritter, D.: Observation of periodic waves in a pulse-coupled neural network. Opt. Lett. **18**, 1253–1255 (1993)
12. Gu, X.D., Guo, S.D., Yu, D.H.: A new approach for automated image segmentation based on unit-linking PCNN. In: Proceedings of IEEE International Conference on Machine Learning and Cybernetics, IEEE ICMLC 2002 Proceedings, Beijing, China, pp. 175–178 (2002)

13. Kuntimad, G., Ranganath, H.S.: Perfect image segmentation using pulse coupled neural networks. IEEE Trans. Neural Netw. **10**(3), 591–598 (1999)
14. Johnson, J.L.: Pulse-coupled neural nets: translation, rotation, scale, distortion and intensity signal invariance for images. Appl. Opt. **33**(26), 6239–6253 (1994)
15. Gu, X.D., Yu, D.H., Zhang, L.M.: Image shadow removal using pulse coupled neural network. IEEE Trans. Neural Netw. **16**(3), 692–698 (2005)
16. Gu, X.: Feature extraction using unit-linking pulse coupled neural network and its applications. Neural Process. Lett. **27**(1), 25–41 (2008)
17. Henriques, J., Caseiro, R., Martins, P.: High-speed tracking with kernelized correlation filters. IEEE Trans. Pattern Anal. Mach. Intell. **37**(3), 583–596 (2015)

Quaternion Spike Neural Networks

Luis Lechuga-Gutiérrez and Eduardo Bayro-Corrochano[(✉)]

Departamento de Control Automático,
Centro de Investigación y de Estudios Avanzados, Unidad Guagalajara,
Zapopan, Jalisco, Mexico
{lrlechuga,edb}@gdl.cinvestav.mx

Abstract. This work presents a new type of Spike Neural Networks (SNN) developed in the quaternion algebra framework. This new neural structure based on SNN is developed using the quaternion algebra. The training algorithm was extended adjusting the weights according to the quaternion multiplication rule, which allows accurate results with a decreased network complexity with respect to the real SNN. The experimental part shows a good performance for robot manipulator control.

Keywords: Spike neural networks · Quaternion · Spikeprop

1 Introduction

Adaptive signal processing is a fast growing area in signal processing and the third generation artificial neural networks have been an important part of the area of Artificial Intelligence and they have been extensively used in many different engineering tasks. Based on how close these networks are to the functioning and architecture of a biological neural network, in one of his seminal papers Wolfgang divides the artificial neural networks into three generations [1].

The first generation of artificial neural networks with a threshold neuron model was used by McCulloch–Pitts in 1943. Their model, also known as the perceptron, had only two states ('high' or 'low'), based on the weighted sum of the input signals and a step function defining the activation. On the other hand, the second generation of artificial neural networks introduces the concept of learning and the possibility of continuous neuron's states. The second generation is able to solve problems that the first cannot, so providing more computational power [2]. Finally, *Spiking Neural Networks* (SNN) have been named third generation artificial neural networks, so that the information is handled in a space–time way; in other words, the shape of the pulse generated by any neuron is identical to that of any other, and the firing time and the firing location are the only variables containing information [3].

2 Third Generation Artificial Neural Networks

The mathematical models representing the functioning of the third generation artificial neurons can be divided in two major categories; namely, behavior and

© Springer International Publishing Switzerland 2016
L. Cheng et al. (Eds.): ISNN 2016, LNCS 9719, pp. 640–646, 2016.
DOI: 10.1007/978-3-319-40663-3_73

threshold models [3]. The abstraction level is directly related with the computational properties of the corresponding model and depends on the inclusion
or not of biological aspects, such as cell's ions channels, in order to describe
the neuron's behavior. The behavior models include biological aspects, while the
threshold ones do not and they assume a special behavior of the firing signal considering a low threshold of voltage, such that the neuron's firing occurs when the
membrane's potential exceeds the given threshold value. The Hodgkin–Huxley's
model is one example of behavior model, while the Perfect Integrate and Fire
(PIF) model is the most simple example of threshold model, which is discussed
in this paper.

Threshold and Firing Models. The Perfect Integrate and Fire (PIF) and
Leaky Integrate and Fire (LIF) models are included as examples of integrate
and fire models and they have been extensively implemented in several research
papers [4], because of their simple implementation and low computational cost.
    In fact, the solution of the ordinary differential equation describing the neural
system is given as follows:

$$V(t) = V_r e^{-\frac{1}{RC}(t-t_0)} + \frac{1}{C} \int_{t_0}^{t} I(\tau) e^{-\frac{1}{RC}(t-\tau)} d\tau \tag{1}$$

where $V(t)$ is the difference of potential between the two sides of the circuit's
branches.

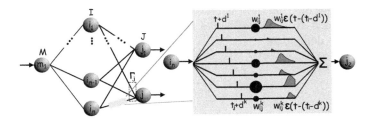

Fig. 1. Feedforward spiking neural network and its respective multiple delayed synaptic
terminals.

2.1   Learning Method

The networks architecture for this method necessarily involves a feedforward
network of spiking neurons with multiple delayed synaptic terminals Fig. 1. Formally, a neuron j, having a set Γj of immediate predecessors, receives a set of
spikes with firing times t_i, $i \in \Gamma_j$. Any neuron generates at most one spike during the simulation interval, and fires when the internal state variable reaches a
threshold θ. The dynamics of the internal state variable $X_j(t)$ are determined

by the impinging spikes it is shown in Fig. 1, whose impact is described by the spike-response function $\epsilon(t)$ weighted by the synaptic efficacy ("weight") w_{ij}:

$$x_j(t) = \sum_{i \in \Gamma_j} w_{ij}\epsilon(t - t_i) \tag{2}$$

Extending the Eq. (2) to include multiple synapses per connection, the state variable x_j of neuron j receiving input from all neurons i can then be described as the weighted sum of the pre-synaptic contributions

$$x_j(t) = \sum_{i \in \Gamma_j} \sum_{k=1}^{m} w_{ij}^k y_i^k(t) \tag{3}$$

where w_{ij}^k denotes the weight associated with synaptic terminal k. The firing time t_j of neuron j is determined as the first time when the state variable crosses the threshold $\theta : x_j(t) \geq \theta$. Thus, the firing time t_j is a non-linear function of the state variable $x_j : t_j = t_j(x_j)$. The threshold θ is constant and equal for all neurons in the network.

Error Backpropagation. For this method called Spike-Propagation [5] the error-backpropagation is derived, of the same way of derivation by Rumelhart et al. The target of the algorithm is to learn a set of firing times, denoted t_j^d, at the output neurons $j \in J$ for a given set of patterns $P[t_1...t_h]$, where $P[t_1...t_h]$ defines a single input pattern described by single spike times for each neuron $h \in H$. In this learning method is chosen the least mean squares error-function as error-function. Given desired spike times t_j^d and actual firing times t_j^a, this error-function is defined by

$$E = \frac{1}{2} \sum_{j \in J} (t_j^a - t_j^d)^2 \tag{4}$$

For error backpropagation, each synaptic terminal is treated as a separate connection k with weight w_{ij}^k. Hence, for a backprop rule, we need to calculate the changes in the weights for each neuron-synapse:

$$\Delta w_{ij}^k = -\eta \frac{\partial E}{\partial w_{ij}^k} \tag{5}$$

When we combine these results, the Eq. (5) evaluates to

$$\Delta w_{ij}^k = -\eta \frac{y_i^k(t_j^a)(t_j^d - t_j^a)}{\sum_{i,l} w_{ij}^l (\partial y_i^l(t)/\partial t(t_j^a))} \tag{6}$$

3   Quaternion Spike Neural Networks

Quaternion algebra (H) was invented by W.R. Hamilton in 1843 in order to extend the properties of complex numbers to 3D space. A quaternion can in fact be defined as a complex number with three imaginary parts [6]. In this section a new Spike Neural structure defined in Quaternion Algebra is introduced and a suitable learning algorithm for such a structure is also reported.

Let us define an QSNN (Quaternion Spike Neural Network) as a multilayer Spike Neural Network in which inputs and outputs values, weights and biases are quaternions. The learning rule algorithm has been also developed as a quaternion expression.

$$\Delta W_{ml}^n = -\eta \frac{\partial E}{\partial W_{ml}^n} \tag{7}$$

Thus the error of Eq. (7) now belongs to the quaternion algebra ($E \in H$), and W_{ml}^n is a quaternionic weights vector of the n delay, between the presynaptic neuron m and the postsynaptic neuron l, so the Eq. (7) extends as

$$\frac{\partial E}{\partial W_{ml}^n} = \frac{\partial E}{\partial W r_{ml}^n} + \frac{\partial E}{\partial W i_{ml}^n}i + \frac{\partial E}{\partial W j_{ml}^n}j + \frac{\partial E}{\partial W k_{ml}^n}k \tag{8}$$

Using the chain rule in (8) for each element, we obtain

$$\frac{\partial E}{\partial W r_{ml}^n} = \frac{\partial E}{\partial tr_{ml}^n}\frac{\partial tr_{ml}^n}{\partial F_{ml}^n}\frac{\partial F_{ml}^n}{\partial W r_{ml}^n} + \frac{\partial E}{\partial ti_{ml}^n}\frac{\partial ti_{ml}^n}{\partial F_{ml}^n}\frac{\partial F_{ml}^n}{\partial W r_{ml}^n} + \frac{\partial E}{\partial tj_{ml}^n}\frac{\partial tj_{ml}^n}{\partial F_{ml}^n}\frac{\partial F_{ml}^n}{\partial W r_{ml}^n} + \frac{\partial E}{\partial tk_{ml}^n}\frac{\partial tk_{ml}^n}{\partial F_{ml}^n}\frac{\partial F_{ml}^n}{\partial W r_{ml}^n}$$

$$\frac{\partial E}{\partial W i_{ml}^n} = \frac{\partial E}{\partial tr_{ml}^n}\frac{\partial tr_{ml}^n}{\partial F_{ml}^n}\frac{\partial F_{ml}^n}{\partial W i_{ml}^n} + \frac{\partial E}{\partial ti_{ml}^n}\frac{\partial ti_{ml}^n}{\partial F_{ml}^n}\frac{\partial F_{ml}^n}{\partial W i_{ml}^n} + \frac{\partial E}{\partial tj_{ml}^n}\frac{\partial tj_{ml}^n}{\partial F_{ml}^n}\frac{\partial F_{ml}^n}{\partial W i_{ml}^n} + \frac{\partial E}{\partial tk_{ml}^n}\frac{\partial tk_{ml}^n}{\partial F_{ml}^n}\frac{\partial F_{ml}^n}{\partial W i_{ml}^n}$$

$$\frac{\partial E}{\partial W j_{ml}^n} = \frac{\partial E}{\partial tr_{ml}^n}\frac{\partial tr_{ml}^n}{\partial F_{ml}^n}\frac{\partial F_{ml}^n}{\partial W j_{ml}^n} + \frac{\partial E}{\partial ti_{ml}^n}\frac{\partial ti_{ml}^n}{\partial F_{ml}^n}\frac{\partial F_{ml}^n}{\partial W j_{ml}^n} + \frac{\partial E}{\partial tj_{ml}^n}\frac{\partial tj_{ml}^n}{\partial F_{ml}^n}\frac{\partial F_{ml}^n}{\partial W j_{ml}^n} + \frac{\partial E}{\partial tk_{ml}^n}\frac{\partial tk_{ml}^n}{\partial F_{ml}^n}\frac{\partial F_{ml}^n}{\partial W j_{ml}^n}$$

$$\frac{\partial E}{\partial W k_{ml}^n} = \frac{\partial E}{\partial tr_{ml}^n}\frac{\partial tr_{ml}^n}{\partial F_{ml}^n}\frac{\partial F_{ml}^n}{\partial W k_{ml}^n} + \frac{\partial E}{\partial ti_{ml}^n}\frac{\partial ti_{ml}^n}{\partial F_{ml}^n}\frac{\partial F_{ml}^n}{\partial W k_{ml}^n} + \frac{\partial E}{\partial tj_{ml}^n}\frac{\partial tj_{ml}^n}{\partial F_{ml}^n}\frac{\partial F_{ml}^n}{\partial W k_{ml}^n} + \frac{\partial E}{\partial tk_{ml}^n}\frac{\partial tk_{ml}^n}{\partial F_{ml}^n}\frac{\partial F_{ml}^n}{\partial W k_{ml}^n} \tag{9}$$

Developing each of the partial derivatives and using the dot product and the cross product between quaternions is achieved regroup the Eq. (9) in, it is achieved a more compact equation

$$\frac{\partial E}{\partial W_{ml}^n} = \left(E \odot \frac{1}{\sum_l W \frac{\partial F_{ml}^n}{\partial tr_{ml}^n}} \right) \otimes t_l^* \tag{10}$$

4   QSNN Vs SNN

In this section QSNN neural networks are compared to SNNs, using the same parameters for the neuronal models. For both cases, a LIF neuronal model is used, with a value of resistance equal to 730 Ohms, a capacitance of 20 $\mu Farads$, a theta $= -30$ mv as threshold, a learning rate of 0.1, the weights were initialized arbitrarily at 1.75, and 10 delays of 50 ms and a time window of 1000 ms were used.

The input vector to the neuron SNN is: $IN = [518\ 534\ 477\ 580]$ while the input vector to the neuron QSNN is: $QIN = [518\ 534i\ 477j\ 580k]$ and 6 tests within the range of 0 to 1000 are conducted to test the efficiency of QSNN against SNN. The results obtained are shown in Fig. 2. It can be seen in all cases as spike neural networks based on quaternions converge faster than conventional neural networks spike. This is because when working with a quaternionic product between the input vector (quaternion) and the weight vector (quaternion) all elements of both vectors are correlated in contrast to a product between two vectors (Fig. 2).

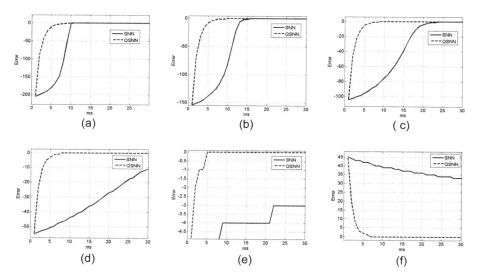

Fig. 2. The comparison between the SNN (solid line) and QSNN (dashed line), under six different setpoints are displayed: in the subfigure (a) the setpoint is 680 ms, (b) 630 ms, (c) 580 ms, (d) 530 ms, (e) 480 ms and the set point for the figure (f) was 430 ms.

5   Kinematic Control of a Manipulator of 6 DOF Using QSNN

As described in the previous section QSNN require fewer iterations to converge to the desired point, compared against SNN, so it was decided to evaluate its performance for kinematic control of a manipulator of 6 degrees of freedom.

In this section the kinematic control of a 6 Degree Of Freedom (DOF) manipulator is presented, using six neurons in parallel, each neuron controlling a degree of freedom. All weights were initialized arbitrarily in 1, so that all the neurons emit the first pulse into the same instant (which decoded it is equal to 93 degrees), this positions the end effector in $[x, y, z] = [-9.976, 4.971, 9.516]$. Manipulator parameters are: Length of link 1, 2 and 3 equal to 10 cm, 10 cm and 5 cm respectively. The initial angles are issued by the response of the presynaptic neurons when weights are initialized. In Fig. (3–a) it is possible to observe the evolution of the angles proposed by neurons to evolve from the initial configuration (initializing all presynaptic weight to 1, all the neurons emitting the first pulse in the same instant of time) of the manipulator to one of the settings required to position the end effector in the setpoint ($[x,y,z] = [12 \ 12 \ 10]$).

5.1   Kinematic Control SNN

This subsection the kinematic control of the Sect. 5 was replicated, in order to make a comparison between QSNN and SNN. In Fig. 4 the kinematic control of

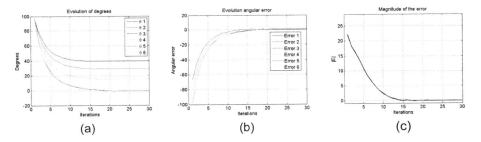

Fig. 3. Figure (a) showns the evolution of the 6 angles, in figure (b) the evolution of the error of each of the 6 degrees of freedom it is shown, and in figure (c) the evolution of magnitude of the error, this magnitude is between the setpoint and the end effector of the manipulator.

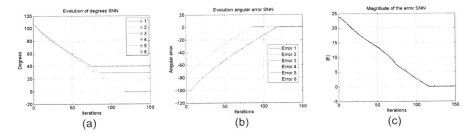

Fig. 4. In figure (a) is shown the evolution of the 6 angles, in figure (b) the evolution of the error of each of the 6 degrees of freedom it is shown, and in figure (c) the evolution of magnitude of the error, this magnitude is between the setpoint and the end effector of the manipulator.

manipulators test was doubled but this time with SNN, resulting considerable increase in the number of iterations required for get the necessary configuration, and the manipulator end effector arrives to the set point. In addition the evolution of the trajectory of each degree of freedom is generated in a milder form in QSNN (Fig. 3), compared to SNN (Fig. 4) this is an advantage if you later want design a trajectory tracking control. This lies in the close relationship that naturally possess imaginary axes with rotations (Fig. 4).

6   Conclusions

It was found that the quaternionic spike neural networks in conjunction with the with the quaternion gradient descendent, working on a hypercomplex algebra outperform to the Spike neural networks. Furthermore they are ideal for applications in the area of control in manipulators with angular movements, having a better performance compared to SNN. Improving the behavior of spikes neural networks.

References

1. Wolfgang, M.: Networks of spiking neurons: the third generation of neural network models. IEEE Trans. Neural Netw. **9**, 1659–1671 (1997)
2. Haykin, S.: Neural Networks: A Comprehensive Foundation. Prentice-Hall, New York (1999)
3. Wulfram, G., Werner, M.: Spiking Neuron Models, Single Neurons, Population, Plasticity. MIT Press, Cambridge (2002)
4. Van Rullen, R., Delorme, A., Gautrais, J., Thorpe, S.: (SpikeNET): a simulator for modeling large networks of integrate and fire neurons. Neurocomput. **22**, 989–996 (1999)
5. Bohte, S., Kok, J., Han, L.: Error-backpropagation in temporally encoded networks of spiking neurons. Neurocomput. **48**, 17–37 (2002)
6. Eduardo, B.C.: TGeometric Computing, for Wavelets Transforms, Robot Vision, Learning, Control and Action. Springer, London (2010)

Vector-Matrix Models of Pulse Neuron for Digital Signal Processing

Vladimir Bondarev[✉]

Sevastopol State University, Sevastopol, Russia
bondarev@sevsu.ru

Abstract. The models of multi-input pulse neuron in generalized vector-matrix form are proposed in order to solve digital signal processing problems. Non-recursive and recursive digital models are considered. Nonrecursive models use the description of linear systems in a convolution form and the input signal is presented as a sequential or a parallel binary vector. Recursive models are based on the description of linear systems in the time domain and use an impulse response and a state space approaches. A learning rule for the mentioned models of a pulse neuron is derived to solve a problem of signal reconstruction and adaptive noise suppression. Results of a computer simulation are presented.

Keywords: Pulse neuron · Vector-matrix model · Adaptive signal processing · Learning rule

1 Introduction

Different models of artificial neural networks are widely used in digital signal processing [1]. In the last decade much attention is paid to the pulsed neural networks [2]. Models of such networks are biologically inspired and based on pulse coding of information. The analysis of properties of pulsed neural networks for digital signal processing is the important task in the field of development of adaptive real-time systems.

Various models of pulse neurons were studied in [3–5]. The simplified representation of internal processes of pulse neuron is reduced to the summation of the post-synaptic potentials caused by input pulses and to the emitting of the output pulse if the potential of the neuron soma exceeds some threshold. In this sense the dynamics of pulse neural networks corresponds to the dynamics of systems with pulse frequency modulation [6, 7].

Various applications of the pulse neuron models for signal processing were considered in [2, 8–10]. The multi-input model of the pulse neuron (PN) for adaptive filtering was studied in [11]. Discrete expressions in a special form for the internal state computation of the PN and the learning rule for the reconstruction of the useful signal were also considered in [11].

In this paper, we develop this approach further. We will focus on the generalization of the model of multi-input PN, its representation in the vector-matrix recursive and non-recursive forms which are widely used for digital signal processing.

© Springer International Publishing Switzerland 2016
L. Cheng et al. (Eds.): ISNN 2016, LNCS 9719, pp. 647–656, 2016.
DOI: 10.1007/978-3-319-40663-3_74

2 Problem Formulation

A multi-input pulse neuron model in which the linear filters to simulate responses to the input pulses are applied is presented in the Fig. 1. In contrast to the models mentioned in the introduction this model has no matching of the potential of the neuron soma to the threshold. This simplification is acceptable because the threshold matching corresponds to the quantification of a neuron output and provides a transformation of a continuous process to a discrete process [12]. This additional transformation can be easily derived if necessary.

It is assumed that PN inputs (see Fig. 1) are bipolar frequency modulated pulse sequences $u_i(t)$. Input signal will be represented either in the form of a sequence of pulses arriving at a single channel input (sequential scheme) or in the form of pulses arriving simultaneously at all PN inputs during time interval Δt (parallel scheme, see vector **b** in Fig. 1).

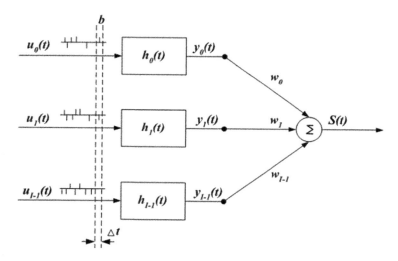

Fig. 1. Pulse neuron

Input pulse sequences $u_i(t)$ cause a linear filter reaction defined by its pulse response $h_i(t)$. These filter reactions correspond to excitatory or inhibitory postsynaptic potentials, depending on the sign of the input pulse. Filter reactions $y_i(t)$ are weighted with synaptic weights w_i and summed to form a neuron state $s(t)$.

Input pulse sequences are represented as the sum of the signed Dirac delta functions [3, 6, 7]:

$$u_i(t) = \sum_j \lambda_i^j \delta(t - t_i^j), t_i^j \le t, \tag{1}$$

where t is current time, t_i^j is a time corresponding to the occurrence of the pulse j at input i, λ_i^j is the sign of the input pulse. Since the filter response under zero initial conditions is determined by a convolution integral, the reaction of every filter to the input (1) can be presented as [3, 7]:

$$y_i(t) = \sum_j \lambda_i^j h_i(t - t_i^j), \tag{2}$$

where $h_i(t)$ are pulse responses of linear filters. Then the state $s(t)$ of the pulse neuron is determined as follows

$$s(t) = \sum_{i=0}^{I-1} w_i y_i(t), \tag{3}$$

where I is the number of PN inputs, w_i is the weighting factor of input i.

Expressions (2) and (3) are shown in the continuous form. To calculate values of $s(t)$ with digital processors a time sampling of (2) and (3) must be performed. Let the values of $s(t)$ are calculated at discrete times $t_n = n\Delta t$, where Δt is a time sampling step.

We will derive the necessary discrete expressions in vector-matrix form which is widely used in a case of solving various adaptive digital signal processing problems [13]. This will extend known methods for adaptive signals processing with the pulse neural networks. Depending on the chosen description of linear filters nonrecursive and recursive PN models will be considered.

3   Nonrecursive Models

3.1   Sequential Input Model

Let pulse responses $h_i(t)$ be characterized by finite duration T and $t - T \le t_i^j \le t$. In accordance with (2) $y_i(t)$ represents the sum of pulse response samples at time moments when pulses arrive at the filter input.

Let an input sequence of pulses of a channel i for the time moment n be a binary vector $\mathbf{b}_i^{\mathrm{T}}(n)$ whose elements are defined as follows

$$b_i(n - k) = \begin{cases} \lambda_i^j, & t_n - (k+1)\Delta t < t_i^j \le t_n - k\Delta t \\ 0, & otherwise \end{cases}, \tag{4}$$

where $k = 0, 1, \ldots, K-1$ and $K = [T/\Delta t]$.

The reaction of each filter to an input pulse sequence in discrete time domain can be written as

$$y_i(n) = \mathbf{b}_i^{\mathrm{T}} \mathbf{h}_i, \tag{5}$$

where $\mathbf{h}_i = (h_i(0), h_i(1), \ldots, h_i(K-1))^{\mathrm{T}}$ denotes the impulse response vector.

Consider the matrices $\mathbf{B}$ and $\mathbf{H}$, where the rows of matrix $\mathbf{B}$ are binary input vectors (4) of every PN channel, and the columns of matrix $\mathbf{H}$ are impulse response vectors $\mathbf{h}_i$. Then output reactions of all filters can be represented by the vector containing elements of the main diagonal of the matrix product

$$\mathbf{y}(n) = diag(\mathbf{BH}), \tag{6}$$

where $\mathbf{y}(n) = (y_0(n), y_1(n), \ldots, y_{I-1}(n))^{\mathrm{T}}$. If the impulse responses of all channels are the same, then $\mathbf{y}(n) = \mathbf{Bh}$.

Hence the generalized model of the PN for the sequential scheme can be written as a vector-matrix expression

$$s(n) = \mathbf{w}^{\mathrm{T}} diag(\mathbf{BH}). \tag{7}$$

In case of identical pulse responses of channels $s(n) = \mathbf{w}^{\mathrm{T}} \mathbf{Bh}$, where $\mathbf{w}^{\mathrm{T}} = (w_0, w_1, \ldots, w_{I-1})$ are the PN weighting vector.

3.2   Parallel Input Model

Let $\mathbf{b}(n)$ be a column vector whose elements correspond to the sign of the pulse at an input i at time moment n:

$$b_i(n) = \begin{cases} \lambda_i^j, & t_n - \Delta t < t_i^j \le t_n \\ 0, & otherwise \end{cases}. \tag{8}$$

In this case, the reaction of filters to the input (8) can be written as the convolution of vectors

$$\mathbf{y}(n) = \sum_{k=0}^{K-1} \mathbf{b}(n-k) . \times \mathbf{h}(k), \tag{9}$$

where $\mathbf{y}(n)$ is a vector of filters output signals, $\mathbf{h}(k) = (h_0(k), h_1(k), \ldots, h_{I-1}(k))$ is a vector, whose elements correspond to parallel samples of impulse responses of all channels, $\mathbf{b}(n-k)$ is the delayed input binary vector, symbols $.\times$ denotes the element-by-element product of vectors. Then

$$s(n) = \mathbf{w}^{\mathrm{T}} \mathbf{y}(n), \tag{10}$$

where vector $\mathbf{y}(n)$ is defined by the formula (9).

If we use matrices $\mathbf{B}$ and $\mathbf{H}$, then columns of the matrix $\mathbf{B}$ will be parallel binary vectors $\mathbf{b}(n)$ and rows of the matrix $\mathbf{H}$ will be vectors of parallel samples of channel impulse responses $\mathbf{h}(k)$. Therefore in addition to the model (9)–(10) a generalized model of a PN for the parallel scheme can be described by the same vector-matrix expression (7).

4 Recursive Models

4.1 Impulse Response Model

Nonrecursive PN models described above are characterized by finite impulse response, which ensures the stability of linear filters. However, in order to store all impulse response samples and shifted input binary vectors a significant amount of memory is required.

It is possible to eliminate the specified disadvantage if we use the recursive implementation (2). In case of known transfer function of the linear filter its pulse response can be written in the form of the following sum [14]:

$$h(t) = \sum_{q=1}^{Q} r_q \exp(\alpha_q t), \tag{11}$$

where r_q denotes residue of transfer function $H(p)$ in the pole α_q. Substituting expression (11) into (2), we find that

$$y_i(t_n) = \sum_{q=1}^{Q} \sum_{j=1}^{n} \lambda_i^j r_{iq} \exp(\alpha_{iq}(t_n - t_i^j)), \tag{12}$$

where $t_i^j \le t_n$. We will take into account that $t_n = t_{n-1} + \Delta t$, and also that only one pulse arrives at the filter input during time interval $(t_{n-1}, t_n]$, i.e. $b_i(n) = \lambda_i^n$ or $b_i(n) = 0$. Then after simple transformations the expression (12) can be written as follows

$$y_i(t_n) = \sum_{q=1}^{Q} \sum_{j=1}^{n-1} \lambda_i^j r_{iq} \exp(\alpha_{iq}(t_{n-1} - t_i^j)) \exp(\alpha_{iq} \Delta t) + b_i(n) r_{iq}. \tag{13}$$

Denoting by

$$x_{iq}(t_{n-1}) = \sum_{j=1}^{n-1} \lambda_i^j r_{iq} \exp(\alpha_{iq}(t_{n-1} - t_i^j)), \tag{14}$$

we derived expressions to calculate of the filter reaction

$$y_i(t_n) = \sum_{q=1}^{Q} x_{iq}(t_n) \tag{15}$$

$$x_{iq}(t_n) = x_{iq}(t_{n-1}) \exp(\alpha_{iq}\Delta t) + b_i(n)r_{iq}. \tag{16}$$

Expression (16) can be rewritten in the vector form as follows:

$$\mathbf{x}_i(n) = \mathbf{x}_i(n-1). \times \exp(\boldsymbol{\alpha}_i\Delta t) + b_i(n)\mathbf{r}_i, \tag{17}$$

where $\mathbf{x}_i, \boldsymbol{\alpha}_i, \mathbf{r}_i$ are Q-dimensional vectors. Then according to (15) the output signal value for the channel i will be equal to the sum of all elements of the vector $\mathbf{x}_i(n)$, i.e.

$$y_i(n) = \mathbf{1}^{\mathrm{T}}\mathbf{x}_i(n), \tag{18}$$

where $\mathbf{1}^{\mathrm{T}}$ is a unit vector of suitable size.

Expression (17) defines the recursive vector form of the filter implementation. The state $s(n)$ of the PN is determined by the formula (10).

4.2   State-Space Model

Other possibility to build recursive models of the PN is based on state-space method. State-space representation of a linear continuous time-invariant system is written in the following form [14]:

$$\dot{\mathbf{x}}_i(t) = \mathbf{A}\mathbf{x}_i(t) + \mathbf{G}u_i(t) \tag{19}$$

$$y_i(t) = \mathbf{C}^{\mathrm{T}}\mathbf{x}_i(t), \tag{20}$$

where $\mathbf{x}_i(t)$ is a n-dimensional state vector, $\mathbf{A}, \mathbf{G}, \mathbf{C}^{\mathrm{T}}$ are matrices of respective dimensions defined by the differential equation that describes the system.

If the state Eq. (19) is integrated over the discretization interval $(t_{n-1}, t_n]$ and either one pulse or none arrives at the filter input during this interval, then the following state-space equations in the recursive discrete form can be derived

$$\mathbf{x}_i(n) = e^{\mathbf{A}\Delta t}\mathbf{x}_i(n-1) + b_i(n)\mathbf{G} \tag{21}$$

$$y_i(n) = \mathbf{C}^{\mathrm{T}}\mathbf{x}_i(n). \tag{22}$$

Expressions (21) and (22) determine the reaction of one channel of the PN in generalized recursive vector-matrix form. Thus the output state $s(t)$ of the PN is defined by the expression (10).

5 Computer Simulation

We will consider application of non-recursive (7), (9)–(10) and recursive (17)–(18), (21)–(22) PN models for adaptive filtering. As an example we will consider the problem of an original signal reconstruction from input sequence of pulses and suppression of additive noise. The general scheme of this problem is illustrated in Fig. 2. On the scheme the adaptive filter (AF) is realized as a cascade connection of the pulse-frequency converter (PFC) and the PN.

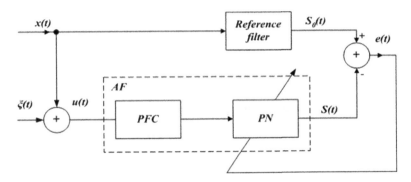

Fig. 2. Adaptive filter based on PN

Let the input signal of the AF will be equal to

$$u(t) = x(t) + \xi(t), \tag{23}$$

where $x(t)$ is the original signal, $\xi(t)$ is a stochastic stationary noise uncorrelated with $x(t)$. The PFC will convert $u(t)$ to pulse sequence (1). The output of the PN is presented by the signal $s(t)$ in accordance with (3). The reference filter defines the desirable linear operator for the original signal conversion.

The challenge now is to adjust weighting vector $\mathbf{w} = (w_0, w_1, \ldots, w_{I-1})$ providing the minimum of the functional

$$J(\mathbf{w}) = E\{[s_0(t) - s(t)]^2\}, \tag{24}$$

where E is mathematical expectation, $s_0(t)$ is the reference filter output signal.

We will obtain the estimation of the gradient of the functional (24) by vector $\mathbf{w}$. Taking (10) into account, we receive

$$\nabla J(\mathbf{w}) = -2E\{e(n)\mathbf{y}(n)\}. \tag{25}$$

Applying the method of the quickest descent and executing stochastic approximation, we receive the simple learning rule to find the vector $\mathbf{w}$. This rule corresponds to the known Widrow-Hoff rule [13]:

$$\mathbf{w}_n = \mathbf{w}_{n-1} + \mu(n)e(n)\mathbf{y}(n), \tag{26}$$

where $\mu(n)$ is a learning rate and $\mathbf{y}(n)$ is determined by (6), (9), (18) or (20).

Let the PN filters have the following pulse responses

$$h_i(t) = (\exp(-\frac{t - \Delta_i}{\tau_m}) - \exp(-\frac{t - \Delta_i}{\tau_s}))H(t - \Delta_i), \tag{27}$$

where τ_m, τ_s are time constants, Δ_i is the time delay, H is the Heaviside step function. The pulse response (27) is often used for the description of postsynaptic potentials [3].

We will consider a simple case when pulse responses of filters have identical form, but are time shifted for $\Delta_i = i\Delta t$.

Then the PN state $s(n)$ in case of nonrecursive filters can be determined directly by the models (7) or (9)–(10).

To apply the recursive models (17)–(18) or (21)–(22) it is necessary to execute preliminary transformations. In accordance with (11) the low frequency filter with first order transfer function $H(p) = 1/(p + \alpha)$ has a pulse response $h(t) = \exp(-\alpha t)$. Then the recursive model (17)–(18) for the pulse response (27) can be represented by means of the difference between reactions of two first order filters:

$$\mathbf{x}_1(n) = a_1 \mathbf{x}_1(n - 1) + \mathbf{b}(n) \tag{28}$$

$$\mathbf{x}_2(n) = a_2 \mathbf{x}_2(n - 1) + \mathbf{b}(n) \tag{29}$$

$$\mathbf{y}(n) = \mathbf{x}_1(n) - \mathbf{x}_2(n), \tag{30}$$

where $a_1 = \exp(-\Delta t/\tau_m)$, $a_2 = \exp(-\Delta t/\tau_s)$, $\mathbf{x}_1(n)$ and $\mathbf{x}_2(n)$ are vectors of the size $I - 1$. The same recursive model can be derived by using expressions (21)–(22).

An AF simulation was carried out by means of nonrecursive (7), (9)–(10), (27) and recursive (28)–(30) models of the PN under the learning rule (26).

The parameters of the models were as follows: $I = 200$, $K = 32$, $\Delta t = 1$ ms, $\tau_m = 5$ ms, $\tau_s = 2$ ms. The original signal was a harmonic oscillation with the frequency of 20 Hz, and the additive disturbance was presented as a white noise with the normal distribution and the variance equal to unit. Transmission ratio of the reference filter was equal to 1. Because in this case $s_0(t) = x(t)$, then the purpose of the PN was the reconstruction of the original signal from its mixture with the additive noise based on sequence of input pulses.

The initial value of the learning rate was equal to 0.5 and decreased in proportion to the $1/\sqrt{n}$. As the PFC the model of the integral bipolar pulse-frequency modulator was used [6, 10].

Results of the simulation showed that the AF constructed on the base of offered nonrecursive and recursive PN models and using the learning rule (26) performs well reconstructing of the original signal and suppresses noise. If the signal-to-noise ratio for an input signal was equal −3 dB, then for an output signal it reached value of 14.5 dB, i.e. the signal-to-noise ratio as a result of the filtering improved by 17.5 dB. Also the values of the PN state $s(n)$ calculated on the basis of specified nonrecursive and the recursive PN models were identical. This confirms the correctness of the derived formulas. With the specified simulation parameters, computation time of PN output state values on the basis of the recursive models was nearly 4.5 times less, than on the basis of nonrecursive models.

6   Conclusion

Presented recursive and non-recursive vector-matrix PN models are quite common and are suited for digital implementation. Recursive models compared to non-recursive models require less processing time (time complexity) and less memory (space complexity) during implementation but they require more detailed stability and quantization error analysis. Non-recursive models show more time and space complexity but can provide learning rules not only for synaptic coefficients but also for pulse responses of every PN channel [11]. Thus digital filters with complex frequency response curve can be implemented using non-recursive models of the PN. Non-recursive models are stable and do not require complex error analysis due to limited duration of the pulse response.

Considered adaptive filtering scheme base on the PN is versatile and can be used to solve different adaptive signal processing problems. Presented PN learning rule for the original signal reconstruction from pulse sequence and for additive noise suppression corresponds to Widrow-Hoff rule, and allows to evaluate quasioptimal values of synaptic coefficients on the basis of derived models. Developed models are also oriented to be applied by analogy of other known learning rules for adaptive filtering, particularly for unsupervised learning rules.

Further directions of this research are the estimation of computation efficiency of proposed PN models and the implementation of additional learning rules as well as their development considering the impulse representation not only for input, but also for output signals.

References

1. Hen, H.Y., Jenq-Neng, H.: Introduction to neural networks for signal processing. In: Hen, H. Y., Jenq-Neng, H. (eds.) Handbook of Neural Network Signal Processing, pp. 1–30. CRC Press, Boca Raton (2002)
2. Maass, W.: Paradigms for computing with spiking neurons. In: van Hemmen, J.L., Cowan, J.D., Domany, E. (eds.) Models of Neural Networks. Early Vision and Attention, vol. 4, pp. 373–402. Springer, New York (2002)
3. Gerstner, W., Kistler, W.M.: Spiking Neuron Models – Single Neurons, Populations, Plasticity. Cambridge University Press, Cambridge (2002)
4. Natschlaeger, T., Ruf, B.: Spatial and temporal pattern analysis via spiking neurons. Netw.: Comput. Neural Syst. 9, 319–332 (1998)
5. Ponulak, F., Kasiński, A.: Introduction to spiking neural networks: information processing, learning and applications. Acta Neurobiol. Exp. 71, 409–433 (2011)

6. Popkov, Y., Ashimov, A.A., Asaubaev, K.: Statistical Theory of Automatic Systems with Dynamic Pulse-Frequency Modulation. Nauka Publ., Moscow (1988). (in Russian)
7. Bondarev, V.N.: On system identification using pulse-frequency modulated signals. Research report 88-E-195, Eindhoven University of Technology, Eindhoven (1988)
8. Wei, D., Harris, J.G.: Signal reconstruction from spiking neuron models. In: Proceedings of the 2004 International Symposium on Circuits and Systems (ISCAS 2004), vol. 5, pp. 353–356. IEEE Press (2004)
9. Bondarev, V.N.: Adaptive pulse-frequency modeling aimed at digital signal processing problems. Vestn. SevGTU **18**, 46–51 (1999). (in Russian)
10. Bondarev, V.N.: Adaptive synthesis of pulse-frequency digital nonrecursive filters. Sbornik nauchnyih trudov AVMS im. P.S. Nahimova **4**, 80–85 (2012). (in Russian)
11. Bondarev, V.N.: Application of digital model of pulse neuron for the adaptive signal filtration. In: Proceedings of the XVII All-Russian Scientific and Technical Conference "Neuroinformatics-2015", Part II, pp. 169–177. NIYaU MIFI Publ., Moscow (2015). (in Russian)
12. Bruckstein, A.M., Zeevi, Y.Y.: Analysis of "Integrate-to-Threshold" neural coding schemes. Biol. Cybern. **34**, 63–79 (1979)
13. Widrow, B., Stearns, S.D.: Adaptive Signal Processing. Prentice-Hall, Englewood Cliffs (1985)
14. Nise, N.S.: Control Systems Engineering, 6th edn. Wiley, Hoboken (2011)

About $\Sigma\Pi$-neuron Models of Aggregating Type

Zaur Shibzukhov[1,2]($\boxtimes$) and Denis Cherednikov[2]

[1] Institute of Applied Mathematics and Automation, Nalchik, Russia
szport@gmail.com
[2] Moscow Pedagogical State University, Moscow, Russia
denis.cherednikov@gmail.com

Abstract. A new class of the artificial neuron models is described in this work. These models are based on assumptions: (1) contributions of synapses are summing with the help of certain aggregation operation; (2) contribution of synaptic clusters are computed with the help of another aggregation operation on the set of simple synapses. These models include a big part of known functional models of neurons. The generalization of the $\Sigma\Pi$-neuron model on the basis of aggregation operations is presented. Correctness of the $\Sigma\Pi$-neuron model of aggregating type under easily verifiable conditions is also presented.

Keywords: Neural network · Neuron model · Aggregation function · Sigma-Pi neuron

1 Introduction

The classic approach to neural networks represents them as network of neurons. Each neuron in the network has multiple inputs from other neurons or inputs of all net and only single output, which can divide into several branches for transferring to other neurons.

Neurons are complicated elements of the human brain. They have branched *dendritic system*, which receives signals from other neurons through synapses. Synapses can be simple with single input and complicated with multiple inputs. They may also occur spatially localized synaptic clusters that form the dendritic zone in which information processing is performed almost independently of other dendritic zones. Synapses within the cluster also have an influence on each other. Synapses are transform their inputs into a positive or negative contribution to the total potential of the neuron.

Dendritic system of the neuron aggregates contributions and forms the *total potential* of the neuron. The output signal is generated on the base of the value of total potential of the neuron.

L. Cheng et al. (Eds.): ISNN 2016, LNCS 9719, pp. 657–664, 2016.
DOI: 10.1007/978-3-319-40663-3_75

2   General Functional Neuron Model of Aggregating Type

Let's a segment $\mathbf{X} \subset \mathbb{R}$ (for example, $[0,1]$ or $[-1,1]$) defines a set of all values, transmitted between neurons; a segment $\mathbf{U} \subseteq \mathbb{R}$ defines a set of values for the total potential of the neuron.

Neuron model of aggregating type is built on the bases of application of aggregation functions [1,2]. It contains several structural layers: *bottom layer of synapse, intermediate layer of synaptic clusters* and *top layer of neuron*.

Synapses could be simple and complicated. *Simple synapse* have single input. It's transformation function is scalar function depending on parameter:

$$u = \mathrm{syn}(x, w),$$

where u – contribution of the synapse, $x \in \mathbf{X}$ – input of the synapse, $w \in \mathbf{W}$ – parameter of synaptic weight. The classic model of simple synapse is the linear model: $\mathrm{syn}(x, w) = wx$. Many of non-linear models could be formulated in the following form:

$$u = \mathrm{syn}(x, w) = w\varphi(x - a),$$

where $\varphi(s)$ is monotone non-decreasing function such that $\varphi(s) = 0 \Leftrightarrow s \leqslant 0$. *Complicated synapse* have multiple inputs. It's transformation function is a function of several arguments depending of single or several parameters of synaptic weight:

$$u = \mathrm{syn}\{x_i \colon i \in \mathbf{i} \mid w_1, \ldots, w_m\},$$

where Δu is contribution of the complicated synapse, x_i is it's i-th unput, $\mathbf{i} \subseteq \{1, \ldots, n\}$ is the set of indexes of inputs, $w_1, \ldots, w_m$ are synaptic weights.

Synaptic cluster brings together a group of spatially localized synapses. It could be considered as non-linear aggregator of contributions of synapses:

$$s = \mathrm{Agg}_C\{u_j \colon j \in \mathbf{j}\},$$

where s is contribution of the synaptic cluster, u_j is contribution of the synapse, $\mathbf{j} \subset \{1, \ldots, N\}$ set of indexes of synapses of the synaptic cluster, Agg_C is aggregation function for calculating contributions of the synaptic clusters.

All contributions of synaptic clusters are aggregated on the top layer. *Total aggregated potential* of the neuron is evaluated as

$$s = \mathrm{Agg}\{s_1, \ldots, s_m\},$$

where $s_1, \ldots, s_m$ are contribution of synapric clusters; Agg is aggregation function for evaluation of total potential neuron.

Output function $\mathrm{out} \colon \mathbf{U} \to \mathbf{X}$ defines the rule for transformation of total potetial of the neuron into it's output y:

$$y = \mathrm{out}(s).$$

3 Single-layered Neuron Model of Agregating Type

in classic model of neuron the contributions of synapses are aggregated via summation:

$$y = \text{out}\Big(\theta + \sum_{j=1}^{m} u_j\Big), \tag{1}$$

where u_i is contribution of j-th synapse, θ is the constant.

The operation of arithmetic summation has nonlinear generalization which saves all it's algebraic properties. This is g-summing [3,4]:

$$\Big(\theta + \sum_{i=1}^{n} u_i\Big)_g = g^{-1}\big(g(\theta) + \sum_{i=1}^{n} g(u_i)\big),$$

where g is monotone inversible function монотонная обратимая функция and $g(0) = 0$. The Acẑel theorem [5] explains why it is g-summation: any continous and associative aggregation function on a segment is a sort of g-summing. Much of known aggregation functions are g-summings. For example:

$g(u)$	$\Big(\theta + \sum\limits_{j=1}^{m} u_j\Big)_g$
$u^{\langle p\rangle} = \text{sign}u\|u\|^p$	$\big(\theta^{\langle p\rangle} + \sum\limits_{j=1}^{m} u_j^{\langle p\rangle}\big)^{\langle 1/p\rangle}$
e^{pu}	$\dfrac{1}{p}\ln\big(e^{p\theta} + \sum\limits_{j=1}^{m} e^{pu_j}\big)$
$\begin{cases} \ln\|u\| + i\pi[\![u < 0]\!], & u \neq 0 \\ 0, & u = 0 \end{cases}$	$\theta \prod\limits_{j=1}^{m} u_j$

The model of neuron when g-summing is used for aggregation of contributions of synapses loks like

$$y = \text{out}_g\Big(g(\theta) + \sum_{j=1}^{m} g(u_j)\Big),$$

where $\text{out}_g(s) = \text{out}(g^{-1}(s))$.

Note that for any summing-like operation $\tilde{+}$ there are two key properties:

$\Sigma 1)$ $a \,\tilde{+}\, 0 = 0 \,\tilde{+}\, a = a;$

$\Sigma 2)$ the equation $a \,\tilde{+}\, x = b$ и $x \,\tilde{+}\, a = b$ must have solutions for any permissiable pairs of a and b.

There are many nonsymmetrical aggregation functions that have these properties, for example:

$$\Big(\theta + \sum_{i=1}^{n} u_i\Big)_{g_0,g_1,\ldots,g_m} = g^{-1}\big(g_0(\theta) + \sum_{j=1}^{m} g_j(u_j)\big),$$

where $g = g_0 + g_1 + \cdots + g_m$, $g_0, g_1, \ldots, g_m$ are monotone inversible functions, which are all increasing or decreasing at the same time. Correspondent model of neuron looks like

$$y = \mathrm{out}_g\Big(g_0(\theta) + \sum_{j=1}^{m} g_j(u_j)\Big),$$

where $\mathrm{out}_g(s) = \mathrm{out}(g^{-1}(s))$.

Thus in general neuron model may have the following form:

$$y = h\Big(\theta + \sum_{j=1}^{m} g_j(u_j)\Big).$$

In general case the single layered model of aggragting neuron may have the following form:

$$y = \mathrm{out} \circ \mathrm{Agg}\{\theta, u_1, \ldots, u_m\},$$

where Agg is aggregation summing-like function.

4   The $\Sigma\Pi$-neuron Model

$\Sigma\Pi$-neuron model is the example of neuron model of aggregating type with synaptic clusters. This model of a neuron was first proposed in [6] and has been used as a model that reflects the local interaction of simple clusters in synaptic synapses [7]. It was shown in [8,9] that the $\Sigma\Pi$-neuron model is more than adequate to meet the information processing processes. Advantage of this model is its relative simplicity and expressiveness. It is shown in [10] that adequate functional model of neurons of the cerebral cortex is obtained when the input $\Sigma\Pi$-neurons signals are pre-mapped using the radial functions. A more detailed analysis of models with complex synapses and synoptic clusters can be found in [11].

In the $\Sigma\Pi$-neuron model synaptic clusters are modeled using aggregating function of product:

$$S(u_1, \ldots, u_m) = \prod_{i=1}^{m} u_i,$$

where $u_1, \ldots, u_m$ are contributions of simple synapses included in synaptic cluster. Total potential of neuron is eavluated as arithmetical sum of contributions of synapses:

$$y = \mathrm{out}\Big(\theta + \sum_{k=1}^{m} \prod_{i \in \mathbf{i}_k} u_{ki}\Big),$$

where $\mathbf{i}_k \subseteq \{1, \ldots, n\}$, u_{ki} is contribution of i-th simple synapse in k-th synaptic cluster. When simple synapses has linear model ($u = wx$) transformation function has the following form:

$$y = \mathrm{out}\Big(\theta + \sum_{k=1}^{m} w_k \prod_{i \in \mathbf{i}_k} x_i\Big), \tag{2}$$

where $w_k = \prod\limits_{i\in\mathbf{i}_k} w_{ki}$. This $\Sigma\Pi$-neuron model is correct for the class of logic functions, i.e. any partial defined logical function $f\colon \mathbf{X} \to \{0,1\}$, where $\mathbf{X} \subseteq \{0,1\}^n$, can be represented by some $\Sigma\Pi$-neuron (2).

The $\Sigma\Pi$-neuron model with simple synapses with transformation function $u = w\varphi(x-a)$ has the following form:

$$y = \mathrm{out}\Big(\theta + \sum_{k=1}^{m} w_k \prod_{i\in\mathbf{i}_k} g_i(x_i - a_{ki})\Big),$$

where $g_i(x) = 0 \Leftrightarrow x \leqslant 0$. It where proved constructively in [12] that this model at simply verified common conditions can correctly express discrete functions defined on the finite subsets of $\mathbb{R}^n$. The constructive learning process is accompanied by a process of minimizing complexity, so in the synaptic cluster participated in the smallest possible number of simple synapses.

This constructive procedure also has one important property: it allow to build *collections* of different $\Sigma\Pi$-neurons on the base of the same training set, which all propduce correct outputs on that training set. This makes it possible to use them with confidence as basic algorithms in boosting-like procedures or in procedures of weighted voting. It is possible because you can always build a $\Sigma\Pi$-neuron, which does not make mistakes at *not less than half* of trainig samples, using as a training *no more than half* of the training samples.

5   The Neuron Model with Aggregating Type of Synaptic Clusters

The $\Sigma\Pi$-model is the example of the neuron model:

$$y = \mathrm{out}\Big(\theta + \sum_{k=1}^{m} u_k\Big),$$

where u_k is contribution of k-th synaptic cluster that is composed of simple synapses:

$$u_k = \mathrm{Agg}_\Pi\{u_i\colon i \in \mathbf{i}_k\},$$

where Agg_Π is an aggregation function, that satisfies the requirement:

$$\mathrm{Agg}_\Pi\{u_i\colon i \in \mathbf{i}\} = 0 \Leftrightarrow \exists i \in \mathbf{i}\colon u_i = 0.$$

The example class of such aggregation functions is generalize product operation and they are called as *quasi-product* functions:

$$\Big(\prod_{i\in\mathbf{i}} u_j\Big)_h = \Big(\prod_{i\in\mathbf{i}} \mathrm{sign}\, u_i\Big) h^{-1}\Big(\prod_{i\in\mathbf{i}} h(|u_i|)\Big),$$

where h is inversible monotone function $\mathbb{R}_+ \to \mathbb{R}_+$, such that $h(0) = 0$.

$h(u)$	$\left(\prod\limits_{i\in\mathbf{i}} u_j\right)_h$
$e^u - 1$	$\left(\prod\limits_{i\in\mathbf{i}} \operatorname{sign} u_i\right) \ln\left(1 + \prod\limits_{i\in\mathbf{i}}(e^{|u_i|} - 1)\right)$

Another example is aggregation functions of the following form:

$$\left(\prod_{i\in\mathbf{i}} \operatorname{sign} u_i\right)\left(h^{-1}\left(\prod_{i\in\mathbf{i}} h(1 + |u_i|)\right) - 1\right),$$

where h is inversible monotone function $\mathbb{R}_+ \to \mathbb{R}_+$, such that $h(1) = 1$.

So the neuron model with synaptic clusters of aggregating type has the following form:

$$y = \operatorname{out}\left(\theta + \sum_{k=1}^{m} \operatorname{Agg}_\Pi\{u_{ki} : i \in \mathbf{i}_k\}\right). \tag{3}$$

The abilities of this model express the following proposition. Let out is *correct output function*, i.e. for any $y \in \mathbf{Y}$ there exists value s, such that $y = \operatorname{out}(s)$.

Let transformation fuction of the simle synapse is such that

(1) $\operatorname{syn}(x - a, w) = 0$ only when $x \leqslant a$ (or $x < a$);
(2) $\operatorname{syn}(x - a, w)$ is monotone on x and w;
(3) for all $x > a$ (or $x \geqslant a$) and all u there exists parameter w, such that $\operatorname{syn}(x - a, w) = u$.

The tipycal example: $\operatorname{syn}(x-a, w) = w\varphi(x-a)$, where φ is monotone increasing fincion and $\varphi(x) = 0 \Leftrightarrow x \leqslant 0$ or $\varphi(x) = 0 \Leftrightarrow x < 0$. In these assumptions the following theorem is true [13].

Theorem. *For any finite and consistent training set it can be built constructively an aggregating neuron*

$$\operatorname{agn}(\mathbf{x}) = out\left(\theta + \sum_{k=1}^{m} \operatorname{Agg}_\Pi\{u_{ki} : i \in \mathbf{i}_k\}\right),$$

such that $y = \operatorname{agn}(\mathbf{x})$ *for any pairs* $\langle \mathbf{x}, y \rangle$ *from the training set.*

In general case the model (3) of aggregating neuron the theorem can be generalised if we replace the arithmetic addition by symmetrical aggregation function of summing type:

$$\begin{aligned} y &= \operatorname{out} \circ \operatorname{Agg}_\Sigma\{\theta, s_1, \dots, s_m\}, \\ s_k &= \operatorname{Agg}_\Pi\{u_{ki} : i \in \mathbf{i}_k\}, \quad k = \overline{1, m}, \end{aligned} \tag{4}$$

where $\operatorname{Agg}_\Sigma$ is aggregation function for summing of contributions of synaptic clusters, which is satisfies requirements $\Sigma 1$–$\Sigma 2$. In that assumptions the full analoge of the above theorem retains it's truth.

6 Conclusion

In this work it was described new classes of the neuron models of aggregating type. They include majority of known functional models of neurons: neurons of perceptron type, $\Sigma\Pi$-neurons, radial neurons. They greatly extend the range of types of the neuron models on the basis of which it would be possible to build neural networks. For one subclass of such models of aggregating type that extend the $\Sigma\Pi$-neuron model it was proved the ability to correctly represent functions defined on discrete sets.

In further work the abilities of the neuron models of aggregating type to approximate wide range of functions will be investigated both empirically and theoretically. Because the $\Sigma\Pi$-neuron models and their extensions are shown good abilities to represent wide range of functions [12,14,15] then the same ability to be expected from the neuron models of aggregating type in general case. Also still have to investigate the effect from application of the different classes of aggregation functions for modelling of synaptic clusters and the process of summation of all contributions to the total potential of the neuron on the quality and complexity of approximating functions.

Acknowledgements. This work is supported by grant RFBR 15-01-03381 and by research program of ONIT RAS.

References

1. Mesiar, R., Komornikova, M., Kolesarova, A., Calvo, T.: Aggregation functions: A revision. In: Bustince, H., Herrera, F., Montero, J. (eds.) Fuzzy Sets and Their Extensions: Representation. Aggregation and Models. Springer, Berlin (2008)
2. Grabich, M., Marichal, J.-L., Pap, E.: Aggregation Functions. Series: Encyclopedia of Mathematics and its Applications. Cambridge University Press, New York (2009)
3. Pap, E.: g-calculus. Univ. u Novom Sadu Zb. Rad. Prirod.-Mat. Fak. Ser. Mat. **23**, 145–150 (1993)
4. Pap, E.: Preprint ESI 1448. Vienna **23**, 145–156 (2004)
5. Aczél, J.: Lectures on Functional Equations and their Applications. Academic Press, New York (1966)
6. Feldman, J.A., Ballard, D.H.: Connectionist models and their properties. Cogn. Sci. **6**, 205–254 (1982)
7. Rumelhart, D.E., Hinton, G., Williams, R.: Learning internal representation by error propagation. Parallel Distrib. Process. **1**, 318–362 (1986)
8. Mel, B.W.: The Sigma-Pi Column: A Model of Associative Learning in Cerebral Neocortex. California institute of technology. cns memo no. 6: Technical report Pasadena, California 91125 (1990)
9. Mel, B.W.: The Sigma-Pi model neuron: roles of the dendritic tree in associative learning. Soc. NeuroSci. Abstr. **16**, 205 (1990)
10. Mel, B.W., Koch, C.: Sigma-Pi learning: on radial basis functions and cortical associative learning. In: Touretzk, D.S. (ed.) Advances in Neural Information Processing Systems, vol. 2, pp. 474–481. Morgan Kaufmann, San Mateo, CA (2000)

11. Mel, B.W.: Why have dendrites? a computational perspective. In: Stuart, G., Spruston, N., Hausse, M. (eds.) Dendrites 2nd edn, Oxford University Press (2007)
12. Sibzukhov, Z.M.: Constructive Learning Methods of $\Sigma\Pi$-Neural Networks. - M: Nauka. (in Russian) (2006)
13. Shibzukhov, Z.M., Cherednikov, D.: About neuron models of aggregating types. Mach. Learn. Data Anal. **1**, 1706–1716 (2015). http://jmlda.org/?page_id=35 (in Russian)
14. Shibzukhov, Z.M.: Recurrent method of constructive learning of some networks $\Sigma\Pi$-Neurons and $\Sigma\Pi$ neural modules. J. Comput. Math. Math. Phisycs. **43**, 1298–1310 (2003)
15. Shibzukhov, Z.M.: On constructive and well-behaved classes of algebraic $\Sigma\Pi$-algorithms. Doklady Math. **81**, 490–492 (2010)

Conversion from Rate Code to Temporal Code – Crucial Role of Inhibition

Mikhail V. Kiselev[✉]

Chuvash State University, Cheboksary, Russia
mkiselev@megaputer.ru

Abstract. This study is an attempt to answer the question – what kind of spiking neural networks could efficiently transform rate-coded input signal into temporally coded form – specific activity of neuronal groups with strictly fixed temporal delays between spikes emitted by different neurons in every group. Since theoretical approach to the solution of this problem appears to be very hard or impossible the network configurations performing this task efficiently were found by means of genetic algorithm. Exploration of their structure showed that while excitatory neurons form the groups with stereotypical firing patterns, the inhibitory neurons of the network make these patterns specific for different rate-coded stimuli and, thus, play the key role in conversion of rate-coded input signal to temporal code.

Keywords: Spiking neural network · Rate code · Temporal code · Polychronization · Chaotic neural network · Neural network self-organization

1   Introduction

Understanding the mechanisms of information representation and processing in nervous system is a mandatory point on the road to comprehension of brain functioning as a whole. Computer simulation is a commonly used methodology for reaching this goal, and spiking neural networks (SNNs) form a class of most biologically plausible neural network models. They imitate one of the most distinctive features of living neural networks – that information is represented in them in the form of sequences of spikes, very short pulses of constant amplitude. Although, there exist many possible schemes of how information can be encoded in spike sequences, they can be grouped in two large classes – rate code and temporal code. The first group uses the mean firing frequency measured in a certain time window for one or several neurons as an informational parameter. The exact firing time of an individual neuron is not important in this approach. But it plays the key role in the second group, where the sequences of spikes emitted by certain neurons with strictly fixed relative delays are considered as information units. These groups of neurons are called polychronous neuronal groups or PNGs [1]. It is known that rate code is widespread in the nervous system, especially in the peripheral nervous system. At the same time, temporal code has many advantages from point of view of computational power, informational capacity, noise/fault tolerance etc. Besides, there are experimental evidences that the temporal coding is really used for information

© Springer International Publishing Switzerland 2016
L. Cheng et al. (Eds.): ISNN 2016, LNCS 9719, pp. 665–672, 2016.
DOI: 10.1007/978-3-319-40663-3_76

representation and processing in the brain cortex [2, 3]. It can serve as a basis for realization of working memory mechanism [4].

Temporal coding was explored extensively in the last decade, however, there is an important problem related to it which remains almost unaddressed so far. Namely, it is not clear how a temporally encoded signal can appear from rate-coded sources. For example, it is demonstrated in [4] how PNGs storing information about recent stimuli can arise in the chaotic SNN. However, it should be noted that these stimuli were already represented as sequences of spikes strictly fixed in time (as in more recent research on SNN training [5, 6]). The aim of the present study is to understand what kind of SNN structures could implement rate/temporal re-coding. Until now, works considering this kind of conversion remain surprisingly few. A possible role of ensembles of hippocampal pyramidal neurons in this process was considered in [7]. General approach to solution of this problem was presented in [1] – but without specification of concrete conditions under which this conversion could be realized and without an experimental evidence for its realization. An attempt to suggest a solution of this problem was made in the recent article [8]. It was demonstrated, that conversion between these two coding schemes could be performed under certain conditions by a homogenous chaotic neural network. However, in [8] PNG response to stimulus was expressed in activity of at most third part of its neurons – so that reactions to the same pattern might imply activities of completely different PNG subsets. Although, from the formal point of view, PNG activity in [8] really contains information about stimuli presented, using this information for further processing is hardly possible. It motivated me to further research the results of which are presented here.

Similar to many other problems related to SNNs, applicability of theoretical analysis to this problem is very limited. For this reason, our main tool in this research was numeric optimization based on genetic algorithm.

The paper is organized in the following way. First, we describe briefly the used models of neurons and network, and the simulation of rate-coded input signal. Further, we describe the network parameter optimization procedure using genetic algorithm (GA) and the optimization criterion. Next section is devoted to structure and properties of the networks – winners in GA. The results are summarized in the Conclusion.

2   The Models of the Neurons and the Network

Since degree of neurophysiological realism was not a number one factor in this study, the simplest functional model of neuron – leaky integrate-and-fire neuron was used. We utilized current-based model of excitatory synapses and conductance-based model of inhibitory synapses, namely, the simplest delta synapse models in which the effect of presynaptic spike on membrane potential u is instantaneous and has zero duration. Thus, presynaptic spike coming to an excitatory synapse with the weight w_i increments u by w_i while spike coming to an inhibitory synapse with the weight w_a decreases u by the value $(u-E_I)(1 - \exp(-w_a))$, where E_I – inhibitory reversal potential. Besides, u permanently decays with the time constant τ to its rest value (equal to 0). If u reaches a threshold value after incoming excitatory spike, the neuron fires, u is reset to 0 and neuron becomes unable to fire again during the time interval τ_R.

Spikes propagation from neuron to neuron takes non-zero time. For excitatory connections this time is randomly selected in the range 1–30 ms. All connections of inhibitory neurons are very fast −1 ms. Network is chaotic in the sense that the probability that two given neurons are connected depends only on their types (excitatory or inhibitory). A time step in our simulation equaled to 1 ms so that excitatory synaptic connections could have $N_D = 30$ different delays.

Excitatory synapses are plastic in our model. We use plasticity rule based on the standard STDP model. However, the classic STDP rule can easily lead to instability in recurrent SNNs – active closed circuits of excitatory neurons grow and become more and more active due to unlimited growth of their connection weights caused by STDP. This constant activity can suppress completely network reaction to any external stimuli. To fight this negative effect, homeostatic amendments to the standard STDP model are introduced. Namely, the value of long-term synaptic potentiation (LTP) is made dependent on postsynaptic neuron activity. Very active neurons lose their ability to strengthen their synapses. It is realized by introducing a new component of neuron state - the variable η. If neuron does not fire for a long time period it equals to 1. When neuron fires, η is decreased by the value $\Delta\eta^-$. After that it grows with the constant speed $\Delta\eta^+$ until reaches 1 again. Synaptic weight values have lower and upper bounds w_{min} and w_{max}. It follows from the fact that the plasticity law is applied to the value of the so called *synaptic resource* W [9] instead of the weight itself. The relationship between these two variables can be expressed as

$$w = w_{min} + \frac{(w_{max} - w_{min})\max(W, 0)}{w_{max} - w_{min} + \max(W, 0)}. \tag{1}$$

Therefore, when W runs from $-\infty$ to $+\infty$, w runs from w_{min} to w_{max}. While in the starting network state weights of all connections of the same type were initialized by the same values, excitatory weights in the process of network evolution scattered in the range $[w_{min}, w_{max})$ due to synaptic plasticity thus providing the network with some intrinsic structure.

Neurons in the network are linked with other neurons and with *input nodes* which serve as a source of external signal. These input nodes send spikes to excitatory and inhibitory neurons via excitatory synapses. The external signal includes two components – noise and rate-coded stimuli. Noise component consists of spikes randomly generated with a certain constant mean frequency. Each stimulus corresponds to some subset of input nodes. These subsets have the same size and may intersect. Regularly, with 200 ms period, spike frequency on one of these subsets becomes very high (300 Hz) during a short time interval (30 ms). Thus, it is a typical example of rate-coded signal. The number of different stimuli in these experiments was 10.

3 Genetic Algorithm

Thus, we want to build a network which would respond to the appearance of certain stimulus in input signal by activity of some PNGs specific for this stimulus. For reaching this goal we used an optimization technique based on genetic algorithm. Several procedures for finding PNGs have been proposed [6, 10]. However, all of them require significant computation time even in case of the moderate size networks used in our research (with $N_N = 1000$ excitatory neurons). For this reason we had to use a simplified optimization criterion. PNG activity consists of spike pairs. If a PNG becomes active after presentation of a certain stimulus then the respective spike pairs will appear after this stimulus significantly more often than after the other stimuli. Our optimization criterion is based on amount of spike pairs specific for different stimuli. It is calculated in the following way. Let $n(i, j, a, p)$ be the number of cases when after presentation of the stimulus p (and before presentation of succeeding stimulus) the i-th neuron fires and a ms after that the j-th neuron fires. We will denote such a spike pair as $e_2(i, j, a)$. Only excitatory neurons are considered here. For each i, j, and a, the stimulus $s(i, j, a)$ corresponding to the maximum value of $n(i, j, a, p)$ is determined. Let us denote this maximum value as $\hat{n}(i, j, a)$. Let N_P be the number of different stimuli, $N(i, j, a)$ – sum of $n(i, j, a, p)$ for all p. Then the value

$$P(i, j, a) = \sum_{k=\hat{n}(i,j,a)}^{N(i,j,a)} b\left(k; N(i, j, a), \frac{1}{N_P}\right), \qquad (2)$$

where $b(.; N, P)$ is binomial distribution, expresses the probability that the given value of $\hat{n}(i, j, a)$ may be observed even in case when $e_2(i, j, a)$ has equal chances to appear after any stimulus. If

$$P(i, j, a) < \frac{0.03}{N_P N_N N_N N_D}, \qquad (3)$$

then we can say that $e_2(i, j, a)$ is specific for the stimulus $s(i, j, a)$. In (3) we take into account the number of independent statistical hypotheses tested. The number of all pairs <i, j> entering all specific $e_2(i, j, a)$ (for any stimulus) is taken as a value of optimization criterion. This procedure admits an efficient realization on massively parallel architectures like GPU[1].

The aim of the genetic algorithm (GA) was to find the combination of 35 network parameters which would maximize the above mentioned value. The set of parameters included quantity of inhibitory neurons, quantities of connections of all kinds (input nodes -> excitatory, input nodes -> inhibitory, excitatory -> excitatory, etc.), weights of these connections (or weight value ranges and initial weight value – for plastic connections), time constants of neurons, parameters entering plasticity rule Each network evolved 1000 s, the last 200 s period was used to calculate the value of

---

[1] 2 NVIDIA GTX TITAN-X cards were used in this work.

optimization criterion. Population size was equal to 300; mutation rate per individual was set to 0.3; probability of selection of the given individual for crossover was proportional to individual's rank. Elitism level was set to 0.1.

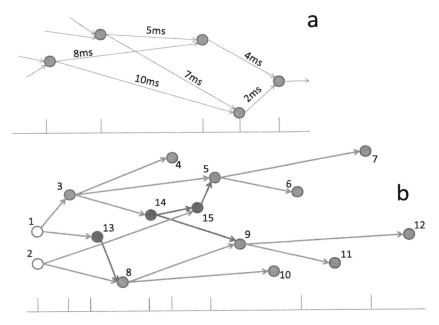

Fig. 1. Structures of selective PNGs. Blue color corresponds to excitation, red color – to inhibition. Neurons are placed in the order of their (possible) firing – from left to right. a. Traditional structure of PNG proposed in [1]. b. PNGs formed by the tree-like system of strong synaptic connections and inhibition-based competition. (Color figure online)

4   Results

5 GA runs have been carried out. The number of specific neuron pairs was great in all GA winners ranging from 93173 to 103559. The sets of specific $e_2(i, j, a)$ for all winners were used to determine PNGs specific for different stimuli. This procedure resembles agglomerative clustering algorithm when we build growing PNGs, adding specific spike pairs $e_2(i, j, a)$ one after one in such a way that would not lead to decrease of PNG selectivity. It is described in full detail in [11].

SNNs with selective PNGs were found in all experiments. Mean PNG size varied in the range 5.14–16. Distribution of selective PNG sizes for all stimuli and for all experiments is shown on Fig. 2. We see that small and moderate size (<10) PNGs dominate, however large PNGs with more than 100 neurons are observed as well. To evaluate selectivity of the PNGs we measured a mean number of spikes in all PNG activations after the stimulus recognized and after all the other stimuli. Great ratio of these two values indicates strong selective PNG group reaction to the stimulus recognized.

Range of this parameter was [35.5–197.22]. Thus, chaotic SNN can really learn with help of synaptic plasticity to convert rate-coded stimuli to temporal encoding.

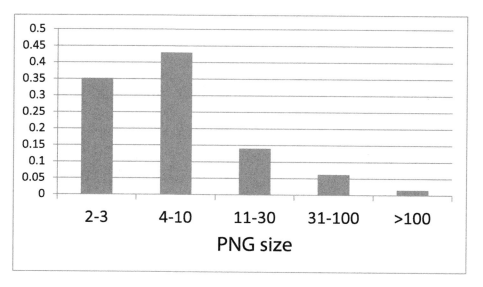

Fig. 2. Distribution of selective PNG sizes for all stimuli and for all experiments.

The detailed analysis of structure of the emerging PNGs gave even more interesting results. Izhikevich and other researchers in their pioneering works on polychronization assumed the following scheme of PNG activity [1]. Every neuron entering a PNG fires if spikes come simultaneously to several its synapses with great weights. Firing chain becomes possible because of fine tuning of synaptic delays in intra-PNG connections (Fig. 1a). PNG activation may be very selective because it is caused by specific combination of firing times of its triggering neurons. Inhibition plays a rather passive role in this process – inhibitory neurons just prevent unlimited excitation growth in the network. Structure of PNGs in our experiments was completely different. All neurons entering the PNGs have only one very high synaptic weight. Presynaptic spike coming to this synapse forces the neuron to fire. PNGs consist of neurons connected by branching tree of these strong links (Fig. 1b). Naturally, firing pattern of neurons in this kind of PNG is very stable. However, this PNG mechanism, by itself, is not selective at all – every firing by its root neurons causes activation of the whole PNG. And here, the key role of inhibitory neurons is elicited. Let us consider simplified examples of two PNGs including neurons 3–7 and 8–12, reacting to stimuli corresponding to input nodes 1 and 2, respectively. Input nodes and PNG neurons have connections with inhibitory neurons (##13, 14 on Fig. 1b). These inhibitory neurons are activated together with the first PNG and stop development of activation of the second PNG. Moreover, they suppress activity of other inhibitory neurons (#15) which could block activation of the first PNG. Thus, inhibitory neurons realize competition between various PNGs making them selective. Certainly, this picture is possible only under conditions of very specific structure of connections between different groups of excitatory and inhibitory neurons. A valuable result of the present research is demonstration that this structure can arise automatically from initial fluctuations of synaptic connections as a result of synaptic plasticity – driven network evolution.

5 Conclusion

Let us summarize the results obtained in this study (which is a part of an ongoing research project).

- In order to determine SNN configurations suitable for the conversion of external rate-coded signal to temporal code, genetic algorithm was used. This approach reached its goal – SNNs which react to rate-coded stimulus by highly specific activity of their polychronous neuronal groups have been found.
- Structure of these PNGs differs significantly from PNGs proposed by the author of this concept, E. Izhikevich. Namely, the key role in their selective activation is played by inhibitory neurons and, in particular, by specific configuration of their pre- and post-synaptic connections creating competitive environment for numerous PNGs existing in the SNN.
- These structures are formed automatically as a result of self-organization process driven by STDP-based synaptic plasticity. Their formation takes reasonable time (hundreds of stimulus presentation cycles).

It remains to say about further goals of the present project. The reported results need a thorough exploration. The influence of many factors should be studied – SNN size, noise, number of different stimuli, their similarity etc. It would be interesting to pass from the conversion of rate-coded stimuli to their memorizing, for example, using working memory mechanism proposed in [4]. These are steps in the general direction of the project – the creation of SNN-based adaptive automated control systems.

References

1. Izhikevich, E.: Polychronization: computation with spikes. Neural Comput. **18**, 245–282 (2006)
2. Ikegaya, Y., Aaron, G., Cossart, R., Aronov, D., Lampl, I., Ferster, D., et al.: Synfire chains and cortical songs: temporal modules of cortical activity. Science **304**, 559–564 (2004)
3. Butts, D.A., Weng, C., Jin, J., Yeh, C.I., Lesica, N.A., Alonso, J.M., Stanley, G.B.: Temporal precision in the neural code and the timescales of natural vision. Nature **449**, 92–95 (2007)
4. Szatmary, B., Izhikevich, E.: Spike-timing theory of working memory. PLoS Comput. Biol. **6**, e1000879 (2010)
5. Gardner, B., Sporea, I., Grüning, A.: Learning spatiotemporally encoded pattern; transformations in structured spiking neural networks. Neural Comput. **27**, 2548–2586 (2015)
6. Guise, M., Knott, A., Benuskova, L.: Bayesian model of polychronicity. Neural Comput. **26**, 1–22 (2014)
7. Mehta, M.R., Lee, A.K., Wilson, M.A.: Role of experience and oscillations in transforming a rate code into a temporal code. Nature **417**, 741–746 (2002)

8. Kiselev, M.: Homogenous chaotic network serving as a rate/population code to temporal code converter. Comput. Intell. Neurosci. **2014**, 476580 (2014). doi:10.1155/2014/476580
9. Kiselev, M.: Self-organization process in large spiking neural networks leading to formation of working memory mechanism. In: Rojas, I., Joya, G., Gabestany, J. (eds.) IWANN 2013, Part I. LNCS, vol. 7902, pp. 510–517. Springer, Heidelberg (2013)
10. Martinez, R., Paugam-Moisy, H.: Algorithms for structural and dynamical polychronous groups detection. In: Alippi, C., Polycarpou, M., Panayiotou, C., Ellinas, G. (eds.) ICANN 2009, Part II. LNCS, vol. 5769, pp. 75–84. Springer, Heidelberg (2009)
11. Kiselev, M.: Rate coding vs. temporal coding – is optimum between? Accepted for Publication in Proceedings of IJCNN-2016 (2016)

Analysis of Oscillations
in the Brain During Sensory Stimulation:
Cross-Frequency Relations

Elena Astasheva[1](✉), Maksim Astashev[2,3],
and Valentina Kitchigina[1,3]

[1] Institute of Theoretical and Experimental Biophysics,
Russian Academy of Sciences, Pushchino, Moscow District, Russia
litgara@rambler.ru, vkitchigina@gmail.com
[2] Institute of Cell Biophysics, Russian Academy of Sciences, Puchshino,
Moscow District, Russia
astashev@yandex.ru
[3] Pushchino State Institute of Life Sciences, Pushchino, Moscow District, Russia

Abstract. Oscillations are necessary for the execution of many cognitive functions. Little attention has been given to the problem of how various types of oscillations are synchronized during information processing. We analyzed the power and correlation characteristics of rhythmic activities in theta, alpha, and gamma ranges when animals were in two behavioral states: quiet wakefulness and the processing of sensory signals. Eight brain areas (cortical and subcortical) were examined. During sensory stimulation, a strengthening of oscillations of all ranges was revealed in many structures. The main result of this study is that cross-frequency relations in all oscillatory bands in cortical and subcortical structures sharply increased during the processing of sensory information.

Keywords: Waking guinea pigs · Cortical and subcortical structures · Oscillations · Cross-frequency analysis · Sensory stimulation · Cross-correlation

1 Introduction

Brain processes depend on interactions between neuronal groups. Oscillations providing the synchronization and dynamic coordination of functionally linked neurons create favorable conditions for cognition [7, 27]. They are necessary for the execution of many cognitive functions, such as space orientation, selective attention, decision-making, and memory [2, 3, 13, 24–26].

Low-frequency (delta, theta, alpha) and high-frequency oscillations (gamma and ripples) have been intensively studied (reviewed in [1, 2]), and recently it has been specified that these different rhythmic activities participate in cognition [20, 25]. One of important oscillations in the brain is the theta rhythm: it correlates with learning and memory processes, especially in human [4, 9, 11, 14, 23]. The alpha rhythm (10–12 Hz) is associated with the ability to perceive external incentives and to react to them: dependence of reaction time and perception of signals on its phase was found in a number of works [6, 11, 15]. The gamma rhythm is often considered as a physiological

© Springer International Publishing Switzerland 2016
L. Cheng et al. (Eds.): ISNN 2016, LNCS 9719, pp. 673–680, 2016.
DOI: 10.1007/978-3-319-40663-3_77

fingerprint of attention [8, 12]. There is also evidence that, whereas fast gamma oscillations are involved in the maintenance of relevant information, the slow alpha rhythm is implicated in suppressing irrelevant information in memory [10, 16, 19] and attention tasks [18, 22].

Despite increasing interest in rhythmic processes, the problems of how different types of oscillations can be synchronized during the processing of sensory signals and how changes in the expression of rhythmic activity affect the correlations between brain structures have been insufficiently studied so far.

The main objective of this study was the analysis of power and correlation characteristics of rhythmic activity in the theta, alpha, and gamma ranges when animals were in two behavior states: quiet wakefulness and the processing of sensory signals.

2   Methods

2.1   Animals and Surgery

All animal experimentations were approved by the local ethics committee following guidelines that are in accordance with the Directive (2010/63/EU) of the European Parliament and of the Council. All efforts were made to minimize the number of animals used and their suffering.

Experiments were performed on waking guinea pigs (males and females, weight from 450 to 540 g were used). A week prior to experiments, the animals undergoing surgery and monopolar electrodes (nichrom, 0.15 mm) were implanted to the eight structures by using stereotaxic coordinates [17] The structures were: (1) entorhinal cortex (EC), (2) supramamillary nucleus (SM), (3) central nucleus of amygdala (CA), (4) medial septal nucleus (MS), (5) dentate gyrus (DG), (6) hippocampus (Hip, CA1 field), (7) frontal cortex (FC), (8) lateral septal nucleus (LS).

Oscillatory activity was recorded in three frequency ranges – theta (4–8 Hz), alpha (10–12 Hz) and gamma (30–80 Hz), in two different functional animal states: during quiet wakefulness and during processing of sensory signal (flash of light, 0.3 Hz).

The animals were slightly restricted in their movements in a ventilated box. The box was large enough to make animals feel comfortable in their natural sitting position during experiments, which lasted about half an hour.

2.2   Recordings and Protocol

Baselines for LFPs were recorded at the beginning of each experiment over a period of five days. Then sensory stimuli were delivered. The recordings were obtained for 7 min before stimulation (control) and then for 7 min during stimulation.

For the LFP recording we used tool measuring amplifiers, which have high-impedance inputs (based on OPA376), therefore the resistance of the electrode, the environment and connecting wires could not take into account. In our system only the voltage at the point of electrode placement was measured.

2.3   Data Processing

To recording and processing field potentials, an 8-channel installation was applied, including: (1) personal computer (MS Windows XP), (2) universal board ADC/DAC data collection L791 (LCard, Moscow, Russia), (3) the software package LGraph2 (LCard, Moscow, Russia) or WinEDR (Strathclyde Electrophysiology Software, University of Strathclyde, Glasgow, UK, (4) a DIY buffer repeater on a OPA376. The record files were processed manually in order to eliminate artifacts in them: the random effects of physical activity of the experimental animal, electrical interference from outside electrical equipment and so forth. To obtain the inter-channel correlations and visualisation, a continuous wavelet analysis with complex Morlet wavelet has been developed. The cross-correlation function was also calculated for the detection of rhythm interrelations; its maximum (with allowance for the possible phase shift) was taken to be the cross-correlation coefficient (Ccr).

2.4   Statistical Analysis

The differences between groups ("control" and "sensory stimulation") were determined using one-way analysis of variance (ANOVA) with the post-hoc Bonferroni test. All calculations were carried out using the Origin7.0 SR program (OriginLab Corporation, USA).

3   Results

The LFP recordings and the subsequent wavelet analysis showed that oscillations of all bands, theta (4–8), alpha (10–12), and gamma (30–80 Hz) occur in all cortical and subcortical areas investigated. During sensory stimulation, a strengthening of oscillations in all ranges in many structures was revealed (Figs. 1 and 2, Table 1).

Expression of rhythmic activity was slightly varied in different structures. Mean coefficients of wavelet power for different types of oscillations in eight structures are shown in Table 1. The analysis showed that the most distinct effect of sensory stimulation on theta oscillations was in the EC, where the theta power increased from 0.005 to 0.02 n.u. (P < 0.001). The increase in alpha oscillations was strongest in the EC and MS (from 0.006 to 0.015 n.u. (P < 0.001)), and somewhat less in the LS (from 0.005 to 0.1 n.u. (P < 0.001)). Changes in gamma power during stimulation were not unidirectional: it can increase (EC, SM, MS), decrease (DG, Hip, FC, LS), or not change (CA).

The main result of our study was that cross-frequency relations (Ccr) in all ranges of oscillations in all brain structures investigated sharply increased during the processing of sensory information (Table 2).

The analysis showed that the most distinct effect of sensory stimulation on theta oscillations was in the EC, where the theta power increased from 0.005 to 0.02 n.u. ($P < 0.001$). The increase in alpha oscillations was strongest in the EC and MS (from 0.006 to 0.015 n.u. ($P < 0.001$)), and somewhat less in the LS (from 0.005 to 0.1 n.u.

Table 1. Mean coefficients of wavelet power for different oscillations in eight structures (mean ± S.E.M.). The power (normalized units) for alpha, theta, and gamma oscillations are shown for two states of animals, quiet wakefulness (control) and the processing of sensory signal. The effect is significant at **$P < 0.01$ or ***$P < 0.001$ versus initial activity. Designation of structures as in Methods.

Structure	State	Theta power	Alpha power	Gamma power
EC	control	0,005 ± 0,002	0,006 ± 0,002	0,005 ± 0,002
	stim	0,02 ± 0,01 ***	0,015 ± 0,01 ***	0,006 ± 0,003 ***
SM	control	0,008 ± 0,004	0,008 ± 0,003	0,004 ± 0,002
	stim	0,014 ± 0,009 ***	0,014 ± 0,01 ***	0,006 ± 0,003 ***
CA	control	0,007 ± 0,004	0,008 ± 0,003	0,006 ± 0,002
	stim	0,014 ± 0,009 ***	0,016 ± 0.01 ***	0,006 ± 0,003
MS	control	0,006 ± 0,002	0,006 ± 0,002	0,004 ± 0,002
	stim	0,014 ± 0,009 ***	0,015 ± 0,01 ***	0,006 ± 0,003 ***
DG	control	0,008 ± 0,004	0,009 ± 0,003	0,006 ± 0,002
	stim	0,008 ± 0,005	0,011 ± 0,008 *	0,005 ± 0,002 ***
Hip	control	0,006 ± 0,004	0,009 ± 0,004	0,006 ± 0,003
	stim	0,008 ± 0,003**	0,009 ± 0,007	0,003 ± 0,001 ***
FC	control	0,006 ± 0,004	0,006 ± 0,003	0,005 ± 0,002
	stim	0,008 ± 0,003**	0,009 ± 0,007 ***	0,003 ± 0,001 ***
LS	control	0,006 ± 0,002	0,005 ± 0,002	0,004 ± 0,002
	stim	0,007 ± 0,005	0,01 ± 0,006***	0,003 ± 0,001 ***

Table 2. Cross-frequency relations (Ccr) in cortical and subcortical structures in two different functional states of animal: quiet wakefulness (control) and the processing of sensory signals (stimulation).

Rhythm	State	EC	SMN	CA	MS	DG	Hip	FC	LS
Theta-alpha	control	0,02	0,02	-0,05	0,05	0,05	0,12	0,03	0,06
	stim	0,38	0,4	0,34	0,35	0,4	0,41	0,35	0,39
Theta-gamma	control	-0,03	0,1	-0,005	0,14	0,13	0,04	-0,05	-0,08
	stim	0,32	0,29	0,26	0,24	0,18	0,18	0,24	0,34
Alpha-gamma	control	-0,18	0,07	0,04	0,009	0,03	0,13	-0,19	0,02
	stim	0,47	0,4	0,46	0,42	0,33	0,42	0,39	0,43

($P < 0.001$)). Changes in gamma power during stimulation were not unidirectional: it can increase (EC, SM, MS), decrease (DG, Hip, FC, LS), or not change (CA).

The main result of our study was that cross-frequency relations (Ccr) in all ranges of oscillations in all brain structures investigated sharply increased during the processing of sensory information (Table 2).

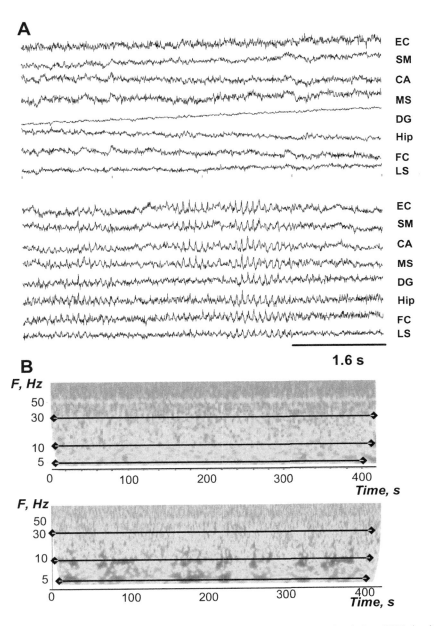

Fig. 1. Typical changes of activities in brain structures during sensory stimulation. LFPs in eight structures (A) and the results of wavelet analysis of the hippocampal activity (B) are shown. In A and B, initial (above) and evoked (below) activities are demonstrated. Arrows indicate the fragments in theta, alpha and gamma bands that were selected for further analysis. Designations of structures in A as in Methods.

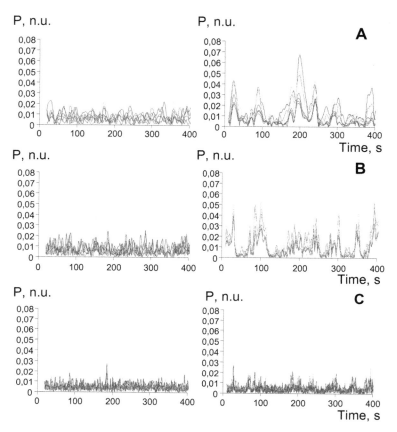

Fig. 2. Time profile of LFPs in eight brain structures of one and the same animal recorded in a control experiment (quiet wakefulness) (on the left) and during sensory stimulation (flash of light, 3 Hz) (on the right) in theta (A), alpha (B), and gamma (C) ranges. On the abscissa, time, s; on the ordinate, the power of wavelet coefficients in normalized units (n.u.).

4   Conclusions

In this study we showed for the first time that theta, alpha, and gamma oscillations occur in various cortical and subcortical areas and that they sharply increase during sensory stimulation in many structures, thought not everywhere. The main important result of our work was that cross-frequency relations in all oscillatory ranges studied sharply increased during the processing of sensory signals. This fact testifies that the synchronization of rhythmic activities in various structures promotes information processing in the brain. The findings of our study correspond to the data of other authors obtained in experiments on animals in working memory, matching, and attention tasks [5, 18, 21].

References

1. Buzsáki, G.: Rhythms of the Brain. Oxford University Press, New York (2006)
2. Buzsáki, G., Wang, X.J.: Mechanisms of gamma oscillations. Ann. Rev. Neurosci. **35**, 203–225 (2012)
3. Buzsáki, G., Watson, B.O.: Brain rhythms and neural syntax: implications for efficient coding of cognitive content and neuropsychiatric disease. Dialogues Clin. Neurosci. **14**(4), 345–367 (2012)
4. Cantero, J.L., Atienza, M., Stickgold, R., Kahana, M.J., Madsen, J.R., Kocsis, B.: Sleep-dependent theta oscillations in the human hippocampus and neocortex. J. Neurosci. **23** (34), 10897–10903 (2003)
5. Chik, D.: Theta-alpha cross-frequency synchronization facilitates working memory control – a modeling study. SpringerPlus **2**(1), 1–10 (2013)
6. Dustman, R.E., Beck, E.C.: Phase of alpha brain waves, reaction time and visually evoked potentials. Electroencephalogr. Clin. Neurophysiol. **18**, 433–440 (1965)
7. Fries, P.: A mechanism for cognitive dynamics: neuronal communication through neuronal coherence. Trends Cogn. Sci. **9**(10), 474–480 (2005)
8. Fries, P.: Neuronal gamma-band synchronization as a fundamental process in cortical computation. Ann. Rev. Neurosci. **32**, 209–224 (2009)
9. Grunwald, M., Busse, F., Hensel, A., Kruggel, F., Riedel-Heller, S., Wolf, H., Arendt, T., Gertz, H.J.: Correlation between cortical theta activity and hippocampal volumes in health, mild cognitive impairment, and mild dementia. J. Clin. Neurophysiol. **18**, 178–184 (2001)
10. Jensen, O., Gelfand, J., Kounios, J., Lisman, J.E.: Oscillations in the alpha band (9–12 Hz) increase with memory load during retention in a short-term memory task. Cereb. Cortex **12** (8), 877–882 (2002)
11. Jensen, O., Kaiser, J., Lachaux, J.P.: Human gamma-frequency oscillations associated with attention and memory. Trends Neurosci. **30**, 7317–7324 (2007)
12. Jutras, M.J., Fries, P., Buffalo, E.A.: Gamma-band synchronization in the macaque hippocampus and memory formation. J. Neurosci. **29**(40), 12521–12531 (2009)
13. Laurent, F., Brotons-Mas, J.R., Cid, E., Lopez-Pigozzi, D., Valero, M., Gal, B., de la Prida, L.M.: Proximodistal structure of theta coordination in the dorsal hippocampus of epileptic rats. J. Neurosci. **35**(11), 4760–4775 (2015)
14. Mitchell, D.J., McNaughton, N., Flanagan, D., Kirk, I.J.: Frontal-midline theta from the perspective of hippocampal "theta". Prog. Neurobiol. **86**, 156–185 (2008)
15. Nunn, C.M., Osselton, J.W.: The influence of the EEG alpha rhythm on the perception of visual stimuli. Psychophysiology **11**(3), 294–303 (1974)
16. Palva, S., Linkenkaer-Hansen, K., Näätänen, R., Palva, J.M.: Early neural correlates of conscious somatosensory perception. J. Neurosci. **25**(21), 5248–5258 (2005)
17. Rapisarda, C., Bacchelli, B.: The brain of the guinea pig in stereotaxic coordinates. Arch. Sci. Biol. **61**, 1–37 (1977)
18. Sauseng, P., Klimesch, W., Gruber, W.R., Birbaumer, N.: Cross-frequency phase synchronization: a brain mechanism of memory matching and attention. Neuroimage **40** (1), 308–317 (2008)
19. Sauseng, P., Klimesch, W., Heise, K.F., Gruber, W.R., Holz, E., Karim, A.A., Glennon, M., Gerloff, C., Birbaumer, N., Hummel, F.C.: Brain oscillatory substrates of visual short-term memory capacity. Curr. Biol. **19**(21), 1846–1852 (2009)
20. Sridharan, D., Knudsen, E.I.: Gamma oscillations in the midbrain spatial attention network: linking circuits to function. Curr. Opin. Neurobiol. **31**, 189–198 (2015)

21. Schack, B., Klimesch, W., Sauseng, P.: Phase synchronization between theta and upper alpha oscillations in a working memory task. Int. J. Psychophysiol. **57**(2), 105–114 (2005)
22. Sridharan, D., Knudsen, E.I.: Gamma oscillations in the midbrain spatial attention network: linking circuits to function. Curr. Opin. Neurobiol. **31**, 189–198 (2015)
23. Tesche, C.D., Karhu, J.: Theta oscillations index human hippocampal activation during a working memory task. Proc. Natl. Acad. Sci. U.S.A. **97**, 919–924 (2000)
24. Vinogradova, O.S.: Expression, control, and probable functional significance of the neuronal theta-rhythm. Prog. Neurobiol. **45**(6), 523–583 (1995)
25. Wang, X.J.: Neurophysiological and computational principles of cortical rhythms in cognition. Physiol. Rev. **90**(3), 1195–1268 (2010)
26. Wang, Y., Romani, S., Lustig, B., Leonardo, A., Pastalkova, E.: Theta sequences are essential for internally generated hippocampal firing fields. Nat. Neurosci. **18**(2), 282–288 (2015)
27. Womelsdorf, T., Schoffelen, J.M., Oostenveld, R., Singer, W., Desimone, R., Engel, A.K., Fries, P.: Modulation of neuronal interactions through neuronal synchronization. Science **316**(5831), 1609–1612 (2007)

Memristor-Based Neuromorphic System with Content Addressable Memory Structure

Yidong Zhu[1,2], Xiao Wang[1,2], Tingwen Huang[3], and Zhigang Zeng[1,2(✉)]

[1] School of Automation, Huazhong University of Science and Technology,
Wuhan 430074, China
zgzeng@hust.edu.cn
[2] Key Laboratory of Image Processing and Intelligent Control of Education Ministry
of China, Wuhan 430074, China
[3] Texas A&M University at Qatar, Doha, Qatar

Abstract. By mimicking the complex biological systems, neuromorphic system is more efficient and less energy-efficient than the traditional Von Neumann architecture. Due to the similarity between memristor and biological synapse, many research efforts have been investigated in utilizing the latest discovered memristor as synapse. This paper improves the original network circuit based on memristor and content addressable memory structure and extends the existing results in the literature. The competition network circuit includes input layer, synapse and output layer. The synapse is made up of two memristors which store information and judge whether input and storage data are same. The output layer consists of subtractor which processes match and mismatch voltage to recognize pattern and the winner-take-all circuit to find out of which storage pattern is the closest to input pattern. The circuit design about read/write framework and working principle are discussed in detail. Finally, the system has been trained and recognizes these 5×6 pixel digit images from 0 to 9 successfully.

Keywords: Memristor · Neural network · Neuromorphic · Pattern recognition

1 Introduction

In recent years, the concept of memristor, originally proposed by Leon Chua in 1971 [1], has generated renewed interest since it has been experimentally found in 2008 by HP Lab [2]. Memristors and memristive devices are basically resistor with varying resistances, which depend on the history of the current through the memristor. Moreover, memristor-based memories can achieve high integration density of $100 \ Gb/cm^2$ [3]. Because of these advantages, they have been thought as a potential candidate for synapses on neuromorphic computing systems and many works have been done in this field. For instance, the memristor was used to represent the connection weight of two neurons utilizing memristor-based synaptic neuron in [4,5]. In [6,7], neural synapse was comprised of transistor

© Springer International Publishing Switzerland 2016
L. Cheng et al. (Eds.): ISNN 2016, LNCS 9719, pp. 681–690, 2016.
DOI: 10.1007/978-3-319-40663-3_78

and memristor and the neural network demonstrates fundamental properties including associative learning and pulse coincidence detection.

On the other hand, memristor-based neural network with basic learning rule called winner-take-all(WTA) has attract much extensive concern. Ebong et al. designed memristor-based neural network for position detection through the WTA learning rule [8]. In [9], a neuromorphic system which consists of memristor array and CMOS neuron uses same learning rule for visual pattern recognition. In [10], neuromorphic crossbar circuit based on memristors recognize speech through this rule. In studying memristor-based neural network with the WTA learning rule, we found that most designers use only one memristor as synapse to realize system where may exist leakage current problem which likely cause a certain error of calculation result about Hamming distance between input and storage pattern particularly in the large-scale neural network. However these networks can conquer the problem by iterating, this iterative training may spend huge time. In [11], authors propose a design based on memristor and content addressable memory(MCAM) architecture which overcomes leakage current shortcoming on crossbar, but the paper only consider how closely input vector match storage vector and leave out of consideration about the percentage of mismatch which also play a crucial role in reducing the judgment error.

In this paper, a improvement circuit which can obtain the degree of mismatch is given based on network circuit in [11]. The remainder of this paper is organized as follows. Memristor type used in the paper and attributes are described is presented in Sect. 2. Section 3 details the network circuit structure and how to write memristance to store pattern and identify pattern. Then, an character recognize application using the novel network circuit is discussed and simulation results is analyzed in Sect. 4. Section 5 concludes this paper.

2 Memristor

In network circuit design, we can use either analog memristor of which memristance can be changed gradually such as HP memristor [2], or binary memristors based on the filamentary-switching mechanism of which memristance only can be changed abruptly and have either a high resistance state(HRS) or low resistance(LRS) [12,13]. Since our circuit only need binary feature in character recognizition and resistance change of binary memristor is faster than analog memristor, we consider binary memristor as neural network synapse in our design and choose a kind of binary memristor model suggested in [13] for simulation. In this model, the memristance M is described by the following equation:

$$I = M^{-1}V_M \tag{1}$$

$$\frac{dM}{dt} = f(V_M)W(M, V_M) \tag{2}$$

$$f(V_M) = \beta \left(V_M - 0.5[|V_M + V_t| - |V_M - V_t|] \right) \tag{3}$$

$$W(M, V_M) = \theta(V_M)\theta(R_{off} - x) + \theta(-V_M)\theta(x - R_{on}) \tag{4}$$

where $f(V_C)$ is a function model of threshold memristor which illustrates memristance would be changed only when $|V_M| > V_t$, and β is a positive constant characterizing the rate of memristance change, $W(M, V_M)$ is a window function which limits memristance change between R_{on} and R_{off}.

3   Proposed Synapse and Architecture

3.1   Competition Network

A 3×3 neuromorphic system based on FM block circuit is shown in Fig. 1. In this structure, WE_i is the write enable signal. By enabling this line these patterns data which need to be stored can access to memristors in write operation. In addition, line ML_i and NL_i sum up match and mismatch current respectively. The output voltage V_{out} calculated by subtraction circuit shows the degree of similarity between storage pattern and recognizition pattern.

3.2   Full Match Block

Based on motivations and inspired by [11], the improving structure of FM block is shown in Fig. 2. The FM block consists of three modules, there modules corresponding to each point are as follows.

Storage Cell. Here we combine two memristors $ME1$ and $ME2$ together to built storage cell, and store data using a complementary memristor resistance type, for instance, memristors $ME1$ and $ME2$ store R_{off} and R_{on} respectively which represents logic 0, correspondingly $ME1$ and $ME2$ store opposite resistance respectively as logic 1. Through a series of appropriate operation the block will generate a match/mismatch output. Firstly, recognizition pattern data should be complementary and expressed as line voltage SX_i and SY_i. According to complementary resistance type of memristor, we stipulate that applying line SX_i, SY_i with low level V_{RL} and high level V_{RH} respectively represents logic 0, to the contrary, logic 1 is represented by applying line SX_i, SY_i with opposite voltage respectively. Secondly, the formula of the cell output voltage V_A deduced by resistance voltage divider is as follows.

$$V_A = \begin{cases} (V_{SX_i} - V_{SY_i})\dfrac{R_{ME2}}{R_{ME1} + R_{ME2}} & V_{SX_i} = V_{RH} \\ (V_{SY_i} - V_{SX_i})\dfrac{R_{ME1}}{R_{ME1} + R_{ME2}} & V_{SX_i} = V_{RL} \end{cases} \tag{5}$$

Finally, if storage data matches recognizition pattern data when R_{off} is much larger than R_{on}, the output voltage V_A approximately equals V_{RH}. Conversely, voltage V_A approximately reaches V_{RL}. According to the output, we can find out whether two pattern match.

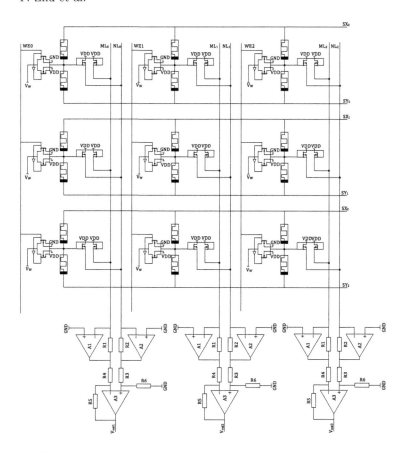

Fig. 1. 3×3 competition network system based on FM block

Transmission Gate. The transmission gate consists of a inverter and two transistors which can achieve two way signals transmission and low resistance which is difficult to meet by only one transistor. The gate is used in write operation to guarantee that storage resistance status of two memristor can change to complementary state according to input.

Voltage to Current Converter. The converter in this design is realized by NMOS and PMOS which can convert voltage to current. The equations in [11] describing their I-V characters are shown below:

(1) For NMOS, if $V_{DS} > V_{GS} - V_T$

$$I_D = \frac{\mu C_{ox}}{2} \frac{W}{L} (V_{GS} - V_T)^2 \tag{6}$$

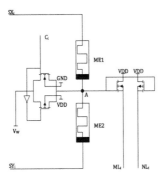

Fig. 2. Design schematic of Full Match block

(2) For PMOS, if $V_{SD} > V_{SG} - |V_{th}|$

$$I_D = \frac{\mu C_{ox}}{2} \frac{W}{L} (V_{SG} - |V_T|)^2 \tag{7}$$

where μ, C_{ox} are fixed parameters under the constant environmental condition, W and L are the sizes and V_T is the threshold voltage of transistor.

3.3 Write Operation

As discussed above, memristor can change its state by applying appropriate voltage. Therefore, by use of this feature, write operation controls voltage to shift memristor state to store pattern data. First of all, applying the same voltage to line SX_i and SY_i and this voltage is decided by the information you want to store. If logic 1 should be written into the selected storage cell, high level V_{WH} will be applied on line SX_i and SY_i. Otherwise, voltage of line SX_i and SY_i is set to low level V_{WL}. Afterwards, enabling write enable signal WE_i select which column is written. Since enable signal turns the transmission gate on, V_W pulls the point A to V_M. Eventually, voltage applying on two memristors is larger than threshold value V_{th} and voltage direction is opposite to each other, so memristances of $ME1$ and $ME2$ are changed and complementary. For unselected cells, their states remain unaltered due to same voltage in line SX_i and SY_i.

3.4 Read Operation

When pattern data is stored on memristor synapse, input pattern can be recognized through reading operation. To carry out the operation, input vectors would drive all the signal lines SX_i and SY_i while this drive voltage is smaller than memristor threshold voltage V_{th} to prevent storage data from changing.

The match current flowing through line ML_i will be summed and converted into voltage V_{A1} through amplifier $A1$, and the voltage size represents the degree of similarity which can be written as follows:

$$V_{A1} = -\sum_{i=1}^{n_1} \left(\frac{\mu C_{ox}}{2} \frac{W_i}{L_i} (V_A - V_{T_i})^2 \right) R_1 \approx -\sum_{i=1}^{n_1} \left(\frac{\mu C_{ox}}{2} \frac{W_i}{L_i} (V_{RH} - V_{T_i})^2 \right) R_1 \quad (8)$$

For the same reason, The mismatch current through line NL_i will be summed and converted into voltage V_{A2} through amplifier $A2$ and this voltage shows the degree of dissimilarity which can be written as follows:

$$V_{SA} = VDD - V_A = VDD - V_{RL} \tag{9}$$

$$V_{A2} = -\sum_{i=1}^{n_2} \left(\frac{\mu C_{ox}}{2} \frac{W_i}{L_i} (V_{SA} - V_{T_i})^2 \right) R_2 = -\sum_{i=1}^{n_2} \left(\frac{\mu C_{ox}}{2} \frac{W_i}{L_i} (VDD - V_{RL} - V_{T_i})^2 \right) R_2 \quad (10)$$

where μ, C_{ox}, W_i, L_i and V_{T_i} are parameters of transistor, and n_1 and n_2 are the number of match and mismatch blocks respectively, V_A is FM block output voltage which can be calculated by (5).

It is unilateral to judge how close is the recognizition pattern to storage pattern only through match or mismatch voltage, so this work takes the two aspects into account to choose the comprehensive result. The final output voltage V_{out} is calculated by subtracting mismatch voltage from match voltage and described based on (8) and (10) as:

$$\begin{aligned}
V_{out} =& R_5 \left(\frac{V_{A2}}{R_3} - \frac{V_{A1}}{R_4} \right) = \frac{R_5 R_1}{R_4} \sum_{i=1}^{n_1} \left(\frac{\mu C_{ox}}{2} \frac{W_i}{L_i} (V_{RH} - V_{T_i})^2 \right) \\
& - \frac{R_5 R_2}{R_3} \sum_{i=1}^{n_2} \left(\frac{\mu C_{ox}}{2} \frac{W_i}{L_i} (VDD - V_{RL} - V_{T_i})^2 \right)
\end{aligned} \tag{11}$$

Under same production processes and technologies, transistor parameters can are set up to equivalent value in same column. So based on the above assumptions (11) can be simplified as follow:

$$V_{out} = \alpha_1 n_1 - \alpha_2 n_2 \tag{12}$$

where α_1 and α_2 are constant parameters which decide the degree of influence that the number of match and mismatch blocks make to the output voltage and they are shown below:

$$\alpha_1 = \frac{R_5 R_1}{R_4} \frac{\mu C_{ox}}{2} \frac{W}{L} (V_{RH} - V_T)^2$$

$$\alpha_2 = \frac{R_5 R_2}{R_3} \frac{\mu C_{ox}}{2} \frac{W}{L} (VDD - V_{RL} - V_T)^2$$

4 Character Recognition Simulation

In this section, character recognition simulation using above circuit is performed and analysed. To execute this operation for competitive network, the WTA circuit which make the output node with the largest output value emerges as the winning node and inhibits all other nodes need to be designed. In [14] puts forward a high-precision VLSI WTA circuit and the simplified basic circuit is shown in Fig. 3. This circuit can meet the requirement about processing output with simple structure and high precision, so this WTA circuit is used to our simulation.

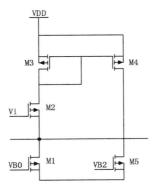

Fig. 3. Schematic diagram of the WTA circuit

Resistance range of resistor R_i on amplifier is crucial for the success of our simulation because unreasonable resistance may cause operational amplifier reaches saturation state or difference between match and mismatch output voltage is smaller than the WTA circuit minimum judgment precision, so discussion about resistance range is as follow.

To prevent operational amplifier from reaching to saturation, resistance should be limited. When all data stored on synapses match input data, maximum output of the amplifier $A1$ is obtained without saturation and shown as follow:

$$V_{A1max} = V_{sat} = -\sum_{i=1}^{n} \left(\frac{\mu C_{ox}}{2} \frac{W_i}{L_i} (V_{RH} - V_{T_i})^2 \right) R_1 \tag{13}$$

$$R_{1max} < \frac{|V_{sat}|}{\sum_{i=1}^{n} \left(\frac{\mu C_{ox}}{2} \frac{W_i}{L_i} (V_{RH} - V_{T_i})^2 \right)} \tag{14}$$

In a similar way, maximum resistance R_{2max} is expressed as follows.

$$R_{2max} < \frac{|V_{sat}|}{\sum_{i=1}^{n} \left(\frac{\mu C_{ox}}{2} \frac{W_i}{L_i} (VDD - V_{RL} - V_{T_i})^2 \right)} \tag{15}$$

In order to guard against the WTA circuit fails to decide which one neuron to win when two storage pattern are similar and only one data is different, resistors are limited as follow:

$$V_{WTA} < V_{diff} = \frac{R_5 R_1}{R_4}\left(\frac{\mu C_{ox}}{2}\frac{W_i}{L_i}(V_{RH}-V_{T_i})^2\right) + \frac{R_5 R_2}{R_3}\left(\frac{\mu C_{ox}}{2}\frac{W_j}{L_j}(VDD-V_{RL}-V_{T_j})^2\right) \quad (16)$$

where V_{diff} represents the smallest difference in two storage pattern and V_{WTA} represents minimum judgment precision of the WTA circuit.

Figure 4 shows the 5×6 pixel input images that need to be recognize. For these images, by using logic 0 and 1, white and black grid are represented respectively. These logic value need to be transformed into voltage for write and read operation in this network circuit. So images need to be scanned line by line from left to right to convert two-dimensional image into one-dimensional input, such as the expression form of digit 0 image is exhibited as follows (17).

$$q_0 = (0,1,1,1,0,1,0,0,0,1,1,0,0,0,1,1,0,0,0,1,1,0,0,0,1,0,1,1,1,0)^T \quad (17)$$

Before network executes character recognition function, these images should be learned by network which make these one-dimensional inputs store at network corresponding synapses according to write operation rule discussed above, therefore this meant that a $30 \times 10 \times 2$ memristor array was required to test this network. In the end, each column of neural synapses store input images logic value vectors which represent digit image respectively, hence, each column output shows that how close is the recognized pattern to storage pattern.

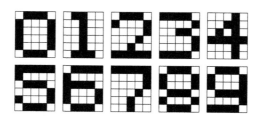

Fig. 4. Ten digit images which are used as input to the network for training

In real hardware conditions, noise can't be ignored, hence the simulation use not only image 1, but also testing image 1 with random noise. Figure 5 shows how noise was added to image 1 to generate noisy images. The original images for number 1 are shown in the left column, and random noise mask images are shown in the middle column. Final resulting noisy images used for test is shown in the right column where all pixels which have noisy image mask are flipped.

Figure 6 shows state of the neurons in this competition network during testing image 1. From the simulation results, the neuron node that represent image 1 has eventually become a winner and the others are failed and suppressed in this competition, so the network correctly identify this image no matter whether there is a random noise.

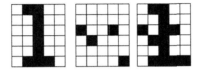

Fig. 5. Input image 1 in the left column is shown. Images with random pixels as black are shown in the middle column to produce noisy image mask. Resulting noisy images used for testing are shown in the right column where all pixels which have noisy image mask are flipped.

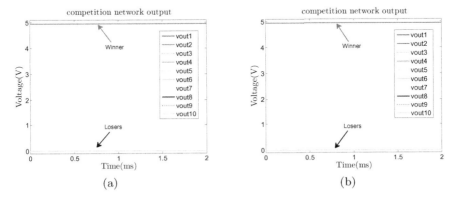

Fig. 6. Only one node represent image 1 achieve victory and other node are suppressed in this competition network based on WTA circuit. (a) use digit image 1 without random noise. (b) use digit image 1 with random noise (Color figure online).

5   Concluding Remarks

In this paper, a novel hybrid CMOS/Memristor competition network is presented. It contains synapse consisting of two memristor which store information and distinguish whether the input and storage data are the same, converter which generates not only match current, but also mismatch current, and output layer established by amplifier which is applied to process match and mismatch current. Finally, resistance range of resistors on amplifier are discussed and the recognizition test is performed successfully no matter whether the input contains random noise or not.

Acknowledgments. This work was supported by Huawei Innovation Research Program (HIRP) under Grant YB2015080050, the Science and Technology Support Program of Hubei Province under Grant 2015BHE013, the Program for Science and Technology in Wuhan of China under Grant 2014010101010004, the Program for Changjiang Scholars and Innovative Research Team in University of China under Grant IRT1245 and the National Priority Research Project NPRP 7-1482-1-278 funded by Qatar National Research Fund.

References

1. Chua, L.O.: Memristor-The missing circuit element. IEEE Trans. Circuit Theor. **18**, 507–519 (1971)
2. Strukov, D.B., Snider, G.S., Stewart, D.R., Williams, R.S.: The missing memristor found. Nature **453**, 80–83 (2008)
3. Ho, Y., Huang, G.M., Li, P.: Nonvolatile memristor memory: Device characteristics and design implications. In: Proceedings of the 2009 International Conference on Computer-Aided Design, pp. 485–490. ACM, New York (2009)
4. Linares-Barranco, B., Serrano-Gotarredona, T.: Memristance can explain spike-time-dependent-plasticity in neural synapses. In: Nature Precedings (2009)
5. Jo, S.H., Chang, T., Ebong, I., Bhadviya, B.B., Mazumder, P., Lu, W.: Nanoscale memristor device as synapse in neuromorphic systems. Nano Lett. **10**, 1297–1301 (2010)
6. Cantley, K.D., Subramaniam, A., Stiegler, H.J., Chapman, R.A., Vogel, E.M.: Hebbian learning in spiking neural networks with nanocrystalline silicon TFTs and memristive synapses. IEEE Trans. Nanotechnol. **10**, 1066–1073 (2011)
7. Cantley, K.D., Subramaniam, A., Stiegler, H.J., Chapman, R.A., Vogel, E.M.: Neural Learning circuits utilizing nano-crystalline silicon transistors and memristors. IEEE Trans. Neural Netw. Learn. Syst. **23**, 565–573 (2012)
8. Ebong, I., Mazumder, P.: CMOS and memristor-based neural network design for position detection. Proc. IEEE **100**(6), 2050–2060 (2012)
9. Chu, M., Kim, B., Park, S., Hwang, H., Jeon, M., Lee, B.H.: Neuromorphic hardware system for visual pattern recognition with memristor array and CMOS neuron. IEEE Trans. Ind. Electron. **62**, 2410–2419 (2015)
10. Truong, S.N., Ham, S.J., Min, K.S.: Neuromorphic crossbar circuit with nanoscale filamentary-switching binary memristors for speech recognition. Nanoscale Res. Lett. **9**, 1–9 (2014)
11. Yang, Y., Mathew, J., Pradhan, D.K.: Matching in memristor based auto-associative memory with application to pattern recognition. In: 12th International Conference on Signal Processing, pp. 1463–1468. IEEE Press, New York (2014)
12. Suri, M., Bichler, O., Querlioz, D., Palma, G., Vianello, E., Vuillaume, D., Gamrat, C., DeSalvo, B.: CBRAM devices as binary synapses for low-power stochastic neuromorphic systems: Auditory (cochlea) and visual (retina) cognitive processing applications. In: 2012 IEEE International Conference on Electron Devices Meeting, pp. 10.3.1–10.3.4. IEEE Press, New York (2012)
13. Biolek, D., Ventra, D.M., Pershin, Y.V.: Reliable SPICE simulations of memristors, memcapacitors and meminductors. arXiv preprint (2013). arXiv:1307.2717
14. Choi, J., Sheu, B.J.: A high precision VLSI winner-take-all circuit for self-organizing neural networks. Solid-State Circ. **28**, 576–594 (1993)

Detailed Structure of the Cortical Magnetic Response to Words

V.L. Vvedensky[1,2(✉)] and A. Yu. Nikolayeva[2]

[1] Kurchatov Institute, Moscow, Russia
VictorLvo@yandex.ru
[2] Moscow State University of Psychology and Education, Moscow, Russia

Abstract. We measured magnetic cortical response to words and analyzed its variability in individual trials. Considerable variations of the amplitudes of different components of the response in different trials suggest hopping of active spot from one point to another within a cortical area responsible for a certain processing stage. This behavior is similar to that observed in experiments with voluntary movement. We believe that only a small fraction of cortical area involved in a certain neural function is active during any particular trial. Next time another spot in the area will be activated. This is the reason of high variability of magnetic signals which are extremely sensitive to the position of the active spot on the folded cortical surface.

Keywords: Cortical response to word · Multi-channel MEG · Single-trial correlation with the average · Cortical networks

1 Introduction

Signals generated by the brain in response to external stimuli are commonly of small amplitude and "buried" in the noise-like ongoing activity, not related to the process under study. Many repetitions of the stimulus (trials) are used to accumulate enough data to extract the signal of interest. This is not natural for complex stimuli like words, which are usually used in combinations, but still multi-channel magnetic measurements (MEG) provide accurate responses to the spoken word consisting of several peaks with different time delays in different cortical areas in both hemi-spheres (Fig. 1). We see that the responses vary considerably between subjects. This is in accord with intra-operative mapping of cortical areas involved in language which shows considerable variability across patients and language tasks [1–3]. MEG data analysis is usually performed individually for each subject, allowing the detection of small differences invisible in grand-averages [4, 5]. Averaging is commonly used to reveal generality of the response across subjects, stimuli and tasks, though often it is difficult to decompose the conglomeration of peaks in the averaged traces into distinct components. This decomposition is very important for untangling the dynamic interaction of the different processes contributing to speech understanding [6]. Traditional accounts distinguish several stages in the processing stream: acoustic, phonetic, phonemic and word recognition, which are believed to follow each other. However, several experiments indicate

© Springer International Publishing Switzerland 2016
L. Cheng et al. (Eds.): ISNN 2016, LNCS 9719, pp. 691–697, 2016.
DOI: 10.1007/978-3-319-40663-3_79

early access to lexical information well before 400 ms after the word onset (see [6] for a summary). The words seem to be perceived as a whole as suggested by the experiments with the written text that has been jumbled [7]. In a stream of repeated objects (words) only few characteristic features are essential for correct recognition. The rich set of magnetic signals provides an opportunity to identify these features. Typically, word–specific changes in the amplitude of the MEG evoked response in a certain time window are reported. Alternatively, latency variation for a certain peak can be associated with the decision time change in the course of cortical computations. Experimental data contain greater amount of reliably measured values which can be used to identify distinctly different phases of the word perception. Our goal is to analyze magnetic cortical responses on a single-trial basis in order to reveal fine details of the processes supporting perception of words.

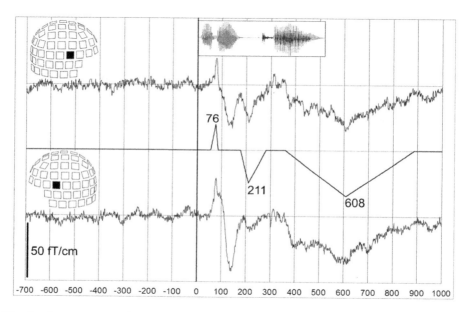

Fig. 1. Average magnetic responses to word in the symmetric points over right and left hemispheres. Sensor positions, where maximal magnetic field gradient is observed, are indicated on the helmet. Audio signal of the perceived word is shown at the top, it starts at time zero, scale in milliseconds. Central trace shows templates of three characteristic triangular peaks: 76 ms, 211 ms and 608 ms.

2   Methods

Four subjects, 2 males and 2 females (age 20–25), participated in 30-min tasks consisting of passive listening of common Russian words (РУКА, ЖЕНА, ЛЕТО, РУДА, КАРА, ЖЕЛЕ, means - HAND, WIFE, SUMMER, ORE, RETRIBUTION, JELLY) spoken by a native female speaker. These words are evenly distributed over the frequency list of Russian language. The words, about 500 ms long, were presented in random order

150 times each with a randomly jittered inter stimulus interval of 2 s ± 250 ms. No special instruction, except listening to the words, was given. None of the participants had known neurological or psychiatric disorders. The study was approved by the local ethics committee of the Moscow University of Psychology and Education (MSUPE) and was conducted following the ethical principles regarding human experimentation (Helsinki Declaration). Magnetic brain responses were recorded with the sampling rate of 1000 Hz using a helmet-shaped whole head magnetometer (Elekta Neuromag MEG system, 306 channels) in the MSUPE. Native hardware filters with band pass 0.1–330 Hz were used and no additional filtering was applied to avoid distortion of the signal shape. During the measurement the participants were sitting in the magnetically shielded room. We processed signals from 204 channels measuring planar gradients of the magnetic field generated by the brain, since the main contribution to the gradient signal comes from the area located under the sensor. General data processing and calculation of the field patterns were done using the Brainstorm software [8], the shape of the cortical surface was reconstructed using Free Surfer software [9].

3   Results

All our subjects displayed magnetic responses to words generally similar to those reported in literature [4, 5, 10]. Here we analyze in detail measurements for a single subject with the purpose to describe cortical processes which generate characteristic response signals. In a single channel one can extract clearly distinguishable nearly triangular peaks with quite different duration, as shown in Fig. 1. The array of sensors provides information on the spatial distribution of the magnetic field over the head. Patterns of the radial component of the field for the instants of maximum field amplitude are shown in Fig. 2. They are dipolar, indicating that localized cortical areas are active during these brain events. The magnetic field maps imply different patches of the cortex active during each event. Calculated positions of these pools of neurons fit into the Lateral sulcus of the brain in both hemispheres. The shape of the brain is reconstructed from the subject's MRI-scan. View from the right is shown in Fig. 2.

Since the average signals display regular shape, we can estimate the amplitude of the signal of that shape in each trial. We used this approach for the study of the magnetic signals of the brain related to the voluntary movement [11, 12]. In spite of the difference between the processes under study (voluntary movement and word perception) the variability of the recorded magnetic signals was considerable in both cases and the similarity between the processes was obvious. Amplitudes of the peak with definite time delay in different trials were evenly distributed between maximum and minimum values, which were of opposite sign. Average value was non-zero. Such a behavior is usually explained as a superposition of a weak nearly stable signal, related to a robust cortical process, and high irrelevant noise.

Distributions of the amplitudes for the peaks, shown in Fig. 3 A and B seem to indicate large irrelevant noise contribution, though careful analysis suggests another explanation. Our magnetic measurements reveal neural activity inside the fissures, as shown in Fig. 2. Direct measurements with subdural cortical electrodes (ECoG) on

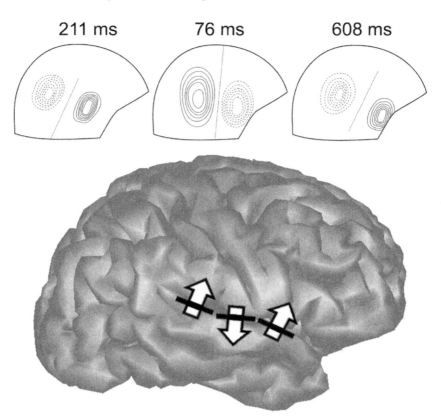

Fig. 2. Distributions of the magnetic field, displayed as isomagnetic charts, for the three peaks (76 ms, 211 ms, 608 ms) shown in Fig. 1. Positions of active pools of neurons for each time instant are shown as arrows on the cortical surface reconstructed from the MRI-scan. These maps and positions are calculated for the average of 150 trials.

subjects with pharmacologically intractable epilepsy have revealed extended areas of cortex activated during experiments with speech perception and production [3, 13]. They surround the Lateral fissure. Subdural grids of electrodes measure electric signals on the outer surface of the brain. Amplitude of the magnetic signal from the brain depends strongly on the position of the generating source on the cortical surface – whether is it inside a fissure or on a gyral region [14]. Functional role of each temporal component of the brain response to word is highly debated [6], though there is not much reason to expect extremely high variability in the processes performing auditory or linguistic tasks. We believe that during each trial, a certain stage of stimulus processing recruits populations of neurons with slightly different location on the cortical surface. Figure 3C illustrates how maximum and zero amplitudes of the magnetic field can be recorded during virtually the same cortical process. Variations of the amplitude of the magnetic field are akin to the flickering of the pendant light in the wind.

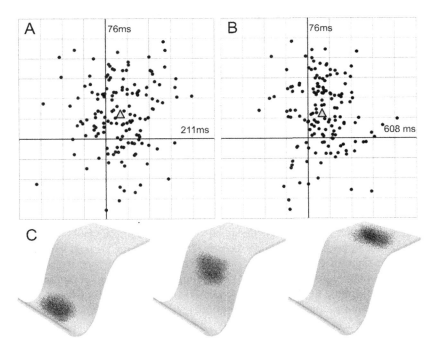

Fig. 3. Comparison of relative amplitudes of three peaks in 150 individual trials, represented as points. Triangles show average values. (A) Amplitude of the peak 76 ms versus peak 211 ms. (B) Amplitude of the peak 76 ms versus peak 608 ms. (C) Sketches demonstrate different possible positions of active pool of neurons on the cortical surface in the fissure. Central position is optimal for generating maximum magnetic field, while other two produce no magnetic signal.

This evident drawback of the magnetic technique is, from the other point of view, the manifestation of extremely high sensitivity of MEG to the slight displacements of the active zone along the cortex. The area capable to perform a definite linguistic (or another) task is much larger than the area activated during any particular trial. The spot jumps from one place to another with each new repetition of the stimulus. This can be seen in Fig. 2. Average isomagnetic charts for the peaks 211 ms and 608 ms have the shape, which implies scatter of source positions in individual trials along the Lateral fissure. This behavior of electrically active cells in the organism is common - muscular activity is organized in a similar way. During long muscular contraction some motor units can become inactive, being substituted by other ones. Such an organization of the neural process makes it immune to local malfunction (due to fatigue, for example). If the word is misunderstood, it can be recognized next time after repetition, using unaffected portion of the cortex. During some trials active spots corresponding to a certain component of the response happen to occupy optimal positions for generation of magnetic field. In these cases the shape of the raw recorded signal is close to the average one, though of much higher amplitude. This was observed earlier for voluntary movement [11].

Figure 3 A, B shows that the neural populations active during the three characteristic peaks change locations independently on each other. These groups of neurons support distinctly different stages of word processing. For this subject these components in this cortical area can be easily separated. More complex shapes of the response were recorded for other subjects where amplitude variations for some peaks correlate in individual trials. In these cases clearly separate peaks in the trace correspond to the same cortical event, and should be treated as a single process. Analysis of the amplitude and field pattern in individual trials is a useful tool for separation of the magnetic brain response into distinct stages of word processing. Each of these stages can be associated with various conditions of the experiment.

4   Conclusions

We conclude that in the set of magnetic signals from different trials most reliable information is provided by high-amplitude records. The spot on the cortical surface active during certain stage of the processing can change location from trial to trial. During many trials the pool of neurons, reliably performing certain cortical computation, does not produce measurable field, being located in the "magnetically mute" area. This fact should be taken into account on the stage of interpretation of the experiment. In order to improve data processing one has to apply grouping approach [12, 15], paying more attention to the trials containing relevant information and disregarding those, which simply overload the processing system.

Acknowledgments. We are grateful to T.A. Stroganova and A.V. Butorina who made high quality measurements. Supported by the Grant of the Russian Fund for Basic Research 15-29-03814-ofi_m.

References

1. Roux, F.-E., Lubrano, V., Lauwers-Cances, V., Trémoulet, M., Mascott, ChR, Démonet, J.-F.: Intra-operative mapping of cortical areas involved in reading in mono and bilingual patients. Brain **127**, 1796–1810 (2004)
2. FitzGerald, D.B., Cosgrove, G.R., Ronner, S., Jiang, H., Buchbinder, B.R., Belliveau, J.W., Rosen, B.R., Benson, R.R.: Location of language in the cortex: a comparison between functional MR imaging and electrocortical stimulation. Am. J. Neuroradiol. **18**, 1529–1539 (1997)
3. Llorens, A., Trebuchon, A., Liegeois-Chauvel, C., Alario, F.-X.: Intra-cranial recordings of brain activity during language production. Front. Psychol. 2, Article 375, 12 (2011)
4. Travis, K.E., Leonard, M.K., Chan, A.M., Torres, C., Sizemore, M.L., Qu, Z., Eskandar, E., Dale, A.M., Elman, J.L., Cash, S.S., Halgren, E.: Independence of early speech processing from word meaning. Cereb. Cortex **10**, 2370–2379 (2013)
5. Pylkkänen, L., Marantz, A.: Tracking the time course of word recognition with MEG. Trends Cogn. Sci. **7**(5), 187–189 (2003)
6. Almeida, D., Poeppel, D.: Word-specific repetition effects revealed by MEG and the implications for lexical access. Brain Lang. **127**(3), 497–509 (2013)

7. Grainger, J., Whitney, C.: Does the huamn mnid raed wrods as a wlohe? Trends Cogn. Sci. **8**(2), 58–59 (2004)
8. Tadel, F., Baillet, S., Mosher, J.C., Pantazis, D., Leahy, R.M.: Brainstorm: a user-friendly application for MEG/EEG analysis. Comput. Intell. Neurosci. **2011**, Article ID 879716, 13 p. (2011). http://neuroimage.usc.edu/brainstorm/
9. Dale, A.M., Fischl, B., Sereno, M.I.: Cortical surface-based analysis. I. Segmentation and surface reconstruction. NeuroImage **9**, pp. 179–194 (1999). http://surfer.nmr.mgh.harvard.edu/
10. Helenius, P., Salmelin, R., Service, E., Connolly, J.F.: Distinct time courses of word and context comprehension in the left temporal cortex. Brain **121**, 1133–1142 (1998)
11. Vvedensky, V.L.: Individual trial-to-trial variability of different components of neuromagnetic signals associated with self-paced finger movements. Neurosci. Lett. **21**(569), 94–98 (2014)
12. Vvedensky, V.L., Prokofyev, A.O.: Timing of cortical events preceding voluntary movement. Neural Comput. **28**, 286–304 (2016)
13. Cogan, G.B., Thesen, Th, Carlson, Ch., Doyle, W., Devinsky, O., Pesaran, B.: Sensory–motor transformations for speech occur bilaterally. Nature **507**, 94–98 (2014)
14. Ahlfors, S.P., Han, J., John, W., Belliveau, J.W., Hämäläinen, M.S.: Sensitivity of MEG and EEG to source orientation. Brain Topogr. **23**, 227–232 (2010)
15. Laskaris, N.A., Fotopoulos, S., Ioannides, A.A.: Mining information from event-related recordings. IEEE Signal Process. Mag. **21**(3), 66–77 (2004)

Pre-coding & Testing Technique for Interfacing Neural Networks Associative Memory

Fayçal Saffih[1($\boxtimes$)], Wan Abdulllah[2], and Zainol Ibrahim[2]

[1] Department of Electrical Engineering,
United Arab Emirates University, Al-Ain, UAE
faycals@uaeu.ac.ae
[2] Department of Physics, University of Malaya,
Kuala Lumpur, Malaysia
{wat,drzai}@um.edu.my

Abstract. Associative memories are one of the most important field of artificial neural networks (ANN) [1–3]. They have been used as content addressable memories (CAM) which enable them to perform pattern recognition [4, 5] as well as mitigating wireless communications noise [6]. Few applications have been investigated [7] to interface the associative memory in order to make it practical CAM memory. Through this vision scope, the present paper is considered as an attempt for the implementation of such interface by suggesting the Pre-coding and Testing Technique (PTT). Furthermore, it will be shown that the suggested technique influences the retrieval capability of associative memories in particular the Bidirectional Associative Memory which is chosen as an application CAM for the validation and assessment of the proposed technique.

Keywords: Artificial intelligence · Artificial neural networks · Bidirectional associative memories · Content addressable memories · Tele-communication noise mitigation · Pattern recognition

1 Introduction

Artificial neural networks are models of parallel-processing interconnected units called neurons dedicated to emulate biological neural networks such as the human brain. Their parallel processing as well as their rich interconnectivity makes them robust against any failure on the global network processing affecting any processing element. Based on the different interconnection configurations and their learning rules, artificial neural networks are classified to realize dedicated tasks. Feed-forward networks are dedicated for classification tasks using the back-propagation learning-rule [8] whereas, feed-back networks like the Hopfield network [9], and its derivatives such as the Bidirectional Associative Memory (BAM) [8, 10], are dedicated for optimization tasks, such as recalling stored memory using their content address-ability.

© Springer International Publishing Switzerland 2016
L. Cheng et al. (Eds.): ISNN 2016, LNCS 9719, pp. 698–705, 2016.
DOI: 10.1007/978-3-319-40663-3_80

The technique used for the present paper scope to differentiate the spurious states of the neural network from those corresponding to stored configurations which are both stable, is based on the encoding-decoding mechanism similarly used by the error-correcting theory [8]. The spurious states are stable configurations of the neural network but different from the stored ones which are stable configurations too. The cyclic encoding strategy has been chosen as the encoding scheme because of its simplicity and its cost-effective hardware implementation.

2   Pre-coding and Testing Technique (PTT)

The idea of differentiating the spurious states from the stored configurations but in another context was introduced by Fontanary and Koberle [9]. They suggested modifying the Hebbian learning rule slightly to check the input state whether it is within the basin of attraction of some memory state. In contrary, the PTT technique let the neural network user distinguish the stored stable state from the spurious one, when the network reaches its stable state. The PTT functionality is described in the following. The candidate binary-word memory to be stored is encoded before sending it to the neural network to learn it. The generated code using the cyclic encoder is concatenated with its corresponding vector generator (candidate binary-word), to form a memory-word, which is sent to the neural network to store it by learning its neurons. We use in the present paper the bi-directional associative memory neural network described later to implement the PTT technique. After the learning and processing phases (or the single learning-processing phase [9]), the BAM network gets its stable state configuration. Then the cyclic encoder encodes the corresponding code-generator vector of the stable state to calculate its corresponding code. After the comparison between the generated code and the corresponding code digits of the stable state, one could state whether the stable state correspond to one of the stored memories or not. Personnaz et al. [10] suggested the use of an encoder inside a neural network, when trying to implement a Hopfield net by exploiting the ring systolic architecture. The difference between their encoding approach and the present technique, is the use of the resultant stable state and its reverse to determine whether it corresponds to a stored memory or not as it may be.

3   Bidirectional Associative Memory

The Bidirectional Associative Memory (BAM) is a Hopfield derivative network, which inherits the two important properties of its original network, the minimization of the computational energy and the stability of the network configuration response. It is composed of two layers in which each neuron of one layer is interconnected with all neurons of the other layer. The BAM belongs to the family of heteroassociative dynamic associative memory (HDAM) proposed by Okajima [11] in 1987.

The BAM network is to be taught using a learning rule, to recall a stored memory starting from a distorted input memory.

4 PTT Application Results

Before going inside the details of the simulation it is worthy to mention that the PTT technique is not always successful in distinguishing correctly the stored memory from the spurious images and some error of the PTT technique is observed. We carry out numerical simulation of the PTT technique implemented on a BAM network composed of 24 neurons connected to 32 neurons. The generator polynomial used by the cyclic encoder is the Gaulois Field [8] primitive function. The generator polynomial g generates a 5-bit-code for the 27-bit-vector generator representing the memory. Then the BAM network synaptic strengths are learned using the Hebbian rule associating the 32-bit-generated memories (27-bit-vector memory concatenated with its generated 5-bit-code) of the first layer with the 24-bit memory of the other layer. Random stimulus configurations, with a given Hamming distance to a stored memory, are generated and input to the BAM network to test its recalling performance using PTT technique introduced earlier. This analysis is performed by comparing the BAM stable state and the stored memories then this result is compared with the PTT result. If the PTT affirm the precedent comparison, the technique is said to be successful, otherwise, it has failed. The Hamming distance representing the number of bits differentiating the input configuration from one of the stored memories is used as a parameter of PTT performance analysis discussed in the next conclusion section. Figures 1, 2, 3, 4, 5 and 6 show the PTT measurement of the BAM correct recalling success rate and the PTT

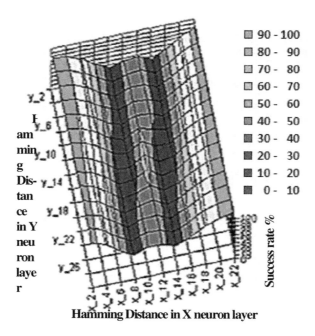

Fig. 1. PTT Success rate measurement (4 stored memories)

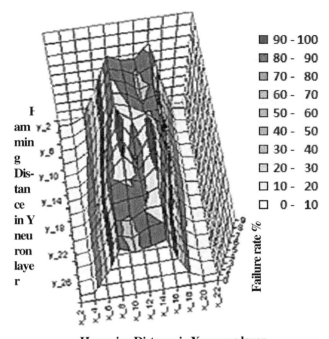

Fig. 2. PTT failure rate measurement (4 stored memories)

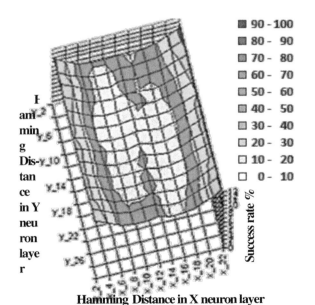

Fig. 3. PTT Success rate measurement (15 stored memories)

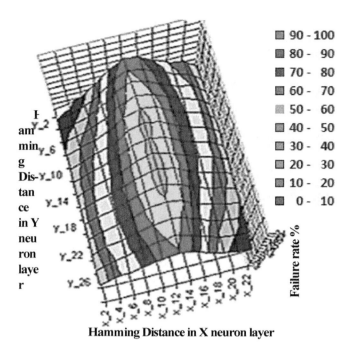

Fig. 4. PTT failure rate measurement (15 stored memories).

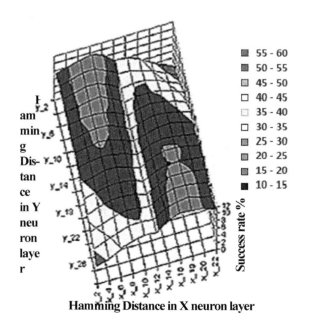

Fig. 5. PTT Success rate measurement (20 stored memories)

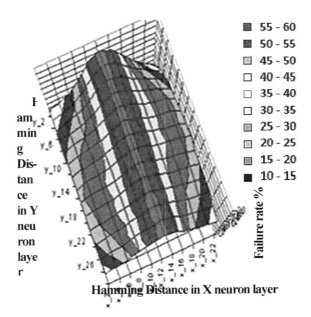

Fig. 6. PTT failure rate measurement (20 stored memories)

correct recalling failure rate. The axes 'x' and 'y' represent the Hamming distance of the input stimulus pattern to BAM network through its layers 'X' and 'Y'. The Hamming distance is shown by the concatenation of an integer number representing the Hamming distance in both axes indices. For example, x_8 represents an 8-bit Hamming distant of the input stimulus pattern to an 'X' layer stored memory.

The graphs show that when the BAM correct recalling measurement by the PTT is high the failure rate of this PTT measurement is low and vice-versa. Besides, the correct recalling measurement as well as the correct-recalling failure measurement are more dependent to the Hamming distance of the input pattern of the 'X' layer rather than the 'Y' layer. This could be explained by the fact that the layer of the minimum number of neurons influences the recalling performance of BAM network more than the other layer. This is an intrinsic property of BAM networks [15, 16].

5   Conclusion

From the results of the simulation above, one would conclude the following:

1. The PTT has a very low probability of error when the Hamming distance between the input vector and any of the stored memories is small.
2. When the Hamming distance grows the probability of recalling stored memories goes smaller and the error of the PTT correct-recalling goes higher.

3. As the number of the learned memories goes higher the probability of the correct recall becomes smaller and conversely the probability of PTT technique correct-recalling error gets higher.

Finally, the PTT for interfacing the BAM network has a good estimation (low error) for the detection of the stable states from the spurious ones for the small Hamming distances. As this distance increases the PTT gradually loses its power of detection and consequently the interfacing technique loses its effectiveness. The same observation is also observed when the number of the stored memories increases to achieve the capacity of the neural network which is the BAM in our case [17].

In addition, PTT correct-recalling success-rate follow the recalling capability BAM networks and indeed it enhances the recalling power of stored memories in a similar way to the neural pre-coding implementation mentioned in [6].

References

1. Anderson, J.A.: A simple neural network generating interactive memory. Math. Biosci. **14**, 197–220 (1972)
2. Nakano, K.: Associatron: a model of associative memory. IEEE Trans. Syst. Man Cybern. **SMS-2**, 380–388 (1972)
3. Yanai, H., Sawada, Y.: Associative memory network composed of neurons with hysteric property. Neural Netw. **3**, 223–228 (1990)
4. Wang, Y., et al.: Bidirectional associative memory. In: Hassoun, M. (ed.), Associative Neural Memories: Theory and Implementation, p. 107. Oxford University Press, New York (1993)
5. Daniel, L., et al.: Biological plausibility of artificial neural networks: learning by non-hebbian synapses. In: Hassoun, M. (ed.) Associative Neural Memories: Theory and Implementation, p. 31. Oxford University Press, New York (1993)
6. Salavati, A.H., Raj Kumar, K., Shokrollahi, A., Gerstner, W.: Neural pre-coding increases the pattern retrieval capacity of Hopfield and Bidirectional Associative Memories. In: 2011 IEEE International Symposium on Information Theory Proceedings (ISIT), July 31 – August 5 2011, pp. 850–854 (2011)
7. Personnaz, L., et al.: Silicon integration of learning algorithms and other auto adaptive properties in a digital feedback network, VLSI design of neural networks. In: Ramacher, U., Ruckert, U. (eds.) Kluwer academic publisher (1991)
8. Rhee, M.Y.: Error Correcting Coding Theory. McGraw-Hill, New York (1989)
9. Fontanary, J.F., Koberle, R.: Neural Networks with transparent memory. J. Phys. A: Math. Gen. **21**, 259–262 (1988)
10. Personnaz, L., et al.: Specification and implementation of a digital hopfield-type associative memory with on chip learning. IEEE Trans. Neural Netw. **3**(4), 529–539 (1992)
11. Okajima, K., Tanaka, S., Fujiwara, S.: A heteroassociative memory network with feedback connection. In: Caudill, M., Butler, C. (eds.) Proceedings of the IEEE First International Conference on Neural Networks, vol. 2, pp. 711–718 (1987)
12. Hassoun, M.H.: Adaptive Dynamic heteroassociative neural memories for pattern classification. In: Proceedings of the SPIE, Optical Pattern Recognition, pp. 75–83 (1989)
13. Hassoun, M.H.: Dynamic heteroassociative neural memories. Neural Netw. **2**(4), 275–287 (1989)

14. Leung, C.S., Cheung, K.F.: Householder encoding for discrete bidirectional associative memory. Proc. IJCNN **1**, 237–241 (1991)
15. Kosko, B.: Adaptive bidirectional associative memories. Appl. Opt. **26**, 4947–4960 (1987)
16. Kosko, B.: Bidirectional associative memories. IEEE Trans. Syst. Man Cybern. **SMC-18**, 49–60 (1988)
17. Mishchenko, K.: Capacity of bidirectional associative memory. Contemp. Eng. Sci. **8**, 825–833 (2015)

The Peculiarities of Perceptual Set in Sensorimotor Illusions

Valeria Karpinskaia[1]($\boxtimes$), Vsevolod Lyakhovetskii[2],
Viktor Allakhverdov[1], and Yuri Shilov[3]

[1] St. Petersburg State University, St. Petersburg, Russia
v.karpinskaya@spbu.ru, vimiall@gmail.com
[2] RAS Institute of Physiology, St. Petersburg, Russia
v_la2002@mail.ru
[3] Samara State Aerospace University, Samara, Russia
sheloves@ssu.samara.ru

Abstract. The effects of perceptual set in sensorimotor domain were studied with the help of Müller-Lyer and Ponzo illusions. The results lead us to conclude that the strengths of Ponzo and Müller-Lyer illusions are influenced by distinct characteristics of visual processing at different levels of visual system. The results of the experiments also demonstrate the existence of illusory perceptual sets in the sensorimotor domain. The type of the illusory perceptual set was found to depend on hemispheric-specific mental representations, which are activated by the movements of the right and the left hand.

Keywords: Müller-Lyer · Ponzo · Perceptual set · Left and right hand

1 Introduction

Following Uznadze and Piage there have been many studies investigating the influence of perceptual set but few of them have used visual illusions. Moreover, those that have done so have only used the Müller-Lyer illusion [6, 7]. However, it is worth noting that different levels of visual system are likely to be responsible for the appearance of different types of illusions, according to many researchers (especially [1]). As a consequence, it has been possible to divide the visual illusions onto the distinct subclasses according to their classification and that is why it is important to study not only the Müller-Lyer illusion but also illusions from other classes such as the Ponzo illusion. In addition, it is important to study the contribution of different hemispheric-specific subsystems of mental representations for their possible influence on different visual illusions. These hemispheric-specific subsystems can be selectively activated when the participant uses his or her right or left hand for sensorimotor measurements [2–4]. The aims of our study were (1) to investigate the possibility of inducing an illusory perceptual set in sensorimotor domain; (2) to investigate possible distinctions between the perceptual sets that are created by the movements of the right hand or the left hand.

© Springer International Publishing Switzerland 2016
L. Cheng et al. (Eds.): ISNN 2016, LNCS 9719, pp. 706–711, 2016.
DOI: 10.1007/978-3-319-40663-3_81

2 Methods

We recorded the extent of our participants' hand movements along the horizontal lines in the Ponzo and Müller-Lyer illusions in two conditions: (i) immediate reproduction and (ii) memorization. Four types of stimuli were used: the shaft flanked by outward-pointing (<>) and inward-pointing (><) arrows evoking the Müller-Lyer illusion (Fig. 1a, b); the shaft flanked by straight cuts (neutral stimuli, Fig. 1c) and the shaft without any flanks in Ponzo illusion (Fig. 1d).

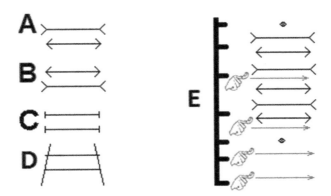

Fig. 1. Stimuli and methods

Forty different stimuli were presented in random order on a monitor screen located 60 cm in front of the participant. The shaft length was either 5 cm, 6.6 cm, 8.3 cm or 11.6 cm. During the presentation of each stimulus, the participant moved his/her hand (either the left hand or the right hand) across the touch screen located directly in front of the monitor, first along the upper shaft and then along the lower shaft (from right to left or from left to right). This constituted the 'memorization' phase. After the participant had finished his/her movements, the stimulus disappeared and participant was asked to reproduce the length of two shafts using his/her hand movements on the touch screen. This constituted the 'reproduction' phase (Fig. 1e). The experiment was performed without feedback. Each participant undertook two tasks with twenty presentations in each (for right and left hand). In the first task, the eyes of the volunteer were open during the 'reproduction' phase; in the second task, the eyes were closed during the 'reproduction' phase. We recorded the start and the end points of the participant's hand movements. The coordinates of the points were used to calculate the strength of the illusion: this was the difference between the reproduced lengths of the upper and the lower shafts. Two groups of right-handed participants participated in experiment. The dominant hand was determined by [5]. Group R ten participants first performed the task using the right hand and then using the left hand. In group L, vice versa. The group R participants first performed the task using the right hand (with eyes open), before performing the task using the right hand (with eyes closed). After that the participants performed the task using the left hand (with eyes open), before performing the task using the left hand (with eyes closed). The group L participants first performed the task using the left hand (with eyes open), before performing the task using the left hand (with eyes closed). After that the participants performed the task using the right hand (with eyes opened), before performing the task using the right hand (with eyes closed).

3 Results and Discussion

Analysis of the results showed that the participants experienced a significant Müller-Lyer illusion during the 'memorization' phase despite the fact that they could see both their hands and the lines as they traced out their lengths. However, there was no comparable effect with the Ponzo illusion in the 'memorization' phase (Table 1).

Table 1. Strength of the Müller-Lyer and Ponzo illusions, cm. Memorization phase.

	Left hand, eyes open	Left hand, eyes closed	Right hand, eyes open	Right hand, eyes closed
Group R				
Müller-Lyer illusion (upper shaft appears longer)	0.74*	0.73*	0.66*	0.50*
Müller-Lyer illusion (bottom shaft appears longer)	0.64*	0.76*	0.46*	0.80*
Ponzo illusion	0.20	0.17	0.00	-0.04
Neutral stimuli	0.26	0.12	-0.06	-0.05
Group L				
Müller-Lyer illusion (upper shaft appears longer)	0.49*	0.44*	1.08*	0.72*
Müller-Lyer illusion (bottom shaft appears longer)	0.24	0.56*	0.61*	0.64*
Ponzo illusion	-0.12	-0.15	0.30	-0.07
Neutral stimuli	0.33	-0.03	-0.28	-0.10

During the 'reproduction' phase, at least one of the variants of the Müller-Lyer and the Ponzo illusions (Fig. 1 a-e) produced an illusion in all four tasks for the group R participants. Similarly, the group L participants experienced at least one of the variants of Müller-Lyer illusion in all four tasks, but the Ponzo illusion was only evident in the task "left hand (eyes open)", "right hand (eyes closed)" (Table 2).

This leads us to conclude that Ponzo and Müller-Lyer illusions are influenced by hemispheric-specific characteristics of visual processing at the different levels of visual system. We also found that at memorization phase only the group R participants showed significant positive correlations between the strength of Müller-Lyer illusion and Ponzo illusion (Table 3).

Table 2. Strength of the Müller-Lyer and Ponzo illusions, cm. Reproduction phase.

	Left hand, eyes open	Left hand, eyes closed	Right hand, eyes open	Right hand, eyes closed
Group R				
Müller-Lyer illusion (upper shaft appears longer)	0.47*	1.19*	0.56*	1.19*
Müller-Lyer illusion (bottom shaft appears longer)	0.67*	0.15	0.68*	-0.09
Ponzo illusion	0.25*	0.67*	0.23*	0.70*
Neutral stimuli	0.14	0.73*	0.04	0.61*
Group L				
Müller-Lyer illusion (upper shaft appears longer)	0.20	0.73*	0.73*	1.07*
Müller-Lyer illusion (bottom shaft appears longer)	0.65*	0.92*	0.70*	0.26
Ponzo illusion	0.43*	0.27	0.16	0.62*
Neutral stimuli	-0.06	-0.46*	0.09	0.21

Table 3. Correlations between strengths of the Müller-Lyer and Ponzo illusions. Memorization phase.

	Left hand, eyes open	Left hand, eyes closed	Right hand, eyes open	Right hand, eyes closed
Group R				
Müller-Lyer illusion (upper shaft appears longer)/Müller-Lyer illusion (bottom shaft appears shorter)	0.46	-0.06	0.25	-0.19
Müller-Lyer illusion (upper shaft appears longer)/Ponzo	0.57	0.73*	0.51	0.69*
Müller-Lyer illusion (bottom shaft appears longer)/Ponzo	0.23	-0.03	0.32	-0.19
Group L				
Müller-Lyer illusion (upper shaft appears longer)/Müller-Lyer illusion (bottom shaft appears shorter)	-0.06	-0.65*	0.10	-0.68*
Müller-Lyer illusion (upper shaft appears longer)/Ponzo	0.92*	0.46	0.78*	0.77*
Müller-Lyer illusion (bottom shaft appears longer)/Ponzo	-0.01	0.10	0.09	-0.77*

Table 4. Correlations between strengths of the Müller-Lyer and Ponzo illusions. Reproduction phase.

	Left hand, eyes open	Left hand, eyes closed	Right hand, eyes open	Right hand, eyes closed
Group R				
Müller-Lyer illusion (upper shaft appears longer)/Müller-Lyer illusion (bottom shaft appears shorter)	0.52	-0.18	0.48	0.43
Müller-Lyer illusion (upper shaft appears longer)/Ponzo	0.45	0.81*	0.83*	0.51
Müller-Lyer illusion (bottom shaft appears longer)/Ponzo	0.17	-0.26	0.56	0.26
Group L				
Müller-Lyer illusion (upper shaft appears longer)/Müller-Lyer illusion (bottom shaft appears shorter)	0.58	0.27	0.58	0.14
Müller-Lyer illusion (upper shaft appears longer)/Ponzo	0.74*	0.68	0.75*	0.50
Müller-Lyer illusion (bottom shaft appears longer)/Ponzo	0.59	-0.19	0.50	-0.11

For the group L participants, we found both positive correlations and statistically significant negative correlations between the strength of the Müller-Lyer illusion and Ponzo illusion, and between two types of Müller-Lyer illusion. At reproduction phase there are fewer significant correlations for both groups, all of them are positive (Table 4). It is possible that these differences in the correlations could cause a different type of perceptual set (assimilative or contrast).

Overall, some 50% of the stimuli produced the illusion of overestimation of the upper shaft and only 25% of the stimuli produced the illusion of overestimation of the lower shaft. This asymmetry of stimuli was responsible for the emergence of a perceptual set, based on the illusion. As a consequence, the participants also made systematic errors during the measurement of the shafts of neutral stimuli (Fig. 1c). This type of set was different for the R and L groups. A contrast perceptual set was found in the group L participants for the "left hand (eyes closed)" task where the participants overestimated the lower shafts of the neutral stimuli. An assimilative perceptual set was found in the group R participants for the "right or left hand (eyes closed)" tasks where the participants overestimated the upper shaft of the neutral stimuli (Table 2).

4   Conclusions

These results reveal the existence of an illusory perceptual set in the sensorimotor domain. Moreover, the type of the illusory perceptual set appears to depend on hemispheric-specific mental representations, which are activated by the movements of the right and the left hand.

Acknowledgements.   Financial support: Russian Humanitarian Scientific Fund 16-36-01008.

References

1. Coren, S., Girgus, J.S., Erlichman, H., Hakstian, A.R.: An empirical taxonomy of visual illusions. Percept. Psychophysics **20**, 129–137 (1976)
2. Karpinskaia, V., Lyakhovetskii, V.: The differences in the sensorimotor estimation of the illusions ponzo and müller-lyer. Experimentalnaia Psichologija **7**, 3 (2014). (in Russian)
3. Karpinskaia, V., Lyakhovetskii, V.: The sensorimotor evaluation of perceptual illusions. Procedia Soc. Behav. Sci. **86**, 323–327 (2013)
4. Kosslyn, S.M., Behrmann, M., Jeannerod, M.: The cognitive neuroscience of mental imagery. Neuropsychologia **33**, 1335–1344 (1995)
5. Oldfield, R.C.: The assessment and analysis of handedness: the edinburgh inventory. Neuropsychologia **9**, 97–113 (1971)
6. Pollack, R.H.: Simultaneous and successive presentation of elements of the Müeller-Lyer figure and chronological age. Percept. Motor Skills **19**, 303–310 (1964)
7. Valerjev, P., Gulan, T.: The role of context in Müller-Lyer illusion: the case of negative Müller-Lyer illusion. Rev. Psychol. **20**, 29–36 (2013)

Generalized Truth Values: From Logic to the Applications in Cognitive Sciences

Oleg Grigoriev[✉]

Fuculty of Philosophy, Lomonosov Moscow State University, Moscow, Russia
grig@philos.msu.ru

Abstract. The aim of this report is to present a new family of generalized truth values which contain ontological, epistemic and vagueness dimensions and form the ordered structures known as a lattices in mathematics. This set may serve as a basis for constructing different logical systems on the one hand and modeling some epistemic situations and cognitive processes on the other. In particular I sketch some interpretations of special logical operations, classical seminegations, in the field of study a phenomenon of dissociation in neuropsychology.

Keywords: Generalized truth values · Multi-dementional epistemic logic · Epistemic situation · Vagueness · Analytic tableaux · Dissociation

1 Introduction

I would like to start with the pure logical motivation for introducing and discussing a special kind of what nowadays is commonly known as *generalized truth values* in studies of logical semantics. But the ultimate goal of current presentation is to sketch some prominent application of this apparatus entirely outside the logic itself (Sect. 6).

One of the sources for invention of non-classical logics in 20th century was a revision of a nature of the objects which can serve as semantic values for linguistic construction when studying them from a logical point of view. In particular the well known principle of bi-valence underlying classical logic was substantively redeemed. It was admitted for sentences to be assigned not only one of the two classical truth values. This approach finally led to the formation of the whole branch of modern logic – many valued logic.

Among of the most fruitful ideas for semantic constructions in many valued logics is that to consider truth values as complex entities having inner structure.

2 Truth and False Having Parts

Let us start with the observation of a construction producing a very special kind of generalized truth values proposed by D. Zaitsev and O. Grigoriev in [3,6].

© Springer International Publishing Switzerland 2016
L. Cheng et al. (Eds.): ISNN 2016, LNCS 9719, pp. 712–719, 2016.
DOI: 10.1007/978-3-319-40663-3_82

The main idea of this approach is to admit even to the basic abstract semantic objects, "truth" and "false", to have components in their inner structure. That is they are not atomic anymore. Starting from purely logical and epistemological position we might single out *ontological* and *epistemic* aspects of truth values. This obviously correlates to the basic conceptions of truth widely accepted in logic. Put it roughly, ontological aspect of truth reflects classical correspondence theory while epistemic one relates to the coherence theory.

Realizing the idea just described we still take as the initial semantic entities usual classical truth values, "truth" and "false", but split each into two components: "ontological" end "epistemic" ones. Thus we separately say about ontological components of "truth" and "false" (designated hereafter as "t" and "f" respectively) and epistemic component of these values (designated by 1 and 0). To construct compound, or generalized, truth values we make use of direct product operation on the sets $\{t, f\}$ and $\{1, 0\}$ thus obtaining the set $4^G = \{\langle f, 0 \rangle, \langle t, 0 \rangle, \langle f, 1 \rangle, \langle t, 1 \rangle\}$. This set can be ordered in a natural way and its elements comprise the structure, four element lattice as represented in the Fig. 1. Vector shows the direction of order.

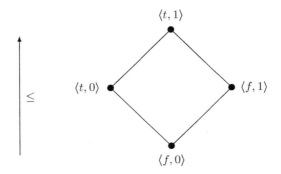

Fig. 1. Lattice $FOUR^{t,1}$

Binary lattice operations $\sqcap$ and $\sqcup$ on the $FOUR^{t,1}$ structure could be defined in a usual way, according with the order relation, but I do not need them in the context of the present article.

The most interesting part of the current construction – is a way of defining unary complementation-like operations. In [3,6] two very specific unary operations on $FOUR^{t,1}$ have been defined, so called *classical semi-complementations*. For the detailed account of their properties see [7]. I would like to stress here only one and the most important feature of these semi-complementations: each of them affect only its own component of a truth value and leaves untouched another. Thus we can use the notations $\sim_t$ and $\sim_1$ emphasizing the relation between operation its "working" component. Table 1 represents a complete definition of these operations.

Table 1. Definition of $\sim_t$ and $\sim_1$ on the lattice $FOUR^{t,1}$

	$\sim_t$	$\sim_1$
$\langle t,1 \rangle$	$\langle f,1 \rangle$	$\langle t,0 \rangle$
$\langle t,0 \rangle$	$\langle f,0 \rangle$	$\langle t,1 \rangle$
$\langle f,0 \rangle$	$\langle t,0 \rangle$	$\langle f,1 \rangle$
$\langle f,1 \rangle$	$\langle t,1 \rangle$	$\langle f,0 \rangle$

One can notice the "back and forth" nature of the operations. Each of them is like a shuttle between two adjacent values.

3   Adding a Dimension of Vagueness

The main drawback of the system of truth values described in the previous section consists in certainty of each parts of a value. Historically generalized truth values initially intended to deal with uncertain or overdetermined information (see classical paper [1]). Thus, we need a *vagueness* component to be incorporated.

The understanding of phenomenon of vagueness heavily relies on the notion of meaning. Philosopher and logician Kit Fine in [5] points out that in most general sense vagueness is a deficiency of meaning. Every linguistic item capable to have a meaning also capable to be vague. K. Fine also distinguishes extensional and intensional vagueness. The idea of extensional vagueness is closely related to the notion of generalized truth value as it presupposes the situation of truth-valuational gaps (or sometimes gluts). One of the way to cope with vagueness within the complex truth-values approach is the usage of specific indefinite truth value. In the context of truth values described we just add "vagueness dimension" as a new component. I denote this new component as 'u'. So, for example, the value $\langle t, u \rangle$ should mean that "it is true ontologically, but vague epistemically".

The problem arises with ontological component. While being vague epistemically seems natural, the ontological vagueness does not make sense for the first sight. Indeed, ontological value reflects the "objective" state of affairs and completely independent from anyone's knowledge. The sentence could be true ontologically even nobody knows the value of the corresponding component. If one respects this strong point of view, only epistemic component admits vagueness and we have six elements lattice of truth values as depicted in Fig. 2 on the left side.

The situation can be relaxed if we slightly change the understanding of what does it mean to be ontologically true. Instead of taking the ultimate concept of ontological truth, we can think of this property as being consistent with some dominant theory or knowledge base. This is quite reasonable because there is not too much sense in practical usage of the "objective" truth value. Thus not only the individual agent could be vague about the truth value of a sentence but the theory as well. This gives us an opportunity to extend the set of truth values

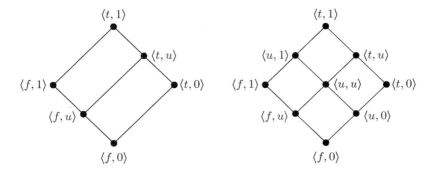

Fig. 2. Lattices SIX^u and $NINE^u$

to the structure shown on the right side of Fig. 2. This understanding however makes very important shift. Epistemic dimension here refers to the knowledge of some single rational agent and the whole construction fits best for designing epistemic logic of a single rational agent.

Again, there are standard lattice operations, ⊓ and ⊔, on both of these structures but I omit their detailed definition due to the lack of space, for details see, for instance, [7]. But it is worth to give a table definition for semi-complementations at least in the case of $NINE^u$.

Table 2. Definition of $\sim_t$ and $\sim_1$ on the lattice $NINE^u$

	$\sim_t$	$\sim_1$
$\langle t,1 \rangle$	$\langle f,1 \rangle$	$\langle t,0 \rangle$
$\langle t,u \rangle$	$\langle f,u \rangle$	$\langle t,u \rangle$
$\langle t,0 \rangle$	$\langle f,0 \rangle$	$\langle t,1 \rangle$
$\langle u,1 \rangle$	$\langle u,1 \rangle$	$\langle u,0 \rangle$
$\langle u,u \rangle$	$\langle u,u \rangle$	$\langle u,u \rangle$
$\langle u,0 \rangle$	$\langle u,0 \rangle$	$\langle u,1 \rangle$
$\langle f,1 \rangle$	$\langle t,1 \rangle$	$\langle f,0 \rangle$
$\langle f,u \rangle$	$\langle t,u \rangle$	$\langle f,u \rangle$
$\langle f,0 \rangle$	$\langle t,0 \rangle$	$\langle f,1 \rangle$

Note that there are situation when truth values behave as fixed points, namely when "working" component is u.

4   Logical Toolkit

Now I turn to the logical part. Let us design first an appropriate formal language. The most natural choice is to use some variant of propositional language which focuses on the structure of compound sentences composed from atomic ones. So let $\mathcal{L}$ denote the language which consists of a denumerable set of propositional variables, named Var, set of logical symbols (propositional connectives): $\{\wedge, \vee, \neg_t, \neg_1\}$ and parentheses as technical symbols. The rules for construction formulae in $\mathcal{L}$ (orjust formulae in this context) are usual with a slight modification for negations: if A is a formula, then $\neg_t A$ and $\neg_1 A$ are also formulae. Let For stand for a set of formulae.

To connect logical language and semantic structure we define interpretation of formulae via function v. For elements of Var a valuation function maps Var to the set of lattice elements, that is $v\colon Var \mapsto NINE^u$. Next complex formulae evaluated by extending v such that $v(A \wedge B) = v(A) \sqcap v(B)$ and $v(A \vee B) = v(A) \sqcup v(B)$. As for negations, we have $v(\neg_t A) = \sim_t (v(A))$ and $v(\neg_1 A) = \sim_1 (v(A))$. Note the correspondence between propositional connectives and lattice operations!

Now the crucial relation to be defined is a *logical consequence relation* or entailment ("$A \vDash B$" abbreviates "from A logically follows B"). It is a relation that captures logically correct reasoning forms and thus of the most importance for logic. There are several interesting ways to define entailment but the most simple and natural just makes use of the lattice order relation and valuation function: $A \vDash B \Leftrightarrow v(A) \leqslant v(B)$.

One of the main purpose in constructing logic – is to find a formalization of a syntactical analogue for $\vDash$, that is to present some logical formalism, describing behavior of corresponding syntactical relation $\vdash$.

5   Reasoning with Analytic Tableaux

The main advantage of *analytic-tableaux* calculi is their close relation to semantic truth conditions for the formulae. Moreover, analytic-tableaux formalisms usually serve as the basis for computer aided decision algorithms for the formalized logical theories. I present here one of the simplest and most natural version of such a system for the logic of $NINE^u$.

For our purposes I will adopt apparatus of so called prefixed analytic tableaux where each formula has a prefix, or sign, which is a metalinguistic symbol corresponding to a component of a supposed truth value of a formula. Thus let $Sn = \{t, f, 1, 0, u_t, u_1\}$ to be the set of signs and denote as γ its arbitrary element. A set of *signed formulae* is formed by means of product $Sn \times For$. For simplicity we write γA instead of $\langle \gamma, A \rangle$ for $\gamma \in Sn$ and formula A. Informally γA means that γ component is actually present (or 'active') in its position inside assigned truth values. For example tA signals that value of A is in the set $\{\langle t, 1 \rangle, \langle t, u \rangle, \langle t, 0 \rangle\}$. Note that I did not make a distinction between two sorts of vagueness – t and 1 ones, because inside the truth value its position makes things clear. But now I must state it explicitly. So I have two corresponding marks: u_t and u_1.

We shall use an expressions of the form "$X; \gamma A$" to represent a set of signed formulae along with a distinguished formula γA. Thus X is a finite set (possibly empty) of signed formulae and expression "$X; \gamma A$" actually means $X \cup \{\gamma A\}$.

A *tableau rule* could be characterized from set-theoretical standpoint as a pair of which the first component is a finite set of signed formulae and the second is a $\bot$ symbol or consist of one or two new finite sets of signed formulae. Now, following a long-time tradition, we represent a tableau rule as a fraction and use such an intuitive notions as '(set of)formula(e) above(below) a line of a rule'.

A *configuration* is a collection of finite sets of signed formulae. A *tableau* is a sequence of configurations $C_0, C_1, C_2 \ldots$ where each C_i for $i > 0$ is obtained from C_{i-1} by application of one of the rules from T^u calculus.

A tableau is *closed* if it has a final configuration each set of which is equal to $\{\bot\}$.

A *tableau for a sequence* $A \vdash B$ is a tableau which starts with a configuration composed from the sets of the form $\{\gamma A, \gamma^< B\}$ for all possible γ and $\gamma^<$ (see table below). A sequence $A \vdash B$ is called T^u-*proved* if there is a closed tableau for $A \vdash B$.

The set of tableau rules listed below comprises the tableau *calculus* T^u. Due to lack of space I cannot provide the proofs of metatheorems here but correctness of rules can be easily seen from properties of lattice $NINE^u$. To reduce space, all possible closure conditions summarized in one $[\bot]$ rule indicating the presence of non-compatible signs of formulae in the same set. Table 3 shows for each sign γ set of its incompatibles, $\gamma^* s$, as well as set of its position predecessors, $\gamma^< s$, needed to setup an initial configuration.

Table 3. Incompatibles and position predecessors of γ

γ	set of γ^*s	set of $\gamma^<$s
t	$\{u_t, f\}$	$\{u_t, f\}$
1	$\{u_1, 0\}$	$\{u_1, 0\}$
u_t	$\{t, f\}$	$\{f\}$
u_1	$\{1, 0\}$	$\{0\}$

Note. A collection of rules below is just a half of the whole family. Actually 1-Rules are easily generated form t-Rules replacing t by 1 and f by 0 everywhere in rules definitions. Likewise 0-Rules are obtained from f-Rules by the same replacement. Finally u_1-Rules are produced from u_t-Rules substituting 1 for t and 0 for f simultaneously.

t-Rules:

$$[t\neg_1] \; \frac{X; t\neg_1 A}{X; tA} \quad [t\neg_t] \; \frac{X; t\neg_t A}{X; fA} \quad [t\wedge] \; \frac{X; t(A \wedge B)}{X; tA, tB} \quad [t\vee] \; \frac{X; t(A \vee B)}{X; tA \mid X; tB}$$

f-Rules:

$$[f\neg_1] \; \frac{X; f\neg_1 A}{X; fA} \quad [f\neg_t] \; \frac{X; f\neg_t A}{X; tA} \quad [f\wedge] \; \frac{X; f(A \wedge B)}{X; fA \mid X; fB} \quad [f\vee] \; \frac{X; f(A \vee B)}{X; fA, fB}$$

u_t-*Rules and* $\perp$-*Rule:*

$$[u_t\neg_1]\ \frac{X; u_t\neg_1 A}{X; u_t A} \qquad\qquad [u_t\wedge]\ \frac{X; u_t(A \wedge B)}{X; u_t A, u_t B \mid X; u_t A, tB \mid X; tA, u_t B}$$

$$[u_t\neg_t]\ \frac{X; u_t\neg_t A}{X; u_t A} \qquad\qquad [u_t\vee]\ \frac{X; u_t(A \vee B)}{X; u_t A, u_t B \mid X; u_t A, fB \mid X; fA, u_t B}$$

$$[\perp]\ \frac{X; \gamma A, \gamma^* A}{\perp}$$

6   Towards the Application of Seminegations

The novelty of current contribution is the set of specific generalized truth values having ontological, epistemic and vagueness dimensions. Evidently a plethora of situations could be found admitting modeling in terms of SIX^u or $NINE^u$. I sketch just one a bit unexpected example concerning a phenomenon of *dissociation* in psychological research.

In a collection [2, P.1] I found an incredible quotation: "In many instances, the outcome of the dissociation issue is that the dissociated items constitute the two sides (if not the several) of a single object". But it is almost exactly the same idea as we put behind the notion of a compound, two-folded truth value!

A notion of dissociation is one of the central in neuropsychology. According to good introductory source [8, Ch5.], "dissociation can be defined as the situation where one cognitive function is preserved and the other one is damaged." Let us imagine two different epistemic situations.

1. *Single agent situation.* Consider a rational agent a with two distinguished cognitive abilities, A and B. For example, one ability is responsible for recognition of round objects while another for square, or one for vertical another for horizontal movement. Let also A be intact while B is broken. Thus we can single out corresponding parts within a truth value: $\langle t, 1_a \rangle$ means in this context that the sentence is true both "ontologically" (which might suppose a knowledge base information) and from a's point of view. This likely means that ability A is responsible for truth assignment. But what if B should be used for evaluation? Evidently a makes mistake or just unaware of the concrete value: $\langle t, 0_a \rangle$ or $\langle t, u_a \rangle$ correspondingly. More interestingly: what kind of negation a would use? My hypothesis that type of negation depends on evaluation of non-negative sentence. A-evaluation essentially induces $\neg_t$ usage while B-evaluation forces $\neg_{1_a}$ which behaves differently.

One more interesting remark. As we know from Table 2 in some situations truth value become fixed points under relevant operations. Say, $\sim_1(\langle t, u \rangle)$ gives $\langle t, u \rangle$ again. This is quite reasonable: how a can know the value of negated sentence if he evaluates the non-negative sentence as u? The unexpected answer – he *can*. The underlying mechanism is another kind of dissociation between conscious experience and nonconscious information processing related to the same function [2,8]. Roughly speaking a is vague about the value of some sentence then this may mean vagueness on the conscious level but some nonconscious structures (called "zombies" in [2]) evaluate this sentence correctly. But how could we know about the existence of this kind of dissociation? Via correct negation! Suppose a assigns the value $\langle t, u \rangle$ for some sentence α but $\neg_{1_a} \alpha$ is evaluated as $\langle f, 0 \rangle$. This means that a negates α correctly exactly because he "knows" the value of α but on the nonconscious level. For logical apparatus this gives even more negation-like operations. In particular just described negation which leads from vagueness to definiteness.

2. *Two agents situation.* In this context another notion from neuropsychology comes into play: *double dissociation.* Now we have two agents a and b and two abilities A and B such that a have intact A but damaged B while b have B and broken A. Double dissociation is interesting in many respects. In particular it helps to answer the question on interdependency of two abilities A and B. The picture seems to be more complex in view of conscious/nonconscious dissociation because some functions of an ability could be hidden in nonconscious ground. In general the mechanism of negations described above could help to reveal these hidden parts.

References

1. Belnap, N.D.: In: Ryle, G. (ed.) How a Computer Should Think. Contemporary Aspects of Philosophy. Oriel Press, London (1977)
2. Dissociation, B.: Interaction between dissociated implicit and explicit processes. In: Rossetti, Y., Revonsuo, A. (eds.) Advances in consciousness research. J. Benjaminis B.V., pp. 30–55 (2000)
3. Grigoriev, O.M.: Bipartite truth and semi-negations. In: Proceedings of 7-th International conference 'Smirnov readings in logic', June 22–24, Moscow (2011)
4. Grigoriev, O.M.: Two Formalisms for a Logic of Generalized Truth Values. Bull. Symbolic Logic **21**, 71–72 (2015)
5. Fine, K.: Vagueness. Truth and Logic. Synthese **30**, 265–300 (1975)
6. Zaitsev, D.V., Grigoriev, O.M.: Two kinds of truth - one logic. Logical Invest. **17**, 121–139 (2011)
7. Zaitsev, D., Shramko, Y.: Bi-facial truth: a case for generalized truth values. Studia Logica **101**, 1299–1318 (2013)
8. Revonsuo, A.: Consciousness: The Science of Subjectivity. Psychology Press, New York (2010)
9. Shramko, Y., Wansing, H.: Some useful sixteen-valued logics: how a computer network should think. J. Phil. Logic **34**, 121–153 (2005)

Modeling of Cognitive Evolution: Agent-Based Investigations in Cognitive Science

Vladimir G. Red'ko[1,2(✉)]

[1] Scientific Research Institute for System Analysis
of the Russian Academy of Sciences, Moscow, Russia
vgredko@gmail.com
[2] National Research Nuclear University MEPhI
(Moscow Engineering Physics Institute), Moscow, Russia

Abstract. The new direction of investigation, namely, modeling of cognitive evolution is described. The cognitive evolution is evolution of animal cognitive abilities. Investigation of cognitive evolution is based on models of autonomous agents. The sketch program for future investigations of cognitive evolution is proposed. Initial models, which were developed in accordance with the sketch program, are characterized. In particular, the model of agents that have several natural needs, models of agent movement in mazes, accumulation of knowledge, and formation of predictions, and the model of plan formation of rather complex behavior are described.

Keywords: Modeling of cognitive evolution · Autonomous agents · Animal cognitive abilities · Evolutionary origin of human intelligence

1 Introduction

The current work describes the new direction of investigation, namely, modeling of cognitive evolution. Cognitive evolution is the evolution of cognitive abilities of biological organisms. The important result of cognitive evolution is human thinking that is used at scientific cognition. This work develops our previous approaches to the problem of modeling of cognitive evolution [1–3].

It should be noted that modeling of cognitive evolution is directly connected with the following fundamental scientific problems:

- How did human thinking origin in the process of biological evolution?
- Why human formal thinking is applicable to cognition of the nature? In particular, why human formal logical inferences (which seem unrelated to external world) that are used in mathematical proofs are applicable to cognition of the real nature?

There is powerful background for modeling of cognitive evolution in several modern areas of computer science and in biological studies of the cognitive abilities of animals.

In addition, modeling of cognitive evolution in future should have a number of interdisciplinary connections: (1) with the foundations of science, foundations of

© Springer International Publishing Switzerland 2016
L. Cheng et al. (Eds.): ISNN 2016, LNCS 9719, pp. 720–730, 2016.
DOI: 10.1007/978-3-319-40663-3_83

mathematics, (2) with the cognitive science, (3) with the biological studies, (4) with the scientific foundations of artificial intelligence.

There are several branches of investigations in the field of the computer science, which are close to modeling of cognitive evolution: "Simulation of Adaptive Behavior" [4], "Artificial Life" [5, 6], "Cognitive Architectures" [7, 8], "Biologically Inspired Cognitive Architectures" [9], scientific foundations of Artificial Intelligence.

All these branches include models of autonomous agents. Mathematical and computer investigations of autonomous cognitive agents consider models of artificial organisms that are similar to biological organisms. Agents (modeled organisms) can adapt to changing environment [4], have ability to perform the reinforcement learning (RL) [10] and hierarchical RL [11, 12]. The autonomous agents can have simple mental abilities [13]. Recent works on autonomous agents consider investigations of antici-pation and prediction abilities of agents [14, 15].

In addition to models of autonomous agents, very interesting experimental inves-tigations of cognitive, mental abilities of biological organisms are conducted currently. In particular, the cognitive and creative properties of New Caledonian crows were investigated [16–18]. Evolutionary roots of human cognitive abilities were analyzed on the base of investigations of mental features of biological organisms.

It should be underlined that the current work is directly aimed to the new area of investigation: modeling of cognitive evolution. We propose the sketch program for future modeling of cognitive evolution and describe initial models of investigations of cognitive evolution. These models are based on autonomous cognitive agents; the models are in accordance with the sketch program.

2 Sketch Program for Future Researches of Cognitive Evolution

Future investigations of cognitive evolution can include the following steps.

(A) *Modeling of adaptive behavior of autonomous agents that have several natural needs: food, reproduction, safety.* Such modeling can be simulations of a natural behavior of simple modeled organisms.

(B) *Investigation of the transition from the physical level of information processing in nervous system of animals to the level of generalized "notions".* Such transition can be considered as emergence of "notions" in animal minds. Usage of notions leads to essential reduction both the needed memory and the time of information processing, therefore it should be evolutionary advantageous.

(C) *Investigations of processes of generating causal relations in animal memory.* Storing relationships between the cause and the effect and the adequate use of these relationships is one of key properties of active cognition of regularities of the external world by animals. This allows to predict events in the external world and to use adequately these predictions.

(D) *Investigations of "logic conclusions" in animal minds.* Actually, at classical conditioning, animals do a "logic conclusion": "If the conditioned stimulus takes place and the conditioned stimulus results in the unconditioned one, then the unconditioned stimulus is expected". Such conclusion is similar to the simple logical conclusion in mathematical deductions: "If the statement A takes place and the statement B is the consequence of the statement A, then the statement B takes place". It is important to understand, how systems of these conclusions operate, to what extent this "animal logic" is similar to ours, human logic.

The listed items outline steps of possible investigations from simplest forms of adaptive behavior to logical rules that are used in mathematical deductions. Following these steps, we began corresponding modeling (see below).

3   Models of Autonomous Cognitive Agents

3.1   Model of Agents with Natural Needs

The computer model of the simple agents, which have needs of (1) food, (2) reproduction, and (3) safety (Step A of the sketch program), was designed and analyzed [19, 20]. The model demonstrated certain cycles of behavior; the needs of food, security, and reproduction are sequentially satisfied in these cycles.

3.2   Model of Formation of Heuristics and Generalized Notions by Self-learning Agent

One of the most important cognitive properties of living organisms is the formation of generalized notions. Using notions leads to a reduction in the required memory and the processing time. However, how do the notions emerge? Can we imagine the processes of formation of notions by means of computer simulation? We outline a computer model, in which the autonomous agent alone produces generalizations and forms notions [21]. This model corresponds to the step B of the sketch program of modeling of cognitive evolution.

Model Description. The behavior of the autonomous agent in the two-dimensional cellular environment in discrete time is considered. Each time moment t, the agent executes one of the following five actions: eating food, moving forward into the adjacent cell, turning to right or to left, to rest. Portions of food are randomly placed into a half of the cells of the cellular world. The agent has a resource $R(t)$, which increases at eating food and decreases, when the agent executes other actions. The agent eats the entire portion of food in the cell at the action "eating food"; then a new portion of food is added randomly into a some free cell of the world.

The control system of the agent determines the choice of the agent action. The control system of the agent is a set of rules:

$$\mathbf{S}_k \rightarrow A_k, \tag{1}$$

where $\mathbf{S}_k$ is the situation, A_k is the action, k is the index of the rule. Each rule has its own weight W_k, the weights of the rules are modified at agent learning. The components of the vector $\mathbf{S}_k$ take the values 0 or 1; they correspond to the presence (1) or absence (0) of a portion of food in a certain cell in the "field of vision" of the agent. The field of vision includes four cells: the cell, in which the agent is, the cell ahead from the agent and the two cells to the right and left from the agent.

Each step of time, the agent selects and executes a certain action, and learns. Selecting the action is as follows. If the control system of the agent includes some rules, for which all components of the vector S_k coincide with the components of the current situation $S(t)$, then with probability $1-\varepsilon$, the rule with maximal weight W_k is selected from these rules. The agent executes the action A_k corresponding to this selected rule. The agent executes a random action with probability ε. A random action is executed also, if in the control system of the agent there are no rules, for which $S_k = S(t)$. If a randomly selected action A is absent in the agent control system, then a new rule $S(t) \rightarrow A$ is added to the set of agent rules. The initial weight W of this rule is 0. The annealing method [22] was used at the computer simulation: at the initial time moments, when rules have not yet formed, the value ε was close to 1, then during next time moments the value of ε exponentially was reduced to zero, the typical time of ε reduction was 1000 time moments.

Rule weights W_k are modified by means of the reinforcement learning [10]. Changes of the agent resource R were rewards at this learning. The reinforcement learning increases the weights of the rules, the use of which leads to an increase in the agent resource.

Simulation Results. The computer simulation demonstrates that the self-learning agent itself generates the following heuristics. The agent executes the action *eating food*, if there is food in the cell containing the agent (irrespectively of presence of food in the other cells of the field of vision of the agent). The agent executes the action *moving forward*, if there is no food in the agent cell, and there is food in the forward cell. The agent executes the action *turning to right or left*, if there is no food in the agent cell and in the forward cell, but there is food in the right or left cell from the agent. The frequency of the action *to rest* is negligible small. Simulation demonstrates that the agent resource increases.

Additionally, the computer program included the averaging procedure. Namely, the average number of executions of certain actions for a given situation S was calculated. This allows the agent to create notions that characterize the external environment. The averaging procedure results in creation of the following agent notions: *there is food in my cell, there is food in the forward cell,* and *there is food in the right or left cell.* These notions are important to the agent, since they are directly related with its actions and with the increase of its resource. Therefore, the agent itself is able to generate autonomously the generalized notions, characterizing the sensory information.

3.3   Models of Fish Exploratory Behavior in Mazes

We designed and analyzed models of cognitive behavior of fish in the course of maze exploration. The models are inspired by the biological experiment on zebrafish, Danio rerio, in mazes [23]. Zebrafish behavior in that experiment was studied in an unfamiliar (for fish) environment, namely in mazes of two types: the four-arm cross-shaped maze (Fig. 1) and the more complex maze with 11 arms.

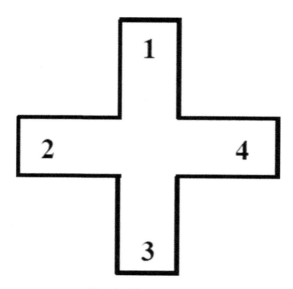

Fig. 1. The cross maze

Our models simulate movements of an agent (modeled zebrafish) in mazes. The time t is discrete, $t = 0,1,2\ldots$ At each step of time, the agent leaves an arm and enters into another one. The models were studied by means of computer simulation.

Model of Knowledge Acquisition. The model of knowledge acquisition (the model 1) describes the agent movements in the cross maze. The model assumes that the agent moves according to the two most frequent types of movements: the type 1 (the movement of an agent between adjacent arms) and the type 2 (the movement between the opposite arms). According to the biological experiment [23], the movement of type 1 is preferable as compared with the movement of type 2. We also take into account the possibility of random movements.

In order to take into account transitions between different types, we introduce the following model parameters: transition probabilities between the two types P_{ij}, i, $j = 1,2$. For example, P_{12} is the probability of transition from the first type to the second one. We also introduce the probabilities of transitions from a random type of movement (indexed "0") to types 1 and 2, P_{00}, P_{01}, P_{02}, as well as the probability of transition to a random movement P_{10} and P_{20} ($P_{00} + P_{01} + P_{02} = 1$, $P_{10} + P_{11} + P_{12} = 1$, $P_{20} + P_{21} + P_{22} = 1$). If there are several possible arms for movement in accordance with the described method, a next arm is selected randomly.

Acquisition of the knowledge about the arms was modeled as follows. It was assumed that the agent has a certain knowledge K_i about each arm, $0 \leq K_i \leq 1$, $i = 1,2,3,4$. Initial values of knowledge K_i are 0. When the agent visits i-th arm, the value K_i becomes equal to 1. Additionally, all values K_i slightly decrease with time: at any time moment t, values K_i are multiplied by the factor d_K ($0 < d_K < 1$, $1 - d_K < < 1$).

A special tendency to enter into those arms, which the agent did not visited for a long time, was introduced as follows. The agent at the time step t considers knowledge K_i about all four arms, and at the next step $t + 1$ the agent with a certain probability P_{choice} moves into the arm, which has a minimal value of K_i.

The parameters of computer simulation were as follows: the factor of knowledge decreasing was $d_K = 0.9$, the transition probabilities P_{ij} were: $P_{00} = 0.4$, $P_{01} = 0.4$, $P_{02} = 0.2$; $P_{10} = 0.1$, $P_{11} = 0.8$, $P_{12} = 0.1$; $P_{20} = 0.2$, $P_{21} = 0.6$, $P_{22} = 0.2$, the probability of movement into the arm with minimal knowledge P_{choice} was variable.

Figure 2 shows the dynamics of the sum of knowledge K_{SUM} about all four arms for different values of probability P_{choice}.

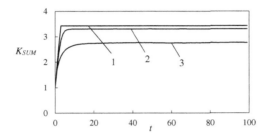

Fig. 2. The dependence of the sum K_{SUM} of knowledge for all arms in the cross maze on the time t for different values of P_{choice} (averaged over 10000 different starting random seeds). (1) $P_{choice} = 1$. (2) $P_{choice} = 0.5$. (3) $P_{choice} = 0$.

Model of Predictions of Future Situations. The computer model that describes the formation of predictions of future situations in the cross maze by the agent (the model 2) has been developed and studied. Predictions are characterized quantitatively by the values of assurance of these predictions A_S. Namely, for the given initial situation S_t and the action A_t, the assurance of the prediction of the next situation S_{t+1} is characterized by the value A_S ($0 \leq A_S \leq 1$). Each situation corresponds to a particular arm. When the agent leaves an arm, there are three possible actions A_t: (1) to turn into the right arm, (2) to go into the opposite arm, (3) to turn into the left arm. The agent predicts the next situation S_{t+1}. Thus, the assurances A_S characterize all possible chains $\{S_t, A_t\} \rightarrow S_{t+1}$.

To some extent, the considered predictions are similar to the formation of simple acceptor of results of action in the theory of functional systems by P.K. Anokhin [24].

The behavior of the agent in the model 2 is as follows. Initially, at $t = 0$, all assurances A_S equal 0; then the agent makes some predictions, and the assurances A_S are changing. At any time moment t ($t > 0$) after the action A_t, the agent predicts the next situation $S_{pr,t+1}$ for the time step $t + 1$. If the assurance for this prediction is small: $A_S(S_t, A_t, S_{pr,t+1}) < T_A$ (T_A is a certain threshold value), then the agent at time moment

$t + 1$ returns to that arm, which it visited at time t-1. If the assurance for this prediction is rather large $(A_S(S_t,A_t,S_{pr,t+1}) > T_A)$, then the agent choose randomly one of the possible actions A_{t+1} and moves randomly into some new arm. Thus, the agent explores the maze.

The described mode of the agent's movement can be considered as an heuristics which corresponds to the opposite trends: (1) the desire to predict the results of the behavior reliably (in this case the low assurance of the correct prediction increases after repetitions of movements), and (2) the search for a new, unpredictable situation (this corresponds to performing random actions at the high current assurance).

The agent makes predictions in the following manner. For the given current situation S_t and action A_t, the agent checked the assurances $A_S(S_t,A_t,S_{pos,t+1})$ for predictions of all possible next situations $S_{pos,t+1}$. Then the agent determines the situation $S_{max,t+1}$ with the maximal value $A_S(S_t,A_t,S_{pos,t+1})$: $S_{pr,t+1} = \arg\ \max\{A_S(S_t,A_t,S_{pos,t+1})\}$, and predicts this situation $S_{pr,t+1}$. Only if all values $A_S(S_t,A_t,S_{pos,t+1})$ are too small, the agent predicts the next situation randomly.

The values of assurances A_S are adjusted as follows. At the time step $t + 1$, the agent checks the prediction that it has made at the time step t. If the prediction of the next situation is correct (the predicted situation $S_{pr,t+1}$ coincides with the real situation S_{t+1}: $S_{pr,t+1} = S_{t+1}$), then the assurance of this prediction is increased according to the expression:

$$\Delta A_S(S_t, A_t, S_{pr,t+1}) = d_I\left[1 - A_S(S_t, A_t, S_{pr,t+1})\right], \tag{2}$$

where S_t and A_t are the situation and the action in the moment t, $S_{pr,t+1}$ is the predicted situation that the agent expects, d_I is the factor of increasing of assurances $(0 < d_I < 1)$.

If the prediction is wrong $(S_{pr,t+1} \neq S_{t+1})$, then the assurance of this prediction decreases:

$$\Delta A_S(S_t, A_t, S_{pr,t+1}) = -d_D A_S\left[(S_t, A_t, S_{pr,t+1})\right], \tag{3}$$

where d_D is the factor of decreasing of assurances $(0 < d_D < 1)$.

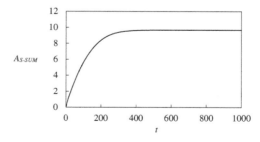

Fig. 3. The dependence of the summarized assurance $A_{S\text{-}SUM}$ on the time t for the whole cross maze (averaged over 10000 different starting random seeds)

In addition, it is assumed that all assurances values are multiplied by the factor d_A ($0 < d_A < 1$, $1-d_A < < 1$) at any time step; i.e. all assurances $A_S(S_t,A_t,S_{t+1})$ are slightly reduced.

The parameters of computer simulation were as follows: the threshold for estimations of assurance values was $T_A = 0.9$, the factors of increasing and decreasing of assurances were $d_I = d_D = 0.3$, the factor of reduction of all assurances was $d_A = 0.995$.

The main results of simulation for this model are represented in Fig. 3, which shows the dependence of the sum of all assurances for the whole cross maze A_{S-SUM} on the time t.

Hypothetical Model of Planning of Movement in Rather Complex Maze. The model 3 assumes that after certain period of maze exploration, the fish is able to form some generalized notions that characterize the essential places (situations) in rather complex maze. The agent has knowledges about situations and reliable predictions (with large A_S) about results of possible actions. The model 3 describes the process of forming the plan of movement to the goal situation, which was not visited for a long time (such situation has minimal value of knowledge K_i). The agent creates a plan of movement from some starting situation to the goal situation as follows.

Using the table of reliable predictions, the agent begins to analyze such situations and actions that result in the goal situation. Then the agent analyzes situations and actions that result in the pre-goal situations, and so on. Thus, *the agent begins from the goal situation* and analyzes consecutively possible ways to reach this situation. The agent also takes into account the distance from the considered situation S_t to the goal situation; this distance is the number of actions needed to reach the goal situation from the considered situation S_t.

Then the agent creates a simple *knowledge database* that is a table. Each row of this table includes the following information: (1) the current situation, (2) the action that reduces the distance between the current situation and the goal situation, (3) the next situation that is the result of the action and (4) distances between the current/next situations and the goal situation.

Finally, using this knowledge database, the agent forms a plan of movement from *the starting situation to the goal situation*. In this process of plan formation, the agent consecutively selects actions, which reduce the distance between the considered situations and the goal situation.

Thus, the models of accumulation of knowledge, prediction of results of actions, and planning of movement towards the goal situation have been developed and investigated. These models in several aspects correspond to the step C of the sketch program of modeling of cognitive evolution.

The similar model of planning by New Caledonian crows is outlined below.

3.4   Model of Plan Formation by New Caledonian Crows

The model is based on the biological experiment on New Caledonian (NC) crows [17]. In that work, NC crows were preliminary trained to execute particular elements of a rather complex behavior. After the preliminary training, the crows should solve the three-stage problem that includes the following particular elements:

(1) to pull up a short stick tied to the end of a string and to release this stick,
(2) to extract a long stick from a barred toolbox by means of the short stick, and
(3) to extract the food from a deep hole by means of the long stick.

It was impossible (a) to extract the food from the deep hole by means of the short stick and the bill, and (b) to extract the long stick from the barred toolbox by means of the bill. Therefore, in order to reach the food, the crow had to execute the ordered chain of sequential actions $1 \rightarrow 2 \rightarrow 3$.

The work [25] describes the simple formalization of planning of the complex goal-directed behavior by NC crows on the base of particular elements of this behavior. The process of planning is very similar to the model of planning by fish (see above). This process includes:

(1) The analysis of way to reach the goal situation (to extract the food).
(2) Guessing the needed predictions.
(3) The estimation of the distance between the considered situation and the goal situation.
(4) The formation of the knowledge database that characterizes situations, actions, results of actions, distances between the considered situations and the goal situation.
(5) The formation of the plan of actions in accordance with the knowledge database.

See [25] for details.

4 Conclusion

Thus, simple models of autonomous cognitive agents have been developed. These models characterize the following cognitive features of agents: (1) formation of generalized notions, (2) generation of predictions of future situations, and (3) formation of plans of rather complex behavior.

It should be underlined that the presented models characterized only initial steps of modeling of cognitive evolution. Nevertheless, investigation of cognitive evolution is directly connected with the fundamental scientific problems; there is powerful background for modeling of cognitive evolution in computer science and in biological researches; modeling of cognitive evolution in future should have a number of interdisciplinary connections. See [3] for details. Thus, modeling of cognitive evolution is the perspective direction of future investigations.

Acknowledgements. This work was supported by the Russian Science Foundation, Grant No. 15-11-30014.

References

1. Red'ko, V.G.: Evolution of cognition: towards the theory of origin of human logic. Found. Sci. **5**, 323–338 (2000)
2. Red'ko, V.G.: The natural way to artificial intelligence. In: Goertzel, B., Pennachin, C. (eds). Artificial General Intelligence, pp. 327–351. Springer, Berlin, Heidelberg, New York (2007)
3. Red'ko, V.G.: Modeling of cognitive evolution: perspective direction of interdisciplinary investigation. Procedia Comput. Sci. **71**, 215–220 (2015). http://www.sciencedirect.com/science/article/pii/S1877050915036686
4. Meyer, J.-A., Wilson, S.W. (eds): From animals to animats. In: Proceedings of the First International Conference on Simulation of Adaptive Behavior. MIT Press, Cambridge (1991)
5. Langton, C.G. (ed.): Artificial Life: The Proceedings of an Interdisciplinary Workshop on the Synthesis and Simulation of Living Systems. Addison-Wesley, Redwood City CA (1989)
6. Langton, C.G., Taylor, C., Farmer, J.D., Rasmussen, S. (eds.): Artificial Life II: Proceedings of the Second Artificial Life Workshop. Addison-Wesley, Redwood City CA (1992)
7. Langley, P., Laird, J.E., Rogers, S.: Cognitive architectures: research issues and challenges. Cogn. Syst. Res. **10**, 141–160 (2009)
8. Laird, L.E.: The Soar Cognitive Architecture. The MIT Press, Cambridge (2012)
9. Samsonovich, A.V.: On a roadmap for the BICA challenge. Biologically Inspired Cogn. Architectures **1**, 100–107 (2012)
10. Sutton, R., Barto, A.: Reinforcement Learning: An Introduction. MIT Press, Cambridge (1998)
11. Dietterich, T.G.: Hierarchical reinforcement learning with the MAXQ value function decomposition. J. Artif. Intell. Res. **13**, 227–303 (2000)
12. Schmidhuber, J.: Formal theory of creativity, fun, and intrinsic motivation (1990-2010). IEEE Trans. Auton. Mental Dev. **2**, 230–247 (2010)
13. Vernon, D., Metta, G., Sandini, G.: A survey of artificial cognitive systems: implications for the autonomous development of mental capabilities in computational agents. IEEE Trans. Evol. Comput. **11**, 151–180 (2007)
14. Butz, M.V., Sigaud, O., Pezzulo, G., Baldassarre, G.: Anticipations, brains, individual and social behavior: an introduction to anticipatory systems. In: Butz, M.V., Sigaud, O., Pezzulo, G., Baldassarre, G. (eds.) ABiALS 2006. LNCS (LNAI), vol. 4520, pp. 1–18. Springer, Heidelberg (2007)
15. Georgeon, O., Marshall, J.: Demonstrating sensemaking emergence in artificial agents: a method and an example. Int. J. Mach. Conscious. **5**, 131–144 (2013)
16. Weir, A.A.S., Chappell, J., Kacelnik, A.: Shaping of hooks in New Caledonian crows. Science **297**, 981 (2002)
17. Taylor, A.H., Elliffe, D., Hunt, G.R., Gray, R.D.: Complex cognition and behavioural innovation in New Caledonian crows. Proc. Royal Soc. B **277**, 2637–2643 (2010)
18. Taylor, A.H., Gray, R.D.: Is there a link between the crafting of tools and the evolution of cognition? Wiley Interdisc. Rev. Cogn. Sci. **5**, 693–703 (2014)
19. Red'ko, V.G., Koval', A.G.: Evolutionary approach to investigations of cognitive systems. In: Samsonovich, A.V., Johannsdottir, K.R. (eds.): Biologically Inspired Cognitive Architectures 2011. Proceedings of Second Annual Meeting of the BICA Society, pp. 296–301. IOS Press, Amsterdam (2011)
20. Koval', A.G., Red'ko, V.G.: The behavior of model organisms that have natural needs and motivations. Math. Biol. Bioinf. **7**, 266–273 (2012). (In Russian)

21. Beskhlebnova, G.A., Red'ko, V.G.: Model of formation of generalized notions by autonomous agents. In: Proceedings of the Fourth International Conference on Cognitive Science. Tomsk: TSU. vol. 1, pp. 174–175 (2010). (In Russian)
22. Kirkpatrick, S., Gelatt, C.D., Vecchi, M.P.: Optimization by simulated annealing. Science **220**, 671–680 (1983)
23. Red'ko, V.G., Nepomnyashchikh, V.A., Osipova, E.A.: Models of fish exploratory behavior in mazes. Biologically Inspired Cogn. Architectures **13**, 9–16 (2015)
24. Anokhin, P.K.: Biology and Neurophysiology of the Conditioned Reflex and its Role in Adaptive Behavior. Pergamon Press, Oxford (1974)
25. Redko, V.G., Nepomnyashchikh, V.A.: Model of plan formation by New Caledonian crows. Procedia Comput. Sci. **71**, 248–253 (2015). http://www.sciencedirect.com/science/article/pii/S1877050915036820

Usage of Language Particularities for Semantic Map Construction: Affixes in Russian Language

Anita Balandina, Artyom Chernyshov[(✉)], Valentin Klimov,
and Anastasiya Kostkina

National Research Nuclear University MEPhI (Moscow Engineering Physics Institute),
Moscow, Russian Federation
anita.balandina@gmail.com, zexirius@gmail.com,
vvklimov@mephi.ru, anakost@bk.ru

Abstract. This paper is devoted to a method of creation the semantic map for Russian language. We consider different approaches for cognitive maps construction which were made for other languages and compare them to the developing algorithm. We also show the main features of Russian words structure and highlight the important of them for the further usage in the concept of semantic map. We introduce the set-theoretical model of the Russian words which will be used in further researches.

Keywords: Semantic maps · Weak semantic maps · Russian language · Affixes

1 Introduction

In the present time word semantics is given special attention in the Web and in other fields related with automated text recognition. For instance, improving search in the Web (semantic search), where the user's queries are uniquely interpreted by the searcher, or sentiment analysis (also opinion mining), whose purpose is to determine the attitude of a speaker or a writer with respect to some topic of the overall contextual polarity (e.g. positive or negative) of a document [1]. This concept could be included into many different web-applications, as a rule in social networks, blogs and other resources of personalization, which the semantic search and sentiment analysis are applied in. In any case, automated evaluation of word semantics means using various scales, or dimensions, that characterize the word meaning [2]. These tools are also known as semantic maps, cognitive maps, semantic spaces.

The most popular approaches in this field are based on usage of vector space model. It means that concepts, words or documents (representations) can be associated with vectors in an abstract multidimensional vector space. There are also other approaches the sense of which consists in using manifolds of more complex topology and geometry. In summary, cognitive map can be defined as a topological or metric space, the topology and/or the geometry of which reflect semantic characteristics and relations among a set of representations (such as words or word senses) embedded in this space.

Besides, there are other approaches, for example, named "weak semantic cognitive mapping" [3] that is not based on the idea of "dissimilarity". The idea of the current

© Springer International Publishing Switzerland 2016
L. Cheng et al. (Eds.): ISNN 2016, LNCS 9719, pp. 731–738, 2016.
DOI: 10.1007/978-3-319-40663-3_84

approach consists in using such notion as "opposite relations" and doesn't take into account individual semantic characteristics of representations given a priori. Only relations, but not semantic features, are given as input. As a result, semantic dimensions of the map that are not predefined to emerge naturally, starting from a randomly generated initial distribution of words in an abstract space with no a priori given semantics and following [3]. The main relation that authors use in this approach is the relation "synonymy-antonymy" for representations. Mainly the approach "week semantic cognitive mapping" was applied to the fixed quantity of English words.

According to different researches, in present time the most of these approaches are used for words of English language. Therefore, we have the designs of cognitive maps for English language and French language with defined metric for space of words [4–6]. In this connection we discover the opportunity of creation the cognitive map of Russian language, where each word will have certain space coordinates and will be easily identified.

The current approach differs from the approaches based on usage of vector space model and has its own specificity because of the definite features of Russian language. The idea of the approach is not based on the notion of "dissimilarity" and we start from using of relations along representations, for example, "synonymy-antonymy". One might say that our approach has similar features with "weak semantic cognitive mapping", however, the approach that we suggest is not applied to English language and other languages as far as consideration must be given to morphological multiplicity generated by Russian language and, hence, more wide semantics of words. In the next paragraphs we consider the existing methods of constructing cognitive maps in detail, analogues of our approach and explain the essence of the approach as well as the future trends of it.

2   Different Approaches of Semantic Maps Creation

Latent semantic analysis (LSA, also known as Latent semantic indexing, LSI) has arguably received the most attention of all the semantic space model implementations [2, 7]. The principles of factor analysis, in particular the identification of latent semantic relationships of the studied phenomena or objects, are the basis of latent semantic analysis. This method is used to retrieve the context-dependent meanings of lexical units by means of statistical processing of a large corpus of texts in the classification or clustering documents. LSA represents the documents and individual words in a so-called "semantic space", in which all subsequent comparisons are performed.

The algorithm of LSA starts by computing a word × document frequency matrix from a large corpus of text, and in result it constitutes very sparse matrix. The row entries (vectors of words) correspond to the frequency of each word in a particular document, and are normalized using an entropy function [2]. In next step the reduction of dimensionality of the sparse matrix is achieved by using singular value decomposition (SVD; a mathematical technique closely related to eigenvector decomposition and factor analysis). It is compression step which is used to capture important regularities in the patterns of word co-occurrences while ignoring smaller variations that may be caused by idiosyncrasies in the word usage in individual documents. In other words, SVD brings out latent semantic relationships between words, even if they have never co-occurred in the same document [2, 8]. The result of condensing the matrix in this way is those words which occur in similar documents. LSA has the essential disadvantages and the one of them is related to the significant reduction in computation speed by increasing the size of input data (for example, in SVD-transformation) [9]. Besides, the current method has been criticized for being prone to the influence of strong orthographic word frequency effects (despite its entropy based normalization) on vectors and vector distances [10], and as being a "bag of words" model, ignoring statistical information inherent in word transitions within documents [11].

Latent Dirichlet Allocation (LDA) is a generative topic model for explaining the observations by using the implicit groups to provide an explanation of why some of the data are similar. LDA was first presented by Bleem David, Andrew Ng and Michael Jordan in 2002 and it was a graphical model for detecting topics [12]. If the observations are the words collected in texts, it argues that each text is a mixture of a small amount of topics and that the appearance of each word is related to the one of the topics of the document. In summary, in LDA, each document may be considered as a mixture of various topics. A topic is not strongly defined, neither semantically nor epistemologically, and is identified on the basis of supervised labeling and (manual) pruning on the basis of their likelihood of co-occurrence. A lexical word may occur in several topics with a different probability, however, with a different typical set of neighboring words in each topic [13].

Weak semantic cognitive map is a kind of semantic spaces that provide metrics for universal semantic dimensions. The idea of weak semantic cognitive mapping [14] is essentially differs from LSA and LDA approaches, above-mentioned in this paragraph and based on the notion "dissimilarity", and consists of separating representations (e.g. words) along as many as possible clearly identified, mutually independent, universal semantic dimensions that make sense for all domains of knowledge. Word-to-word relations of antonymy, synonymy, hyponymy and hypernymy are used in construction of the maps. The emergent semantic map dimensions are: "valence" (the first principal component of the weak semantic map derived from synonym-antonym relations), "arousal" (the second principal component), and "abstractness", or ontological generality (the semantic dimension derived from hypernym-hyponym relations) [15].

So, as we can see there are plenty of different approaches and algorithms of the semantic maps construction, however all of them have a significant disadvantage – none of them seem suitable for its usage for construction the semantic map of Russian language. In the following paragraph this problem will be explained in detail.

3   Features of the New Approach

Previously there were considered mostly English kinds of semantic maps. For Russian language for the moment there is no fully working algorithms for creating semantic maps. The closest decision was offered by so called Russian WordNet team, which decided to create a Russian analogy of the English WordNet version. For this moment their task is half done, but anyway their progress is very small. It is connected to some difficulties mostly caused by translation of synsets (sets of synonymic words).

The other researches in this area weren't so successful, even in comparison to Russian WordNet. For this moment, the semantic map area is quite unexplored. In this connection we started our research to try to solve this problem.

During our following researches we are going to introduce a fundamentally new approach of creating semantic maps for Russian language. There we highlighted the most common problems, which we can face during our research and the methods which could help us to avoid them.

1. Variety of Russian words. There is huge amount of Russian words which meaning is quite close and the only difference between them are affixes. In Russian language these words are called «same root». To solve this problem, it was decided to extract the root of each word and place it on the map. As the addition of some affixes to the root as a rule (except some affixes) can create definite appraisal of the word, therefore using the special functions and algorithms we will be able to find the positions of the new words without placing it on the semantic map. This approach will help to reduce the semantic map size and difficulty without losing its expressive power.

2. Complicated morphological structure of Russian word. Each word in Russian language, in the simplest case, consists of the root, sets of suffixes and prefixes, postfixes and inflexions. However, there are plenty of exceptions of this rule, or the words, including quite big amount of affixes. To handle of these words, we have to use huge dictionaries and be well versed in the area of morphology.

3. Variety of synonyms and antonyms. The expressiveness of Russian language is higher than for English languages, therefore there are plenty of synonyms and antonyms of each word. To simplify the semantic map and make the relation of synonymy and antonymy more formalized it was decided to add special relationships of the words. These relationships will be defined through the special coefficients. For example, is two words are synonyms, the coefficient of synonymy between them will be 1, if they are antonyms it will be -1, in other case it will be 0. Adding this relation will make the map easier for analysis and usage.

4. Lack of analogies of the semantic maps for Russian languages. As it was already mentioned, for this moment there no fully working algorithm for semantic map construction.

During the development of our approach we tried to solve all the problems listed above, so the method considers all the specialties of the Russian languages, it solves the problem with variety of the words, and the main feature of the approach is the opportunity to avoid the difficulty of the Russian words structure, and even its usage to reduce computational complexity.

4   Specialties of Russian Words Construction

In Russian language there are some different methods of word construction. Their classification is given on the Fig. 1.

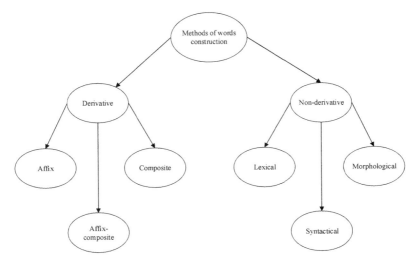

Fig. 1. The classification of word construction methods.

For the emotional color shaping of words in the Russian language we normally use suffixes. Prefixes may also be used to increase or decrease the meaning of the words, but less than suffixes. Therefore, during our research, we will consider only derivative methods of word construction as other methods will not help us to define appraisals. Affix classification is given on the Fig. 2.

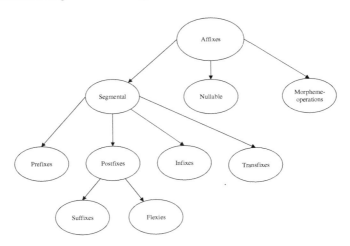

Fig. 2. The classification of affixes.

Each affix may specify a certain emotional color. If the prefix or suffix does not affect the meaning of the word, but simply forms a new part of speech, we will consider that it sets a zero emotional color. In the table below we represent the classification of the most common Russian suffixes and show whether they can attach the appraisal to the word or not. Therefore, in the process of the text analyzation, special algorithms will be able to determine the emotions of the whole text.

5 The Representation of the Word Structure

In Russian language, every word can be represented as concatenation of three sets (ordering elements in concatenation may depend on the particular case)

$$W = < R, a_1, \ldots, a_n, b_1, \ldots, b_m >, m, n \geq 0, \tag{1}$$

where

- W – described word,
- R – root of the word,
- $\{a_i\}$ – nullable set of affixes (suffixes, prefixes, postfixes)
- $\{b_i\}$ – nullable set of other elements (ending and so on).

Generally, removing an affix from any word converts it into another word, which is also available in the lexicon. Thus, each affix determines its relation type on the set of the words. If each word have some coordinates on semantic map then each affix can be associated with a vector of a typical shift, caused by adding affix to the word. These vectors can be considered as lattice generators.

This approach has several advantages. Firstly, it is a perceptible reduction of computational complexity, since there are about a hundred different affixes in Russian language. Thus, instead of positioning each word form to a semantic map we only need to arrange the root of words. Another obvious advantage of this method is a better allocation of emotional color and style of speech or text, based on the analysis using affixes that will help in further research of modeling the process of human thought and activity.

On the other hand, use of affixes for semantic maps constructing associated with some difficulties. For example, the shift vector for the same affix may vary for different words by main principal components compared to traditional construction of weak semantic maps.

Moreover, the relatively small number of words associated with affixes that can evaluate semantic shift of the coordinates. If we throw out the card all the words that are not related by affixes, the map will be very small. And in the dictionary there are a plenty of words that are so unrelated, that effect of affix relationship on semantic map will be imperceptible. Despite these difficulties, the use of affixes can sometimes be useful as assistant for main semantic map construction method.

To determinate the capability of application of new approach we offer to use affix association coefficient of semantic map, which is defined as:

$$K = \frac{N_a}{N}, \tag{2}$$

where

- K – affix association coefficient
- N_a – quantity of words connected by affix relationship,
- N - quantity of words in whole semantic map

Further explorations imply developing the calculating algorithm which will help us to create semantic map, and

As the example of the affixes influence to the words, let's consider the word "kot" ("cat"), let it has (x,y) coordinates on the semantic map. Adding the affix "ik" to the root will create a new word "kotik". The meaning of the word hasn't changed, but this suffix made it more soft, as if the speaker wanted to show, that this cat is very nice. After this modification of the word, it will have new space coordinates on the semantic map $(x+\Delta x, y+\Delta y)$. This shift $(\Delta x, \Delta y)$ is defined by the affix "ik". So, adding the affix has generated the new semantic space with new appraisal of the word.

6   Conclusion

The rapid growth of interest in the area of automated proceeding and recognition texts extracted from different documents or the Internet or its usage by automated systems with considering its emotional color, shows us the necessity of designing the semantic cognitive map for Russian language. We considered the most popular semantic maps, designed for different languages and found out that there is now fully completed cognitive map for Russian language.

The basis of our approach is usage of affixes to define the emotional colors of the words. We represented the set-theoretical model of the Russian word considering all the Russian morphemes. By the means of this model we can describe each Russian word.

Using the special algorithms and the described model we can create a fully working semantic map for Russian language. It will have many advantages including, for example, its calculating power.

Our further plans of this research include the developing the calculating algorithm which will help us to create semantic map. We are also going to test in on different dictionaries and check its calculating power.

Acknowledgements.   The funding for this research was provided by the Russian Science Foundation, Grant RSF 15-11-30014.

References

1. Pang, B., Lee, L.: Opinion Mining and Sentiment analysis. Now Publishers Inc., (Chapter 4.1.2 Subjectivity Detection and Opinion Identification) (2008)
2. Keith, J., Westbury, C., Goldman, J.: Performance impact of stop lists and morphological decomposition on word-word corpus-based semantic space models. Behav. Res. Methods **47**, 666–684 (2015)

3. Samsonovich, A., Ascoli, G.: Augmenting weak semantic cognitive maps with an "abstractness" dimension. Hindawi Publishing Corporation Comput. Int. Neurosci. **2013**, 3 (2013)

4. Ploux, S., Ji, H.: A Model for Matching Semantic Maps between Languages (French/English, English / French). Computational Linguistics, Massachusetts Institute of Technology Press (MIT Press), vol. 29, 155–178 (2003)

5. Cysouw, M.: Semantic maps as metrics on meaning. Linguist. Discovery **8**, 70–95 (2010)

6. Haspelmath, M.: The geometry of grammatical meaning: semantic maps and cross-linguistic comparison. New Psychol. Lang. **2**, 211–242 (2003)

7. Landauer, T., Dumais, S.: A solution to plato's problem: the latent semantic analysis theory of acquisition, induction, and representation of knowledge. Psychol. Rev. **104**, 211–240 (1997)

8. Buckeridge, A., Sutcliffe, R.: Disambiguating noun compounds with latent semantic indexing. In: Second International Workshop on Computational Terminology (COMPUTERM 2002), within the 19th International Conference on Computational Linguistics (COLING 2002), Taipei, Taiwan, 14, pp. 1–7 (2002)

9. Deerwester, S., Dumais, S., Furnas, F., Landauer, T., Harshman, R.: Indexing by latent semantic analysis. J. Am. Soc. Inf. Sci. **41**, 391–407 (1990)

10. Rohde, D., Gonnerman, L., Plaut, D.: An improved model of semantic similarity based on lexical co-occurence. Commun. ACM **8**, 627–633 (2006)

11. Perfetti, C.: The limits of co-occurrence: tools and theories in language research. Discourse Process. **25**, 363–377 (1998)

12. Blei, D., Ng, A., Jordan, M.: Latent dirichlet allocation. J. Mach. Learn. Res. **3**, 993–1022 (2003)

13. Neelam Sushma, N., Kondareddy, G.V.: Reputation-based trust evaluation by extracting user's assessment. Int. J. Techn. Res. Appl. **3**, 5–16 (2016)

14. Samsonovich, A.V., Ascoli, G.A.: Principal semantic components of language and the measurement of meaning. PLoS ONE **5**, 1–17 (2010)

15. Samsonovich, A.V.: Semantic cross-correlation as a measure of social interaction. Biologically Inspired Cogn. Architectures **7**, 1–8 (2014)

Author Index

Abdulllah, Wan   698
Aliseychik, Anton   292
Allakhverdov, Viktor   706
Astashev, Maksim   673
Astasheva, Elena   673

Bagherpour, Solmaz   490
Bakhshiev, A.V.   317
Balandina, Anita   731
Bao, Chunhua   401
Bao, Lanying   401
Baranov, Sergey   185
Basov, O.O.   497
Bayro-Corrochano, Eduardo   640
Benderskaya, Elena   444
Bizin, M.M.   497
Blagoveshchenskaya, Ekaterina A.   513
Bondarev, Vladimir   647
Bozhokin, Sergey V.   49
Budkina, Elena M.   277
Budkov, V. Yu.   497

Cao, Jinde   284
Chang, Zheng   30
Chen, He   242
Chen, Weineng   603
Chen, Xiangyong   284
Chen, Zonggan   603
Cheng, Chunying   401
Cheng, Long   233, 338
Cheng, Ming   233
Cherednikov, Denis   657
Chernykh, German A.   108
Chernyshov, Artyom   731
Chistyakova, Tamara   565
Cong, Fengyu   30

Dashkina, Aleksandra I.   513
de Carvalho, Francisco de A.T.   393
Deguchi, Toshinori   409
Deng, Lu   233
Dmitrieva, Ludmila A.   108
Dorogov, A. Yu.   204
Duan, Haibin   356

Er, Meng Joo   426, 474
Evdokimenkov, Veniamin   365
Evnevich, Elena L.   40
Ezhov, Alexandr A.   375

Fang, Yongchun   242, 338
Ferreira, Marcelo R.P.   393
Fu, Li   346

Ghasemi, A.   454
Gorbachenko, Vladimir I.   310
Grigoriev, Oleg   712
Grodetskiy, Yuri   302
Gu, Xiaodong   631
Gundelakh, F.V.   317
Guo, Liuxiao   260
Guo, Ping   540, 555
Gurakov, Mihail   620

Han, Min   21
Hartill, Bruce   328
Hodashinskiy, Ilya   620
Hou, Yuqing   92
Hou, Zeng-Guang   233
Hu, Xiaoguang   346, 356
Huang, Huifen   3
Huang, Tingwen   681

Ibrahim, Zainol   698
Ishii, Naohiro   409

Ji, Peng   555
Jin, Cong   595

Karpinskaia, Valeria   706
Karpov, Alexey A.   115, 418
Kawaguchi, Masashi   409
Kaya, Heysem   115
Kheidorov, Igor E.   613
Khomenko, Yulia   100
Khromov, Andrei G.   375
Kim, Roman   365
Kipyatkova, Irina   418

Kiselev, Mikhail V.   665
Kitchigina, Valentina   673
Klimov, Valentin   731
Kohlert, Christian   565
Kolbin, Ilya S.   583
Kostkina, Anastasiya   731
Kostyuchenko, Evgeny   620
Kotenko, Igor   521
Koval, Alexandra   100
Kovalets, Pavel E.   613
Krasilshchikov, Mikhail   365
Krivonosov, Egor   620
Kudashev, Oleg   82
Kuperin, Yuri A.   108
Kuznetsov, Evgenii B.   277

Lavrentyeva, Galina   82
Lazovskaya, Tatiana V.   277, 310, 513
Lechuga-Gutiérrez, Luis   640
Lee, Eric Wai Ming   385
Leonov, Sergey S.   277
Levonevskiy, Dmitriy K.   40
Li, Bing   143
Li, Dan   356
Li, Feng   3
Li, Hong   30
Li, Michael M.   434
Li, Michelle Ching Wa   385
Li, Wenjie   66
Li, Zhijun   250
Li, Zhouhong   211
Liu, Derong   269
Liu, Fan   426
Liu, Hang   631
Liu, Jia   260
Liu, Jinan   595
Liu, Xiaoli   250
Liu, Xing   223
Liu, Youyi   30
Liu, Zhenwei   152
Long, Shujun   143
Lyakhovetskii, Vsevolod   706

Malychina, Galina   74
Malykhina, Galina   302
Mescheryakov, Roman   620
Moradzadeh, M.   454
Motienko, A.I.   497
Mugica, Francisco   490

Nagornova, Zhanna   100
Navleva, Angelina A.   108
Nebot, Àngela   490
Nikolayeva, A. Yu.   691
Niu, Haisha   152
Novikov, Andrei   444
Novoselov, Sergey   82

Orlov, Igor   292
Osipov, Vasiliy   177

Pang, Shaoning   328
Pavlovsky, Vladimir   292
Pekhovsky, Timur   82
Peng, Zhouhua   223
Perets, Dmitry   100
Petukhov, Alexandr Y.   56
Podoprosvetov, Alexey   292
Polevaya, Sofia A.   56
Popa, Călin-Adrian   127
Pratama, Mahardhika   426

Qin, Sitian   160
Qiu, Jianlong   284
Qiu, Li   260

Razygraev, Alexander   565
Red'ko, Vladimir G.   720
Reviznikov, Dmitry L.   583
Ribeiro, Bernardete   531
Ristainiemi, Tapani   30
Ryabukhina, Viktoria V.   513

Saenko, Igor   521
Saffih, Fayçal   698
Saitov, I.A.   497
Saitov, S.I.   497
Salah, Albert Ali   115
Sarrafzadeh, Abdolhossein   328
Sasaki, Hiroshi   409
Savchenko, Andrey V.   505
Sebrjakov, German   365
Shayeghi, H.   454
Shemyakina, Natalia   100
Shemyakina, Tatiana A.   547
Shi, Wen   603
Shibzukhov, Zaur   657
Shilov, Yuri   706
Shishova, Marina   292

Silva, Catarina   531
Simões, Eduardo C.   393
Simonchik, Konstantin   82
Skorik, Fadey   521
Smolin, Vladimir   292, 573
Song, Qiankun   168
Sonkin, Konstantin   100
Soroka, Alexander M.   613
Stankevich, Lev   100
Su, Chun-Yi   250
Sun, Ning   242
Sun, Zhanquan   3
Suslova, Irina B.   49

Tan, Manchun   135
Tan, Yongyi   160
Tarasenko, F.D.   482
Tarkhov, Dmitriy A.   277, 310, 482, 513, 547
Tarkov, Mikhail S.   196
Terentyeva, Svetlana S.   375
Teterin, Mikhail   565
Tomyshev, Maxim   620

Vasilyev, Alexander N.   277, 310, 547, 583
Venkatesan, Rajasekar   474
Vorobiev, Vladimir I.   40
Vvedensky, V.L.   691

Wang, Changming   66
Wang, Dan   223
Wang, Fuqiang   160
Wang, Jun   21
Wang, Lingling   346
Wang, Ning   426, 474
Wang, Ruohui   12

Wang, Xiao   681
Wang, Zhanshan   152
Wei, Qinglai   269

Xin, Xin   555
Xu, Desheng   135
Xue, Ruijie   338

Yakovenko, Anton   74
Yalavarthi, Vijaya Krishna   474
Yang, Chengdong   284
Yang, Tiantian   30
Yang, Yimin   338
Yang, Yongqing   260
Yin, Qian   540
Yu, Hongnian   233
Yu, Jian   540

Zaitsev, Dmitry   465
Zaitseva, Natalia   465
Zeng, Zhigang   681
Zhan, Zhihui   603
Zhang, Chengkun   21
Zhang, Guofeng   346, 356
Zhang, Jun   603
Zhang, Xuebo   338
Zhang, Yingwei   152
Zhang, Yong   426
Zhao, Jing   328
Zhao, Zhenjiang   168
Zhou, Jiongru   66
Zhou, Tiantong   66
Zhu, Yidong   681
Zhukov, Maxim V.   310
Zou, Ling   66

Printed in the United States
By Bookmasters